ANNUAL REVIEW OF PLANT PHYSIOLOGY

EDITORIAL COMMITTEE (1986)

ANNUAL REVIEW OF PLANT PHYSIOLOGY

VOLUME 37, 1986

WINSLOW R. BRIGGS, *Editor*

Carnegie Institution of Washington, Stanford, California

RUSSELL L. JONES, *Associate Editor*

University of California, Berkeley

VIRGINIA WALBOT, *Associate Editor*

Stanford University

ANNUAL REVIEWS INC.   4139 EL CAMINO WAY   PALO ALTO, CALIFORNIA 94306 USA

ANNUAL REVIEWS INC.
Palo Alto, California, USA

International Standard Serial Number: 0066–4294
International Standard Book Number: 0–8243–0637–6
Library of Congress Catalog Card Number: A–51–1660

Annual Review and publication titles are registered trademarks of Annual Reviews Inc.

Annual Reviews Inc. and the Editors of its publications assume no responsibility for the statements expressed by the contributors to this *Review*.

Typesetting by Kachina Typesetting Inc., Tempe, Arizona; John Olson, President
Typesetting coordinator, Janis Hoffman

PRINTED AND BOUND IN THE UNITED STATES OF AMERICA

ANNUAL REVIEWS INC. is a nonprofit scientific publisher established to promote the advancement of the sciences. Beginning in 1932 with the *Annual Review of Biochemistry*, the Company has pursued as its principal function the publication of high quality, reasonably priced *Annual Review* volumes. The volumes are organized by Editors and Editorial Committees who invite qualified authors to contribute critical articles reviewing significant developments within each major discipline. The Editor-in-Chief invites those interested in serving as future Editorial Committee members to communicate directly with him. Annual Reviews Inc. is administered by a Board of Directors, whose members serve without compensation.

For the convenience of readers, a detachable order form/envelope is bound into the back of this volume.

CONTENTS

ARTICLES IN OTHER *ANNUAL REVIEWS* OF INTEREST TO
PLANT PHYSIOLOGISTS

From the *Annual Review of Biochemistry,* Volume 55 (1986):

Reactive Oxygen Intermediates in Biochemistry, *Ali Naqui, Britton Chance, and Enrique Cadenas*

Genetics of Mitochondrial Biogenesis, *Alexander Tzagoloff and Alan M. Myers*

Carbohydrate-Binding Proteins: Tertiary Structures and Protein-Sugar Interactions, *Florante A. Ouiocho*

Eukaryotic DNA Replication, *Judith L. Campbell*

Transcription of Cloned Eukaryotic Ribosomal RNA Genes, *Barbara Sollner-Webb and John Tower*

The Transport of Proteins into Chloroplasts, *Gregory W. Schmidt and Michael L. Mishkind*

From the *Annual Review of Biophysics and Biophysical Chemistry,* Volume 15 (1986):

Halorhodopsin: A Light-Driven Chloride Ion Pump, *Janos K. Lanyi*

Electrostatic Interactions in Membranes and Proteins, *Barry H. Honig, Wayne L. Hubbell, and Ross F. Flewelling*

Chromosome Classification and Purification Using Flow Cytometry and Sorting, *J. W. Gray and R. G. Langlois*

Recent Advances in Planar Phospholipid Bilayer Techniques for Monitoring Ion Channels, *Roberto Coronado*

Identifying Nonpolar Transbilayer Helices in Amino Acid Sequences of Membrane Proteins, *D. M. Engelman, T. A. Steitz, and A. Goldman*

Applications of NMR to Studies of Tissue Metabolism, *M. J. Avison, H. P. Hetherington, and R. G. Shulman*

Recombinant Lipoproteins: Implications for Structure and Assembly of Native Lipoproteins, *David Atkinson and Donald M. Small*

The Role of the Nuclear Matrix in the Organization and Function of DNA, *William G. Nelson, Kenneth J. Pienta, Evelyn R. Barrack, and Donald S. Coffey*

From the *Annual Review of Ecology and Systematics,* Volume 16 (1985):

Ecology of Kelp Communities, *Paul K. Dayton*

Resource Limitation in Plants—An Economic Analogy, *Arnold J. Bloom, F. Stuart Chapin, III, and Harold A. Mooney*

From the *Annual Review of Cell Biology*, Volume 1 (1985):

Protein Localization and Membrane Traffic in Yeast, *Randy Schekman*

Microtubule Organizing Centers, *B. R. Brinkley*

Chromosome Segregation in Mitosis and Meiosis, *Andrew W. Murray and Jack W. Szostak*

Nonmuscle Actin-Binding Proteins, *T. P. Stossel, C. Chaponnier, R. M. Ezzell, J. H. Hartwig, P. A. Janmey, D. J. Kwiatkowski, S. E. Lind, D. B. Smith, F. S. Southwick, H. L. Yin, K. S. Zaner*

Using Recombinant DNA Techniques to Study Protein Targeting in the Eucaryotic Cell, *Henrik Garoff*

Progress in Unraveling Pathways of Golgi Traffic, *Marilyn Gist Farquhar*

Biogenesis of Peroxisomes, *P. B. Lazarow and Y. Fujiki*

From the *Annual Review of Genetics*, Volume 19 (1985):

Maize *Adh1, Michael Freeling and D. Clark Bennett*

Comparative Organization of Chloroplast Genomes, *Jeffrey D. Palmer*

From the *Annual Review of Microbiology*, Volume 39 (1985):

Carbon Metabolism in *Rhizobium* Species, *Mark Stowers*

Plant Virus Satellites, *R. I. B. Francki*

Plant and Fungal Protein and Glycoprotein Toxins Inhibiting Eukaryote Protein Synthesis, *A. Jiménez and D. Vázquez*

From the *Annual Review of Phytopathology*, Volume 23 (1985):

The Molecular Genetics of Plant Pathogenic Bacteria and Their Plasmids, *Nickolas J. Panopoulos and Richard C. Peet*

Transposon Mutagenesis and Its Potential for Studying Virulence Genes in Plant Pathogens, *Dallice Mills*

Monoclonal Antibodies in Plant Disease Research, *Edward L. Halk and Solke H. De Boer*

Of Special Interest...
TROPICAL BIOLOGY

We are pleased to call your attention to this year's issue of the *Annual Review of Ecology and Systematics* (Volume 17), emphasizing tropical biology. Due in November, it will include:

Nutrition and Nutritional Relationships of *Pinus radiata*: *J. Turner and M. Lambert, Forest Commission, NSW, Australia*

Nutrient Cycling in Moist Tropical Forest: *P.M. Vitousek and R.L. Sanford, Stanford Univ., Stanford, CA*

Ecology of Tropical Dry Forest: *P.G. Murphy, Michigan State Univ., East Lansing, MI; A.E. Lugo, Southern Forest Experiment Station, USDA Forest Service, Rio Piedras, Puerto Rico*

Biology of Birds of Paradise and Bower Birds: *J. Diamond, Univ. California School of Medicine, Los Angeles, CA*

Ecology of African Grazing and Browsing Mammals: *S.J. McNaughton and N.J. Georgiadis, Syracuse Univ., Syracuse, NY*

Perspectives for an Etiology of Stand-Level Dieback: *D. Mueller-Dombois, Univ. Hawaii at Manoa, Honolulu, HI*

The Role of Competition in Plant Communities in Arid and Semiarid Regions: *N. Fowler, Univ. Texas, Austin, TX*

The Ecology of Tropical Arboviruses: *T. Yuill, Univ. Wisconsin, Madison, WI*

Pollination Systems in the Humid Tropics: *K.S. Bawa, Univ. Massachusetts, Boston, MA*

Designing Agricultural Ecosystems for the Humid Tropics: *J.J. Ewel, Stanford Univ., Stanford, CA*

The Future of Tropical Ecology: *D.H. Janzen, Univ. Pennsylvania, Philadelphia, PA*

Ecology of Coralline Algal Crusts: *R. Steneck, Univ. Maine at Orono, Walpole, ME*

The Socioecology of Primate Groups: *J. Terborgh, Princeton Univ., Princeton, NJ; C.H. Janson, State Univ. New York, Stony Brook, NY*

Competition and Resource Partitioning in the Guild of Neotropical Frugivorous Mammals: *N. Smythe, Smithsonian Tropical Research Institute, APO Miami, FL*

Soil-Plant Relationships in the Tropics: *D.J. Lathwell and T.L. Grove, Cornell Univ., Ithaca, NY*

Ann. Rev. Plant Physiol. 1986. 37:1–22

CONFESSIONS OF A HABITUAL SKEPTIC

Norman E. Good

Department of Botany and Plant Pathology, Michigan State University, East Lansing, Michigan 48824

CONTENTS

When I was invited by the Editorial Committee of the *Annual Review of Plant Physiology* to prepare these memoirs, I was pleased and flattered. At last I had an opportunity to leave a legacy in the literature, limited only by the laws of libel. As it turned out, writing these memoirs has proved extraordinarily difficult. Heretofore my literary ramblings have been constrained by the subject matter and disciplined by editors. Now I must try to record the diffuse speculations of a lifetime without benefit of external restraint.

This article is only marginally autobiographical. It is primarily the history of my ideas on biological topics. It is only incidentally a brief account of a life that has been fairly uneventful except in terms of my exposure to the influence of outstanding intellects.

THE MAKING OF AN AGNOSTIC

I lived the first three decades of my life on a farm on the outskirts of the city of Brantford in the province of Ontario in Canada. Brantford is of some historical

1

interest. Originally it was the new capitol of the Mohawk Nation of the Iroquois Federation when those allies of the British were invited to leave the United States after the American Revolution. The famous Joseph Brant, chief of the Mohawks, brought his people up the Grand River from Lake Erie to the ford at the headwaters of navigation, thereafter known as Brant's ford. In view of past and future events, there is some irony in the fact that the British granted to Brant and his people the lands along the river "from its mouth to its source, for them and their posterity to enjoy forever." These same lands had already been occupied by these same Iroquois for centuries. Their early occupation of the area blocked an otherwise easy spread of French settlements along the St. Lawrence river into the rich heartland of America. Indeed, the military prowess of the Iroquois and their occupation of my homeland are important reasons why this article is being written in English instead of French. It is also noteworthy that the "forever" mentioned in the treaty holds a record for brevity of eternity. Once the French political power in America was broken, the Iroquois ceased to be useful and were elbowed aside.

My paternal great-grandfather obtained a grant of the same lands from a forgetful British Crown in 1837. The framed original of this not so original deed still hangs in the stairwell of the Georgian house he built a few miles north of the main Mohawk settlement. One hundred and twenty-five years later, my wife and I added to our family of four children two Iroquois babies. This belated attempt to redress the wrong done their ancestors has been very satisfying, but I am not sure how much good it has done the remnant of the Six Nations.

More to the point of this narrative, there is in the old house an extensive library stocked with scientific and literary classics, written and printed over a period of 300 years. Among the books are two immense volumes of Joseph Black's lecture notes, published in the 18th century. It is probably not an exaggeration to attribute the birth of chemistry to Black. He developed ways of working with gases as chemical reactants, paving the way for Lavoisier; he discovered carbon dioxide; and he was the first to work with heat as a quantity (rather than with temperature which is an intensity). Not much of this leaps out from the books, however, because the language of chemistry that we use now has been invented since Black's time. There is also a treatise on electricity written and published about the time of Franklin's experiments by an author I do not now recall, a treatise remarkable for clarity and rigor of reasoning. Then there is a copy of Lyell's *Principles of Geology* published well before the appearance of Darwin's *Origin of Species*. When I looked into Lyell's book, I was astonished to find Darwin's principle already clearly and explicitly enunciated with respect to geology, namely that the things we see now can be explained in terms of processes now operating and need not be attributed to cataclysmic events. Only much later did I discover that Darwin made a point of acknowledging his debt to Lyell, the father of modern geology. In the library are many other texts of comparable importance to the history of science. Obviously, some of my ancestors had scientific interests in a day when science was mostly in the hands of laymen. If there are genes for an interest in things physical, chemical, and biological, I certainly have a right to a share.

I do not know why, but the area around Brantford attracted other immigrants with scientific or literary aspirations. The farm next to ours was and still is occupied by Carlyles, great-grandnephews and nieces and heirs of Thomas Carlyle. (Much as I love some of my Carlyle cousins, I still cannot read the outpourings of the senior Thomas). However, the best known among these immigrants was a certain Professor Bell from the University of London who sought the "wholesome" air of the New World to cure his tubercular sons. Unfortunately, in this he was only partly successful. His younger son shared his years of early manhood with my father's uncles and aunts. The senior Professor Bell was an "elocutionist," which in those days meant a specialist in speech. Among his many accomplishments was a truly phonetic alphabet, totally ignored by the public but much lauded by George Bernard Shaw. I well remember my father's old aunt telling a story that may have grown somewhat over the many intervening years. Apparently, the Professor had a dog that he had trained to emit a constant low growl, something not difficult to do. He then inserted his fingers into the dog's mouth and, by appropriate manipulations, made the dog utter comprehensible English words! Dr. Bell's surviving son followed in his father's footsteps; he also studied the physics of speech. But, in anticipation of modern techniques, he replaced the dog with instruments, converting the sounds of speech into electrical analogs and then converting these analogs back into sound. The trick was so successful that it became widely used for long-distance transmission of speech and for the recording of sounds. Unfortunately, none of my ancestors had the foresight to invest in the technique.

It is not altogether surprising that my father, steeped in such traditions, should have elected to study the then new discipline of physical chemistry at the University of Toronto. He was a natural scholar with a remarkable combination of manual, mathematical, and verbal skills. He could (and did) shoe horses with shoes he made in a forge he had built with his own hands. He laid bricks with as much skill as a professional. His intellectual accomplishments were formidable. He was a candidate for a Rhodes scholarship when these were new, but lost out because of his utter boredom with sports. I believe that he was at one time on the Board of Governors of the University of Toronto. Whenever I tested him with lists of little-used words, he was never at a loss. He looked up telephone numbers in the directory only once. He invariably solved the more difficult mathematical school problems of his children, more than 40 years after he had studied mathematics himself, and often with proofs that had teachers shaking their heads in amazement. He was physically very powerful, short and immensely broad, with muscular coordination that would have made him a good athlete if he had cared. He kept his great strength to old age. For instance, he was 79 years old before we began to relieve him of the most arduous farm work.

I suppose that my father's breadth of interests and his versatility were a kind of professional undoing. In any event, he early abandoned the study of the physical sciences (including aeronautics at a time when only birds could fly) for political science and economics. He was a minor authority on monetary theory and he was briefly in Canadian federal politics. When I was young the house was filled with economists and politicians but never scientists. However, a prophet is not without honor save in his own family, and my admiration for my father is moderated by the fact that he was not easy to live with. Although I did envy him his intellect and his skills and I admired his disinterested dedication to worthy social causes, my adulation is well within bounds. His understanding of people, including himself, was notoriously defective. He had a theoretical dedication to agriculture as a way of life, not as a business, and he did not hesitate to sacrifice his children on that altar. Although he was very hard working and energetic, he never took any responsibility for the boring routine work of raising livestock or the other repetitive aspects of farming that we dubbed "the chores." He was much too committed to his worthy causes to forego any of them in the interest of milking the cows twice daily. As a consequence, whichever of his sons must milk the cows had, perforce, a very erratic education.

The fact that I was often that son has long since ceased to rankle, if it ever did. I never was sure what I wanted to do with my life anyway. This uncertainty is still with me and will probably continue with me until time makes the question irrelevant. Besides, there were some benefits from the enforced absences from school. The intermittent nature of my formal education gave me long periods to consolidate my ideas and to clarify my understanding of basic concepts. (Nothing clarifies concepts like the dropping away of minutiae.) Another consequence of having my education interrupted all through the war years is that, when I did return to university, my colleagues and fellow graduate students were largely war veterans, an exceptionally mature and dedicated lot. Therefore, I have no fundamental professional regrets, although I did feel some nostalgia for the sparkle of a dark-browed Jeanie and a romance that never was.

There were three interruptions in my education totaling nine years. I did not graduate from the University of Toronto until 1948 when I was 31 years old. These interludes and the reason for them had profound and sometimes contradictory effects. I came to feel that school was a luxury pursuit and not a serious part of life. To this day, I cannot think of academic efforts as real work. (Some of my more candid colleagues might say that said frame of mind is obvious.) Deep inside me is a feeling that work is done with a plow and tractor, or a fork, or an axe. Consequently, I am always a little puzzled by the fact that I am being paid to play in the laboratory. Thus I am doomed forever to be, in my own mind, a dilettante, but being a dilettante has its advantages. I may be opinionated, dogmatic, and blunt to a point of incivility, but I am not too involved emotionally in my arguments. It is only a game.

Another consequence of my erratic schooling, intervals of reading being interrupted by long hours of repetitive physical work with no one to talk to, is that I early began to separate concepts from the words commonly used to express the concepts—just as an illiterate deaf-mute must. This mental trick puts a very different complexion on the contemplation of science. It can also make communication difficult when at last there is someone to communicate with. Thus having an idea and putting it into words can be quite separate processes in my mind. It may be that this mental trick is the most valuable residue from the breaks in my formal education. One of the pitfalls of sciences and indeed of all rational thought is the tendency to create words for concepts, words that then take on a "reality" in their own right, independent of the concept. Words separated from the parent concepts tend to usurp the role of concepts and to create "ideas" that are actually meaningless. Thus, if we are not careful, science can become a sort of litany, full of sound and fury, signifying nothing. This predilection for ideas over words was reinforced by my first chemistry teacher at the university, F. W. Kenrick, who was dedicated to but one cause. He spent several months lecturing us on the definition of a solution and on the definition of the terms used in the definition! In almost every lecture he implored us: "Know what you are talking about." In the idea section that follows, I am going to try to apply this criterion to a few of the dogmas of modern biology, but I am reconciled to the fact that there will be some annoyed readers.

After I graduated from the University of Toronto, I attended the California Institute of Technology. In those days G. W. Beadle of *Neurospora*-gene-enzyme fame, was head of the Biology Division, Linus Pauling was head of the Chemistry Division, and Max Delbrück was teaching us to use bacteriophages as biological tools. The whole Institute was in the birth pangs of so-called molecular (actually macromolecular) biology. Much of this passed me by because I was involved with other things, but the intellectual ferment could not fail to have an effect. My own research, under the direction of Herschell Mitchell, dealt with the chemistry and metabolism of small molecules, notably amino acids, but I did use *Neurospora* mutants as tools and some of my work did have genetic implications.

My first postdoctoral position involved a complete change of research area. I helped Allan Brown at the University of Minnesota in his task of differentiating respiration from concurrent photosynthesis, difficult because the two opposite processes both involve molecular oxygen and carbon dioxide. We succeeded by using isotopically labeled oxygen as the substrate for respiration, since the oxygen produced from water by photosynthesis is, of course, unlabeled. The problem was of considerable importance in those days because the great Otto Warburg had decreed that photosynthesis requires an immense uptake of oxygen, reconsuming three quarters of the photosynthetically produced oxygen. We succeeded in showing that Warburg was wrong, as we now know from many other kinds of experiments.

I have sometimes been asked how we came to miss the phenomenon of photorespiration since our instrument was ideally suited for detecting it. The simple answer is that we did not miss photorespiration at all. We observed it and adjusted conditions to eliminate it, using low concentrations of oxygen and high concentrations of carbon dioxide. We chose these conditions specifically to avoid a phenomenon that was already well known but irrelevant to our concerns. Photorespiration had been described many years before by Warburg, who called it "photocombustion."

My exposure to photosynthesis research with Brown introduced me to a new area in which I have dabbled in an intermittent and desultory way ever since. In late 1952, I joined Robin Hill in Cambridge, where Sanger was already determining the structure of insulin and Watson and Crick were theorizing on the structure of DNA. Again I let most of the excitement of the new molecular biology pass me by. Again my work was directed toward different goals. Robin and I developed a useful technique, since much employed, in which chloroplast lamellae reduce substances like flavins and viologens which then in turn reduce oxygen. Thus they are at the same time acceptors of electrons from chloroplast lamellae and catalysts of oxygen reduction by the lamellae. One of these substances, methylviologen, has since become popular as a herbicide under the alternative name, paraquat.

I cannot leave the account of this period of my life without acknowledging my appreciation of Robin Hill. He is an unusual person in many respects, among which is an unusual sensitivity and understanding coupled to an unusual lack of articulateness. The wisdom of his comments (which is often very great) only penetrates the average skull after a long delay, from hours to years. Withal, Robin is one of the most discerning and original scientists I have ever encountered. His contributions to our understanding of photosynthesis have been enormous. As early as 1953, he was speculating on the possibility of two sequential light reactions in photosynthesis. It is scandalous that the Nobel Prize Committee has continued to overlook him. Perhaps the oversight has occurred because he keeps a low profile, publishes rather infrequently, and abhors self-advertisement.

In 1954, I returned to Canada to take up a research position with the federal Department of Agriculture, in a new laboratory dedicated to fundamental research on biological problems. There, for about two years, I worked with the late Wolf Andreae on indoleacetic acid (auxin) metabolism in plants. We showed that a major product of the metabolism of exogenous indoleacetic acid is indoleacetylaspartic acid, an observation that may or may not have physiological importance. Then I went back to studying photosynthesis, especially photosynthetic electron transport coupled to ATP formation. I confirmed Jagendorf's observation that ammonium salts uncouple the two linked processes, and I extended the observation to a large number of aliphatic amines.

In the course of these studies on chloroplast photophosphorylation, I observed that many anions could also have uncoupling effects, especially at high concentrations, whereas "inner" salts such as glycine and other amino acids had no such effect. Therefore, it seemed to me that photophosphorylation in vitro might be laboring under a handicap because all of the buffers we used to control pH of necessity contained anions. So I made a series of amino acid buffers, a wide range of aminocarboxylic and aminosulfonic acids with pKa's appropriate for biological research. They have proved useful, not only in photosynthesis research but also in general biochemical research, in tissue culture, and in medicine.

When I came to Michigan State University in 1962, I had the good fortune to work with Seikichi Izawa. He is responsible for any reputation I may have for constructive and reliable research in photosynthesis. For 12 years he put up with me and my indolence while he trained some outstanding graduate students. For laboratory skills, hard work, scientific imagination, and ability to pick promising research topics, he is exceptional. The research done in the Izawa years in our laboratory is too diverse to catalog here. It was both in the mainstream of photosynthesis research and out of it; although we were looking at much discussed aspects of chloroplast electron transport and photophosphorylation, our conclusions were sometimes considered aberrant. Among our observations that have been much noted and well accepted are the in vitro unstacking and stacking of chloroplast lamellae and the role of lipophilic strong oxidants as acceptors of electrons directly from photosystem II. (Indophenol dyes which are reputed to accept electrons from photosystem II actually accept electrons primarily from photosystem I.)

Although I left a Department of Agriculture laboratory in Canada to come to an academic position at Michigan State University, it was only at Michigan State that I began to interact with agronomists. When the energy concerns of the 1970s dragged me kicking and screaming from my laboratory, I was forced to give some thought to where my skills and knowledge could be applied in the interest of growing better crops. In some ways this has been a recipe for schizophrenia. My analysis of the factors involved in crop yield (2) convinced me that my professional discipline had little to offer, but all of my instincts were crying out to me to involve myself with the cultivation of the soil or with subjects pertinent to the cultivation of the soil. Now that society has begun to return to its characteristic euphoria after its brief bout with hysteria in the 1970s, we scientists are being allowed to drift back to our various ivory towers. Meanwhile, I have made a personal compromise that has no rational justification but nevertheless soothes my soul. As I approach retirement, I find myself taking more and more satisfaction from growing things and working in the fields. No doubt this is a manifestation of a second childhood, of waning intellectual powers. Sometimes my more sedate colleagues raise their eyebrows when I arrive at the department in tattered and not too clean bib overalls—my badge of reverse snobbery. I play the lottery of growing thousands of seedling grapes and apples, not for hope of gain but for the joy of wielding a hoe in the still of the evening or the early morning. Besides, grapes taste much better than golf balls.

Now I close this brief biographical sketch and move on to more important things, a history of my ideas right and wrong in the context of evolving concepts in biology.

THE WATER RELATIONS OF PLANTS

When I returned to the University of Toronto after the war, I studied biology with an emphasis on botany. One of the subjects that intrigued me was the water economy of plants. It is obvious that terrestrial plants have a major problem in obtaining and conserving water; they must take up carbon dioxide out of an oftentimes very dry atmosphere without at the same time and by the same pathway losing an inordinate amount of water. Moreover, plants grow in large part by cellular expansion, that is by the uptake of water. For these reasons the water relations of plants have always played a major role in studies of plant physiology.

Unfortunately, the literature on plant reactions involving water has usually been couched in language that has done more to confuse than to clarify. Indeed, I found the terminology with its "osmosis" and "suction pressure deficits" so confusing that I had to stop, wipe my mental slate clean, and ask "but what is really going on?" When I did that, I found the deliberations of the experts simple and almost self-evident. Let me summarize without recourse to any of the conventional jargon:

1. Thermal agitation of the molecules of any fluid causes each molecule to move in a random way, that is to say equally in all directions. There is thus a net movement of molecules down gradients of concentration. This is obvious from experience and it is also conceptually obvious. If there are two molecules of water at point A and only one molecule of water at point B, the random motion of individual water molecules will cause more molecules of water to move from A toward B than from B toward A. There will be a net movement of water from A to B.

2. The effective "concentration" of water depends on the actual concentration but also on the presence of other substances that may bind water molecules or otherwise modify their ability to do things. It also depends on any differences in pressure on the water. This general ability to do things is referred to as the potential activity or sometimes simply as the "potential" or the "activity." I do not believe that any other terminology is necessary if we want to discuss the passive movement of water from regions of higher water potential to regions of lower water potential. This subject is practically important but conceptually trivial.

Botanists have tended to treat water relations in terms of the spontaneous water movements that I have been describing, but I doubt that the answers to many of the problems lie there. The really interesting thing about the water relations of living things is that not all is passive and the wedding of active and passive processes presents difficult problems. For instance, roots of many plants seem able to take up water *against* a potential gradient and to extrude almost pure water from cut stems in defiance of the forces for equilibration that we have been discussing. Clearly, such defiance of equilibration requires metabolic energy to do the necessary uphill work. On the other hand, the roots seem quite easily permeated by water (as they must be). Water moves out of them quite freely when the external water potential is lowered sufficiently and they take up water rapidly when the internal water potential is lowered sufficiently. There is a major problem in biological energetics here. Must we postulate the coexistence of a pump and rapid back-leak?

Many years ago Levitt (5) made a most important calculation. He showed that the entire respiration of roots could not produce enough energy to drive such a leaky system if some of the reported root pressures actually existed. Therefore, he concluded, quite logically and quite wrongly, that such great water disequilibria could not exist. The problem with his conclusion is that huge water disequilibria *do* exist in some biological systems and, indeed, ubiquitous disequilibria of all sorts of substances are characteristic of living things. Are we to postulate the maintenance of such disequilibria against back-leaks? For instance, considerable metabolic energy must be expended by the salt-glands of mangroves or the kidneys of desert animals, both of which excrete almost pure salt, just to separate the salts from the water. By Levitt's argument, these processes are made insupportable by the amounts of energy required to maintain the water concentration differences against the leaks implied by measured "permeability" to water. Since the costly processes do occur, I am forced to conclude that the pumps and back-leaks cannot be allowed to coexist. We must look very carefully at measurements of the conductivities of membranes to water and to all other substances. What are we really measuring? A "pump" can run in either direction depending on the demands on it, but in neither direction does it operate in a manner analogous to diffusion. The "pump" can operate as a "gate" or as a conduction channel without providing a pathway for diffusion. These considerations are necessarily vague but nevertheless extremely important.

A strange misconception that I have heard voiced over the years is that "pumps" concentrating solutes of various kinds do exist in plants but not "pumps" concentrating water. However, a massive solute pump is *ipso facto* a water pump, since separating solutes from water is the same thing as separating water from solutes. It all depends on which side of the membrane we place the observer.

I am forced to the somewhat revolutionary conclusion that the "permeabilities" of membranes as measured are totally misinterpreted. Movements in the direction of concentration (i.e. activity) differences must not be automatically equated with passive diffusion. Consider the unidirectional transport of auxins down their concentration gradients, which must owe nothing to ordinary diffusion. Perhaps if we study membrane function more carefully, we will find that all metabolites, even water, can be carried by "gated" processes, although I am not sure what "gated" means in mechanistic terms.

To reiterate: we must beware of arguments based on fancied "permeabilities" when there is as yet no evidence at all that the life processes studied are the typical first-order reactions of passive equilibration by diffusion. Diffusion and life are antithetical, even if life processes must be carried out in the context of diffusion-imposed stresses. Concentrating things against gradients, which is the essence of life, is inevitably energy-consuming anyway, but maintaining a vast multitude of concentration differences in the face of an equal multitude of rapid back-diffusion processes is unthinkable in view of the fact that living things are often fairly efficient engines. The downhill transport of substances we see may be pumps running backward and not diffusive leaks at all. Pumps running backward may be conserving energy, not wasting it. Understanding this aspect of membrane function is probably the most challenging and most important task in all of physiology.

PROTON PUMPS

After my undergraduate days I left the consideration of membrane function for many years. Only when it became apparent that electron transport in chloroplasts was linked to ATP formation through transmembrane phenomena did my attention return to the topic. It had long been known that concentration differences could be created during the hydrolysis of ATP and indeed this was not surprising. The creation of concentration differences requires an input of metabolic energy and ATP was known to be a ubiquitous reservoir of cellular energy. However, it was left to my colleague of Cambridge days, Peter Mitchell, to point out that the reverse process could be used in the generation of

ATP during electron transport. Thus he showed that electron transport in mitochondria creates ion activity differences across membranes, and André Jagendorf showed that electron transport in chloroplast lamellae does the same thing. It was also shown that these ion activity differences can drive ATP synthesis, as Mitchell had postulated. For this insight Peter Mitchell was awarded a well-deserved Nobel Prize.

Unfortunately, the language introduced to describe the phenomena and to explain the mechanisms has been murky at best and misleading at worst. The name of the process, "chemiosmosis," which has grasped the imagination of plant and animal physiologists alike, may have some historical justification in terms of a now long-dead version. It seems harmless. It is nevertheless nonsensical; "osmotic" refers to the effect of solutes on the potential activity of solvents, and no one now suggests that ATP is made by a local decrease in the potential of water.

Yet this is an unimportant matter compared to the very misleading term "proton pump," a term that conjures up a process having no meaning in the context of the chemistry of aqueous systems. Being meaningless, it blocks all rational analysis of events. Moving hydrogen ions from one place to another has no effect on the concentration of hydrogen ions! This fact is counterintuitive and nonchemists have difficulty in grasping it. However, I cannot understand why so many biochemists and biophysicists persist in the use of the misleading language (or at least tolerate it). Somehow they seem convinced that the term "proton pump" is a useful allegory, that we are, in any case, looking at a chicken-egg situation. It is not so. Electrical neutrality must be preserved (except at the surfaces of charged membranes) and therefore the sum of the cationic charges must equal the sum of the anionic charges at any place where we can make measurements. It follows that the concentration of hydrogen ions is prescribed absolutely by the difference between C^+ and A^-, where C^+ represents the sum of the cationic charges other than H^+ and A^- represents the sum of the anionic charges other than OH^-. Movements of H^+ are irrelevant, as we see when we note that concentration of hydrogen ions actually *decreases* toward the negative pole during electrophoresis in spite of the fact that hydrogen ions are moving *into* that region. This apparent conundrum can be resolved when we realize that the concentrations of H^+ and OH^- adjust, using the almost infinite reservoir of these ions in the slightly dissociated water, and that H^+ can be consumed or produced in reactions involving other weak acids. Thus, as Stewart has pointed out (7), the concentration of hydrogen ions is a dependent variable, a result not a cause. To assign phenomena to a "proton pump" is to center attention on a partial reaction, only one part of an equation and a dependent part at that. We must be prepared to write the entire reaction sequence if we want to gain an insight into what is actually occurring.

It is interesting to speculate on the origin of the unfortunate emphasis on "proton pumps." I suspect that a minor contribution may have come from phonetics, the tendency of the English language to alliteration, a holdover from medieval Germanic poetry where rhymes were at the beginning of words rather than at the end. The more important reason is recent history. The energy-conserving ion movements were first detected as pH changes simply because pH meters are available and cheap. Furthermore, very minute discrepancies in the movements of strong ions, cations vs anions, can be detected with a pH meter. For instance, if one has a 10 millimolar concentration of KCl and the concentration of K^+ is precisely equal to the concentration of Cl^-, the pH will be exactly 7.0. If one now removes 0.01% of the Cl^- or introduces 0.01% of additional K^+, there will be a *tenfold decrease* in the concentration of H^+ and the pH will be 8.0. The sensitivity of the measurement, however, does not lend any credibility to the movement of hydrogen ions as the cause of anything. Quite the contrary. I fear that further progress in the understanding of ATP synthesis (or indeed progress in the understanding of any transmembrane phenomenon attributed to "proton pumps") is mired in most unfortunate terminology.

The "electrogenic proton pumps" so dear to the hearts of the chemiosmoticists are not only nonsense, they are nonsense said backward. It would seem that the primary event that leads to the conservation of energy in ATP is a charge separation across a membrane. In photophosphorylation the reaction-center pigment seems to be oriented in such a way that the excitation transfers an electron across a dielectric layer. In oxidative phosphorylation, the same thing seems to be accomplished by placing catalysts of different partial reactions of the overall oxidation of substrates on different sides of a dielectric. Indeed, Mitchell has very wisely pointed out that the situation is analogous to the situation in a fuel cell (6). The charge separation is quickly followed by a migration of cations and anions in appropriate directions under the influence of the electric field, perhaps principally by Mg^{2+} migrations in vivo although ADP, P_i, or even malate, bicarbonate, and carbonate are candidates. In the suspensions of chloroplast lamellae usually used in phosphorylation studies, Mg^{2+} and Cl^- are the major ions migrating in opposite directions (4). Note again that the migrations of H^+ and OH^- are irrelevant. The consequent imbalance of cations and anions, which is far too large to be tolerated, is almost entirely obliterated by changes in the numbers of cationic and anionic charges on the membrane proteins. Thus the anionic charges left behind when Mg^{2+} migrates out of the membrane disappear when carboxylate anions are protonated to become carboxyl groups (anionic $-COO^-$ becomes neutral $-COOH$) or neutral amino groups ($-NH_2$) become cationic ($-NH_3^+$).

The change in pH, which originally alerted us to the phenomenon, is only a minor tertiary effect of a quite different initial cause, a minor process detected only because the concentration of hydrogen ions is such a sensitive indicator of strong ion imbalance. However, it is quite possible that the important end result for photophosphorylation, the "energization" of the membrane that makes ATP formation possible, represents a region of the membrane protonated in the manner described. If so, the protonated region must be oriented with respect to the sides of the membrane because the "sidedness" of the situation is preserved and acids stored on the inside of lamellar vesicles can be used to sustain phosphorylation.

It is time that we stop drawing cartoons with little H^+ arrows traversing membranes and think instead of what the measurements of pH changes really mean in terms of balanced equations. Also, it is time to stop thinking naively about the membrane "permeability" to hydrogen ions. In fact, as pointed out above, measurements of "permeability" to anything must be interpreted with great caution.

The above description of events is, of course, a very incomplete picture of how ATP is made, but it is not as incomplete as the conventional "chemiosmotic" picture. Moreover, the above description has the merit of being real and plausible chemistry without any appeal to magic proton pumps. It is not hypothetical at all. Rather, it is a description of the chemical stoichiometries that almost inevitably follow from the observations.

PHOTOSYNTHESIS AND CROP YIELD

When I moved to Michigan State University in 1962, I had over ten years of somewhat intermittent research in photosynthesis behind me. Therefore, it is not surprising that I was occasionally asked by production-minded agronomists for advice on methods for measuring photosynthesis. My response was to ask questions in return: What do you want to know and why do you want to know it? If their answer involved estimates of crop yield, I made myself unpopular by the only honest response possible, to wit: weigh the plant. As concern for energy sources mounted during the 1970s, this flip but appropriate answer became less and less appreciated, and the pressure on photosynthesis researchers to contribute to the practical problems of society mounted. I was even prevailed upon to coauthor a chapter of a book on the factors determining plant growth and crop yield (2). My conclusions were not altogether popular with my colleagues, who looked on the then current concerns of society as professionally providential.

It is a truism to say that crop yield is the result of photosynthesis and the translocation of the photosynthate into the economically important parts of the plant. It does not follow, however, that either the biochemistry of photosynthesis or the mechanism of translocation is an important limiting factor determining crop yield. The syllogism that, since plants grow entirely by photosynthesis, they must accumulate biomass to the extent that photosynthesis exceeds respiration is true but dreadfully misleading. One trouble lies in the distinction between actual photosynthesis and potential photosynthetic capacity. Moreover, the rate of growth of a plant depends almost equally on the rate of net photosynthesis and on the rate of reinvestment of the photosynthate in new photosynthetic machinery. In fact, the growth of a plant can be approximated by the following equation:

$$P_\tau = P_0\epsilon^{\alpha R\tau}$$

where P_0 is the size at the beginning of the observation, P_τ is the size at time τ, ϵ is 2.71828---, α is the proportion of the photosynthate reinvested in more photosynthetic machinery, R is the rate of photosynthesis per unit of photosynthetic machinery, and τ is the time of measurement. Furthermore, R can be dependent on α because reinvestment of photosynthate provides a sink for the products, and this in itself often increases the proportion of the photosynthetic potential of the machinery actually used. But α is an expression of the *growth* potential of the organism, not an expression of the photosynthesis potential at all. Similarly, the relative growth rates of the various plant parts (and hence the harvest index) are expressions of differences in the growth potentials of these parts and almost certainly do not reflect limitations on the translocation pathway.

For these reasons, it is particularly important that photosynthesis research aimed at increasing crop yield be broadly interpreted in terms of constraints on photosynthesis and growth and not be over-concerned with the process of photosynthesis itself. For instance, it is obvious that the growth of plants, and therefore net photosynthesis, is greatly enhanced by available nitrogen. Yet nitrogen is not directly one of the reactants in photosynthesis. Nitrogen affects photosynthesis through the growth that builds new photosynthetic machinery. Too great a preoccupation with the details of photosynthesis, if it detracts from studies of stress, external deficiencies, and genetic determinants of growth, is a sure recipe for frustrating those geneticists and agronomists who would like guidance from plant physiologists.

This is not to denigrate photosynthesis research or any other search for knowledge for its own sake. We cannot afford to smother fundamental research, research not directed toward the solution of specific practical problems. It must never be forgotten that we run out of practical solutions very quickly when we run out of understanding. No matter how practical a problem, it must be approached from the standpoint of analysis and understanding. Unfortunately, the traditional category of research called "fundamental" is sometimes equated to no application (and by implication, useless) while "applied" is equated to description and testing, with no real need to understand. Either attitude is disastrous if we really are interested in solving problems.

GENES AND ENZYMES

When I went to Caltech in 1948, the dogma of one-gene-one-enzyme was at the height of its popularity, but I would have none of it. The concept seemed simplistic and the evidence minuscule. In hindsight, it would appear that I was both right and wrong; my objections were largely valid, but I underestimated the potential value of any theory, right or wrong, sophisticated or simplistic, if it directs research into useful channels. The alchemists discovered chemistry in their search for the philosopher's stone.

In order to appreciate my objections to the gene-enzyme dogma, one must realize that the gene as we knew it then was not a sequence of bases in DNA coding for a specific protein. Rather it was a lesion, something wrong with a region of a chromosome that interfered with some aspect of biological function or form. In fact, I used to bait my geneticist friends by pointing out that the gene was only a mathematical abstraction born of segregation ratios! Further, I used to say that enzymes, like everything else in the cell, must be susceptible to genetic modification. In that case, where was the evidence for any primacy of a gene-enzyme relationship? If one could modify a specific bristle on an insect by a gene, why should another gene not modify the conformation of an enzyme? The counterargument that, lacking a specific gene-enzyme relationship, it should be possible to modify an enzyme by several different unlinked genes, did not seem compelling. If one could find only one gene to modify a bristle, did that argue for a unique and direct gene-bristle relationship? Besides, the factual base for the gene-enzyme argument was woefully weak.

When I was a student, many genetic changes were attributed to the absence of particular enzymes in metabolic pathways, but as I remember it, an actual absence of the enzymes had only been demonstrated in two cases. In all other cases, the enzyme deficiencies were inferred from nutritional requirements of the mutants; inferences require less work than observations. My lab-mate Francis Haskins (3) worked out an elaborate and substantially correct scheme for the pathway of synthesis of tryptophan and niacin on the basis of a series of *Neurospora* mutants. Almost as an afterthought, and because he was thorough, he decided to do appropriate genetic studies, whereupon he discovered that the various mutants apparently blocked by deficiencies in quite different enzymes were, in fact, all alleles! Probably modern geneticists can interpret this observation to their satisfaction, but it was not comforting then in terms of the original gene-enzyme hypothesis.

The thing I failed to appreciate when I was exposed to the one-gene-one-enzyme dogma was the fact that, if it was true in even one instance, the phenomenon offered a wonderful way of looking at a gene-protein relationship. In other words, it offered an opportunity to follow the transfer of genetic information from DNA (which was even then accepted as the primary genetic material) to a protein, and proteins made up most of the structures and catalyzed most of the functions of cells.

Another factor contributing to my skepticism was the fact that I was then deeply immersed in studying the chemistry and metabolism of small molecules and I underestimated the potential of techniques for working with macromolecules. Consequently, my chance of studying the biology of macromolecules ("molecular biology") was missed.

I regret now that I did not keep up with the discipline of molecular biology as it was unfolding. Modern genetics is an incomparable tool for the study of a host of problems in biology. On the other hand, I worry lest the discipline of molecular biology become unduly parochial, lest molecular biologists approach the problems of plant physiology with the same naiveté that plant physiologists sometimes use in their approach to the problems of plant productivity. Unfortunately, the applications of genetics to plant physiology cannot be less naive than the physiological assumptions on which they are predicated. The nature of the lore of molecular biology as it stands today is also something of an impediment in making physiological or anatomical modifications of plants by the techniques of genetic engineering. It is not yet obvious how DNA-protein manipulations can answer some of the most important questions in biology, especially questions about the organization and regulation in eukaryote cells and in the tissues of multicellular organisms. How close are we to understanding the timing of the expression of genes, when they are expressed at all? Cut off my hand and I am forever handless but it is not so with my close cousin the salamander. Why? What tells a sieve tube cell to be a sieve tube cell? The list of problems of development yet to be solved is endless, yet such questions bear immediately on the genetic control of physiological functions.

In this case I am not pessimistic, however, and I hope that molecular biologist will continue to attack problems of regulation and differentiation. Already some of the changes in cells associated with specialized forms and functions have been correlated to DNA changes, sometimes reversible and sometimes irreversible. If differentiation and the multiplicity of functions of cell types is ever to be explained, it will be molecular biologists who do the explaining.

WHITHER BIOLOGY?

The impact of modern genetics on our perceptions of taxonomy and evolution is just beginning to be felt. Again I am at a loss for hard facts and again I regret my inattention to the science growing up around studies of nucleic acids and proteins. The little information I have acquired by diffusion has whetted my appetite for more, and I am excited by implications I can barely assess. Perhaps I will study these things in my next career.

Originally we knew genes primarily as malfunctions in development, by the consequent absence of some discernible structure or metabolic reaction. As more and more mutations blocking metabolic pathways were studied, a higher and higher proportion of mutant genes could be described in terms of inactive or less active enzymes. This state of affairs was desirable from the standpoint of the then current primitive analysis of gene action, but it posed something of a problem for those primarily interested in evolution and in the genetic variability that must provide the basis for natural selection. It was difficult to picture evolution in terms of losses of function when evolution is so obviously and so necessarily associated with changes improving functions. Of course, part of the dilemma lay in the nature of the mutations we felt obliged to study, the all-or-none effects of genetic disasters. It is not difficult to imagine small changes that would occasionally make small incremental improvements in gene function, but such small changes were too difficult to study when we were still incapable of refined genetic analysis.

Mutations, great or small, are presumably random in nature. However, their consequences in terms of survival of the organism must nearly always be bad, if only because the preexisting gene products had already been designed for particular functions by ages of natural selection. In other words, random changes, when the number of such possible changes is almost limitless and the changes are in every conceivable direction, should almost always be disadvantageous over a wide range of conditions. One would expect that improvement by such a lottery would be slow indeed, so slow that we might not entertain the concept of evolution by mutagenesis at all if we were not impressed with the idea that evolutionary time is very long and that populations subject to natural selection are very large. But are the pertinent populations really large and do selection acts occur with great frequency? The answer is clearly yes when we consider many unicellular organisms, but it is certainly no when we consider many long-lived multicellular organisms.

Mutagenesis and selection must play by some remarkable rules in such multicellular organisms; consider as an example the elephant. Once the great majority of the cells has given up the ability to form gametes, mutations in them became irrelevant to evolution; the changes cannot be transmitted to progeny. If a mutation occurs that makes for better bones or brawn, it can only be selected when the individual elephant possessing it survives better. Thus selection in the germ line usually involves a characteristic not expressed in the germ line! Mutations that can be selected must occur very early in the development of the individual, before the cells are committed to their various specialized roles as sterile workers. Presumably the majority of such mutations in animals takes place in the germ lines of preceding generations. Furthermore, the world population of elephants has probably always been small and the same can be said for many organisms. The numbers are particularly unpromising when we consider that selection of improvements must be delayed for many years as the elephant matures; thus there can only be one selection act in 20 or 30 years, and the selection involves only one cell per generation, the fertilized ovum. If we accept the conventional picture of evolution by mutagenesis followed by natural selection, the problems are formidable. The entire world population of elephants would seem to be evolving with one cell selection each 20 years from a few thousand cells!

The numbers problem with higher plants is much less formidable. The selection unit is less well defined and the "germ line" is neither as clearly defined nor as protected from the environment as in animals. Indeed, each growing point of a plant can be looked upon as an "individual" producing either gametes or specialized cells for its own betterment. Moreover, the specialized cells behind the growing point (and genetically related to the growing point) almost immediately begin to contribute to the well-being of the growing point and thereby influence the production of and survival of gametes. Therefore, many mutations, though they affect only the specialized functions of nonreproducing cells, can nevertheless be promptly selected and transmitted to future generations through the better production of gametes and the better development of the propagules. Of course, some of the genetic changes in the "germ line" (in the special reproductive meristems) affect whole plant func-tions, and in these instances selection is again delayed for a whole plant generation. The evolutionary problems of elephants and oak trees are alike with regard to such changes.

Modern studies of molecular biology may provide a hint for a partial resolution of this numbers problem. It is now apparent that genetic information can sometimes be transferred among very different organisms by processes that have nothing to do with meiosis or the regular transfer of genes to progeny. Thus, in principle at least, all genes in all living things may be accessible to all other living things. Such lateral transfers of parts of genes, genes, or clusters of genes would tremendously expand the pool of diversity available to each organism. The evolution of elephants might be accelerated remarkably if preselection of some of the genes for elephants could be accomplished in rabbits! There are so many implications of lateral gene transfer that I cannot resist devoting a few paragraphs to a layman's speculations.

First, let us consider the frequency with which lateral gene transfer occurs. The transfer of genes among some prokaryotes is commonplace on the scale of evolutionary change, a regular laboratory exercise. The transfer of genes from prokaryotes to eukaryotes is less routine, but it is nevertheless well known. The best documented case is the crown gall disease of dicotyledonous plants, where bacteria transform the normal organized tissues of plants into tumors by transfer of genes from the bacterium to the host, genes that provide the plant with an unregulated supply of its own hormones. Other genes can be transferred along with the tumor-producing genes, and there is no reason why these other genes cannot also be transferred without the genes for tumor formation. The latter cryptic transformations could be happening with appreciable frequency all over the plant and animal kingdoms without our knowledge. None of the known mechanisms for lateral gene transfer can yet be assessed as potential factors in evolution. And who knows how many unknown mechanisms exist? The transfer of genetic information from eukaryotes to prokaryotes is less well documented, but it can be inferred in one or two instances (1).

However, I do not believe that the way to estimate the evolutionary importance of lateral gene transfer is by estimating the frequency of such transfers. Rather, we must ask ourselves how frequently such events are used by evolution when they have occurred and how effective they are if used. The importance of lateral gene transfer will only be assessed when the appropriate taxonomy of genes has been developed. Such studies are already in progress, but they have not been carried far enough to give us any general answers. For instance, the plant-derived globin in the leghemoglobin of nitrogen-fixing nodules (and the intron-interrupted DNA coding for the globin) are almost identical to the DNA and globin responsible for our hemoglobin. There are several ways in which this apparent homology could be explained, but in view of the known processes of lateral gene transfer, it seems to me that the most satisfactory explanation involves acceptance of the apparent homology as real, as a manifestation of the actual relationship of our oxygen-carrying system to the system in beans.

How is chaos to be avoided if genes migrate willy-nilly among organisms? Obviously chaos does occur occasionally; witness crown gall, some other cancers, and many other types of disorder. But disorders can be selected against and "foreign" intruders excluded by mechanisms known and unknown. I am reminded of the analogy of the introduction of foreign genes and foreign species into a plant community. No matter how abundant the propagules from species foreign to a forest community, the foreign genes and foreign species almost never gain a foothold unless there is a major disturbance in the community. I suggest that genes in a genome may share some of the characteristics of species in a biological community. The fact that ecologists cannot fully explain the stability of communities does not alter the fact that the stability is very real and very great. Perhaps we should spend more time training gene ecologists to study the ecology of the genome. If this concept of the genome as a relatively stable unit has validity, we may have to be less sanguine about attempts at genetic engineering.

It is even possible that the picture of lateral gene transfer and the concept of species as a community of genes may provide us with a clue to the reason for very different rates of evolution in different organisms and an explanation of "living fossils." Suppose a species develops some unusual activity, some critical metabolic function conducted in an unusual way. Then virtually any transfer of genetic material from another species might be disadvantageous or even lethal. I have no particular function in mind, but for the sake of argument let us consider a ridiculous extreme. Suppose that some species had discovered how to make and use D-amino acids instead of L-amino acids. Such an organism would not be able to use any genetic information not originating in itself and its evolution would, of necessity, be restricted to its private game of 20-sided dice.

The importance of horizontal gene transfers in evolution cannot long remain uncertain as more and more is learned of the relationships of genes among different organisms. It may be that all controversy will be taken out of the question before this article is published. Meanwhile, it seems to me that the topic is one of the most intriguing in biology. It is a topic that should bring together the interests of taxonomists, ecologists, and geneticists in a manner long overdue.

EPILOGUE

The period of my scientific training was so critical to the development of modern biology that it has already been described by many authors. Since I have no hope of rivaling Watson's *Double Helix,* I am tempted to refrain entirely from discussions of well-known personalities or of well-known events. On the other hand, such a course implies a graceless lack of appreciation of many colleagues, when there is no lack of appreciation in my heart. Forgive me one and all. At least you will be spared the odium of association with the more unorthodox of these discussions. However, I do want to acknowledge my appreciation of a thoroughly satisfactory family, biological and academic. It is a source of pride to me that two of my sons are already launched on successful scientific careers of their own and that several of my graduate students have achieved distinction. Strangely, I am also proud of the fact that I am known to my students and family alike as "The Old Goat." I am proud of the appellation because I feel that dignity implies a lack of self-confidence and reverence implies an element of fear. I hope that neither insecurity on my part nor fear on their part played any part in our relationships.

I also hope that I do not emerge from this account of my ideas as being only negative-minded, glorying in iconoclasm and the discomfiture of the doctrinaire. I am aware of the dangers of negativism for its own sake or even of justified skepticism if it stifles imagination; witness my objection to the one-gene-one-enzyme theory. Nevertheless, I do not apologize for negativism. I think that there is merit in some of the criticisms of dogma that I have presented here. Perhaps different ways of looking at things may get some mental flywheels off dead center, even if some of the different ways prove absurd. Moreover, no theory can be considered safe to use until it has been thoroughly tested by the attentions of professional doubters.

Many of the developments that have occurred during my career have pleased me and many have distressed me. Perhaps the most pleasing has been the changing role of women in science. When I was at Caltech there were no girls at all among the students, undergraduate or graduate. There were a very few female postdoctoral associates, but female students were specifically barred. An illustration of the change is ever before me. I have on my desk two photographs of international gatherings of photosynthesis researchers. The first was taken at Gatlinburg, Tennessee, in 1952. In that august assembly of 75 scientists there were only two women. The second was taken at Ventura, California, in 1983. In the latter assembly there were 23 women out of a total of about 140, quite an improvement in 31 years but not enough improvement; the disparity is still too large. However, the greatest cause for rejoicing is the change in attitude among the women themselves. In the early 1950s few of the male scientists I knew had any reservation about women in science, but many of the women seemed insecure. Now the level of confidence of women in science is justifiably high. It is good to have the missing half of the population accepting themselves in science at last. Welcome!

On the other hand, I worry about quite a number of educational problems. The university and high school educational systems are too chaotic and un- structured. Part of the trouble arises from a doctrinaire belief that early educa- tion should remain a regional or even a community affair, as though reading, writing, arithmetic, and quantum mechanics varied from state to state. Also, in the name of a "liberal" education quite nonsensical things are often taught, things that may be harmless in themselves but must not serve as alternatives to rigorous training. Languages, including the English language, are inadequately taught, if they are taught at all. Mathematics, the ultimate abstraction, is too often taught as an abstraction only, with the result that students rarely learn the mathematics of processes. For instance, few of my students realize that linear (arithmetic) processes do not occur since they imply reactions that do not change the conditions of the reaction, a contradiction. How many students appreciate that an exponential (logarithmic) plot implies that the process is proportional to the amount of something remaining?

The competitive grant system has been, on balance, a wonderful introduc- tion. One need only compare institutionally supported science with grant supported science. The former is often riddled with local politics. Un- fortunately, as the competitive grant system matures it is also subject to political pressures, the politics of sciencemanship. Scientists come to be evaluated by the amount of money they spend or by the number of papers they publish. Versatility on the part of scientists also suffers if money is too freely available.

It is more "economical" to buy things than to make them, which is all very well if one plods well-trodden paths, but where then are the new techniques? The economy argument can even become ridiculous: one of my associates took me to task for repairing a piece of glass equipment on the ground that my time was too valuable. While I enjoyed the mistaken estimate of my general usefulness, I found the argument unconvincing. I made the repair in about one-tenth of the time it would have taken to visit the glassblower. On the other hand, modern electronic equipment, sensors, separation devices, etc, are so accurate but so specialized that purchase of costly equipment is almost necessary. However, I hope that equipment will gradually become more and more modular, so that research versatility and inventiveness will not vanish from the face of the earth. Training students in the principles and practices of instrumentation must not be neglected simply because instrument companies are also in the business. I do not want instrument companies directing my research.

I cannot bear to close this account without recording an anecdote that should be part of the lore of plant physiology. This tale, which I know to be true, has to do with the origin of the term "phytotron," a term that is widely accepted and used although I suspect that it is bad Greek. Shortly before the newly built controlled climate greenhouse at Caltech was formally opened in 1949, James Bonner and Sam Wildman repaired to the local Greasy Spoon for morning coffee. Quoth James (whose aptness of phraseology is legendary): "They should call the place a phytotron, where they bombard spinach with high speed carrots." I am glad to have the opportunity to set the record straight on this important matter, which has been much and erroneously discussed. I fear that the facts may have been suppressed by influential biologists to whom science is a sacred cow and not a joyful lark.

Literature Cited

1. Bannister, J. V., Parker, M. W. 1975. The presence of ccpper/zinc dismutase in the bacterium *Photobacterium leiognathi*. A likely case of gene transfer from eukaryotes to prokaryotes. *Proc. Natl. Acad. Sci. USA* 82:149–52

2. Good, N. E., Bell, D. H. 1980. Photosynthesis, plant productivity and crop yield. In *The Biology of Crop Productivity*, ed. P. S. Carlson, pp. 3–51. New York: Academic

3. Haskins, F. A., Mitchell, H. K. 1952. An example of the influence of modifying genes in *Neurospora*. *Am. Nat.* 86:231–37

4. Hind, G., Nakatani, H. Y., Izawa, S. 1974. Light-dependent redistribution of ions in suspensions of chloroplast thylakoid membranes. *Proc. Natl. Acad. Sci. USA* 71:1484–88

5. Levitt, J. 1947. The thermodynamics of active (non-osmotic) water absorption. *Plant Physiol.* 22:514–25

6. Mitchell, P. 1968. *Chemiosmotic Coupling and Energy Transduction.* Printed for Glynn Research Ltd. by Hall Graphics, Plymouth, England

7. Stewart, P. A. 1983. Modern quantitative acid-base chemistry. *Can. J. Physiol. Pharmacol.* 61:1444–64

Ann. Rev. Plant Physiol. 1986. 37:23–47

CELLULAR POLARITY

E. Schnepf

Zellenlehre, Fakultät für Biologie, Universität Heidelberg, D-6900 Heidelberg, Federal Republic of Germany

CONTENTS

INTRODUCTION

"Little that cells do is as intriguing or baffling as their ability to locate within themselves most cellular components with exquisite precision. Cells do this both during maintenance and growth, locating replacements with nicety, and during development, producing a closely choreographed rearrangement of many or all the cellular components. The control of this choreography is virtually a mystery" (33).

Many environmental factors influence cells and organisms polarly, such as light, gravity, and adjacent cells. Correspondingly, most cells and organisms have an intrinsic polar organization. Time also has a direction; therefore developmental processes often are vectorial and result in polar growth. Even various cell components have a polar structure (39): microtubules, flagella, actin, and cellulose fibrils, not to mention the macromolecules. This review is

23

0066-4294/86/0601-0023$02.00

focused on plant cell polarity. It presents some of the new observations made since the last article (1978) of this series (107) (for further reviews on special aspects of cell polarity see 9, 10, 137, 142, 151, 175, 178; and of multicellular systems see 34, 36, 39).

What can be visualized or measured in polarly organized cells are phenomena, i.e. structural or functional manifestations of the polarity. Cell polarity in a narrow sense means heteropolarity: the cell has an axis with two different poles. This axis need not necessarily be identical with the longitudinal axis of the cell. In some unicellular algae the polar axis (Golgi apparatus–nucleus–pyrenoids) is perpendicular to the equipolar longitudinal axis (46). Generally a polar axis is firmly fixed. It cannot be reversed easily, so the few exceptions are of special interest. It reorients gradually, however, as a response to environmental factors, e.g. during growth of filamentous cells. The establishment of a new polar axis is nevertheless a widespread phenomenon and occurs when a filament branches.

POLAR GROWTH OF CELLS

The most obvious example of cell heteropolarity is polar growth, tip growth (151), though there are cells such as *Micrasterias* that grow multipolarly (74, 75).

Polar Cell Wall Formation

Tip growth implies local formation of cellulose or other fibrillar wall constituents and local exocytosis of secretory Golgi vesicles by which pectins and other matrix substances are incorporated into the wall. It depends on appropriate stress and strains. If the equilibrium between the deposition of wall material and the expansion of the cell is disturbed, thickened walls are formed in the cell tip or conversely the cell bursts apically. Deviations in the precise location of these processes cause irregularities in cell diameter and shape.

When cellulose is synthesized polarly, the synthase complexes within the plasma membrane can be expected to be polarly distributed. There is an increasing amount of evidence that these complexes are partly represented by "rosettes," consisting of six particles. They can be observed in the protoplasmic fracture face (PF) of the plasma membrane in freeze-fractured cells, provided that chemical prefixations and other pretreatments have been avoided. The rosettes are concentrated in the very tip of *Adiantum* protonema filaments, i.e. in the growing region (172). They are nearly lacking in more basal parts of the cell.

A similar gradient in distribution was found in *Funaria* caulonema cells (123). The number of rosettes corresponds with growth rate. That was demonstrated by comparing quickly elongating caulonema and slowly growing chloronema cells and by experiments in which growth was inhibited by cultivation in darkness (140). Colchicine induces subapical bulging instead of apical elongation. It likewise causes a rather irregular distribution of rosettes, often with a subapical maximum of frequency. Mild plasmolysis is restricted to the cell apex. It stops growth; the rosettes disappear within a few minutes (140). The rather strict correlation between growth rate and rosette frequency allowed some calculations on the kinetics of glucan synthesis (about 1800 glycosidic bonds per glucan chain per minute) and of cellulose elementary fibril production in the tip region. The calculations suggest that the rosettes are short-lived, presumably producing only one cellulose elementary fibril each (123, 140).

It is probable that the rosettes are incorporated polarly into the plasma membrane by exocytotic Golgi vesicles, as was shown in *Micrasterias* (75). During cellulose synthesis the rosettes move laterally within the plasma membrane. This migration does not seem to be directed (123) by, for example, peripheral microtubules, which are largely lacking in the very tip (137); the microfibrils are deposited in a dispersed fashion.

Pectins and similar wall constituents are synthesized in dictyosomes. Polymerization and methylation of the polysaccharides continue within the Golgi vesicles (151). The polar exocytosis of these substances implies a polarly oriented membrane flow (72), the more so because dictyosomes are not found in the very tip of fast-growing cells. In this region Golgi vesicles accumulate to form a "clear cap" or "tip body" before they extrude their contents. The transport of Golgi vesicles in plant cells is generally believed to be driven by actin. The microtubule system does not seem to be involved directly, for colchicine does not inhibit tip growth in pollen tubes (53) or in certain fungal hyphae (138) (for exceptions, see below). In contrast, cytochalasin stops it in pollen tubes (53) and in developing *Fucus* zygotes (12) but only reduces growth rate in moss caulonema cells (135). Other transport mechanisms such as electrophoresis may also play a role (107). The vesicles fuse with the plasma membrane predominantly, but not exclusively in the apical dome (136). It is, therefore, questionable whether recognition processes between vesicle membrane and plasma membrane control local growth.

The polar exocytosis of wall material incorporates the membranes of the Golgi vesicles into the plasma membrane; it supplies the latter with the membrane material needed for cell expansion but also with specific constituents like cellulose synthase complexes (see above) and ion pumps and channels (12). Generally the membrane area incorporated by exocytosis is considerably higher than the area required for extension (102, 151). In consequence, membrane material must be retrieved, either by endocytotic processes or by the recycling of membrane constituents in molecular or micellar form, unchanged or after partial degradation.

There are no obvious structures that might be involved in the endocytotic internalization of membranes, with the exception of coated pits in the plasma membrane. They might develop into coated vesicles and, perhaps after having lost the clathrin coat, re-fuse with elements of the endomembrane system, e.g. with Golgi cisternae or vesicles. Indeed, coated pits are found predominantly in the apical region but only in low number in basal areas of the plasmalemma of *Funaria* caulonema tip cells (136). It is, however, questionable whether the amount of membrane recycling via coated pits is sufficient to retrieve the excess of membranes. The number of coated pits and coated vesicles is low in pollen tubes (102, 151). It must be mentioned here that in animal cells the number of coated pits is rather constant whereas the number of coated vesicles varies with endocytotic activity (84). Presumably, the coated pits function predominantly in receptor-mediated uptake of substances.

In tip-growing cells, elongation rate and the maintenance of wall thickness and of cell diameter are strictly correlated. It is largely unknown how the constancy of the cell diameter is controlled. In some cases, e.g. in *Vaucheria* (70) and in *Onoclea* gametophytes (20), the diameter of the apex is influenced by light.

Growing tips have a tendency to burst. It was therefore concluded that local cell wall synthesis and wall lysis are delicately balanced by simultaneous extrusion of wall precursors, synthases and lytic enzymes (5). Alternatively, it was suggested that the wall is initially plastic, i.e. in the apical dome, but more rigid (because of gradual cross-linking of the microfibrils) in the lateral parts of the wall (180). Moreover, the wall is strengthened here by additionally deposited microfibrils.

Picton & Steer (101, 103) suggest that a microfilamentous network within the cytoplasm also stabilizes the tip region of pollen tubes against internal osmotic pressure. This idea is circumstantially supported for some cell types by experimental results. In growing fungal hyphae the cell diameter increases when cytochalasin is applied (1, 25). In other cell types, microtubules seem to control cellulose microfibril deposition in the tip region and the cell diameter during elongation (71, 81). Disturbances in the equilibrium of calcium ions also cause changes in cell diameter, often in connection with other growth irregularities (115, 125, 135). Ca^{2+} controls the cytoskeletal system and oriented exocytosis (118). However, polarly growing cells do not react uniformly upon substances that might act upon the cytoskeleton.

Regular tip growth is related to the mechanical characteristics of the elongating cell. These in turn depend on the geometry of the cell and determine the tensional stresses in meridional and latitudinal direction (47, 151). The coincidence of maximal extension rates and minimal stress at the apex of a growing *Chara* rhizoid indicates that there is a feedback between the extension pattern and the turgor (47). During growth, the cell must be informed on shape and dimensions. It should be able not only to have positional information, i.e. to measure distances and directions, but also to measure turgor pressure and stresses (151).

When an elongating cell meets a firm obstacle, the direction of growth changes. It can be assumed that the existence of the barrier is recognized by variations in the pattern of pressure and tension in the apex. The plasma membrane functions as a sensor (176). The high sensitivity of cellulose fibril formation against turgor reduction, indicated by the sensitivity of the particle rosettes in the plasma membrane, may be one of the keys to understanding these processes.

Little is known about how a cell terminates elongation. Pollen tubes or fungal hyphae are able to grow nearly infinitely. They stop growing when they have reached their destination. Other cell types, such as root hairs, have a limited growth. This problem raises the question of how a cell measures distances, volumes, or developmental time (depending on physical time and on synthetic rates). In multicellular filaments, the cell length seems to be regulated by rhythmical divisions of the nucleus, which maintains a distinct distance from the tip. When, however, in *Funaria* caulonemata the diameter of the cell is increased by the ionophore A 23187, the cells become shorter while the volume remains constant (135).

A special case of polar growth is spore or pollen germination and the formation of a side branch in uniseriate filaments. In contrast to the growing tip, a preexisting rigid cell wall is present here. When the new tip develops, generally a new wall layer is formed. It expands while the old cellulosic wall is not thinned out but ruptured mechanically by the internal pressure of the growing new tip. This is demonstrated by the structure of the rim of the pierced old wall (6, 133, 139). Lytic enzymes facilitate wall penetration (31, 77, 156). They are locally released by exocytosis (96) and that, again, raises the problem of vectorial transport and locally restricted membrane fusion.

Orientation of Polar Growth by Environmental Factors

Environmental stimuli (light, gravity, chemical substances, etc) can orient the direction of polar growth. The cell responds either by bowing, i.e. by differential growth of opposite cell wall flanks in the tip zone while the growth center maintains its apical position (148, 151), or by bulging, i.e. by the displacement of the growth center from the apex to a subapical site (28, 151). Bending results in a curved cell, bulging in a sharp angle.

Chara rhizoids respond by bowing when the direction of gravity is changed (153). The very tip contains a group of statoliths, vesicles containing crystallites of $BaSO_4$. They have a slightly subapical position (147). When the cell is treated with cytochalasin B, they sediment onto the plasma membrane in the apical dome (48). Tip growth then stops. When a rhizoid is treated for 1 min with cytochalasin B, the statolith vesicles return to their normal position within about 30 min and the cell then continues to elongate. Obviously, actin filaments participate in locating the statoliths. If the rhizoids are turned from a vertical into a horizontal position, the statoliths sediment onto the physically lower flank. After about 10 min, the cell begins to bow as the upper flank expands faster than the lower one. After 2–3 h, the graviresponsal curvature is completed; the statoliths are then in their normal position (152).

Graviresponse of polar growth in *Chara* rhizoids follows a simple feedback principle (151): As a consequence of the asymmetric distribution of the statoliths, the exocytosis of Golgi vesicles is inhibited at the lower flank. Most Golgi vesicles incorporate their contents as well as their membrane into the wall and the plasma membrane, respectively, at the upper flank. Therefore this flank grows more than the lower one, the cell bows until the statoliths are equally distributed again. The delicate correlation between cellulose fibril formation, pectin incorporation, and wall expansion is remarkable.

Gravitropism is also observed, e.g. in moss protonemata (133). It is possible but not proved that amyloplasts in the tip region (135) function as statoliths here.

In filamentous green algae, protonemata of mosses and ferns, and also in hyphae of various fungi, polar growth is oriented by light. The tip responds phototropically and polarotropically, i.e. it perceives the vibration plane of the light. Phytochrome and/or blue light receptors (42, 66) and occasionally also chlorophyll (70) are the photoreceptors.

In *Adiantum* protonema cells, the phototropic response is determined by phytochrome. The Pr and Pfr forms are in different dichroic orientations at the cell flank (69). The polarotropic orientation of growth in primary protonemata of *Physcomitrella* depends on the wavelength and the photon flux rate of monochromatic light. The cells elongate parallel to the electrical vector of plane-polarized light in blue light, and at higher flux rates of red light. They are positively phototropic in far-red light and in red light of low flux rates, but grow perpendicular to the incident light in high flux rates of red light (66).

Early phototropic responses include rapid changes in the membrane potential (108). In *Onoclea* gametophytes a blue light photoreceptor seems to act on cell expansion via an auxin–hydrogen ion system but phytochrome via an ethylene–calcium ion system (20).

In *Ceratodon* protonemata, light is perceived by the phytochrome system. It is transduced into changes of bioelectric potentials, of phospholipid metabolism and of ion translocation; calcium ions accumulate at the illuminated flank. The growth responses become manifest in an apical swelling and subapical bulging, culminating in phototropically oriented growth (42). Ionophores like monensin and nigericin inhibit growth. In the presence of these drugs, unilaterally illuminated cells to not reorient their growth. Nevertheless, the stimulus is perceived. When the inhibitors are removed, the cells react in darkness on the previous irradiation with phototropic bending. This memory effect can last several hours. Phototropic mutants grow normally and show the usual cell organization but cannot perceive the direction of the light (43).

A similar memory effect has been observed in rhizoids of *Boergesenia*. When the bending after irradiation is inhibited by low temperature, rewarming in darkness induces the phototropical response (60).

Chemotropism is common in polarly growing cells. Pollen tubes reach the egg cell, sexually determined hyphae of fungi reach the corresponding organ of the sexual partner, and parasitic fungi invade a leaf through stomata, guided by generally unknown substances (151). Little is known about the receptors of the stimuli, their position (the plasma membrane?), or the transduction of the stimuli.

Cytoplasmic Polarity in Cells with Tip Growth

DISTRIBUTION AND DIFFERENTIATION OF CELL ORGANELLES   Interrelated with tip growth, cell organelles are polarly distributed and in some cases even have different structure in different areas of the cell. The cells of *Acetabularia* (142, 161) and of the moss caulonema (68, 135, 137) have been studied extensively in this respect. A few of those organelles or cell constituents that are directly involved in cell wall formation were mentioned in the previous sections.

The increasing stability of the cell wall toward the cell base is a consequence of additionally deposited cellulose (26, 162). Callose lines the wall of pollen tubes, except at the growing tip (101). Presumably it makes the wall impermeable so that the entry of ions, especially calcium ions, is thought to be restricted to the cell apex. As a result, exocytosis occurs preferentially in the apical dome (101). This idea cannot, however, be applied to other tip-growing cells in which callose is deposited only when growth is disturbed.

From attempts to reverse the polar axis, e.g. by centrifugation, it was concluded that the plasma membrane plays a very prominent role in polarity. Polar differentiation of the plasma membrane is manifest not only in the distribution of particle rosettes but also in the distribution of other intramembranous particles, the function of which is unknown (166, 172).

A polarization of the plasma membrane is not restricted to growing cells. In animal gland cells, exocytosis, receptors of hormones and secretagogues, and budding of viruses are selectively located either in the apical or in the basolateral area (83). In these cells the desmosomes restrict the lateral diffusion of membrane components. In tip-growing plant cells, other mechanisms must be effective.

The perhaps most striking polar structure in cells with tip growth is the clear cap or tip body (151). It consists mainly of an apical accumulation of exocytotic

Golgi vesicles, together with endoplasmic reticulum (ER) elements. The tip body is big in cells with a high growth rate, e.g. in pollen tubes (115, 116), root hairs (146), moss caulonema cells (135), and hyphae of various fungi (138), but it is small or even lacking in slowly elongating cells, e.g. in moss chloronema cells (139). It disappears when elongation is experimentally affected (57, 115, 135) or ceases during development (97). Fungal hyphae often contain within the apical accumulation of exocytotic vesicles a "Spitzenkörper." It consists of a well-delineated aggregate of vesicles that are presumed to originate from the ER (98).

The tip body is a zone where most other cell organelles are excluded, especially the bigger ones. They migrate into the apical dome when the tip body has disappeared. In cells with cytoplasmic streaming (such as pollen tubes, root hairs, and *Chara* rhizoids) the cyclosis does not include the clear cap. In the tip body of *Chara* rhizoids individual exocytotic vesicles have been observed by dark-field microscopy of living cells. They show a saltatory movement with a net flux toward the apex. At the wall they sometimes rebound until they disappear (151).

Dictyosomes, ER, mitochondria, and sometimes lipid bodies each represent the subapical zone, whereas plastids, if present, follow more basally (67, 68, 115, 135). In *Funaria* caulonema tip cells the dictyosomes have more cisternae, up to nine, in the apical region where they are especially active, but less, only four to six, in the basal region of the cell. The mitochondria are longer in the proximal region than they are in the distal region (135). Analyses with fluorochromes reveal a gradient in membrane distribution, with the maximum in the tip region in pollen tubes (105, 120) and in *Funaria* caulonema tip cells (132).

The proximal part of a tip-growing cell is usually filled with a single big vacuole. It enlarges during growth. Smaller vacuoles extend into the apical region. They increase in size proximally and seem to fuse with the basal one (67, 68, 135).

Generally, the nucleus migrates forward during growth to maintain a defined distance from the tip, not only in cells that divide like protonema cells (67, 135) or fungal hyphae but also in the unicellular *Chara* rhizoids (151). Exceptions to this rule include root hairs and *Acetabularia,* in which the nucleus may have or has a basal position, respectively. The distance between the nucleus and the growing apex increases gradually in *Physcomitrium* protonemata (67) and in the infection hyphae of the phycomycete *Lagenisma* (138); it also increases immediately before cell division in *Adiantum* protonemata (171).

The chloroplasts may be not only unevenly distributed but also polarly differentiated. In moss caulonema cells they are spherical or constricted in the distal but elongate in the proximal region (67, 135). They divide preferentially in the apical areas. During the cell cycle their distribution changes somewhat in

Physcomitrium. In the *Funaria* caulonema tip cells, the starch content of the chloroplasts increases from base to tip (135). That might be related to the sugar concentration in the cytosol. The tip region can be supposed to contain relatively high concentrations of sugar. It is the center of polysaccharide synthesis at the plasma membrane and at dictyosomes and Golgi vesicles, which in both cases is connected with a transmembrane transport of the sugars. A group of chloroplasts with especially large starch grains may also function as statoliths. When they are dislocated or when growth is slowed down they lose most of the starch (135).

In *Acetabularia* (55, 142, 161), the structure, activity, and DNA content of the chloroplasts depend on their location. In the apical part of the stalk they are small, elongate, divide rapidly, and have well-developed grana but almost no starch. In the middle part they are less elongate, larger, and contain starch. In the rhizoid near the nucleus, they resemble those at the tip. In other parts of the rhizoid they are spherical and have a high starch content but a reduced thylakoid system. Chloroplasts devoid of DNA are by far more frequent in the basal than in the apical parts (85).

Microfilaments seem to exist all over the cytoplasm in most tip-growing cells, as indicated by cytoplasmic streaming but also evidenced directly (143, 145, 157). Presumably they are concentrated in the very tip (12). In root hairs, they occur in bundles as well as in individual microfilaments (143) and are, in basal regions, largely oriented in the direction of the cytoplasmic streaming.

Microtubules have mainly a peripheral position in root hairs (144, 158). They are also found in the endoplasm in protonema cells (106, 135, 139) or fungal hyphae (45, 89, 138), often in close association with the nucleus. They also occur in the very tip (56, 58), though presumably not in every cell type. They seem to be especially labile during fixation here. In the just developing germ tube of *Funaria* they are initially lacking (139). The just initiated side branch and the apical dome of a caulonema tip cell contain only very few microtubules (133, 135, 137).

PHYSIOLOGICAL GRADIENTS   Corresponding to the gradients in growth and organelle distribution, there exist cytochemical gradients and gradients in pH and in affinity for dyes (42, 159, 160). In the giant *Acetabularia* cell, polysomes and protein synthesis are unevenly distributed, reaching a maximum around the nucleus and at the tip of the growing stalk (87, 142, 161). The ribosomes migrate from the base to the apex at a rate of about 2–4 mm per day. Relatively high concentrations of poly(A) RNA in the tip region are reported in pollen tubes as well (110).

Analyses of pollen tubes by proton-induced X-ray emission showed a rather equal distribution of sulfur, potassium, chlorine, and iron (113) but a distinct tip-to-base gradient for zinc (27, 113), phosphorus (113, 120), and especially calcium (113, 120, 121). This method detects all atoms of the respective elements, be they free, bound, or stored. The distribution especially of the free ions was, however, artificially changed in these analyses because the specimens had to be dried and in some cases were fixed chemically.

A calcium gradient was visualized also by chlorotetracycline in pollen tubes and other cells with tip growth (42, 105, 114, 116, 120). Chlorotetracycline fluoresces intensely when it is complexed with calcium and bound at a membrane. ^{45}Ca autoradiography likewise demonstrated that the calcium concentration in pollen tubes is highest in the tip (61). It should be mentioned here that calmodulin-binding phenothiazines reveal a gradient in calmodulin (44).

Ionic currents generally enter tip-growing cells at the apex and leave them laterally, behind the tip. Electrical polarity has been demonstrated repeatedly, e.g. in pollen tubes (179), protonemata (18, 109), fungal hyphae (3, 54, 78, 79), germinating fucoid zygotes (12, 100), and various other algae (11, 168, 169) (for survey, see 175, 178).

The current is generated by a passive flow of mainly K^+ and Ca^{2+} into the tip and of protons, which are actively pumped out basally. Though calcium ions comprise only a small portion of the transcellular current (78, 100, 177) they seem to be especially important for cell polarity. Their relative immobility enables them to function in control (126), perhaps via fixed electrical charges (178).

Usually, currents and growth are closely related. Darkened *Acetabularia* cells stop growing, lose their electrical polarity, and cytoplasmic streaming ceases after 2–3 weeks (11). When they are reilluminated, electrical polarity and cytoplasmic streaming recover in a few minutes, somewhat irregularly at first. Recovery is firmly established after 2–4 h. Growth continues after about 30 h.

Inwardly moving currents predict the site of branch emergence in *Achlya* (79) and *Vaucheria* (178) and precede germination in fucoid zygotes (105).

The ionic currents and their direction are supposed to be driven by polarly distributed ion pumps and channels in the plasma membrane (12, 177, 178). Proton pumps may likewise be involved in the control of polar growth (41, 78). ATP-driven proton pumping is one of the initial events of growth response to blue light in the tips of the filamentous *Onoclea* gametophytes (21).

POLARITY IN NONELONGATING CELLS

Only a few arbitrarily selected examples of polarity in nonelongating cells are described briefly here. Receptor cells have recently been studied extensively. In *Chara* rhizoids the site of graviperception and of growth response are closely related, spatially and functionally (151, see above). The statocytes in root caps only perceive the gravistimulus and then transduce the information, at least in part by electrical currents, to the elongation zone of the root (8). The statocytes have a strong structural polarity. A complex of ER cisternae is situated at the distal pole and amyloplasts sediment upon it as a result of gravity (51, 149). Graviperception depends on an undisturbed polar arrangement of cell organelles (150).

When they are displaced by mild centrifugation they re-migrate into their normal position within a few hours (150), which indicates that their positioning is somehow controlled by the plasma membrane. Microtubules do not seem to be involved directly in graviperception (50); rather, they presumably stabilize the polar arrangement of the ER (49).

Well known for their polar organization are megasporocytes. Polarity becomes manifest in organelle distribution, aniline blue wall fluorescence, and location of plasmodesmata (128). It is an expression of a regulated development of the gametophyte in relation to the developing surrounding tissue (181) and is important for fertilization.

INDUCTION AND CONTROL OF CELL POLARITY

Nucleus and Cytoskeleton

When multicellular filaments like protonemata develop, the nucleus in the tip cell must move. During *Funaria* spore germination a short outgrowth develops before the nucleus begins to migrate (139; for similar observations in *Fucus* zygotes, see 100). Later the nucleus becomes associated with microtubules and is transported into the outgrowth. The translocation can be inhibited by colchicine. Then the germ tube does not elongate further but branches irregularly and new outgrowths are formed. Auxin has similar depolarizing effects. In normal development the nucleus divides within the outgrowth. One of the daughter nuclei re-migrates into the spore and later into a newly developing outgrowth. The site where this second germ tube is formed, and the direction in which it elongates, is determined by the position of the nucleus, which in turn depends on a functional microtubule system, as shown by experiments with D_2O (139).

On the contrary, the position of the nuclei influences the further development of the outgrowth and orients the second outgrowth and thereby the establishment of a new polar axis. Microtubules are tools in this interplay (136, 137).

A similar feedback system controls side branch formation of *Funaria* caulonema cells (133, 134, 137). Here the position of the nucleus determines the site of outgrowth formation in a subapical cell. The position of the nucleus is controlled by environmental and by endogenous factors. Under suitable conditions, i.e. cultivation on vertically oriented agar plates, the nucleus of the subapical cell initially resides upon the physically lower flank of the cell. It seems to function like a statolith. It keeps a certain distance from the distal cross wall, which is determined by unknown endogenous factors, as is the angle at which the outgrowth branches off. The cell begins to bulge opposite the nucleus, somewhat distally (see also 67). Then the nucleus becomes associated with microtubules, it migrates toward the outgrowth, and divides at its base; the primary side branch cell is cut off. The migration of the nucleus and also of the chloroplasts into the outgrowth depends on microtubules. It is inhibited by colchicine, as is further development of the outgrowth.

In the apical cell of the *Funaria* caulonema likewise a feedback between plasma membrane and nucleus controls polar growth and microtubules and perhaps microfilaments participate therein (135–137). Under normal conditions, the elongating tip determines the position of the nucleus. If the nucleus is centrifuged toward the basal end of the cell, it returns to its former position, as do all other cell organelles, and elongation continues after a few hours when the normal polar arrangement is restored. If the re-migration of the nucleus is inhibited by colchicine, a new outgrowth arises somewhat distally from the nucleus. As in the case of side branch formation, it does not develop further because the nucleus remains immobile. Under normal conditions the polarly differentiated plasma membrane (that part of the cell which together with the cell wall is not displaced during centrifugation) controls cell polarity and the position of the nucleus. An irreversibly displaced nucleus, in turn, repolarizes the plasma membrane and induces the formation of a new tip. When a premitotic nucleus is centrifuged toward the proximal cell pole and divides there before it can re-migrate, a new tip even at the cell base arises that elongates with reversed polarity (135–137; for similar observations in fern protonemata, see 94). Colchicine inhibits normal tip growth. The cell does not elongate; it forms a bulge subapically that does not develop further (135). These growth irregularities may also be caused by an inhibition of nuclear migration, i.e. of the feedback control between the nucleus and the apical plasma membrane.

On the other hand, there are unicellular tip-growing cells like pollen tubes (32), root hairs (151), and infection hyphae of the phycomycete, *Lagenisma* (138), that are not much affected by antimicrotubule drugs. Obviously elongation and the position of the nucleus are less strictly coordinated here than in multicellular filaments in which the single cells have an equal length, and the nucleus therefore must have a distinct position when it divides.

This assumption is supported by observations in *Schizophyllum* hyphae (112) and in germinating fern spores. In the latter cells, a feedback control seems to exist similar to that in *Funaria*. Initially there is no morphological polarization in the spores (6). It becomes manifest by the migration of the nucleus from its original central position to the proximal face and then to one end of the spore in *Onoclea* (164; for other ferns, see 111). It is associated then with microtubules. Antimicrotubule drugs but also membrane-active agents affect nuclear movement (6, 93, 163). Presumably the migration is driven by microtubules and oriented by membranes. In the next developmental step the cell divides unequally; the small cell becomes a rhizoid.

When the spore is centrifuged at an early stage, the nucleus resumes its normal migration and the division is unequal. The polarity of the cell is stable during this phase. When it is centrifuged just before or during mitosis, the cell divides symmetrically (7, 125). Polarization is also severely disturbed when the movement of the nucleus is inhibited by antimicrotubule agents or when cell division is disturbed by caffeine (92, 163, 165). Cytochalasin B has no effect (158). Obviously, actin filaments are less important in these processes (for *Funaria*, see 135).

In moss sporogenesis microtubules likewise participate in the transport of nucleus and plastids and thus contribute to the establishment of sporocyte polarity (14, 15). However, it is perhaps an oversimplification to assume that microtubules function in polar development generally and exclusively via the transport of nuclei and chloroplasts. In the siphonalean green alga *Bryopsis*, various antimicrotubule agents cease cytoplasmic streaming within about 15 min. Cell elongation stops within some hours. In a few days new growth points are generated over the cell flanks. It is suggested that the growth anomalies here are triggered by the arrest of cytoplasmic streaming (95).

Occasionally antimicrotubule agents inhibit the transport of exocytotic vesicles toward the cell apex and thus alter growth (57, 71). Here, as well as perhaps in cells in which microtubules participate in the orientation of cellulose microfibrils (for recent surviews, see 36, 81), microtubules influence the manifestation of polarity rather than control it. The same is true for microfilaments (163). Antimicrofilament agents slow down or stop elongation (53, 135), reduce inward currents at cell poles (12), and cause growth irregularities (25), perhaps by inhibiting cytoplasmic streaming and polar vesicle transport, but they do not alter polarity. A microfilament network is believed to stabilize pollen tube tips against internal osmotic pressure (101, 103) and seems to be involved in the fixation of the polar axis in fucoid zygotes (12, 107).

Plasma Membrane and Calcium

The stability of polarity in centrifugation experiments has long been recognized. It was recently shown again in root hair development (99) and again demonstrates the importance of the plasma membrane. However, in those cells in which the position of the nucleus does not influence cell polarity, the plasma membrane does not control polarity absolutely but does so together with other cell components, among which, in normal plant cells, the cell wall is essential (9, 107). When a polarized cell is plasmolyzed, polarity is frequently lost or severely disturbed (9, 73). Protoplasts from previously polarized cells are nonpolar initially. Cell wall formation precedes the fixation of a new polar axis (16, 17, 59, 65). Not all cells require a close contact between plasma membrane and cell wall to maintain the polar differentiation of the former. In *Micrasterias* the specific pattern of the plasma membrane controls multipolar growth, and is maintained even in mildly plasmolyzed cells (74, 75). The apical deposition of wall material in plasmolyzed root hairs (141) and protonema cells (140) likewise indicates that the polar axis is still present here. In genuinely naked cells such as flagellates, the polarity of the plasma membrane is stabilized by the cytoskeleton; in animal epithelial cells the contact with adjacent cells also contributes to stability.

The unequal distribution of ion pumps and channels within the plasma membrane manifests its polar structure and is perhaps one of the first and fundamental steps in cell polarization. Calcium channels seem to play a prominent role. The following model (175) is based mainly on observations made in fucoid zygotes. They are initially nonpolarized. Ion pumps and channels, especially calcium channels, are randomly distributed in the plasma membrane. Polarizing environmental factors (or, if they are lacking, perhaps even chance) lead to a somewhat uneven distribution, to a labile polarity. As a consequence, a current is driven through the cell. By positive feedback the pumps and channels become concentrated on opposite cell poles, with the calcium channels at the presumptive outgrowth: the currents become stronger, are self-amplified (63).

To explain this feedback system, three different mechanisms have been suggested, none of which need exclude the others.

1. Self electrophoresis. The pumps and channels themselves or their transporting entirities are assumed to be charged. They move through the cytoplasm, driven by the electrical field generated by the transcellular current. Opposite charges lead to incorporation at opposite poles of the plasma membranes (63, 107, 182).

2. Lateral electrophoresis. The transcellular current creates an electrical field not only across the cytoplasm but also along the plasma membrane. If pumps and channels are oppositely charged, they will redistribute and cap by lateral movement in the plasma membrane (62, 63, 127).

3. Contractility. Golgi vesicles incorporate calcium channels into the plasma membrane. When the secretion is slightly asymmetric, unequally distributed calcium channels will cause a calcium gradient within the cytoplasm. Calcium ions control the organization of the microfilament system and its activity. As a result, exocytotic Golgi vesicles, the membrane of which contains new calcium channels, are preferentially incorporated at one pole (12, 107).

These models are developed for polarizing fucoid zygotes but may be applied to other cells as well. Each explains the early occurrence of endogenous currents. The "contractile" model is mainly based on the observation that F-actin becomes concentrated at the presumptive outgrowth by the time the position of the polar axis is fixed, and that cytochalasin B blocks the polar transport of exocytotic Golgi vesicles, reduces the inward currents, and reversibly inhibits polarization (12). It does not disorganize existing polarity.

Each model is also compatible with the observation that electrical fields or forced local entry of ions polarize nonpolar cells such as fucoid zygotes, spores and pollen grains (18, 64, 175), and moss protoplasts (17), or even reverse polar particle distribution in the plasma membrane (13). Nevertheless, it is not understood completely why germinating pollen grains and elongating pollen tubes grow toward an electrode but do not prefer either the positive or the negative one (154, 175, 178). Likewise a strong homogeneous magnetic field (14 Tesla) induces pollen tubes to grow either parallel or antiparallel to the direction of the field (154, 155).

Generally, polarization seems to start from a very labile distribution and orientation of molecules. The key for the understanding of polarization is that inequalities are enhanced and stabilized by positive feedback. Magnetotropism, for example, may be caused by an accumulation of membrane proteins in membrane areas that are oriented perpendicular to the field. They accumulate there because an α helix preferentially lies parallel to the field (154; D. Sperber, personal communication). Inner cytoplasmic structures (154) such as the actin filament system may also be oriented by the field.

Light or other signals are believed to be perceived by specific receptors in the plasma membrane. The perception then influences the distribution or function of membrane pumps and channels. It has long been known that light polarizes cells (for review, see 107, 175). Newly detected is the polarizing effect on moss protoplasts (16, 65).

It is widely accepted that calcium channels play an important role in the establishment and control of polarity, though the observations vary in detail, and the question of primary and secondary effects can often not be answered with certainty. The occurrence of calcium gradients, as mentioned above, is an early phenomenon in polarization (105, 120). Calcium is necessary for spore germination (174). Light increases the uptake of calcium in protonema cells (42). Forced local calcium entry, by application of the calcium ionophore A 23187, polarizes fucoid zygotes (126) and *Funaria* spores (18); the monovalent cation ionophore valinomycin has no effect (126).

Nifedipine blocks calcium channels, which may be voltage gated (124), and thus it modulates internal calcium. When treated with nifedipine, pollen grains form broad protuberances instead of tube-like outgrowths and the tubes assume an ameboid shape (119). Calcium ion fluxes, electrical currents, and voltage-gated calcium channels are other examples of a kind of feedback system in polarity.

Experiments with various inhibitors (which, however, may not be strictly specific in any case) demonstrate that calcium ions play a prominent role not only in cell polarization but also as secondary messengers in the control of polar growth. When the delicate internal equilibrium of calcium is disturbed, growth rate is reduced and anomalies result. The calcium ionophore A 23187 stops tip growth but not cytoplasmic streaming in pollen tubes, disorients local wall secretion, and affects polar organelle distribution (52, 101, 115), whereas the broad-range cationophore X-537 A inhibits elongation rather unspecifically (117).

In *Funaria* caulonema cells, however, growth is only slowed down and wall deposition is made somewhat more irregular after treatment with A 23187, while the organelle distribution is maintained for a relatively long time (135). When calcium ions are complexed with chlorotetracycline, the walls of pollen tubes are irregularly thickened, the polar zonation of organelles is disorganized, and the growing tubes reverse toward the pollen grains (118).

Agents that affect calmodulin (104, 105) or the cyclic nucleotide phosphodiesterase (and in consequence regular calcium distribution, 122) disturb germination and tip growth of pollen tubes and of moss caulonema cells. Similar effects are caused by substances that are supposed to interfere with the internal sequestration of Ca^{2+} by mitochondria or the endoplasmic reticulum (104) or with the movement of calcium away from membrane sites (105). Control of the cytoskeleton, of exocytosis (104, 118), and of tip extension (101, 104) are among the most prominent functions of Ca^{2+} in polar growth.

Generally, the phenomena of transcellular electrical currents, locally restricted calcium fluxes, and polar growth are closely coupled. However, the currents remain when elongation and cytoplasmic streaming are blocked by cytochalasin B (179). They are reduced but still exist in cytochalasin-treated fucoid zygotes (12). During branching of *Achlya* hyphae the currents at the original tip decrease or even reverse transiently, although the elongation rate does not change (79). Sporadically, outward currents occur instead of inward ones in growing tips of *Trichoderma* hyphae (54). Ion fluxes differentially affect tip growth and polar accumulation of organelles in *Griffithsia,* as shown by experiments with ionophores (169). Currents without growth and growth without currents have also been observed in this red alga (168).

Obviously there is no direct connection between exocytosis and currents. It may be rather that the currents are important in the establishment of the polar axis (168). Once fixed, polarity is maintained and occasionally even manifested by polar growth, even when the ion fluxes do not run. The polar distribution of membrane channels and pumps is stable and does not depend on them. Accordingly, it is difficult to reverse a fixed polar axis.

Blade pieces of the coenocytic alga *Caulerpa* regenerate rhizoids at the original basal end and blades at the apical end, regardless of inversion and centrifugation. Only rhizome segments do not regenerate strictly polarly (88). Even anucleate pieces of *Acetabularia* stalks maintain polarity (142). Inverse grafts develop two caps distally when they are connected at the basal poles.

When they are joined at the apical poles, they grow out at the fusion site and form a new cap there (86).

Virtually nothing is known about how a cell maintains polarity or the polar differentiation of the plasma membrane. Feedback mechanisms with currents, the cell wall, and cytoskeletal elements may stabilize the uneven distribution of components of the plasma membrane and of organelles. As is demonstrated by the experiments mentioned above, a reversion can be induced when further control elements are involved, e.g. the position of the nucleus.

Generally, the reversion of a polar axis is preceded by a longer period of depolarization. Isolated cells of higher plants lose cell polarity and reestablish it, if at all, only after several cell cycles, having first formed a callus.

Correlative Control in Multicellular Systems

The fact that isolation causes depolarization reveals that cell polarity in multicellular systems is controlled not only by intracellular and environmental factors in a narrow sense but also by the adjacent cells, the tissue, and the plant body as a whole. The single cells are interrelated. They influence each other to form the organism (10). During development, new polar axes may be established by homeogenetic induction (173).

One manifestation of these interactions is the polarly oriented cell divisions, which are among the fundamental steps in organogenesis. They may be strictly predetermined as for example in *Azolla* roots (38). Because these developmental processes were reviewed recently in this series by Green (36) and Gunning and Hardham (39), they are not treated in this chapter. Interested readers are referred to further review articles (10, 40, 82). It should only be noted here that the orientation of cortical microtubule arrays is related to the plane of division (37, 80) and that oriented cell divisions are essential for polar development in fern protonemata (19, 91, 163, 165).

The control of polar development by cell interactions is most obvious in apical cells. The apical cell of a multicellular filament retains the polar axis when it is isolated from the subapical cell, but the elongation rate is reduced and the pattern of cell divisions changed (22, 76). When the apical cell is removed surgically, the subapical cell assumes the character of an apical cell. It is transformed directly only in exceptional cases, e.g. in the brown alga *Sphacelaria*. This requires that the subapical cell be connected with further basal cells for a longer time (23). Generally, a redifferentiation of a subapical cell is preceded by one or more cell divisions (22, 23, 76). The new tip then represents a kind of a new side branch rather than being a direct replacement of the former tip.

When apical growth is disturbed in protonemata, the filaments change their developmental pattern; moss caulonemata become chloronematic (76, 137). Apical dominance in multicellular filaments has been demonstrated repeatedly

(10, 29, 30, 137). Obviously, apical and subapical cells control each other mutually.

Phytohormones seem to function as messengers in these control systems (10). High auxin concentrations disorient polar development in protonemata (139). The existence of a special growth regulator, rhodomorphin, was demonstrated by Waaland (167, 170). Rhodomorphin controls polar development in the red alga, *Griffithsia*. When the large multinucleate cells of this alga are isolated, they regenerate a new rhizoid at the basal and a new shoot cell at the apical pole. When two cells are placed into an empty *Nitella* cell in tandem arrangement, each cell forms a rhizoid but only the apically situated cell develops a new shoot cell. The rhizoid of this cell elicits in the basally situated cell a special outgrowth, which then fuses with the rhizoid. It was possible to induce this development with an extract of rhizoids and to show that the fusion hormone is species specific.

It is well known that auxin is involved in the control of polar cell differentiation in higher plants. Sachs (129–131) demonstrated that xylem vessels form in response of a flux of auxin rather than being induced by variations of hormone concentration. There is a positive feedback between signal flux and the differentiation of the cells into facilitated channels for this same flux. Wounding, grafting, and external application of auxin determine the course of xylem strands and even induce the formation of rings of tracheids. These results support the suggestion that auxin fluxes control the polar orientation of differentiation whereas auxin concentrations influence the rate of vessel differentiation and their number and size (2).

Factors or mechanisms other than auxin fluxes may also be involved in the control of polar development in the whole plant. It can likewise be assumed that they are self-enhancing by positive feedback (129). Electrical currents play a role not only in controlling tip-growing cells but also in transferring information from the root cap to the elongating zone after gravistimulation (8).

CONCLUSIONS AND CONCEPTIONS

Polarity, the "choreographed rearrangement of cellular components during development" (33) (see Introduction), is an intrinsic character of cells. It is based upon the interrelationship of various cell components and on feedback mechanisms, not on simple cause-and-effect relations. In a few cases, the role of the individual participants in the choreography becomes clearer to researchers. At the same time it becomes obvious that the search for a single maitre de ballet puts one on the wrong track. Within the network of interrelationships, the plasma membrane with its ion channels and ion pumps has a central position.

The electrical currents, the cell wall, the cytoskeleton, the position of the nucleus, and other cell components are not only controlled by the plasma membrane, they also react to it. In multicellular systems, the individual cell influences the development of the adjacent tissue—as can easily be seen when the cell has been damaged—but it is also controlled mutually by the surrounding cells and by the whole plant body. The old question of whether the cells form the plant or the plant forms the cells (4) originates from wrong assumptions.

Polarity and the realization of polarity require positional information. It is still largely unknown what the signals are primarily and how they are perceived and transduced, i.e. how the "positional effect" can be understood. Membranes, cytoskeleton, electrical currents, ions (especially Ca^{2+}), and signalling substances like auxin in multicellular plants can be supposed to be involved somehow in polar development. Various models postulate two diffusible morphogens, the rate of synthesis of each being inversely related to the concentration of the other (173), or mechanisms of autocatalytic short-range activation in conjunction with long-range inhibition (35, 90). The existence of such substances has not yet been demonstrated in plants. These concepts and other mathematical models (24) may stimulate polarity research in the near future.

Literature Cited

1. Allen, E. D., Aiuto, R., Sussman, A. S. 1980. Effects of cytochalasins on *Neurospora crassa*. I. Growth and ultrastructure. *Protoplasma* 102:63–75
2. Aloni, R., Zimmermann, M. H. 1983. The control of vessel size and density along the plant axis. A new hypothesis. *Differentiation* 24:203–8
3. Armbruster, B. L., Weisenseel, M. H. 1983. Ionic currents traverse growing hyphae and sporangia of the mycelial water mold *Achlya debaryana*. *Protoplasma* 115:65–69
4. Barlow, P. W. 1982. "The plant forms cells, not cells the plant": the origin of de Bary's aphorism. *Ann. Bot.* 49:269–71
5. Bartnicki-Garcia, S., Lippman, E. 1972. The bursting tendency of hyphal tips of fungi: presumptive evidence for a delicate balance between wall synthesis and wall lysis in apical growth. *J. Gen. Microbiol.* 73:487–500
6. Bassel, A. R., Kuehnert, C. C., Miller, J. H. 1981. Nuclear migration and asymmetric cell division in *Onoclea sensibilis* spores: an ultrastructural and cytochemical study. *Am. J. Bot.* 68:350–60
7. Bassel, A. R., Miller, J. H. 1982. The effects of centrifugation on asymmetric cell division and differentiation of fern spores. *Ann. Bot.* 50:185–98
8. Behrens, H. M., Weisenseel, M. H., Sievers, A. 1982. Rapid changes in the pattern of electric current around the root tip of *Lepidium sativum* L. following gravistimulation. *Plant Physiol.* 70:1079–83
9. Bentrup, F. W. 1984. Cellular polarity. In *Encyclopedia of Plant Physiology: Cellular Interactions* (NS), ed. H.-F. Linskens, J. Heslop-Harrison, 17:473–90. Berlin/Heidelberg: Springer-Verlag. 743 pp.
10. Bopp, M. 1984. Cell pattern and differentiation in bryophytes. In *Positional Controls in Plant Development*, ed. P. W. Barlow, D. J. Carr, pp. 157–91. Cambridge: Cambridge Univ. Press. 502 pp.
11. Borghi, H., Puiseux-Dao, S., Durand, M., Dazy, A. C. 1983. Morphogenesis, bioelectrical polarity and intracellular streaming in a giant cell, *Acetabularia mediterranea*: studies on their recovery after prolonged dark period. *Plant Sci. Lett.* 31:75–86
12. Brawley, S. H., Robinson, K. R. 1985. Cytochalasin treatment disrupts the endogenous currents associated with cell polarization in fucoid zygotes: studies of the role of F-actin in embryogenesis. *J. Cell Biol.* 100:1173–84
13. Brower, D. L., McIntosh, J. R. 1980. The effects of applied electric fields on *Micrasterias*. I. Morphogenesis and the pattern of cell wall deposition. *J. Cell Sci.* 42:261–77

14. Brown, R. C., Lemmon, B. E. 1982. Ultrastructure of meiosis in the moss *Rhynchostegium serrulatum*. I. Prophasic microtubules and spindle dynamics. *Protoplasma* 110:23–33

15. Brown, R. C., Lemmon, B. E. 1982. Ultrastructure of sporogenesis in the moss, *Amblystegium riparium*. I. Meiosis and cytokinesis. *Am. J. Bot.* 69:1096–1107

16. Burgess, J., Linstead, P. J. 1981. Studies on the growth and development of protoplasts of the moss, *Physcomitrella patens*, and its control by light. *Planta* 151:331–38

17. Burgess, J., Linstead, P. J. 1982. Cell-wall differentiation during growth of electrically polarized protoplasts of *Physcomitrella*. *Planta* 156:241–48

18. Chen, T.-H., Jaffe, L. F. 1979. Forced calcium entry and polarized growth of *Funaria* spores. *Planta* 144:401–6

19. Cooke, T. J., Paolillo, D. J. Jr. 1980. The control of the orientation of cell divisions in fern gametophytes. *Am. J. Bot.* 67:1320–33

20. Cooke, T. J., Racusen, R. H. 1982. Cell expansion in the filamentous gametophyte of the fern *Onoclea sensibilis* L. *Planta* 155:449–58

21. Cooke, T. J., Racusen, R. H., Briggs, W. R. 1983. Initial events in the tip-swelling response of the filamentous gametophyte of *Onoclea sensibilis* L. to blue light. *Planta* 159:300–7

22. Ducreux, G. 1983. Isolement expérimental des cellules terminales de l'apex de *Sphacelaria cirrosa* (Roth) C. Agardh (Sphacelariales, Phéophycées) et analyse comparée de leur potentialités morphogénétiques. *Phycologia* 22:415–29

23. Ducreux, G. 1984. Experimental modification of the morphogenetic behavior of the isolated sub-apical cell of the apex of *Sphacelaria cirrosa* (Phaeophyceae). *J. Phycol.* 20:447–54

24. Edelstein, L. 1982. A molecular switching mechanism in differentiation. Induction of polarity in cambial cells. *Differentiation* 23:1–9

25. El Mougith, A., Dargent, R., Touze-Soulet, J.-M. 1984. Effect of cytochalasin A on growth and ultrastructure of *Mucor mucedo* L. *Biol. Cell.* 52:181–90

26. Emons, A. M. C., Wolters-Arts, A. M. C. 1983. Cortical microtubules and microfibril deposition in the cell wall of root hairs of *Equisetum hyemale*. *Protoplasma* 117:68–81

27. Ender, C., Li, M. Q., Martin, B., Povh, B., Nobiling, R., et al. 1983. Demonstration of polar zinc distribution in pollen tubes of *Lilium longiflorum* with the Heidelberg proton microprobe. *Protoplasma* 116:201–3

28. Etzold, H. 1965. Der Polarotropismus und Phototropismus der Chloronema von *Dryopteris filix-mas* (L.) Schott. *Planta* 64:254–80

29. Faivre-Baron, M. 1978. Étude des mécanismes de corrélations d'inhibition chez le jeune gamétophyte du *Gymnogramme calomelanos* L. *Biochem. Physiol. Pflanz.* 172:79–91

30. Faivre-Baron, M. 1980. Action de deux inhibiteurs de la synthèse protéique sur la morphologie prothallienne d'une Fougère. *Flora* 169:467–75

31. Fèvre, M. 1979. Glucanases, glucan synthases and wall growth in *Saprolegnia monoica*. In *Fungal Walls and Hyphal Growth*, ed. J. H. Burnett, A. P. J. Trinci, pp. 225–63. Cambridge: Cambridge Univ. Press. 418 pp.

32. Franke, W. W., Herth, W., VanDer-Woude, W. J., Morré, D. J. 1972. Tubular and filamentous structures in pollen tubes: possible involvement as guide elements in protoplasmic streaming and vectorial migration of secretory vesicles. *Planta* 105:317–41

33. Fulton, A. B. 1980. Calcium ions, electrical currents and the choreography of (some) eucaryotic cells. *Cell* 22:5–6

34. Furuya, M. 1984. Cell division patterns in multicellular plants. *Ann. Rev. Plant Physiol.* 35:349–73

35. Gierer, M. 1981. Physik der biologischen Gestaltbildung. *Naturwissenschaften* 68:245–51

36. Green, P. B. 1980. Organogenesis—a biophysical view. *Ann. Rev. Plant Physiol.* 31:51–82

37. Green, P. B., Lang, J. M. 1981. Toward a biophysical theory of organogenesis: birefringence observations on regenerating leaves in the succulent, *Graptopetalum paraguayense* E. Walther. *Planta* 151:413–26

38. Gunning, B. E. S. 1981. Microtubules and cytomorphogenesis in a developing organ: the root primordium of *Azolla pinnata*. In *Cytomorphogenesis in Plants*. *Cell Biology Monogr.* ed. O. Kiermayer, 8:301–25. Wien/New York: Springer-Verlag. 439 pp.

39. Gunning, B. E. S., Hardham, A. R. 1982. Microtubules. *Ann. Rev. Plant Physiol.* 33:651–98

40. Hardham, A. R. 1982. Regulation of polarity in tissues and organs. In *The Cytoskeleton in Plant Growth and Development*, ed. C. W. Lloyd, pp. 377–441. London/New York: Academic. 457 pp.

41. Harold, R. L., Harold, F. M. 1980. Oriented growth of *Blastocladiella emersonii* in gradients of ionophores and inhibitors. *J. Bacteriol.* 144:1159–67

42. Hartmann, E. 1984. Influence of light on phototropic bending of moss protonemata of *Ceratodon purpureus* (Hedw.) Brid. *J. Hattori Bot. Lab.* 55:87–98

43. Hartmann, E. 1984. Die Bedeutung von Calzium bei der phototropischen Reaktion von Moosprotonemata. *Mittbd. Bot. Tagg. Wien,* p. 47 (Abstr.)

44. Hausser, I., Herth, W., Reiss, H.-D. 1984. Calmodulin in tip-growing plant cells, visualized by fluorescing calmodulin-binding phenothiazines. *Planta* 162:33–39

45. Heath, I. B., Heath, M. C. 1978. Microtubules and organelle movements in the rust fungus *Uromyces phaseoli* var. *vignae. Cytobiologie* 16:393–411

46. Hegewald, E., Schnepf, E. 1985. Zur Struktur und Taxonomie spindelförmiger Chlorellales (Chlorophyta): *Schroederia, Pseudoschroederia* gen. nov., *Closteriopsis. Algol. Stud.* In press

47. Hejnowicz, Z., Heinemann, B., Sievers, A. 1977. Tip growth: patterns of growth rate and stress in the *Chara* rhizoid. *Z. Pflanzenphysiol.* 81:409–24

48. Hejnowicz, Z., Sievers, A. 1981. Regulation of the position of statoliths in *Chara* rhizoids. *Protoplasma* 108:117–37

49. Hensel, W. 1984. A role of microtubules in the polarity of statocytes from roots of *Lepidium sativum* L. *Planta* 162:404–14

50. Hensel, W. 1984. Microtubules in statocytes from roots of cress (*Lepidium sativum* L.). *Protoplasma* 119:121–34

51. Hensel, W., Sievers, A. 1981. Induction of gravity-dependent plasmatic responses in root statocytes by short time contact between amyloplasts and the distal endoplasmic reticulum complex. *Planta* 153:303–7

52. Herth, W. 1978. Ionophore A 23 187 stops tip growth, but not cytoplasmic streaming, in pollen tubes of *Lilium longiflorum. Protoplasma* 96:275–82

53. Herth, W., Franke, W. W., VanDerWoude, W. J. 1972. Cytochalasin stops tip growth in plants. *Naturwissenschaften* 59:38–39

54. Horwitz, B. A., Weisenseel, M. H., Dorn, A., Gressel, J. 1984. Electric currents around growing *Trichoderma* hyphae, before and after photoinduction of conidiation. *Plant Physiol.* 74:912–16

55. Hoursiangou-Neubrun, D., Dubacq, J. P., Puiseux-Dao, S. 1977. Heterogeneity of the plastid population and chloroplast differentiation in *Acetabularia mediterranea.* In *Progress in Acetabularia Research,* ed. C. L. F. Woodcock, pp. 175–94. New York/London: Academic. 341 pp.

56. Howard, R. J. 1981. Ultrastructural analysis of hyphal tip cell growth in fungi: Spitzenkörper, cytoskeleton and endomembranes after freeze-substitution. *J. Cell Sci.* 48:89–103

57. Howard, R. J., Aist, J. R. 1977. Effects of MBC on hyphal tip organization, growth, and mitosis of *Fusarium acuminatum,* and their antagonism by D_2O. *Protoplasma* 92:195–210

58. Howard, R. J., Aist, J. R. 1979. Hyphal tip cell ultrastructure of the fungus *Fusarium:* improved preservation by freeze-substitution. *J. Ultrastruct. Res.* 66:224–34

59. Ishizawa, K., Enomoto, S., Wada, S. 1979. Germination and photo-induction of polarity in the spherical cells regenerated from protoplasm fragments of *Boergesenia forbesii. Bot. Mag. Tokyo* 92:173–86

60. Ishizawa, K., Wada, S. 1979. Growth and phototropic bending in *Boergesenia* rhizoid. *Plant Cell Physiol.* 20:973–82

61. Jaffe, L. A., Weisenseel, M. H., Jaffe, L. F. 1975. Calcium accumulations within the growing tips of pollen tubes. *J. Cell Biol.* 67:488–92

62. Jaffe, L. F. 1977. Electrophoresis along cell membranes. *Nature* 265:600–2

63. Jaffe, L. F. 1981. The role of ion currents in establishing developmental gradients. In *International Cell Biology 1980–1981,* ed. H. G. Schweiger, pp. 507–11. Berlin/Heidelberg/New York: Springer-Verlag. 1033 pp.

64. Jaffe, L. F., Nuccitelli, R. 1977. Electrical controls of development. *Ann. Rev. Biophys. Bioeng.* 6:445–76

65. Jenkins, G. I., Cove, D. J. 1983. Light requirements for regeneration of protoplasts of the moss *Physcomitrella patens. Planta* 157:39–45

66. Jenkins, G. I., Cove, D. J. 1983. Phototropism and polarotropism of primary chloronemata of the moss *Physcomitrella patens:* responses of the wild type. *Planta* 158:357–64

67. Jensen, L. C. W. 1981. Division, growth, and branch formation in protonema of the moss *Physcomitrium turbinatum:* studies of sequential cytological changes in living cells. *Protoplasma* 107:301–17

68. Jensen, L. C. W., Jensen, C. G. 1984. Fine structure of protonemal apical cells of the moss *Physcomitrium turbinatum. Protoplasma* 122:1–10

44   SCHNEPF

69. Kadota, A., Wada, M., Furuya, M. 1982. Phytochrome-mediated phototropism and different dichroic orientation of Pr and Pfr in protonemata of the fern *Adiantum capillus veneris* L. *Photochem. Photobiol.* 35:533–36
70. Kataoka, H. 1981. Expansion of *Vaucheria* cell apex caused by blue or red light. *Plant Cell Physiol.* 22:583–95
71. Kataoka, H. 1982. Colchicine-induced expansion of *Vaucheria* cell apex. Alteration from isotropic to transversally anisotropic growth. *Bot. Mag. Tokyo* 95:317–30
72. Katsaros, C., Galatis, B., Mitrakos, K. 1983. Fine structural studies on the interphase and dividing apical cells of *Sphacelaria tribuloides* (Phaeophyta). *J. Phycol.* 19:16–30
73. Kesseler, H. 1960. Morphologische und zellphysiologische Untersuchungen an *Chaetomorpha linum*. *Helgol. Wiss. Meeresunters.* 7:114–24
74. Kiermayer, O. 1964. Untersuchungen über die Morphogenese und Zellwandbildung bei *Micrasterias denticulata* Bréb. *Protoplasma* 59:76–132
75. Kiermayer, O. 1981. Cytoplasmic basis of morphogenesis in *Micrasterias*. See Ref. 38, pp. 147–90
76. Knoop, B. 1973. Untersuchungen zum Regenerationsmechanismus bei *Funaria hygrometrica* Sibth. I. Die Auslösung der Caulonemaregeneration. *Z. Pflanzenphysiol.* 70:22–33
77. Knox, R. B., Heslop-Harrison, J. 1971. Pollen-wall proteins: electron-microscopic localization of acid phosphatase in the intine of *Crocus vernus*. *J. Cell Sci.* 8:727–33
78. Kropf, D. L., Caldwell, J. H., Gow, N. A. R., Harold, F. M. 1984. Transcellular ion currents in the water mold *Achlya*. Amino acid proton symport as a mechanism of current entry. *J. Cell Biol.* 99:486–96
79. Kropf, D. L., Lupa, M. D. A., Caldwell, J. H., Harold, F. M. 1983. Cell polarity: endogenous ion currents precede and predict branching in the water mold *Achlya*. *Science* 220:1385–87
80. Lang Selker, J. M., Green, P. B. 1984. Organogenesis in *Graptopetalum paraguayense* E. Walther: shifts in orientation of cortical microtubule arrays are associated with periclinal divisions. *Planta* 160:289–97
81. Lloyd, C. W. 1984. Toward a dynamic helical model for the influence of microtubules on wall patterns in plants. *Int. Rev. Cytol.* 86:1–51
82. Lloyd, C. W., Barlow, P. W. 1982. The co-ordination of cell division and elonga-

tion: the role of the cytoskeleton. See Ref. 40, pp. 203–28
83. Lombardi, T., Montesano, R., Wohlwend, A., Amherdt, M., Vassalli, J.-D., Orci, L. 1985. Evidence for polarization of plasma membrane domains in pancreatic endocrine cells. *Nature* 313:694–96
84. Lubinski, J., Huet, C. 1984. Independent variation in the number of coated pits and of coated vesicles in cultured fibroblasts. *Biol. Cell.* 52:119–28
85. Lüttke, A. 1981. Heterogeneity of chloroplasts in *Acetabularia mediterranea*. Heterogeneous distribution and morphology of chloroplast DNA. *Exp. Cell Res.* 131:483–88
86. Lüttke, A. 1983. Polarity of *Acetabularia mediterranea*: stability in the anucleate state. *Ann. Bot.* 52:905–13
87. Lüttke, A., Grawe, F. 1984. The pattern of protein synthesis in *Acetabularia mediterranea*. *Ann. Bot.* 54:103–10
88. Matilsky, M. B., Jacobs, W. P. 1983. Regeneration in the coenocytic marine alga *Caulerpa*, with respect to gravity. *Am. J. Bot.* 70:635–38
89. McKerracher, L. J., Heath, I. B. 1985. Microtubules around migrating nuclei in conventionally-fixed and freeze-substituted cells. *Protoplasma* 125:162–72
90. Meinhardt, H. 1979. Eine Theorie der Steuerung der räumlichen Zelldifferenzierung. *Biol. Zeit* 9:33–39
91. Miller, J. H. 1980. Orientation of the plane of cell division in fern gametophytes: the roles of cell shape and stress. *Am. J. Bot.* 67:534–42
92. Miller, J. H., Bassel, A. R. 1980. Effects of caffeine on germination and differentiation in spores of the fern, *Onoclea sensibilis*. *Physiol. Plant.* 50:213–20
93. Miller, J. H., Greany, R. H. 1976. Rhizoid differentiation in fern spores: experimental manipulation. *Science* 193:687–89
94. Mineyuki, Y., Furuya, M. 1980. Effect of centrifugation on the development and timing of premitotic positioning of the nucleus in *Adiantum* protonemata. *Dev. Growth Differ.* 22:867–74
95. Mizukami, M., Wada, S. 1983. Morphological anomalies induced by antimicrotubule agents in *Bryopsis plumosa*. *Protoplasma* 114:151–62
96. Mullins, J. T. 1979. A freeze-fracture study of hormone-induced branching in the fungus *Achlya*. *Tissue Cell* 11:585–95
97. Najim, L., Turian, G. 1979. Conidiogenous loss of structuro-functional polarity in the hyphal tips of *Sclerotinia fructigena*. *Eur. J. Cell Biol.* 20:24–27

98. Najim, L., Turian, G. 1979. Ultrastructure de l'hyphe végétatif de *Sclerotinia fructigena. Can. J. Bot.* 57:1299–
1313
99. Nakazawa, S., Yamazaki, Y. 1982.
Cellular polarity in root epidermis of
Gibasis geniculata. Naturwissenschaften
69:396–97
100. Nuccitelli, R. 1978. Ooplasmic segregation and secretion in the *Pelvetia* egg is
accompanied by a membrane-generated
electrical current. *Dev. Biol.* 62:13–33
101. Picton, J. M., Steer, M. W. 1982. A
model for the mechanism of tip extension
in pollen tubes. *J. Theor. Biol.* 98:15–
20
102. Picton, J. M., Steer, M. W. 1983. Membrane recycling and the control of secretory activity in pollen tubes. *J. Cell Sci.*
63:303–10
103. Picton, J. M., Steer, M. W. 1983. Evidence for the role of Ca^{2+} ions in tip
extension in pollen tubes. *Protoplasma*
115:11–17
104. Picton, J. M., Steer, M. W. 1985. The
effects of ruthenium red, lanthanum,
fluorescein isothiocyanate and trifluoperazine on vesicle transport, vesicle fusion
and tip extension in pollen tubes. *Planta*
163:20–26
105. Polito, V. S. 1983. Membrane-associated calcium during pollen grain germination: a microfluorometric analysis. *Protoplasma* 117:226–32
106. Powell, A. J., Lloyd, C. W., Slabas, A.
R., Cove, D. J. 1980. Demonstration of
the microtubular cytoskeleton of the
moss, *Physcomitrella patens,* using antibodies against mammalian brain tubulin.
Plant Sci. Lett. 18:401–4
107. Quatrano, R. S. 1978. Development of
cell polarity. *Ann. Rev. Plant Physiol.*
29:487–510
108. Racusen, R. H., Cooke, T. J. 1982. Electrical changes in the apical cell of the fern
gametophyte during irradiation with pho-
tomorphogenetically active light. *Plant
Physiol.* 70:331–34
109. Racusen, R. H., Cooke, T. J. 1984. Extracellular measurements of an electrical
dipole in the apical cell of the fern gametophyte. *Plant Physiol.* 75:131 (Abstr.)
110. Raghavan, V. 1981. Distribution of poly-
(A)-containing RNA during normal pollen development and during induced pollen embryogenesis in *Hyoscyamus niger.*
J. Cell Biol. 89:593–606
111. Raghavan, V., Huckaby, C. S. 1980. A
comparative study of cell division patterns during germination of spores of *An-
emia, Lygodium* and *Mohria* (Schi-
zaeaceae). *Am. J. Bot.* 67:653–63
112. Raudaskoski, M. 1980. Griseofulvin-

113. Reiss, H.-D., Grime, G. W., Li, M. Q.,
Takacs, J., Watt, F. 1985. Distribution
of elements in the lily pollen–tube tip,
determined with the Oxford scanning
microprobe. *Protoplasma* 126:147–52
114. Reiss, H.-D., Herth, W. 1978. Visualization of the Ca^{2+}-gradient in growing
pollen tubes of *Lilium longiflorum* with
chlorotetracycline fluorescence. *Protoplasma* 97:373–77
115. Reiss, H.-D., Herth, W. 1979. Calcium
ionophore A 23 187 affects localized wall
secretion in the tip region of pollen tubes
of *Lilium longiflorum. Planta* 145:225–
32
116. Reiss, H.-D., Herth, W. 1979. Calcium
gradients in tip growing plant cells
visualized by chlorotetracycline fluorescence. *Planta* 146:615–21
117. Reiss, H.-D., Herth, W. 1980. Effects of
the broad-range ionophore X 537 A on
pollen tubes of *Lilium longiflorum. Planta* 147:295–301
118. Reiss, H.-D., Herth, W. 1982. Dis-
oriented growth of pollen tubes of *Lilium
longiflorum* Thunb. induced by prolonged treatment with the calcium-
chelating antibiotic, chlorotetracycline.
Planta 156:218–25
119. Reiss, H.-D., Herth, W. 1985. Nifedi-
pine-sensitive calcium channels are involved in polar growth of lily pollen
tubes. *J. Cell Sci.* 76:247–54
120. Reiss, H.-D., Herth, W., Nobiling, R.
1985. Development of membrane- and
calcium-gradients during pollen germination in *Lilium longiflorum. Planta*
163:84–90
121. Reiss, H.-D., Herth, W., Schnepf, E.,
Nobiling, R. 1983. The tip-to-base calcium gradient in pollen tubes of *Lilium
longiflorum* measured by proton-induced
X-ray emission (PIXE). *Protoplasma*
115:153–59
122. Reiss, H.-D., Schnepf, E. 1983.
Papaverine effects on development of
Funaria caulonema filaments. *Z. Pflanzenphysiol.* 110:339–54
123. Reiss, H.-D., Schnepf, E., Herth, W.
1984. The plasma membrane of the
Funaria caulonema tip cell: morphology
and distribution of particle rosettes and
the kinetics of cellulose synthesis. *Planta*
160:428–35
124. Reuter, H. 1983. Calcium channel modulation by neuro-transmitters, enzymes
and drugs. *Nature* 301:569–74
125. Robinson, A. I., Miller, J. H., Helfrich,
R., Downing, M. 1984. Metal-binding

induced alterations in site of dividing
nuclei and structure of septa in a dikaryon
of *Schizophyllum commune. Protoplasma* 103:323–31

sites in germinating fern spores (*Onoclea sensibilis*). *Protoplasma* 120:1–11

126. Robinson, K. R., Cone, R. 1980. Polarization of fucoid eggs by a calcium ionophore gradient. *Science* 207:77–78

127. Robinson, K. R., Jaffe, L. F. 1975. Polarizing fucoid eggs drive a calcium current through themselves. *Science* 187:70–72

128. Russell, S. D. 1979. Fine structure of megagametophyte development in *Zea mays. Can. J. Bot.* 57:1093–1110

129. Sachs, T. 1981. The control of the patterned differentiation of vascular tissues. *Adv. Bot. Res.* 9:152–262

130. Sachs, T. 1981. Polarity changes and tissue organization in plants. In *International Cell Biology 1980–1981*, ed. H. G. Schweiger, pp. 489–96. Berlin/Heidelberg/New York: Springer-Verlag. 1033 pp.

131. Sachs, T., Cohen, D. 1982. Circular vessels and the control of vascular differentiation in plants. *Differentiation* 21:22–26

132. Saunders, M. J., Hepler, P. K. 1981. Localization of membrane-associated calcium following cytokinin treatment in *Funaria* using chlorotetracycline. *Planta* 152:272–81

133. Schmiedel, G., Schnepf, E. 1979. Side branch formation and orientation in the caulonema of the moss, *Funaria hygrometrica:* normal development and fine structure. *Protoplasma* 100:367–83

134. Schmiedel, G., Schnepf, E. 1979. Side branch formation and orientation in the caulonema of the moss, *Funaria hygrometrica:* experiments with inhibitors and with centrifugation. *Protoplasma* 101:47–59

135. Schmiedel, G., Schnepf, E. 1980. Polarity and growth of caulonema tip cells of the moss, *Funaria hygrometrica. Planta* 147:405–13

136. Schnepf, E. 1981. Polarity and gradients in tip growing plant cells. See Ref. 130, pp. 485–88

137. Schnepf, E. 1982. Morphogenesis in moss protonemata. See Ref. 40, pp. 321–44, 405–41

138. Schnepf, E., Heinzmann, J. 1980. Nuclear movement, tip growth and colchicine effects in *Lagenisma coscinodisci* Drebes (Oomycetes, Lagenidiales). *Biochem. Physiol. Pflanz.* 175:67–76

139. Schnepf, E., Hrdina, B., Lehne, A. 1982. Spore germination, development of the microtubule system and protonema cell morphogenesis in the moss, *Funaria hygrometrica:* effects of inhibitors and of growth substances. *Biochem. Physiol. Pflanz.* 177:461–82

140. Schnepf, E., Witte, O., Rudolph, U., Deichgräber, G., Reiss, H.-D. 1985. Tip cell growth and the frequency and distribution of particle rosettes in the plasmalemma: experimental studies in *Funaria* protonema cells. *Protoplasma* 127:222–29

141. Schröter, K., Sievers, A. 1971. Wirkung der Turgorreduktion auf den Golgi-Apparat und die Bildung der Zellwand bei Wurzelhaaren. *Protoplasma* 72:203–11

142. Schweiger, H. G., Berger, S. 1981. Pattern formation in *Acetabularia*. See Ref. 38, pp. 119–45

143. Seagull, R. W., Heath, I. B. 1979. The effect of tannic acid on the in vivo preservation of microfilaments. *Eur. J. Cell Biol.* 20:184–88

144. Seagull, R. W., Heath, I. B. 1980. The organization of cortical microtubule arrays in the radish root hair. *Protoplasma* 103:205–29

145. Seagull, R. W., Heath, I. B. 1980. The differential effects of cytochalasin B on microfilament populations and cytoplasmic streaming. *Protoplasma* 103:231–40

146. Sievers, A. 1963. Beteiligung des Golgi-Apparates bei der Bildung der Zellwand von Wurzelhaaren. *Protoplasma* 56:188–92

147. Sievers, A. 1967. Elektronenmikroskopische Untersuchungen zur geotropischen Reaktion. II. Die polare Organisation des normal wachsenden Rhizoids von *Chara foetida. Protoplasma* 64:225–53

148. Sievers, A., Heinemann, B., Rodriguez-Garcia, M. I. 1979. Nachweis des subapikalen differentiellen Flankenwachstums im *Chara*-Rhizoid während der Graviresponse. *Z. Pflanzenphysiol.* 91:435–42

149. Sievers, A., Hensel, W. 1982. The nature of graviperception. In *Plant Growth Substances*, ed. P. F. Wareing, pp. 497–506. London/New York: Academic. 683 pp.

150. Sievers, A., Heyder-Caspers, L. 1983. The effect of centrifugal accelerations on the polarity of statocytes and on the graviperception of cress roots. *Planta* 157:64–70

151. Sievers, A., Schnepf, E. 1981. Morphogenesis and polarity of tubular cells with tip growth. See Ref. 38, pp. 263–99

152. Sievers, A., Schröter, K. 1971. Versuch einer Kausalanalyse der geotropischen Reaktionskette im *Chara*-Rhizoid. *Planta* 96:339–53

153. Sievers, A., Volkmann, D. 1979. Gravitropism in single cells. In *Encyclo-*

pedia of Plant Physiology: Physiology of Movements (NS), ed. W. Haupt, M. E. Feinleib, 7:567–72. Berlin/Heidelberg: Springer-Verlag. 731 pp.

154. Sperber, D. 1984. Das Wachstum pflanzlicher Zellen und Organe im magnetischen und elektrischen Feld. Konstanzer Dissertationen Nr. 51. Konstanz: Hartung-Gorre. 55 pp.

155. Sperber, D., Dransfeld, K., Maret, G., Weisenseel, M. H. 1981. Oriented growth of pollen tubes in strong magnetic fields. Naturwissenschaften 68:40–41

156. Thomas, D. des S., Mullins, J. T. 1967. Role of enzymatic wall softening in plant morphogenesis: hormonal induction in Achlya. Science 156:84–85

157. Traas, J. A. 1984. Visualization of the membrane bound cytoskeleton and coated pits of plant cells by means of dry cleaving. Protoplasma 119:212–18

158. Traas, J. A., Braat, P., Derksen, J. W. 1984. Changes in microtubule arrays during the differentiation of cortical root cells of Raphanus sativus. Eur. J. Cell Biol. 34:229–38

159. Turian, G. 1978. The "Spitzenkörper," centre of the reducing power in the growing hyphal apices of two septomycetous fungi. Experientia 34:1277–79

160. Turian, G. 1979. Cytochemical gradients and mitochondrial exclusion in the apices of vegetative hyphae. Experientia 35: 1164–66

161. Vanden Driessche, T. 1983. Spatial and temporal organization of Acetabularia: combinatorial conditions for morphogenesis. In Endocytobiology II, ed. H. E. A. Schenk, W. Schwemmler, pp. 329–40. Berlin: de Gruyter. 1071 pp.

162. Vermeulen, C. A., Wessels, J. G. H. 1984. Ultrastructural differences between wall apices of growing and nongrowing hyphae of Schizophyllum commune. Protoplasma 120:123–31

163. Vogelmann, T. C., Bassel, A. R., Miller, J. H. 1981. Effects of microtubule-inhibitors on nuclear migration and rhizoid differentiation in germinating fern spores (Onoclea sensibilis). Protoplasma 109:295–316

164. Vogelmann, T. C., Miller, J. H. 1980. Nuclear migration in germinating spores of Onoclea sensibilis: the path and kinetics of movement. Am. J. Bot. 67:648–52

165. Vogelmann, T. C., Miller, J. H. 1981. The effect of methanol on spore germination and rhizoid differentiation in Onoclea sensibilis. Am. J. Bot. 68:1177–83

166. Volkmann, D. 1984. The plasma membrane of growing root hairs is composed of zones of local differentiation. Planta 162:392–403

167. Waaland, S. D. 1975. Evidence for a species-specific cell fusion hormone in red algae. Protoplasma 86:253–61

168. Waaland, S. D., Lucas, W. J. 1984. An investigation of the role of transcellular ion currents in morphogenesis of Griffithsia pacifica Kylin. Protoplasma 123:184–91

169. Waaland, S. D., Lucas, W. J. 1984. Control of intracellular localization in a red alga. Plant Physiol. 75:132 (Abstr.)

170. Waaland, S. D., Watson, B. A. 1980. Isolation of a cell-fusion hormone from Griffithsia pacifica Kylin, a red alga. Planta 149:493–97

171. Wada, M., Mineyuki, Y., Kadota, A., Furuya, M. 1980. The changes of nuclear position and distribution of circumferentially aligned cortical microtubules during the progression of cell cycle in Adiantum protonemata. Bot. Mag. Tokyo 93:237–45

172. Wada, M., Staehelin, L. A. 1981. Freeze-fracture observations on the plasma membrane, the cell wall and the cuticle of growing protonemata of Adiantum capillus-veneris L. Planta 151:462–68

173. Warren Wilson, J. 1980. A control system for initiating and maintaining polarity. Ann. Bot. 46:701–11

174. Wayne, R., Hepler, P. K. 1984. The role of calcium ions in phytochrome-mediated germination of spores of Onoclea sensibilis. Planta 160:12–20

175. Weisenseel, M. H. 1979. Induction of polarity. See Ref. 153, pp. 485–505

176. Weisenseel, M. H. 1980. Polaritätsinduktion und polares Wachstum pflanzlicher Zellen und Organe. Biol. Zeit 10:39–44

177. Weisenseel, M. H., Jaffe, L. F. 1976. The major growth current through lily pollen tubes enters as K^+ and leaves as H^+. Planta 133:1–7

178. Weisenseel, M. H., Kicherer, R. M. 1981. Ionic currents as control mechanism in cytomorphogenesis. See Ref. 38, pp. 379–400

179. Weisenseel, M. H., Nuccitelli, R., Jaffe, L. F. 1975. Large electrical currents traverse growing pollen tubes. J. Cell Biol. 66:556–67

180. Wessels, J. G. H., Sietsma, J. H. 1981. Cell wall synthesis and hyphal morphogenesis. A new model for apical growth. In Cell Walls '81, ed. D. G. Robinson, H. Quader, pp. 135–42. Stuttgart: Wiss. Verlagsges. 297 pp.

181. Willemse, M. T. M. 1981. Polarity during megasporogenesis and megagametogenesis. Phytomorphology 31:124–34

182. Woodruff, R. I., Telfer, W. H. 1980. Electrophoresis of proteins in intercellular bridges. Nature 286:84–86

Ann. Rev. Plant Physiol. 1986. 37:49–72

PLANT CHEMILUMINESCENCE[1,2]

F. B. Abeles

Appalachian Fruit Research Station, United States Department of Agriculture, Route 2, Box 45, Kearneysville, West Virginia 25430

CONTENTS

[1]The US Government has the right to retain a nonexclusive royalty-free license in and to any copyright covering this paper.

[2]Abbreviations: ATP, adenosine triphosphate; BHA, butylated hydroxyanisole; BHT, butylated hydroxytoluene; CL, chemiluminescence; DABCO, 1,4-diaza-bicyclo-(2,2,2)-octane; DNA, deoxyribonucleic acid; DTBQ, diterthydroquinone; EDTA, ethylene diaminetetraacetic acid; NADH, reduced nicotinamide adenine dinucleotide; NADPH, reduced nicotinamide adenine dinucleotide phosphate; PMN, polymorphonuclear leukocytes; RNase, ribonuclease; SHAM, salicylhydroxamic acid; SOD, superoxide anion dismutase.

INTRODUCTION

Luminescence is a general term indicating light-producing processes. There are three kinds of luminescence: fluorescence, bioluminescence, and chemiluminescence (CL). In all cases, light is produced when molecules in an electronically excited state decay to a stable ground state. The quantum yield (number of photons emitted divided by the total number of molecules reacting in the process) is reduced when electrons move to the lower ground state through nonvisible or radiationless transfers such as the production of heat. Light is the source of energy for fluorescence, and chemical reactions such as oxidation are the source of energy for bioluminescence and chemiluminescence. While chlorophyll is the usual light-absorbing pigment for fluorescence, other unsaturated or ring compounds can be involved in the process. Bioluminescence is not known to occur in vascular plants. There are two features that distinguish bioluminescence from chemiluminescence. Bioluminescence is an enzymatically mediated process and has high quantum yield. The quantum yield for bioluminescence varies from 100% to 1%, while the quantum yield for chemiluminescence is orders of magnitude lower. Chemiluminescence, while occurring in most, if not all, living cells, is also a property of a number of oxidative reactions. The page on which this is written is chemiluminescent and is probably producing 100 or so photons a minute (83).

Bioluminescence

Bioluminescence, also called phosphorescence, is a form of CL that can occur in the body or excretion of certain living organisms. Bioluminescence occurs in fungi *(Lampteromices japonicus)* bacteria *(Photobacterium fisheri)*, protozoans *(Gonyaulux* sp.) coelenterates, annelids, mollusks, crustaceans, millepedes, insects, hemicordates, and vertebrates (62). These organisms are inhabitants of water, soil, and air and use the light they produce for breeding or food-gathering purposes.

The luciferin-luciferase system is the most prevalent bioluminescent system, while other organisms use photoproteins. The luciferins can be long-chained fatty aldehydes, sesquiterpenes, and other complex organic compounds. Many of these compounds are accumulated via a food chain relationship, for example, the feeding of fish on the crustacean *Cypridena*. The energy for light production is derived from the oxidation of a specific substrate which either yields light directly or transfers its energy to another fluorescent compound. The oxidation reactions can yield CO_2 or the conversion of an aldehyde to a carboxylic acid.

Dioxetanes are intermediates in many of these reactions. Their emission maxima can range from 459 nm (blue) in teleost fish to 582 nm (yellow) for the firefly. Most, however, are about 500 nm (green) (109, 133).

Fluorescence

Photoactivated light production has been called delayed light emission, photo-induced luminescence, and fluorescence. This delayed light production occurs after light-generated photosynthetic or other high energy intermediates recombine in the dark. Unlike CL, sufficient light is produced so that an image intensification system can be used to take pictures (phytoluminographs) of the plant (45). Fluorescence has been used to characterize the heat resistance of leaves from fruit trees (76) and to evaluate the maturity and freshness of apricots (31).

Fluorescence of nonphotosynthetic tissue is evident in the sensitive systems used to measure CL. Fluorescence from roots (137), seedlings (94), and nonliving organic substances such as dried milk (122) have been reported. Fluorescence can cause artifacts in CL studies. Most materials have to be kept in the dark for 10 min or more (94), so that photoexcited molecules can decay to a ground state before CL measurements are started.

Chemiluminescence

While it is common knowledge that plants use photons for photosynthesis and photomorphogensis, it is not as evident that they also produce them by CL. Some of the other terms used in the literature to describe the same phenomenon are: ultraweak or superweak bioluminescence, ultraweak photon emission, ultraweak or superslight chemiluminescence, mitogenetic radiation, and spontaneous luminescence. A working definition of CL is that it is an oxidatively driven production of light, has an emission spectrum that ranges from 200 nm to 700 nm, an intensity of 10 to 1000 photons s^{-1} cm^{-2}, and a quantum efficiency of 10^{-3} to 10^{-14} photons per activated molecule.

In the 1920s, Gurwitsch and coworkers were investigating the mechanisms controlling cell division in onion root tips. They were interested in learning whether rapidly dividing cells produced radiation that was capable of promoting cell division in neighboring cells (55). They tested the idea that dividing cells produced a "mitogenetic" radiation, supposedly in the UV part of the spectrum, by using tips from one set of onion roots as an emitter and the nondividing portion of another root as a receptor. The mitogenetic radiation-emitting root was sheathed in a metal tube with its tip exposed at the open end. This emitter was placed next to nondividing parts of other roots and changes in the rate of cell division in the receptor root measured. Subsequent workers tested the idea that mitogenic radiation was a form of high energy ionizing radiation by using Geiger type ionization tubes. They failed to observe the production of ionizing radiation from plants (107, 117). Grebe et al (54) reported that they observed the production of UV light from *Helianthus* leaves. However, except for a report that yeast produce light in the 250 nm to 380 nm range (69), no confirmation for the production of UV light by plants has appeared. In all cases in which it has been measured, the emission spectrum of CL is now known to be in the green to yellow portion of the spectrum (see below).

With the advent of sensitive photomultiplier tubes, Colli et al (33) demonstrated that etiolated seedlings of wheat, beans, lentils, and corn produced light. The amount of light increased during the germination process and was about tenfold higher in roots than in stems or seeds. The emission spectrum was broad, starting in the violet region at 400 nm and increasing to a maximum at 550 nm in the green. Because their photomultiplier was insensitive to far red light, they were unable to record light production above 600 nm. They also reported that light production was oxygen dependent, and bean tissue wounded by cutting it into sections produced twice as much light as intact plants.

Since these early observations, other investigators have observed CL from all parts of plants. CL has been observed from seeds (17, 22, 87), stems (1, 3, 103), roots (1, 39, 103, 104, 110, 129, 131), and fruits (44). In general, roots produced ten or more times the amount of light than stems (1, 103).

A number of reviews on the early studies of CL in plants have appeared (13, 34, 41, 94, 102). The more generalized topic of light production in other organisms including bacteria, fungi, insects, and fish has also been summarized (102, 109, 111, 133, 134). Because of the importance of the leukocyte in medical research, the CL of these cells has been studied extensively (5, 115). Some reviews have focused on the chemical nature of the substances involved in the production of light, including luminegenic probes, and the chemical processes involved in their activation (14, 16, 71, 81, 134). Other reviews have dealt with the oxidation of cells (18), oxidation in foods (70), the use of CL in analytical chemistry (59, 71, 83, 134), and the instrumentation utilized in this area of research (111, 133).

ACTIVATED OXYGEN

Activated oxygen is the driving force in CL since little or no light is produced when cells or tissues are placed in an anaerobic environment (see below). Much has been written on the chemistry of activated oxygen, and the summary that follows is only intended to serve as an outline of the terms and definitions used in this area of research. The excellent review by Korycka-Dahl & Richardson (70), and the reviews cited therein, were used as the source of the information summarized below.

Singlet Oxygen

Activated oxygen occurs in four forms. The normal unactivated ground state of oxygen is called triplet oxygen ($^3O_2$). The neutral oxygen atom has eight electrons. Two are tightly bound in the ls orbital and play no role in oxygen chemistry. The remaining six are valence electrons and are in either the 2s or 2p orbitals. In the O_2 molecule there are 12 valence electrons. If these electrons are added sequentially to molecular orbitals following the Pauli exclusion principal and Hund's rule, the triplet state of oxygen has two unpaired electrons with the same (↑ ↑ parallel) spin in each of the two highest orbitals. When placed in a magnetic field this molecule exists in three states because the two unpaired electrons generate a magnetic field. The singlet state ($^1O_2$) is one in which the 2p electrons are in the same orbital and have opposing or antiparallel (↑ ↓) spin and no magnetic moment. The energy of $^1O_2$ is 22.4 kcal above the $^3O_2$ ground state. $^1O_2$ is not a free radical but an electrophilic species seeking electrons to fill the vacant molecular orbital. While the reaction of oxygen with a double-bonded substrate such as a lipid requires energy of activation, $^1O_2$ reacts readily with molecules with a high density of electrons. $^1O_2$ has a lifetime in the order of microseconds and will produce light when it decays to the triplet ground state. When a single molecule decays to the ground state it produces light in the infrared region at 1269 nm in what is referred to as monomol emission. When two molecules of $^1O_2$ combine and return to the ground state the light produced is called a dimol emission. Most of the light produced is in the far red and red region of the spectrum at 703 nm and 634 nm (65), with a small amount in the blue and green portion of the spectrum at 460 and 510 nm (10).

Superoxide Anion (O_2^{*-})

The univalent reduction of O_2 yields O_2^{*-} [$O_2 + e^- \rightarrow O_2^{*-}$]; O_2^{*-} is more stable than $^1O_2$ since its decay rate is 100 molecules per mole per second (16). The additional electron makes this molecule both a strong nucleophile and free radical. O_2^{*-} is the conjugate base of hydroperoxy acid which has a pK of 4.9.

Under physiological conditions most of the acid is in the form of the conjugate base $[O_2*^- + H^+ \leftrightarrows O_2H*]$.

Hydrogen Peroxide (H_2O_2)

Adding an additional electron to O_2*^- results in the formation of H_2O_2 or O_2^{-2}. This electron increases the oxygen to oxygen bond distance from 1.21 Å in O_2 to 1.49 Å for H_2O_2 and decreases the heat of dissociation from 118 kcal/mole to 35 kcal. H_2O_2 is considered to be stable under physiological conditions of pH and temperature and in the absence of scavengers such as catalase and peroxidase. Levels of H_2O_2 in fruits of passion fruit have been estimated at 100 μM (89).

Hydroxy Radical (OH*)

The last step in this oxygen reduction series is the formation of OH* $[H_2O_2 + e^- \rightarrow OH^- + OH*]$. OH* is a strong oxidizing agent and reacts readily with most organic compounds found in the cell. Because of this its lifetime in the cell is probably short.

Production of Activated Oxygen

The production of various forms of activated oxygen by physical, chemical, and biochemical means has been reviewed by Korycka-Dahl & Richardson (70). This subject is complicated by the fact that various forms of activated oxygen react among themselves and the exact form of activated oxygen in any particular system is always in doubt. For example, $^1O_2$ is the product of xanthine oxidase, aldehyde oxidase, galactose oxidase, dihydroorotate oxidase, NADPH-cytochrome reductase, and ferredoxin- and NADPH-reductase (16). In the presence of light, O_2*^- is a byproduct of photosynthesis. H_2O_2 is the product of other oxidases including D-amino acid oxidase, uricase, α-hydroxy acid oxidase, and glucose oxidase. O_2*^- can dismutate spontaneously or with the aid of superoxide dismutase to form $^1O_2$ and H_2O_2 respectively. A disproportionation of H_2O_2 can give rise to O_2*^-. A reaction (Haber-Weiss) between O_2*^- and H_2O_2 can form $^1O_2$ and OH*. $^1O_2$ can also be formed by the reaction of O_2*^- with radical cations. To complicate matters further, other reactions that have been proposed include the formation of $^1O_2$ and hydroxide ions from OH* and O_2*^-.

Scavengers and Inhibitors

The use of activated oxygen scavengers or inhibitors figure prominently in CL research. They are used as probes to evaluate the role of specific metabolic pathways or forms of activated oxygen in the chemiluminescent process. However, because of problems with specificity, penetration, and the fact that the cell contains its own scavengers (enzymes and substances such as carbohydrates, nucleic acids, and amino acids), the results obtained are often contradictory and difficult to interpret.

AZIDE   Azide is both a $^1O_2$ scavenger and an inhibitor of myeloperoxidase (10, 72, 135). Azide may also scavenge OH*. It was reported to reduce CL from Fenton's reagent ($FeSO_4$ + H_2O_2) which generates OH*.

Azide was reported to be an inhibitor of CL produced by: the addition of NADPH to renal microsomes (108), the reaction between submitochondrial particles and organic hydroperoxides (25), and the oxidation of arachidonic acid (29). In the production of CL from roots however, azide was a weak or less effective inhibitor than cyanide (1, 29, 103). These results suggested that $^1O_2$ was not an important source of CL in root tissue.

DABCO   DABCO is a $^1O_2$ scavenger that was not oxidized by $^1O_2$ (88). It has been used in numerous studies as a $^1O_2$ probe or scavenger. DABCO was found to decrease CL from a reaction mixture containing bovine heart submitochondrial particles and organic hydroperoxides (26), CL from a lipid peroxidation system (108), and the CL from soybean seeds (22). However, it had the opposite effect and stimulated CL in numerous other systems thought to generate $^1O_2$ (22, 27, 29, 50, 72).

DIPHENYLISOBENZOFURAN, DIMETHYLFURAN   Diphenylisobenzofuran has been described as a $^1O_2$ trap (16). The related dimethylfuran was shown to have no effect on CL from soybean seeds (22), but it did inhibit CL in a ferricytochrome c-peroxide system (27).

DEUTERIUM OXIDE   Deuterium oxide has been used as a probe in the sense that it prolongs the half life of $^1O_2$. It did not increase the CL from soybean seeds (22).

SUPEROXIDE ANION DISMUTASE (SOD)   SOD has been used to evaluate the role of superoxide anion in various processes by virtue of its ability to convert O_2^{*-} to H_2O_2. Since it is a macromolecule, its use is limited to in vitro systems or systems in which the source of O_2^{*-} is external to the cell. SOD has been shown to inhibit CL of bacteria (101), and luminol-enhanced CL of chloroplasts (120). SOD was found to have a small effect on soybean root CL (103). SOD has also been used to demonstrate the production of O_2^{*-} by PMNs (30). Penicillamine has also been used as a O_2^{*-} probe since it has SOD-like activity (13a, 139).

SULFITE   Sulfite has been proposed as a scavenger for O_2^{*-} produced by xanthine oxidase (11).

BENZOATE   Sodium benzoate is a purported OH* trap (6, 9). It was found to reduce CL from PMNs but had no effect on alveolar macrophages (6). Benzoate was also found to have little or no effect on CL by *Listeria monocytogenes* (101) or NADPH-induced CL from renal microsomes (108).

MANNITOL   Mannitol has been used to inhibit CL from Fentons reagent (9). It, along with sodium formate and thiourea, are OH* radical scavengers (57). Mannitol was found to have no effect on the CL from roots (103).

ASCORBIC ACID   Ascorbate is thought to play a role in reducing levels of activated oxygen in the cell. It is oxidized by both H_2O_2 and O_2*^- to form dehydroascorbic acid. The later reacts with glutathione in the presence of dehydroascorbate reductase to form oxidized glutathione (47). The oxidized glutathione is subsequently reduced by NADPH (58). Ascorbic acid has been shown to reduce CL from wounded soybean roots (103) and the horseradish peroxidase system (118).

TIRON   Tiron is an iron chelator and has been used as an inhibitor of peroxidase. Tiron was found to reduce CL from soybean roots (103). Other iron chelators, EDTA and 1,10-phenanthroline were found to inhibit NADPH induced renal microsomal CL (108).

CATALASE   Catalase was found to reduce the CL from soybean roots (103) and the production of light by water-alcohol fractions from roots (92). Catalase was used to inhibit CL production by a xanthine oxidase-luminol mixture (11). This enzyme has also been used to demonstrate the production of H_2O_2 by PMNs (30).

SALICYLHYDROXAMIC ACID (SHAM)   SHAM, an inhibitor of lipoxygenase, has been found to inhibit the rise in CL associated with soybean germination but not the endogenous CL found in unimbibed seeds (22).

CYANIDE   Cyanide has been employed as a general metabolic inhibitor. CL from pea (1) and soybean (103) was reported to be inhibited by cyanide. However, others have either failed to observe an effect (19) or have reported a promotive effect (48, 77).

DETERGENTS   Triton X-100, Zwitterdent, and Lubrol were reported to increase CL from soybean roots (104). Para-chloromercuribenzoate was reported as an inhibitor of NAD(P)H oxidase (120).

Physiological Roles of Activated Oxygen

A number of roles have been proposed for the various forms of activated oxygen produced by the cell. They include: mediators of cellular damage, promoters of senescence, metabolite oxidation, and biocidal barriers against potential pathogens.

The herbicide paraquat (methyl viologen) is toxic only in the presence of light. In the illuminated chloroplast, paraquat is reduced to a radical that forms O_2^{*-}, and subsequently other forms of activated oxygen such as H_2O_2 (60, 140). The toxicity of this herbicide is a dramatic demonstration of the necessity of controlling the production, and disposal of, activated oxygen in the cell. The photodynamic action of paraquat has a natural analog in the phenomenon known as sunscald, where plants are damaged by a combination of heat and light. In such conditions the damaged photosynthetic apparatus produces an excess of O_2^{*-} that cannot be detoxified by endogenous SOD and carotenoids present in the chloroplast (97). Catalase plays an important role in maintaining H_2O_2 below toxic levels. This has been demonstrated by the report that a lethal barley mutant is deficient in catalase. Death of the leaves is caused by the accumulation of H_2O_2 generated by peroxisomes and released into the cell (64).

The observation that an inhibitor of catalase is produced when tropical plants are chilled lends support to the idea that H_2O_2 may be important in chilling damage. However, chilling did not increase levels of H_2O_2 in the tissue. H_2O_2 has been reported to enhance the senescence of pears (23) and cereals (85). H_2O_2 was also found to increase in cotyledons as they age from an initial level of 7 mM to 18 mM after 10 days (46). In leaves, the effect of H_2O_2 was to increase the levels of RNase and protease and decrease levels of chlorophyll and protein. Antioxidants such as ascorbic acid, L-cysteine, reduced glutathione, and ferrous sulphate reduced the peroxide-mediated senescence (85). It has been suggested that leaf senescence may be a consequence of cumulative membrane deterioration resulting from increasing levels of lipid peroxidation controlled in part by a decline in the activity of SOD and catalase as the leaf ages (36). As carnation flowers age, the lipid fluidity of microsomal membranes decreases, and this effect appears to be the consequence of unregulated super-oxide production (80, 121). Additional support for the idea that lipid peroxidation plays a role in the final stages of plant senescence stems from the observation that a variety of lipoxygenase inhibitors were found to delay the symptoms of senescence in carnations (12).

Lignin is a polymer of hydroxycinnamyl alcohols such as coniferyl, sinapyl, and p-coumaryl alcohol. The polymerization is initiated by peroxidase and H_2O_2 (56). The H_2O_2 used in this reaction can be produced by the reduction of oxygen by NAD(P)H in the presence of a cell wall-bound peroxidase or flavin coenzymes (116).

H_2O_2 and peroxidase (myeloperoxidase) play an important role in the biocidal activity of PMNs (66, 67, 82). Activation of the PMN by phagocytosis results in the formation of O_2^{*-}, H_2O_2, OH*, and perhaps $^1O_2$ as a result of glucose oxidation by the hexose monophosphate shunt (5, 6, 115, 138). Allen et al (7) were the first to report that a burst of CL was associated with the phagocytic activity of PMNs. The CL is associated with a membrane-spanning NAD(P)H oxidase located in the plasma membrane. The enzyme is cyanide-resistant but p-chloromercuribenzoate-sensitive (120). Glucose metabolism supplies the NADPH since inhibitors such as deoxy-D-glucose prevented CL and its effect was reversed by glucose (106). Patients with chronic granulomatous disease have defective PMNs which do not destroy the bacteria they engulf. These cells also fail to produce CL when activated by phagocytosis (138). The emission spectrum of PMNs and the myeloperoxidase-H_2O_2-Cl system have similar broad peaks with a maximum near 570 nm (8, 30). As discussed below, chemiluminegenic probes such as luminol have been useful in studying the role of activated oxygen in PMN phagocytosis (6).

Peroxidases are located in both cell walls and peroxisomes of plant cells (24). This situation is analogous to the PMN, where peroxidases are secreted into the phagosome. This has led workers to suggest that plant peroxidases, and H_2O_2 produced by stressed cells, may play a role in biocidal activity of wounded or infected tissue (63, 73, 125–127).

NATURE OF CHEMILUMINESCENT MOLECULES

Lipids

Lipid oxidation via lipoxygenase probably plays an important role in CL (93). Unsaturated fatty acids produce CL with a quantum yield of 10^{-11} for 18:3 and 22:6 acids (2) and 10^{-12} for linolenic acid (114). The activation energy was reported to be 10 kcal mole^{-1} (114). The emission spectrum of oxidized lipids has a 550 nm peak resulting from the production of excited ketones or aldehydes, and a broad band between 600 nm and 650 nm suggesting the production of $^1O_2$ (2). A number of investigators have pointed out that there are similarities between the emission spectra of oxidizing lipids and that of the CL observed from plants (79), rat liver homogenates (84, 136), and microsomal lipid peroxidation systems (86).

In addition to CL, lipid peroxidation produces ethylene, ethane (18), and long-lived toxic aldehydes such as 4-hydroxynonenal (28). The 4-hydroxynonenal in turn induced ethane production and CL when applied to rat liver cells.

Lipids produce CL because of enzymatic and nonenzymatic reactions. Unimbibed seeds produce light associated with nonenzymatic reactions (20). After 4 h of imbibition, and presumably initiation of germination, CL increased in a linear fashion. SHAM, a lipoxidase inhibitor, reduced the CL of imbibed but not dry seeds. BHA, a general scavenger of free radicals and common food additive, blocked both reactions. According to Boveris et al (20), approximately 20% of the oxygen consumption of soybean seeds resulted from lipid oxidation. While most research supports the hypothesis that lipid oxidation plays a key role in CL, there are two reports that treatments which decreased the levels of lipid in tissue increased CL from corn roots (37, 87).

Lipoxygenase

Boveris et al (17) have provided an in-depth study of the role of lipoxygenase (linoleate:oxygen oxidoreductase, EC 1.12.11.12) as a source of CL. This iron-containing enzyme catalyzes the oxidation of linoleic acid by molecular oxygen, yielding hydroperoxides. The lipoxygenase-catalyzed oxidation of linoleate was a source of low-level chemiluminescence. The ratio of CL to conjugated diene present was about 2.5 to 3.3×10^{-14} photons s^{-1} per diene molecule (20). As with intact plants, no CL was observed in a nitrogen gas phase.

The lipoxygenase reaction was sensitive to SHAM (80% inhibition at 0.5 mM), the free radical trap 2,5-di-t-butylquinol (85% inhibition at 0.1 mM), and insensitive to cyanide. DABCO, a dimol emission enhancer in aqueous systems, did not affect the first CL peak but did enhance the second peak by about 45%. However, this interpretation is difficult to understand since DABCO is a singlet oxygen trap and should remove this species of activated oxygen from the system (88). Cyanide, up to 1 mM, did not inhibit CL or oxygen uptake which agrees with the observation that lipoxygenase is not sensitive to cyanide (17).

In a subsequent study, lipoxygenase was inhibited by antioxidant quinols such as BHA, BHT, DTBQ, propyl gallate, and hydroxamic acids. In a cell-free system, these compounds caused a 75% inhibition of oxygen uptake (BHA was most active at 10 μM). The other compounds were also effective but required larger amounts. BHA was also the most effective in blocking oxygen uptake by soybean axis (100 M BHA caused a 28% inhibition) (20). The emission spectrum in the visible region had maxima at 450 nm and 550 nm and a shoulder around 630 nm. The 630 nm band supports the idea of singlet oxygen involvement (17).

CL is associated with a respiratory burst in phagocytic cells originating from a membrane-spanning NAD(P)H oxidase located within the plasma membrane. Membrane phospholipids are converted into arachidonic acid by phospholipase A2. This enzyme was inhibited by the Upjohn Company compound U3585 and prostaglandin E1. The arachidonic acid is then oxidized by lipoxygenase to form leukotrienes. The leukotrienes are potential mediators of inflammation. The lipoxygenase was inhibited by ETYA and NDGA (abbreviations not defined in original paper). All of these lipoxygenase inhibitors, U3585, prostaglandin E1, ETYA, and NDGA inhibited CL. An alternative pathway from arachidonic acid, catalyzed by cyclooxygenase, which yields prostaglandins, was susceptible to the inhibitory action of acetylsalicylic acid (aspirin) and indomethacin. These inhibitors had no effect on CL which suggests that cyclooxygenase is not involved in leukocyte CL (106). The authors concluded that lipoxygenase is involved in CL through the decomposition of unstable hydroperoxy intermediates generated in the formation of leukotrienes.

However, conflicting results were obtained by Cadenas et al (29). In their experiments, CL was produced by the addition of arachidonic acid to a suspension of ram vesicular gland microsomes or purified prostaglandin synthase (cyclooxygenase). Maximal CL was observed 15 to 30 s after the addition of arachidonic acid. The emission spectrum showed two peaks at 634 nm and 703 nm which is indicative of $^1O_2$ dimol emission. CL was enhanced by DABCO and inhibited by azide, indomethacin, acetylsalicylic acid, and beta-carotene.

CL from rat liver homogenates increased threefold following oral administration of autooxidized linseed oil. This liver CL was quenched in vitro by the following free radical scavengers: BHT, d-α-tocopherol and 2,5-diphenylfuran. Negligible inhibition was observed following addition of SOD, catalase, and D-mannitol. The emission spectrum had peaks at 530 nm, 585 nm, and 635 nm and had the features associated with lipid peroxidation and the presence of $^1O_2$ (84).

Dioxetanes

Dioxetanes are thought to play a key role in the production of light from lipids and other unsaturated hydrocarbons. Details of their formation and subsequent breakdown to form light have been reviewed (14, 81). Dioxetanes are formed by the reaction of dienes with $^1O_2$ to yield a four-membered ring consisting of two carbon atoms joined to two oxygen atoms. These compounds break down to form activated carbonyls which in turn decay to yield ketones and light in the range of 350 nm to 480 nm (25). Shimonura (109) has published the emission spectra of a number of luciferins whose light production is mediated via dioxetanes. In general, the light produced is in the green (500 nm) portion of the spectrum.

Carbonate Radicals

Light can also be generated from carbonate radicals (CO_3^{*-}). These radicals are produced in a reaction between carbonate ions and OH^*. The carbonate radicals thus formed break down to form CO_2, H_2O, and light (61). This mechanism was thought to be involved in the observation that 0.2 M carbonate promoted the CL of the bacteria *Listeria monocytogenes* (101).

Nucleic Acids

DNA has been proposed as a source of CL. Rattemeyer et al (99) provided evidence that chromatin may be a source of photons. In some respects, this work is an extension of the concept of mitogenetic radiation, where as discussed earlier, it was thought that nuclei produced radiation capable of stimulating cell division in neighboring cells. CL of cucumber roots was increased after treatment with ethidium bromide. Ethidium bromide intercalates between DNA base pairs. Increasing the concentration of ethidium bromide causes unwinding of the supercoils followed by reformation of the coils but in an opposite direction. The authors noted a similarity between the dose response curve for the effect of ethidium bromide on CL and the action of ethidium bromide on sedimentation of DNA. As the concentration of ethidium bromide applied to roots increased, CL first increased and then decreased (99).

EMISSION SPECTRA

CL emission spectra have been reported from activated oxygen systems, enzymatic systems, plants, PMNs, and microorganisms. The emission spectra from bioluminescent systems, which involve a specific luciferase and have a higher quantum yield, have been described (34, 133). Because of the small amount of light produced, it is difficult to establish CL emission spectra. The use of filters to measure light production over a portion of the spectrum results in a further reduction in the amount of light available for measurement. The spectral sensitivity of photomultipliers used in these studies are variable and are often insensitive to red and far red light. For purposes of this review the colors of the spectrum in nanometers are defined as: violet <440, blue 440–500, green 500–550, yellow 550–590, red 590–650, and far red >650. In general, chemiluminescent emission spectra are broad Gaussian curves.

Singlet Oxygen

The decay of $^1O_2$ produces light. During monomol emission, when a single molecule is involved, the conversion of $^1O_2$ to $^3O_2$ yields far red light at 1260 nm and 1320 nm (25). There are no reports of the production of light with these wavelengths from plants. The second reaction is between two molecules of $^1O_2$ or dimol emission. The reaction involves two $^1O_2$ molecules or the reaction between a $^1O_2$ and $^3O_2$. The light produced is red (634 nm and 703 nm). Lesser amounts of blue and green light (460 nm and 510 nm) are also produced (10, 18, 25, 61, 65, 72).

Dioxetane

Light is also produced from dioxetane or activated carbonyl groups as they decay from activated to stable states. Most of the light produced by dioxetanes is green (500 nm), though it can vary from 459 nm to 582 nm (109). Activated carbonyl groups produce light with shorter wavelengths, primarily violet, (350 nm to 480 nm) (18, 25). In general, short wavelength light (violet to green) is associated with the deactivation of oxygen attached to a carbon atom while longer wavelength red light (630 nm and higher) is a reflection of the decay of activated oxygen itself.

Plants

For the most part, plant CL emission spectra show a maximum in the green portion of the spectrum. The following maxima have been reported. Soybean, 580 nm (22) and 460 nm to 620 nm (21); bean, 550 nm (33); peas and wheat, 550 nm (79); cucumber, 500 nm and 700 nm (102); pumpkin roots, 530 nm (129); and barley roots, 500 nm (131). The production of light in the UV region, or mitogenic radiation, as it is called, has been reported from *Helianthus* (54). UV light, 250 nm to 380 nm, production by yeast has also been reported (69). Another fungus whose light production extends to the UV region is *Entomophthora virulenta* (113). The peak of the PMN CL emission spectrum is similar to that observed for plants, 570 nm (8) and 575 nm (30).

Enzymatic Reactions

The emission spectra of a number of enzymatic systems have been studied because they were thought to be the source of CL from intact cells. For lipoxygenases, or lipid oxidation, there is a blue peak at 450 nm, a yellow component around 550 nm to 580 nm, and a red one near 640 nm (2, 21, 29, 84, 86, 114, 136). Myeloperoxidase, the peroxidase from PMNs, produces a peak at 570 nm (8). A mixture of protein, H_2O_2, and OCl^- produced light at 475 nm and 625 nm (10). Other peroxidative systems produced primarily blue light, xanthine oxidase, 435 nm (61), and cytochrome c + H_2O_2, 460 nm (112).

From the above data, it is difficult to assign the source of plant CL to a single specific light-generating system. However, based on the emission spectra studies presently available, lipoxidation, and to a lesser extent peroxidation, are prime candidates as the enzyme systems responsible for CL. The role of $^1O_2$ dimol emission appears to be minimal.

CHEMILUMINIGENIC PROBES

Chemiluminigenic substrate probes (hereafter probes) are exogenously applied chemicals that produce light after reacting with various forms of activated oxygen in the cell. Good probes are nontoxic, have a high (0.01) quantum efficiency yield, are soluble (10^{-3} to 10^{-4} M) in physiological media, and have a known reactivity with species of activated oxygen. The advantage of using probes in studies of activated oxygen is that less biological material is needed and CL is not dependent upon a supply of endogenous chemiluminigenic substrates. Probes are used where the purpose of the study is to evaluate solely the production of activated oxygen. Otherwise, CL is dependent upon the activation of endogenous chemiluminigenic substances as well as the production of activated oxygen. Five compounds are commonly used as probes at the present time. They are luminol (5-amino-2,3-dihydro-1,4-phthalazine dione), lucigenin (10,10'-dimethyl-9,9'-biacridinium dinitrate), lophine (2,4,5-triphenylimidazole), oxalate esters [bis (2,4,6-trichlorophenyl) oxalate], and pyrogallol. The following references can be consulted for a description of the structures, properties, and use of these probes in analytical systems (71, 123, 124, 134).

The most frequently used probe is luminol. It has been used to measure activated oxygen in PMNs (5, 6, 105, 135, 138), mitochondria (75), chloroplasts (119), and roots (1, 92). Luminol and lucigenin were also used to demonstrate the production of oxygen radicals by xanthine oxidase (124).

Probes measure different species of activated oxygen. Luminol will react with O_2^{*-}, OCl^-, I^-, MnO_4-, H_2O_2, and perborate (71). Kricka & Thorpe (71) also pointed out that the process is catalyzed by peroxidase, haemin, $Fe(CN)^{6-}$, Ni^{2+}, or Cr^{3+}. The production of light with luminol involves a process of dioxygenation while lucigenin undergoes reductive dioxygenation (5). The fact that these two probes undergo different reactions during activation was clearly demonstrated by Seim (105). He found that 90% of the effectiveness of luminol was lost as PMNs remained in culture over a four-day period, while lucigenin lost only 15% of its activity. In another study, luminol provided a faster response time with PMNs, and its light production was inhibited by azide but SOD had no effect. On the other hand, lucigenin had a slower response time, azide had no effect, and SOD was inhibitory (135).

CHEMILUMINESCENCE BY MITOCHONDRIA AND ENZYMES

Mitochondria

Mitochondria have been shown to produce CL. Beefheart mitochondria (15) produced CL as a result of their generation of O_2^{*-} and H_2O_2. Additional CL was produced if acetaldehyde, ATP, or antimycin A was added to the system. The CL was thought to be a result of the reaction between acetaldehyde with O_2^{*-} and H_2O_2 which formed metastable intermediates. Bean mitochondria have also been shown to produce light (51). In this system, maximum light production was associated with an active oxidative phosphorylation system. The CL of rat liver mitochondria was shown to be enhanced by the addition of a derivative of luminol (75).

Lipoxygenase

The role of lipoxygenase in CL has been covered earlier in this review.

Peroxidase

Peroxidases, both plant and animal, have been shown to produce CL (1, 8, 42, 44, 53, 118). Horseradish peroxidase and H_2O_2 generate dioxetanes from aldehydes which then produce excited aldehydes and formic acid. The emission spectrum is similar to acetone phosphorescence (32). A direct reaction between haemoproteins (such as peroxidase) and H_2O_2 will also produce light (112).

NADPH Oxidase

NADPH oxidase, a flavoprotein widely distributed in mitochondria, plasma membranes, endoplasmic reticulum, and peroxisomes, generates O_2*^- and OH* (5, 35, 91, 98, 100, 116, 120, 132). The oxidation of NADPH by pea homogenates was shown to be a source of CL (1).

Xanthine Oxidase

Xanthine oxidase, utilizing acetaldehyde as a substrate, has been shown to be a source of CL (61, 101).

FACTORS CONTROLLING CL IN PLANTS

Oxygen

Oxygen, or activated oxygen, is the key component of CL. This is shown by the numerous reports that CL is absent or low when plants are incubated in a a N_2 gas phase (19, 38, 48, 128, 131). A half-maximal rate of CL was observed at 3% O_2 (137). Barley root CL was reported to increase in a linear fashion from 0% to 20% O_2; thereafter the rate of increase was slower, but still linear up to a concentration of 100% (131).

Wounding

A number of reports have shown that wounding plants mechanically (103, 104), with disease (78), with chemicals such as formaldehyde (68), actinomycin D (94), by electrical (38) or by thermal stress (4), results in increased CL. Salin & Bridges (104) postulated that increased CL resulted from elevated peroxidative activity following cellular disorganization.

Temperature

Temperature has a number of effects on CL. It governs the rate at which O_2 or activated O_2 reacts with lipids and other unsaturated substrates. Temperature also controls the rate at which activated O_2 is formed from O_2 by oxidases and other enzymes. Finally, high and low temperature extremes have lethal effects on cells which can be monitored by the additional CL produced by the damaged cells.

The CL of soybean seeds was shown to consist of two reactions by studying the temperature dependancy of CL. One reaction had an activation energy of 20 kJ mole^{-1} and the second, 68 kJ mole^{-1}. The first reaction was thought to be by a lipoxidase and the second by fatty acid oxidation (22). However, in this system, similar curves were observed with dry and imbibed seeds, and the production of activated O_2 by the tissue itself was probably not a part of the reaction. Other workers have also noted a discontinuity of CL production curves at 35°C (4, 22), which is indicative of the contribution of two or more processes in CL. On the other hand, other workers report smooth logarithmic curves from temperature dependancy studies, for example in a study of the temperature dependance of CL from living and heat-killed barley roots (131). Though the two curves both showed a logarithmic increase in CL as a function of temperature, the living roots produced more light than the denatured ones. However, when the temperature reached a lethal temperature of 40°C the CL production of living roots superimposed upon the CL of killed roots. Similar results have been observed in other studies (4, 49).

The observation that killing cells at low or high temperatures causes a change in the CL temperature dependency curves has been used in a number of studies designed to measure either the frost hardiness (3, 74, 95, 96) or heat tolerance (40, 76) in horticultural plants. One explanation for the increased CL associated with freezing cells is that freezing water produces latent heat and this energy can produce the additional CL observed (43).

Magnetic Fields

Magnetic fields were reported to influence CL as well as respiration and growth of bean seedlings (39). Low magnetic fields (62 Oe) that promoted respiration and growth by 10% increased CL by 40%. Higher magnetic fields decreased respiration, growth, and CL.

Metal Ions

An increase in CL is also produced when *Elodea canadensis* cells are killed with formaldehyde. When the plants were grown in the presence of excess Cu^+ the CL burst was inhibited; on the other hand, Hg^{2+} increased CL production (68). CL and respiration of intact wheat roots was decreased by a 0.2 mM solution of $Co(NO_3)$ (110).

pH

The production of CL from barley roots was increased when the roots were incubated at pH 9.2 in comparison to roots suspended at pH 7.3 (48). A lower pH was maximal for CL production from the homogenates of rye roots (7.5) (53) and lentils (8.0) (33).

Radiation

CL of pea sprouts was reduced after the plants were irradiated by X rays or UV light (130). X rays also promoted CL from mitochondria isolated from beans (52).

Hormones

Indoleacetic acid reduced baseline levels of soybean root CL (104). However, upon wounding there was an enhancement of light emission in IAA treated roots (104) or when it was applied, along with H_2O_2, to strawberry receptacles (44). In the latter case it acted as probe and replaced pyrogallol which had a similar effect. In a more defined system, indoleacetic acid and peroxidase produced CL, indole-3-carboxaldehyde, and CO_2 (32, 41). The action of indoleacetic acid in these systems is one of a CL probe as opposed to its normal hormonal capacity.

CONCLUSIONS

CL can be observed as a phenomenon and used as a technique. As a phenomenon, we have not progressed much beyond the original observations of Colli et al (33). We know that plants, like other forms of organic matter, produce light in the presence of O_2. The amount of light can be increased by wounding and stress which probably causes a decompartmentalization of the cell. While we know that O_2^{*-} or H_2O_2 are probably the most important forms of activated O_2 in the process, and that activated ketones and dioxetanes are important light-emitting substances, the overall process needs to be sorted out. The technical problems associated with obtaining precise emission spectra make it difficult to identify the relative contributions of lipid oxidation as opposed to the oxidation of other substances, or direct light production from $^1O_2$. As a phenomenon, the CL story is complex and there is always the concern that its meaning may be trivial. However, as a technique, CL may provide useful information on the production and levels of activated O_2 in the cell. CL may be used in much the same way as luciferase and luciferin are used to measure ATP. Additional work is needed to understand what forms of activated O_2 cause chemiluminigenic probes to produce light in plants. In addition, new probes should be tested for their usefulness in plant systems. Progress in this area will result in a better understanding of the role of activated O_2 in growth, development, senescence, and disease resistance in plants.

Literature Cited

1. Abeles, F. B., Leather, G. R., Forrence, L. E. 1978. Plant chemiluminescence. *Plant Physiol.* 62:696–98
2. Adamson, A. W., Slawson, V. 1976. Chemilumnescent autooxidation of unsaturated fatty acid films. Linolenic acid and docosahexenoic acid on silica gel. *Colloid Interface Sci.* 5:193–201
3. Agaverdiyev, A. S., Doskoch, Y. E., Tarusov, B. N. 1965. Effect of low temperatures on the ultraweak luminescence of plants. *Biofizika* 10:832–36
4. Agaverdiyev, A. S., Tarusov, B. N. 1965. Ultraweak chemiluminescence of the stems of wheat in relation to temperature. *Biofizika* 10:351–52
5. Allen, R. C. 1982. Biochemiexcitation: Chemiluminescence and the study of biological oxygenation reactions. In *Chemical and Biological Generation of Excited States*, ed. W. Adams, G. Cilento, pp. 1–35. New York: Academic
6. Allen, R. C., Loose, L. D. 1976. Phagocytic activation of a luminol-dependent chemiluminescence in rabbit alveolar and peritoneal macrophages. *Biochem. Biophys. Res. Commun.* 69:245–52
7. Allen, R. C., Stjernholm, R. L., Steele, R. H. 1972. Evidence for the generation of an electronic excitation state(s) in human polymorphonuclear leukocytes and its participation in bactericidal activity. *Biochem. Biophys. Res. Commun.* 47:679–84
8. Andersen, B. R., Brendzel, A. M., Lint, T. F. 1977. Chemiluminescence spectra of human myeloperoxidase and polymorphonuclear leukocytes. *Infect. Immunol.* 17:62–66
9. Andersen, B. R., Harvath, L. 1979. Light generation with Fenton's reagent. Its relationship to granulocyte chemiluminescence. *Biochim. Biophys. Acta* 584:164–73
10. Andersen, B. R., Lint, T. F., Brendzel, A. M. 1978. Chemically shifted singlet oxygen spectrum. *Biochim. Biophys. Acta* 542:527–36
11. Arneson, R. M. 1970. Substrate-induced chemiluminescence of xanthine oxidase and aldehyde oxidase. *Arch. Biochem. Biophys.* 136:352–60
12. Baker, J. E., Wang, C. Y., Terlizzi, D. E. 1985. Delay of senescence in carnations by pyrazon, phenidone analogues, and tiron. *HortScience* 20:121–22
13. Baremboin, G. M., Domanski, A. N., Turoverov, K. K. 1966. Chemilumines-

cence of cells and organisms. In *Luminescence of Biopolymers and Cells*, pp. 114–42. New York: Plenum
13a. Birker, P. J. M. W., Freeman, H. C. 1977. Structure, properties, and function of a copper(I)-copper(II) complex of d-penicillamine - pentathallium(I) Mu - 8-chlorododeca(d - penicillaminato) - octacuprate(1)-hexacupra te (II) normal-hydrate. *J. Am. Chem. Soc.* 99:6890–99
14. Bogan, D. J. 1982. Gas-phase dioxetane chemiluminescence. See Ref. 5, pp. 37–83
15. Boh, E. E., Baricos, W. H., Bernofsky, C., Steele, R. H. 1982. Mitochondrial chemiluminescence elicited by acetaldehyde. *J. Bioenerg. Biomembr.* 14:115–33
16. Bors, W., Saran, M., Lengfelder, E., Spottl, R., Michel, C. 1974. The relevance of the superoxide anion radical in biological systems. *Curr. Top. Radiat. Res. Q.* 9:247–309
17. Boveris, A., Cadenas, E., Chance, B. 1980. Low level chemiluminescence of the lipoxygenase reaction. *Photobiochem. Photobiophys.* 1:175–82
18. Boveris, A., Cadenas, E., Chance, B. 1981. Ultraweak chemiluminescence: A sensitive assay for oxidative radical reactions. *Fed. Proc.* 40:195–98
19. Boveris, A., Cadenas, E., Reiter, R., Filipkowski, M., Nakase, Y., et al. 1980. Organ chemiluminescence: Noninvasive assay for oxidative radical reactions. *Proc. Natl. Acad. Sci. USA* 77:347–51
20. Boveris, A., Puntarulo, S. A., Roy, A. H., Sanchez, R. A. 1984. Spontaneous chemiluminescence of soybean embryonic axes during imbibition. *Plant Physiol.* 76:447–51
21. Boveris, A., Sanchez, R. A., Varsavsky, A. I., Cadenas, E. 1980. Spontaneous chemiluminescence of soybean seeds. *FEBS Lett.* 113:29–32
22. Boveris, A., Varsavsky, A. I., Da Silva, S. G., Sanchez, R. A. 1983. Chemiluminescence of soybean seeds: Spectral analysis, temperature dependence and effect of inhibitors. *Photochem. Photobiol.* 38:99–104
23. Brennan, T., Frenkel, C. 1977. Involvement of hydrogen peroxide in the regulation of senescence in pear. *Plant Physiol.* 59:411–16
24. Brinkman, F. G., Sminia, T. 1977. Histochemical location of catalase in peroxisomes and of peroxidase in cell walls

and Golgi bodies of cells in differentiating potato tuber tissue. *Z. Pflanzenphysiol.* 84:407–12

25. Cadenas, E., Arad, I. D., Boveris, A., Fisher, A. B., Chance, B. 1980. Partial spectral analysis of the hydroperoxide-induced chemiluminescence of the perfused lung. *FEBS Lett.* 111:413–18

26. Cadenas, E., Boveris, A., Chance, B. 1980. Low-level chemiluminescence of bovine heart submitochondrial particles. *Biochem. J.* 186:659–67

27. Cadenas, E., Boveris, A., Chance, B. 1980. Low-level chemiluminescence of hydroperoxide-supplemented cytochrome c. *Biochem. J.* 187:131–40

28. Cadenas, E., Muller, A., Brigelius, R., Sies, H., Lang, J., et al. 1984. Low-level chemiluminescence, alkane production and glutathione depletion in isolated hepatocytes caused by a diffusible product of lipid peroxidation, 4-hydroxynonenal. In *Oxygen Radicals in Chemistry and Biology. Proc. 3rd Int. Conf.*, ed. W. Bors, M. Saran, D. Tait, pp. 318–24. Berlin/New York: de Gruyter

29. Cadenas, E., Sies, H., Nastainczyk, W., Ullrich, V. 1983. Singlet oxygen formation detected by low-level chemiluminescence during enzymatic reduction of prostaglandin G2 to H2. *Hoppe-Seyler's Z. Physiol. Chem.* 364:519–28

30. Cheson, B. D., Christensen, R. L., Sperling, R., Kohler, B. E., Babior, B. M. 1976. The origin of the chemiluminescence of phagocytosing granulocytes. *J. Clin. Invest.* 58:789–96

31. Chuma, Y., Nakaji, K., Ohura, M. 1982. Maturity and freshness evaluation of Japanese apricots by means of delayed light emission. *J. Fac. Agric. Kyushu Univ.* 27:21–31

32. Cilento, G. 1982. Electronic excitation in dark biological processes. See Ref. 5, pp. 277–307

33. Colli, L., Facchini, U., Guidotti, G., Lonati, R. D., Orsenigo, M., et al. 1955. Further measurements on the bioluminescence of the seedlings. *Experientia* 11:479–81

34. Cormier, M. J., Hercules, D. M., Lee, J. 1973. *Chemiluminescence and Bioluminescence.* New York: Plenum

35. Curnutte, J. T., Kipnes, R. S., Babior, B. M. 1975. Defect in pyridine nucleotide dependent superoxide production by a particulate fraction from the granulocytes of patients with chronic granulomatous disease. *N. Engl. J. Med.* 293:628–32

36. Dhindsa, R. S., Plumb-Dhindsa, P., Thorpe, T. A. 1981. Leaf senescence: correlated with increased levels of membrane permeability and lipid peroxidation, and decreased levels of superoxide dismutase and catalase. *J. Exp. Bot.* 32:93–101

37. Doskoch, Y. E., Gorev, A. S. 1972. Role of antioxidants of a lipid and phenol nature in the genesis of quanta of superweak spontaneous chemiluminescence in plants. *Dokl. Akad. Nauk SSSR* 207:992–95 (In Russian)

38. Doskoch, Y. E., Kovrizhnykh, V. V., Tarusov, B. N. 1974. The effect of electronic potential on the spontaneous hyperweak chemiluminescence of plants. *Dokl. Akad. Nauk SSSR* 218:1451–53 (In Russian)

39. Doskoch, Y. E., Strekova, V. Y., Tarakanova, G. A., Tarusov, B. N. 1969. Spontaneous ultraweak chemiluminescence of plants in relation to a change of their biological activity in a permanent-magnetic field. *Sov. Plant Physiol.* 16:222–26 (From Russian)

40. Doskoch, Y. E., Yakovlev, A. P., Tarusov, B. N. 1969. Spontaneous ultraweak chemiluminescence of inbred lines and interline plant hybrids. *Biofizika* 14:561–63

41. Duran, N. 1982. Singlet oxygen in biological processes. See Ref. 5, pp. 345–69

42. Duran, N., Zinner, K., De Baptista, R. C., Vidigal, C. C. C., Cilento, G. 1976. Chemiluminescence from the oxidation of auxin derivatives. *Photochem. Photobiol.* 24:383–88

43. Dzhanumov, D. A., Veselovskii, V. A., Tarusov, B. N., Marenkov, V. S., Pogosyan, S. I. 1971. Temperature resistance of plants studied by methods of spontaneous and photoinduced chemiluminescence. *Sov. Plant Physiol.* 18:496–500 (From Russian)

44. Dzieciol, U., Antoszewski, R. 1969. Peroxidase activity of strawberry receptacle as determined by chemiluminescence. *Biol. Plant.* 11:457–64

45. Ellenson, J. L., Raba, R. M. 1983. Gas exchange and phytoluminography of single red kidney bean leaves during periods of induced stomatal oscillations. *Plant Physiol.* 72:90–95

46. Ferguson, B., Watkins, C. B., Harman, J. E. 1983. Inhibition by calcium of senescence of detached cucumber cotyledons. *Plant Physiol.* 71:182–86

47. Foyer, C. H., Halliwell, B. 1977. Purification and properties of dehydroascorbate reductase from spinach leaves. *Phytochemistry* 16:1347–50

48. Gasanov, R. A., Mamedov, T. G., Tarusov, B. N. 1963. Spontaneous induction of luminescence in plants under aerobic

and anaerobic conditions. *Dokl. Akad. Nauk SSSR* 150:913–15 (From Russian)
49. Gasanov, R. A., Mamedov, T. G., Tarusov, B. N. 1963. Relationship between the very weak chemiluminescence and the heat resistance of plant organisms. *Dokl. Akad. Nauk SSSR* 153:947–49 (From Russian)
50. Goda, K., Chu, J., Kimura, T., Schaap, A. P. 1973. Cytochrome *c* enhancement of singlet molecular oxygen production by the NADPH-dependent adrenodoxin reductase-adrenodoxin system: The role of singlet oxygen in damaging adrenal mitochondrial membranes. *Biochem. Biophys. Res. Commun.* 52:1300–6
51. Gorlanov, N. A., Churmasov, A. V. 1974. Influence of x-rays on the ultraweak chemiluminescence of the mitochondria of kidney bean plants. *Radiobiology* 14:185–87 (From Russian)
52. Gorlanov, N. A., Kokorev, Yu. M. 1973. Influence of gamma irradiation of seeds on the ultraweak chemiluminescence and antioxidant activity of wheat, corn, and buckwheat seedlings. *Radiobiology* 13:201–5 (From Russian)
53. Grabiec, S., Bogdanski, K., Marczukajtis, A. 1968. Variability of chemiluminescence intensity of grain tissues in rye of various periods of storage durations. *Bull. Acad. Pol. Sci. Ser. Sci. Biol.* 16:761–64
54. Grebe, L., Krost, A., Peukert, L. 1937. Versuche zum physikalischen Nachweis der mitogenetischen Strahlung. *Strahlentherapie* 60:575–81
55. Gurwitsch, A. G., Gurwitsch, L. D. 1959. *Die Mitogenetische Strahlung.* Jena: Fischer Verlag
56. Halliwell, B. 1978. Lignin synthesis: The generation of hydrogen peroxide and superoxide by horseradish peroxidase and its stimulation by manganese (II) and phenols. *Planta* 140:81–88
57. Halliwell, B. 1978. Superoxide-dependent formation of hydroxyl radicals in the presence of iron chelates. *FEBS Lett.* 92:321–26
58. Halliwell, B., Foyer, C. H. 1978. Properties and physiological function of a glutathione reductase purified from spinach leaves by affinity chromatography. *Planta* 139:9–17
59. Hara, T., Toriyama, M., Tsukagoshi, K. 1983. Determination of a small amount of biological constituent by use of chemiluminescence. I. The flow-injection analysis of protein. *Bull. Chem. Soc. Jpn.* 56:1382–87
60. Hassan, H. M., Fridovich, I. 1979. Paraquat and *Escherichia coli*. Mechanism of

production of extracellular superoxide radical. *J. Biol. Chem.* 254:10846–852
61. Henry, J. P., Michelson, A. M. 1979. Light emissions involving the superoxide anion. In *Biochemical and Medical Aspects of Active Oxygen,* ed. O. Hayaishi, K. Asada, pp. 135–51. Baltimore: Univ. Park Press
62. Herring, P. J., ed. 1978. *Bioluminescence in Action.* New York: Academic. 461 pp.
63. Jacob, A. A., Low, I. E., Paul, B. B., Strauss, R. R., Sbarra, A. J. 1972. Mycoplasmacidal activity of peroxidase-H_2O_2-halide systems. *Infect. Immunol.* 5:127–31
64. Kendall, A. C., Keys, A. J., Turner, J. C., Lea, P. J., Miflin, B. J. 1983. The isolation and characterization of a catalase-deficient mutant of barley (*Hordeum vulgare* L.). *Planta* 159:505–11
65. Khan, A. U., Kasha, M. 1964. Rotational structure in the chemiluminescence spectrum of molecular oxygen in aqueous systems. *Nature* 204:241–43
66. Klebanoff, S. J. 1967. Iodination of bacteria: A bactericidal mechanism. *J. Exp. Med.* 126:1063–78
67. Klebanoff, S. J. 1968. Myeloperoxidase-halide-hydrogen peroxide antibacterial system. *J. Bactiol.* 95:2131–38
68. Kochetov, Yu. V., Tarusov, B. N. 1976. Ultraweak chemiluminescence of the leaves of aquatic plants. *Biophysics* 20:551–54
69. Konev, S. V., Lyskova, T. I., Nisenbaum, G. D. 1966. Very ultraweak bioluminescence of cells in the ultraviolet region of the spectrum and its biological role. *Biofizika* 11:361–63
70. Korycka-Dahl, M. B., Richardson, T. 1978. Activated oxygen species and oxidation of food constituents. *CRC Crit. Rev. Food Sci. Nutr.* 10:209–41
71. Kricka, L. J., Thorpe, G. H. G. 1983. Chemilumnescent and bioluminescent methods in analytical chemistry. *Analyst* 108:1274–96
72. Krinsky, N. I. 1984. Biology and photobiology of singlet oxygen. See Ref. 28, pp. 453–64
73. Lehrer, R. I. 1969. Antifungal effects of peroxidase systems. *J. Bacteriol.* 99:361–65
74. Limberger, G. E. 1974. Spontaneous and photoinduced chemiluminescence in apple trees differing in frost resistance. In *Biofizicheskie I Fiziologo-Biokhimi Cheskie,* pp. 62–70. Issledovanii Plodovykh I Iagodnykh Kul'tur. (In Russian)
75. Lippman, R. D. 1980. Chemilumnescent measurement of free radicals and

antioxidant molecular-protection inside living rat-mitochondria. *Exp. Gerontol.* 15:339–51

76. Lishchuk, A. I., Il'nitskii, O. A. 1976. Application of the methods of photoinduced chemiluminescence for studying heat resistance of fruit trees. *S-kh. Biol.* 11:933–34 (In Russian)

77. Lloyd, D., Boveris, A., Reiter, R., Filipkowski, M., Chance, B. 1979. Chemiluminescence of *Acanthamoeba castellanii. Biochem. J.* 184:149–56

78. Mamaev, A. T., Tarusov, B. N., Kutlaev, B. N. 1975. Chemiluminescence of grapevine, normal and infected by phylloxera, downy mildew and chlorosis; testing for resistance to diseases and pests in plant breeding. In *Problemy Sel's-kokhoziaistvennoi,* ed. G. Dobrovol'skii, pp. 384–88. V.Nauki v Moskovskom Universitete. (In Russian)

79. Mamedov, T. G., Popov, G. A., Konev, V. V. 1972. Spectral composition of superweak biochemiluminescence of plants and some lipids. *Trans. Moscow Obshchest. Ispyt. Prir.* 39:177–79 (In Russian)

80. Mayak, S., Legge, R. H., Thompson, J. E. 1983. Superoxide radical production by microsomal membranes from senescing carnation flowers: and effect on membrane fluidity. *Phytochemistry* 22:1375–80

81. McCapra, F. 1978. The chemistry of bioluminescence. In *Bioluminescence in Action,* ed. P. J. Herring, pp. 49–73. New York: Academic

82. McRipley, R. J., Sbarra, A. J. 1967. Role of the phagocyte in host-parasite interactions. XII. Hydrogen peroxide-myeloperoxidase bactericidal system in the phagocyte. *J. Bacteriol.* 94:1425–30

83. Mendenhall, G. D. 1977. Analytical applications of chemiluminescence. *Angew. Chem. Int. Ed. Engl.* 16:225–32

84. Miyazawa, T., Kaneda, T., Takyu, C., Inaba, H. 1983. Characteristics of tissue ultraweak chemiluminescence in rats fed with autooxidized linseed oil. *J. Nutr. Sci. Vitaminol.* 29:53–64

85. Mondal, R., Choudhuri, M. A. 1981. Role of hydrogen peroxide in senescence of excised leaves of rice and maize. *Biochem. Physiol. Pflanzen.* 176:700–9

86. Nakano, M., Noguchi, T., Sugioka, K., Fukuyama, H., Sato, M., et al. 1975. Spectroscopic evidence for the generation of singlet oxygen in the reduced nicotinamide adenine dinucleotide phosphate-dependent microsomal lipid peroxidation system. *J. Biol. Chem.* 250: 2404–6

87. Nenko, N. I., Krasnook, N. P., Fisenko, V. Y. 1981. Intensity of superweak luminescence and changes in the lipid complex of maize germinating seed embryo roots induced by (5H)-furanon-2. *Biochem. Biophys. Cult. Plants* 13: 400–4 (In Russian)

88. Ouannes, C., Wilson, T. 1968 Quenching of singlet oxygen by tertiary aliphatic amines. Effect of DABCO. *J. Am. Chem. Soc.* 90:6527–28

89. Patterson, B. D., Payne, L. A., Chen, Y-Z., Graham, D. 1984. An inhibitor of catalase induced by cold in chilling sensitive plants. *Plant Physiol.* 76:1014–18

90. Deleted in proof

91. Pederson, T. C., Aust, S. D. 1972. NADPH-dependent lipid peroxidation catalyzed by purified NADPH-cytochrome *c* reductase from rat liver microsomes. *Biochem. Biophys. Res. Commun.* 48:789–95

92. Pogosyan, S. I., Aver'yanov, A. A., Merzlyak, M. N., Veselovskii, V. A. 1978. Extracellular chemiluminescence of plant roots. *Dokl. Akad. Nauk SSSR* 239:974–76 (In Russian)

93. Pogosyan, S. I., Koretskaya, T. F., Veselovskii, V. A. 1974. Substrates and inhibitors of superweak luminescence of plant tissues. *Trans. Moscow Soc. Nat.* 50:99–103 (In Russian)

94. Popp, F. A., Ruth, B., Bahr, W., Bohm, J., Grass, P., et al. 1981. Emission of visible and ultraviolet radiation by active biological systems. *Collect. Phenomol.* 3:187–214

95. Prilutskii, A. V., Doskoch, Y. E., Tarusov, B. N. 1973. Evaluation of frost-resistance in mulberry tree based on flashes of spontaneous ultralow chemiluminescence. *S-kh. Biol.* 8:841–46 (In Russian)

96. Prilutskii, A. V., Doskoch, Y. E., Tarusov, B. N. 1974. Specificity of spontaneous ultraweak chemiluminescence and some morpho-physiological symptoms of winter hardiness in mulberry trees as affected by the conditions of the thermoperiod. *S-kh. Biol.* 9:397–404 (In Russian)

97. Rabinowitch, H. D., Sklan, D., Budowski, P. 1982. Photooxidative damage in the ripening tomato fruit. Protective role of superoxide dismutase. *Physiol. Plant.* 54:369–74

98. Ramasarma, T., Swaroop, A., MacKellar, W., Crane, F. L. 1981. Generation of hydrogen peroxide on oxidation of NADH by hepatic plasma membranes. *J. Bioenerg. Biomem.* 13:241–53

99. Rattemeyer, M., Popp, F. A., Nagl, W. 1981. Evidence of photon emission from DNA in living cells. *Naturwissenchaften* 68:572–73

100. Rich, P. R., Boveris, A., Bonner, W. D. Jr., Moore, A. L. 1976. Hydrogen peroxide generation by the alternate oxidase of higher plants. *Biochem. Biophys. Res. Commun.* 71:695–703
101. Roth, J. A., Kaeberle, M. L. 1980. Chemiluminescence by *Listeria* monocytogenes. *J. Bacteriol.* 144:752–57
102. Ruth, B. 1979. Experimental investigations on ultraweak photon emission. In *Electromagnetic Bio-Information,* ed. Popp, F. A., pp. 107–22. Vienna: Urban & Schuwizenberg
103. Salin, M. L., Bridges, S. M. 1981. Chemiluminescence in wounded root tissue. *Plant Physiol.* 67:43–46
104. Salin, M. L., Bridges, S. M. 1983. Chemiluminescence in soybean root tissue: Effect of various substrates and inhibitors. *Photobiochem. Photobiophys.* 6:57–64
105. Seim, S. 1983. Role of myeloperoxidase in the luminol-dependent chemiluminescence response of phagocytosing human monocytes. *Acta Pathol. Microbiol. Immunol. Scand. Sect. C* 91:123–28
106. Semadeni, B., Weidemann, M. J., Peterhans, E. 1984. Biological aspects of chemiluminescence induced by sendai virus in mouse spleen cells. In *Nonsegmented Negative Strand Viruses,* pp. 451–58. New York: Academic
107. Seyfert, F. 1932. Uber den physikalischen Nachweis von mitogenetischen strahlen *Jahrb. Wiss. Bot.* 76:747–64 (In German)
108. Shah, S. V., Cruz, F. C., Baricos, W. H. 1983. NADPH-induced chemiluminescence and lipid peroxidation in kidney microsomes. *Kidney Int.* 23:691–98
109. Shimonura, O. 1982. Mechanism of bioluminescence. See Ref. 5, pp. 249–76
110. Shkliaev, Iu. N., Nikolaevskii, V. S. 1975. The effect of non-root supplement of cobalt on the chemiluminescence and respiration of the wheat root. *Agrokhimiia* 3:120–22 (In Russian)
111. Slawinska, D., Slawinski, J. 1983. Biological chemiluminescence. *Photochem. Photobiol.* 37:709–15
112. Slawinski, J., Galezowski, W., Elbanowski, M. 1981. Chemiluminescence in the reaction of cytochrome *c* with hydrogen peroxide. *Biochim. Biophys. Acta* 637:130–37
113. Slawinski, J., Majchrowicz, I., Grabikowski, E. 1981. Ultraweak luminescence from germinating resting spores of *Entomophthora virulenta. Acta Mycol.* 17:131–39
114. Slawson, V., Adamson, A. W. 1976. Chemilumnescent autooxidation of lin-

olenic acid films on silica gel. *Lipids* 11:472–77
115. Spitznagel, J. K. 1977. Bacteriocidal mechanisms of the granulocyte. In *The Granulocyte: Functions and Clinical Utilization,* pp. 103–31. New York: Liss
116. Stich, K., Ebermann, R. 1984. Investigation of hydrogen peroxide formation in plants. *Phytochemistry* 23:2719–22
117. Strum, B. 1935. Eine neue Ausfuhrungsform des Geigerschen Spitzensahlers zur Lichtmessung und Beispiele fur seine Anwending. *Z. Physik* 94:85–103 (In German)
118. Szczodrowska, B., Wlodarczyk-Graetzer, M., Slawinski, J. 1972. Utilization of chemiluminescence for the determination of peroxidase activity in plant extracts. II. Kinetics and mechanism of reaction of peroxidase oxidase oxidation or purpurogallin. *Zesz. Nauk. Wyzsz. Szk. Roln. Szczecinic* 9:455–68 (In Polish)
119. Takahama, U., Nishimura, M. 1977. Light-induced chemiluminescence of luminol in spinach chloroplast fragments: Reaction of O_2^- with electron transfer components. *Plant Cell Physiol.* 18:1139–48
120. Takanaka, K., O'Brien, P. J. 1975. Mechanisms of H_2O_2 formation by leukocytes. Evidence for a plasma membrane location. *Arch. Biochem. Biophys.* 169:428–35
121. Thompson, J. E., Mayak, S., Shinitzky, M., Halevy, A. H. 1982. Acceleration of membrane senescence in cut carnation flowers by treatment with ethylene. *Plant Physiol.* 69:859–63
122. Timms, R. E., Roupas, P., Rogers, W. P. 1982. Determination of oxidative deterioration of milk powder and reconstituted milk by measurement of chemiluminescence. *J. Dairy Sci.* 49:649–54
123. Totter, J. R. 1975. Light production in alkaline mixtures of reducing agents and dimethylbiacridylium nitrate. *Photochem. Photobiol.* 22:203–11
124. Totter, J. R., de Dugros, E. C., Riveiro, C. 1960. The use of chemiluminescent compounds as possible indicators of radical production during xanthine oxidase action. *J. Biol. Chem.* 235:1839–42
125. Urs, N. V. R., Dunleavy, J. M. 1974. Bactericidal activity of horseradish peroxidase on *Xanthomonas phaseoli* var. Sojensis. *Phytopathology* 64:542–45
126. Urs, N. V. R., Dunleavy, J. M. 1974. The function of peroxidase in resistance of soybean to bacterial pustule. *Crop Sci.* 14:740–44

127. Urs, N. V. R., Hill, J. H. 1978. Inactivation of southern bean mosaic virus by a horseradish peroxidase-hydrogen peroxide system. *Phytopathol. Z.* 91:365–68

128. Vartapetian, B. B., Agapova, L. P., Aver'yanov, A. A., Veselovskii, V. A. 1974. A study of molecular oxygen transport from overground parts to the roots of plants by the methods of chemiluminescence. *Sov. Plant Physiol.* 21:406–8 (From Russian)

129. Vartapetian, B. B., Agapova, L. P., Aver'yanov, A. A., Veselovskii, V. A. 1974. New approach to study of oxygen transport in plants using chemiluminescent method. *Nature* 249:69

130. Veselova, T. V., Veselovskii, V. A. 1971. Effect of ultraviolet and x-irradiation on superslight chemiluminescence of pea sprouts. *Radiobiologiia* 11:627–30 (In Russian)

131. Veselovskii, V. A., Sekamova, E. N., Tarusov, V. N. 1963. Mechanism of ultraweak spontaneous luminescence of organisms. *Biofizika* 8:125–27

132. Wakeyama, H., Takeshige, K., Takayanagi, R., Minakami, S. 1982. Superoxide-forming NADPH oxidase preparation of pig polymorphonuclear leukocytes. *Biochem. J.* 205:593–601

133. Wampler, J. E. 1978. Measurements and physical characteristics of luminescence. See Ref. 61, pp. 1–48

134. Whitehead, T. P., Kricka, L. J., Carter, T. J. N., Thorpe, G. H. G. 1979. Analytical luminescence: Its potential in the clinical laboratory. *Clin. Chem.* 25: 1531–46

135. Williams, A. J., Cole, P. J. 1981. The onset of polymorphonuclear leukocyte membrane stimulated metabolic activity. *Immunology* 43:733–39

136. Wright, J. R., Rumbaugh, R. C., Colby, H. D., Miles, P. R. 1979. The relationship between chemiluminescence and lipid peroxidation in rat hepatic microsomes. *Arch. Biochem. Biophys.* 192: 344–51

137. Yafarova, I. O., Veselovskii, V. A. 1969. Study of the kinetic patterns of the luminescence of the root system of seedlings. *Biofizika* 14:364–66

138. Yanai, M., Quie, P. G. 1981. Chemiluminescence by polymorphonuclear leukocytes adhering to surfaces. *Infect. Immunol.* 32:1181–86

139. Youngman, R. J., Dodge, A. D. 1979. Mechanism of paraquat action: Inhibition of the herbicidal effect by a copper chelate with superoxide dismutating activity. *Z. Naturforsch. Teil C* 34:1032–35

140. Youngman, R. J., Elstner, E. F. 1981. Oxygen species in paraquat toxicity: The crypto-OH radical. *FEBS Lett.* 129:265–68

Ann. Rev. Plant Physiol. 1986. 37:73–92

ORGANIZATION OF THE ENDOMEMBRANE SYSTEM

N. Harris

Department of Botany, University of Durham, Durham DH1 3LE, United Kingdom

CONTENTS

INTRODUCTION

The major components of the endomembrane system of higher plant cells are generally considered to include the endoplasmic reticulum (ER), the Golgi apparatus (GA), and their transition elements, with the various vesicles, vacuoles, and other organelles derived from them. The nuclear envelope (NE) (some authors consider only the outer membrane) is also included in most discussions as is the plasmalemma (PL), although this is not strictly an endomembrane. The membranes of the semiautonomous organelles, the mitochondria and plastids, are usually excluded from general consideration of the endomembrane system.

0066-4294/86/0601-0073$02.00

It is now more than a decade since Morré and Mollenhauer developed the endomembrane "concept" in a review of the endomembrane system of higher plant cells (63). In this they attempted to describe not merely the individual structures of the endomembrane system but to correlate structure with function in a developmental, and sometimes physical, continuum which involves the various components of the endomembrane system. Since the development of the endomembrane concept, increasing emphasis has been placed on studies of structure/function relationships. Significant advances have been made in methodology to aid this approach. However, there remain tantalizing gaps in our knowledge that are difficult to fill either by examination of a series of static images from electron microscopy or by analysis of specific organelle-enriched fractions. It is thus true that while there is increasing evidence to support the endomembrane concept in many biological tissues, there are also examples from careful studies of higher plant tissues which suggest that such a concept is not universally applicable.

This review considers some of the advances made in the study of the three-dimensional, structural interrelationships of the components of the endomembrane system and how these are associated with both intracellular compartmentalization and extracellular secretion.

METHODS

The detailed form of the endomembrane system within higher plant cells has been revealed by electron microscopy. This technique usually involves chemical fixation of tissues and their subsequent embedding and thin sectioning. Assuming that the original fixation is acceptable, two problems remain: (*a*) cells are three-dimensional, but at approximately 80nm, sections are essentially two-dimensional; and (*b*) the dynamic events of living tissues can only be inferred from a series of static images. Recent advances in both microscopical and biochemical techniques and the development of techniques that bridge these disciplines have added to our understanding of structure/function aspects of the endomembrane system.

Microscopical Studies

Although serial sectioning can lead to three-dimensional reconstruction, this can be both difficult and time consuming and the results are not easily presented or interpreted. The examination of selectively stained thick sections, initially by high voltage electron microscopy (HVEM), has given dramatic three-dimensional images of the endomembrane system. Originally osmium impregnation methods were used for selective staining of parts of the endomembrane system, although more recently complexes of zinc iodide and osmium tetroxide (ZIO) (e.g. 34, 41, 54) or zinc oxide and potassium ferri- (43) or ferro-cyanide (84) (OsFeCN) have been used (see 34, 40). These complexes can act as selective stains for the ER, Golgi cisternae, nuclear envelope, etc, although the precise nature of the staining reaction is unclear. Although originally developed to capitalize upon the increased beam penetration available in HVEM, thick section studies have also been applied to conventional (C)TEM (see 34).

While such EM methods have given structural information about the interrelationships of the components of the endomembrane system, cytochemical techniques have added to our understanding of the associated biochemical compartmentations and changes therein. Enzyme cytochemistry and autoradiography continue to be useful, but the developments in EM immunocytochemistry have perhaps been the most important. Specific proteins can be localized at an ultrastructural level as well as at a histological level, using monospecific antibodies that are raised either in a polyclonal or monoclonal system. The bound monospecific antibodies are located with a "marker" visible under either optical or electron microscopy. For histological and cytological work, fluorescent markers are becoming increasingly important, while at an EM level, the earlier developments with ferritin-labeled secondary antibodies (3) are now being replaced by the use of gold colloids coated with either Protein A (81) or a suitable secondary antibody.

Other microscopical techniques of increasing value in study of the endomembrane system include the use of image enhancement techniques and fluorescent probes (e.g. 38) to study the dynamic changes in living tissues and the use of in situ hybridization at both optical and electron microscopy levels (e.g. 37a).

Biochemical Studies

Many of the functional properties of the endomembrane components have been determined largely by their biochemical composition and capabilities after isolation. There are, however, obvious problems in separating fractions that are similar in composition and may even have some limited physical interconnection. The ER, for example, is found in both "rough" and "smooth" forms, depending upon the presence or absence of associated polyribosomes, and it is composed of both cisternal (usually rough) and tubular elements that are interconnected. During extraction both convert to microsomal forms, although with Mg^{2+} in the extraction medium the respective rough and smooth microsomal fractions may be separated by density gradient centrifugation. With the correct level of Mg^{2+} the ribosomes remain attached and the rough fraction has a density of 1.15–1.18 g.cm^{-3}. The separation is complicated by the presence of Golgi components with a density around 1.15 g.cm^{-3}, the mitochondrial fraction at 1.18–1.2 g.cm^{-3}, and sometimes a broken chloroplast fraction at 1.14–1.17 g.cm^{-3}. Similarly, extraction of the smooth microsomal fraction, or stripped rough ER in the presence of low Mg^{2+} concentrations, is complicated by other fractions with similar densities. Despite such problems, recent advances in experimental technique have resulted in improved fractionation of, for example, intact vacuoles from protoplasts and also Golgi-enriched fractions. The latter may now be prepared without stabilizing fixatives, and the fraction therefore retains more of the characteristic biochemical activities. These and other improvements are discussed below in relation to structure/function aspects of the components of the endomembrane system.

COMPONENTS OF THE ENDOMEMBRANE SYSTEM

Endoplasmic Reticulum

In its various forms the endoplasmic reticulum is the most abundant membrane within the higher plant cell. It is, however, not an organelle of static form or amount but varies, in both respects with cell type, stage of differentiation and predominant metabolic activity. The ER is the source of most of the membrane synthesized within the cell and is also the site of many of the proteins and lipids that are subsequently destined either for intracellular deposition or extracellular secretion. ER structure and function have been the subjects of a number of recent reviews (e.g. 16).

Plant cells exhibiting high rates of synthesis of protein, either for intracellular accumulation within vacuolar protein bodies or for external secretion, generally contain abundant rough cisternal ER. The cisternae are interconnected by smooth tubular ER. Cells involved in transport or secretion of lipids or sugars have an extensive network of tubular ER. Changes in the metabolic rate or activity during cell development are accompanied by changes in ER form, often with the development of intermediate forms that may have associated polyribosomes.

Although the cisternal and tubular nature of ER was known from conventional thin-section electron microscopy, the use of thicker sections has proved particularly useful in illustrating its three-dimensional structure within the cells. This approach has the advantage of readily distinguishing between tubules and vesicles which generally show similar profiles in conventional thin sections. This is important when considering the transport of material from the ER to other components of the endomembrane system. The interconnection of cisternae by tubular elements allows for compartmentalization and transport within the ER of a cell and may also allow for an "intrasymplastic" transport between cells where tubular ER is in close association with plasmodesmata (see below). Transport from the ER, however, is less clearly defined. Thin sections show dilations at the periphery of the cisternae, and it was generally assumed that these represented the start of vesiculation in the first stage of transport from the ER. Examination of thick sections of seed cotyledon parenchyma cells undergoing rapid storage protein synthesis revealed that such dilations were not forming vesicles but merely sections through the junctions of the cisternal and tubular elements, the dilation arising from the greater diameter of the tubular ER than the trans-lumen distance of the cisternal ER (33). Examination of thin sections alone can give a misleading impression of the structural organization of the ER. Thin sections of mung bean (*Vigna radiata* L.) storage parenchyma indicated that during seed germination there was an apparent increase in cisternal ER that was associated with the synthesis of the degradative enzymes involved in the mobilization of the storage reserves, and an increase in the ER marker enzyme NADH-cytochrome *c* reductase was also recorded, although extraction of total ER showed a decline by more than 50%. These apparently contradictary findings were clarified by the quantitative examination of selectively stained thick sections which showed that while there was an increase in rough cisternal ER, the form most readily visualized in thin sections, there was a dramatic reduction of the tubular ER network (37). Biochemical analysis showed that there was not merely a breakdown of the pre-existing tubular ER and a transition of some to a cisternal form but that there was also biogenesis of new ER components (28).

The thick section approach has also been used to study the changes in ER form and abundance that occur during cytokinesis. Both ZIO (41) and OsFeCN (42) have been used as selective stains in studies of endomembrane transitions. These have described the break-up of the nuclear envelope during prophase and the rearrangements of the ER. These rearrangements involve accumulation of cisternal ER at both poles while the nuclear spindle is penetrated by tubular ER. Conventional uranyl acetate and lead citrate contrast staining of such thick sections can also point out the distribution of spindle microtubules. These are associated with the polar cisternal ER aggregates, which may act as "anchors" during anaphase chromosomal migration, and also with the spindle tubular ER. Microtubule structure is affected by Ca^{2+} concentration. Using potassium antimonate to determine Ca^{2+} distribution, Wick & Hepler (100) demonstrated the potential of both polar and spindle ER to sequester Ca^{2+} and thus play a regulatory role in chromosomal movement.

The significance of Ca^{2+} in the regulation of cellular activities is becoming increasingly apparent, with regulation of Ca^{2+} concentration accomplished either through intracellular compartmentalization or transport into the apoplast. Initial evidence suggests that the ER may have a role in regulating cytosolic Ca^{2+} levels (11, 73, 89).

The Golgi Apparatus

The Golgi apparatus (GA) is composed of the dictyosomes and their vesicles, which are of different types. It is a major component of the endomembrane system of both plant and animal cells, regulating and directing the intracellular deposition and extracellular secretion of both its own and ER products, which it can also modify. It is also the cause of some controversy because its function and turnover are a central feature of the endomembrane concept. There is disagreement about the definition of the Golgi apparatus, depending upon either "the recognition of a common architecture" as a starting point (63) or the belief that "generalisations of this type are misleading and certainly do not reflect the great variability and function of this organelle" (79). This apparent polarization is, however, an overstatement as both schools share a common realization that there are basic central features but that there is also variation in the structure which can often be associated with known or presumed function. This section deals with the basic structure of the Golgi apparatus; the transitions between the ER and GA and the vesicular movements from the GA are dealt with below when considering functional roles.

The dictyosomes consist of stacks of cisternae although the precise form of the cisternae and their number vary both between tissues and within tissues at different stages of cell development. Higher plant dictyosomes show considerable polarity (86) with an apparent "forming" face, also referred to as a *cis* or proximal face and a maturing (*trans* or distal) face. Most botanical microscopists are now adopting the terms *cis* and *trans,* which are widely used by animal cell biologists. It should be noted, however, that animal cells do not normally show the same degree of polarity across the dictyosome stack as is seen in plant cells.

In higher plants the outermost cisterna of the *cis* face often has a reticulate form, although the extent of the fenestration and the presence or absence of central plate-like regions seem to vary not only with tissue but also with investigator. Viewing transverse (i.e. perpendicular to the stack) and planar thin sections of maize root-cap dictyosomes, Robinson (77) found an apparent central plate in *cis* face cisternae of "root cap" cells but came to substantially different conclusions than Mollenhauer and Morré (60, 63), whose micrographs also showed an apparent plate region in the *cis* dictyosome. Robinson also concluded that there was no evidence of dictyosome polarity (cf his micrographs). This was contradicted by Shannon et al (86), who, when viewing sections in the plane of the stack, noted an extensive, fenestrated *cis* face in ZIO fixed root tip cells. The holes seen in the fenestrated face are less than 0.05 μm in diameter, and in thin (0.08 μm) sections it is perhaps unreasonable to expect to see such a network defined clearly in either transverse or planar orientation; in the latter the section thickness will, on average, contain two overlapping cisternae.

The central region of the dictyosome contains 2–5 cisternae that show a central plate region in both transverse and planar sections. These cisternae have a peripheral reticulum with associated vesicles. The *trans* face is characterized by cisternae in which the intracisternal lumen is smaller and which show fenestrations across the cisternae that are visible in both thin and thick sections.

The *trans* face has numerous associated vesicles that vary in size, form, and apparent content, depending upon the cell type and predominant biosynthetic activity.

The role of the intercisternal elements is unclear, although suggestions include (*a*) a possible role in cross-linking of cisternae to stabilize the dictyosome stack, and (*b*) conflicting ideas concerning a possible physical constraint that may regulate vesicle production at the periphery of the cisternae (50, 58).

The obvious polarity, in general form, from *cis* to *trans* side is also reflected in membrane thicknesses and staining characteristics which show a gradient from the ER to *cis* to *trans* to plasmalemma and also implies endomembrane flow. The intramembrane particle distribution and cytochemical and enzyme cytochemical tests also illustrate various aspects of biochemical polarity across the plant dictyosome and have been reviewed in detail (79). The distribution of associated polyribosomes also varies with the dictyosome polarity (62). In general, plant dictyosomes show a greater degree of polarization than animal cell dictyosomes, but there is considerable variation in both overall form and detailed structure of plant dictyosomes that apparently is related to their primary role in the transport of either predominantly carbohydrates, proteins, or lysosomal enzymes (49). Numerous studies of Golgi activity have employed a range of potential or suspected inhibitors of membrane flow and secretory activity; these include the wide use of the sodium ionophore monensin (e.g. 19, 21, 61, 78, 88) and the potassium ionophore nigericin (21, 78) and also cytochalasin (53, 87). Such papers have led to numerous models for Golgi functioning, although it is recognized that, for example, monensin has a general effect on the ionic status of the cell rather than acting as a specific inhibitor of Golgi function, and that the effects of the inhibitors and chelators may be indirect by affecting other aspects of cell dynamics.

Direct connections exist between the ER and the GA. These are most clearly illustated in thick section studies of tissues in which some form of lysosomal activity is taking place. They are associated with the formation of a structurally characteristic GERL (Golgi-endoplasmic reticulum-lysosome). This has been described in vacuolating root tip cells (55, 59), seed storage parenchyma tissue in which controlled storage-protein hydrolysis occurs within vacuolar protein bodies (39), and in ligule cells in *Isoetes* (51). The reticulate membrane pattern seen in thick, selectively stained sections of such tissues is similar to that of the "partially coated reticulum" described below.

Direct connections between ER and GA are not always apparent. How, in such cases, is the dictyosome "replenished"? In some instances transition vesicles pass between the ER and GA, but where these are not present (e.g. maize root cap cells) alternative mechanisms have been proposed. These include suggestions that (*a*) the cisternae are permanent structures with only the peripheral vesicles being cycled (24), and (*b*) membrane is synthesized at the *cis* face (80).

While there is debate concerning the formation of and transport to the *cis* face, there is agreement that the dictyosomes produce a range of vesicles that either may transport sequestered material to the plasmalemma for external secretion or generate compartmentalization within the cell.

Vesicles and Vacuoles

There is wide variation in the extent and form of plant cell vacuoles and vesicles. Their structure, development, and function have been the subjects of a number of extensive reviews (e.g. 56), and the compartmentalization aspects are considered elsewhere in this volume (7a). As vesicles and vacuoles are an integral part of the endomembrane system, some aspect of their associated structural and biochemical transitions and their roles in secretion and deposition are discussed below. The relationship of coated (clathrin) vescicles with the plasmalemma, dictyosomes, and the "partially coated reticulum" are considered together in the next section.

Plasmalemma and Plasmodesmata

Although not strictly an endo membrane, the plasmalemma is usually considered in discussion of the endomembrane system because of its relationship to both exo- and endo-cytosis. The plasmalemma is the thickest of the cell membranes and often shows a differential staining of its cytoplasmic and paramural faces.

The plasmalemma has important functions in ion transport and the synthesis and deposition of cell wall components including the cellulose microfibrils. It is usually continuous between adjacent cells via the plasmodesmata, which consist of a plasmalemma tubule enclosing a desmotubule that is derived from the (tubular) ER trapped during cytokinesis in the forming cell plate. The network of tubular ER present at the developing cell plate may also help to locate and stabilize the aggregating vesicles into a single plane (94).

There is no doubt that plasmodesmata are involved in symplastic transport, but the detailed structure of the plasmodesmata and their relationship to intercellular symplastic transport is debated. Different interpretations of the staining patterns suggest that (*a*) it is possible to consider the desmotubule as an open channel with luminal continuity between ER of adjacent cells, or (*b*) the desmotubule is a solid structure derived from the ER, with symplastic transport occurring only in the cytosolic compartment between the desmotubule and the tubule of the plasmalemma. Recent reviews tend to support the latter (32), although it may be that, as with other elements of the endomembrane system, the structure/function relationships of plasmodesmata vary between cells and physiological states.

The involvement of the plasmalemma and endomembrane system in endocytosis is also a matter of some debate. While it is generally accepted that vesicular transport from the Golgi apparatus to the plasmalemma is involved in extracellular secretion, and can involve coated vesicles (e.g. 26), endocytosis (pinocytosis) as a general feature of plant cells is disputed. Although it has been suggested that endocytosis is unlikely to be a normal feature of plant cell activity, in specialized cases endocytosis is apparent—for example, in protoplast uptake of cationized ferritin (45, 91).

Coated vesicles are also sometimes found associated with the GA and can give rise to the characteristic "partially coated reticulum" (57), although this is not always associated with the GA (72). Recent purification of a coated vesicle fraction and its biochemical characterization (57) suggests that the partially coated reticulum may be involved in the secretion of peroxidase and possibly its sorting from other components passing through the GA (41). It is tempting to speculate that the presence of coated pits or vesicles at the plasmalemma and the partially coated reticulum indicates membrane flow between the two sites: it is usually assumed by botanists to be from the GA to the plasmalemma. The uptake of cationized ferritin, however, shows that endocytosis can occur via coated pits, and it may be that there are two separate populations of coated vesicles undertaking distinctly different roles within the cell (27).

Nuclear Envelope

There are striking similarities between the biochemical and structural characteristics of the membrane of the nuclear envelope and the ER. Some authors believe the nuclear envelope consists of an inner and an outer membrane, but because of a direct connection at the nuclear pores, most writers consider the membrane in its entirety. Ultrastructural and biochemical aspects of the plant nucleus, including the nuclear envelope, have been reviewed in detail (48). However, with regard to the organization of the endomembrane system, the relationship between the nuclear envelope and the ER should be considered.

Close structural associations are seen between the ER and the nuclear envelope, although the less frequent use of potassium permanganate as a fixative has resulted in fewer reports of apparent direct connections between the two types of membrane in higher plant cells. The NE has NADH-cytochrome c reductase activity, but unlike the ER, the membrane also has a high level of strongly bound nucleic acids. The outer membrane may have associated polyribosomes while the inner membrane has associated chromatin. The nuclear pores are complex, probably octagonal, structures whose distribution and frequency within the nuclear envelope vary with nuclear activity. The precise relationships of pore distribution to nuclear function are not understood, and there are conflicting reports of apparent relationships between nuclear pore distribution and adjacent endomembrane components (33, 85). Nuclear pore structures are maintained within ER-like membranes during mitotic breakup of the nuclear envelope (41), with probable interconnection of the pores by residual NE laminal proteins.

Lipid Bodies and Microbodies

Microbodies and lipid bodies are usually considered in discussions of the endomembrane system because of their ontogenic relationship with the ER. Microbodies are respiratory organelles bounded by a single membrane that has some similarity to the ER membrane. Microbodies are found in close association with the organelle that supplies substrate, and they are characterized by metabolic pathways associated with flavin-linked oxidases. Although involved in essentially catabolic reactions, with the production of H_2O_2, which is removed by the catalase activity present, microbodies have membranes that apparently are freely permeable to most small molecules that may then be used in biosynthetic pathways. Leaf peroxisomes, in which glycolate oxidase is the primary flavin oxidase, increase during leaf greening whereas glyoxysomes, which undertake the glyoxylate cycle, are associated with the mobilization of lipids during seed germination. The structure and activity of plant microbodies have been dealt with in a number of recent comprehensive reviews (6, 92). The role of the endomembrane system in the synthesis of lipid bodies during seed development is described below.

There is a close relationship between the ER, glyoxysomes, and lipid bodies during seed germination. There is discussion as to whether the glyoxysomal proteins are cotranslationally synthesized at the ER or whether they are synthesized within the cytosol and subsequently incorporated. Recent evidence, reviewed by Trelease (93), suggests that the glyoxysomal enzymes are synthesized within the cytosol whereas the membrane may be derived from the ER. Two contrasting ultrastructural studies of glyoxysome formation are based on watermelon and cotton seed germination. In watermelon (98) an increasing population of pleiomorphic glyoxosomes was found, while in cotton there was no increase in numbers but only in size of glyoxysomes that had developed during seed maturation from ER membrane and cytosolically synthesized enzymes (52).

It is evident that the endomembrane system is composed of a wide variety of organelles with different functions within the plant cell. There are numerous examples of close interrelationships between the components at both a structural and functional level, and it was from such examples that the "Endomembrane Concept" was developed. Some aspects of component transitions are considered below under either intracellular deposition or extracellular secretion. The structural and functional aspects of each example are considered together since division would be a rather artificial imposition on systems that are dynamic and usually sequential.

FUNCTIONAL TRANSITIONS

The endomembrane system plays a major role in the biogenesis of cytoplasmic organelles, the deposition of material within them, and the biosynthesis and transport of material destined for extracellular secretion.

Intracellular Deposition

Two examples of the role of the endomembrane system in intracellular deposition are considered. The first involves the debate about the formation of lipid reserves in developing seeds and the second considers some of the recent advances in work on storage protein deposition in seeds.

LIPID BODY FORMATION   Lipids, often in the form of triacylglycerols, are deposited in the storage reserve tissues of a wide range of seeds in organelles referred to as spherosomes, oleosomes, or lipid bodies. Some authors distinguish between small spherosomes that are present in nonstorage tissues and larger oleosomes or lipid bodies that are restricted to storage tissues, while others question such a distinction, referring to all types as either oleosomes (e.g. 7) or lipid bodies (97). It is generally acknowledged that the lipid deposits are bounded by a phospholipid-protein monolayer that is probably derived from the ER.

Unlike animal tissues, in which lipid deposits have a cytosolic synthesis, storage lipid synthesis in plants is associated with the plastids and the ER. The plastids, which lie in close proximity to rough cisternal ER, synthesize the fatty acids that are incorporated into triacylglycerides at the ER (90). From the ultrastructural evidence of attachment of ER and lipid droplets it has been suggested that the lipids are accumulated between the lipid bilayer of the ER membrane, resulting in the formation of the "half-unit" membrane around lipid bodies (97). An alternative scheme proposes that lipid bodies develop in the cytoplasm close to the surface of the plastids, become encircled by ER, and acquire a surface coat of lipid protein synthetically derived, but quite distinct from, the ER (6).

PROTEIN BODY FORMATION   The protein reserves of seeds are deposited in membrane-bound organelles now commonly termed "protein bodies." A distinction should be made between protein deposition in monocots and dicots. In cereal grains the storage proteins are found predominantly in subaleurone and outer layers of the starchy endosperm and are hydrolyzed during germination from nonliving tissues in a relatively uncontrolled manner prior to peptide uptake at the embryo scutellum. By contrast, legume seeds contain substantial reserves of protein in protein bodies within the cotyledon storage parenchyma tissues, from which mobilization within living tissues is regulated.

Cereal protein bodies   Storage proteins are classified by their solubility, with the alcohol-soluble prolamins being the predominant form in wheat, barley, and rye, and globulins (soluble in salt solutions) predominant in oats. Ultrastructural examination of the development of protein body formation in endosperm tissue of barley (14) and wheat (4, 70) indicates that the prolamins are initially deposited within a membrane-bound compartment and subsequently released into the cytosol where they may continue to aggregate additional protein and membranous material before forming a matrix during seed desiccation. Whether the surrounding membrane, when present, is ER or tonoplast is as yet unclear, although if the latter, then some form of transport from the ER (the site of synthesis) to vacuole (the site of deposition) would be required. Thick section studies of developing wheat endosperm indicate the presence of numerous dictyosomes closely associated with both cisternal ER and forming protein bodies (71). In maize there is ultrastructural and biochemical evidence to suggest that (*a*) polyribosomes are directly associated with the protein body membrane (12) and that (*b*) rough microsomal preparations are capable of synthesizing the storage protein zein (13) with cleavage of the preprotein occurring cotranslationally. Oat globulins are also synthesized by ER-associated polyribosomes although high molecular weight proteins are probably transported to the protein bodies and cleaved into the smaller α and β subunits (1). Rice has the additional feature of having various distinguishable types of protein bodies, each thought to contain predominantly different storage proteins (5). All types of rice protein bodies, however, are derived from the ER, although transport from the Golgi apparatus may be involved in the accumulation of some types of protein deposits (68).

Legume seed protein bodies   The major legume storage proteins are globulins and are often glycoproteins containing low levels of N-acetylglucosamine and mannose. Using pulse-chase autoradiography of labeled amino acids applied to cotyledon storage parenchyma tissue, it was found that the label first accumulated at the rough ER and subsequently in the protein bodies (2). That the amino-acid incorporation was into storage protein polypeptides was shown later (18). During legume seed development the onset of the phase of storage protein deposition is accompanied by major changes in the form of the vacuoles of the storage cells. The large central vacuole is replaced by numerous small vacuoles, although the precise manner of this transformation is not certain; it may be by subdivision of the original vacuole (23) or by its replacement with numerous newly derived smaller vacuoles or a combination of the two (36). The problem may be difficult to resolve by conventional EM techniques as these involve a sequential study of a dynamic organelle that is highly sensitive to fixative concentrations, the optimum for which may be varying with changes in the organelle structure. Morphometric analysis shows that there is a massive increase in vacuolar surface area (22) as well as a substantial increase in the amount of rough cisternal ER associated with the onset of storage protein synthesis (10).

Immunocytochemical studies have been used to follow the deposition of legume storage proteins and have confirmed the role of the ER (20) as well as showing that different storage proteins are accumulated together within the protein bodies. A possible role in the transport of the storage proteins from ER to protein bodies has been suggested for the Golgi apparatus. This was originally based on circumstantial, ultrastructural evidence (24, 36), although recent biochemical work has shown that the oligosaccharide side-chains of some storage proteins are posttranslationally modified by glycosyl transferases in the Golgi apparatus (95). Chrispeels (17) was also able to show, using pulse-trace experiments with tritiated fucose as substrate, that phytohemagglutinin (PHA), a major storage protein in *Phaseolus vulgaris* seeds, was fucosylated within the Golgi apparatus and transferred by vesicles to the developing protein bodies. The wider role of the Golgi apparatus in posttranslational modification and transport of storage proteins has recently been the subject of an excellent review (19). Immunocytochemical studies apparently confirm a role for the Golgi apparatus in the transport of some storage proteins (20, 21).

A promising recent innovation uses immunocytochemical localization in sections cut from frozen tissues which are resin-embedded on the grid after staining; this can give apparently excellent preservation of both tissue structure and antigenicity while also allowing good contrast staining. The method has been used to show localization of phaseolin and PHA in the ER and Golgi complex of developing bean cotyledons (29, 30). Using sections of resin-embedded tissue, preliminary evidence suggests that there may be sorting of the storage proteins within the dictyosome stack, with nonglycosylated storage proteins being secreted from the mid region while glycosylated proteins continue through the *trans* aspect (35).

Extracellular Secretion

Three particular aspects of the extracellular secretion from plant cells are considered and discussed in relation to the organization and role of the endomembrane system. They include the secretion of proteins, polysaccharides, and root cap slime. There are specific structural characteristics associated with the endomembrane system of the cells involved in each type of secretion, and emphasis on the differences and similarities lies at the center of the discussion concerning the general applicability of the Endomembrane Concept.

PROTEINS   The characteristic feature of protein-secreting cells is the presence of numerous rough cisternal ER elements. Particular emphasis has been paid to the synthesis and secretion of α-amylase in cells of the aleurone layer of germinating barley. Synthesis and secretion, which are initiated by gibberellic acid, are accompanied by proliferation of the rough cER. Recent work has demonstrated that the ER is the site of α-amylase synthesis (46, 47), although it is still uncertain whether the secretory vesicles that carry the α-amylase to the plasmalemma are derived from the ER or the Golgi apparatus. This contrasts with the evidence of intracellular protein deposition (or secretion into vacuoles) where there is now clear evidence for the role of the Golgi apparatus (see above).

POLYSACCHARIDES   The cell wall cellulose microfibrils, and the matrix of pectins and hemicelluloses in which they are embedded, are the extracellular polysaccharides of the plant cell. Cellulose microfibrils in higher plants are synthesized by arrays of intramembranous particles (IMPs) in the plasmalemma (64) while the site of synthesis of the matrix polysaccharides is probably the Golgi apparatus (74).

The relationship between the microtubules of the cytoskeleton and the plasmalemmal IMPs, which form "rosettes" on the PF face and "terminal globules" on the EF face, indicates that the two are closely linked in the deposition of the cell wall (44). The precise role of the cytoskeletal elements in both the direction of vesicles from the endomembrane system to the plasmalemma and the constraints that the microtubules may impose upon the direction of the forming fibrils remain to be elucidated. Cytochemical and biochemical evidence suggests that wall polysaccharides are accumulated in Golgi vesicles prior to extracellular depostion (65).

ROOT CAP SLIME   Maize root cap cells have been used extensively as a test system for examining secretory activity and the role of the endomembrane system (59, 77, 78, 86) and also as a possible bioassay tissue for the evaluation of some inhibitors of Golgi activity (88). The latter report is of particular significance as it raises serious doubts about interpretation of many of the previous inhibitor studies, including the use of monensin, and suggests that "there is no conclusive evidence that any specific inhibitors of Golgi (dictyosome) activity have yet been discovered." Slime, which is similar to the cell wall pectin fraction (101), is produced by the outer root cap cells that are characterized by their abundant ER, large population of hypertrophied dictyosomes, and numerous small vacuoles. Synthesis of slime polysaccharide is initiated in the ER (cf wall matrix polysaccharides) with intermediate transfer from nucleotide sugars to lipid before addition to glycoprotein and subsequent polymerization in the Golgi apparatus prior to secretion (66). Concentration of slime occurs in the secretory vesicles that are connected to the dictyosome cisternae by peripheral tubules; on release from the dictyosome the vesicles change shape, becoming more spherical, and migrate to the plasmalemma where fusion of the vesicle membrane with the plasmalemma results in release of the slime to the apoplast (83). The results from a freeze-fracture study of the endomembrane system and the plasmalemma of slime-secreting cells of *Lepidium sativum* L. (cress) suggest that the Golgi apparatus regulates the molecular composition of the plasma membrane by selecting specific membrane components prior to exocytosis of the Golgi-derived slime vesicles (96), as well as supporting the concept of membrane flow during secretion events in root cap cells.

OTHER SECRETIONS   Studies of the role of the endomembrane system in extracellular secretion include work on salt glands, nectaries, ligules, and digestive glands where there is conflicting evidence on transport through the GA (49, 76, 86). Studies of the secretory apparatus of salt glands of *Cynodon* (69) illustrate differences between secreting and nonsecreting cells, while work on *Tamarix aphylla* L. (9) and *Frankenia pauciflora* (67) come to different conclusions regarding the role of the characteristic interfacial apparatus. In *Cynodon* the changes in the partitioning membranes during secretion led to an increase in the extracellular space (which appears to be an intracellular tubular network in thin section) and an enhancement of the surface area to volume ratio which may facilitate apoplast/symplast transfer in a manner similar to the more permanent arrangements found in transfer cells. A similar extensive tubular network is also found in some nectar-secreting cells, for example, the trichomes of cotton (25) and *Abutilon* (31). Although we do not yet know whether such networks have continuous direct access to the apoplast, calculations from nectar secretion rates suggest that nectar secretion cannot be maintained by granulocrine secretion or membrane transport (75). The secretory papillae of the extrafloral nectaries of cotton also contain conspicuous ER and transport the prenectar along a symplastic route as in *Abutilon* (31). It is not clear, however, if transport from cell to cell is via the ER/desmotubule lumen, in an "intra-symplastic" pathway, or via the space between the desmotubule and the plasmalemma.

Ligules (the membranous extension of the adaxial epidermis of some leaves) have a high rate of extracellular secretion. Both protein and polysaccharide are secreted from cells with an extensive endomembrane system of both tubular and cisternal ER and dictyosomes (15, 51) with some dictyosome-dictyosome and dictyosome-ER interconnection (51).

THE ENDOMEMBRANE CONCEPT

Since the inception of the Endomembrane Concept it has been subjected to some criticism (e.g. 77, 80, 99). The principal support for the concept comes from evidence of direct interconnections involved in both structural and functional transitions (usually in protein secretions) and from careful immunocytochemical studies of transport pathways during protein synthesis and deposition. There is, however, inconclusive evidence regarding ER-GA interaction in polysaccharide synthesis and secretion, particularly with regard to dictyosome turnover; is it by membrane renewal from the ER (in line with the Endomembrane Concept) or by self assembly within the dictyosomes? Does lack of evidence for ER-GA membrane flow in this case really mean that the Endomembrane Concept is invalid? I think not. Rather, it indicates that the concept may not be universal in its applicability to endomembrane transitions within higher plant cells but is a (valuable) framework in which to consider the secretory pathways.

CONCLUSIONS

It is clear that there are many unresolved questions concerning the endomembrane system of higher plant cells. High on this list are matters concerned with the central controlling role of the Golgi apparatus in the modification and directional secretion of the various products passing through it. We still know little of how the basic membrane turnover is accomplished within the dictyosomes, as well as whether there is any recycling, for segregation or purification, of vesicles, membrane components, etc, as proposed for animal cell systems (e.g. 82). Immunocytochemical localization of specific membrane proteins may prove a powerful tool in such work, although early plant studies have pointed to some potential problems of monoclonal antibodies (8). Other important problems concern the general interaction of the cytoskeleton with the endomembrane system and its role in the movement of vesicles. We still have little information on such movements—are vesicles directed to their "target" by the transport process or moved (actively or passively) around the cell until they "recognize" their target?

The answers to such problems will only be obtained by a multidisciplinary approach, and I would suggest, perhaps with some prejudice, that it is likely that the role of (sophisticated) cytochemistry will prove of particular importance. It seems unlikely that a common "model system" for all endomembrane transitions will be found. Instead, the range of mechanisms employed to synthesize, segregate, and secrete will be almost as diverse as the types of products that are being processed, sometimes simultaneously, in plant cells.

ACKNOWLEDGMENTS

Although too numerous to mention individually, I must thank the many participants at the 3rd International Meeting on Botanical Microscopy (July 1985, York) who willingly discussed their latest views on plant cell structure/function relationships. I should also like to thank Nigel Chaffey for his comments and help and Denis for her continuing patience and support.

Literature Cited

1. Adeli, K., Altosaar, I. 1983. Role of the endoplasmic reticulum in the biosynthesis of oat globulin precursors. *Plant Physiol.* 73:949–55
2. Bailey, C. J., Cobb, A., Boulter, D. 1970. A cotyledon slice system for the electron autoradiographic study of the synthesis and intracellular transport of seed storage proteins of *Vicia faba*. *Planta* 95:103–18
3. Baumgartner, B., Tokuyasu, K. T., Chrispeels, M. J. 1980. Immunocytochemical localization of reserve proteins in the endoplasmic reticulum of developing bean *(Phaseolus vulgaris)* cotyledons. *Planta* 150:419–25
4. Bechtel, D. B., Gaines, R. L., Pomeranz, Y. 1982. Protein secretion in wheat endosperm—formation of the matrix protein. *Cereal Chem.* 59:336–43
5. Bechtel, D. B., Juliano, B. O. 1980. Formation of protein bodies in the starchy endosperm of rice (*Oryza sativa* L.): a reinvestigation. *Ann. Bot.* 45:503–9
6. Beevers, H. 1979. Microbodies in higher plants. *Ann. Rev. Plant Physiol.* 30:159–93
7. Bergfeld, R., Hong, Y.-N., Kühn, T., Schopfer, P. 1978. Formation of oleosomes (storage lipid bodies) during embryogenesis and their breakdown during seedling development in cotyledons of *Sinapis alba* L. *Planta* 143:297–307
7a. Boller, T., Wiemken, A. 1986. Dynamics of vacuolar compartmentation. *Ann. Rev. Plant Physiol.* 37:137–64
8. Bolwell, G. P., Northcote, D. H. 1984. Demonstration of a common antigenic site on endo-membrane proteins of *Phaseolus vulgaris* by a rat monoclonal antibody. *Planta* 162:139–46
9. Bosabalidis, A. M., Thomson, W. W. 1984. Ultrastructural differentiation of an unusual structure lining the anticlinal walls of the inner secretory cells in *Tamarix* salt glands. *Bot. Gaz.* 145:427–35
10. Briarty, L. G. 1978. The mechanisms of protein body development in legumes and cereals. In *Plant Proteins*, ed. G. Norton, pp. 81–106. London: Butterworth
11. Buckhout, T. J. 1984. Characterisation of Ca^{2+} transport in purified endoplasmic reticulum membrane vesicles from *Lepidium sativum* roots. *Plant Physiol.* 76:962–67
12. Burr, B., Burr, F. A. 1976. Zein synthesis in maize endosperm by polyribosomes attached to protein bodies. *Proc. Natl. Acad. Sci. USA* 73:515–19
13. Burr, F. A., Burr, B. 1981. *In vitro* uptake and processing of prezein and other maize preproteins by maize membranes. *J. Cell Biol.* 90:427–34
14. Cameron-Mills, V., von Wettstein, D. 1980. Protein body formation in the developing barley endosperm. *Carlsberg Res. Commun.* 45:577–94
15. Chaffey, N. J. 1985. Structure and function in the grass ligule: The membranous ligule of *Lolium temulentum* L. as a secretory organ. *Protoplasma* 127:128–32
16. Chrispeels, M. J. 1980. The endoplasmic reticulum. In *The Biochemistry of Plants*, ed. N. E. Tolbert, 1:390–412. New York: Academic. 705 pp.
17. Chrispeels, M. J. 1983. The Golgi apparatus mediates the transport of phytohemagglutinin to the protein bodies in bean cotyledons. *Planta* 158:140–51
18. Chrispeels, M. J. 1984. Biosynthesis, processing and transport of storage proteins and lectins in cotyledons of developing legume seeds. *Philos. Trans. R. Soc. London Ser. B* 304:309–22
19. Chrispeels, M. J. 1985. The role of the Golgi apparatus in the transport and posttranslational modification of vacuolar (protein body) proteins. In *Oxford Surveys of Plant Molecular and Cell Biology*, ed. B. J. Miflin, 2:43–68. Oxford: Clarendon
20. Craig, S., Goodchild, D. J. 1984. Periodate-acid treatment of sections permits on-grid immunogold localisation of pea seed vicilin in ER and Golgi. *Protoplasma* 122:35–44
21. Craig, S., Goodchild, D. J. 1984. Golgi-mediated vicilin accumulation in pea cotyledon cells is re-directed by monensin and nigericin. *Protoplasma* 122:91–97
22. Craig, S., Goodchild, D. J., Hardham, A. R. 1979. Structural aspects of protein accumulation in developing pea cotyledons. I. Qualitative and quantitative changes in parenchyma cell vacuoles. *Aust. J. Plant Physiol.* 6:81–98
23. Craig, S., Goodchild, D. J., Miller, C. 1980. Structural aspects of protein accumulation in developing pea cotyledons. II. Three-dimensional reconstructions of vacuoles and protein bodies from serial sections. *Aust. J. Plant Physiol.* 7:329–37
24. Dieckert, J. W., Dieckert, M. C. 1976. The chemistry and cell biology of the vacuolar proteins of seeds. *J. Food Sci.* 41:475–82
25. Eleftheriou, E. P., Hall, J. L. 1983. The extrafloral nectaries of cotton. I. Fine

structure of the secretory papillae. *J. Exp. Bot.* 34:103–19

26. Fowke, L. C., Griffing, L. R., Mersey, B. G., van der Valk, P. 1983. Protoplasts for studies of the plasma membrane and associated organelles. *Protoplasts 1983* (Lecture Proc., 6th Int. Protoplast Symp., Basel). *Experientia Suppl.* 46:101–10

27. Fowke, L. C., Tanchak, M. A., Griffing, L. R., Mersey, B. G. 1985. Plant coated vesicles. *Proc. R. Microsc. Soc.* 20:10BM

28. Gilkes, N. R., Chrispeels, M. J. 1980. The endoplasmic reticulum of mungbean cotyledons. Biosynthesis during seedling growth. *Planta* 149:361–69

29. Greenwood, J. S., Chrispeels, M. J. 1985. Immunocytochemical localisation of phaseolin and phytohemagglutinin in the endoplasmic reticulum and Golgi complex of developing bean cotyledons. *Planta* 164:295–302

30. Greenwood, J. S., Keller, G. A., Chrispeels, M. J. 1984. Localisation of phytohemagglutinin in the embryogenic axis of *Phaseolus vulgaris* with ultrathin cryosections embedded in plastic after indirect immunolabelling. *Planta* 162:548–55

31. Gunning, B. E. S., Hughes, J. E. 1976. Quantitative assessment of symplastic transport of pre-nectar into the trichomes of *Abutilon* nectaries. *Aust. J. Plant Physiol.* 3:619–37

32. Gunning, B. E. S., Overall, R. L. 1983. Plasmodesmata and cell-to-cell transport in plants. *BioScience* 33:260–65

33. Harris, N. 1979. Endoplasmic reticulum in developing seeds of *Vicia faba*. A high voltage electron microscopic study. *Planta* 146:63–69

34. Harris, N. 1983. The uses of high voltages and thick sections in botanical electron microscopy. In *New Frontiers in Food Microstructure*, ed. D. B. Bechtel, pp. 287–316. Minnesota: Am. Assoc. Cereal Chem. 392 pp.

35. Harris, N. 1984. Immunocytochemical localisation of the intra-cellular secretion of the major legume storage proteins via the Golgi apparatus. *Proc. R. Microsc. Soc.* 19:S16

36. Harris, N., Boulter, D. 1976. Protein body formation in cotyledons of developing cowpea *(Vigna unguiculata)* seeds. *Ann. Bot.* 40:739–44

37. Harris, N., Chrispeels, M. J. 1980. The endoplasmic reticulum of mungbean cotyledons: Quantitative morphology of cisternal and tubular ER during seedling growth. *Planta* 148:293–303

37a. Harris, N., Croy, R. R. D. 1986. Localization of mRNA for pea legumin: *in situ* hybridisation using a biotinylated cDNA probe. *Protoplasma*. In press

38. Harris, N., Gates, P. 1984. Fluorescence microscopy of the endomembrane system of living plant cells. *Plant Cell Environ.* 7:699–703

39. Harris, N., Oparka, K. J. 1982. Connections between dictyosomes, ER and GERL in cotyledons of mungbean (*Vigna radiata* L.). *Protoplasma* 114:93–102

40. Hawes, C. R. 1982. Applications of high voltage electron microscopy to botanical ultrastructure. *Micron* 12:227–57

41. Hawes, C. R., Juniper, B. E., Horne, J. C. 1981. Low and high voltage electron microscopy of mitosis and cytokinesis in maize roots. *Planta* 152:397–407

42. Hepler, P. K. 1980. Membranes in the mitotic apparatus of barley cells. *J. Cell Biol.* 86:490–99

43. Hepler, P. K. 1981. The structure of the endoplasmic reticulum revealed by osmium tetroxide-potassium ferricyanide staining. *Eur. J. Cell Biol.* 26:102–10

44. Herth, W. 1985. Plasma-membrane rosettes involved in localised wall thickening during xylem vessel formation of *Lepidium sativum* L. *Planta* 164:12–21

45. Joachim, S., Robinson, D. G. 1984. Endocytosis of cationic ferritin by bean leaf protoplasts. *Eur. J. Cell Biol.* 34:212–16

46. Jones, R. L. 1980. Quantitative and qualitative changes in the endoplasmic reticulum of barley aleurone layers. *Planta* 150:70–81

47. Jones, R. L., Jacobsen, J. V. 1982. The role of the endoplasmic reticulum in the synthesis and transport of α-amylase in barley aleurone layers. *Planta* 156:421–32

48. Jordan, E. G., Timmis, J. N., Trewavas, A. J. 1980. The plant nucleus. See Ref. 16 pp. 489–588

49. Juniper, B. E., Hawes, C. R., Horne, J. C. 1982. The relationship between the dictyosomes and the form of endoplasmic reticulum in plant cells with different export programmes. *Bot. Gaz.* 143:135–45

50. Kristen, U. 1978. Ultrastructure and a possible function of the intercisternal elements in dictyosomes. *Planta* 138:29–33

51. Kristen, U. 1980. Endoplasmic reticulum-dictyosome interconnections in ligula cells of *Isoetes lacustris*. *Eur. J. Cell Biol.* 23:16–21

52. Kunce, C. M., Trelease, R. N., Doman, D. C. 1984. Ontogeny of glyoxysomes in maturing and germinating cotton seeds—a morphometric analysis. *Planta* 161:156–64

53. Lockhausen, J., Kristen, U. 1983.

Effects of cytochalasin B on the mitochondria and the Golgi apparatus in ovary gland cells. *Z. Pflanzenphysiol.* 110: 191–99

54. Marty, F. 1973. Sites réactifs a l'iodure de zinc tetroxide d'osmium dans cellules de la racine d'*Euphorbia characias*. *C. R. Acad. Sci. Ser. D* 227:2681–84

55. Marty, F. 1978. Cytochemical studies on GERL, provacuoles and vacuoles in root meristematic cells of Euphorbia. *Proc. Natl. Acad. Sci. USA* 75:852–56

56. Marty, F., Branton, D., Leigh, R. A. 1980. Plant vacuoles. See Ref. 16, pp. 625–59

57. Mersey, B. G., Griffing, L. R., Rennie, P. J., Fowke, L. C. 1985. The isolation of coated vesicles from protoplasts of soybean. *Planta* 163:317–27

58. Mollenhauer, H. H., Morré, D. J. 1975. A possible role for intercisternal elements in the formation of secretory vesicles in plant Golgi apparatus. *J. Cell Sci.* 19:231–37

59. Mollenhauer, H. H., Morré, D. J. 1976. Transition elements between endoplasmic reticulum and Golgi apparatus in plant cells. *Cytobiology* 13:297–306

60. Mollenhauer, H. H., Morré, D. J. 1980. The Golgi apparatus. See Ref. 16, pp. 437–88

61. Morré, D. J., Boss, W. F., Grimes, H., Mollenhauer, H. H. 1983. Kinetics of Golgi apparatus membrane flux following monensin treatment of embryogenic carrot cells. *Eur. J. Cell Biol.* 30:25–32

62. Morré, D. J., Boss, W. F., Mollenhauer, H. H. 1984. Distribution of Golgi apparatus-associated polyribosomes across the polarity axis of dictyosomes of wild carrot (*Daucus carota* L.). *Protoplasma* 123:221–25

63. Morré, D. J., Mollenhauer, H. H. 1974. The Endomembrane Concept: a functional integration of the endoplasmic reticulum and Golgi apparatus. In *Dynamic Aspects of Plant Ultrastructure*, ed. A. W. Robards, pp. 84–137. London: McGraw-Hill

64. Mueller, S. C., Brown, R. M. 1980. Evidence for an intramembrane complex component associated with a cellulose microfibril synthesising complex in higher plants. *J. Cell Biol.* 84:315–26

65. Northcote, D. H. 1979. The involvement of the Golgi apparatus in the biosynthesis and secretion of glycoproteins and polysaccharides. *Biomembranes* 10:51–76

66. Northcote, D. H. 1982. The synthesis and transport of some plant glycoproteins. *Philos. Trans. R. Soc. London Ser. B* 300:195–206

67. Olesen, P. 1979. Ultrastructural observations on the cuticular envelope in salt glands of *Frankenia pauciflora*. *Protoplasma* 99:1–9

68. Oparka, K. J., Harris, N. 1982. Rice protein body formation: All types are initiated by dilation of the ER. *Planta* 154:184–88

69. Oross, J. W., Thomson, W. W. 1984. The ultrastructure of *Cynodon* salt glands: secreting and non-secreting. *Eur. J. Cell Biol.* 34:287–91

70. Parker, M. L. 1982. Protein accumulation in developing endosperm of a high protein line of *Triticum dicoccoides*. *Plant Cell Environ.* 5:37–43

71. Parker, M. L., Hawes, C. R. 1982. The Golgi apparatus in developing endosperm of wheat *Triticum aestivum*. *Planta* 154:277–83

72. Pesacreta, T. J., Lucas, W. J. 1985. Presence of a partially coated reticulum in angiosperms. *Protoplasma* 125:173–84

73. Picton, J. M., Steer, M. W. 1985. The effects of ruthenium red, lanthanum, fluorescein isothiocyanate and trifluoperazine on vesicle transport, vesicle fusion and tip extension in pollen tubes. *Planta* 163:20–26

74. Ray, P. M., Eisinger, W. R., Robinson, D. G. 1976. Organelles involved in cell wall polysaccharide formation and transport in pea cells. *Ber. Dtsch. Bot. Ges.* 89:121–46

75. Robards, A. W., Stark, M. 1985. The pathway of sugar movement through the Abutilon nectary gland. *Proc. R. Microsc. Soc.* 20:22BM

76. Robins, R. J., Juniper, B. E. 1980. The secretory cycle of *Dionaea muscipula*. I The fine structure and effect of stimulation on the fine structure of the digestive gland cells. *New Phytol.* 86:279–96

77. Robinson, D. G. 1980. Dictyosome-endoplasmic reticulum associations in higher plant cells? A serial section analysis. *Eur. J. Cell Biol.* 23:22–26

78. Robinson, D. G. 1981. The ionic sensitivity of secretion-associated organelles in root cap cells of maize. *Eur. J. Cell Biol.* 23:267–72

79. Robinson, D. G. 1985. *Plant membranes. Endo- and plasma- membranes of plant cells*. New York: Wiley. 331 pp.

80. Robinson, D. G., Kristen, U. 1982. Membrane flow via the Golgi apparatus of higher plant cells. *Int. Rev. Cytol.* 77:89–127

81. Roth, J. 1984. The protein A-gold technique for antigen localisation in tissue sections by light and electron microscopy. In *Immunolabelling for Electron Microscopy*, ed. J. M. Polak, I. M.

Varndell, pp. 113–22. Amsterdam: Elsevier 370 pp.

82. Rothman, J. E. 1981. The Golgi apparatus: two organelles in tandem. *Science* 213:1212–19

83. Rougier, M. 1981. Secretory activity of the root cap. In *Encyclopedia of Plant Physiology: Plant Carbohydrates II* (NS), ed. W. Tannar, F. A. Loewus, 3B:542–74. Berlin: Springer-Verlag. 766 pp.

84. Schnepf, E., Hausman, K., Herth, W. 1982. The osmium tetroxide-potassium ferrocyanide (OsFeCN) staining technique for electron microscopy. A critical evaluation using ciliates, mosses and higher plants. *Histochemie* 76:261–71

85. Severs, N. J., Jordan, E. G., Williamson, D. H. 1976. Nuclear pore absence from areas of close association between nucleus and vacuole in synchronous yeast cultures. *J. Ultrastruct. Res.* 54:374–87

86. Shannon, T. M., Henry, Y., Picton, J. M., Steer, M. W. 1982. Polarity in higher plant dictyosomes. *Protoplasma* 112:189–95

87. Shannon, T. M., Picton, J. M., Steer, M. W. 1984. The inhibition of dictyosome vesicle formation in higher plant cells by cytochalasin D. *Eur. J. Cell Biol.* 33:144–47

88. Shannon, T. M., Steer, M. W. 1984. The root cap as a test system for the evaluation of Golgi inhibitors. II. Effect of potential inhibitors on slime droplet formation and structure of the secretory system. *J. Exp. Bot.* 35:1708–14

89. Sievers, A., Behrens, H. M., Buckhout, T. J., Gradmann, D. 1984. Can a Ca^{2+} pump in the endoplasmic reticulum of *Lepidium* roots be the trigger for rapid changes in membrane potential after gravistimulation? *Z. Pflanzenphysiol.* 114:195–200

90. Stumpf, P. K. 1980. Biogenesis of saturated and unsaturated fatty acids. In *The Biochemistry of Plants*, ed. P. K.

Stumpf, 4:177–204. New York: Academic

91. Tanchak, M. A., Griffing, L. R., Mersey, B. G., Fowke, L. C. 1984. Endocytosis of cationised ferritin by coated vesicles of soybean protoplasts. *Planta* 162:481–86

92. Tolbert, N. E. 1980. Microbodies—peroxisomes and glyoxysomes. See Ref. 16, pp. 359–88

93. Trelease, R. N. 1984. Biogenesis of glyoxysomes. *Ann. Rev. Plant Physiol.* 35:321–47

94. Vigil, E. L. 1984. Role of ER in mitosis and cytokinesis of meristematic cells in cotton apices. *J. Histochem. Cytochem.* 31:1062–68

95. Vitale, A., Chrispeels, M. J. 1984. Transient N-acetylglucosamine in the biosynthesis of phytohemagglutinin: Attachment in the Golgi apparatus and removal in protein bodies. *J. Cell Biol.* 99:133–40

96. Volkmann, D. 1981. Structural differentiation of membranes involved in the secretion of polysaccharide slime by root cap cells. *Planta* 151:180–88

97. Wanner, G., Formanek, H., Theimer, R. R. 1981. The ontogeny of lipid bodies (spherosomes) in plant cells. Ultrastructural evidence. *Planta* 151:109–23

98. Wanner, G., Vigil, E. L., Theimer, R. R. 1982. Ontogeny of microbodies (glyoxysomes) in cotyledons of dark-grown watermelon (*Citrullus vulgaris* Schrad.) seedlings: Ultrastructural evidence. *Planta* 156:314–25

99. Whaley, W. G., Dauwalder, M. 1979. The Golgi apparatus, the plasmamembrane and functional integration. *Int. Rev. Cytol.* 58:199–245

100. Wick, S. M., Hepler, P. K. 1980. Localisation of Ca^{2+} containing antimonate precipitates during mitosis. *J. Cell Biol.* 86:500–13

101. Wright, K., Northcote, D. H. 1974. The relationship of root cap slimes and pectins. *Biochem. J.* 139:525–34

Ann. Rev. Plant Physiol. 1986. 37:93–136
Copyright © 1986 by Annual Reviews Inc. All rights reserved

PHOTOREGULATION OF THE COMPOSITION, FUNCTION, AND STRUCTURE OF THYLAKOID MEMBRANES

Jan M. Anderson

CSIRO, Division of Plant Industry, GPO Box 1600, Canberra A.C.T. 2601, Australia

CONTENTS

INTRODUCTION

The development and growth of plants is directly and dramatically influenced both by the light intensity and light quality the plant receives during growth.

93

Light is essential, not only as the driving force of photosynthesis, but also as a trigger and a modulator of morphogenic responses. Plants have devised unique mechanisms to capture light for photosynthesis, convert solar energy to stored chemical energy as ATP and NADPH, transform CO_2 to carbohydrates, drive protein transport, and synthesize macromolecules in the chloroplast, as well as to adapt to changes in their environment. Such mechanisms are also unique for the environment, the plant species, and the genes. Without light there is no photosynthesis. Thus, it is not surprising that plants have evolved a novel set of responses to light that is a crucial factor of their environment. Higher plants have at least three photoreceptors: protochlorophyllide, phytochrome, and the blue light receptors. In angiosperms, the conversion of protochlorophyllide to chlorophyllide is strictly light dependent; hence it is an essential factor for the accumulation of the Chl-proteins (35, 63, 96). Phytochrome exists as two photointerconvertible forms, absorbing red (Pr) or far-red (Pfr) light. The photoconversion of Pr to Pfr in vivo induces many morphogenic responses, whereas the conversion of Pfr to Pr cancels the induction of these responses; Pfr is considered to be the active form (171, 197, 207). The blue light receptors are also important regulators, especially in lower plants and algae (57, 210, 211).

Plants adapt to their particular light environment, its intensity, quality, and duration, by modulating the composition of their thylakoid membranes in a coordinated, integrated manner in order to make the best use of the available quantum flux. Modulations in the relative proportions of light-harvesting pigments, PSII and PSI reaction centers, electron carriers, and ATP synthetase will in turn influence thylakoid function and structure.

Plants must cope with contrasting terrestrial light climates that may vary by at least two orders of magnitude. Also, the daily quantum flux available will vary seasonally, diurnally, and spatially (not only within the plant canopy at the macrolevel, but also across a leaf). It is well known that plants growing in shaded habitats are capable of only low rates of photosynthesis which saturate at very low light intensities. In contrast, plants grown in open sunny habitats have much greater capacities for photosynthesis which saturate at much higher light intensities (47, 48, 50, 52, 144, 152, 239). Nevertheless, the quantum yields measured as the amount of CO_2 fixed per mole of photosynthetically active quanta absorbed are rather similar (48, 50). Plants growing in extremely shaded habitats (e.g. the forest floor of a tropical rainforest in Queensland, Australia) receive only a daily quantum flux in the 400–700 nm waveband of 1.1% on an overcast day of the quantum flux above the canopy and 0.4% on a clear day (49). About 60% of the photosynthetically useful radiation is in the form of sunflecks, very short intermittent bursts of direct sunlight. The remaining diffuse irradiance that reaches the forest floor is greatly enriched in far-red, enriched in green, and depleted in red and blue light because of strong absorption by the chlorophylls and carotenoids of higher leaves in the canopy (49).

The spectral distribution of the diffuse irradiance has a much greater far-red/red ratio than sunlight, which may have ecological significance, and has an influence on the phytochrome response (109, 215). While it is difficult to determine the relative amounts of green and far-red light in the diffuse irradiance available for photosynthesis, it seems that there will be more light available for PSI (far-red plus green) than for PSII (green only).

Plants are classified as sun or shade species depending on their ability to adapt to light intensity, which results from genetic adaptation to the prevailing light environment of their native habitat (48, 50). However, it has been established that the photosynthetic capacities of plants are also influenced by the light intensity or quality plants receive during growth. Single genotypes grown under high white light resemble sun plants, and those grown under low white light intensity resemble shade plants (48, 50, 52, 144, 152, 239). Considerations of the effect of light adaptation have been mainly limited to higher plants in this review rather than including the algae. The organization of the protein complexes along thylakoid membranes, strategies of light-harvesting, and patterns of membrane appression in most algae are not necessarily the same as those of higher plants. Moreover, consideration of the photoregulation of thylakoid membranes during development is outside the scope of this review, although it is an important aspect that has been reviewed elsewhere (27, 136).

I shall discuss briefly the function and molecular organization of higher plant thylakoids, then deal with the long-term light adaptation of the composition and function of the light-harvesting assemblies of PSII and PSI in response to light quantity, light quality, and the duration of irradiance, and the effects of all these on thylakoid membrane structure. Next I discuss the effects of light regulation on modulations of the electron transport complexes and ATP synthetase and their relation to the function of thylakoid membranes. In addition to these long-term adaptive changes that result from the complex interactions of light on photosynthesis and the expression of the plant genomes, thylakoid membranes possess short-term, adaptive changes that permit flexible and dynamic alterations in the distribution of the excitation energy between the photosystems. These adaptations help to maintain balanced electron transport under fluctuating light conditions and also limit photoinhibition. Finally, I highlight the complex interactions of light with the plant genomes and the resulting diverse effects of light regulation on pigment and protein syntheses.

MOLECULAR ORGANIZATION OF THYLAKOID PROTEINS AND LIPIDS

Most of the thylakoid proteins are organized into five intrinsic membrane-spanning complexes; together with closely associated extrinsic proteins, these perform light-harvesting, electron transport, proton translocation, and ATP-synthesizing functions of thylakoid membranes (Figure 1). The function and molecular organization of the thylakoid proteins and lipids are considered briefly below: recent reviews include those by Murphy (176), Haehnel (98), Anderson (15), Barber (31), Staehelin & Arntzen (216), Cox & Olsen (71), and Kaplan & Arntzen (119).

The PSII complex that mediates noncyclic electron transport from water to the plastoquinone pool includes 5 proteins (176, 190). Two of these noncovalently bind the inner antenna chl a and β-carotene molecules of PSII. The reaction center apoprotein (51–47 kilodaltons) also binds P680, phaeophytin, and Q_A (the first quinone-type electron acceptor of PSII), and the core antenna apoprotein (44 kDa) also has a putative quinone-binding site. The PSII complex also includes the Q_B, herbicide-binding protein (32 kDa) (127), the D2 protein (34 kDa) which has a putative quinone binding site, cytochrome b-559 (9 kDa) present as a heterodimer, and 4 Mn and 4 Cl atoms. The extrinsic proteins of the oxygen-evolving complex, designated by their apparent molecular masses as the 33, 24, and 17 kDa proteins, are exposed at the lumenal thylakoid surface (19, 175). Three less characterized polypeptides (24, 22, and 10 kDa) appear also to be associated with PSII complex (19).

The cytochrome b/f complex, which oxidizes plastoquinone and reduces plastocyanin, accepts electrons from PSII complex via the plastoquinone pool, acts as a proton pump, and catalyzes the electrogenic Q cycle (98). This complex has 5 subunits: cytochrome f (34 kDa), cytochrome b-563 (23 kDa) which has two hemes, the Fe-S Rieske redox center protein (20 kDa) and the 17 kDa protein [probable stoichiometry of $1:2:1:2$ (176)], and a 5 kDa polypeptide of unknown function.

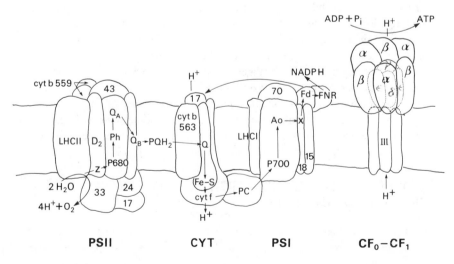

Figure 1   Scheme for the organization of thylakoid protein complexes of thylakoid membranes.

Plastocyanin (11 kDa) transfers electrons from the cytochrome b/f complex to PSI complex which forms NADPH. PSI complex also participates in cyclic photophosphorylation with the cytochrome b/f complex via ferredoxin (98, 176). PSI complex contains the β-carotene-P700-chl a-proteins of PSI (68 kDa) which perform the initial charge separation from P700 $\rightarrow$ phaeophytin $\rightarrow$ Q_A (98). Recently, the complete amino acid sequences of the two PSI reaction center polypeptides of maize have been deduced from a photogene-containing segment of maize cDNA (84). The apoproteins have much higher molecular weights of 83.2 and 85 kDa (84); each probably binds 20 chl a molecules. The later PSI electron acceptors (X $\rightarrow$ center B $\rightarrow$ center A) appear to be Fe-S centers associated with 18 and 15 kDa polypeptides, and several other polypeptides of lower M_R are also associated with PSI complex (176). The extrinsic polypeptides, ferredoxin (11 kDa) with an Fe_2–S_2 center and ferredoxin-NADP-reductase (37 kDa), appear to form a ternary complex with $NADP^+$ independently of attachment to PSI complex (33).

The proton gradient generated by electron transfer through both photosystems is used by ATP synthetase to make ATP. ATP synthetase consists of an extrinsic coupling factor, CF_1, which is composed of 5 subunits: 3 α (59 kDa), 3 β (56 kDa), 1 γ (37 kDa), 1 δ (17.5 kDa), and 2 ξ (13 kDa), and an intrinsic membrane-spanning, proton-conducting sector, CF_O, which has 3 subunits: subunit 1 (15 kDa), subunit II (12.5 kDa), and subunit III (8 kDa) with a probable stoichiometry of 1 : 2 : 10–12 (218).

In addition, chloroplasts that possess chl b have specific chl a/b-proteins that are associated with PSII (LHCII) and PSI (LHCI). LHCII is the main thylakoid component accounting for more than half the total chlorophyll and intrinsic protein of most plant thylakoids (225). LHCII consists of multiple apoproteins of 29–25 kDa (39, 156, 225), but most plants have two main apoproteins (27 and 25 kDa), each of which binds an average of 7–14 pigment molecules in a consensus stoichiometry of 4 chl a: 3 chl b: 2 xanthophylls (225). Another chl a/b-protein complex has an apoprotein (29 kDa) that binds more chl a than chl b (61, 62). As nuclear DNA contains a multigene family for the LHCII apoproteins (69, 74, 227, 230), this may account for the multiple apoproteins of LHCII that have yet to be biochemically characterized and whose function and location in the complete LHCII-PSII complex assembly is unknown. Specific PSI chl a/b-proteins (chl a/chl b $\sim$ 3.6) consisting of four or more polypeptides of 25–20 kDa are associated with a "native" PSI complex which contains one P700 per 110–180 chlorophyll molecules (104, 225). Less is known about these minor chl a/b-proteins of LHCI, but they exhibit a very strong degree of immunological cross-reactivity with LHCII, suggesting there may be structural homologies between the two classes of nuclear-encoded light-harvesting complexes (P. K. Evans, J. M. Anderson, unpublished observations).

The acyl lipids that form 25–30% of the mass of plant thylakoid membranes act as a fluid matrix for the functional supramolecular complexes and also allow diffusional processes to occur. They appear to be fairly uniformly distributed between higher plant species, although changes in lipid composition and acyl chain saturation occur during development. The five classes of acyl lipids are monogalactosyldiacylglycerol, digalactosyldiacylglycerol, sulphoquinovosyldiacylglycerol, phosphatidylglycerol, and phosphatidylcholine. Important characteristics of the acyl lipid composition of thylakoid membranes are that the main acyl lipid, monogalactosyldiacylglycerol, which comprises up to 50% of the total acyl lipid, does not form lamellar bilayer structures in aqueous dispersion but rather forms cylindrical inverted micelles. Second, the uncharged glycolipids make up some 70% of the total lipid with phospholipids rarely exceeding 20%. Third, the acyl residues are mainly unsaturated (85–90%) with 18:3 linolenic acid predominating (176, 198).

Discrete, functionally distinct electron transport complexes and ATP synthetase, whose subunits are encoded by the nuclear or chloroplast genomes (105), pose problems for their coordinated synthesis and assembly. Since the polypeptide composition of each complex must remain constant, with the exception of variable amounts of LHCII and LHCI around core PSII and PSI complexes, respectively, there must be a tight, highly coordinated regulation of the chloroplast- and nuclear-encoded subunits of each complex. In principal, however, there need not necessarily be a fixed and constant stoichiometry between thylakoid electron transport complexes that are functional-structural entities (13, 14, 162). Moreover, lateral heterogeneity in the distribution of complexes means also that the pigment beds of PSII and PSI may be stoichiometrically independent of each other.

Lateral Asymmetry of Thylakoid Complexes and Lipids

Developing or mutant thylakoids containing only core PSII and PSI complexes, cytochrome b/f complex, and ATP synthetase are unstacked (13, 14). In the light-induced development of higher plant thylakoids, there is a step-wise assembly of functional complexes. Following the synthesis and assembly of the chl a/b-proteins of LHCII and LHCI and their subsequent interaction with core PSII and PSI complexes, there is a lateral rearrangement of thylakoid intrinsic protein complexes and concomitant membrane appression. This results in the thylakoid membrane network of chl b-containing chloroplasts being structurally differentiated into appressed regions (grana partitions) which are interconnected with the nonappressed regions (mainly stroma thylakoids); the inner surface of these flattened thylakoid vesicles enclose the lumenal space that is continuous between appressed and nonappressed regions.

Both morphological and biochemical studies have shown that the appressed and nonappressed regions have distinct composition, function, and structural organization. Long-term, light-regulated environmental changes result in variations in the extent of membrane appression, and hence of the relative stoichiometries of thylakoid complexes and thylakoid function. Recent evidence for lateral asymmetry in the distribution of thylakoid supramolecular protein complexes has been reviewed extensively (15, 16, 31, 98, 176, 216). ATP synthetase is located exclusively in nonappressed regions (169). The location of other thylakoid complexes has been suggested by membrane fractionation studies, particularly those using the aqueous polymer, two-phase partition method (cf 4) which allows right-side-out vesicles derived mainly from nonappressed membrane regions to be separated from inside-out vesicles enriched in appressed membranes (4). Most PSII-LHCII complexes are located in appressed regions (PSIIα) (calculated to be 70–80% of total PSII chl in spinach (21)), but some PSII (PSIIβ) with less LHCII is present in the nonappressed stroma thylakoids. The extent of PSI lateral heterogeneity is not resolved; on a chl basis, it was calculated that 80–90% of the total chl of PSI was in stroma thylakoids (21), whereas light-induced measurements of P680 or P700 indicated a more extreme lateral heterogeneity (98). While membrane fractionation studies first suggested that the cytochrome b/f complex was uniformly located between both regions (12, 70), later other researchers argued for exclusion of this complex from appressed regions (31, 89). Recently, immunocytochemical techniques provided direct evidence for the location of two intrinsic thylakoid complexes, cytochrome b/f complex (10, 92) and PSI complex (233), in both appressed and nonappressed membrane regions, and PSII-LHCII complex mainly, but not exclusively, in appressed regions (93, 232).

Significantly, this lateral heterogeneity of the distribution of thylakoid protein complexes is matched by lateral asymmetry of the acyl lipids (94, 177), even though the appressed and nonappressed regions are continuous. On a protein basis, the appressed regions are depleted in both galactolipid classes and sulphoquinovosyldiacyglycerol but have similar amounts of phosphatidylglycerol (94, 176, 177). Murphy (176) calculates that lipids occupy only 24% of the appressed membrane area and 50% in nonappressed membranes.

Molecular Mechanisms of Thylakoid Stacking

The extent of membrane appression varies according to the developmental stage of the chloroplast, long-term changes caused by environmental conditions (particularly light), and short-term, dynamic changes induced by reversible phosphorylation of a specific subpopulation of LHCII, as discussed later. There is compelling evidence that part of LHCII is an essential factor for membrane appression. This tentative identification was first made on the basis of developmental or mutant studies where the extent of membrane stacking was correlated with the amount of LHCII (see 51). Direct evidence was provided by studies with the reconstitution of LHCII into thylakoid lipids or membranes lacking LHCII. LHCII proteoliposomes interacted with each other to form structures reminiscent of grana stacks (205). The insertion of exogenous LHCII into thylakoids from intermittent light-grown plants that have no LHCII and only limited membrane appression resulted in significant membrane appression (72). Further, it has been demonstrated by many researchers (see 15, 216) that membrane appression is prevented by limited proteolysis of thylakoid membranes or LHCII proteoliposomes since proteases remove a stromal surface-exposed peptide fragment (~ 1.5 kDa) from LHCII. The essential segment of the LHCII N-terminal region cleaved by proteases is (Lys-Arg)-Ser-Thr-Thr-Lys-Lys (173, 206). The 3 or 4 positively charged amino acids decrease the overall negative surface charge of thylakoids and aid in membrane appression by decreasing the electrostatic repulsive forces between surfaces of adjacent membranes (30). It has also been suggested that the essential surface-exposed LHCII segment contributes to charge interaction with LHCII on an adjacent membrane and may also contribute to van der Waals' attractive forces (174). In addition to the 3–4 positively charged amino acids, LHCII has 3 stroma-exposed carboxyl groups (206). Descriptions of the physical forces such as van der Waals' interactions, electrostatic repulsion, and hydration forces are reviewed elsewhere (29, 30, 226). Barber (29, 30) argues for a lateral charge diffusion model for membrane appression where those complexes with high net negative charge will migrate away, and those with lower surface charge will congregate in appressed areas. Murphy (176) suggests that the extreme segregation of PSII-LHCII complexes may arise as a secondary consequence of the intermembrane LHCII-LHCII associations. If so, the reason for PSII in appressed regions is simply that it is dragged along by the considerable LHCII antenna surrounding the core PSII complex (176). Conditions that lead to the cessation of intermembrane LHCII-LHCII interactions and the detachment of LHCII from core PSII during protein phosphorylation lead to the exclusion of both units from the appressed region. Conversely, dephosphorylation leads to reassociation of LHCII with PSII, and LHCII-LHCII interactions between adjacent membranes result in an increase in the area of appressed regions.

LONG-TERM PHOTOREGULATION

Pigment Composition

LIGHT INTENSITY   The marked adaptation of the pigment composition and content of higher plant thylakoids from plants grown under different light intensities is well established (13, 47, 48, 50, 52, 144, 152, 167, 239). Obligate shade plants have much more chlorophyll per chloroplast (17) and lower chl a/chl b ratios (~ 2.0–2.4) compared to sun plants (~ 2.8–3.6) (48, 50, 144, 167). Leaves from individual trees exhibit a similar gradient between sun-exposed and shade leaves (144, 148). With individual plants grown under controlled intensity conditions, those adapted to high irradiance have higher chl *a*/chl *b* ratios and less chlorophyll per chloroplast (138, 139, 145, 148, 150, 167, 242). Recently it has been shown that the chl *a*/chl *b* ratios of pea thylakoids from plants grown at 8 light intensities are not linear with light irradiance but decrease more rapidly the lower the light intensity (139) (Figure 2a). Indeed, adaptation occurs across bifacial leaves (220, 221) with spinach thylakoids adjacent to the upper- and lower-most surfaces having very different chl *a*/chl *b* ratios of 3.5 and 2.6, respectively (220). The carotenoid composition also varies, with thylakoids from sun and high light intensity plants having a higher proportion of β-carotene and less xanthophylls, especially lutein and neoxanthin, compared to shade and low light intensity plants (144, 148, 150). Since β-carotene is mainly associated with the core chl *a*-proteins, and lutein and neoxanthin with LHCII (56, 79, 153), this accounts for the finding that the xanthophyll/β-carotene and chl *a*/β-carotene ratios are higher in the low light or shade adaptation.

LIGHT QUALITY   In attempts to identify the effects of the blue light receptor and phytochrome, plants have also been grown under different light qualities of equal irradiance. Comparisons of barley (60, 147, 149), mustard (240), bean (22, 67), soybean (75), corn (78, 80), and *Atriplex triangularis* (141) grown in blue, white, or red light of equal irradiance show that blue-light thylakoids have higher chl *a*/chl *b* ratios and less chlorophyll per chloroplast than red-light chloroplasts. From these adaptive changes in pigment composition, thylakoid structure, and function, Lichtenthaler proposed that blue light induces a sun-type adaptation and red light a shade-type adaptation (60, 144, 147, 149). However, both the plant species and the level of irradiance used are very important. Strangely, thylakoids from peas grown in continuous red, white, and blue light of equal irradiance had no differences in pigment or chlorophyll-protein composition (141). Startlingly different results were obtained recently with a shade plant (142). Leaves of the fern *Asplenium australasicum* grown in red light were pale green compared to those grown in blue or white light and had a 2.5-fold decrease in total chl/frond area or threefold decrease in total chl/fresh weight (142). Thylakoids of red-light *Asplenium* had a chl *a*/chl *b* ratio of 2.62 compared to 2.37 in blue-light thylakoids, relatively less LHCII and more PSI chl, with a PSII/PSI chl content ratio of 1.6 compared to 2.1 in the blue-light thylakoids, and only slight differences in the ratios of appressed to nonappressed thylakoid membranes (142). These results demonstrate that the red-light adaptation of the shade plant *Asplenium* mimicked the sun adaptation and blue-light adaptation mimicked the shade adaptation, the reverse situation to that of other plants thus far investigated (6, 22, 60, 75, 78, 80, 147, 149, 240). Since shade plants grow in light habitats enriched in far-red and green and depleted in blue and red (49), their levels of both phytochrome and blue light receptors will probably vary markedly from those of sun plants; this may

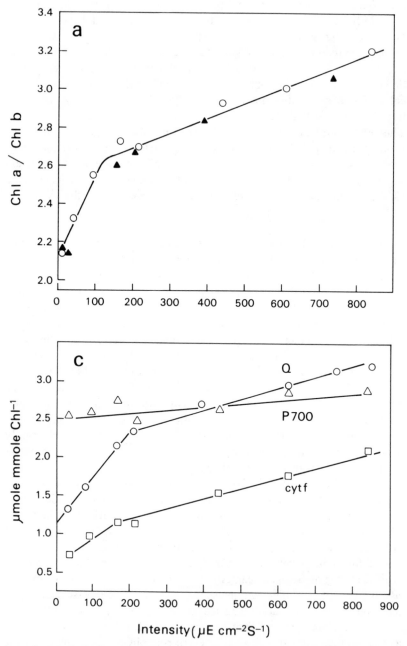

Figure 2   Effect of light intensity on the composition and function of pea thylakoids isolated from peas grown under 7 different light intensities (139, 140). (a) chl *a*/chl *b* ratios (b) amounts of Q (PSII complex), cyt *f* (cytochrome *b/f* complex), and P700 (PSI complex); (c) PSII and PSI electron transport; (d) CF₁ activity.

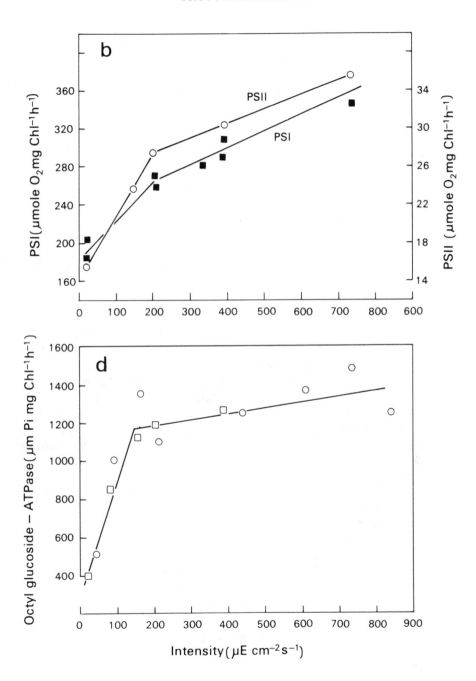

account for the remarkable differences seen in the blue- and red-light adaptation of *Asplenium*.

In this context, the recent studies of Humbeck et al (115, 116) are relevant. *Scenedesmus* grown under blue light had a greater chlorophyll content/cell (chl *a*/chl *b* ratio of 3.1) and more LHCII, while that grown in red light had a lower chlorophyll content (chl *a*/chl *b* ratio of 5.0) (115). Thus, the blue-light adapted *Scenedesmus* mimicked low intensity or shade adaptation, and the red light mimicked high intensity or sun adaptation. Humbeck & Senger (116) suggest that these differences in red and blue light adaptation in green algae might be explained by phylogenetic adaptation to their environment. With increasing depth of water, far-red and red light is attenuated and transmission is restricted to the blue-green region (133). Clearly the notion that the photosynthetic apparatus of sun plants is identical with that of high intensity or blue light-adapted plants, and that of shade plants with low intensity or red-light plants, is no longer tenable.

A shade habitat combines low intensity with a very low red/far-red ratio (49); this may alter the degree of effectiveness of phytochrome (109, 215). Hence other light quality studies have explored the effects of either far-red enriched fluorescent (+FR) and far-red deficient incandescent (−FR) illumination, or red (R) and far-red (FR) irradiance of equal photon fluence rates on thylakoid composition, structure, and function. With peas (90, 162, 167) and corn (78, 80) there were few differences observed in the chl *a*/chl *b* ratios: peas, +FR, 2.7 and −FR, 2.9 (162); corn mesophyll, R, 3.4 and FR, 3.2 (78); corn bundle sheath, R. 4.5 and FR, 4.8 (78). In developing soybean, however, the effect was more dramatic, with a chl *a*/chl *b* ratio of 2.45 with FR compared to 2.92 with R (77). Often the light climates enriched or depleted in far-red will have other quality differences in the 400–700 nm region so it is impossible to make comparisons. The complexity of light quantity and quality relations is evident also in a recent study of corn mesophyll and bundle sheath chloroplasts (80). Under high-intensity blue, white, and red light sources, which establish very different Pfr/P_{total} ratios, and hence phytochrome effectiveness (109, 215), there was no direct correlation between pigment content and Pfr/P_{total}. In a comparison of low red light to low red plus high far-red light, the synthesis of PSI complex (measured as the 68 kDa apoprotein of PSI/total protein) was directly related to intensity, whereas PSII complex (measured as the 51 and 47 kDa apoproteins of PSII/total protein) was rather constant, resulting in an increase of PSI/PSII reaction center ratios with increasing red irradiance levels (80).

Recently it has been established that the variations in chl *a*/chl *b* ratios and xanthophyll/β-carotene ratios observed in plants adapted to varying light conditions reflect modulations in the proportions of chl-protein complexes. Since most of the chl *b* and xanthophylls are associated with LHCII, and LHCII is mainly located in the appressed grana partitions (21), one would predict high amounts of LHCII in shade, low light, and some red and far-red enriched-adapted thylakoids, consistent with their higher ratios of appressed to nonappressed membranes. With the introduction of improved mild SDS polyacrylamide gel electrophoresis procedures in 1978 (see 225), 7–10 chl-protein complexes can be resolved with low free pigment ($\sim$ 10%), and qualitative comparisons of chl-protein composition are feasible. LHCII is enriched in thylakoids from chloroplasts of shade plants compared to sun plants (11, 65), in low intensity compared to high intensity light (138, 139, 146, 150, 179), in soybeans grown in far-red compared to red light (77), and in *Atriplex triangularis* grown in red compared to blue light (142), consistent with higher ratios of appressed/nonappressed thylakoid membranes.

Concomitant with relative increases in the amount of total chl associated with the chl *a/b*-proteins of LHCII and LHCI, there is a decrease in that of the core chl *a*-proteins of PSII (CPa) and PSI (CP1). It is impossible to get an accurate estimate of the amount of CP1 alone by gel electrophoresis methods, as the bands of lowest mobility (CP1a[1], CP1a[2]) contain CP1 plus some LHCI; this method allows only a qualitative estimate of the total chlorophyll associated with PSI. For PSII core complexes, either one or two bands are resolvable: one band (CPa) (as in 11) contains both PSII chl *a*-proteins plus some tightly bound chl *a/b*-proteins of PSII, or two bands (61, 62, 155, 156) which may have lost some chlorophyll. Nevertheless, either gel procedure is suitable for relative comparison of PSII complexes. In all light-adaptation studies thus far examined by mild gel electrophoresis, an increase in LHCII and total PSII chl is paralleled by a decrease in total PSI chlorophylls consistent with the proposed lateral heterogeneity in their distribution (21).

It has been suggested that an increased amount of chlorophyll associated with PSII in shade plants may help to redress the energy imbalance of shade light that favors excitation of PSI (14, 48). Additionally, increasing the chlorophyll *b* and lutein content might increase the proportion of blue-green light captured by PSII (13). Recently, Evans (82) has calculated that alterations to the relative proportions of chlorophyll-protein complexes occurring in shade adaptation do not enhance the utilization of light of different spectral qualities, in contrast to Anderson's proposal (13).

Thylakoid Structure

There are well-characterized changes in chloroplast ultrastructure from plants adapted to sun and shade (17, 167), high intensity and low intensity (52, 148, 150, 152), blue and red (7, 60, 147, 152), and far-red deficient and far-red enriched (78, 80, 162) light. Shade, low light, some red light, and far-red enriched chloroplasts have more thylakoid membranes and less stromal volume compared to sun, high light, blue light, and far-red deficient light-adapted chloroplasts. This is most evident in obligate shade plants (17, 52, 167) where the thylakoids occupy almost the entire chloroplast, significantly reducing the relative stromal volume. There appears to be a direct correlation between chlorophyll content and amount of thylakoid membranes. Where the available light is limiting, light capture is increased by an enhanced chlorophyll content and hence thylakoid membrane content.

In addition to the high thylakoid/stromal volume ratio of shade and low intensity light chloroplasts, their thylakoids have broader grana stacks, many more thylakoids per granal stack, and more grana stacks relative to sun chloroplasts. An index of grana stacking is conveniently measured as the ratio of length of appressed membranes to the length of nonappressed membranes: e.g. spinach or pea thylakoids with 50–60% appressed membranes have a ratio of 1.0–1.5, and this ratio may increase to 4–5 in obligate shade leaves (17). The ratio of appressed/nonappressed thylakoid membranes is also greater in far-red enriched than in far-red deficient-adapted pea (167) and soybean (77) thylakoids grown under equal photon fluence rates. In any plant species, there is a rough correlation between the relative extent of membrane appression and chl a/chl b ratio. This is to be expected, since chl b is located mainly in LHCII and LHCII is distributed preferentially in the grana partitions (21).

Effect of Light Regulation on the Size of PSII and PSI Units

Since PSII and LHCII are mainly located in appressed membranes, and PSI mainly in the nonappressed regions, the modulations in chl-protein content resulting from long-term light adaptation will influence not only thylakoid structure but also the light-harvesting assemblies of PSII and PSI. Originally the photosynthetic unit was defined functionally by Emerson and Arnold as the number of chl molecules cooperating in light absorption for the evolution of one O_2 molecule: later it was defined either as the total chl molecules per P700 (47, 52) or cytochrome f (239), since it was assumed that there was a fixed stoichiometry of P680:cytochrome f:P700 of 1:1:1. With the recent recognition that plants have varying amounts of PSII and PSI reaction centers (97, 165, 167) and varying antenna sizes of PSII and PSI (165, 167, 223) it is necessary to determine the antenna sizes of both PSII (total PSII chl/P680) (157) and PSI (total PSI chl/P700) individually.

Unfortunately, it is difficult to measure either the concentrations of reaction centers or sizes of PSI and PSII units. A very qualitative method relies on measuring the "apparent" total chl associated with PSII and PSI by determination of the amounts of chl associated with chl-proteins by gel electrophoresis (11, 14). While this method only gives an "apparent" total chl content of individual photosystems, it shows the amount of chl a/b-antenna proteins compared to core chl a-proteins. The number of antenna molecules transferring excitation energy to P680 or P700 can be determined by kinetic analysis (165, 167, 223), but few investigators have the required sophisticated machines. Oxygen flash yields (73, 118) or optical difference spectroscopy can be used to determine concentrations of P700 or P680. Initially Jursinic (118) found that the amount of C^{14}-atrazine bound to the Q_B protein agreed with the estimation of P680 determined by oxygen flash yields. Recently, however, Dennenberg & Jursinic (73) suggest that the light-induced measurements of Q_A at 325 nm in the presence of ferricyanide used to estimate P680 (162, 164, 165, 167) overestimate the amount of Q_A and hence the amount of P680. Whitmarsh & Ort (237, 238) challenge the notion of varying stoichiometries of P680 and P700 in sun plants as they found constant molar ratios of P680: cyt f: P700 for spinach (both summer and winter), tomato, pigweed, and radish. They suggest that the light-induced absorbance change at 325 nm attributable to Q_A reduction significantly overestimates the amount of P680 (238). While there are no reliable and simple methods yet to determine P680 and P700, most investigators advocate flexibility in the amounts of P680 relative to P700 under different light growth conditions.

Before discussing the antenna sizes of PSII and PSI, we must consider another complication resulting from the heterogeneity of PSII. The concept of the heterogeneity of PSII was introduced in order to interpret the biphasic nature of primary PSII activity which had been measured as the fluorescence induction rise (166, 168). Melis et al propose that PSII occurs as two structural-functional entities, termed PSIIα and PSIIβ, which differ in both their effective light-harvesting chl antenna sizes (166, 168) and location in the membranes. PSIIα has the larger antenna unit size because of a full complement of chl a/b-proteins of LHCII, and it is located in appressed thylakoids (18, 164), whereas PSIIβ, which has a smaller complement of LHCII, is located in the nonappressed stroma thylakoids (18, 163, 164, 223). Alternatively, it has been suggested that apparent PSII heterogeneity is the result of incomplete blockage by DCMU of some PSII units (108), or arises as a consequence of connectivity between PSII units (55, 128). Recently Percival et al (191) showed that the contribution of PSIIα and PSIIβ to the fluorescence induction curves depends on the ratio of PSII to LHCII, with PSIIβ characteristics being generated either by disconnecting LHCII from PSII or by preferentially exciting PSII relative to LHCII. Thus, the presence of PSIIβ centers in the nonappressed thylakoid regions may reflect in part the decoupling of the specific population of phosphorylated LHCII from PSIIα centers and the subsequent diffusion of the LHCII-depleted PSIIα and the free phosphorylated LHCII from the appressed to the nonappressed regions. In any event, the ratios of PSIIα/PSIIβ vary under different growth conditions (162, 165, 167, 191). If these interpretations are correct, it is important to establish the functional and structural heterogeneity of the chl a/b-proteins of LHCII.

EFFECT OF LIGHT QUANTITY   There is agreement that an increase in light irradiance during plant growth causes substantial increases in PSII reaction centers and slight increases in PSI reaction centers on a chl basis (138–140, 167). The Q/P700 ratios for peas adapted to 7 light intensities over a tenfold range increased from 1.2 to 2.3 at the highest intensity (140). As the light intensity during plant growth increased, there was a substantial decrease in the "apparent" antenna size of PSII units and little decrease in that of PSI (140). However, in another study, fluorescence induction measurements indicated no differences in the antenna size of PSII in light-adapted pea chloroplasts (167). By contrast, Hodges & Barber (107) conclude that a low intensity light pea leaf had a higher PSIIα/PSIIβ ratio than a high intensity light pea leaf consistent with a greater overall PSII/PSI content, lower chl a/chl b ratio, and greater appressed to nonappressed thylakoid membrane area of the low intensity light pea leaf. In tomato, Hodges (106) found an increase in PSIIα following transfer of the plants from high to low light. Lichtenthaler et al (150) conclude that low-light radish had a larger PSII unit than high-light radish, consistent with the higher LHCII/CPa ratio for the low-light plants. Adaptation to light irradiance clearly alters the LHCII/CPa ratio (138, 139, 150), with low intensity light peas having a ratio of 8.61 compared to 3.55 in high intensity light peas (139). Since the relative amount of CPa (gel electrophoresis) correlates roughly with the amount of Q (measured as atrazine binding) in a number of studies (138–140, 142), it appears that LHCII/CPa ratios may indicate the apparent size of PSII units.

EFFECT OF LIGHT QUALITY   Assessment of the stoichiometries of PSII and PSI reaction centers and the sizes of PSII and PSI units in shade plants are meager. In the earliest study (52), fluorescence induction rise measurements indicated that the total chl/Q ratios of *Alocasia* and spinach were similar. Recently, Melis & Harvey (167) found high Q/P700 ratios of 2.4–3.8 for three shade plants compared to values of 1.8–2.2 for a variety of sun plants. These high Q/P700 ratios of shade plants resulted from a slight decrease in P700 and a marked increase in Q on a chl basis (167). In contrast, Chu & Anderson (65) found much less Q in *Alocasia* (measured as atrazine binding) which had a Q/P700 ratio of ~ 1.0 compared to values of 1.6–1.8 for peas or spinach. *Alocasia* had a 1.6-fold larger PSII unit and a slightly smaller PSI unit than spinach (65). The lower relative amounts of CPa found in the shade plants (11, 65) are consistent with less total P680 or Q per unit chl. Lichtenthaler et al (144, 148) also suggest that the shade leaves of beech had more antenna chlorophylls and a larger PSII unit than sun leaves. By contrast, Melis (90, 162) observed that the size of PSIIα for the shade fern *Polystichum* was 210, smaller than the PSIIα unit of spinach (~ 350), while that of PSI was also lower.

Comparisons between peas grown in far-red enriched and far-red depleted light show marked differences in the stoichiometries of PSII and PSI reaction centers (162, 167); the Q/P700 ratios were 1.8 and 2.8 in the far-red depleted and far-red enriched peas, respectively (162). In contrast, the sizes of PSIIα, PSIIβ, and PSI units were similar (162). The greater ratio of PSII/PSI reaction centers in plants growing with light predominantly absorbed by PSI (far-red light) is consistent with the higher PSII/PSI reaction center ratios observed in shade plants (90, 162, 167). These few results obtained illustrate the uncertainties of comparing data from various plants in different laboratories using different experimental procedures. Melis (162, 167) concludes that light quality does not substantially affect the size of PSII and PSI units in shade plants, whereas Chu & Anderson (65) and Lichtenthaler et al (144, 148) suggest it may do so.

It is clear that the amounts of LHCII relative to core CPa, and LHCI to CP1 are not constant, with shade (11, 65, 144, 148) and low-light (138, 139, 150) adaptation leading to increased amounts of LHCII and decreased amounts of CPa. This variation apparently indicates an increase in the effective PSII antenna size in shade and low-light adapted plants relative to that of sun and high-light adapted plants. This point is not resolved since Melis et al (162, 167) suggest that there is little adaptation either by light quality or irradiance in the antenna size of PSIIα and PSIIβ. In other studies of far-red plus red and red light soybeans (77) and corn (78, 80), there was an enrichment in the apoproteins of CPa compared to those of CP1, and hence a greater PSII/PSI reaction center ratio in the far-red plus red light. Eskins et al (80) conclude that although far-red light influences leaf and plant size, it is apparently red irradiance that influences the PSII/PSI reaction center ratio.

The marked modulations observed in the proportions of chl-protein complexes and hence PSII and PSI reaction centers, and the relative number of chl molecules associated with each, strongly support the concept of highly adaptive thylakoids in which the stoichiometries of the chl-protein complexes are adjusted and optimized in response to light conditions during growth. There is insufficient data available yet to make valid generalizations on the light adaptation of either PSII or PSI unit sizes or P680/P700 ratios. Different plant species appear to respond differently, and whether the plants are being grown under continuous or light-dark regimes may also be of importance. Melis (162) suggests that the main strategy in the sun-shade and high-low light adaptation is to change the relative number of PSIIα, PSIIβ, and PSI units without alteration in actual photosynthetic unit size. Other studies (106, 107, 138, 140, 144, 148, 203) suggest that shade plants and low-light plants have larger PSII units with lower chl a/chl b and higher LHCII/CPa ratios than sun and high-light plants. The apparently larger PSII units may be the result of an increase in the number of PSIIα units compared to PSIIβ units. More research and better methods are needed to resolve this conflict.

Thylakoid Function

As discussed, plants respond to light by modulation of the relative amounts of chl-protein complexes and hence stoichiometries of PSII and PSI reaction centers and unit sizes so as to harvest light energy effectively and attain maximum photosynthetic yields. Significantly, plants also modulate the relative amounts of other electron carriers such as the cytochromes, plastoquinone, plastocyanin, and ferredoxin. The early studies of light quantity adaptation (47, 52, 95, 203, 241, 242) established that the relative amounts of P700 are not greatly affected by light intensity, whereas there are marked increases in the amounts of other electron transport carriers on a chl basis. Adaptation to light intensity of a single genotype results in significant increases in the electron transport capacity of both PSII and PSI with increasing irradiance received during plant growth (52, 138, 140, 167, 241). With peas grown in varying light intensities, the changes in both PSII and PSI electron transport were marked and nonlinear (140; Figure 2). In line with these closely coordinated changes in the rates of electron transport of both photosystems, the amounts of Q, PQ, and cytochrome f all changed in a similar way (Figure 2), but the amount of P700 was not greatly altered. There is a linear relationship between the amounts of Q (measured as atrazine binding), cytochrome f, and CF_1 (140), suggesting irradiance modulates these components in much the same way. Since each is an invariant component of the PSII complex, the cytochrome b/f complex, and ATP synthetase, respectively, this suggests that the biosynthesis and assembly of these three complexes are well coordinated in peas and regulated by similar mechanisms during growth. On the other hand, there must be different mechanisms of regulation for PSI complex, LHCII, and LHCI. These results indicate that the chl-proteins must be photoregulated by different mechanisms. This would be expected since the apoproteins of LHCII and LHCI are nuclear encoded and those of core chl a-proteins of PSII and PSI are chloroplast encoded (105). However, there appears also to be a differential regulation of PSII chl a-proteins versus PSI chl a-proteins that is partly influenced by the level of irradiance.

In the sun-shade adaptation, the capacities of both PSII and PSI electron transport are markedly lower in shade than sun plants (17, 47, 48, 50, 52, 167). In the shade plant *Alocasia,* there was a fourteenfold decrease in PSII electron transport and a sixfold decrease in PSI electron transport compared to high-light *Atriplex* (52). Shade plant chloroplasts have only ⅓ to ½ the amounts of cytochromes b-559, b-563, and f, and plastoquinone on a chl basis relative to sun plants (52, 95, 203). Thus, the much greater capacity of sun plants for electron transport per mole of chl relative to shade plants is related partly to the marked changes in the amounts of the cytochrome b/f complex and the mobile electron carriers plastoquinone, plastocyanin, and ferredoxin.

Comparison of several sun plants grown in continuous blue or red light shows that the blue-light thylakoids had higher PSII activities and lower chl/plastoquinone and chl/cyt f ratios, consistent with their higher chl a/chl b ratios and greater photosynthetic capacities (6, 7, 22, 60, 141, 147, 149, 240). In *Atriplex triangularis* (141), blue-light thylakoids have increased proportions of Q, plastoquinone, cytochromes b-559, f, and b-563, and P700 relative to red-light thylakoids resulting in greater rates of PSI, PSII, and whole chain electron transport. Adaptation resulted in a fourfold faster rate of photosynthesis of blue-light compared to red-light leaves which correlates with changes in composition of the thylakoid membranes (141). In the shade plant *Asplenium,* there were higher levels of Q, cytochrome f, and CF_1 activity in thylakoids from red-light ferns, but lower levels of P700 and plastoquinone, indicating that the plant species is important in red and blue light quality adaptation (142).

The light regulation of ATP synthetase has seldom been determined in light quantity or quality adaptation studies. Berzborn et al (43) demonstrated that the concentration of CF_1 was two times higher on a chl basis in spinach grown under high light intensity compared to low light intensity. The observed increase in CF_1, and hence ATP synthetase, correlated with a relative increase in the amount of nonappressed thylakoid membranes (40% in low light compared to 60% in high light) consistent with the exclusive location of ATP synthetase in nonappressed membranes (169). With peas grown under seven light intensities (140), there was an increase in the octylglucoside-activated Mg^{2+} ATP synthetase as the light intensity during growth increased (Figure 2), with the rates of increase being more marked at the lower light intensities. The response of CF_1 activity to light intensity parallels those of chl a/chl b ratios, chl/Q, and chl/cyt f ratios (140). Indeed, there is a linear relationship between cytochrome f and ATP synthetase, suggesting these two complexes are directly influenced by light irradiance (140). Similarly, *Atriplex* grown in blue light had more CF_1 per unit chl than did red-light *Atriplex,* in line with the increased capacity of electron transport of both PSII and PSI and CO_2 fixation of the blue-light plants (144).

Photoregulation Across Leaves

Bifacial leaves receive most of their light at their upper surfaces (88). While the palisade tissue is illuminated with strong white light, the spongy tissue receives lower intensity light enriched in green and far-red because of the strong absorption of light by chlorophyll and carotenoids (222). Thus, the light conditions of the palisade and spongy tissue are qualitatively similar to those of sun and shade plant leaves. Indications of differentiation of the photosynthetic properties of the chloroplasts of palisade and spongy tissue were demonstrated by comparison of the fluorescence curves of the upper and lower surfaces of bifacial leaves which resembled those of sun and shade leaves, respectively (126, 208).

The first comparisons of the photosynthetic properties of chloroplasts isolated from the two types of tissue from *Camellia japonica* and spinach were made by Terashima (220), who developed a method to cut sections parallel to the leaf surface. With *Camellia japonica*, the palisade chloroplasts (chl *a*/chl *b* ratio of 3.2) had higher amounts of P700, PSII polypeptides, and PQ, and lower amounts of LHCII compared to spongy tissue chloroplasts (chl *a*/chl *b* ratio of 2.6) (221). Maximal rates of both PSII and PSI electron transport under saturating light were greater in the palisade than in spongy tissue chloroplasts (221). Similar results were found with spinach palisade chloroplasts having a twofold higher overall electron transport than spongy tissue chloroplasts, explained in part by their higher contents of PQ, cyt *f*, plastocyanin, ferrodoxin, and ferredoxin NADP reductase on a chl basis (220). The higher overall electron transport rate of palisade chloroplasts (chl *a*/chl *b* ratio of 3.41) was also accompanied by a threefold higher CO_2 fixation rate than that of spongy tissue chloroplasts (chl *a*/chl *b* ratio of 3.04). Palisade chloroplasts had fewer thylakoid membranes, smaller granal stacks, and more stroma compared to spongy tissue chloroplasts; these ultrastructural differences are also analogous to those observed for sun and shade plant chloroplasts, respectively (17, 52).

The photosynthetic and ultrastuctural properties of the spinach chloroplasts from 10 sections segmented parallel to the leaf surface changed continuously from the upper to the lower surface, irrespective of the differentiation into palisade and spongy tissue (220). The chl *a*/chl *b* ratio was 3.5 in fraction 1 at the upper surface and 2.6 in fraction 10 at the lower surface. The cytochrome *f* and ribulose bisphosphate carboxylase content also decreased continuously, as did PSII and PSI electron transport activity. Moreover, the ratio of appressed to nonappressed thylakoids and the thylakoid to stroma volume ratio both continuously increased as a function of depth from the upper leaf surface. After a fully expanded spinach leaf had been inverted for 7 days and received more light on its lower surface, there were inverted profiles of chl *a*/chl *b* ratios and PSII electron transport activity, showing that mature chloroplasts adapt to changed light environment (220). These elegant experiments of Terashima (220–222) reveal for the first time a vertical gradient in the photosynthetic properties of chloroplasts across bifacial leaves, resulting from dynamic adjustments of each chloroplast to its respective light environment, in essentially the same manner as in sun and shade plant chloroplasts.

Effect of Photosynthetic Inhibitors

Analogous effects to those observed in shade adaptation are found when plants are grown in the presence of low concentrations of PSII inhibitors. This effect of inhibitors has been demonstrated for monuron (124), fluometuron (235), methabenzthiazuron (83), bentazon (151, 160, 161), and San 9785 (137). The chl a/chl b ratios of inhibitor-treated thylakoids are lower than those of control thylakoids, and the ratios of appressed to nonappressed membranes are greatly increased. In all cases, the grana stacks are wider than usual, and often there are many more thylakoids per granum. For example, in barley chloroplasts, although the chlorophyll level per plastid was unaltered by San 9785 treatment, the chl a/chl b ratio decreased to 2.7 compared to 3.5 in control thylakoids as a result of the San 9785-treated thylakoids having more LHCII and less PSI chl (137). With the various inhibitors causing partial inhibition of PSII, the plants respond by enhancing PSII relative to PSI so as to compensate for loss of PSII activity, independent of the herbicide used. It appears that the adaptive responses to partial inhibition of PSII are analogous to those of shade adaptation. Since shade plants receive greater excitation of PSI than PSII, they also respond by increasing PSII relative to PSI to maximize light absorption. With the PSII inhibitors it is the decrease of electron transport that apparently triggers the increase in LHCII, suggesting that photosynthesis itself exerts a positive feedback control.

SUMMARY   Plants respond to light by marked adaptation of the composition of their thylakoid membranes, and hence their function and structure. These dynamic adaptations of the photosynthetic assemblies of PSII and PSI, electron transport complexes, and ATP synthetase are influenced by the light quality, light quantity, and the duration of illumination in a complex but coordinated fashion. This adaptation involves an integrated adjustment of the various thylakoid processes to make the best use of the available quantum flux. This light adaptation involves a *multifaceted* response rather than a *single* response, such as altering the concentration of one unique rate-limiting component. Adaptation involves a variety of strategies that include alterations in the relative number of pigment molecules and their composition in both PSII and PSI, the relative number of PSIIα, PSIIβ, and PSI units, and the amounts of electron transport complexes and ATP synthetase. Apparently the photoregulation of gene expression occurs at light intensities greater than that thought to saturate the phytochrome-mediated regulation of chloroplast development (122). These substantial structural, organizational, and functional changes in light-adapted thylakoids suggest also that photosynthesis itself is involved in positive feedback control mechanisms that promote the accumulation of photosynthetic components (e.g. an increase of PSII relative to PSI in the presence of PSII inhibitors or shade light).

SHORT-TERM PHOTOREGULATION

In the previous section, I have considered that the extensive flexibility in the light-regulated expression of the plant genomes ensures that chloroplasts possess the necessary amounts of thylakoid and stromal components to maintain optimal rates of NADPH and ATP production and consumption for the particular environment of the leaf. Chloroplasts also possess important strategies that permit short-term, rapidly reversible, adaptive responses to light, as I will now discuss briefly.

Interaction Between Thylakoids and Stroma

It is important to recognize that thylakoids can generate variable ATP/NADPH ratios by variations in noncyclic electron flow from O_2 to NADPH, and in PSI cycling events, as well as in variations of both H^+/e and H^+/ATP stoichiometries (110). These variations may result from both developmental and environmental changes. For example, the proportion of energy going into carbohydrate metabolism relative to other pathways such as the biosyntheses of macromolecules or ATP-driven transport systems will depend on the developmental stage of the leaf and also on the plant species. Further, light itself is one of the environmental factors that causes variations in stoichiometries of NADPH and ATP, which in turn markedly influence carbon assimilation (110). Several enzymes of the carbon reduction pathway are also light regulated (59). Additionally, light-induced changes in the stromal pH, Mg^{2+}, and substrate concentrations influence the activities of several enzymes of the carbon assimilation pathway (59, 110). Conversely, the turnover of ATP and NADPH by carbon assimilation influences the redox state and the phosphorylation potential of the chloroplasts. In turn, these factors affect thylakoid function and organization as expressed, for example, in the activation state of CF_1 (170) or in the activity of the thylakoid protein kinases (8, 9, 39, 110). Further, phosphorylation of thylakoid PSII proteins by the light-activated kinase reversibly decreases the light-saturated rate of PSII electron transport (113). Light certainly plays a complex role in the short-term regulation of the tightly coupled processes of electron transport and carbon assimilation.

Protein Phosphorylation

Higher plants (64) and green algae (54) possess a reversible, relatively rapid response to detect and correct imbalances in the rates of excitation of the photosystems resulting from changes in the spectral quality and intensity of ambient light. The basis of this fine adjustment of the relative excitation of PSII and PSI is thought to involve a kinase/phosphatase-mediated phosphorylation/ dephosphorylation of part of LHCII. This topic has been reviewed extensively (8, 9, 32, 40, 41, 102, 216).

Both experimental observations and theoretical considerations some of which are outlined below, suggest that this regulatory effect results in part from protein phosphorylation causing changes in membrane organization that permit a decreased interaction between LHCII and PSII. Upon overexcitation of PSII, the PQ pool is reduced, the protein kinase is activated, and part of LHCII is phosphorylated. Following phosphorylation the increase of net negative charge of LHCII in adjacent membranes ensures that the membrane adhesion will be decreased. The phosphorylated LHCII apparently detaches itself from the PSII complexes, and both phosphorylated LHCII and core PSII complexes are then free to migrate laterally to the nonappressed regions and possibly interact with PSI complexes. Conversely, overexcitation of PSI leads to oxidation of the PQ pool, inactivation of the kinase, dephosphorylation of the LHCII units by a phosphatase, and then migration of the dephosphorylated LHCII relocated around core PSII complexes back to the stacked membrane areas.

Although several proteins of PSII are phosphorylated in the light, including the chl a-binding 44 kDa protein, several proteins in the 34–30 kDa region, and a 10 kDa protein, LHCII is the dominant protein phosphorylated by the light-dependent kinase (41, 188, 189, 217). The protein kinase that is activated by high light intensity appears to be mainly controlled by the redox state of the plastoquinone pool (8, 9, 39, 110). Kinase activity may also be controlled by the ATP/ADP ratio, with ADP apparently acting as an inhibitor even when the plastoquinone pool is reduced (28, 112). The site of phosphorylation of LHCII is thought to be the threonine residue(s) near the N-terminus of LHCII which is located at the outer stroma-facing thylakoid surface (39, 173). Following this light-mediated phosphorylation of part of LHCII, there is a limited decrease in thylakoid stacking in peas (44, 131, 219) and barley (212). Phosphorylation also induces the removal of some LHCII units from PSII core complexes, allowing both complexes to migrate to the nonappressed, PSI-rich thylakoid regions. This has been measured as an increase in LHCII (20, 129, 135, 189) and PSII centers (128) in nonappressed membranes. Freeze-fracture analysis also reveals a movement of the 8 nm freeze-fracture particles, thought to be LHCII, from appressed to nonappressed regions (131, 212). However, Haworth & Melis (103) were unable to measure any increase in the antenna size of PSI upon phosphorylation. As yet there is no direct evidence to show that the phosphorylated LHCII actually transfers energy to PSI. Horton & Black (111) suggest that phosphorylation of LHCII leads merely to a decrease in the rate of transfer of excitation energy to PSII reaction centers.

It should be noted that no more than 25% of LHCII is estimated to be phosphorylated in vitro (216). Recently, it has been shown that the main apoproteins of LHCII exhibit different degrees of phosphorylation and lateral migration. Specific phosphate incorporation was higher in the 25 kDa apoprotein compared to the 27 kDa apoprotein of LHCII, and after phosphorylation there was a greater increase of the 25 kDa apoprotein in nonappressed thylakoids compared to the 27 kDa apoprotein (134). Of the four chl a/b-apoproteins of PSII resolved by Machold (155, 156), only two (chl a/b-AP2a and chl a/b-AP2b) were phosphorylated in *Vicia faba* and *Hordeum vulgare* (O. Machold, personal communication). After phosphorylation, these two apoproteins had much higher specific radioactivities in stromal compared to granal thylakoids. These results suggest that a specific subpopulation of LHCII is available for phosphorylation. This may be a reflection of the structural heterogeneity of the polypeptide composition of LHCII that is coded for by multifamily nuclear genes (69, 74, 227, 230). It is also possible that only those LHCII located at the outer edges of the granal stacks are accessible to the kinase. Murphy (176) suggests that the random phosphorylation of the LHCII pool in the middle of the granal stacks of the protein-dense appressed regions might be severely restricted. Further, the termination of LHCII phosphorylation might be achieved fairly soon by deactivation of the kinase, since the detachment of some LHCII units from PSII complexes would already reduce the amount of excitation energy reaching PSII reaction centers, lead to an alteration in the redox state of PQ, and initiate phosphatase activation. If so, the phosphorylation process might be self-limiting (176).

Reversible phosphorylation of LHCII may also function in the maintenance of high efficiency during changing contributions of cyclic phosphorylation to total phosphorylation (8). Increased cyclic phosphorylation by PSI would cause an imbalance in noncyclic electron transport in favor of PSII which might also trigger LHCII phosphorylation (8).

The diversity in the molecular structure and localization of light-harvesting complexes in algae (46, 133) leads to very different strategies for the regulation of light energy distribution between the photosystems in blue-green and red algae. With increasing depth of water, the light intensity is drastically diminished, and the spectrum is enriched in green and greatly depleted in far-red, then red and blue light (133); this favors excitation of PSII rather than PSI, which is the reverse situation to that of shaded habitats. Cyanobacteria and red algae have no lateral heterogeneity in the localization of PSII and PSI. Instead, PSII and PSI complexes are uniformally distributed along unstacked thylakoid membranes with the phycobiliproteins (not LHCII) assembled as phycobilisomes attached to the outer thylakoid surface (133). In red algae the changes in distribution of excitation energy between the photosystems occur in milliseconds rather than minutes as for higher plants, and the millisecond alterations do not involve phosphorylation (45, 46). Biggins et al (46) provide a model for the mechanism of excitation energy transfer in blue-green and red algae. Under conditions of excitation by mainly PSII light, the phycobilisomes will transfer energy to PSII with a high probability of excitation transfer from PSII to PSI. With preferential excitation of PSI, the proton gradient set up by cyclic phosphorylation and/or changes in the redox state of components cause a change in charge distribution of the photosystems, resulting in conformational changes that ensure the photosystems are separated. Higher plants require a different strategy, because in direct light there is a comparable excitation of both photosystems, and in shaded conditions it is PSI rather than PSII that is preferentially excited (49). If PSII and PSI were in close proximity in higher plants, much of the light energy received by LHCII could be transferred to PSI. Lateral segregation of much of the PSII and PSI pigment beds and reaction centers limits the possibility of spillover and avoids overexcitation of PSI (13, 29).

Photoinhibition

Both long-term and short-term adaptive changes permit chloroplasts to balance excitation rates, electron transfer, and carbon assimilation under a wide range of light conditions. However, prolonged exposure of plants to light intensities in excess of those found in natural growth conditions causes a marked decline in photosynthetic capacity (oxygen evolution and CO_2 fixation) termed photoinhibition (48, 186, 193). Although it is well documented that photoinhibitory damage selectively inhibits PSII (193), the molecular mechanisms of this inactivation are not established.

Arntzen et al (26, 130) have proposed an hypothesis for the primary site of photoinhibition that involves the Q_B herbicide-binding protein of PSII (127). The chloroplast gene *(psbA)* encoding the Q_B protein is transcribed and translated on 70S chloroplast ribosomes to give a 34.5 kDa precursor peptide (53). Following the excision of a small peptide sequence from the precursor protein, the processed Q_B protein is bound to the thylakoid membrane in a nonfunctional pool. When a site becomes available on a Q_B protein-depleted PSII complex (formed during photoinhibition), integration occurs from the Q_B protein pool to reform a functional PSII complex (26). Excessive high light intensity damage to the Q_B protein apparently causes it to be removed from PSII complex by a highly efficient, intrinsic membrane protease (185), thereby inhibiting electron transfer from PSII complex to the cytochrome *b/f* complex. The proposed selective removal of the Q_B protein following photoinhibition in isolated pea thylakoids (185) and *Chlamydomonas* cells (130, 184) requires de novo synthesis of the Q_B protein and its integration into PSII complex for recovery of photosynthetic competence. It is suggested that the quinone anion causes photodamage to the Q_B protein, either directly or indirectly, by reactions with oxygen to form oxygen radicals at the quinone-binding site (26). An alternative hypothesis proposed by Cleland et al (66, 67) suggests that the primary site of damage is closer to the reaction center and involves Q_A rather than Q_B. Evidence for this proposal has been obtained recently with spinach chloroplasts, where photoinhibition caused a decrease in the extent of the absorbance changes of PSII reaction centers, phaeophytin and Q_A (67).

Several of the PSII proteins bind herbicides and have putative quinone-binding sites. Q_A is assumed to be bound to the 51 kDa apoprotein of PSII, with P680, phaeophytin, and Q_A positioned across the membrane to allow for rapid charge transfer (98, 176). The 44 kDa chl a-protein of PSII also binds phenol-type herbicides, presumably at the binding site believed to be common to the 44 kDa and Q_B proteins (231). Thus the quinone-binding domains of the 51 and 44 kDa apoproteins and Q_B protein must be adjacent to each other in order for efficient electron transfer from Q_A to Q_B to occur. Hence photoinhibition may cause inactivation at both the Q_a and Q_B sites.

In any event, the Q_B protein is rapidly turned over in the light (53, 127, 158) with the rate of turnover being proportional to light intensity, and fifty- to eightyfold faster than the turnover rates of other thylakoid proteins under high light intensity (158, 184, 236). Thus, under high light intensities capable of saturating photosynthesis, a major component of thylakoid protein synthesis must be involved in the formation of Q_B protein. This is reflected by the high levels of mRNA for the Q_B protein in chloroplasts (53). Arntzen et al (26) suggest that the high rates of turnover of Q_B protein in the light is a normal consequence of its function as a quinone protein in electron transfer. Photoinhibition occurs when the rate of inactivation and subsequent removal of the Q_B protein exceeds the rate of its resynthesis. Regardless of the primary site and mechanism of inactivation in PSII, further damage can then occur, including inactivation of the reaction centers and photodestruction of pigments. The inhibitory effect of protein phosphorylation on the maximum capacity of PSII electron transfer may afford partial protection against irreversible damage to PSII (113). In addition, excess light may also inhibit electron transport through PSI. The mechanism of the photoinhibition of PSI is unknown, but oxygen is required (26). Also, the activities of carbon cycle enzymes will quickly decline (194) and the capacity for carbon assimilation will be severely limited.

Plants adapted to low light intensities are particularly prone to photoinhibition (195, 196), partly because of their limited electron transport capacity. Faced with high intensities, this adaptation of lower electron transport capacity is potentially catastrophic since the plastoquinone pool will become reduced and lead to photoinhibition. Horton (110) suggests that the adaptation of the light-harvesting systems of both photosystems is a balance between efficiency in one extreme condition and disaster in the other. The light-regulated, short-term mechanism of phosphorylation of part of the LHCII pool could widen the light intensity range over which chloroplasts function (110). Photophosphorylation of other PSII proteins and the concomitant decrease in maximum PSII electron transport capacity may also decrease photoinhibition (113, 114).

MECHANISMS OF LIGHT REGULATION

Photoregulation of Pigment Synthesis

It is scarcely surprising that most photosynthetic organisms have developed mechanisms for the light regulation of pigment synthesis since the pigments are not needed in the dark. As discussed, chloroplasts from plants grown at lower light intensities or in the shade have high thylakoid to stroma volume ratios together with low chl a/chl b ratios, signifying more LHCII and LHCI, and of necessity less chl a-proteins of PSII and PSI. Conversely, high light intensity or sun plant chloroplasts have lower thylakoid/stroma volume ratios and higher chl a/chl b ratios because of more chl a-proteins of PSII and PSI and less LHCII and LHCI. These modulations in amounts of pigment-protein complexes strongly suggest that there must be different mechanisms for the photoregulation of chl a/b-proteins and chl a-proteins.

CHLOROPHYLL BIOSYNTHESIS   Two sites of light regulation are known for chlorophyll biosynthesis (35, 63). First, the level of synthesis of δ-aminolaevulinic acid, a key precursor, appears to be indirectly controlled by phytochrome (171). Second, in dark-grown, etiolated angiosperms the reduction of protochlorophyllide to chlorophyllide by the enzyme protochlorophyllide reductase is strictly light dependent (96). However, gymnosperms and most algae are able to synthesize chlorophyll in darkness (35, 96). In developing angiosperms, a ternary complex of NADPH-enzyme-protochlorophyllide is converted to NADP-enzyme-chlorophyllide only after excitation of the bound protochlorophyllide molecule (96). Consequently, continuous chlorophyllide synthesis requires essentially continuous illumination, and the final amount of chlorophyll in thylakoid membranes depends on the irradiance the chloroplast receives during growth. Recently, Adamson et al (1–3) have demonstrated that angiosperms are also capable of dark synthesis of chlorophyll; however, light is required for the initial obligatory chlorophyll biosynthesis that precedes dark synthesis. In angiosperms, there must also be a light-independent protochlorophyllide reductase, or perhaps one or more of the multiple pathways of chlorophyll biosynthesis postulated by Rebeiz (200) are light independent.

The final stages of chl a and chl b biosynthesis that appear to be light independent are complex and poorly understood (35, 63). Further, both chl a and chl b, as well as their immediate precursors, appear to be chemically heterogeneous (200, 243). Position 4 of the tetrapyrrole complex may have an ethyl group (as in chl a or chl b) or a vinyl group, and the isocyclic ring may also be esterified. Rebeiz et al (200, 243) propose multiple pathways for both chl a (6 pathways) and chl b biosynthesis. While the details of these pathways remain to be established, undoubtedly chlorophyll formation is more complicated than originally thought. This complexity in chemical structures and biosynthesis of chlorophylls is paralleled by an in vivo spectral heterogeneity of chlorophylls evident in the many spectral forms of chl a and chl b (51, 58). The in vivo spectral forms may arise in part from the chemical heterogeneity of the pigments as well as from their particular protein environments.

Another major unanswered question concerns the final stages of chl b

biosynthesis. Although not proven, it is generally accepted that chl b is formed from chl a (chlide $a \rightarrowtail$ chl $a \rightarrowtail$ chl b) and light is not directly required for the conversion. However, Oelze-Karow & Mohr (182) postulate that chlide b might be the direct precursor for chl b (chlide $a \rightarrowtail$ chlide $b \rightarrowtail$ chl b). While the synthesis of chl b in mustard seedlings appears to be indirectly mediated by phytochrome (181, 183), in wheat there is no obligatory requirement for either Pfr or threshold amounts of chlide (125), and the dark formation of both chl a and chl b in *Tradescantia* leaves was inhibited if the leaves were irradiated with red light before the dark treatment (2).

Another complication that may occur in chlorophyll synthesis in developing plastids is demonstrated by the reorganization of PSII units in the dark. Following short light treatments (8 h) of etiolated bean plants subsequently transferred to darkness (24 h), there was a dramatic increase of the chl a/chl b ratio from 4.4 to 18.1, a marked decrease in LHCII and grana stacking, with little change in chlorophyll content (5, 25). Apparently the chl a and chl b molecules were released from LHCII, chl b transformed to chl a, and then incorporated into PSII chl a-proteins.

The presence of multiple pathways for chl a and b biosynthesis, which may or may not be light regulated, may differ in both developmental stages of the leaves and plant species, which, together with the apparent turnover of certain chlorophyll-proteins in the dark, makes for an exceedingly complex situation. Nevertheless, it is vital to elucidate what factors directly control the relative amounts of chl a and chl b in chloroplasts.

CAROTENOID BIOSYNTHESIS   The biosynthetic pathways for carotenoids within the chloroplast are identical with that of phytol up to the formation of the C_{20} compound geranylgeranyldiphosphate (GGPP) (202). Condensation of two molecules of GGPP yields the colorless phytoene, which in turn leads to the various carotenoids by a number of dehydrogenation and cyclization steps. The light control of carotenoid biosynthesis has been reviewed extensively (99, 199). Although light clearly stimulates carotenoid synthesis, strict photoregulation is restricted to some fungi and bacteria (202). In higher plants, β-carotene and the xanthophylls are present in low amounts in etiolated leaves, so light control must be considered as a modulation rather than an induction. Since no light control has been detected for phytol formation, it is probable that photoregulation of carotenoid formation needs to operate at a point beyond GGPP formation (202). Rüdiger & Benz (202) suggest that the control could be effected by the availability of the thylakoid apoproteins to which the carotenoids must be bound.

Accumulation of carotenoids in the absence of chlorophyll is promoted to some extent by phytochrome, but substantial amounts of carotenoids appear to be limited by the availability of chlorophyll (87, 180). Conversely, when carotenoid synthesis is blocked by the herbicide SAN 9789 (86), chlorophyll synthesis and membrane assembly are still possible. This strategy implies that carotenoids (required as protectors against photooxidation of chlorophylls) become available only when the chlorophylls are being actively synthesized; however, in mustard seedling cotyledons the carotenoids are always synthesized ahead of chlorophyll (87). Frosch & Mohr (87) propose that a "push-pull" regulation controls the accumulation of carotenoids. The push regulation is thought to result from phytochrome-mediated regulation of enzyme levels prior to the pool of free carotenoids, while the pull regulation results from the formation of functional carotenoid-chlorophyll-protein complexes that would drain the free chlide pool. It is more complicated than this, however, because the different carotenoids are variously distributed between the chl-proteins, with β-carotene mainly associated with chl a-proteins, and lutein and neoxanthin with the chl a/b-proteins (56, 79, 153).

In summary, we do not know how chlorophyll and carotenoid syntheses are regulated by light in a coordinated manner during chloroplast development, nor do we know how plant cells adjust their final amounts of chl a, chl b, and carotenoids. It is generally assumed, however, that chlorophyll and carotenoids are not produced in excess of the apoproteins of the pigment-protein complexes. Regulation of the total chl a, chl b, and carotenoid content may also be governed by the intracellular concentrations of one or more photosynthetic products.

Light Regulation of Gene Expression

The effects of light on the regulation of gene expression of the thylakoid membrane components are extremely complex. Obviously some regulation must occur at virtually every stage in the synthesis of each component to ensure their coordinated synthesis, as well as delivery of the nuclear- and chloroplast-encoded polypeptides of each thylakoid complex in the appropriate stoichiometric amounts at the proper times. There must also be the means for selective regulation to allow plants to respond to environmental changes, particularly light. Although a full consideration of the light regulation of gene expression is outside the scope of this review, some points arising from the fast expanding research in this area are presented. Extensive coverage is available in the reviews of Tobin & Silverthorne (229), Herrmann et al (105), Harpster & Apel (100), Jenkins et al (117), and Lamb & Lawton (132).

The biogenesis of angiosperm thylakoid membranes is under strict light control. Mature chloroplasts differentiate from either rudimentry proplastids present in light-grown plants or from etioplasts which develop in plants maintained in darkness. Etioplasts possess few of the thylakoid polypeptides, they have no chlorophyll nor pigment proteins, and photosynthesis is impossible. The dramatic changes observed during the light-induced transformation of etioplasts to chloroplasts are controlled by at least three photoreceptors: protochlorophyllide (35, 63, 96, 120), phytochrome (171, 207), and the blue-light receptors (57, 210, 211).

While the molecular mechanisms of the interactions between the photoreceptors and gene expression are only dimly perceived as yet, a unique feature of membrane biogenesis is the need for the coordinated expression of both nuclear and chloroplast genes. Light control during thylakoid biogenesis could occur at several levels during transcription, mRNA processing, translation, protein transport and processing, and assembly into functional complexes. It is already clear that transcriptional changes are only part of the regulatory process, and there must be a two-way interaction between the chloroplast and the other cellular compartments (229).

One of the cardinal points of light control of nuclear-encoded thylakoid proteins is at the level of transcription (100, 105, 117, 229). For example, dark-grown barley (23) and *Lemna* (228) have very low levels of LHCII mRNA. As the levels of LHCII mRNA are greatly increased by brief red light treatment, and suppressed by immediate far-red light treatment, phytochrome controls this response. Moreover, it has now been shown that this phytochrome-induced increase in the steady state level of LHCII mRNA is the result of increases in the rates of transcription rather than changes in the stability of the transcripts (172, 213). It must be stressed, however, that there is considerable variability among different species in the relative levels of nuclear mRNAs found in dark-grown leaves (105, 229). While the levels of mRNA are low in most species examined (42, 91, 178), etiolated pea seedlings contain substantial LHCII mRNA prior to light treatment (117, 224). Although light-grown peas contain greater amounts of the transcript than etiolated peas, there is a substantial time lag before the levels of LHCII mRNA increase (117, 224). The effects of light on the accumulation of LHCII mRNA in peas (224) and maize (178) may be related to a general stimulation of chloroplast development.

Interestingly, the mRNAs may be expressed differently in various tissues: LHCII mRNA levels present in roots did not increase upon illumination (224). Since some proteins, including LHCII, are encoded by small multigene families (69, 74, 227, 230), certain genes could be strictly photoregulated while others might be active in both the light and dark. In any event, it is likely that individual gene families will exhibit varying degrees of tissue-, cell-, and temporal-specific expression among gene members (100).

Kaufman et al (121–123) have demonstrated that there is a differential response of nuclear genes to light intensity. Phytochrome regulation occurs at two fluence ratios of red light (121, 122): the usual low fluence (LF) response has a threshold of ∼ 1 µmol m^{-2}s^{-1} and is reversed by red light, while the very low fluence (VLF) response has a threshold of 10^{-4} mol m^{-2} s^{-1} and is not reversed by red light. In peas, the small subunit of ribulose bisphosphate carboxylase transcripts have only a LF response, but LHCII transcripts have both a VLF and a LF response (122). Comparison of 20 cDNA clones corresponding to nuclear-encoded transcripts that respond to light in pea buds suggests at least five groups of coordinately regulated genes based on the characteristics of their phytochrome responses (121, 123). Significantly, those transcripts with only a LF red response are also regulated by the blue light photoreceptors (123). These important results demonstrate how light intensity and quality, together with varying amounts of photoreceptors in particular plants, allow different levels of gene expression that cause the marked modulations of thylakoid components.

In addition to promoting the rate of transcription of certain nuclear mRNAs, phytochrome may also cause a decrease in the rate of transcription of such nuclear genes as protochlorophyllide reductase (34) and phytochrome (68, 187), though regulation at the level of mRNA degradation is not excluded. In both cases, the levels of mRNA decline rapidly in dark-grown plants following red light treatment (34, 68, 187).

Changes in the proportions of chloroplast-encoded thylakoid proteins during light-induced development have also been correlated with corresponding alterations in the levels of chloroplast mRNAs (100, 105, 117, 132, 229). Marked increases in the transcript levels of the gene coding for the Q_B protein in maize (36), mustard (154), pea (224), but not in mung bean (224), were found in light-treated compared to etiolated seedlings. In some cases, phytochrome was shown to be the photoreceptor regulating the level of Q_B protein mRNA (154, 224). Genes whose transcripts increase appreciably in abundance during chloroplast biogenesis are termed photogenes. DNA/RNA hybridization techniques have allowed the localization of photogenes within the chloroplast genomes of maize (201), barley (192), and mustard (154). Rodermel & Bogorad (201) demonstrated that the photogenes were scattered over $\sim$ 19% of the maize chloroplast genome and these genes were transcribed either polycistronically or individually. In maize, the levels of the transcript pools reached maximal levels at different times during the illumination period (early, middle, or late) and then declined drastically, except for Q_B protein mRNA which increased continuously (201).

Herrmann et al (105), in studies with spinach and *Oenothera,* emphasize that there is no shutdown of the transcription of any genes in the dark; differences in genes are found only in the degree of increases of mRNA levels during illumination. While the genes encoding ATP synthetase and the cytochrome *b/f* complex appeared to be coordinately expressed with both mRNA levels and protein levels increasing in parallel during illumination, the protein subunits of PSI complex appeared sequentially (105). Herrmann et al (105) suggest that light controls are also exerted at the post-transcriptional level. Increased transcription of some chloroplast genes may be a general function of chloroplast developmental processes rather than a specific light regulation of chloroplast transcript levels. When mature plants of *Spirodela oligorrhiza* were maintained heterotrophically in the dark, the level of the Q_B protein synthesis was drastically curtailed, although the level of Q_B protein mRNA was unaltered compared to that of light-grown plants (85). Fromm et al (85) conclude that light-regulation of the Q_B protein occurs at the translational level, and their results support the idea that the effect of light on chloroplast mRNA may be mediated by its effect on chloroplast development rather than as a specific regulator of chloroplast transcript levels.

Finally, the mechanisms coordinating the interaction of nuclear and chloroplast genomes need to be elucidated. Does phytochrome directly regulate the formation of chloroplast-encoded transcripts, or do the observed effects result indirectly from phytochrome control of nuclear genes? Rodermel & Bogorad (201) suggest that the level of transcription of the nuclear gene for the small subunit of ribulose bisphosphate carboxylase may be regulated by events in the chloroplast. Indirect evidence in favor of control of nuclear processes by chloroplast-derived factors is also seen in the studies of Taylor et al using pigment-deficient maize mutants (101, 159, 178). If chloroplast development and carotenoid biosynthesis were prevented by herbicide treatment (100) or by a mutation in carotenoid biosynthesis (101, 159, 178), LHCII mRNA levels were very low and did not increase upon illumination. Elimination of functional plastids can lead to specific suppression of the nuclear genes for LHCII. Surprisingly, the levels of LHCII mRNA were normal in chl-deficient mutants (101).

Photoregulation of Chlorophyll-Proteins

As discussed, the pigments and proteins of pigment-protein complexes are under different photoregulatory mechanisms, with chlorophyll synthesis requiring continuous illumination to excite successive protochlorophyllide molecules bound to protochlorophyllide reductase, and phytochrome and the blue light receptors regulating gene transcription, as well as translational and post-translational events.

Little is known yet about the assembly of the pigment-proteins in thylakoid membranes. However, the stage of chloroplast development seems to be important and indicates once again the requirement of post-transcriptional events (39, 100, 105, 117, 178). For LHCII, it is established that plants may accumulate significant levels of LHCII mRNA but yet have little or no LHCII. This is so for etiolated plants after exposure to brief red light treatments (23, 228), plants grown in intermittent light that have chl *a* only (38, 214, 234), and chl *b*-less barley mutants (24, 37, 209). While the presence of chl *a* and chl *b* are certainly necessary for the accumulation of LHCII, it is not established whether other points of photocontrol allow translation of LHCII mRNA only when both chl a and chl b are being synthesized. Alternatively, LHCII apoproteins may be synthesized only to be degraded in the absence of chl a and chl b). Since preexisting LHCII is rapidly degraded in immature pea leaves (38), bean leaves (5, 25), and radish cotyledons (146), Bennett (39) favors the idea of post-translational control whereby the LHCII apoproteins are inserted into thylakoid membranes but are rapidly degraded if chl a and chl b are not available. However, Slovin & Tobin (214) propose that light controls the translation of LHCII mRNA via a mechanism that does not employ phytochrome.

Recent evidence for post-transcriptional control of LHCII comes from studies with a chlorophyll b-deficient wheat mutant that has only limited LHCII (76). Application of chloramphenicol (to suppress chloroplast 70S ribosomes) to etiolated mutant seedlings during greening causes an increase in the chl a/chl b ratio, LHCII concentration, and membrane stacking (76). This suggests that a chloroplast-encoded protein mediates post-transcriptional events for the nuclear-encoded LHCII. In a nuclear maize mutant that fails to accumulate core PSII complex, there is an accelerated turnover of the chloroplast-encoded 47 kDa, chl a-protein, and the Q_B protein (143), suggesting that the nucleus plays a direct or indirect role in the stabilization of these chloroplast-encoded PSII complex proteins. Key unanswered questions for the light regulation of chl-proteins include: (a) characterization of the factors controlling the amount of total chlorophyll synthesized, particularly the relative proportions of chl a versus chl b; (b) the role of carotenoids in chl-protein assembly; (c) the binding of pigments to the apoproteins; (d) the stabilization of chl-proteins in mature thylakoids; and (e) relative modulations of chl a-proteins versus chl a/b-proteins.

CONCLUDING REMARKS

Light adaptation of the thylakoid membranes in plants grown at low light intensity or shade appears to be a question of the most economical use of light. In order to capture the meager, perhaps limiting light, a substantial part of the photosynthetic capacity of the chloroplast needs to be invested in the synthesis and maintenance of the light-harvesting assemblies of both photosystems. With the increased synthesis of chlorophylls and carotenoids and their apoproteins that constitute a major portion of thylakoid proteins, the amount of thylakoid membranes greatly increases. Indeed, in obligate shade plant chloroplasts the thylakoid membranes almost fill the entire chloroplast volume, leaving little room for the enzymes of carbon assimilation. As the maximal irradiance is very low, the capacity for NADPH and ATP formation can be reduced drastically, and in turn CO_2 fixation is very low. Conversely, adaptation of plants to direct sunlight or high light intensity results in the increased synthesis of PSII and PSI core complexes, the cytochrome b/f complex, and ATP synthetase at the expense of light-harvesting chl a/b-proteins in order to utilize the high levels of irradiance and maintain optimal electron transport capacity and high photosynthetic capacity.

The ability of plants to adapt their photosynthetic apparatus to different light environments by modulation of chl a/b-proteins versus electron transport complexes and ATP synthetase is effective only within certain limits. At low irradiance, there are increased amounts of thylakoid membranes with greater areas of membrane appression and much more LHCII and LHCI relative to core PSII and PSI complexes. The extra pigments needed for light-harvesting are inserted as chl a/b-proteins rather than antenna chl a-proteins, perhaps because they do not include any P680 or P700. Although there are two chl a-proteins in each core PSII or PSI complex, they appear always to be in the same stoichiometry, so larger light-harvesting units cannot be assembled by increasing the antenna chl a-protein relative to the reaction center chl a-protein. Significantly, Evans (81) estimates that CPa, CP1, and LHCII require 67, 35, and 25 mol nitrogen per mol chl, respectively. Investment in LHCII relative to CPa and CP1 is consistent with the nitrogen costs for each complex and the maintenance of light-harvesting capacity in preference to electron transport capacity (82). At low irradiance, thylakoids apparently have larger photosynthetic units with more chl a/b-proteins, less chl a-proteins, and fewer reaction centers to maximize light-harvesting, while at high irradiance there are more reaction centers, with more chl a-proteins, less chl a/b-proteins, and smaller antenna to minimize the deleterious effects of photoinhibition. This modulating effect that appears to be more dramatic in PSII than in PSI needs further study.

It is already evident that the mechanisms of light regulation for the expression of particular genes may not operate universally in all plant species, nor in all cell types, and not at each stage of development. Light being essential for photosynthesis will itself also influence the pools of redox intermediate metabolites, NADPH and ATP; this in turn will certainly play a part in the regulation of gene expression. The multiplicity of mechanisms and complex diversity of light regulation reflect the varying requirements of plants in their perception of light. Further studies are needed to elucidate the coordinated regulation of thylakoid protein complexes, which involve nuclear- and chloroplast-encoded proteins and pigment biosynthetic pathways, and the role of light in gene expression.

ACKNOWLEDGMENTS

I am very grateful for discussions with Drs. Denis Murphy, Ta-Yan Leong, John R. Evans, Fred Chow, and Jack Barrett; my thanks also to Mrs. Denese McCann for typing the manuscript.

Literature Cited

1. Adamson, H., Griffiths, T., Packer, N., Sutherland, M. 1985. Light-independent accumulation of chlorophyll a and b and protochlorophyllide in green barley *Hordeum vulgare* L. *Physiol. Plant.* 64:345–52

2. Adamson, H., Hiller, R. G. 1981. Chlorophyll synthesis in the dark in angiosperms. In *Photosynthesis. Chloroplast Development*, ed. G. Akoyunoglou, 5:213–21. Philadelphia, Pa: Balaban Int. Sci. Serv. 1018 pp.

3. Adamson, H., Packer, N., Gregory, J. 1985. Chloroplast development and the synthesis of chlorophyll and protochlorophyllide in *Zostera* transferred to darkness. *Planta* 165:469–76

4. Akerlund, H.-E., Andersson, B. 1983. Quantitative separation of spinach thylakoids into photosystem II-enriched inside-out vesicles and photosystem I-enriched right-side-out vesicles. *Biochim. Biophys. Acta* 725:34–40

5. Akoyunoglou, A., Akoyunoglou, G. 1984. Mechanism of thylakoid reorganization during chloroplast development in higher plants. *Isr. J. Bot.* 33:149–62

6. Akoyunoglou, G., Anni, H. 1984. Blue light effect on chloroplast development in higher plants. See Ref. 211, pp. 397–406

7. Akoyunoglou, G., Anni, H., Kalosakas, K. 1980. The effect of light quality and the mode of illumination on chloroplast development in etiolated bean leaves. See Ref. 210, pp. 473–84

8. Allen, J. F. 1983. Regulation of photosynthetic phosphorylation. *CRC Crit. Rev. Plant Sci.* 1:1–22

9. Allen, J. F., Bennett, J., Steinback, K. E., Arntzen, C. J. 1981. Chloroplast protein phosphorylation couples plastoquinone redox state to distribution of excitation energy between photosystems. *Nature* 291:21–25

10. Allred, D. R., Staehelin, L. A. 1985. Lateral distribution of the cytochrome b_6/f and coupling factor ATP synthetase complexes of chloroplast thylakoid membranes. *Plant Physiol.* 78:199–202

11. Anderson, J. M. 1980. Chlorophyll-protein complexes of higher plant thylakoids: Distribution, stoichiometry and organization in the photosynthetic unit. *FEBS Lett.* 117:327–31

12. Anderson, J. M. 1982. Distribution of the cytochromes of spinach chloroplasts between the appressed membranes of grana stacks and stroma-exposed thylakoid regions. *FEBS Lett.* 138:62–66

13. Anderson, J. M. 1982. The role of chlorophyll-protein complexes in the function and structure of chloroplast thylakoids. *Mol. Cell Biochem.* 46:161–72

14. Anderson, J. M. 1982. The significance of grana stacking in chlorophyll b-containing chloroplasts. *Photobiochem. Photobiophys.* 3:225–41

15. Anderson, J. M. 1984. Molecular organization of chloroplast thylakoid membranes. In *Advances in Photosynthesis Research*, ed. C. Sybesma, 3:1–10. The Hague: Nijhoff/Dr. W. Junk. 927 pp.

16. Anderson, J. M., Andersson, B. 1982. The architecture of photosynthetic membranes: Lateral and transverse organization. *Trends Biochem. Sci.* 7:288–92

17. Anderson, J. M., Goodchild, D. J., Boardman, N. K. 1973. Composition of the photosystems and chloroplast structure in extreme shade plants. *Biochim. Biophys. Acta* 325:573–85

18. Anderson, J. M., Melis, A. 1983. Localization of different photosystems in separate regions of chloroplast membranes. *Proc. Natl. Acad. Sci. USA* 80:745–49

19. Andersson, B. 1986. Proteins participating in photosynthetic water oxidation. In *Encyclopedia of Plant Physiology: Photosynthesis III* (NS), ed. L. A. Staehelin, C. J. Arntzen. Berlin: Springer-Verlag. In press

20. Andersson, B., Akerlund, H.-E., Jergil, B., Larsson, C. 1982. Differential phosphorylation of the light-harvesting chlorophyll-protein complex in appressed and non-appressed regions of the thylakoid membrane. *FEBS Lett.* 149:181–85

21. Andersson, B., Anderson, J. M. 1980. Lateral heterogeneity in the distribution of chlorophyll-protein complexes of the thylakoid membranes of spinach chloroplasts. *Biochim. Biophys. Acta* 593:427–40

22. Anni, H., Akoyunoglou, G. 1981. The effect of blue and red light on the development of the photosynthetic units during greening of etiolated bean leaves. See Ref. 2, pp. 885–94

23. Apel, K. 1979. Phytochrome-induced appearance of mRNA activity for the apoprotein of the light-harvesting chlorophyll a/b protein of barley (*Hordeum vulgare*). *Eur. J. Biochem.* 97:183–88

24. Apel, K., Kloppstech, K. 1978. The plastid membranes of barley (*Hordeum vulgare*). Light-induced appearance of mRNA coding for the apoprotein of the light-harvesting chlorophyll a/b-protein. *Eur. J. Biochem.* 85:581–88

25. Argyroudi-Akoyunoglou, J. H., Akoyu-noglou, A., Kalosakas, K., Akoyunog-lou, G. 1982. Reorganization of the photosystem II unit in developing thyla-koids of higher plants after transfer to darkness. Changes in chlorophyll b, light-harvesting chlorophyll protein con-tent, and grana stacking. Plant Physiol. 70:1242–48

26. Arntzen, C. J., Kyle, D. J., Wettern, M., Ohad, I. 1984. Photoinhibition: A conse-quence of the accelerated breakdown the apoprotein of the secondary electron acceptor of photosystem II. In Biosynthe-sis of the Photosynthetic Apparatus: Molecular Biology, Development and Regulation, ed. J. P. Thornber, L. A. Staehelin, R. Hallick, pp. 313–24. New York: Liss. 405 pp.

27. Baker, N. R. 1984. Development of chloroplast photochemical functions. In Chloroplast Biogenesis, ed. N. R. Baker, J. Barber, pp. 207–52. Amsterdam: Else-vier. 379 pp.

28. Baker, N. R., Markwell, J. P., Thornber, J. P. 1982. Adenine nucleotide inhibition of phosphorylation of the light-harvest-ing chlorophyll a/b-protein complex. Photochem. Photobiophys. 4:211–17

29. Barber, J. 1980. An explanation for the relationship between salt-induced thyla-koid stacking and the chlorophyll fluores-cence changes associated with changes in spillover of energy from photosystem II to photosystem I. FEBS Lett. 118:1–10

30. Barber, J. 1982. Influence of surface charges on thylakoid structure and func-tion. Ann. Rev. Plant Physiol. 33:261–95

31. Barber, J. 1983. Photosynthetic electron transport in relation to thylakoid mem-brane composition and organization. Plant Cell Environ. 6:311–22

32. Barber, J. 1986. Surface electrical charges and protein phosphorylation. See Ref. 19. In press

33. Batie, C. J., Kamin, H. 1984. Electron transfer by ferredoxin: NADP+ reduc-tase. Rapid-reaction evidence for partici-pation of a ternary complex. J. Biol. Chem. 259:11976–85

34. Batschauer, A., Apel, K. 1984. An in-verse control by phytochrome of the ex-pression of two nuclear genes in barley (Hordeum vulgare L.). Eur. J. Biochem. 143:593–97

35. Beale, S. I. 1984. Biosynthesis of photosynthetic pigments. See Ref. 27, pp. 135–205

36. Bedbrook, J. R., Link, G., Coen, D. M., Bogorad, L., Rich, A. 1978. Maize plas-tid gene expressed during photoregulated

development. Proc. Natl. Acad. Sci. USA 75:3060–64

37. Bellemare, G., Bartlett, S. G., Chua, N.-H. 1982. Biosynthesis of chlorophyll a/b-binding polypeptides in wild type and the Chlorina f2 mutant of barley. J. Biol. Chem. 257:7762–67

38. Bennett, J. 1981. Biosynthesis of the light-harvesting chlorophyll a/b protein polypeptide turnover in darkness. Eur. J. Biochem. 118:61–70

39. Bennett, J. 1983. Regulation of photo-synthesis by reversible phosphorylation of the light-harvesting chlorophyll a/b protein. Biochem. J. 212:1–13

40. Bennett, J. 1983. Regulation of photo-synthesis by chloroplast protein phos-phorylation. Philos. Trans. R. Soc. Lon-don Ser. B, 302:113–25

41. Bennett, J. 1984. Chloroplast protein phosphorylation and the regulation of photosynthesis. Physiol. Plant. 60:583–90

42. Bennett, J., Jenkins, G. I., Hartley, M. R. 1984. Differential regulation of the accumulation of the light-harvesting chlorophyll a/b complex and ribulose bis-phosphate carboxylase/oxygenase in greening pea leaves. J. Cell. Biochem. 25:1–13

43. Berzborn, R. J., Müller, D., Roos, P., Andersson, B. 1981. Significance of dif-ferent quantitative determinations of photosynthetic ATP-synthase CF₁ for heterogeneous Cf₁ distribution and grana formation. In Photosynthesis. Structure and Molecular Organization of the Photosynthetic Membrane, ed. G. Akoyunoglou, 3:107–20. Philadelphia: Balaban Int. Sci. Serv. 1042 pp.

44. Biggins, J. 1982. Thylakoid con-formational changes accompanying membrane protein phosphorylation. Bio-chim. Biophys. Acta 679:479–82

45. Biggins, J., Bruce, D. 1985. Mechanism of the light state transition in photosyn-thesis. III. Kinetics of the state transition in Porphyridium cruentum. Biochim. Biophys. Acta 806:230–36

46. Biggins, J., Campbell, C. L., Bruce, D. 1984. Mechanism of the light state transi-tions in photosynthesis II. Analysis of phosphorylated polypeptides in the red alga, Porphyridium cruentum. Biochim. Biophys. Acta 767:138–44

47. Björkman, O. 1973. Comparative studies on photosynthesis in higher plants. In Photophysiology, ed. A. C. Giese, 8:1–63. New York: Academic

48. Björkman, O. 1981. Responses to differ-ent quantum flux densities. In Encyclope-dia of Plant Physiology: Physiological Plant Ecology I (NS), ed. O. L. Lange,

P. S. Nobel, C. B. Osmond, H. Ziegler, 12A:57–107. Berlin/Heidelberg: Springer-Verlag. 625 pp.

49. Björkman, O., Ludlow, M. M. 1972. Characterization of the light climate on the floor of a Queensland rainforest. *Carnegie Inst. Washington Yearb.* 71:85–94

50. Boardman, N. K. 1977. Comparative photosynthesis of sun and shade plants. *Ann. Rev. Plant Physiol.* 28:355–77

51. Boardman, N. K., Anderson, J. M., Goodchild, D. J. 1978. Chlorophyll-protein complexes and structure of mature and developing chloroplasts. *Curr. Top. Bioenerg.* 8:35–109

52. Boardman, N. K., Björkman, O., Anderson, J. M., Goodchild, D. J., Thorne, S. W. 1974. Photosynthetic adaptation of higher plants to light intensity: Relationship between chloroplast structure, composition of the photosystems and photosynthetic rates. In *Proc. 3rd Int. Congr. Photosynthesis*, ed. M. Avron, pp. 1809–27. Amsterdam: Elsevier. 2194 pp.

53. Bogorad, L. 1981. Chloroplasts. *J. Cell Biol.* 91:256S–70S

54. Bonaventura, C., Myers, J. 1969. Fluorescence and oxygen evolution from *Chlorella pyrenoidosa*. *Biochim. Biophys. Acta* 189:366–83

55. Bowes, J. M., Horton, P. 1982. The effect of redox potential on the kinetics of fluorescence induction in photosystem II particles from *Phormidium laminosum*. *Biochim. Biophys. Acta* 680:127–33

56. Braumann, Th., Weber, G., Grimme, L. H. 1982. Carotenoid and chlorophyll composition of light-harvesting and reaction centre proteins of the thylakoid membrane. Sigmoidicity, energy transfer and the slow phase. *Photobiochem. Photobiophys.* 4:1–8

57. Briggs, W. R., Iino, M. 1983. Blue-light-absorbing photoreceptors in plants. *Philos. Trans. R. Soc. London Ser. B* 303:347–59

58. Brown, J. S. 1977. Spectroscopy of chlorophyll in biological and synthetic systems. *Photochem. Photobiol.* 26:319–26

59. Buchanan, B. B. 1980. Role of light in the regulation of chloroplast enzymes. *Ann. Rev. Plant Physiol.* 31:341–74

60. Buschmann, C., Meier, D., Kleudgen, H. K., Lichtenthaler, H. K. 1978. Regulation of chloroplast development by red and blue light. *Photochem. Photobiol.* 27:195–98

61. Camm, E. L., Green, B. R. 1980. Fractionation of thylakoid membranes with the nonionic detergent octyl-β-D-glucopyranoside. Resolution of chlorophyll-protein complex II into two chlorophyll-protein complexes. *Plant Physiol.* 66:428–32

62. Camm, E. L., Green, B. R. 1983. Relationship between the two minor chlorophyll *a*-protein complexes and the photosystem II reaction centre. *Biochim. Biophys. Acta* 724:291–93

63. Castelfranco, P. A., Beale, S. I. 1983. Chlorophyll biosynthesis: Recent advances and areas of current interest. *Ann. Rev. Plant Physiol.* 34:241–78

64. Chow, W. S., Telfer, A., Chapman, D. J., Barber, J. 1981. State 1 - State 2 transition in leaves and its association with ATP-induced chlorophyll fluorescence quenching. *Biochim. Biophys. Acta* 638:60–68

65. Chu, Z.-X., Anderson, J. M. 1984. Modulation of the light-harvesting assemblies in chloroplasts of a shade plant, *Alocasia macrorrhiza*. *Photobiochem. Photobiophys.* 8:1–10

66. Cleland, R. E., Critchley, C. 1986. Studies on the mechanism of photoinhibition in higher plants. II. Inactivation by high light of photosystem II reaction center function in isolated spinach thylakoids and O$_2$ evolving particles. *Photobiochem. Photobiophys.* In press

67. Cleland, R. E., Melis, A., Neale, P. J. 1986. The mechanism of photoinhibition: Photochemical reaction center inactivation in photosystem II of chloroplasts. *Photosynth. Res.* In press

68. Colbert, J. T., Hershey, H. P., Quail, P. H. 1983. Autoregulatory control of translatable phytochrome mRNA levels. *Proc. Natl. Acad. Sci. USA* 80:2248–52

69. Coruzzi, G., Broglie, R., Cashmore, A., Chua, N.-H. 1983. Nucleotide sequences of two pea cDNA clones encoding the small subunit of ribulose 1,5-bisphosphate carboxylase and the major chlorophyll *a/b*-binding thylakoid polypeptide. *J. Biol. Chem.* 258:1399–1402

70. Cox, R. P., Andersson, B. 1981. Lateral and transverse organization of cytochromes in the chloroplast thylakoid membrane. *Biochem. Biophys. Res. Commun.* 103:1336–42

71. Cox, R. P., Olsen, L. F. 1982. The organization of the electron transport chain in the thylakoid membrane. In *Topics in Photosynthesis*, ed. J. Barber, 4:49–79. Amsterdam: Elsevier. 287 pp.

72. Day, D. A., Ryrie, I. J., Fuad, N. 1984. Investigations of the role of the main light-harvesting chlorophyll-protein complex in thylakoid membranes. Reconstitution of depleted membranes from

intermittent-light-grown plants with the isolated complex. *J. Cell Biol.* 97: 163–72

73. Dennenberg, R. J., Jursinic, P. A. 1985. A comparison of the absorption changes near 325 nm and chlorophyll *a* fluorescence characteristics of the photosystem II acceptors Q_A and Q_{400}. *Biochim. Biophys. Acta* 808:192–200

74. Dunsmuir, P., Smith, S. M., Bedbrook, J. 1983. The major chlorophyll *a/b* binding protein of petunia is composed of several polypeptides encoded by a number of distinct nuclear genes. *J. Mol. Appl. Genet.* 2:285–300

75. Duysen, M., Eskins, K., Dybas, L. 1985. Blue and white light effects on chloroplast development in a soybean mutant. *Photochem. Photobiol.* 41:667–72

76. Duysen, M. E., Freeman, T. P., Williams, N. D., Huckle, L. L. 1985. Chloramphenicol stimulation of light-harvesting chlorophyll protein complex accumulation in a chlorophyll *b* deficient wheat mutant. *Plant Physiol.* 78:531–36

77. Eskins, K., Duysen, M. 1984. Chloroplast structure in normal and pigment-deficient soybeans grown in continuous red or far-red light. *Physiol. Plant.* 61:351–56

78. Eskins, K., Duysen, M., Dybas, L., McCarthy, S. 1985. Light quality effects on corn chloroplast development. *Plant Physiol.* 77:29–34

79. Eskins, K., Duysen, M. E., Olson, L. 1983. Pigment analysis of chloroplast pigment-protein complexes in wheat. *Plant Physiol.* 71:777–79

80. Eskins, K., McCarthy, S., Dybas, L., Duysen, M. 1986. Corn chloroplast development in low fluence red light and in low fluence red light plus far-red light. *Physiol. Plant.* In press

81. Evans, J. R. 1984. *Photosynthesis and nitrogen partitioning in leaves of T. aestivum and related species.* PhD thesis. Australian National Univ., Canberra. 201 pp.

82. Evans, J. R. 1986. A quantitative analysis of light distribution between the two photosystems, considering variation in both the relative amounts of the chlorophyll-protein complexes and the spectral quality of light. *Photobiochem. Photobiophys.* In press

83. Fedtke, C., Deichgräber, G., Schnepf, E. 1977. Herbicide induced changes in wheat chloroplast ultrastructure and chlorophyll *a/b* ratio. *Biochem. Physiol. Pflanz.* 171:307–12

84. Fish, L. E., Kück, U., Bogorad, L. 1985. Two partially homologous adja-

cent light-inducible maize chloroplast genes encoding polypeptides of the P700 chlorophyll *a*-protein complex of photosystem I. *J. Biol. Chem.* 260:1413–21

85. Fromm, H., Devic, M., Fluhr, R., Edelman, M. 1985. Control of *psbA* gene expression in mature *Spirodela* chloroplasts: Light regulation of 32-kd protein synthesis is independent of transcript level. *EMBO J.* 4:291–95

86. Frosch, S., Jabben, M., Bergfeld, R., Kleinig, H., Mohr, H. 1979. Inhibition of carotenoid biosynthesis by the herbicide SAN 9789 and its consequences for the action of phytochrome on plastogenesis. *Planta* 145:497–505

87. Frosch, S., Mohr, H. 1980. Analysis of light-controlled accumulation of carotenoids in mustard (*Sinapis alba* L.) seedlings. *Planta* 148:279–86

88. Fukshansky, L. 1981. Optical properties of plants. In *Plants and the Daylight Spectrum*, ed. H. Smith, pp. 21–40. London/New York: Academic. 508 pp.

89. Ghirardi, M. L., Melis, A. 1983. Localization of photosynthetic electron transport components in mesophyll and bundle sheath chloroplasts of *Zea mays*. *Arch. Biochem. Biophys.* 224:19–28

90. Glick, R. E., McCauley, S. W., Melis, A. 1985. Effect of light quality on chloroplast membrane organization and function. *Planta* 164:487–94

91. Gollmer, I., Apel, K. 1983. The phytochrome-controlled accumulation of mRNA sequences encoding the light-harvesting chlorophyll *a/b* protein of barley (*Hordeum vulgare* L.). *Eur. J. Biochem.* 133:309–13

92. Goodchild, D. J., Anderson, J. M., Andersson, B. 1985. Immunocytochemical localization of the cytochrome *b/f* complex of chloroplast thylakoid membranes. *Cell Biol. Int. Rep.* 9:715–21

93. Goodchild, D. J., Andersson, B., Anderson, J. M. 1985. Immunocytochemical localization of polypeptides associated with the oxygen evolving system of photosynthesis. *Eur. J. Cell Biol.* 36:294–98

94. Gounaris, K., Sundby, C., Andersson, B., Barber, J. 1983. Lateral heterogeneity of polar lipids in the thylakoid membranes of spinach chloroplasts. *FEBS Lett.* 156:170–74

95. Grahl, H., Wild, A. 1975. Studies on the content of P-700 and cytochromes in *Sinapis alba* during growth under two different light intensities. In *Environmental and Biological Control of Photosynthesis*, ed. R. Marcelle, pp. 107–13. The Hague: Dr. W. Junk. 408 pp.

96. Griffiths, W. T., Oliver, R. P. 1984.

Protochlorophyllide reductase-structure, function and regulation. In *Chloroplast Biogenesis*, ed. R. J. Ellis, pp. 245–58. Cambridge: Cambridge Univ. Press. 346 pp.

97. Haehnel, W. 1976. The ratio of the two light reactions and their coupling in chloroplasts. *Biochim. Biophys. Acta* 423:499–509

98. Haehnel, W. 1984. Photosynthetic electron transport in higher plants. *Ann. Rev. Plant Physiol.* 35:659–93

99. Harding, R. W., Shropshire, W. Jr. 1980. Photocontrol of carotenoid biosynthesis. *Ann. Rev. Plant Physiol.* 31: 217–38

100. Harpster, M., Apel, K. 1985. The light-dependent regulation of gene expression during plastid development in higher plants. *Physiol. Plant.* 64:147–52

101. Harpster, M. H., Mayfield, S. P., Taylor, W. C. 1984. Effects of pigment-deficient mutants on the accumulation of photosynthetic proteins in maize. *Plant Mol. Biol.* 3:59–71

102. Haworth, P., Kyle, D. J., Horton, P., Arntzen, C. J. 1982. Chloroplast membrane protein phosphorylation. *Photochem. Photobiol.* 36:743–48

103. Haworth, P., Melis, A. 1983. Phosphorylation of chloroplast thylakoid membrane proteins does not increase the absorption cross-section of photosystem 1. *FEBS Lett.* 160:277–80

104. Haworth, P., Watson, J. L., Arntzen, C. J. 1983. The detection, isolation and characterization of a light-harvesting complex which is specifically associated with photosystem I. *Biochim. Biophys. Acta* 724:151–58

105. Herrmann, R. G., Westhoff, P., Alt, J., Tittgen, J., Nelson, N. 1985. Thylakoid membrane proteins and their genes. In *Molecular Form and Function of the Plant Genome*, ed. L. van Vloten-Doting, G. S. P. Groot, T. C. Hall, pp. 233–56. New York: Plenum

106. Hodges, M. 1984. *Chlorophyll fluorescence and thylakoid membrane organisation*. PhD thesis. Imperial College, London. 325 pp.

107. Hodges, M., Barber, J. 1983. Photosynthetic adaptation of pea plants grown at different light intensities: State 1-State 2 transitions and associated chlorophyll fluorescence changes. *Planta* 157:166–73

108. Hodges, M., Barber, J. 1983. The significance of the kinetic analysis of fluorescence induction in DCMU-inhibited chloroplasts in terms of photosystem 2 connectivity and heterogeneity. *FEBS Lett.* 160:177–81

109. Holmes, M. G., Smith, H. 1977. The function of phytochrome in the natural environment - II. The influence of vegetation canopies on the spectral energy distribution of natural daylight. *Photochem. Photobiol.* 25:539–45

110. Horton, P. 1985. Interactions between electron transfer and carbon assimilation. In *Photosynthetic Mechanisms and the Environment*, ed. J. Barber, N. R. Baker, 6:135–87. Amsterdam: Elsevier Biomedical. In press

111. Horton, P., Black, M. T. 1981. Light-induced redox changes in chloroplast cytochrome *f* after phosphorylation of membrane proteins. *FEBS Lett.* 132:75–77

112. Horton, P., Foyer, C. 1983. Relationships between protein phosphorylation and electron transport in the reconstituted chloroplast system. *Biochem. J.* 210:517–21

113. Horton, P., Lee, P. 1984. Phosphorylation of chloroplast thylakoids decreases the maximum capacity of photosystem-II electron transfer. *Biochim. Biophys. Acta* 767:563–67

114. Horton, P., Lee, P. 1985. Phosphorylation of chloroplast membrane proteins partially protects against photoinhibition. *Planta* 165:37–42

115. Humbeck, K., Schumann, R., Senger, H. 1984. The influence of blue light on the formation of chlorophyll-protein complexes in *Scenesdesmus*. See Ref. 211, pp. 359–65

116. Humbeck, K., Senger, H. 1984. The blue light factor in sun and shade plant adaptation. See Ref. 211, pp. 344–51

117. Jenkins, G. I., Gallagher, T. F., Hartley, M. R., Bennett, J., Ellis, R. J. 1984. Photoregulation of gene expression during chloroplast biogenesis. See Ref. 15, 4:863–72

118. Jursinic, P. 1984. Photosystem II/I stoichiometry by the methods of absorbance change at 325 and 705 nm, ^{14}C-atrazine binding, and oxygen flash yields. See Ref. 15, 1:485–88

119. Kaplan, S., Arntzen, C. J. 1982. Photosynthetic membrane structure and function. In *Photosynthesis*, ed. Govindjee, 1:65–151. New York: Academic. 799 pp.

120. Kasemir, H. 1983. Action of light on chlorophyll(ide) appearance. *Photochem. Photobiol.* 37:701–8

121. Kaufman, L. S., Briggs, W. R., Thompson, W. F. 1985. Phytochrome control of specific mRNA levels in developing pea buds: The presence of both very low fluence and low fluence responses. *Plant Physiol.* 78:388–93

122. Kaufman, L. S., Thompson, W. F., Briggs, W. R. 1984. Different red light requirements for phytochrome-induced accumulation of *cab* RNA and *rbcS* RNA. *Science* 226:1447–49

123. Kaufman, L. S., Watson, J. C., Briggs, W. R., Thompson, W. F. 1985. Photoregulation of nuclear encoded transcripts: Blue light regulation of specific transcript abundance. In *The Molecular Biology of the Photosynthetic Apparatus*, ed. K. Steinback, C. J. Arntzen, L. Bogorad, S. Bonitz. Cold Spring Harbor: Cold Spring Harbor Lab. In press

124. Klein, S., Neuman, J. 1966. The greening of etiolated bean leaves and the development of chloroplast fine structure in absence of photosynthesis. *Plant Cell Physiol.* 7:115–23

125. Klockare, B., Virgin, H. I. 1984. Chlorophyll *b* formation in darkness after phototransformation of protochlorophyllide and during a period of intermittent light. *Isr. J. Bot.* 33:175–83

126. Kulandaivelu, G., Noorudeen, A. M., Sampath, P., Periyanan, S., Raman, K. 1983. Assessment of the photosynthetic electron transport properties of upper and lower leaf sides *in vivo* by fluorometric method. *Photosynthetica* 17:204–9

127. Kyle, D. J. 1985. The 32000 dalton Q_B protein of photosystem II. *Photochem. Photobiol.* 41:107–16

128. Kyle, D. J., Haworth, P., Arntzen, C. J. 1982. Thylakoid membrane protein phosphorylation leads to a decrease in connectivity between photosystem II reaction centers. *Biochim. Biophys. Acta* 680:336–42

129. Kyle, D. J., Kuang, T.-Y., Watson, J. L., Arntzen, C. J. 1984. Movement of a sub-population of the light harvesting complex (LHC_{II}) from grana to stroma lamellae as a consequence of its phosphorylation. *Biochim. Biophys. Acta* 765:89–96

130. Kyle, D. J., Ohad, I., Arntzen, C. J. 1984. Membrane protein damage and repair: Selective loss of a quinone-protein function in chloroplast membranes. *Proc. Natl. Acad. Sci. USA* 81:4070–74

131. Kyle, D. J., Staehelin, L. A., Arntzen, C. J. 1983. Lateral mobility of the light-harvesting complex in chloroplast membranes controls excitation energy distribution in higher plants. *Arch. Biochem. Biophys.* 222:527–41

132. Lamb, C. J., Lawton, M. A. 1983. Photocontrol of gene expression. In *Encyclopedia of Plant Physiology: Photomorphogenesis* (NS), ed. W. Shropshire, Jr., H. Mohr, 16A:213–57. Berlin: Springer-Verlag. 456 pp.

133. Larkum, A. W. D., Barrett, J. 1983. Light-harvesting processes in algae. *Adv. Bot. Res.* 10:1–219

134. Larsson, U. K., Andersson, B. 1985. Different degrees of phosphorylation and lateral mobility of two polypeptides belonging to the light-harvesting complex of Photosystem II. *Biochim. Biphys. Acta* 809:396–402

135. Larsson, U. K., Jergil, B., Andersson, B. 1983. Changes in the lateral distribution of the light-harvesting chlorophyll-*a/b*-protein complex induced by its phosphorylation. *Eur. J. Biochem.* 136:25–29

136. Leech, R. M. 1984. Chloroplast development in angiosperms: Current knowledge and future prospects. See Ref. 27, pp. 1–21

137. Leech, R. M., Walton, C. A., Baker, N. R. 1985. Some effects of 4-chloro-5-(dimethylamino) - 2 -phenyl-3(2H)-pyridazinone (San 9785) on the development of chloroplast thylakoid membranes in *Hordeum vulgare* L. *Planta* 165:277–83

138. Leong, T.-Y., Anderson, J. M. 1983. Changes in composition and function of thylakoid membranes as a result of photosynthetic adaptation of chloroplasts from pea plants grown under different light conditions. *Biochim. Biophys. Acta* 723:391–99

139. Leong, T.-Y., Anderson, J. M. 1984. Adaptation of the thylakoid membranes of pea chloroplasts to light intensities. I. Study on the distribution of chlorophyll-protein complexes. *Photosynth. Res.* 5:105–15

140. Leong, T.-Y., Anderson, J. M. 1984. Adaptation of the thylakoid membranes of pea chloroplasts to light intensities. II. Regulation of electron transport capacities, electron carriers, coupling factor (CF_1) activity and rates of photosynthesis. *Photosynth. Res.* 5:117–28

141. Leong, T.-Y., Anderson, J. M. 1984. Effect of light quality on the composition and function of thylakoid membranes in *Atriplex triangularis. Biochim. Biophys. Acta* 766:533–41

142. Leong, T.-Y., Goodchild, D. J., Anderson, J. M. 1985. Effect of light quality on the composition, function, and structure of photosynthetic thylakoid membranes of *Asplenium australasicum* (Sm.) Hook. *Plant Physiol.* 78:561–67

143. Leto, K. J., Bell, E., McIntosh, L. 1985. Nuclear mutation leads to an accelerated turnover of chloroplast-encoded 48 kd and 34.5 kd polypeptides in thylakoids lacking photosystem II. *EMBO J.* 4:1645–53

144. Lichtenthaler, H. K. 1981. Adaptation of leaves and chloroplasts to high quanta fluence rates. In *Photosynthesis*, ed. G. Akoyunoglou, 6:278–88. Philadelphia: Balaban Int. Sci. Serv. 769 pp.
145. Lichtenthaler, H. K., Burgstahler, R., Buschmann, C., Meier, D., Prenzel, U., Schönthal, A. 1982. Effect of high light and high light stress on composition, function and structure of the photosynthetic apparatus. In *Stress Effects on Photosynthesis*, ed. R. Marcelle, pp. 353–70. The Hague: Dr. W. Junk
146. Lichtenthaler, H. K., Burkard, G., Kuhn, G., Prenzel, U. 1981. Light-induced accumulation and stability of chlorophylls and chlorophyll-proteins during chloroplast development in radish seedlings. *Z. Naturforsch. Teil C* 36: 421–30
147. Lichtenthaler, H. K., Buschmann, C. 1978. Control of chloroplast development by red light, blue light and phytohormones. In *Chloroplast Development*, ed. G. Akoyunoglou, J. H. Argyroudi-Akoyunoglou, pp. 801–16. Amsterdam: Elsevier. 888 pp.
148. Lichtenthaler, H. K., Buschmann, C., Döll, M., Fietz, H.-J., Bach, T., et al. 1981. Photosynthetic activity, chloroplast ultrastructure, and leaf characteristics of high-light and low-light plants and of sun and shade leaves. *Photosynth. Res.* 2:115–41
149. Lichtenthaler, H. K., Buschmann, C., Rahmsdorf, U. 1980. The importance of blue light for the development of suntype chloroplasts. See Ref. 210, pp. 485–94
150. Lichtenthaler, H. K., Kuhn, G., Prenzel, U., Buschmann, C., Meier, D. 1982. Adaptation of chloroplast-ultrastructure and of chlorophyll-protein levels to high-light and low-light growth conditions. *Z. Naturforsch. Teil C* 37:464–75
151. Lichtenthaler, H. K., Kuhn, G., Prenzel, U., Meier, D. 1982. Chlorophyll-protein levels and degree of thylakoid stacking in radish chloroplasts from high-light, low-light and bentazon-treated plants. *Physiol. Plant.* 56:183–88
152. Lichtenthaler, H. K., Meier, D. 1984. Regulation of chloroplast photomorphogenesis by light intensity and light quality. See Ref. 96, pp. 261–81
153. Lichtenthaler, H. K., Prenzel, U., Kuhn, G. 1981. Carotenoid composition of chlorophyll-carotenoid-proteins from radish chloroplasts. *Z. Naturforsch. Teil C* 37:10–12
154. Link, G. 1982. Phytochrome control of plastid mRNA in mustard (*Sinapis alba* L.). *Planta* 154:81–86
155. Machold, O. 1981. Chlorophyll *a/b*-proteins and light-harvesting complex of *Vicia faba* and *Hordeum vulgare*. *Biochem. Physiol. Pflanzen.* 176:805–27
156. Machold, O. 1984. Chlorophyll *a/b*-proteins in their relation to the light-harvesting complex. See Ref. 15, 2:107–14
157. Malkin, S., Armond, P. A., Mooney, H. A., Fork, D. C. 1981. Photosystem II photosynthetic unit sizes from fluorescence induction in leaves. Correlation to photosynthetic capacity. *Plant Physiol.* 67:570–79
158. Mattoo, A. K., Hoffman-Falk, H., Marder, J. B., Edelman, M. 1984. Regulation of protein metabolism: Coupling of photosynthetic electron transport to *in vivo* degradation of the rapidly metabolized 32-kilodalton protein of the chloroplast membranes. *Proc. Natl. Acad. Sci. USA* 81:1380–84
159. Mayfield, S. P., Taylor, W. C. 1984. Carotenoid-deficient maize seedlings fail to accumulate light-harvesting chlorophyll *a/b* binding protein (LHCP) mRNA. *Eur. J. Biochem.* 144:79–84
160. Meier, D., Lichtenthaler, H. K. 1981. Ultrastructural development of chloroplasts in radish seedlings grown at high- and low-light conditions and in the presence of the herbicide bentazon. *Protoplasma* 107:195–207
161. Meier, D., Lichtenthaler, H. K., Burkard, G. 1980. Change of chloroplast ultrastructure in radish seedlings under the influence of the photosystem II-herbicide bentazon. *Z. Naturforsch. Teil C* 35:656–64
162. Melis, A. 1984. Light regulation of photosynthetic membrane structure, organization, and function. *J. Cell Biochem.* 24:271–85
163. Melis, A. 1985. Functional properties of photosystem II_β in spinach chloroplasts. *Biochim. Biophys. Acta* 808:334–42
164. Melis, A., Anderson, J. M. 1983. Structural and functional organization of the photosystems in spinach chloroplasts: Antenna size, relative electron-transport capacity, and chlorophyll composition. *Biochim. Biophys. Acta* 724:473–84
165. Melis, A., Brown, J. S. 1980. Stoichiometry of system I and system II reaction centers and of plastoquinone in different photosynthetic membranes. *Proc. Natl. Acad. Sci. USA* 77:4712–16
166. Melis, A., Duysens, L. N. M. 1979. Biphasic energy conversion kinetics and absorbance difference spectra of photosystem II of chloroplasts. Evidence for two different photosystem II reaction

centers. *Photochem. Photobiol.* 29:373–82

167. Melis, A., Harvey, G. W. 1981. Regulation of photosystem stoichiometry, chlorophyll *a* and chlorophyll *b* content and relation to chloroplast ultrastructure. *Biochim. Biophys. Acta* 637:138–45

168. Melis, A., Homann, P. H. 1976. Heterogeneity of the photochemical centers in system II of chloroplasts. *Photochem. Photobiol.* 23:343–50

169. Miller, K. R., Staehelin, L. A. 1976. Analysis of the thylakoid outer surface. Coupling factor is limited to unstacked membrane regions. *J. Cell Biol.* 68:30–47

170. Mills, J. D., Mitchell, P. 1982. Thiol modulation of CF_0-CF_1 stimulates acid/base-dependent phosphorylation of ADP by broken pea chloroplasts. *FEBS Lett.* 144:63–67

171. Mohr, H. 1984. Phytochrome and chloroplast development. See Ref. 27, pp. 305–47

172. Mösinger, E., Batschauer, A., Schäfer, E., Apel, K. 1985. Phytochrome control of *in vitro* transcription of specific genes in isolated nuclei from barley (*Hordeum vulgare*). *Eur. J. Biochem.* 147:137–42

173. Mullet, J. E. 1983. The amino acid sequence of the polypeptide segment which regulates membrane adhesion (grana stacking) in chloroplasts. *J. Biol. Chem.* 258:9941–48

174. Mullet, J. E., Baldwin, T. O., Arntzen, C. J. 1981. A mechanism for chloroplast thylakoid adhesion mediated by the Chl *a/b* light-harvesting complex. See Ref. 43, pp. 577–82

175. Murata, N., Miyao, M. 1985. Extrinsic membrane proteins in the photosynthetic oxygen-evolving complex. *Trends Biochem. Sci.* 10:122–24

176. Murphy, D. J. 1985. The molecular organisation and function of the photosynthetic membranes of higher plants. *Biochim. Biophys. Acta.* In press

177. Murphy, D. J., Woodrow, I. E. 1983. Lateral heterogeneity in the distribution of thylakoid membrane lipid and protein components and its implications for the molecular organisation of photosynthetic membranes. *Biochim. Biophys. Acta* 725:104–11

178. Nelson, T., Harpster, M. H., Mayfield, S. P., Taylor, W. C. 1984. Light-regulated gene expression during maize leaf development. *J. Cell Biol.* 98:558–64

179. Nurmi, A., Vapaavuori, E. 1982. Chlorophyll-protein complexes in *Salix* sp. 'Aquatica gigantea' under strong and weak light I. Spectral characterization of

the chlorophyll-protein complexes. *Plant Cell Physiol.* 23:785–90

180. Oelmüller, R., Mohr, H. 1985. Carotenoid composition in milo (*Sorghum vulgare*) shoots as affected by phytochrome and chlorophyll. *Planta* 164:390–95

181. Oelze-Karow, H., Kasemir, H., Mohr, H. 1978. Control of chlorophyll *b* formation by phytochrome and a threshold level of chlorophyllide *a*. See Ref. 147, pp. 787–92

182. Oelze-Karow, H., Mohr, H. 1978. Control of chlorophyll *b* biosynthesis by phytochrome. *Photochem. Photobiol.* 27:189–93

183. Oelze-Karow, H., Rösch, H., Mohr, H. 1983. Prevention by phytochrome of photodelay in chlorophyll accumulation. *Photochem. Photobiol.* 37:565–69

184. Ohad, I., Kyle, D. J., Arntzen, C. J. 1984. Membrane protein damage and repair: Removal and replacement of inactivated 32-kilodalton polypeptides in chloroplast membranes. *J. Cell Biol.* 99:481–85

185. Ohad, I., Kyle, D. J., Hirschberg, J. 1985. Light-dependent degradation of the Q_B-protein in isolated pea thylakoids. *EMBO J.* 4:1655–59

186. Osmond, C. B. 1981. Photorespiration and photoinhibition. Some implications for the energetics of photosynthesis. *Biochim. Biophys. Acta* 639:77–98

187. Otto, V., Mösinger, E., Sauter, M., Schäfer, E. 1983. Phytochrome control of its own synthesis in *Sorghum vulgare* and *Avena sativa*. *Photochem. Photobiol.* 38:693–700

188. Owens, G. C., Ohad, I. 1982. Phosphorylation of *Chlamydomonas reinhardi* chloroplast membrane proteins in vivo and in vitro. *J. Cell Biol.* 93:712–18

189. Owens, G. C., Ohad, I. 1983. Changes in thylakoid polypeptide phosphorylation during membrane biogenesis in *Chlamydomonas reinhardii* y-1. *Biochim. Biophys. Acta* 722:234–41

190. Pakrasi, H. B., Arntzen, C. J. 1986. Photosystem II reaction center: Polypeptide subunits and functional cofactors. See Ref. 19

191. Percival, M. P., Webber, A. N., Baker, N. R. 1984. Evidence for the role of the light-harvesting chlorophyll *a/b* protein complex in photosystem II heterogeneity. *Biochim. Biophys. Acta* 767:582–89

192. Poulsen, C. 1983. The barley chloroplast genome: Physical structure and transcriptional activity in vivo. *Carlsberg Res. Commun.* 48:57–80

193. Powles, S. B. 1984. Photoinhibition of

photosynthesis induced by visible light. *Ann. Rev. Plant Physiol.* 35:15–44

194. Powles, S. B., Chapman, K. S. R., Whatley, F. R. 1982. Effect of photoinhibitory treatments on the activity of light-activated enzymes of C_3 and C_4 photosynthetic carbon metabolism. *Plant Physiol.* 69:371–74

195. Powles, S. B., Critchley, C. 1980. Effect of light intensity during growth on photoinhibition of intact attached bean leaflets. *Plant Physiol.* 65:1181–87

196. Powles, S. B., Thorne, S. W. 1981. Effect of high-light treatments in inducing photoinhibition of photosynthesis in intact leaves of low-light grown *Phaseolus vulgaris* and *Lastreopsi microsora. Planta* 152:471–77

197. Quail, P. H. 1983. Rapid action of phytochrome in photomorphogenesis. See Ref. 132, pp. 178–212

198. Quinn, P. J., Williams, W. P. 1983. The structural role of lipids in photosynthetic membranes. *Biochim. Biophys. Acta* 737:223–66

199. Rau, W. 1980. Photoregulation of carotenoid biosynthesis: An example of photomorphogenesis. In *Pigments in Plants*, ed. F.-C. Czygan, pp. 80–103. Stuttgart/New York: Fischer. 447 pp. 2nd ed.

200. Rebeiz, C. A., Wu, S. M., Kuhadja, M., Daniell, H., Perkins, E. J. 1983. Chlorophyll *a* biosynthetic routes and chlorophyll *a* chemical heterogeneity in plants. *Mol. Cell. Biochem.* 57:97–125

201. Rodermel, S. R., Bogorad, L. 1985. Maize plastid photogenes: Mapping and photoregulation of transcript levels during light-induced development. *J. Cell Biol.* 100:463–76

202. Rüdiger, W., Benz, J. 1984. Synthesis of chloroplast pigments. See Ref. 96, pp. 225–44

203. Rühle, W., Wild, A. 1979. Measurements of cytochrome *f* and P-700 in intact leaves of *Sinapis alba* grown under high-light and low-light conditions. *Planta* 146:377–85

204. Deleted in proof

205. Ryrie, I. J., Anderson, J. M., Goodchild, D. J. 1980. The role of light-harvesting chlorophyll *a*/*b*-protein complex in chloroplast membrane stacking. Cation-induced aggregation of reconstituted proteoliposomes. *Eur. J. Biochem.* 107: 345–54

206. Ryrie, I. J., Fuad, N. 1982. Membrane adhesion in reconstituted proteoliposomes containing the light-harvesting chlorophyll *a*/*b*-protein complex: The role of charged surface groups. *Arch. Biochem. Biophys.* 214:475–88

207. Schopfer, P., Apel, K. 1982. Intracellular photomorphogenesis. See Ref. 132, pp. 258–88

208. Schreiber, U., Fink, R., Vidaver, W. 1977. Fluorescence induction in whole leaves: Differentiation between the two leaf sides and adaptation to different light regimes. *Planta* 133:121–29

209. Schwarz, H. P., Kloppstech, K. 1982. Effects of nuclear gene mutations on the structure and function of plastids in pea. The light-harvesting chlorophyll *a*/*b* protein. *Planta* 155:116–23

210. Senger, H., ed. 1980. *The Blue Light Syndrome.* Berlin: Springer-Verlag. 665 pp.

211. Senger, H., ed. 1984. *Blue Light Effects in Biological Systems.* Berlin: Springer-Verlag. 538 pp.

212. Simpson, D. J. 1983. Freeze-fracture studies on barley plastid membranes VII. Structural changes associated with phosphorylation of the light-harvesting complex. *Biochim. Biophys. Acta* 725:113–20

213. Silverthorne, J., Tobin, E. M. 1984. Demonstration of transcriptional regulation of specific genes by phytochrome action. *Proc. Natl. Acad. Sci. USA* 81:1112–16

214. Slovin, J. P., Tobin, E. M. 1982. Synthesis and turnover of the light-harvesting chlorophyll *a*/*b*-protein of *Lemna gibba* grown under intermittant red-light: 1. Possible translational control. *Planta* 154:465–57

215. Smith, H. 1982. Light quality, photoperception, and plant strategy. *Ann. Rev. Plant Physiol.* 33:481–518

216. Staehelin, L. A., Arntzen, C. J. 1983. Regulation of chloroplast membrane function: Protein phosphorylation changes the spatial organization of membrane components. *J. Cell Biol.* 97:1327–37

217. Steinback, K. E., Bose, S., Kyle, D. J. 1982. Phosphorylation of the light-harvesting chlorophyll-protein regulates excitation energy distribution between photosystem II and photosystem I. *Arch. Biochem. Biophys.* 216:356–61

218. Strotmann, H., Bickel-Sandkötter, S. 1984. Structure, function, and regulation of chloroplast ATPase. *Ann. Rev. Plant Physiol.* 35:97–120

219. Telfer, A., Hodges, M., Millner, P. A., Barber, J. 1984. The cation-dependence of the degree of protein phosphorylation-induced unstacking of pea thylakoids. *Biochim. Biophys. Acta* 766:554–62

220. Terashima, I. 1984. *Ecophysiological anatomy of a leaf photosynthetic system.* PhD thesis. Univ. Tokyo, Japan. 89 pp.

221. Terashima, I., Inoue, Y. 1984. Com-

parative photosynthetic properties of palisade tissue chloroplasts and spongy tissue chloroplasts of *Camellia japonica* L.: Functional adjustment of the photosynthetic apparatus to light environment within a leaf. *Plant Cell Physiol.* 25:555–63

222. Terashima, I., Saeki, T. 1983. Light environment within a leaf I. Optical properties of paradermal sections of *Camellia* leaves with special reference to differences in the optical properties of palisade and spongy tissues. *Plant Cell Physiol.* 24:1493–1501

223. Thielen, A. P. G. M., van Gorkom, H. J. 1981. Quantum efficiency and antenna size of photosystems II_α, II_β and I in tobacco chloroplasts. *Biochim. Biophys. Acta* 635:111–20

224. Thompson, W. F., Everett, M., Polans, N. O., Jorgensen, R. A., Palmer, J. D. 1983. Phytochrome control of RNA levels in developing pea and mung-bean leaves. *Planta* 158:487–500

225. Thornber, J. P. 1986. Biochemical characterization and structure of pigment-proteins of photosynthetic organisms. See Ref. 19. In press

226. Thorne, S. W., Duniec, J. T. 1983. The physical principles of energy transduction in chloroplast thylakoid membranes. *Q. Rev. Biophys.* 16:197–278

227. Timko, M. P., Cashmore, A. R. 1983. Nuclear genes encoding the constituent polypeptides of the light-harvesting chlorophyll *a*/*b*-protein complex from pea. In *Plant Molecular Biology*, ed. R. B. Goldberg, pp. 403–12. New York: Liss. 498 pp.

228. Tobin, E. M. 1981. Phytochrome-mediated regulation of messenger RNAs for the small subunit of ribulose 1,5-bisphosphate carboxylase and the light-harvesting chlorophyll *a*/*b*-protein in *Lemna gibba*. *Plant Mol. Biol.* 1:35–51

229. Tobin, E. M., Silverthorne, J. 1985. Light regulation of gene expression in higher plants. *Ann. Rev. Plant Physiol.* 36:569–93

230. Tobin, E. M., Wimpee, C. F., Silverthorne, J., Stiekema, W. J., Neumann, G. N., Thornber, J. P. 1984. Phytochrome regulation of the expression of two nuclear-coded chloroplast proteins. See Ref. 123, pp. 325–34

231. Trebst, A. 1980. Inhibitors in electron flow: Tools for the functional and structural localization of carriers and energy conservation sites. *Methods Enzymol.* 69:675–715

232. Vallon, O., Wollman, F. A., Olive, J. 1985. Distribution of intrinsic and extrinsic subunits of the PSII protein complex between appressed and non-appressed regions of the thylakoid membrane: An immunocytochemical study. *FEBS Lett.* 183:245–50

233. Vaughn, K. C., Vierling, E., Duke, S. O., Alberte, R. S. 1983. Immunocytochemical and cytochemical localization of photosystems I and II. *Plant Physiol.* 73:203–7

234. Viro, M., Kloppstech, K. 1982. Expression of genes for plastid membrane proteins in barley under intermittent light conditions. *Planta* 154:18–23

235. Wergin, W. P., Potter, J. R. 1975. The effects of fluometuron on the ultrastructural development, chlorophyll accumulation and photosynthetic competence in developing velvetleaf seedlings. *Pestic. Biochem. Physiol.* 5:265–79

236. Wettern, M., Ohad, I. 1984. Light-induced turnover of thylakoid polypeptides in *Chlamydomonas reinhardi*. *Isr. J. Bot.* 33:253–63

237. Whitmarsh, J., Ort, D. R. 1984. Quantitative determination of the electron transport complexes in the thylakoid membranes of spinach and several other plant species. See Ref. 15, pp. 231–34

238. Whitmarsh, J., Ort, D. R. 1984. Stoichiometries of electron transport complexes in spinach chloroplasts. *Arch. Biochem. Biophys.* 231:378–89

239. Wild, A. 1979. Physiology of photosynthesis in higher plants. The adaptation of photosynthesis to light intensity and light quality. *Ber. Dtsch. Bot. Ges.* 92:341–64

240. Wild, A., Holzapfel, A. 1980. The effect of blue and red light on the content of chlorophyll, cytochrome *f*, soluble reducing sugars, soluble proteins and the nitrate reductase activity during growth of the primary leaves of *Sinapis alba*. See Ref. 210, pp. 444–51

241. Wild, A., Rühle, W., Grahl, H. 1975. The effect of light intensity during growth of *Sinapis alba* on the electron-transport and the noncyclic photophosphorylation. See Ref. 95, pp. 115–21

242. Wild, A., Wolf, G. 1980. The effect of different light intensities on the frequency and size of stomata, the size of cells, the number, size and chlorophyll content of chloroplasts in the mesophyll and the guard cells during the ontogeny of primary leaves of *Sinapis alba*. *Z. Pflanzenphysiol.* 97:325–42

243. Wu, S.-M., Rebeiz, C. A. 1985. Chloroplast biogenesis. Molecular structure of chlorophyll *b* (E489 F666). *J. Biol. Chem.* 260:3632–34

Ann. Rev. Plant Physiol. 1986. 37:137–64
Copyright © 1986 by Annual Reviews Inc. All rights reserved

DYNAMICS OF VACUOLAR COMPARTMENTATION

Thomas Boller and Andres Wiemken

Botanisches Institut, Abteilung Pflanzenphysiologie, Universität Basel, Hebelstrasse 1, CH-4056 Basel, Switzerland

CONTENTS

INTRODUCTION

The central vacuole, one of the most conspicuous and distinguishing features of mature plant cells, was reviewed in this series for the first time in 1978 (102). At that time, methods for the isolation of vacuoles had been well established for yeast but had only recently become available for higher plants. Since that time, these methods have been refined and modified in a variety of ways (reviewed in 134). Applied to a multitude of plant tissues, they yielded a torrent of information about the properties of the tonoplast and the composition of the vacuolar sap, as summarized in a number of reviews (20, 93, 98, 134, 170, 171).

137

0066-4294/96/0601-0137$02.00

Isolated vacuoles convey a static image of optically homogeneous, motionless spheres with little if any metabolic activity. In accord with this image, the primary function of vacuoles in vivo is concerned with statics of plant form and structure. Plants live on resources that are ubiquitous but generally extremely dilute in the environment (sunlight, CO_2, soil nutrients). To utilize these resources, plants need as much space as possible. The vacuole has a fundamental, static role in this task; vacuoles fill a large space with an inert, metabolically cheap building material, namely water equilibrated osmotically with salts and solutes. Thus vacuoles economically extend the reach of the cytoplasm (183).

In addition, vacuolar compartmentation has many dynamic aspects. Vacuoles change form and structure during growth and differentiation. Proteins and other macromolecules can accumulate in vacuoles and may later be degraded. Ions and metabolites are not simply deposited in the vacuoles but rather remain in continual interplay and exchange with cytoplasmic constituents. Thereby the vacuoles form an internal environment for the cytoplasm, essential to its homeostasis; they fulfill a role analogous to the body fluids of animals (104). These dynamic aspects of vacuolar compartmentation are emphasized in the present review.

DYNAMICS OF VACUOLAR FORM AND STRUCTURE

Ontogeny and Differentiation of Vacuoles

Early ultrastructural studies have indicated that vacuoles are ultimately derived from the endoplasmic reticulum (ER). More recent work has traced the origin of vacuoles to a tubular membrane network at the *trans* face of the Golgi system, called GERL (Golgi-ER-lysosome system) (97, 98). Three-dimensional pictures of thick sections from root meristems, stained with zinc iodide-osmium, show how this network encloses portions of the cytoplasm and how the portions appear to be sealed off by lateral expansion and fusion of the membrane tubules (97). The sequestered cytoplasm is thought to be digested in an autophagic process; the small vacuoles arising expand and fuse to form the large central vacuole (98).

The actual contribution of the Golgi apparatus to the ontogeny of the vacuoles remains unclear. Studies on the formation of protein bodies (considered to be vacuoles specialized in protein storage; see below) hint at two different possibilities. In developing cereal seeds, protein bodies appear to be formed from the ER directly, without involvement of the Golgi apparatus (reviewed in 61). In developing legume seeds, protein bodies are generated by fragmentation of the central vacuole of the developing cotyledon cells (167). Immunocytochemistry (53 and references therein) and pulse-chase experiments combined with cell fractionation (85, 167) indicate that storage proteins are vectorially synthesized on the ER, pass through the Golgi complex, and accumulate in the protein bodies. This indicates a continuous flow of Golgi-derived material and, by inference, of Golgi-derived vesicles to the vacuole. Interestingly, pulse-chase studies in *Phaseolus vulgaris* showed that an integral protein of the protein body membrane was also synthesized in the ER but was transported to the protein bodies much more slowly than the storage proteins accumulating within the protein bodies (89).

Vacuoles can change their form dramatically in the course of development, as seen in a study of guard cell differentiation in *Allium* (117, 118). The vacuoles in these cells autofluoresce and can be observed by fluorescence microscopy in vivo. Guard mother cells contain a spherical vacuole that transiently forms a tubular network similar to the GERL observed at an ultrastuctural level (97). This network—which is continually in motion—is maintained through the course of one cell division and then returns to a spherical structure without apparent autophagic processes. This is reminiscent of the vacuolar dynamics in yeast where large vacuoles cleave into small vesicles and are restored again by fusion during the cell cycle (184). Fragmentation of vacuoles, followed by fusion, occurs also during the growth cycle of cultured plant cells (3, 52).

Similar processes can occur in the absence of cell division in differentiated cells (98). For example, the motor cells in the tertiary pulvinus of *Albizzia* leaves change in a diurnal rhythm from a univacuolate state when fully expanded to a multivacuolate state when the cell volume decreases (28). Although the vacuolar volume undergoes pronounced changes during the course of this rhythm, the cytoplasmic volume remains constant.

Tonoplast and Membrane Flow

The developmental studies cited above indicate that the tonoplast grows by incorporation of membranes derived from the ER and GERL and possibly from the Golgi apparatus. Little is known about tonoplast turnover and about the possibility that membrane flow also occurs in the opposite direction, from the tonoplast to other membrane systems.

Membrane flow from the ER to the central vacuole has been observed in tobacco protoplasts evacuolated by centrifugation (26). The evacuolated protoplasts formed a new central vacuole from the fusion of small vacuoles and ER cisternae, and could no longer be distinguished from untreated control protoplasts after 24 h. It is tempting to speculate that in the untreated protoplasts, membrane flow from the ER to the tonoplast occurred similarly but was balanced by some sort of tonoplast turnover.

Recent studies indicate that the plasmalemma and the tonoplast may be linked by a pathway of membrane flow. Ultrastructural work has shown that isolated protoplasts take up cationic ferritin by endocytosis and discharge it into the endomembrane system, including small vacuoles (65, 150). Apparently, endocytic vesicles derived from the plasmalemma ultimately can fuse with the tonoplast. Studies indirectly showing that impermeant fluorescent markers are transported by pinocytosis from the plasmalemma into the vacuole in yeast cells (89a, 129) have led to the same conclusion.

In this context, it is interesting that rhizobia are taken up by an endocytic process in legume nodules. While the bacteria multiply and differentiate intracellularly, they remain enclosed within the peribacteroid membrane. This membrane has been found to contain antigenic determinants (23) and an ATPase (11a) in common with the plant plasmalemma and the Golgi apparatus. However, it seems related also to the tonoplast since the peribacteroid space contains the enzyme α-mannosidase (105), a marker for plant vacuoles (see below). The similarity of the peribacteroid membrane and the tonoplast is also apparent in an ineffective soybean nodule where the peribacteroid membranes fuse and form vacuoles in which lysis of the bacteroids occurs (182) or in developing nodule cells which degrade infection threads in vacuoles (6).

DYNAMICS OF MACROMOLECULES IN THE VACUOLAR COMPARTMENT

Vacuoles as Lysosomes

Vacuoles isolated from yeast cells have been found to contain a range of hydrolytic enzymes and have therefore been identified as the lysosomes of yeast cells (103, 185). On the basis of ultrastructural and cytochemical studies of higher plants, Matile (101, 102) concluded that vacuoles of higher plants can also be considered lysosomes. Since then, this hypothesis has been verified using isolated vacuoles from a variety of plant tissues; most of the intracellular activity of many typical lysosomal hydrolases is located in the central vacuole (reviewed in 14).

α-Mannosidase has proved to be a marker enzyme for yeast vacuoles where it is bound to the vacuolar membrane (164, 185). α-Mannosidase is also exclusively vacuolar in protoplasts derived from various tissues of higher plants (14, 18, 96, 139). Since it is very stable, easy to assay, present in plant cells in high activity, and almost absent from commercial cell-wall digesting enzymes used to prepare protoplasts, it can be used as a marker for plant vacuoles. In contrast to the yeast enzyme, plant α-mannosidase is not associated with the tonoplast but is a soluble enzyme in the vacuolar sap (18, 169); therefore, it cannot be used as a marker for the tonoplast in cell homogenates. It should be noted that α-mannosidase has also been found in other parts of the cell. Some tissues contain α-mannosidase in the cell wall as well as in the vacuole (18,

163). A small fraction of the enzyme is associated with rough ER in developing cotyledons that actively synthesize α-mannosidase (163). Like α-mannosidase, β-N-acetylglucosaminidase is a soluble, exclusively vacuolar enzyme in protoplasts and is useful as a marker for the vacuolar sap (18, 96).

A focal point of interest has been the localization of proteinases in higher plants. In most tissues investigated, proteolytic activity was primarily vacuolar (18, 29, 59, 84, 96, 112, 113, 152, 178, 187). However, in bean leaves, a highly active proteinase appeared to be present exclusively in the cell wall compartment (165).

Import and Export of Vacuolar Proteins

Our knowledge about the dynamics of vacuolar proteins is derived mostly from the special case of protein bodies in seeds (2, 61). As mentioned above, the storage globulins of legumes are vectorially synthesized and cotranslationally glycosylated in the ER, transported into the Golgi system where further glycosylation takes place, and finally packed into the protein bodies where they are processed by partial deglycosylation and partial proteolysis.

During germination, proteinase is transported into protein bodies. Immunohistochemical studies of mung bean seedlings have shown that the newly formed proteinase first appears in small vesicles (most likely derived from ER) and later in the protein bodies (9), suggesting that import of the proteinase occurs by fusion of the vesicles with the protein body membrane. In endosperm of germinating *Ricinus* seeds, protein bodies fuse and form large vacuoles (112). These vacuoles could be isolated and continued to degrade the storage proteins in vitro (113).

In general, these results are consistent with current views on import of lysosomal proteins in animals (32) and of vacuolar proteins in yeast (141). It remains to be seen to what extent the details of processing and transport are also similar in animals and plants. The typical acid hydrolases in plant vacuoles are generally glycoproteins like the hydrolases of animal lysosomes (46). In animal tissues, the precursors of lysosomal enzymes appear to be modified in the Golgi system so as to obtain a mannose-6-phosphate residue that may function as a recognition marker for further transfer to the lysosomes (32). However, in a recent study of developing pea seeds, no mannose-6-phosphate was found on the glycoproteins (47).

Mature vacuoles generally contain little protein, only about 1–10% of the total cellular protein; these proteins, including the acid hydrolases, might simply be remains of the early developmental phase of the vacuoles where autophagy might have occurred. This static picture is not always correct. Environmental stimuli can lead to a rapid import of specific proteins into mature vacuoles. For example, wounding induces the accumulation of proteinase inhibitors in the vacuoles of tomato leaf cells (110) and of carboxypeptidase inhibitors in the vacuoles of potato leaf cells (62). In bean leaves, ethylene induces the formation of chitinase, an enzyme thought to represent a defense against fungi, that accumulates in the vacuole (19). It is interesting that in vitro translation of the mRNAs for proteinase inhibitor (110) and chitinase (U. Vögeli and T. Boller, unpublished) yield immunoprecipitable translation products larger in size than the mature proteins, indicating that the proteins are modified in vivo on their way to the vacuole.

The dynamics of vacuolar proteins is also apparent in a specialized tissue found in soybean leaves, the paraveinal mesophyll (39, 40, 42). Cells of this tissue possess large vacuoles containing specific glycoproteins, as shown by histochemical stains (39) and immunohistochemistry (42). Ultrastructural evidence implicates the Golgi complex in the production of the glycoproteins (39). Qualitatively, the vacuolar proteins accumulate at anthesis and disappear during pod filling, indicating that they have a temporary storage function (40). In depodded plants, the glycoproteins accumulate rapidly in the vacuoles (42); apparently, removal of the main sink for nitrogen prevents mobilization of the glycoproteins.

Recent experiments have shown a rapid turnover of proteins in the vacuoles of *Acer* cell cultures (30). Vacuoles isolated from cells that had been fed with [^{3}H]leucine for 18 h contained 30% of the total trichloroacetic acid-insoluble radioactivity and degraded half of this material within 6 h in vitro (30).

There is no clear-cut evidence that vacuolar proteins can be released into the cytosol or into the apoplast except by lysis of the whole vacuole. The carnivorous plant *Dionaea* possesses digestive glands that secrete proteinases and other acid hydrolases upon stimulation. It has been suggested that the vacuoles of unstimulated glands store the digestive enzymes (133). A recent histochemical reexamination indicated, however, that acid phosphatase was secreted by way of the dictyosomes, without involvement of the central vacuole (60).

Digestion of Cytoplasmic Components in Vacuoles?

Although autophagic processes appear to occur in the early development of vacuoles (97, 98), the question of whether mature vacuoles degrade cytoplasmic macromolecules remains open, as discussed earlier by Leigh (78). It has been reported that the disappearance of ribulosebisphosphate carboxylase (RBPC), a major leaf protein, is closely correlated with a reduction in the number of chloroplasts per cell as determined in protoplasts from senescing wheat leaves (187). On the basis of this and of electron microscopic pictures showing chloroplasts within the vacuoles of senescing cells, it has been suggested that chloroplasts are taken up into the vacuoles and then broken down (187). More recently, a reexamination of protoplasts from senescing wheat leaves showed that chloroplasts became increasingly fragile during senescence, which made the counting method used in Ref. 187 unreliable (177). When the tissue was fixed with glutaraldehyde before macerating the cells and counting the chloroplasts, only a small reduction in chloroplast number per cell was found in the course of senescence. In contrast, the amount of RBPC per chloroplast decreased by 80%, indicating that the protein was digested without concomitant degradation of whole chloroplasts (177). Thus, chloroplasts are not degraded by autophagy in senescing wheat leaves.

The possibility remains that RBPC is selectively exported to the vacuole and degraded there. Ultrastructural work indicates that the tonoplast interacts with the outer membrane of plastids (119, 166), and thus the vacuole could take up material from the chloroplast. Alternatively, RBPC could be hydrolyzed by proteinases in the chloroplast without any contribution of the vacuolar compartment. Indeed, there are several indications that chloroplasts have their own specific proteolytic system (177).

To summarize, although vacuolar proteinases are highly active in plants, their function in the dynamics of cellular-protein turnover is unknown. In this context, it is interesting to glance over recent work with yeast. Over 95% of the total proteinase activity of yeast is vacuolar (185). Mutants that lack all known vacuolar proteinases have been constructed (1, 188). These mutants show nearly normal vegetative growth but are defective in sporulation and ascus formation (188). Obviously, the vacuolar proteinases are dispensable for a large part of the life cycle of yeast. Proteinases outside the vacuoles might be important during vegetative growth. Such proteinases have been described in the mutants devoid of vacuolar proteinases (1). Their detection in wild type cells would have been exceedingly difficult because of the huge background activity of the vacuolar proteinases.

Vacuoles and Heterophagy

The endocytosis of rhizobia in legume nodules and their eventual degeneration (6, 105) may represent a case of heterophagy (98), although it is unclear whether plant hydrolases or bacterial autolytic enzymes cause the observed degeneration. Interactions of plants with biotrophic fungi (e.g. endomycorrhizal symbionts and obligate pathogens) can be considered examples of incipient heterophagy. The intracellular haustoria formed by these fungi remain enclosed by a plant membrane, the extrahaustorial membrane. The extrahaustorial membrane is continuous with the plant plasmalemma but different in some of its properties (58, 140). It is tempting to speculate that the extrahaustorial membrane can develop into a tonoplast-like membrane, and that the fungal haustorium is thereby enclosed in a vacuole-like organelle within the plant cell.

DYNAMICS OF METABOLITES AND IONS IN THE VACUOLAR COMPARTMENT

Ions and metabolites in the vacuolar compartment can show very different dynamics. In many cases, substances are deposited in the vacuole and never return to the cytoplasm. Such static cases, exemplified by the formation of calcium oxalate crystals in the vacuoles of specialized cells (41), befit the classical image of the plant vacuoles as a substitute for the excretion system of animals. However, vacuolar compartmentation can also be dynamic, ranging from the seasonal change in accumulation and mobilization of carbohydrates in storage organs (71), diurnal fluctuations of vacuolar organic acids in plants with crassulacean acid metabolism (CAM, 159), to the very rapidly changing compartmentation in the vacuoles of opening or closing stomatal complexes (192). Even secondary products are often unexpectedly dynamic and show high rates of turnover (8).

Studies with isolated vacuoles have provided considerable information about the composition of vacuolar sap, about the energization of the tonoplast, and about the transport systems operating in vacuoles (reviewed in 15, 79). Unfortunately, there are few cases in which dynamic processes such as accumulation and mobilization have been reconstituted in experiments with isolated vacuoles. Nevertheless, the data obtained have helped to understand and interpret more indirect work on compartmentation in intact cells or tissues.

Composition of the Vacuolar Sap

Early microscopic studies of cells have shown that vacuoles are the principal sites of accumulation of many colored secondary products such as anthocyanins and betanins; these compounds can be used as vacuolar markers in cell-fractionation studies (see 79). In addition, vacuoles contain many other secondary products and have been shown, by direct isolation, to contain close to 100% of the intracellular alkaloids (34, 35), saponins (162), glucosinolates (86 and references therein), cyanogenic glycosides (138), coumaryl glycosides (114) and other glycosides, e.g. the glycoconjugates of gibberellic acid (45) and of 2,4-D (142). A vast number of the secondary compounds accumulated in the vacuoles are toxic to pathogens, parasites, and herbivores as well as to the plant itself. The glycosides stored in the vacuoles are generally nontoxic but often generate highly toxic aglycones upon hydrolysis. The specific glycosidases catalyzing the corresponding reactions are localized outside the vacuole (86, 114, 162); accordingly, the toxic aglycones are produced only when the compartmentation is disrupted.

Since vacuoles generally constitute more than 90% of the cell volume, it is obvious that the largest part of the osmotically active solutes must be vacuolar (71, 79). Analyses of isolated vacuoles have shown this to be true (70, 72, 73,

81, 96, 109, 168, 190). It would be interesting to know if a given solute has a higher concentration in the vacuole than in the cytoplasm. Unfortunately, with isolated vacuoles, it is analytically nearly impossible to distinguish between 90% vacuolar location—which would signify equal distribution between vacuole and cytoplasm—and 100% vacuolar location (82).

The absence of certain metabolites or ions from isolated vacuoles is also significant (71, 79). Such compounds are either rapidly lost during isolation—a problem largely eliminated by rapid isolation techniques for vacuoles (21, 69)—or specifically excluded from the vacuole. Examples of substances less concentrated in the vacuole than in the cytosol are proline in water-stressed *Nicotiana* leaves (116), proline and glutamate in *Melilotus* leaves (21) and *Acer* cell cultures (3), and sucrose in roots storing gentianose (72) or stachyose (70). Glycinebetaine and proline have been calculated to be 10–100 times more concentrated in the cytoplasm than in the vacuole of beet root cells (80).

Energization of the Tonoplast

Solutes may be accumulated in the vacuoles by active transport against an electrochemical concentration gradient. In general terms, active transport can arise from primary active transport systems, consuming the chemical "energy" of ATP, NADH or other energy-rich compounds, or by secondary active transport systems, coupling the flow of a solute to a preexisting electrochemical gradient of another solute across the membrane (see 128 for a review).

In the plasmalemma of bacteria, fungi, and plants, primary active transport systems generate an electrochemical proton gradient across the membrane (proton motive force, PMF). This PMF is the driving force for a variety of secondary active transport systems; the membrane can be said to be energized by the PMF (128).

Vacuoles have long been known to contain an acidic sap. Therefore, it has been logical to suspect that the tonoplast is also the seat of an active proton pump, fueled by ATP, which acts as a primary active transport system. Considerable effort has been devoted to proving this hypothesis, as summarized in several recent reviews (79, 93, 149).

TONOPLAST ATPase   Investigations of vacuolar ATPase are difficult because vacuoles contain unspecific phosphatases and may be contaminated with other membranes rich in ATPases. Early reports on vacuolar ATPase have not always paid sufficient attention to these potentially interfering activities (see 14). Below we point out that ATPase in vacuolar preparations can be distinguished from unspecific phosphatases by its substrate specificity and from other membrane ATPases by its sensitivity to inhibitors, in particular, by its anion sensitivity. Anion-sensitive ATPase is enriched in purified vacuoles.

Substrate specificity of tonoplast ATPase   Vacuoles generally contain high activities of acid phosphatases that readily hydrolyze ATP (4, 83). In vacuoles and tonoplast membranes isolated from tobacco cell cultures, most of the ATP-splitting activity was due to acid phosphatase (24).

Two techniques have been used to assay specific ATPase in preparations of vacuoles containing acid phosphatase: (*a*) determination of the difference in the rate of ATP-hydrolysis in the presence and absence of Mg^{2+}, since specific ATPases require Mg^{2+}, in contrast to acid phosphatase (18, 173), or (*b*) measurement of ATP hydrolysis in the presence of 0.1 mM molybdate which inhibits most although not all of the acid phosphatase activity but not specific ATPases (4, 33, 83, 146, 158, 175). Vacuolar ATPase defined in this way was membrane-bound, in contrast to acid phosphatase (83, 173), and showed a marked specificity for ATP. ATP was hydrolyzed three- to tenfold more rapidly than other nucleotides (83, 146, 173). Some of these other nucleotides might be hydrolyzed not by ATPase but by other enzymes (see below).

Inhibitor sensitivity of tonoplast ATPase   Inhibitors have been employed to distinguish tonoplast ATPase from ATPases of nonvacuolar origin. The ATPase in preparations of vacuoles was found to be insensitive to vanadate (4, 147, 154, 175), an inhibitor of plasmalemma ATPase, to oligomycin (4, 173, 175) and azide (4, 147, 175), inhibitors of mitochondrial ATPase, and to phlorizin (147), an inhibitor of chloroplast ATPase.

A specific characteristic of ATPase obtained from vacuolar preparations is its anion sensitivity: it is inhibited by high concentrations (50 mM) of nitrate (10, 66, 175) and is stimulated by chloride ions (4, 10, 66, 175). These properties are similar to those of vacuolar ATPase from fungi (22, 161) and to those of the ATPase found in membrane preparations from *Hevea* latex (91, 92). The ATPase present in a light membrane fraction from plant microsomes also has very similar properties; the microsomal anion-sensitive, vanadate-insensitive ATPase is therefore called "tonoplast-type" ATPase (149).

Enrichment of specific ATPase activity in the tonoplast   In studies of enzyme localization, it is important to present a balance sheet for the total and specific activities of the enzyme with data for both the total homogenate and for the purified organelle fractions (121). This has not often been done in studies of ATPases in isolated vacuoles (see 14). Balance sheets for nucleotide-specific ATPase have been provided in studies of isolated vacuoles from tobacco (18), red beet (83), and sugarcane (158). ATPase was found to be enriched in the purified vacuoles. However, it was two to four times less enriched than vacuolar markers. This confirmed the expectation that ATPases occurred in other cellular locations as well.

The question arises whether the anion-sensitive ATPase ("tonoplast-type" ATPase) characteristic of vacuoles is located exclusively on the tonoplast and hence provides the long-sought tonoplast marker. Vacuoles isolated from the mesophyll of CAM plants contain particularly high ATPase activities (4, 146, 147). Smith et al (146) presented balance sheets for total protoplast homogenates and isolated vacuoles; they showed that the vanadate- and azide-insensitive ATPase co-purified with the vacuolar marker, malic acid, and thus provided strong evidence that the "tonoplast-type" ATPase is exclusively vacuolar in this tissue. Interestingly, tonoplast-type ATPase increased about fivefold upon induction of CAM metabolism in *Mesembryanthemum*, a facultative CAM plant (148).

Mesophyll cells of CAM plants contain a hypertrophic vacuole but very little Golgi and ER membranes. Recent work indicates that the anion-sensitive, vanadate- and azide-insensitive ATPase occurs in other elements of the endomembrane system in addition to the tonoplast and, therefore, is not suitable as a tonoplast marker in tissues containing well-developed endomembrane systems. Chanson et al (31) obtained evidence for a nitrate-sensitive MgATPase on Golgi membranes. Bennett et al (10) compared microsomes from total homogenates of red beet with membrane vesicles prepared from isolated vacuoles. They found that nitrate-sensitive ATPase yielded a sharp peak at 1.10 g cm^{-3} in sucrose gradients run with membranes from purified vacuoles (10); the same activity showed a much broader distribution in gradients with membrane vesicles obtained from microsomes (10, 120), indicating that nitrate-sensitive ATPase is present on membranes with densities different from the tonoplast.

OTHER PHOSPHOHYDROLASES IN TONOPLAST PREPARATIONS Mg^{2+}-dependent pyrophosphatase has been found in membranes from isolated vacuoles of red beet root (83, 176) and tulip petals (173). No evidence for an enrichment of this enzyme in vacuolar membranes has been presented; however, its high specific activity in the preparations suggests it is localized in the tonoplast and is not the result of contaminating membranes. In contrast, Thom et al (158) found membrane-bound pyrophosphatase in protoplasts but not in vacuoles. Rea & Poole (125) have demonstrated that the pyrophosphatase in a tonoplast-type membrane fraction from red beet microsomes is a proton pump.

Sugarcane vacuoles contain high activities of membrane-bound GTPase and GDPase (154, 158). These activities are not traceable to ATPase since GTP and GDP do not interfere with the hydrolysis of [γ-^{32}P]ATP (158).

ATP-DEPENDENT PROTON PUMPING IN ISOLATED VACUOLES   The ATPases present in sealed microsomal vesicles have been found to function as electrogenic proton pumps (149). ATP-dependent proton pumping has also been studied in isolated vacuoles. Using the anthocyanin pigments within the vacuoles as pH indicators, Wagner & Lin (172) have shown that MgATP can cause acidification of the vacuolar sap. More recently, Mandala & Taiz (90) examined MgATP-dependent acidification in vacuoles isolated from maize coleoptile protoplasts, using [^{14}C]methylamine uptake into vacuole-derived vesicles as an assay. They found that the ATP-dependent proton pumping responded almost exactly to a range of inhibitors as did the proton-pumping activity in a microsomal preparation derived from the whole tissue. Furthermore, they compared proton-pumping activity and a number of organelle markers in homogenates from protoplasts and vacuoles and found that ATP-dependent methylamine uptake was exclusively vacuolar, like α-mannosidase. Thus, ATP-dependent methylamine uptake in microsomes, described earlier by Mettler et al (106), can be used as a tonoplast marker in corn coleoptiles (90).

The effect of ATP on the electrical membrane potential difference ($\Delta\Psi$) of isolated vacuoles was studied with permeant chemical probes. ATP induced a shift of $\Delta\Psi$ to more positive values inside, indicating that the ATP-dependent proton pump was electrogenic (107, 154, 155). Jochem et al (66) measured $\Delta\Psi$ and pH by impaling isolated vacuoles with microelectrodes. A statistical comparison of measurements with isolated vacuoles in the absence and in the presence of MgATP was performed. The mean $\Delta\Psi$ increased from + 9 to + 18 mV upon addition of ATP, and the mean pH decreased from 5.4 to 4.3. Both changes are consistent with the presence of an inwardly directed, electrogenic proton pump on the tonoplast. Little is known about the regulation of the proton pump. In microsomal vesicles from corn (124) and red beet (9a), it was recently found that the tonoplast-type ATPase was strongly inhibited by ADP (124).

PROTON MOTIVE FORCE IN ISOLATED VACUOLES   The shifts of $\Delta\Psi$ and ΔpH induced by ATP indicate that ATP can establish an electrochemical proton gradient (PMF) in isolated vacuoles and thereby energize the tonoplast. It is noteworthy that a PMF often persists in isolated vacuoles for a considerable time even in the absence of any exogenous energy source. Isolated vacuoles maintain the stored solutes, including organic acids (see above). Therefore, the pH inside the vacuole is often (66, 76, 109, 180) although not always (111) one or two units below that of the external incubation medium. Under the same conditions, the $\Delta\Psi$ across the tonoplast, measured with impaled electrodes (7, 66) or certain chemical probes (155), is close to 0 mV [see (50) for a critical review of the methodology for measurements of the electrical membrane potential in isolated vacuoles]. Hence, the isolated vacuoles maintain an electrochemical proton gradient and remain in a partly energized state. When a protonophore (uncoupler) is added, e.g. CCCP or FCCP, then the PMF collapses, but the ΔpH is still maintained as long as the organic acids remain in the vacuole and the membrane potential approaches a proton diffusion potential (155, 180).

Mechanisms of Transport and Accumulation in Vacuoles

SPECIFIC TRANSPORT SYSTEMS OF THE TONOPLAST   Isolated vacuoles have been found to possess transport systems for a variety of substances. There are specific transport systems for the glucose analog 3-O-methylglucose (55, 156) and for sucrose (36, 68, 75, 156, 186).

A permease for malate has been identified in *Bryophyllum* vacuoles (27). This transport system was quite specific for malate, had a K_m of 1 mM, and was not affected by ATP and uncouplers, indicating that it catalyzed a passive exchange of malate through the tonoplast. A transport system for malate was also found in barley vacuoles (95); there, ATP increased the rate of malate uptake, indicating an active transport mechanism.

Alkaloid transport has been investigated using vacuoles from cell cultures established from numerous alkaloid-containing plants (34, 35). Using very small external concentrations of radioactive alkaloids (0.4 μM), it was shown that rapid transport occurred only with the species-specific alkaloids. For example, nicotine was taken up by *Nicotiana* and *Datura* vacuoles but not by *Catharanthus* vacuoles, while the opposite was true for catharanthine (34). These findings are surprising since nicotine has been successfully used as a membrane-permeant probe to estimate vacuolar pH in protoplasts of *Acer* (77). A transport system for coumaroyl glucosides has been identified in vacuoles of barley mesophyll cells (181). Barley protoplasts do not normally contain coumaroyl glucosides but rapidly accumulate them in the vacuoles when fed with coumarins like scopoletin or esculetin (181).

Transport of inorganic ions has been studied mainly with fractions of microsomal vesicles thought to be derived from the tonoplast. The results indicate that the tonoplast contains transport systems catalyzing Ca^{2+}/H^+ antiport (56, 122), Na^+/H^+ antiport (12), Cl^- transport (11, 13), and NO_3^- transport (13). Isolated barley vacuoles lose more than 50% of the Cl^- and NO_3^- within 30 min of incubation (E. Martinoia, personal communication), indicating that there is an efficient transport system for these ions.

ACTIVE TRANSPORT IN ISOLATED VACUOLES?   Several workers have tried to demonstrate active transport in isolated plant vacuoles (34–36, 55, 75, 95, 181). The criteria generally used to define active transport were (*a*) stimulation by ATP and (*b*) inhibition by ionophores. Thom & Komor (153) obtained evidence that uptake of 3-O-methylglucose in isolated vacuoles was accompanied by a transient acidification of the incubation medium, indicating a H^+-sugar antiport as the mechanism for sugar uptake.

Although the available evidence is consistent with the concept of PMF-driven, secondary active transport systems in the tonoplast, most of the studies have not demonstrated rigorously that transport is active, i.e. that net accumulation occurs. A major obstacle is that isolated vacuoles generally contain large amounts of the substances being investigated. This makes it difficult to establish net accumulation. Radioactive tracer accumulating within the vacuole might reflect either net accumulation or exchange with preexisting unlabeled vacuolar compounds by way of a passive transport system.

Arginine transport in isolated vacuoles provides a good example of this problem (15). Isolated yeast vacuoles contained ca 200 mM arginine in their interior. A highly active arginine transport system catalyzes uptake of [^{14}C]arginine (17). Prelabeling of the vacuoles with [^{3}H]arginine has demonstrated that the [^{14}C]arginine is taken up strictly in exchange against the [^{3}H]arginine already accumulated (17). Competition experiments have shown that [^{14}C]lysine is taken up by the same transport system. Net accumulation of lysine can be observed readily in the absence of any exogenous energy source. However, [^{14}C]lysine uptake is accompanied with a stoichiometric efflux of [^{3}H]arginine in prelabeled vacuoles (T. Boller, unpublished). Hence, lysine "accumulation" is in fact a passive exchange transport of basic amino acids, not active transport. Arginine uptake is strongly stimulated by ATP or NADPH; however, double-labeling experiments have shown that the stimulation of uptake is balanced by a stimulation of efflux, indicating that ATP and NADPH enhance exchange transport but do not cause active transport, i.e. net accumulation (15).

Active, ATP-dependent transport of arginine and other amino acids has been demonstrated in membrane vesicles isolated from yeast vacuoles (115, 136). These vesicles were devoid of the internal pool of arginine. They took up arginine only upon addition of ATP, apparently by an arginine/H$^+$ antiport mechanism; when energization was blocked by the addition of an ionophore, the accumulated arginine leaked out again (115). The active transport system for arginine has about a twentyfold lower affinity and a tenfold lower V_{max} on a protein basis than the passive transport system described earlier (17). Thus, it would be very difficult to obtain evidence for the active transport system in intact vacuoles, where the passive transport system takes up labeled arginine much more rapidly. On the other hand, it obviously proved to be difficult to discover the passive exchange transport system with the isolated tonoplast vesicles. The "arginine/histidine exchange system" described by Sato et al (137), using rather involved experiments, resembles the passive transport system described earlier (17). In conclusion, isolated vacuoles appear to be well suited for detecting passive transport systems whereas membrane vesicles are better suited for detecting active transport systems.

It is likely that tonoplast vesicles will also provide an attractive model system to study active transport in higher plant vacuoles. In fact, secondary active transport of citrate driven by ATP has already been demonstrated in vesicles prepared from membranes of the lutoids from *Hevea* latex (94), often considered to represent a model for the tonoplast (93). Recently, putative tonoplast vesicles isolated from sugarbeet root have been used to provide evidence for a H^+/sucrose antiport system (24a).

TRANPORT BY GROUP TRANSLOCATION?   In bacteria, uptake of sugars is often obligately coupled to phosphorylation. Similar group translocation systems have been proposed for transport into the vacuole.

Many secondary compounds accumulate in the vacuoles as glycosides (see above). It has been postulated that the corresponding glycosyl transferases are associated with the tonoplast and act as group translocators (44). Current evidence does not support this hypothesis. Cell fractionation studies have shown that the glycosyl transferases in various biosynthetic pathways are localized in the cytosol, not in the tonoplast (63, 64, 123, 181). In some cases, flavonoyl glycosides are further derivatized by malonylation (100) and methylation (67). The corresponding transferases have been shown to be soluble and not associated with the vacuole (67, 100).

It has also been hypothesized that accumulation of sucrose in vacuoles depends on a group translocation mechanism (25, 157). The results of a study of sugar uptake in grape pericarp tissue have been interpreted to indicate the existence of a multienzyme complex in the tonoplast, containing all the enzymes needed for synthesis of sucrose from glucose, which also translocates sucrose (25). Recently, isolated vacuoles from sugarcane have been found to take up radioactivity from UDP-[^{14}C]glucose (157). The radioactivity comigrates with sucrose and sucrose phosphate in thin-layer chromatograms. The authors postulate (but do not prove experimentally) that the radioactive compounds are sucrose and sucrose phosphate labeled in both hexoses. They suggest that the five enzymes required to synthesize sucrose from UDP-glucose form a tonoplast-bound multienzyme complex that acts as a group translocator (157). A simpler hypothesis would have been that the tonoplast would contain a transport system for sucrose phosphate but not for sucrose; then, hydrolysis of the transported sucrose phosphate in the vacuole (e.g. by acid phosphatase) would lead to accumulation of sucrose without an active transport system (51). However, isolated sugarcane vacuoles do not take up sucrose phosphate (157).

ACCUMULATION OF VACUOLAR COMPOUNDS BY TRAPPING MECHANISMS   Vacuoles with their acidic pH can accumulate weak bases by an ion-trap mechanism. Compounds such as neutral red, methylamine, or 9-aminoacridine pass the tonoplast as the lipophilic free base. In the vacuoles, they are protonated and trapped as cations for which the membrane is impermeable (111, 180). Alkaloids are also thought to accumulate in vacuoles by an ion-trap mechanism, but recent results on alkaloid uptake in isolated vacuoles contradict this (35). Ions can also be accumulated in vacuoles by precipitation as insoluble salts such as calcium oxalate (41) or by association with charged substances of high molecular weight such as tannins or polyphosphate that act as ion exchangers (102).

Trapping could also occur by conformational changes of the transported substances. Matern et al (99) have proposed that apigenin 7-O-(6-O-malonyl-glucoside), a vacuolar pigment from parsley, is trapped inside the vacuoles by a conformational change induced by lowered pH values. Rataboul et al (123) have provided evidence for a trapping of o-coumaric acid glucoside in vacuoles of sweet clover. This compound is initially synthesized as the *trans* isomer in the cytoplasm but accumulates as the *cis* isomer in the vacuoles. Experiments have shown that tonoplast vesicles prepared from isolated vacuoles are readily permeable to the *trans* form but impermeable to the *cis* form. This suggests that, in vivo, the glycoside enters the vacuole in the *trans* form and is trapped there after its nonenzymatic conversion into the *cis* form.

Homeostatic Functions of Vacuolar Compartmentation

In the following, an attempt is made to understand how the transport systems demonstrated in isolated vacuoles function in cytoplasmic homeostasis, i.e. the maintenance of the delicate dynamic equilibria of the cytoplasm in the face of environmental changes.

CYTOPLASMIC PROTON CONCENTRATION   Cytoplasmic pH is strictly controlled in plants (145). The vacuole, which has proton pumps and often contains large amounts of organic acids, plays an important role in pH homeostasis. Recently, ^{31}P NMR spectroscopy has emerged as a powerful technique for estimating cytoplasmic and vacuolar pH in situ (130). The method relies on a pH-dependent shift of the resonance peak of phosphate (and other phosphory-lated compounds). In plant tissues, generally, two peaks are seen for phosphate, one for phosphate in a more acidic environment taken to represent the vacuole, and a second one in an environment slightly above neutral taken to represent the cytoplasm. The shift of these peaks can be used to estimate vacuolar and cytoplasmic pH, respectively.

Evidence for a role of the vacuole in cytoplasmic pH homeostasis has come from ^{31}P NMR studies of cells incubated in buffers of different pH values. The vacuolar pH decreased with decreasing external pH, while the cytoplasmic pH remained constant (160, 189). This indicates that, at low external pH, an increased influx of protons into the cytoplasm is compensated for by an increased transport of protons into the vacuole. Similarly, in cell cultures subjected to anaerobic stress, which causes excess proton production in the cytoplasm, there is a parallel decrease of vacuolar and cytoplasmic pH (189). In contrast, in root tips, anaerobiosis causes a rapid decline in cytoplasmic pH without a concomitant change in vacuolar pH, indicating that vacuoles do not take up the excess protons produced in the cytoplasm (131, 132). Later, the vacuolar pH increases, indicating that the vacuole releases protons and aggravates cytoplasmic acidosis (131, 132); under severe stress, the tonoplast appears to lose its capacity to retain protons, a condition that might spell death to the cell (131).

OSMOTIC ADJUSTMENT   The most dramatic example of osmotic regulation occurs in guard cells, where the opening mechanism involves rapid intracellular accumulation of K^+ and Cl^- or malate in guard cells. Current models for stomatal opening stress the importance of ion pumps at the plasmalemma (192). It is obvious, however, that ions taken up by the guard cell flow further into the vacuole, since the vacuolar volume increases markedly during stomatal opening. Interestingly, isolated guard cell vacuoles rapidly swell when incubated in the presence of K^+ (143).

In stomata, both osmotic potential and turgor change rapidly in response to environmental stimuli. Many plant cells have the capacity to keep turgor constant by osmotic adjustment even when the water potential in the environment changes. For example, rhythmic day-night variations in leaf water potential can be compensated for by a rhythmic change in the osmotic potential of the cell sap, leaving turgor pressure constant. In various plants, the daily adjustment is equivalent to the accumulation of several 100 mM of solutes, mainly in the form of sugars and amino acids (108). Large amounts of sugars and amino acids have also been found to accumulate in cultured plant cells exposed to water stress (57). Although most of these solutes are vacuolar, it is not known to what extent transport systems at the tonoplast are actively involved in this type of regulation.

Osmotic adjustment is also important in plants subjected to salt stress. Halophytes use abundantly available inorganic ions to build up a high solute potential. These ions must be strictly compartmented in the vacuole to prevent damage to the cytoplasm (191). Studies with isolated vacuoles, e.g. from *Suaeda* (37), will show whether the tonoplast of halophytes is equipped with special pumps for Na^+ and Cl^-. Glycophytes, in contrast, tend to exclude an excess of inorganic ions in the roots; they may need to build up a sufficiently high solute potential by the accumulation of organic molecules in the vacuoles (191; see 135 for a recent example).

ACCUMULATION AND MOBILIZATION OF ORGANIC ACIDS IN CAM PLANTS
The accumulation of large amounts of organic acids during the night and their remobilization during the day provides another impressive example of the dynamics of vacuolar compartmentation (159). Recent studies with isolated vacuoles have verified the existence of a tonoplast ATPase, predicted earlier on the basis of indirect evidence (87). It remains to be shown whether this ATPase

(4, 66, 146, 147) indeed energizes the uptake of malate ions, and how its activity is regulated during the rhythmic changes between accumulation and mobilization of malic acid.

HOMEOSTASIS OF CYTOPLASMIC INORGANIC IONS   The concentration of ions in the cytoplasm is often strictly controlled. An extreme example is Ca^{2+} which is present in the cytoplasm at concentrations in the micromolar range; in contrast, Ca^{2+} generally occurs in the vacuoles in concentrations around 10 mM (88). Furthermore, in many plants, large amounts of Ca^{2+} are deposited as calcium oxalate in the vacuoles of specialized idioblasts (41). Little is known about the role of tonoplast transport systems in cytoplasmic Ca^{2+} homeostasis (88). Recently, a Ca^{2+}-calmodulin-dependent protein kinase activity has been discovered in tonoplast preparations obtained from isolated *Acer* vacuoles (151). The enzyme rapidly phosphorylates a restricted set of tonoplast proteins. It is intriguing to speculate that calcium transport systems involved in Ca^{2+} homeostasis are regulated by reversible phosphorylation.

Another example is regulation of phosphate concentration. Results obtained with ^{31}P NMR spectroscopy indicate that cytoplasmic phosphate concentration is strictly controlled by movement of phosphate in and out of the vacuole. For example, addition of mannose to maize root tips (74) or wheat leaf tissue (179) causes rapid synthesis and accumulation of mannose-6-phosphate in the cytoplasm. Nevertheless, cytoplasmic phosphate concentrations remain constant while the vacuolar phosphate concentration decreases rapidly (74, 179). Similarly, when cultured *Acer* cells are transferred to a phosphate-deficient medium, they appear to lose vacuolar phosphate while maintaining cytoplasmic phosphate concentration (126). When these cells are transferred into a phosphate-rich medium, the vacuolar pool expands without change in the cytoplasmic phosphate concentration (126). Vacuolar phosphate is also accumulated in sucrose-starved cells (127).

INTRACELLULAR PARTITIONING OF NEWLY ASSIMILATED PHOTOSYNTHETIC PRODUCTS   Studies of the intracellular distribution of newly assimilated carbon provide further examples of dynamic vacuolar compartmentation. Tracer flux kinetics (48) and direct microdissection (38) of photosynthetic tissue after pulse-labeling with $^{14}CO_2$ have indicated that newly formed sucrose occurs in two pools, namely, a cytoplasmic pool that is rapidly exported to the phloem, and a vacuolar pool that is slowly exported.

Isolated protoplasts have permitted direct analysis of the role of vacuoles in the distribution of newly fixed carbon. Mesophyll protoplasts were incubated with $H^{14}CO_3^-$ in the light for a short period of time, the vacuoles rapidly isolated, and the distribution of [^{14}C]sucrose and other assimilates between the vacuolar and the extravacuolar compartment measured (5, 16, 69). In barley, newly assimilated sucrose rapidly entered the vacuole (69), whereas newly fixed sucrose was excluded from the vacuole in sweet clover (16) and in spinach (5). Newly formed organic acids rapidly accumulated in the vacuoles of all three species (5, 16, 69).

Intracellular transport studies with protoplasts have a major limitation (15): metabolism and, in particular, regulation of intracellular compartmentation might differ considerably between cells and protoplasts. For example, isolated mesophyll protoplasts do not export the assimilates (16, 69). Therefore, it will be useful to combine studies that allow direct analysis of vacuoles with more indirect techniques such as tracer-flux kinetics or the recently developed method for nonaqueous fractionation (49).

Metabolic Processes Involving the Vacuoles

The static image of vacuoles has led to the notion that vacuoles are metabolically inert (170). Recent data indicate, however, that the tonoplast and the vacuolar sap participate in specific metabolic pathways.

ETHYLENE BIOSYNTHESIS   Guy & Kende (54) have obtained evidence that the last step of ethylene biosynthesis, the conversion of 1-aminocyclopropane-1-carboxylic acid (ACC) to ethylene, takes place in the vacuole. Isolated vacuoles contained most of the ACC present in the protoplasts and converted the stored ACC to ethylene. The ethylene-forming system in isolated vacuoles was similar to that of the intact plant with regard to stereospecificity and inhibitors (54).

SACCHARIDE BIOSYNTHESIS AND MOBILIZATION   The vacuolar compartmentation and transformation of saccharides has been reviewed (71). The first step in the mobilization of stored sucrose in beet root (81) and of stored stachyose in Stachys tubers (70) occurs in the vacuole. Vacuoles isolated from barley leaf tissue (174) and from storage roots of Jerusalem artichoke (43) have been found to contain both the fructans involved in temporary carbohydrate storage and the enzymes necessary to form and mobilize the fructans.

Yamaki (190) has reported that sorbitol oxidase from apple cotyledons is partially located in vacuoles, and has suggested that release of sorbitol from the vacuole where it is stored is coupled to its oxidation to glucose. Unfortunately, the enrichment of sorbitol oxidase in the vacuolar preparation has not been adequately documented (190).

TRANSFORMATION OF SECONDARY METABOLITES   The transformation of O-sinapoyl glucose to O-sinapoyl malate in radish cotyledons during seedling development appears to take place in the vacuoles since isolated vacuoles from this tissue contain both the sinapoyl esters and the enzyme catalyzing their conversion, 1-sinapoylglucose: malate sinapoyltransferase (144). Similarly, in radish leaves, vacuoles contain a series of related hydroxycinnomoyl-malic acid esters (*p*-coumaroyl-, caffeoyl-, and feruloylmalate) and the enzyme(s) catalyzing their formation from the respective acyl glucosides and malic acid (147a).

CONCLUSION

It is clear that the study of isolated vacuoles has greatly advanced our understanding of vacuolar compartmentation. In particular, it has permitted the unambiguous localization of many enzymes in vacuoles and the characterization of numerous transport systems in the tonoplast. Undoubtedly, the regulation of these enzymes and transport systems plays a central role for cytoplasmic homeostasis. Isolated vacuoles provide an excellent system to investigate the regulatory mechanisms involved. By gentle techniques, vacuoles may be isolated essentially without disturbing their membranes and without causing the release of the potentially dangerous secondary compounds and hydrolases stored in their interior. Isolated vacuoles expose the cytoplasmic surface to the incubation medium. Hence, it is easy to apply and remove the cytosolic effectors suspected of playing a role in the regulation of compartmentation. It is one limitation of this experimental system that isolated vacuoles are no longer integrated into the cytoplasm and into the flow of the endomembrane system and lack part of the dynamics of vacuoles in living cells. Nevertheless, work with isolated vacuoles should continue to yield new information about the dynamics of vacuolar compartmentation.

ACKNOWLEDGMENTS

We wish to thank Gilbert Alibert, Enrico Martinoia, and Philippe Matile for helpful comments and Fred Meins, Jr. for critically reading the manuscript.

Literature Cited

1. Achstetter, T., Wolf, D. H. 1985. Proteinases, proteolysis and biological control in the yeast *Saccharomyces cerevisiae*. *Yeast* 1:000–00
2. Akazawa, T., Hara-Nishimura, I. 1985. Topographic aspects of biosynthesis, extracellular secretion, and intracellular storage of proteins in plant cells. *Ann. Rev. Plant Physiol.* 36:441–72
3. Alibert, G., Carrasco, A., Boudet, A. M. 1982. Changes in biochemical composition of vacuoles isolated from *Acer pseudoplatanus* L. during cell culture. *Biochim. Biophys. Acta* 721:22–29
4. Aoki, K., Nishida, K. 1984. ATPase

activity associated with vacuoles and tonoplast vesicles isolated from the CAM plant, *Kalanchoë daigremontiana*. *Physiol. Plant.* 60:21–25
5. Asami, S., Hara-Nishimura, I., Nishimura, M., Akazawa, T. 1985. Translocation of photosynthates into vacuoles in spinach leaf protoplasts. *Plant Physiol.* 77:963–68
6. Bal, A. K. 1985. Vacuolation and infection thread in root nodules of soybean. *Cytobios* 42:41–47
7. Barbier-Brygoo, H., Romieu, C., Grouzis, J. P., Gibrat, R., Grignon, C., Guern, J. 1984. Evidence for the con-

tribution of surface potential to the trans-tonoplast potential difference measured on isolated vacuoles with microelectrodes. *Z. Pflanzenphysiol.* 114:215–19

8. Barz, W., Köster, J. 1981. Turnover and degradation of secondary (natural) products. In *Biochemistry of Plants: Secondary Plant Products,* ed. E. E. Conn, 7:35–84. New York: Academic. 798 pp.

9. Baumgartner, B., Tokuyasu, K. T., Chrispeels, M. J. 1978. Localization of vicilin peptidohydrolase in the cotyledons of mung bean seedlings by immunofluorescence microscopy. *J. Cell Biol.* 79:10–19

9a. Bennett, A. B., O'Neill, S. D., Eilmann, M., Spanswick, R. M. 1985. H⁺-ATPase activity from storage tissue of *Beta vulgaris.* III. Modulation of ATPase activity by reaction substrates and products. *Plant Physiol.* 78:495–99

10. Bennett, A. B., O'Neill, S. D., Spanswick, R. M. 1984. H⁺-ATPase activity from storage tissue of *Beta vulgaris.* I. Identification and characterization of an anion-sensitive H⁺-ATPase. *Plant Physiol.* 74:538–44

11. Bennett, A. B., Spanswick, R. M. 1983. Optical measurements of ΔpH and ΔΨ in corn root membrane vesicles: Kinetic analysis of Cl⁻ effects on a proton-translocating ATPase. *J. Membr. Biol.* 71:95–107

11a. Blumwald, E., Fortin, M. G., Rea, P. A., Verma, D. P. S., Poole, R. J. 1985. Presence of host-plasma membrane type H⁺-ATPase in the membrane envelope enclosing the bacteroids in soybean root nodule. *Plant Physiol.* 78:665–72

12. Blumwald, E., Poole, R. J. 1985. Na⁺/H⁺ antiport in isolated tonoplast vesicles from storage tissue of *Beta vulgaris. Plant Physiol.* 78:163–67

13. Blumwald, E., Poole, R. J. 1985. Nitrate storage and retrieval in *Beta vulgaris:* Effects of nitrate and chloride on proton gradients in tonoplast vesicles. *Proc. Natl. Acad. Sci. USA* 82:3683–87

14. Boller, T. 1982. Enzymatic equipment of plant vacuoles. *Physiol. Veg.* 20:247–57

15. Boller, T. 1985. Intracellular transport of metabolites in protoplasts: Transport between cytosol and vacuole. In *Physiological Properties of Plant Protoplasts,* ed. P. E. Pilet, pp. 76–86. Berlin: Springer-Verlag. 283 pp.

16. Boller, T., Alibert, G. 1983. Photosynthesis in protoplasts from *Melilotus alba:* Distribution of products between vacuole and cytosol. *Z. Pflanzenphysiol.* 110:231–38

17. Boller, T., Dürr, M., Wiemken, A.

1975. Characterization of a specific transport system for arginine in isolated yeast vacuoles. *Eur. J. Biochem.* 54:81–91

18. Boller, T., Kende, H. 1979. Hydrolytic enzymes in the central vacuole of plant cells. *Plant Physiol.* 63:1123–32

19. Boller, T., Vögeli, U. 1984. Vacuolar localization of ethylene-induced chitinase in bean leaves. *Plant Physiol.* 74:442–44

20. Boudet, A. M., Alibert, G., Marigo, G. 1984. Vacuoles and tonoplast in the regulation of cellular metabolism. In *Membranes and Compartmentation of Plant Functions,* ed. A. M. Boudet, G. Marigo, P. Lea, pp. 29–47. Oxford: Oxford Press

21. Boudet, A. M., Canut, H., Alibert, G. 1981. Isolation and characterization of vacuoles from *Melilotus alba* mesophyll. *Plant Physiol.* 68:1354–58

22. Bowman, E. J., Bowman, B. J. 1982. Identification and properties of an ATPase in vacuolar membranes of *Neurospora crassa. J. Bacteriol.* 151:1326–37

23. Brewin, N. J., Robertson, J. G., Wood, E. A., Wells, B., Larkins, A. P., et al. 1985. Monoclonal antibodies to antigens on the peribacteroid membrane from *Rhizobium*-induced root nodules of pea cross-react with plasma membranes and Golgi bodies. *EMBO J.* 4:605–11

24. Briskin, D. P., Leonard, R. T. 1980. Isolation of tonoplast vesicles from tobacco protoplasts. *Plant Physiol.* 66:684–87

24a. Briskin, D. P., Thornley, W. R., Wyse, R. E. 1985. Membrane transport in isolated vesicles from sugarbeet taproot. II. Evidence for a sucrose/H⁺-antiport. *Plant Physiol.* 78:871–75

25. Brown, S. C., Coombe, B. G. 1982. Sugar transport by an enzyme complex at the tonoplast of grape pericarp cells? *Naturwissenschaften* 69:43–45

26. Burgess, J., Lawrence, W. 1985. Studies of the recovery of tobacco mesophyll protoplasts from an evacuolation treatment. *Protoplasma* 126:140–46

27. Buser-Suter, C., Wiemken, A., Matile, P. 1982. A malic acid permease in isolated vacuoles of a crassulacean acid metabolism plant. *Plant Physiol.* 69:456–59

28. Campbell, N. A., Garber, R. C. 1980. Vacuolar reorganization in the motor cells of *Albizzia* during leaf movement. *Planta* 148:251–55

29. Canut, H., Alibert, G., Boudet, A. M. 1985. Proteases of *Melilotus alba* mesophyll protoplasts. I. Intracellular localization. *Plant Sci.* 39:163–69

30. Canut, H., Alibert, G., Boudet, A. M. 1985. Hydrolysis of intracellular proteins in vacuoles isolated from *Acer pseudoplatanus* cells. *Plant Physiol.* 79: In press

31. Chanson, A., McNaughton, E., Taiz, L. 1984. Evidence for a KC1-stimulated, Mg^{2+}-ATPase on the Golgi of corn coleoptiles. *Plant Physiol.* 76:498–507

32. Creek, K. E., Sly, W. S. 1984. The role of the phosphomannosyl receptor in the transport of acid hydrolases to lysosomes. In *Lysosomes in Biology and Pathology*, ed. J. T. Dingle, R. T. Dean, W. S. Sly, 7:63–82. New York: Elsevier

33. D'Auzac, J. 1975. Caractérisation d'une ATP-ase membranaire en présence d'une phosphatase acide dans les lutoïdes du latex d'*Hevea brasiliensis*. *Phytochemistry* 14:671–75

34. Deus-Neumann, B., Zenk, M. H. 1984. A highly selective alkaloid uptake system in vacuoles of higher plants. *Planta* 162:250–60

35. Deus-Neumann, B., Zenk, M. H. 1985. Accumulation of alkaloids in plant vacuoles does not involve an ion-trap mechanism. *Planta* 166

36. Doll, S., Rodier, F., Willenbrink, J. 1979. Accumulation of sucrose in vacuoles isolated from red beet tissue. *Planta* 144:407–11

37. Dracup, M. N. H., Greenway, H. 1985. A procedure for isolating vacuoles from leaves of the halophyte *Suaeda maritima*. *Plant Cell Environ.* 8:149–54

38. Fisher, D. B., Outlaw, W. H. 1979. Sucrose compartmentation in the palisade parenchyma of *Vicia faba* L. *Plant Physiol.* 64:481–83

39. Franceschi, V. R., Giaquinta, R. T. 1983. The paraveinal mesophyll of soybean leaves in relation to assimilate transfer and compartmentation. I. Ultrastructure and histochemistry during vegetative development. *Planta* 157:411–21

40. Franceschi, V. R., Giaquinta, R. T. 1983. The paraveinal mesophyll of soybean leaves in relation to assimilate transfer and compartmentation. II. Structural, metabolic and compartmental changes during reproductive growth. *Planta* 157:422–31

41. Franceschi, V. R., Horner, H. T. 1980. Calcium oxalate crystals in plants. *Bot. Rev.* 46:361–427

42. Franceschi, V. R., Wittenbach, V. A., Giaquinta, R. T. 1983. Paraveinal mesophyll of soybean leaves in relation to assimilate transfer and compartmentation III. Immunohistochemical localization of specific glycopeptides in the

vacuole after depodding. *Plant Physiol.* 72:586–89

43. Frehner, M., Keller, F., Wiemken, A. 1984. Localization of fructan metabolism in the vacuoles isolated from protoplasts of Jerusalem-artichoke tubers (*Helianthus tuberosus* L.). *J. Plant Physiol.* 116:197–208

44. Fritsch, H. J., Grisebach, H. 1975. Biosynthesis of cyanidin in cell cultures of *Haplopappus gracilis*. *Phytochemistry* 14:2437–42

45. Garcia-Martinez, J. L., Ohlrogge, J. B., Rappaport, L. 1981. Differential compartmentation of gibberellin A_1 and its metabolites in vacuoles of cowpea and barley leaves. *Plant Physiol.* 68:865–67

46. Gaudreault, P. R., Beevers, L. 1983. Glycoprotein nature of glycosidases from leaves of *Pisum sativum* L. *J. Exp. Bot.* 34:1145–54

47. Gaudreault, P. R., Beevers, L. 1984. Protein bodies and vacuoles as lysosomes. Investigations into the role of mannose-6-phosphate in intracellular transport of glycosidases in pea cotyledons. *Plant Physiol.* 76:228–32

48. Geiger, D. R., Ploeger, B. J., Fox, T. C., Fondy, B. R. 1983. Sources of sucrose translocated from illuminated sugar beet source leaves. *Plant Physiol.* 72:964–70

49. Gerhardt, R., Heldt, H. W. 1984. Measurement of subcellular metabolite levels in leaves by fractionation of freeze-stopped material in nonaqueous media. *Plant Physiol.* 75:542–47

50. Gibrat, R., Barbier-Brygoo, H., Guern, J., Grignon, C. 1985. Transtonoplast potential difference and surface potential of isolated vacuoles. See Ref. 93, pp. 83–97

51. Glasziou, K. T., Gayler, K. R. 1972. Storage of sugars in stalks of sugar cane. *Bot. Rev.* 38:471–90

52. Glund, K., Tewes, A., Abel, S., Leinhos, V., Walther, R., Reinbothe, H. 1984. Vacuoles from cell suspension cultures of tomato (*Lycopersicon esculentum*)—isolation and characterization. *Z. Pflanzenphysiol.* 113:151–61

53. Greenwood, J., Chrispeels, M. J. 1985. Immunocytochemical localization of phaseolin and phytohemagglutinin in the endoplasmic reticulum and Golgi complex of developing bean cotyledons. *Planta* 164:295–302

54. Guy, M., Kende, H. 1984. Conversion of 1-aminocyclopropane-1-carboxylic acid to ethylene by isolated vacuoles of *Pisum sativum* L. *Planta* 160:281–87

55. Guy, M., Reinhold, L., Michaeli, D.

1979. Direct evidence for a sugar transport mechanism in isolated vacuoles. *Plant Physiol.* 64:61–64

56. Hager, A., Hermsdorf, P. 1981. A H⁺/Ca²⁺ antiporter in membranes of microsomal vesicles from maize coleoptiles, a secondary energized Ca^{2+} pump. *Z. Naturforsch. Teil C* 36:1009–12

57. Handa, S., Bressan, R. A., Handa, A. K., Carpita, N. C., Hasegawa, P. M. 1983. Solutes contributing to osmotic adjustment in cultured plant cells adapted to water stress. *Plant Physiol.* 73:834–43

58. Harder, D. E., Chong, J. 1984. Structure and physiology of haustoria. In *The Cereal Rusts*, ed. W. R. Bushnell, A. P. Roelfs, 1:431–76. Orlando: Academic. 546 pp.

59. Heck, U., Martinoia, E., Matile, P. 1981. Subcellular localization of acid proteinase in barley mesophyll protoplasts. *Planta* 151:198–200

60. Henry, Y., Steer, M. W. 1985. Acid phosphatase localization in the digestive glands of *Dionaea muscipula* Ellis flytraps. *J. Histochem. Cytochem.* 33:339–44

61. Higgins, T. J. V. 1984. Synthesis and regulation of major proteins in seeds. *Ann. Rev. Plant Physiol.* 35:191–221

62. Holländer-Czytko, H., Andersen, J. K., Ryan, C. A. 1985. Vacuolar localization of wound-induced carboxypeptidase inhibitor in potato leaves. *Plant Physiol.* 78:76–79

63. Hrazdina, G., Marx, G. A., Hoch, H. C. 1982. Distribution of secondary plant metabolites and their biosynthetic enzymes in pea (*Pisum sativum* L.) leaves. *Plant Physiol.* 70:745–48

64. Hrazdina, G., Wagner, G. J., Siegelman, H. W. 1978. Subcellular localization of enzymes of anthocyanin biosynthesis in protoplasts. *Phytochemistry* 17:53–56

65. Joachim, S., Robinson, D. G. 1984. Endocytosis of cationic ferritin by bean leaf protoplasts. *Eur. J. Cell. Biol.* 34:212–16

66. Jochem, P., Rona, J. P., Smith, J. A. C., Lüttge, U. 1984. Anion-sensitive ATPase activity and proton transport in isolated vacuoles of species of the CAM genus *Kalanchoë*. *Physiol. Plant.* 62:410–15

67. Jonsson, L. M. V., Donker-Koopman, W. E., Uitslager, P., Schram, A. W. 1983. Subcellular localization of anthocyanin methyltransferase in flowers of *Petunia hybrida*. *Plant Physiol.* 72:287–90

68. Kaiser, G., Heber, U. 1984. Sucrose transport into vacuoles isolated from barley mesophyll protoplasts. *Planta* 161:562–68

69. Kaiser, G., Martinoia, E., Wiemken, A. 1982. Rapid appearance of photosynthetic products in the vacuoles isolated from barley mesophyll protoplasts by a new fast method. *Z. Pflanzenphysiol.* 107:103–13

70. Keller, F., Matile, P. 1985. The role of the vacuole in storage and mobilization of stachyose in tubers of *Stachys sieboldii*. *J. Plant Physiol.* 119:369–80

71. Keller, F., Matile, P., Wiemken, A. 1985. Distribution of saccharides between cytoplasm and vacuole in protoplasts. See Ref. 15, pp. 116–21

72. Keller, F., Wiemken, A. 1982. Differential compartmentation of sucrose and gentianose in the cytosol and vacuoles of storage root protoplasts from *Gentiana lutea* L. *Plant Cell Rep.* 1:274–77

73. Kenyon, W. H., Severson, R. F., Black, C. C. 1985. Maintenance carbon cycle in Crassulacean acid metabolism plant leaves. Source and compartmentation of carbon for nocturnal malate synthesis. *Plant Physiol.* 77:183–89

74. Kime, M. J., Ratcliffe, R. G., Loughman, B. C. 1982. The application of ³¹P nuclear magnetic resonance to higher plant tissue. II. Detection of intracellular changes. *J. Exp. Bot.* 33:670–81

75. Knuth, M. E., Keith, B., Clark, C., Garcia-Martinez, J. L., Rappaport, L. 1983. Stabilization and transport capacity of cowpea and barley vacuoles. *Plant Cell Physiol.* 24:423–32

76. Komor, E., Thom, M., Maretzki, A. 1982. Vacuoles from sugarcane suspension cultures. III. Protonmotive potential difference. *Plant Physiol.* 69:1326–30

77. Kurkdjian, A., Morot-Gaudry, J. F., Wuilleme, S., Lamant, A., Jolivet, E., Guern, J. 1981. Evidence for an action of fusicoccin on the vacuolar pH of *Acer pseudoplatanus* cells in suspension culture. *Plant Sci. Lett.* 23:233–43

78. Leigh, R. A. 1979. Do plant vacuoles degrade cytoplasmic components? *Trends Biochem. Sci.* 4:N37–38

79. Leigh, R. A. 1983. Methods, progress and potential for the use of isolated vacuoles in studies of solute transport in higher plant cells. *Physiol. Plant.* 57:390–96

80. Leigh, R. A., Ahmad, N., Wyn Jones, R. G. 1981. Assessment of glycinebetaine and proline compartmentation by analysis of isolated beet vacuoles. *Planta* 153:34–41

81. Leigh, R. A., ap Rees, T., Fuller, W. A., Banfield, J. 1979. The location of acid

invertase activity and sucrose in the vacuoles of storage roots of beetroot (*Beta vulgaris*). *Biochem. J.* 178:539–47

82. Leigh, R. A., Tomos, A. D. 1983. An attempt to use isolated vacuoles to determine the distribution of sodium and potassium in cells of storage roots of red beet (*Beta vulgaris* L.). *Planta* 159:469–75

83. Leigh, R. A., Walker, R. R. 1980. ATPase and acid phosphatase activities associated with vacuoles isolated from storage roots of red beet (*Beta vulgaris* L.). *Planta* 150:222–29

84. Lin, W., Wittenbach, V. A. 1981. Subcellular localization of proteases in wheat and corn mesophyll protoplasts. *Plant Physiol.* 67:969–72

85. Lord, J. M. 1985. Precursors of ricin and *Ricinus communis* agglutinin. Glycosylation and processing during synthesis and intracellular transport. *Eur. J. Biochem.* 146:411–16

86. Lüthy, B., Matile, P. 1984. The mustard oil bomb: rectified analysis of the subcellular organization of the myrosinase system. *Biochem. Physiol. Pflanzen.* 179:5–12

87. Lüttge, U., Ball, E. 1979. Electrochemical investigation of active malic acid transport at the tonoplast into the vacuoles of the CAM plant *Kalanchoë daigremontiana*. *J. Membr. Biol.* 47:401–22

88. Macklon, A. E. S. 1984. Calcium fluxes at the plasmalemma and tonoplast. *Plant Cell Environ.* 7:407–13

89. Mäder, M., Chrispeels, M. J. 1984. Synthesis of an integral protein of the protein-body membrane in *Phaseolus vulgaris* cotyledons. *Planta* 160:330–40

89a. Makarow, M. 1985. Endocytosis in *Saccharomyces cerevisiae*: Internalization of α-amylase and fluorescent dextran into cells. *EMBO J.* 4:1861–66

90. Mandala, S., Taiz, L. 1985. Proton transport in isolated vacuoles from corn coleoptiles. *Plant Physiol.* 78:104–9

91. Marin, B. 1983. Evidence for an electrogenic adenosine-triphosphatase in *Hevea* tonoplast vesicles. *Planta* 157:324–30

92. Marin, B. 1983. Sensitivity of tonoplast-bound adenosine-triphosphatase from *Hevea* to inhibitors. *Plant Physiol.* 73:973–77

93. Marin, B., ed. 1985. *Biochemistry and Function of Vacuolar Adenosine-Triphosphatase in Fungi and Plants*. Berlin: Springer-Verlag. 259 pp.

94. Marin, B., Smith, J. A. C., Lüttge, U. 1981. The electrochemical proton gradient and its influence on citrate uptake in tonoplast vesicles of *Hevea brasiliensis*. *Planta* 153:486–93

95. Martinoia, E., Flügge, U. I., Kaiser, G., Heber, U., Heldt, H. W. 1985. Energy-dependent uptake of malate into vacuoles isolated from barley mesophyll protoplasts. *Biochim. Biophys. Acta* 806:311–19

96. Martinoia, E., Heck, U., Wiemken, A. 1981. Vacuoles as storage compartments for nitrate in barley leaves. *Nature* 289:292–93

97. Marty, F. 1978. Cytochemical studies on GERL, provacuoles, and vacuoles in root meristematic cells of *Euphorbia*. *Proc. Natl. Acad. Sci. USA* 75:852–56

98. Marty, F., Branton, D., Leigh, R. A. 1980. Plant vacuoles. In *The Biochemistry of Plants: A Comprehensive Treatise*, ed. N. E. Tolbert, 1:625–58. New York: Macmillan. 705 pp.

99. Matern, U., Heller, W., Himmelspach, K. 1983. Conformational changes of apigenin 7-*O*-(6-*O*-malonylglucoside), a vacuolar pigment from parsley, with solvent composition and proton concentration. *Eur. J. Biochem.* 133:439–48

100. Matern, U., Potts, J. R. M., Hahlbrock, K. 1981. Two flavonoid-specific malonyltransferases from cell suspension cultures of *Petroselinum hortense*: Partial purification and some properties of malonyl-coenzyme A:flavone/flavonon-7-*O*-glucoside malonyl-transferase and malonyl-coenzyme A:flavonol-3-*O*-glucoside malonyltransferase. *Arch. Biochem. Biophys.* 208:233–41

101. Matile, P. 1975. *The Lytic Compartment of Plant Cells*. Berlin: Springer-Verlag. 183 pp.

102. Matile, P. 1978. Biochemistry and function of vacuoles. *Ann. Rev. Plant Physiol.* 29:193–213

103. Matile, P., Wiemken, A. 1967. The vacuole as the lysosome of the yeast cell. *Arch. Mikrobiol.* 56:148–55

104. Matile, P., Wiemken, A. 1976. Interactions between cytoplasm and vacuole. In *Encyclopedia of Plant Physiology N S*, ed. C. R. Stocking, U. Heber, 3:255–87. Berlin: Springer-Verlag

105. Mellor, R. B., Dittrich, W., Werner, D. 1984. Soybean root response to infection by *Rhizobium japonicum*: mannoconjugate turnover in effective and ineffective nodules. *Physiol. Plant Pathol.* 24:61–70

106. Mettler, I. J., Mandala, S., Taiz, L. 1982. Characterization of *in vitro* proton pumping by microsomal vesicles isolated from corn coleoptiles. *Plant Physiol.* 70:1738–42

107. Miller, A. J., Brimelow, J. J., John, P. 1984. Membrane-potential changes in vacuoles isolated from storage roots of red beet (*Beta vulgaris* L.). *Planta* 160:59–65

108. Morgan, J. M. 1984. Osmoregulation and water stress in higher plants. *Ann. Rev. Plant Physiol.* 35:299–319

109. Moskowitz, A. H., Hrazdina, G. 1981. Vacuolar contents of fruit subepidermal cells from *Vitis* species. *Plant Physiol.* 68:686–92

110. Nelson, C. E., Ryan, C. A. 1980. *In vitro* synthesis of pre-proteins of vacuolar compartmented proteinase inhibitors that accumulate in leaves of wounded tomato plants. *Proc. Natl. Acad. Sci. USA* 77:1975–79

111. Nishimura, M. 1982. pH in vacuoles isolated from castor bean endosperm. *Plant Physiol.* 70:742–44

112. Nishimura, M., Beevers, H. 1978. Hydrolases in vacuoles from castor bean endosperm. *Plant Physiol.* 62:44–48

113. Nishimura, M., Beevers, H. 1979. Hydrolysis of protein in vacuoles isolated from higher plant tissue. *Nature* 277:412–13

114. Oba, K., Conn, E. E., Canut, H., Boudet, A. M. 1981. Subcellular localization of 2-(β-D-glucosyloxy)-cinnamic acids and the related β-glucosidase in leaves of *Melilotus alba* Desr. *Plant Physiol.* 68:1359–63

115. Ohsumi, Y., Anraku, Y. 1981. Active transport of basic amino acids driven by a proton motive force in vacuolar membrane vesicles of *Saccharomyces cerevisiae*. *J. Biol. Chem.* 256:2079–82

116. Pahlich, E., Kerres, R., Jäger, H. J. 1983. Influence of water stress on the vacuole/extravacuole distribution of proline in protoplasts of *Nicotiana rustica*. *Plant Physiol.* 72:590–91

117. Palevitz, B. A., O'Kane, D. J. 1981. Epifluorescence and video analysis of vacuole motility and development in stomatal cells of *Allium*. *Science* 214:443–45

118. Palevitz, B. A., O'Kane, D. J., Kobres, R. E., Raikhel, N. V. 1981. The vacuole system in stomatal cells of *Allium*. Vacuole movements and changes in morphology in differentiating cells as revealed by epifluorescence, video and electron microscopy. *Protoplasma* 109:23–55

119. Peoples, M. B., Beilharz, V. C., Waters, S. P., Simpson, R. J., Dalling, M. J. 1980. Nitrogen redistribution during grain growth in wheat (*Triticum aestivum* L.). II. Chloroplast senescence and the degradation of ribulose-1,5-bisphosphate carboxylase. *Planta* 149:241–51

120. Poole, R. J., Briskin, D. P., Krátký, Z., Johnstone, R. M. 1984. Density gradient localization of plasma membrane and tonoplast from storage tissue of growing and dormant red beet. Characterization of proton-transport and ATPase in tonoplast vesicles. *Plant Physiol.* 74:549–56

121. Quail, P. H. 1979. Plant cell fractionation. *Ann. Rev. Plant Physiol.* 30:425–84

122. Rasi-Caldogno, F., DeMichelis, M. I., Pugliarello, M. C. 1982. Active transport of Ca^{2+} in membrane vesicles from pea. Evidence for a H^+/Ca^{2+} antiport. *Biochim. Biophys. Acta* 693:287–95

123. Rataboul, P., Alibert, G., Boller, T., Boudet, A. M. 1985. Intracellular transport and vacuolar accumulation of o-coumaric acid glucoside in *Melilotus alba* mesophyll cell protoplasts. *Biochim. Biophys. Acta* 816:25–36

124. Rausch, T., Ziemann-Roth, M., Hilgenberg, W. 1985. ADP is a competitive inhibitor of ATP-dependent H^+ transport in microsomal membranes from *Zea mays* L. coleoptiles. *Plant Physiol.* 77:881–85

125. Rea, P. A., Poole, R. J. 1985. Proton-translocating inorganic pyrophosphatase in red beet (*Beta vulgaris* L.) tonoplast vesicles. *Plant Physiol.* 77:46–52

126. Rébeillé, F., Bligny, R., Martin, J. B., Douce, R. 1983. Relationship between the cytoplasm and the vacuole phosphate pool in *Acer pseudoplatanus* cells. *Arch. Biochem. Biophys.* 225:143–48

127. Rébeillé, F., Bligny, R., Martin, J. B., Douce, R. 1985. Effect of sucrose starvation on sycamore (*Acer pseudoplatanus*) cell carbohydrate and P_i status. *Biochem. J.* 226:679–84

128. Reinhold, L., Kaplan, A. 1984. Membrane transport of sugars and amino acids. *Ann. Rev. Plant Physiol.* 35:45–83

129. Riezman, H. 1985. Endocytosis in yeast: Several of the yeast secretory mutants are defective in endocytosis. *Cell* 40:1001–9

130. Roberts, J. K. M. 1984. Study of plant metabolism in vivo using NMR spectroscopy. *Ann. Rev. Plant Physiol.* 35:375–86

131. Roberts, J. K. M., Callis, J., Jardetzky, O., Walbot, V., Freeling, M. 1984. Cytoplasmic acidosis as a determinant of flooding intolerance in plants. *Proc. Natl. Acad. Sci. USA* 81:6029–33

132. Roberts, J. K. M., Wemmer, D., Ray, P. M., Jardetzky, O. 1982. Regulation of cytoplasmic and vacuolar pH in maize root tips under different experimental conditions. *Plant Physiol.* 69:1344–47

133. Robins, R. J., Juniper, B. E. 1980. The secretory cycle of *Dionaea muscipula*

Ellis. II. Storage and synthesis of the secretory proteins. *New Phytol.* 86:297–311

134. Ryan, C. A., Walker-Simmons, M. 1983. Plant vacuoles. *Methods Enzymol.* 96:580–89

135. Sacher, R. F., Staples, R. C. 1985. Inositol and sugars in adaptation of tomato to salt. *Plant Physiol.* 77:206–10

136. Sato, T., Ohsumi, Y., Anraku, Y. 1984. Substrate specificities of active transport systems for amino acids in vacuolar-membrane vesicles of *Saccharomyces cerevisiae*. Evidence of seven independent proton/amino acid antiport systems. *J. Biol. Chem.* 259:11505–8

137. Sato, T., Ohsumi, Y., Anraku, Y. 1984. An arginine/histidine exchange transport system in vacuolar-membrane vesicles of *Saccharomyces cerevisiae*. *J. Biol. Chem.* 259:11509–11

138. Saunders, J. A., Conn, E. E. 1978. Presence of the cyanogenic glycoside dhurrin in isolated vacuoles from *Sorghum*. *Plant Physiol.* 61:154–57

139. Saunders, J. A., Gillespie, J. M. 1984. Localization and substrate specificity of glycosidases in vacuoles of *Nicotiana rustica*. *Plant Physiol.* 76:885–88

140. Scannerini, S., Bonfante-Fasolo, P. 1983. Comparative ultrastructural analysis of mycorrhizal associations. *Can. J. Bot.* 61:917–43

141. Schekman, R. 1982. The secretory pathway in yeast. *Trends Biochem. Sci.* 7:243–46

142. Schmitt, R., Sandermann, H. 1982. Specific localization of β-D-glucoside conjugates of 2,4-dichlorophenoxyacetic acid in soybean vacuoles. *Z. Naturforsch. Teil C* 37:772–77

143. Schnabl, H., Kottmeier, C. 1984. Determination of malate levels during the swelling of vacuoles isolated from guard-cell protoplasts. *Planta* 161:27–31

144. Sharma, V., Strack, D. 1985. Vacuolar localization of 1-sinapoyl-glucose:L-malate sinapoyltransferase in protoplasts from cotyledons of *Raphanus sativus*. *Planta* 163:563–68

145. Smith, F. A., Raven, J. A. 1979. Intracellular pH and its regulation. *Ann. Rev. Plant Physiol.* 30:289–311

146. Smith, J. A. C., Uribe, E. G., Ball, E., Heuer, S., Lüttge, U. 1984. Characterization of the vacuolar ATPase activity of the crassulacean-acid-metabolism plant *Kalanchoë daigremontiana*. *Eur. J. Biochem.* 141:415–20

147. Smith, J. A. C., Uribe, E. G., Ball, E., Lüttge, U. 1984. ATPase activity associated with isolated vacuoles of the crassulacean acid metabolism plant *Kalanchoë daigremontiana*. *Planta* 162:299–304

147a. Strack, D., Sharma, V. 1985. Vacuolar localization of the enzymatic synthesis of hydroxycinnamic acid esters of malic acid in protoplasts from *Raphanus sativus* leaves. *Physiol. Plant.* 65:45–50

148. Struve, I., Weber, A., Lüttge, U., Ball, E., Smith, J. A. C. 1985. Increased vacuolar ATPase activity correlated with CAM induction in *Mesembryanthemum crystallinum* and *Kalanchoë blossfeldiana* cv. Tom Thumb. *J. Plant Physiol.* 117:451–68

149. Sze, H. 1985. H$^+$-translocating ATPase: Advances using membrane vesicles. *Ann. Rev. Plant Physiol.* 36:175–208

150. Tanchak, M. A., Griffing, L. R., Mersey, B. G., Fowke, L. C. 1984. Endocytosis of cationized ferritin by coated vesicles of soybean protoplasts. *Planta* 162:481–86

151. Teulières, C., Alibert, G., Ranjeva, R. 1985. Reversible phosphorylation of tonoplast proteins involves tonoplast-bound calcium-calmodulin-dependent protein kinase(s) and protein phosphatase(s). *Plant Cell Rep.* 4:199–201

152. Thayer, S. S., Huffaker, R. C. 1984. Vacuolar localization of endoproteinases EP$_1$ and EP$_2$ in barley mesophyll cells. *Plant Physiol.* 75:70–73

153. Thom, M., Komor, E. 1984. H$^+$-sugar antiport as the mechanism of sugar uptake by sugarcane vacuoles. *FEBS Lett.* 173:1–4

154. Thom, M., Komor, E. 1984. Role of the ATPase of sugar-cane vacuoles in energization of the tonoplast. *Eur. J. Biochem.* 138:93–99

155. Thom, M., Komor, E. 1985. Electrogenic proton translocation by the ATPase of sugarcane vacuoles. *Plant Physiol.* 77:329–34

156. Thom, M., Komor, E., Maretzki, A. 1982. Vacuoles from sugarcane suspension cultures. II. Characterization of sugar uptake. *Plant Physiol.* 69:1320–25

157. Thom, M., Maretzki, A. 1985. Group translocation as a mechanism for sucrose transfer into vacuoles from sugarcane cells. *Proc. Natl. Acad. Sci. USA* 82:4697–4701

158. Thom, M., Willenbrink, J., Maretzki, A. 1983. Characteristics of ATPase from sugarcane protoplast and vacuole membranes. *Physiol. Plant.* 58:497–504

159. Ting, I. P. 1985. Crassulacean acid metabolism. *Ann. Rev. Plant Physiol.* 36:595–622

160. Torimitsu, K., Yazaki, Y., Nagasuka, K., Ohta, E., Sakata, M. 1984. Effect of

external pH on the cytoplasmic and vacuolar pHs in mung bean root-tip cells: a ^{31}P nuclear magnetic resonance study. *Plant Cell Physiol.* 25:1403–9

161. Uchida, E., Ohsumi, Y., Anraku, Y. 1985. Purification and properties of H$^+$-translocating, Mg^{2+}-adenosine triphosphatase from vacuolar membranes of *Saccharomyces cerevisiae. J. Biol. Chem.* 260:1090–95

162. Urban, B., Laudenbach, U., Kesselmeier, J. 1983. Saponin distribution in the etiolated leaf tissue and subcellular localization of steroidal saponins in etiolated protoplasts of oat (*Avena sativa* L.). *Protoplasma* 118:121–23

163. Van der Wilden, W., Chrispeels, M. J. 1983. Characterization of the isozymes of α-mannosidase located in the cell wall, protein bodies, and endoplasmic reticulum of *Phaseolus vulgaris* cotyledons. *Plant Physiol.* 71:82–87

164. Van der Wilden, W., Matile, P., Schellenberg, M., Meyer, J., Wiemken, A. 1973. Vacuolar membranes: isolation from yeast cells. *Z. Naturforsch. Teil C* 28:416–21

165. Van der Wilden, W., Segers, J. H. L., Chrispeels, M. J. 1983. Cell walls of *Phaseolus vulgaris* leaves contain the azocoll-digesting proteinase. *Plant Physiol.* 73:576–78

166. Vaughn, K. C., Duke, S. O. 1981. Evaginations from the plastid envelope: A method for transfer of substances from plastid to vacuole. *Cytobios* 32:89–95

167. Vitale, A., Chrispeels, M. J. 1984. Transient *N*-acetylglucosamine in the biosynthesis of phytohemagglutinin: Attachment in the Golgi apparatus and removal in the protein bodies. *J. Cell Biol.* 99:133–40

168. Wagner, G. J. 1979. Content and vacuole/extravacuole distribution of neutral sugars, free amino acids and anthocyanin in protoplasts. *Plant Physiol.* 64:88–93

169. Wagner, G. J. 1981. Enzymic and protein character of tonoplast from *Hippeastrum* vacuoles. *Plant Physiol.* 68:499–503

170. Wagner, G. J. 1982. Compartmentation in plant cells: The role of the vacuole. In *Recent Advances in Phytochemistry*, ed. L. Creasy, G. Hrazdina, 16:1–45. New York: Plenum. 277 pp.

171. Wagner, G. J. 1983. Higher plant vacuoles and tonoplast. In *Isolation of Membranes and Organelles from Plant Cells*, ed. J. L. Hall, A. L. Moore, pp. 83–118. New York: Academic. 317 pp.

172. Wagner, G. J., Lin, W. 1982. An active proton pump of intact vacuoles isolated from *Tulipa* petals. *Biochim. Biophys. Acta* 689:261–66

173. Wagner, G. J., Mulready, P. 1983. Characterization and solubilization of nucleotide-specific, Mg^{2+}-ATPase and Mg^{2+}-pyrophosphatase of tonoplast. *Biochim. Biophys. Acta* 728:267–80

174. Wagner, W., Keller, F., Wiemken, A. 1983. Fructan metabolism in cereals: Induction in leaves and compartmentation in protoplasts and vacuoles. *Z. Pflanzenphysiol.* 112:359–72

175. Walker, R. R., Leigh, R. A. 1981. Characterization of a salt-stimulated ATPase activity associated with vacuoles isolated from storage roots of red beet (*Beta vulgaris* L.). *Planta* 153:140–49

176. Walker, R. R., Leigh, R. A. 1981. Mg^{2+}-dependent, cation-stimulated inorganic pyrophosphatase associated with vacuoles isolated from storage roots of red beet (*Beta vulgaris* L.). *Planta* 153:150–55

177. Wardley, T. M., Bhalla, P. L., Dalling, M. J. 1984. Changes in the number and composition of chloroplasts during senescence of mesophyll cells of attached and detached primary leaves of wheat (*Triticum aestivum* L.). *Plant Physiol.* 75:421–24

178. Waters, S. P., Noble, E. R., Dalling, M. J. 1982. Intracellular localization of peptide hydrolases in wheat (*Triticum aestivum* L.) leaves. *Plant Physiol.* 69:575–79

179. Waterton, J. C., Bridges, I. G., Irving, M. P. 1983. Intracellular compartmentation detected by ^{31}P-NMR in intact photosynthetic wheat-leaf tissue. *Biochim. Biophys. Acta* 763:315–20

180. Weigel, H. J., Weis, E. 1984. Determination of the proton concentration difference across the tonoplast membrane of isolated vacuoles by means of 9-aminoacridine fluorescence. *Plant Sci. Lett.* 33:163–75

181. Werner, C., Matile, P. 1985. Accumulation of coumaroylglucosides in vacuoles of barley mesophyll protoplasts. *J. Plant Physiol.* 118:237–49

182. Werner, D., Mörschel, E., Kort, R., Mellor, R. B., Bassarab, S. 1984. Lysis of bacteroids in the vicinity of the host cell nucleus in an ineffective (fix$^-$) root nodule of soybean (*Glycine max*). *Planta* 162:8–16

183. Wiebe, H. H. 1978. The significance of plant vacuoles. *BioScience* 28:327–31

184. Wiemken, A., Matile, P., Moor, H. 1970. Vacuolar dynamics in synchronously budding yeast. *Arch. Mikrobiol.* 70:89–103

185. Wiemken, A., Schellenberg, M., Urech,

K. 1979. Vacuoles: the sole compartments of digestive enzymes in yeast (*Saccharomyces cerevisiae*)? *Arch. Microbiol.* 123:23–35

186. Willenbrink, J., Doll, S. 1979. Characteristics of the sucrose uptake system of vacuoles isolated from red beet tissue. Kinetics and specificity of the sucrose uptake system. *Planta* 147:159–62

187. Wittenbach, V. A., Lin, W., Hebert, R. R. 1982. Vacuolar localization of proteases and degradation of chloroplasts in mesophyll protoplasts from senescing primary wheat leaves. *Plant Physiol.* 69:98–102

188. Wolf, D. H. 1982. Proteinase action *in vitro* versus proteinase function *in vivo:* Mutants shed light on intracellular proteolysis in yeast. *Trends Biochem. Sci.* 7:35–37

189. Wray, V., Schiel, O., Berlin, J., Witte, L. 1985. Phosphorus-31 nuclear magnetic resonance investigation of the *in vivo* regulation of intracellular pH in cell suspension cultures of *Nicotiana tabacum:* The effects of oxygen supply, nitrogen, and external pH change. *Arch. Biochem. Biophys.* 236:731–40

190. Yamaki, S. 1982. Localization of sorbitol oxidase in vacuoles and other subcellular organelles in apple cotyledons. *Plant Cell Physiol.* 23:891–99

191. Yeo, A. R. 1983. Salinity resistance: physiologies and prices. *Physiol. Plant.* 58:214–22

192. Zeiger, E. 1983. The biology of stomatal guard cells. *Ann. Rev. Plant Physiol.* 34:441–75

Ann. Rev. Plant Physiol. 1986. 37:165–86

CROSS-LINKING OF MATRIX POLYMERS IN THE GROWING CELL WALLS OF ANGIOSPERMS

Stephen C. Fry

Department of Botany, University of Edinburgh, The King's Buildings, Mayfield Road, Edinburgh EH9 3JH, United Kingdom

CONTENTS

THE PROBLEM

The growing plant cell wall is a biphasic structure, consisting of a rigid skeleton of cellulose microfibrils held together (and apart) by a gel-like matrix (48, 54, 57, 60). The matrix makes up about two-thirds of the wall's[1] dry weight and is built up of several noncellulosic polysaccharides and glycoproteins. Most of these polymers, after extraction from the wall, are soluble: they are polyhy-

---

[1]*Note:* the term "wall" refers to "growing cell wall of angiosperms."

0066-4294/86/0601-0165$02.00

droxy, hydrophilic molecules, mostly polyionic, and in the living plant they are highly hydrated. Yet the wall matrix is water-insoluble and very coherent.

When hydrophilic polymers are insoluble, there must be interpolymer cross-links (covalent, noncovalent, or both). Little is known of the cross-links that hold the wall matrix polymers together (43, 49, 57). Part of the difficulty until recently was our incomplete knowledge of their primary structures (i.e. sugar and amino acid sequences). We now know much more about these primary structures (47, 48), and the time is ripe for an attempt to define the cross-links. What are the cross-links, and how can we study their making and breaking? There are many unanswered questions, but we are beginning to make progress.

WALL CROSS-LINKS: GENERAL

The Importance of Wall Matrix Cross-Links

Much interest has been focused on the mode of action of plant growth regulators (PGRs), which often act by altering the *extensibility* of the cell wall (69), probably by altering the wall matrix. The chemical basis of this physical change in the wall is unknown; the quantity and primary structures of the matrix polymers generally show little change after plant growth regulator treatment (69), suggesting that a change in the superstructure is involved, e.g. the making or breaking of interpolymer cross-links. Further progress requires a better knowledge of the chemistry of matrix cross-links.

Matrix cross-links are also likely to determine wall *digestibility* (26). Important situations where walls are partially digested include the laboratory isolation of protoplasts, the rotting of plant litter, the preparation of silage, the commercial production of fruit "juice," the breaching of walls by invading pathogens, the nutrition of herbivores, and the action of dietary fiber in the human colon. Walls differ in their digestibility; this can be seen easily in attempts to isolate protoplasts from different tissues which nevertheless contain very similar wall matrix polymers. Again the source of variability is likely to be superstructural, e.g. interpolymer cross-links. An understanding of such cross-links would further our understanding of the digestibility of cell walls.

Cross-links between walls are responsible for cell *adherence* (8). This is important in plant development (maintaining sister cells in close contact) and in the food industry (cell/cell bonding influences the texture of processed fruit and vegetables). Also, the formation of new interpolymer cross-links is necessary for the bonding of juxtaposed nonsister cells between the stock and scion during grafting.

Extensibility, digestibility, and adherence are important properties of cell walls, dictated by wall cross-links. The wish to predict and manipulate these three properties encourages basic research into the chemistry and metabolism of wall cross-links.

Diversity of Cross-Links

There are many cross-links, both covalent and noncovalent, that could conceivably hold the wall together (Table 1). Particular attention is drawn to the following (Figure 4), which are discussed in more detail later: coupled phenols (28), glycosidic bonds (41), uronoyl ester bonds (15, 28, 43, 45, 74), hydrogen bonds (30), and ionic bonds (14), including Ca^{2+} bridges (38).

Types of Evidence for Cross-Links

Four main types of evidence have been brought forward in support of particular cross-links. These are:

(a) ISOLATION AND CHEMICAL CHARACTERIZATION OF CROSS-LINKED FRAGMENTS   This approach (see Figure 1a) is applicable only to covalent bonds. Cell walls are treated with an agent that breaks the backbone of the polymer but leaves intact the cross-link. A relatively small molecule is then isolated that contains just enough of each backbone to identify the polymer(s) involved, plus the covalent cross-link.

(b) SOLUBILIZATION OF THE POLYMER FROM THE WALL MATRIX BY TREATMENT WITH SPECIFIC CROSS-LINK CLEAVING REAGENTS   One could also accept in evidence a loosening of the cell wall (e.g. as measured in an extensiometer) in response to a sufficiently specific cleaving agent (see Figure 1b). The approach is essentially the converse of method a above. The main difficulty is proof of specificity; the cross-link in question must be broken, but not any other cross-links nor the primary structure of the polymer. A range of possible treatments is given in Table 1. A few of these treatments probably can be used safely—e.g. salts, urea, guanidinium thiocyanate,[2] chelating agents,[2] sugar haptens, and pure esterases—at room temperature and buffered at pH 4–7. Other cleaving reagents have side effects (Table 2).

Another problem is that a polymer may be held in the wall by a combination of different bonds. Multiple treatments would then be necessary for solubilization. Therefore, although successful solubilization by a specific cross-link-breaking reagent indicates a role for that cross-link, failure does not negate it.

(c) SUPPRESSION OF WALL BINDING OF NEWLY SECRETED POLYMER BY AGENTS THAT INHIBIT CROSS-LINK FORMATION   Most of the polysaccharides and glycoproteins of the wall matrix seem to be synthesized intracellularly (see Figure 1c) and secreted through the plasmalemma in soluble form (33). Their subsequent insolubilization in the wall may be a gradual and possibly enzyme-catalyzed step. If so, specific inhibition of this step would

[2]Guanidium and chelating agents are also salts and can thus break ordinary ionic bonds.

Table 1 Some possible cross-links between wall polymers

Cross-link	Possible example	Formation in vivo		Cleaving agents (aq. unless stated)	Comments
		Mechanism	Inhibitors		
(a) Covalent					
Glycosidic	Arabinogalactan · (Gal-Rha) · RG-I	Golgi [and in wall?(1)]	—	Endoglycanases, hot acid, dry HF (51)	Integral part of a single PS?
Ester (uronoyl)	Pectin · (*uronoyl-Glc*) · cellulose	Transesterification by wall enzymes?	—	Esterases, Na₂CO₃, NaOH, methanolic sodium methoxide	Speculation (15, 28, 43, 45, 74)
Ester (feruloyl etc)	Feruloyl · (*Ara*) · pectin	Golgi—feruloyl-CoA?	—	Esterases, NaOH, MeOH/NaOMe (50)	Precursor of cross-links (26)
Phenolic coupling 1. Ether 2. Biphenyl	1. Extensin · (*Tyr-Tyr*) · extensin 2. Pectin · (*Fer-Fer*) · pectin	Wall peroxidase + H₂O₂	Ascorbate, dithiothreitol, NaCN, etc (12, 23)	NaClO₂ at pH 4 and 70°C. (Diferulate also cleaved by NaOH etc)	See Figures 2, 3. Plants have no known enzymes to break these cross-links. Peroxidase may also act to provide H₂O₂ (28).
Ether	PS · (ether) · feruloyl-PS			NaClO₂? BBr₃?(7)	

Disulphide	Cystine (**R**-*SS*-**R**')	Oxidation of Cys in ER?	Dithiothreitol, mercaptoethanol	Cyst(e)ine is rare in wall proteins
(b) Noncovalent				
Hydrogen bond	Hemicellulose · cellulose	Nonenzymic	MMNO(40), KOH [Urea, guanidinium thiocyanate, and heat are not very effective (61)]	Strengthened at low [H_2O], and possibly by wall phenols (20)
Ionic bond	Extensin · pectin	Nonenzymic	Salts (LaCl$_3$ > CaCl$_2$ > NaCl), acids, alkalies	(64)
Calcium bridge	Pectin · (Ca^{2+}) · pectin	May need enzymic removal of Me-esters (71)	Chelating agents, e.g. EGTA, CDTA, EDTA, oxalate, hexametaphosphate; low pH	Special case of ionic bond reinforced by coordination (38)
Hydrophobic interaction	Gelling of pectin (also involves H-bonds)	Nonenzymic	Organic solvents	Possibly also between wall phenols (20)
v.d. Waals bonds	(many)	Nonenzymic	(Reagents that change molecular conformation?)	Nonspecific attraction between interlocking molecules
Lectin bond	Lectin · PS	Nonenzymic	Sugar hapten; denaturation	Special case of several of above

Abbreviations: Ara, arabinose; CoA, coenzyme-A; ER, endoplasmic reticulum; Fer, ferulate; Gal, galactose; Glc, glucose; Me, methyl; MeOH, methanol; NaOMe, sodium methoxide; PS, polysaccharide; RG-I, rhamnogalacturonan-I; Rha, rhamnose; Tyr, tyrosine.

Table 2 Side effects of cross-link cleaving agents

Cross-link cleaved	Treatment used	Other bonds broken						Precautions (where available); comments
		Glycosidic	Ester	Peptide	H-bonds	Ionic	Ca bridge	
Glycosidic	1. Hot acid	H	H	h	c	C	C	Some glycosidic bonds labile to very weak acid, e.g. galacturonate (63), apiose
	2. Endoglycanases	h						"Wrong" glycosidic bonds broken if enzyme incompletely pure or specific
Ester	1. Ice-cold 1M Na_2CO_3 (pH ca 11) (38)	(e)	H		c	C	—	Elimination suppressed by starting reaction cold. Feruloyl esters hydrolyzed only very slowly
	2. 0.5 M NaOH (pH ca 13½) under N_2 (22)	P,e	H	(h)	C	C	C	Elimination suppressed by starting reaction cold. Peeling partly suppressed by $NaBH_4$
	3. Esterases		H					
Coupled phenols	$NaClO_2$, pH 4, 70°C (23, 51, 55)	h	—		—	—	—	Important to include control at same pH and temperature but without $NaClO_2$

Bond type	Treatment						Comments
H-bonds	1. 1-5M KOH, 20°C under N₂	P,E	H	h	C	C	Elimination suppressed by starting reaction cold. Peeling partly suppressed by NaBH$_4$ (40)
	2. MMNO-hydrate, 120°C	h	H	—	C	—	Must be buffered if heated
	3. 8 M urea (5)				C	C	
	4. Guanidinium thiocyanate				C	—	Preferentially solubilizes mannose-rich polysaccharides? (61)
Ionic bonds (64)	25 mM LaCl₃, 100 mM CaCl₂ or 500 mM NaCl at pH 4–7 and 20°C						Could reinforce Ca bridges
Ca bridges	1. Oxalate, EDTA, hexametaphosphate, at pH 3·5–7·5, 70–120°C	E,h	h	—	c		Use EGTA or CDTA
	2. EGTA, CDTA at pH 6·5–7·5, 20°C				c		Shown to effect complete removal of Ca²⁺ from cell walls (38)
Hydrophobic interaction	Organic solvents		*				*Hydrogen bonds strengthened; polymers often precipitated

Abbreviations: C,c, bonds broken competitively; E,e, glycosidic bonds broken by elimination reaction (especially backbone of methylesterified pectins and sugars linked to serine or threonine); H,h, hydrolysis; P,p. "peeling" reaction (stepwise removal of sugars from reducing end of polysaccharide). *Upper case,* serious bond-breaking; *lower case,* minor; *(blank),* bond-breaking unlikely; (—), no data.

make newly secreted polymers stay soluble longer. Possible inhibitors of cross-link formation are given in Table 1. Many are toxic or nonspecific. Some may fail because polymers are held in the wall by more than one type of bond.

Although the formation of noncovalent cross-links seems to be nonenzymic, enzymes may still be required to convert the polymers into a state suitable for such cross-linking. For example, action of pectinesterase may be a prerequisite for the formation of Ca^{2+} bridges between galacturonan molecules (78). Specific inhibitors of pectinesterase would therefore be valuable in the present context.

(*d*) FORMATION OF THE CROSS-LINK USING PURIFIED COMPONENTS IN VITRO   If a cross-link normally forms between two wall polymers in vivo, it should be possible to mimic this process in vitro (see Figure 1d), with the addition of enzymes where appropriate. Formation of a particular cross-link in vitro confirms that such a cross-link could form in vivo, but not that it does. The following account describes recent attempts to apply methods *a* through *d* to the major polymers of the wall matrix.

SPECIFIC WALL CROSS-LINKS

Extensin

Work on extensin illustrates all four methods. Extensin is an insoluble, highly basic glycoprotein, typically making up 1–10% of the wall matrix. It is built of at least two different polypeptide backbones (65), both rich in hydroxyproline residues bearing *short* (mono- to tetrasaccharide) side-chains (48). The proposal (41) that wall polysaccharides are glycosidically attached to extensin probably arose through failure to distinguish between extensin and the soluble, acidic, arabinogalactan-proteins, which bear *long* carbohydrate side-chains (17).

ISOLATION OF CROSS-LINKED FRAGMENTS   During attempts to break the polypeptide backbone of extensin without cleaving covalent cross-links, both acid and alkaline hydrolysis yielded, besides the expected amino acids, an unknown phenolic compound (23). The unknown was shown (by dinitrophenylation with Sanger's reagent) to possess two -NH_2 groups and (by measurement of isoelectric point) two -COOH groups, as well as (by spectrophotometry) one phenolic -OH group. It was shown to be metabolically derived from [^{14}C]tyrosine (but not from the closely related [^{14}C]phenylalanine), and prolonged heating in very strong alkali partially hydrolyzed it to tyrosine. Its mass spectrum (fast-atom-bombardment method) was consistent with an oxidatively coupled dimer of tyrosine [i.e. 2 tyrosines *minus* 2H (23)], and the structure shown in Figure 2a was confirmed by chemical synthesis [oxida-

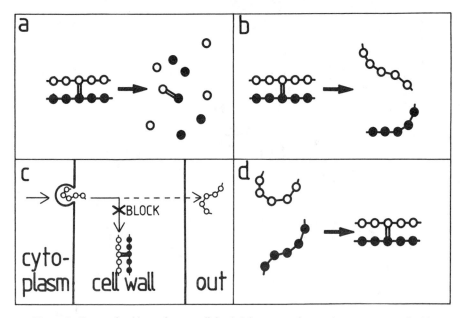

Figure 1 Types of evidence for cross-links (=) between polymers (O-O-O-O, ●-●-●-●). (*a*)
Isolation of cross-linked fragment; (*b*) liberation of polymers by cleaving of cross-link; (*c*) blocking
of wall-binding of newly secreted polymer in vivo; (*d*) formation of cross-link in vitro.

tion of tyrosine with ferricyanide (27)]. The unknown was thus two tyrosine
units joined by a diphenyl-ether bridge and was named *isodityrosine,* an isomer
of the dityrosine (Figure 2b) found in the cuticular protein of insects (3).

The structure of isodityrosine is such that it could cross-link polypeptides. A
plausible reaction for its formation is the peroxidatic coupling of two tyrosine
residues (Figure 2). In addition, an oxidatively coupled trimer of tyrosine has
been found in plant cell walls (25), and this could cross-link up to three
polypeptide chains.

There is evidence that isodityrosine is a component of extensin. First, the
amount of isodityrosine closely paralleled the amount of hydroxyproline in
SDS-washed cell walls of various angiosperm tissues (23) [although it was
apparently absent from *Chlamydomonas* (58)]. Second, a purified tryptic
peptide of extensin contained isodityrosine (16).

To serve a cross-linking role, the isodityrosine bridges of extensin must be
interpolypeptide. The tryptic peptide of extensin contained intramolecular
isodityrosine, the two participating tyrosine units being separated by a single
lysine residue (16). However, this analysis was of the smallest isodityrosine-
containing peptide; it is possible that larger ones will reveal intermolecular
bridges. Indeed, soluble dimers of extensin (possibly caught in the act of

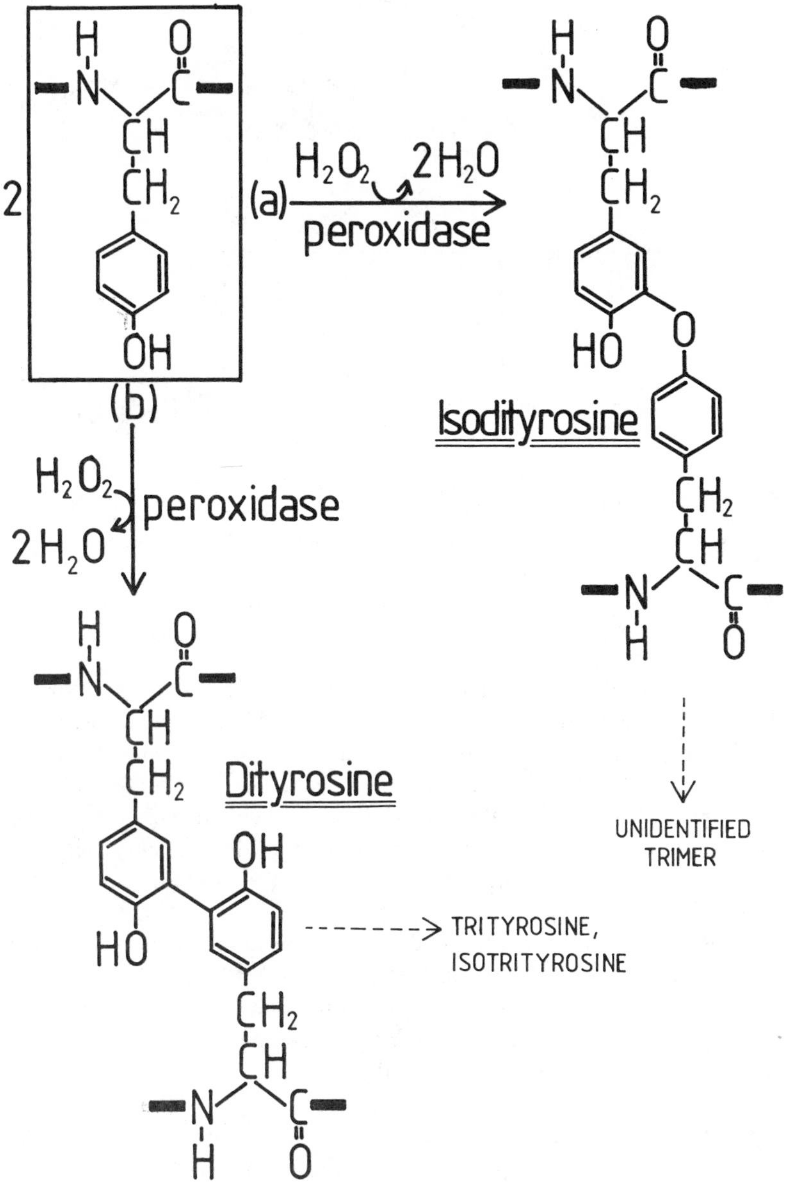

Figure 2   Oxidative coupling of tyrosine residues to produce (*a*) isodityrosine cross-links (in vitro at low pH and in vivo in plant cell walls) or (*b*) dityrosine (in vitro at high pH and in vivo in animals). Coupling to a third tyrosine residue can, in both cases, lead to trimers.

becoming part of the insoluble wall matrix) have been isolated from carrot cell walls, suggesting intermolecular covalent bonding, although it was not established that the bonds were isodityrosine. EM showed that the dimers had cross-links at specific points, usually close to the ends of the monomers (67).

SOLUBILIZATION OF THE POLYMER   Some extensin, mainly the newly laid down molecules, can be extracted from the cell wall with salts ($LaCl_3$ > $CaCl_2$ > NaCl). The proportion extracted varies greatly, e.g. 10–50% for carrot root slices (9, 44), 0–10% for cultured tomato cells (64), 0% for cultured sycamore cells (25), 90% for cultured spinach cells (25), and 10–15% for potato tubers (44). Salt extractability indicates ionic binding, presumably of the cationic extensin to the anionic pectins. The basic amino acids of extensin tend to occur every third residue (65), which (assuming a polyproline-II helix, i.e. with 3 amino acids per turn) puts consecutive positive charges on one face of the polypeptide molecule and 0.94 nm apart, perhaps allowing concerted binding to several -COO$^-$ groups of a single pectin molecule. [Homogalacturonan has a threefold screw axis of repeating unit 1.305 nm (57).]

The extensin not extracted by salt is also insoluble in chaotropic agents, boiling SDS, acidified phenol, reducing agents, and anhydrous hydrogen fluoride (which depolymerizes polysaccharides including cellulose), and is therefore said to be covalently wall-bound (23, 49, 51). This extensin can, however, be solubilized with acidified $NaClO_2$ (chlorite) (51, 55), which concomitantly splits isodityrosine (23) but not peptide bonds (55). Thus, the idea that isodityrosine holds extensin in the wall matrix is supported.

INHIBITION OF CROSS-LINKING IN VIVO   Salt or SDS insolubility is a test for covalent wall binding of newly synthesized (pulse-labeled) extensin. Covalent binding assayed by this method was partially prevented by inhibitors of peroxidase, e.g. dithiothreitol, ascorbate, and cyanide. Formation of isodityrosine was blocked in parallel (12, 23). This indicates that the binding is catalyzed by peroxidase or a similar enzyme and is compatible with the mechanism in Figure 2a.

CROSS-LINKING IN VITRO   If the scheme in Figure 2a is correct for extensin, the reaction should be demonstrable in vitro. Horseradish peroxidase plus H_2O_2 will indeed dimerize free tyrosine; the optimum pH is ~9, but the product is dityrosine, not isodityrosine (31). Dityrosine has not been found in plants. However, at pH 3–5 (which is more typical of the plant cell wall), where the total rate of dimerization is rather low, roughly equal amounts of dityrosine and isodityrosine are formed (S. C. Fry, unpublished). Therefore, it cannot be argued that commercial peroxidase preparations lack a specific isozyme responsible for isodityrosine formation.

Tyrosine residues of soluble proteins can be dimerized in vitro by horseradish peroxidase plus H_2O_2, optimally at pH 9, thereby cross-linking the protein molecules. Again, the dimer formed is dityrosine (2), and extensin is no exception (13). However, tyrosine residues of ionically bound extensin, present in isolated cell walls, were converted to isodityrosine (not dityrosine) when the walls were incubated in pH 6 buffer. The extensin simultaneously became covalently bound, suggesting a cross-linking role for isodityrosine. The process was inhibited by ascorbate, compatible with catalysis by wall-bound peroxidase (13). The action of peroxidase on extensin at low pH in the absence of cell walls has not yet been reported.

Taken together, the results strongly suggest that wall-bound peroxidase can cross-link extensin via isodityrosine bridges. The main block to concluding this unequivocally is the lack of proof that isodityrosine bridges are subtended by tyrosine residues in different extensin molecules.

Pectins

The analysis of cross-linking in pectins is complicated by the diversity of potential cross-links. Pectins are based on a backbone rich in $\alpha(1\rightarrow4)$-linked galacturonate (homogalacturonan). Discrete "hairy" domains (RG-I) occur in which the backbone is also rich in rhamnose residues, many of which bear neutral arabinose- and/or galactose-rich side-chains. Other domains (RG-II) contain at least 12 different monosaccharides. The -COOH groups of some of the galacturonate residues of pectins are methyl esterified. Pectins may also carry acetyl and phenolic groups (48, 52).

Proposed cross-links include Ca^{2+} bridges, other ionic bonds, H-bonds, glycosidic bonds, ester bonds, and phenolic coupling. This diversity means that certain approaches to the detection of cross-links (Figures 1b, 1c) have limited use because the loss of one type of cross-link might not prevent the others holding the pectin in the wall.

ISOLATION OF CROSS-LINKED FRAGMENTS
Phenolic coupling products It seems plausible that pectin molecules are cross-linked by oxidative coupling of their phenolic substituents (Figure 3). Pectins from spinach culture cell walls contained the cinnamate derivatives ferulate and p-coumarate (20, 21), ester-linked through their -COOH groups to pectins (24). There were ~10 feruloyl groups per 1000 pectic sugar residues (the degree of polymerization of RG-I). Partial hydrolysis of the pectin with "Driselase" gave two feruloyl disaccharides, 4-*O*-(6-*O*-feruloyl-β-D-galactopyranosyl)-D-galactose and 3-*O*-(3-*O*-feruloyl-α-L-arabinopyranosyl)-L-arabinose, which accounted for >60% of the wall's ferulate (22). The feruloylated sugar residues were nonreducing ends of pectic side-chains (24), probably exposed domains of the pectin molecule, accessible to extracellular

peroxidases. Kinetic evidence showed that the highly site-specific feruloylation occurs intracellularly on newly synthesized pectin molecules (S.C. Fry, unpublished).

The simplest coupling product of ferulate is diferulate (Figure 3). Spinach walls yielded small amounts of diferulate upon alkaline hydrolysis (S.C. Fry, unpublished), plus much larger amounts of unidentified cinnamate derivatives that had chromatographic properties of oxidatively coupled oligomers (26). The polymers to which these were linked were resistant to enzymic hydrolysis, probably indicating a tightly cross-linked part of the wall.

Thus, some pectins contain potential cross-linking points (ferulate and *p*-coumarate); and cross-links that could have arisen from them have been detected after alkaline hydrolysis. The goal of detecting a pair of pectic fragments, cross-linked by a diferuloyl (or similar) bridge, is yet to be achieved.

Hemicellulose-pectin glycosidic bonds   Fragments isolated from sycamore culture cell walls were reported to be the linkage point between xyloglucan and pectin and formed the basis of a widely adopted wall model (1, 41). For instance, pectinase solubilized pectic fragments (PG-1B; see 41) containing traces of glucose and xylose (or methylxylose). Similarly, pectinase plus cellulase solubilized a fragment (C2) that had properties of pectin but also contained traces of glucose, xylose (or methylxylose), and fucose (or methylfu

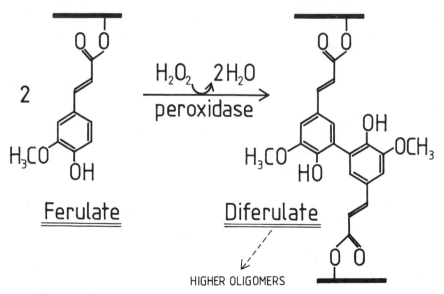

Figure 3   Oxidative coupling of ferulate residues to produce diferulate cross-links. Further coupling probably leads to higher oligomers.

cose) (5). However, current knowledge of wall polymers suggests that the source of the minor sugars in PG-1B and C2 could have been the pectin RG-II rather than xyloglucan (48). A third mixture of fragments (U), solubilized from pectinase-pretreated walls by 8 M urea, contained many different sugars, some characteristic of xyloglucan and others possibly of pectic origin, although the uronic acid was not identified and might have been hemicellulosic glucuronate rather than pectic galacturonate (5). Later work failed to isolate significant quantities of xyloglucan-pectin hybrid fragments (47, 49), but the possibility of a small proportion of glycosidic bonds between hemicellulose and pectin is not excluded (38, 56, 61, 68).

SOLUBILIZATION OF PECTINS   Some of the pectin in plant tissue is soluble in cold water, indicating little or no binding to the cell wall. Further pectin can be solubilized in several ways, but some stays permanently with the α-cellulose (11).

Ca bridges   The presumption of Ca bridges is the basis for the widespread use of chelating agents (e.g. oxalate, hexametaphosphate, EDTA, CDTA, EGTA) to extract pectins. Often the chelating treatments are combined with heating, and the results fail to give real evidence for Ca bridges since heating will cleave pectic backbones regardless of pH (4, 63). Indeed, although pectins are slowly solubilized from sycamore cell walls by hot oxalate at pH 3–6, hot formate (not a chelating agent) at the same pH has the same effect (S.C. Fry, unpublished).

   Good evidence for Ca bridges was obtained by treatment of walls with cold CDTA at neutral pH, which removed essentially all of the Ca and concomitantly solubilized a proportion of the pectin (37). The proportion solubilized was high in fruit, moderate in leafy vegetables, and low in tissue cultures. The pectin not solubilized must have been held by other bonds as well as (or instead of) Ca bridges.

Other ionic bonds   Acidic pectins could bind to basic extensin, but breakage of such bonds with salt does not solubilize much of the wall's pectin nor augment the action of CDTA (37). Nevertheless, it seems possible that ionic bonding to extensin plays a steering role, orientating the pectin in the wall.

Coupled phenols $NaClO_2$, which would cleave cross-linked cinnamate derivatives, e.g. diferulate (Figure 3), solubilized some pectin (~12%) from sycamore culture cell walls and rendered a subsequent EGTA extraction more effective (S.C. Fry, unpublished). Similar results were obtained with kale cell walls (J. M. Barwick and M. C. Jarvis, personal communication). This is consistent with the cross-linking of pectins via phenolic substituents, but does not prove it because $NaClO_2$ breaks the extensin network and could therefore act by releasing pectin from physical entanglement with extensin. It has also been speculated that polysaccharide-bound phenols may be oxidatively coupled to protein-bound tyrosine (53).

Ester bonds   Cold Na_2CO_3, which would hydrolyze ester bonds but cause negligible elimination-degradation, solubilized some CDTA-insoluble wall pectins (39). This suggests that ester bonds help to hold pectin in the wall: these esters could a priori be methylesterified galacturonate residues (unlikely because pectinesterase does not mimic Na_2CO_3), diferuloyl and related bridges (20), or ester bonds between uronic acids and neutral sugars (15, 28, 43, 45, 74). Action of Na_2CO_3 by breaking H-bonds seems unlikely because more powerful chaotropic agents are ineffectual (42). Interpolymeric ester bonds therefore seem probable, and their identification is an important challenge.

INHIBITION OF CROSS-LINKING IN VIVO   Pectin produced by naked protoplasts dissolves into the culture medium, showing that pectin does not readily self-assemble in the absence of an existing cell wall; the pectin is retained after some cellulose (and extensin?) has been deposited during wall regeneration (33). It is not clear whether the cellulose-rich framework traps the pectin by a specific chemical bonding or by physical entanglement.

CROSS-LINKING IN VITRO   Certain pectins will cross-link in vitro in the presence of Ca^{2+} owing to the formation of Ca bridges (6, 72, 78). Pectins with many side-chains (e.g. RG-I) cannot do this, nor can heavily methylesterified homogalacturonans (77). The degree of methylesterification can be controlled (lowered) in the cell wall by pectinesterase, which removes methyl groups nonrandomly so that blocks of contiguous ionized galacturonate residues are generated (71). These acidic blocks will form interpolymer Ca-bridged junction zones. Pectin randomly de-esterified to the same degree (by NaOH) is much less easily cross-linked by Ca^{2+}, since blocks of concerted Ca-bridges cannot form (38, 71).

Some pectins undergo Ca-independent self-aggregation in vitro at low pH and low water activity (38) (exploited in jam making). Suitable conditions possibly occur locally in the inner structure of the wall matrix.

Oxidative coupling of feruloyl-pectins has recently been demonstrated (60a), and a soluble model compound, feruloyl-galactomannan, will gel in vitro with peroxidase + H_2O_2 owing to the formation of diferuloyl cross-links (29). *p*-Coumaroyl-galactomannan also gels, but via unknown cross-links (29).

Hemicelluloses

In dicot primary walls, the major hemicelluloses are neutral xyloglucans and acidic arabinoxylans (48); in monocots they are acidic arabinoxylans and neutral β-(1→3, 1→4)-glucans (74).

ISOLATION OF CROSS-LINKED FRAGMENTS   The evidence for possible pectin-hemicellulose hybrid fragments is discussed above.

Monocot arabinoxylans are substituted with ferulate, p-coumarate, and p-hydroxybenzoate (34, 70). The ferulate is all Z-ferulate in dark-grown plants, but it can photoisomerize (76). Enzymic hydrolysis yielded a feruloyl-arabinofuranosyl-xylose [surprisingly claimed to be β-linked (66)] and mild acid hydrolysis gave feruloyl-arabinose (B. Ahluwalia and S. C. Fry, submitted), showing that the feruloyl residues are on side-chains. That these feruloyl residues can oxidatively couple is suggested by the isolation, from alkaline hydrolysates of grass cell walls, of diferulic acid (Figure 3; 34, 35, 46, 62). It is yet to be shown whether the two -COOH groups of the diferulate were, in the intact wall, esterified to different arabinoxylan molecules and were thus cross-links. The report of carbohydrate fragments containing diferulate (62) is a step toward this goal.

SOLUBILIZATION OF HEMICELLULOSES   Some hemicelluloses are H-bonded to cellulose. Most chaotropic agents, e.g. 8 M urea or guanidinium thiocyanate, solubilize only a small percentage of hemicellulose from walls, although the very strong chaotropic agent 4-methyl-morpholine-N-oxide hydrate (MMNO) at 120°C solubilized hemicellulose and cellulose completely, suggesting H-bonding (40). However, MMNO cleaves a small percentage of glycosidic bonds (40) and most ester bonds (S. C. Fry, unpublished); its effect on peptide bonds and coupled phenols has not been reported. Therefore, solubilization by MMNO does not prove cross-linking by H-bonds and does not exclude the possibility of other (e.g. ester) cross-links. Nevertheless, MMNO is a valuable tool for future studies.

If diferuloyl esters cross-link hemicelluloses, de-esterification should facilitate solubilization. In an attempt to break ester bonds but not H-bonds, methanolic sodium methoxide was applied to grass walls and then washed out with pure methanol; some arabinoxylan thereby became water-extractable, suggesting that ester cross-links had indeed been present (50). Similarly, neutral hydroxylamine solubilized arabinoxylan from wheat endosperm, again suggesting ester cross-links (45). It remains to be seen whether these esters were diferuloyl or uronoyl bonds.

Pretreatment of maize cell walls with NaClO$_2$ allowed the solubilization of arabinoxylan by milder alkali (0.01M KOH) than otherwise (10). This is consistent with intermolecular cross-links involving phenols. 0.01M KOH, under the conditions used, hydrolyzes few wall-bound feruloyl groups (S. C. Fry, unpublished). However, NaClO$_2$ cleaves a few glycosidic bonds (Table 2), and its use requires caution.

The β-(1→3, 1→4)-glucans of barley can be digested by proteases from an apparent molecular weight of 40,000,000 down to a limit of about 1,000,000 daltons. This suggests that about forty 1,000,000-dalton glucan molecules are anchored to a common protein core. The protein lacks hydroxyproline and the glucan-protein bond is unidentified (18).

FORMATION OF CROSS-LINKS IN VITRO   Hemicelluloses can self-assemble via H-bonds in vitro as seen, for instance, in soluble β-(1→3, 1→4)-glucan/arabinoxylan mixtures (74). Similarly, xyloglucan binds to cellulose in vitro (5). The in vitro cross-linking of feruloyl-galactomannan via diferuloyl bridges has been discussed. Another potential cross-link is an ether between the phenolic -OH group of ferulate and an -OH group on a polysaccharide. The peroxidase-catalyzed formation of this sort of cross-link has been suggested (19, 75).

Wall Enzymes and Lectins

The wall contains numerous enzymes (28). Some of these can be eluted with salts and are probably ionically bound to acidic pectins (14). Others cannot be eluted with salt and are said to be covalently bound (28). The nature of the covalent bond is unknown; one possibility is isodityrosine. Coupling of tyrosine residues may even be an enzyme-activating mechanism (73). The wall may also contain lectins. These can be eluted with salt, and the elution is not facilitated by sugars, indicating that ionic bonds hold the lectins in the wall rather than bonds with specific polysaccharide haptens (32).

WALL MODELS AND PHYSICAL ENTANGLEMENT

It would be premature to draw up a precise model of the growing cell wall. Not enough is known about the siting or abundance of cross-links or the conformation (57) or orientation (36, 59, 60) of the polymers. Nevertheless, it seems likely that H-bonds, Ca-bridges, other ionic bonds, coupled phenols, and ester bonds all play a role in building the wall, as summarized in Figure 4. Other cross-links may also be involved.

An important question is the degree of precision with which cross-links are formed. For instance, it is not known whether particular tyrosine residues in extensin are predisposed to form isodityrosine, and, if so, whether it is done by coupling to the equivalent or to different tyrosine residues in other extensin molecules. Neither is it clear whether multiple bonds tend to occur between the same pair of extensin molecules or if a more ramifying network is produced.

It is also unknown whether polymer-polymer interactions are based mainly on specific bonds or whether physical entanglement plays a major role. The latter seems probable owing to the high molecular weight of wall polymers. Physical entanglement could be more subtle if looped molecules occurred; for

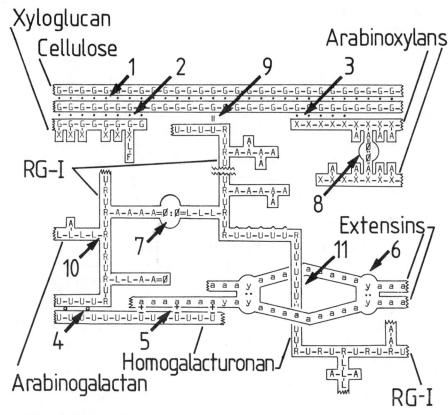

Figure 4 Representative primary structures and possible cross-links of wall polymers. This is not a model of the plant cell wall, and no significance is placed on the chain-length, orientation, conformation, or spacing of the molecules. The diagram illustrates:

(·) Hydrogen-bonds	1. cellulose-cellulose
	2. xyloglucan-cellulose
	3. xylan-cellulose
(○) Calcium bridges	4. homogalacturonan-homogalacturonan
(±) Other ionic bonds	5. extensin-pectin
(:) Coupled phenols	6. extensin-extensin
	7. pectin-pectin
	8. arabinoxylan-arabinoxylan
(=) Ester bonds	9. pectin-cellulose
(−) Glycosidic bonds	10. arabinogalactan-rhamnogalacturonan
(◊) Entanglement	11. pectin-in-extensin

Abbreviations used in the structures are:

A = Arabinose; F = Fucose; G = Glucose; L = Galactose; R = Rhamnose; U = Galacturonic acid; Û = Galacturonic acid methyl ester; a = Amino acid other than tyrosine; y = Tyrosine; y:y = Isodityrosine; Ø = Ferulic acid; Ø:Ø = Diferulic acid.

instance, extensin might form intra- or intermolecular loops via isodityrosine bridges (Figure 4). The loops could envelop microfibrils (16). Alternatively, small (or slender) loops could envelop homogalacturonans, which would then be unable to get out of the noose because the homogalacturonan would be flanked by branched rhamnogalacturonan domains. Formation of loops in extensin (peroxidase-catalyzed coupling of tyrosine residues) would tend to occur in the wall in the vicinity of pectin molecules because pectin ionically binds both peroxidase and extensin, bringing them together for free radical formation. Pectin would also neutralize the positive charges on extensin, facilitating the mutual approach of free radical extensin molecules for coupling. A steering mechanism for the proposed physical entanglement (Figure 4, item 11) can thus be envisaged.

REGULATION OF MAKING AND BREAKING OF CROSS-LINKS

Two enzymes seem particularly promising for studies of cross-link formation: peroxidase, which catalyzes the formation of isodityrosine and diferulate cross-links, and pectinesterase, which produces acidic pectic domains capable of making Ca-bridges and perhaps of acting as a template for isodityrosine formation (see previous paragraph). Pectinesterase possibly also catalyzes transesterification, leading to the formation of uronoyl-sugar ester cross-links (28). It is interesting that both peroxidase and pectinesterase contain or require Ca^{2+} and are inhibited by low pH (13, 78). This is consistent with a regulatory role for these enzymes in making wall cross-links since (a) cross-links lower the extensibility of the wall (20, 21), and (b) Ca^{2+} lowers and H^+ raises wall extensibility (6, 69).

The breaking of cross-links in vivo has been studied very little. Here the situation is complicated by the fact that wall networks can be loosened either by undoing the knots (cross-links) or by cutting the strings (polymers). Enzymes involved could thus include wall-bound esterases, glycanases, or proteases. Progress awaits a clearer definition of the load-bearing polymers and cross-links.

ACKNOWLEDGMENT

I am grateful to the Agricultural and Food Research Council for a grant in support of the previously unpublished work.

Literature Cited

1. Albersheim, P. 1978. Concerning the structure and biosynthesis of the primary cell wall of plants. *Int. Rev. Biochem.* 16:127–50
2. Amadò, R., Aeschbach, R., Neukom, H. 1984. Dityrosine: *in vitro* production and characterization. *Methods Enzymol.* 107: 377–88
3. Andersen, S. O. 1964. The cross-links in resilin identified as dityrosine and trityrosine. *Biochim. Biophys. Acta* 93:213–15
4. Aspinall, G. O., Cottrell, I. W. 1970. Lemon-peel pectin. II. Isolation of homogeneous pectins and examination of some associated polysaccharides. *Can. J. Chem.* 48:1283–89
5. Bauer, W. D., Talmadge, K. W., Keegstra, K., Albersheim, P. 1973. The structure of plant cell walls. II. The hemicellulose of suspension-cultured sycamore cells. *Plant Physiol.* 51:174–87
6. Baydoun, E. A.-H., Brett, C. T. 1984. The effect of pH on the binding of calcium to pea epicotyl cell walls and its implications for the control of cell extension. *J. Exp. Bot.* 35:1820–31
7. Benton, F. L., Dillon, T. E. 1942. The cleavage of ethers with boron bromide. I. Some common ethers. *J. Am. Chem. Soc.* 64:1128–29
8. Blaschek, W., Franz, G. 1983. Influence of growth conditions on the composition of cell wall polysaccharides from cultured tobacco cells. *Plant Cell Rep.* 2: 257–60
9. Brysk, M. M., Chrispeels, M. J. 1972. Isolation and partial characterization of a hydroxyproline-rich cell wall glycoprotein and its cytoplasmic precursor. *Biochim. Biophys. Acta* 257:421–32
10. Carpita, N. C. 1984. Fractionation of hemicelluloses from maize cell walls with increasing concentrations of alkali. *Phytochemistry* 23:1089–93
11. Chambat, G., Joseleau, J.-P., Barnoud, F. 1981. The carbohydrate constituents of the cell wall of suspension cultures of *Rosa glauca. Phytochemistry* 20:241–46
12. Cooper, J. B., Varner, J. E. 1983. Insolubilization of hydroxyproline-rich cell wall glycoprotein in aerated carrot root slices. *Biochem. Biophys. Res. Commun.* 112:161–67
13. Cooper, J. B., Varner, J. E. 1984. Cross-linking of soluble extensin in isolated cell walls. *Plant Physiol.* 76:414–17
14. Crasnier, M., Moustacas, A.-M., Ricard, J. 1985. Electrostatic effects and calcium ion concentration as modulators of bound plant cell wall acid phosphatase. *Eur. J. Biochem.* In press
15. Das, N. N., Das, S. C., Dutt, A. S., Roy,

A. 1981. Lignin-xylan ester linkage in jute fiber *(Corchorus capsularis). Carbohydr. Res.* 94:73–82
16. Epstein, L., Lamport, D. T. A. 1984. An intramolecular linkage involving isodityrosine in extensin. *Phytochemistry* 23: 1241–46
17. Fincher, G. B., Stone, B. A., Clarke, A. E. 1983. Arabinogalactan-proteins: structure, biosynthesis, and function. *Ann. Rev. Plant Physiol.* 34:47–70
18. Forrest, I. S., Wainwright, T. 1977. The mode of binding of β-glucans and pentosans in barley endosperm cell walls. *J. Inst. Brew.* 83:279–86
19. Freudenberg, K., Grion, G. 1959. Beitrag zum Bildungsmechanismus des Lignins und der Lignin-Kohlenhydrat-Bindung. *Chem. Ber.* 92:1355–63
20. Fry, S. C. 1979. Phenolic components of the primary cell wall and their possible role in the hormonal regulation of growth. *Planta* 146:343–51
21. Fry, S. C. 1980. Gibberellin-controlled pectinic acid and protein secretion in growing cells. *Phytochemistry* 19:735–40
22. Fry, S. C. 1982. Phenolic components of the primary cell wall: feruloylated disaccharides of D-galactose and L-arabinose from spinach polysaccharide. *Biochem. J.* 203:493–504
23. Fry, S. C. 1982. Isodityrosine: a new cross-linking amino acid from plant cell-wall glycoprotein. *Biochem. J.* 204:449–55
24. Fry, S. C. 1983. Feruloylated pectins from the primary cell wall: their structure and possible functions. *Planta* 157:111–23
25. Fry, S. C. 1983. Oxidative phenolic coupling reactions cross-link hydroxyproline-rich glycoprotein molecules in the plant cell wall. In *Current Topics in Plant Biochemistry and Physiology,* ed. D. D. Randall, D. G. Blevins, R. L. Larson, B. J. Rapp, 2:59–72. Univ. Missouri Press. 264 pp.
26. Fry, S. C. 1984. Incorporation of [14C]cinnamate into hydrolase-resistant components of the primary cell wall. *Phytochemistry* 23:59–64
27. Fry, S. C. 1984. Isodityrosine—its detection, estimation and chemical synthesis. *Methods Enzymol.* 107:388–97
28. Fry, S. C. 1985. Primary cell wall metabolism. *Oxford Surveys of Plant Molecular and Cell Biology,* ed. B. J. Miflin, 2:1–42. Oxford: Clarendon
29. Geissmann, T., Neukom, H. 1971. Vernetzung von Phenolcarbonsäureestern von Polysacchariden durch oxydative

phenolische Kupplung. *Helv. Chim. Acta* 54:1108–12

30. Grant, G. T., McNab, C., Rees, D. A., Skerrett, R. J. 1969. Seed mucilages as examples of polysaccharide denaturation. *Chem. Commun.* 1969:805–6

31. Gross, A. J., Sizer, I. W. 1959. The oxidation of tyramine, tyrosine and related compounds by peroxidase. *J. Biol. Chem.* 234:1611–14

32. Haass, D., Frey, R., Thiesen, M., Kauss, H. 1981. Partial purification of a hemagglutinin associated with cell walls from hypocotyls of *Vigna radiata*. *Planta* 151:490–96

33. Hanke, D. E., Northcote, D. H. 1974. Cell wall formation by soybean callus protoplasts. *J. Cell. Sci.* 14:29–50

34. Harris, P. J., Hartley, R. D. 1980. Phenolic constituents of the cell walls of monocotyledons. *Biochem. Syst. Ecol.* 8:153–60

35. Hartley, R. D., Jones, E. C. 1976. Diferulic acid as a component of cell walls of *Lolium multiflorum*. *Phytochemistry* 15:1157–60

36. Hayashi, R., Morikawa, H., Nakajima, N., Ichikawa, Y., Senda, M. 1980. Oriented structure of pectic polysaccharides in pea epidermal cell walls. *Plant Cell Physiol.* 21:999–1005

37. Jarvis, M. C. 1982. The proportion of calcium-bound pectin in plant cell walls. *Planta* 154:344–46

38. Jarvis, M. C. 1984. Structure and properties of pectin gels in plant cell walls. *Plant Cell Environ.* 7:153–64

39. Jarvis, M. C., Hall, M. A., Threlfall, D. R., Friend, J. 1981. The polysaccharide structure of potato cell walls: chemical fractionation. *Planta* 152:93–100

40. Joseleau, J.-P., Chambat, G., Chumpitazi-Hermoza, B. 1981. Solubilization of cellulose and other plant structural polysaccharides in 4-methylmorpholine *N*-oxide: an improved method for the study of cell-wall constituents. *Carbohydr. Res.* 90:339–44

41. Keegstra, K., Talmadge, K. W., Bauer, W. D., Albersheim, P. 1973. The structure of plant cell walls. III. A model of the walls of suspension-cultured sycamore cells based on the interconnections of the macromolecular components. *Plant Physiol.* 51:188–96

42. Knee, M. 1978. Properties of polygalacturonate and cell cohesion in apple fruit cortical tissue. *Phytochemistry* 17:1257–60

43. Lamport, D. T. A. 1970. Cell wall metabolism. *Ann. Rev. Plant Physiol.* 21:235–70

44. Lyndon, R. F., Steward, F. C. 1963. The incorporation of ^{14}C-proline into the proteins of growing cells. I. Evidence of synthesis in different proteins and cellular components. *J. Exp. Bot.* 14:42–55

45. Mares, D. J., Stone, B. A. 1973. Studies on wheat endosperm. II. Properties of the wall components and studies on their organization in the wall. *Aust. J. Biol. Sci.* 26:813–30

46. Markwalder, H.-U., Neukom, H. 1976. Diferulic acid as a possible crosslink in hemicelluloses from wheat germ. *Phytochemistry* 15:836–37

47. McNeil, M., Darvill, A. G., Albersheim, P. 1979. The structural polymers of the primary cell walls of dicots. In *Progress in the Chemistry of Organic Natural Products*, ed. W. Herz, H. Grisebach, G. W. Kirby, 37:191–249. Vienna: Springer-Verlag

48. McNeil, M., Darvill, A. G., Fry, S. C., Albersheim, P. 1984. Structure and function of the primary cell walls of higher plants. *Ann. Rev. Biochem.* 53:625–63

49. Monro, J. A., Penny, D., Bailey, R. W. 1976. The organization and growth of primary cell walls of lupin hypocotyl. *Phytochemistry* 15:1193–98

50. Morrison, I. M. 1977. Extraction of hemicelluloses from plant cell-walls with water after preliminary treatment with methanolic sodium methoxide. *Carbohydr. Res.* 57:C4–C6

51. Mort, A. J., Lamport, D. T. A. 1977. Anhydrous hydrogen fluoride deglycosylates glycoproteins. *Anal. Biochem.* 82:289–309

52. Neukom, H., Amadò, R., Pfister, M. 1980. Neuere Erkenntnisse auf dem Gebiete der Pektinstoffe. *Lebensm. Wiss. Technol.* 13:1–6

53. Neukom, H., Markwalder, H.-U. 1978. Oxidative gelation of wheat flour pentosans: A new way of cross-linking polymers. *Cereal Foods World* 23:374–76

54. Northcote, D. H. 1972. Chemistry of the plant cell wall. *Ann. Rev. Plant Physiol.* 23:113–32

55. O'Neill, M. A., Selvendran, R. R. 1980. Glycoproteins from the cell wall of *Phaseolus coccineus*. *Biochem. J.* 187:53–63

56. O'Neill, M. A., Selvendran, R. R. 1985. Hemicellulosic complexes from the cell walls of runner bean (*Phaseolus vulgaris*). *Biochem. J.* 227:475–81

57. Preston, R. D. 1979. Polysaccharide conformation and cell wall function. *Ann. Rev. Plant Physiol.* 30:55–78

58. Roberts, K., Grief, C., Hills, G. J., Shaw, P. 1985. Cell wall glycoproteins: structure and function. *J. Cell Sci. Suppl.* 1: In press

59. Roelofsen, P. A., Kreger, D. K. 1951. The submicroscopic structure of pectin in collenchyma cell walls. *J. Exp. Bot.* 2:332–43

60. Roland, J.-C., Vian, B. 1979. The wall of the growing plant cell: its three dimensional organisation. *Int. Rev. Cytol.* 61:129–66

60a. Rombouts, F. M., Thibault, J. F. 1985. Sugar beet pectin—chemical structure and gelation through oxidative coupling. In *Recent Advances in the Chemistry and Function of Pectins,* ed. M. Fishman, J. J. Jen. Washington, DC: Am. Chem. Soc. In press

61. Selvendran, R. R., Stevens, B. J. H., O'Neill, M. A. 1985. Developments in the isolation and analysis of cell walls from edible plants. In *Biochemistry of Plant Cell Walls,* ed. C. T. Brett, J. R. Hillman, pp. 39–78. Cambridge: Cambridge Univ. Press. 312 pp.

62. Shibuya, N. 1984. Phenolic acids and their carbohydrate esters in rice endosperm cell walls. *Phytochemistry* 23: 2233–37

63. Smidsrød, O., Haug, A., Larson, B. 1966. The influence of pH on the rate of hydrolysis of acidic polysaccharides. *Acta Chem. Scand.* 20:1026–34

64. Smith, J. J., Muldoon, E. P., Lamport, D. T. A. 1984. Isolation of extensin precursors by direct elution of intact tomato cell suspension cultures. *Phytochemistry* 23:1233–40

65. Smith, J. J., Muldoon, E. P., Willard, J. J., Lamport, D. T. A. 1986. Tomato extensin precursors P1 and P2 are highly periodic structures. *Phytochemistry.* In press

66. Smith, M. M., Hartley, R. D. 1983. Occurrence and nature of ferulic acid substitution of cell-wall polysaccharide in graminaceous plants. *Carbohydr. Res.* 118:65–80

67. Stafstrom, J. P., Staehelin, L. A. 1985. Cross-linking patterns in salt-extractable extensin from carrot cell walls. *Plant Physiol.* In press

68. Stevens, B. J. H., Selvendran, R. R. 1984. Hemicellulosic polymers of cabbage leaves. *Phytochemistry* 23:339–47

69. Taiz, L. 1984. Plant cell expansion: Regulation of cell wall mechanical properties. *Ann. Rev. Plant Physiol.* 35:585–657

70. Tanner, G. R., Morrison, I. M. 1983. Phenolic-carbohydrate complexes in the cell walls of *Lolium perenne. Phytochemistry* 22:1433–39

71. Taylor, A. J. 1982. Intramolecular distribution of carboxyl groups in low methoxyl pectins—a review. *Carbohydr. Polymers* 2:9–17

72. Thom, D., Grant, G. T., Morris, E. R., Rees, D. A. 1982. Characterisation of cation binding and gelation of polyuronates by circular dichroism. *Carbohydr. Res.* 100:29–42

73. Tressel, P., Kosman, D. J. 1980. *o,o*-Dityrosine in native and horseradish peroxidase-activated galactose oxidase. *Biochem. Biophys. Res. Commun.* 92:781–86

74. Wada, S., Ray, P. M. 1978. Matrix polysaccharides of oat coleoptile cell walls. *Phytochemistry* 17:923–31

75. Whitmore, F. W. 1976. Binding of ferulic acid to cell walls by peroxidases of *Pinus elliottii. Phytochemistry* 15:375–78

76. Yamamoto, E., Towers, G. H. N. 1985. Cell wall bound ferulic acid in barley seedlings during development and its photoisomerization. *J. Plant Physiol.* 117:441–49

77. Yamaoka, T., Chiba, N. 1983. Changes in the coagulating ability of pectin during the growth of soybean hypocotyls. *Plant Cell Physiol.* 24:1281–90

78. Yamaoka, T., Tsukada, K., Takahashi, H., Yamauchi, N. 1983. Purification of a cell-wall bound pectin-gelatinizing factor and examination of its identity with pectin methylesterase. *Bot. Mag. Tokyo* 96: 139–44

Ann. Rev. Plant Physiol. 1986. 37:187–208

SIDEROPHORES IN RELATION TO PLANT GROWTH AND DISEASE[1]

J. B. Neilands

Department of Biochemistry, University of California, Berkeley, California 94720

Sally A. Leong

Department of Plant Pathology, USDA-ARS, University of Wisconsin, Madison, Wisconsin 53706

CONTENTS

INTRODUCTION

Siderophores (Gr. "iron-bearers") (35) are defined as low molecular weight, virtually Fe(III)-specific ligands produced by microorganisms as scavenging agents in order to combat low iron stress. Nearly all aerobic and facultative anaerobic bacterial species form siderophores, but a few exceptions are known among the *Legionella, Neisseria, Hemophilus, Yersinia, Listeria,* and *Streptococcus.* When chemical and direct nutritional tests for siderophores are negative, one can resort to a bioassay based on reversal of iron limitation imposed by a synthetic chelator, ethylenediamine-di-(*o*-hydroxyphenylacetic acid) (EDDA) (46). All of the known probes available failed to provide evidence in *Legionella* for siderophore synthesis (59). These are exceptions, however, and it is relatively safe to assume that most bacteria and virtually all fungi, with the possible exception of *Saccharomyces cerevisiae,* may be found to form siderophores. Failure to detect a siderophore in the supernatant fluid of iron-starved cultures leaves open the mode of iron assimilation in vitro and in vivo, although in the latter situation siderophores may well be expressed. Also, in a few instances the siderophore is not excreted, but is accumulated within the cell or its envelope. A case in point are the mycobactins from mycobacteria (73).

The biological rationale for siderophore synthesis is founded on the propensity of ionic iron to form, under aerobic conditions at biological pH, oxyhydroxide polymers of the general composition FeOOH (66). In the absence of oxygen, Fe(II) remains soluble throughout the entire range of biological pH, thus obviating the need for siderophores during the anaerobic phase of life. As the atmosphere gradually switched from reducing to oxidizing, the surface iron precipitated in the various forms of ferric hydroxide ($K_{sol} = 10^{-38}$ M), and synthesis of siderophores was necessary for assimilation of a nutritionally essential element which, although abundant, became for all practical purposes unavailable. At this stage of the evolutionary process it became a critical necessity to regulate iron intake since, as we now know, the metal facilitates the conversion of O_2 to the profoundly toxic oxidizing free radicals such as OH· (84).

The purpose of this short review is to direct attention to the relationship between siderophores and plant growth and disease. Siderophores per se have been reviewed extensively in recent years, and the reader is referred to several previous chapters in this series (48–50). Current research on the molecular genetics of siderophore systems in *Escherichia coli* has been surveyed elsewhere (12, 51). Volume 58 of *Structure and Bonding* is devoted entirely to siderophores of microorganisms and plants; the comprehensive review therein by Hider (31) is of special interest. In the summer of 1985, separate meetings on different aspects of iron in biology were held in Nebraska, England, France, and West Germany. The meeting in Tübingen, West Germany, had siderophores as its major focus; the proceedings will be edited by G. Winkelmann, D. vander Helm, and J. B. Neilands and published by Verlag Chemie. The NATO Workshop at Wye College, Kent, is of specific interest in the present context. This symposium was titled: "Iron, Siderophores and Plant Disease." The transactions of this meeting, a gathering of 57 speakers from 11 different countries, are being edited by T. R. Swinburne, East Malling Research Station, Kent, England, and will be published by Plenum Press.

In what follows we have chosen to survey briefly some of the individual siderophores which in one way or another have been connected to plant growth or disease. This will serve as a background for more specialized reviews which are bound to follow in this rapidly expanding field of research.

It is now more than 30 years since siderophore activity was shown to be present in soil (41). Hence it is pertinent to consider the role of these chelating agents in the iron nutrition of the green plant. A considerable body of work with microorganisms pathogenic to humans and animals has provided convincing evidence that iron is a virulence factor (19, 83). All living tissue, regardless of source, must be low in free iron. This is a condition conducive to elaboration of siderophores (21). As a consequence, it is natural to attempt to extend the clinical and animal experience to plants. Thus the focus of our discussion will be on the interaction of siderophores with plants as this relationship applies to growth and disease.

FERRICHROME

The first growth factor for a bacterium was reported in 1912. The compound, which became known as mycobactin, was fully characterized as a hydroxamic acid by Snow (73) and colleagues before its role as a siderophore was recognized. Spurred on by the spectacular success with penicillin, American (78) and Soviet (22) scientists searched for antimetabolites against infectious disease and cancer. One result was an antibiotic, designated grisein and albomycin in the USSR and US, respectively. These were subsequently shown to be identical, iron-hydroxamate complex compounds. While the isolation of mycobactin and grisein-albomycin preceded ferrichrome, the last-named is generally regarded as the prototypical siderophore since it was the first member of this now important and diversified class of natural products to be studied as an iron carrier. Ferrichrome was associated with iron assimilation since, unlike mycobactin, it was obtained as the iron complex and was derived from the smut fungus *Ustilago spherogena,* an organism remarkable for its accumulation of very high levels of cytochrome *c*.

Ferrichrome (Figure 1) is one member of a large family of siderophores that may be described chemically as cyclic peptides containing a tripeptide of N^δ-acyl-N-$^\delta$-hydroxyornithine and a variable combination of glycine, serine, and occasionally alanine. The simplest member of the series is ferrichrome itself, wherein a tripeptide of glycine completes the cyclic moiety and all three acyl functions are acetyl residues. The presence of different acyl moieties attached to the same peptide makes for a very large variation in the number of structural types (17, 33). Probably all members of the ferrichrome series must contain at least one glycine in order for the peptide chain to execute the characteristic hydrogen-bonded β-turn detected by both crystallography and high resolution NMR analysis.

Apart from their early identification as siderophores, the ferrichromes are noteworthy for several additional reasons. They were the first siderophores shown to be overproduced by iron starvation (20), an observation that was promptly extended to other fungi and to bacteria and found to apply to both hydroxamate and catechol-type ligands (21). This enabled production from many different sources of quite reasonable amounts of pure siderophores, an essential consideration in the years before the sensitive spectroscopic character-ization tools, such as NMR and MS, became available. Winkelmann (86) surveyed a large collection of fungi and found all of them to produce one or more ferric hydroxamate-type siderophore; none failed to form ferrichrome.

Ferrichrome serves as a source of iron for tomato and duckweed (*Lemna* sp.) and assays with the siderophore auxotroph *Arthrobacter* JG-9 indicated some uptake of the intact chelate (53). Ferrichrome proved to be a superior source of iron when compared to the standard synthetic chelate EDDA for iron nutrition of oat (*Avena sativa*, var. Victory), and it was suggested that for monocots the natural chelates may be a more efficient source of the metal (60).

Growth tests with *Arthrobacter* JG-9 indicate the presence of a few nM "ferrioxamine B" equivalent in bulk soil and approximately ten times this concentration next to the roots (61). Although *Arthrobacter* does not respond to catechol-type siderophores, it is omnivorous with respect to the hydroxamate-type and will even use synthetic chelates at the proper concentration. Confirma-tion of the presence of ferrichrome-type siderophores in soil was obtained through use of a differential assay with *E. coli* RW193 and its *ton*A derivative AN193, the latter defective in ferrichrome uptake by virtue of the absence of an outer membrane receptor protein (58). Examination of synthetic retro-ferrichrome indicates that uptake of the natural Λ,*cis* coordination isomer is steriospecific (85).

A soil isolate, *Pseudomonas* Fc-1, grows on the ferrichromes as its sole source of C and N. The initial attack appears to be via a peptidase that opens the cyclohexapeptide ring (80).

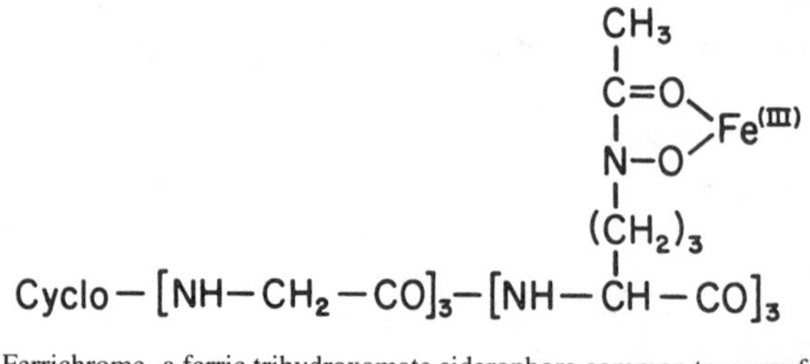

$$\text{Cyclo} - [NH-CH_2-CO]_3 - [NH-CH-CO]_3$$

with pendant group:

CH₃ — C=O, N—O coordinating to Fe(III), N—(CH₂)₃

Figure 1 Ferrichrome, a ferric trihydroxamate siderophore common to many fungal species.

Ferrichrome is possibly the most commonly produced fungal siderophore. It is made by all species of *Penicillium, Aspergillus niger, Ustilago* spp., and other fungi (86). It is also commonly used as a siderophore by bacteria, including *E. coli*, which have not as yet been demonstrated to make this compound (50). There is nothing in the literature to connect ferrichrome to plant pathology, save for its production by *U. sphaerogena* and *U. maydis*, which are parasites on barnyard grass and corn, respectively. A related siderophore, malonichrome (17), has been identified from *Fusarium roseum* and from *F. oxysporum* (J. Leong, personal communication).

The high-affinity iron transport system of *U. maydis* is currently under investigation with the goal of defining the role of siderophore-mediated iron assimilation in disease (S. A. Leong, personal communication). The fungus produces two siderophores, ferrichrome and ferrichrome A. Mutants defective in the production of one or both siderophores have been isolated by chemical mutagenesis. One class of mutants, unable to produce either siderophore, can be fully complemented by supplementation of culture media with N^δ-hydroxyl-L-ornithine, suggesting that ferrichrome and ferrichrome A share a common biosynthetic step, namely, the N^δ-hydroxylation of L-ornithine. Pathogenicity tests employing these siderophoreless mutants should establish whether or not these compounds are important for plant parasitism in this system.

In a recent survey of opportunistic and systemic fungi pathogenic to humans, evidence for the presence of siderophores of the hydroxamate type was found in *Absidia crymbifera, Aspergillus niger, Rhizopus arrhizus, Rhizopus oryzae, Blastomyces dermatitidis, Histoplasma capsulatum, Sporothrix schenickii, Candida albicans,* and *Trichophyton mentagrophytes*. This study represents the first application of periodate oxidation as a method for detection of natural hydroxamic acids (27).

RHODOTORULIC ACID

The siderophore-active dihydroxamic acid, rhodotorulic acid (Figure 2), is formed by *Rhodotorula pilimanae* ATCC26423, a basidiomycetous yeast, and by related fungi (3). It can be regarded as a dipeptide fragment of ferrichrome

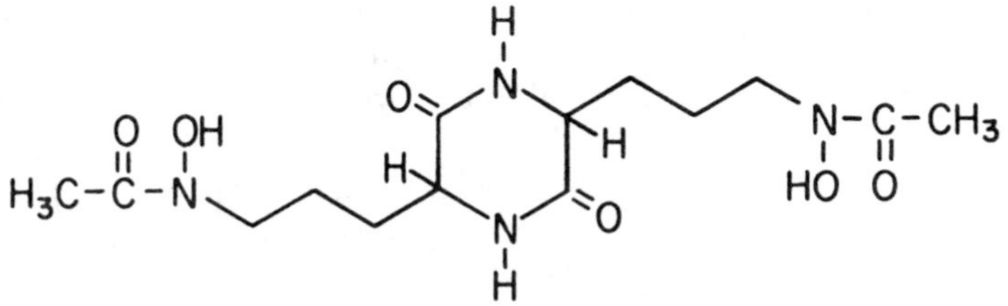

Figure 2   Rhodotorulic acid, a siderophore from basidiomycetous yeasts and related fungi.

and has been shown to form a chelate containing 2/3 atom of Fe(III). Since rhodotorulic acid is formed in very large quantities by *R. pilimanae,* a number of practical experiments with this siderophore are possible.

Trace amounts of rhodotorulic acid and other siderophores gave typical "Green Islands" when applied to excised bean leaves (4). A biochemical explanation for this dramatic result, which was not observed with the synthetic chelator EDTA, is yet to be disclosed. It may be related to the ability of siderophores to bind iron in such a way that generation of oxygen radicals is prevented, an effect that has been observed with deferriferrioxamine B.

A role for siderophores in the weathering of rocks is supported by the finding (2) that powdered silicate standards (200 mesh) shaken 3 weeks in phosphate buffer containing 0.01 M rhodotorulic acid released substantial amounts of Fe and, surprisingly, Mg. Rocks in arid regions appear to acquire a typical reddish orange or darker lacquer known as "desert varnish." Recently, microcolonial fungi have been found associated with such coatings (74).

The advantages of using rhodotorulic acid as a model for investigation of the molecular genetics of biosynthesis of a eukaryotic siderophore have been enumerated and supported experimentally by isolation of mutants (40).

Dimerumic acid, from *Fursarium dimerum,* is a siderophore structurally identical to rhodotorulic acid save for the substitution of the two acetyl groups with residues of trans-Δ^2-anhydromevalonic acid. It has been isolated from iron-stressed cultures of *Verticillium dahliae,* but no information is available to implicate it as a virulence factor in this soil-borne phytopathogen (30).

FERRIOXAMINES

The mesylate salt of deferriferrioxamine B is available commercially from Ciba-Geigy under the trade name Desferal. It is employed for deferration therapy in cases of transfusion-induced siderosis but is of limited value because it is not orally effective. Siderophores of the ferrioxamine family have been reported in over 80% of all actinomycetal strains investigated, with ferrioxamine class B representing the predominant component. It follows that siderophores of the ferrioxamine class can be expected to occur generally in soil and to be stabilized at some overall steady state concentration established by the rate of synthesis and degradation. All members of this large class of hydroxamate siderophores are comprised of repeating units of 1-amino-ω-N-hydroxyaminoalkane (pentane or butane) alternated with succinic or acetic acids as acyl functions. Both linear and cyclic members are known (Figure 3). It has recently been reported that higher plant NADH:nitrate reductase (EC 1.6.6.1), which is known to reduce ferricyanide, cytochrome *c,* and ferric citrate in addition to nitrate, is also capable of electron transfer to ferrioxamine B (15). The authors propose that siderophores may furnish iron to higher plants and that nitrate reductase in the roots may function as a reductase for the complexed Fe(III).

FUSARININES

The fusarinines (fusigens) are linear and cyclic hydroxamic acids composed of units of N^δ-hydroxyornithine bearing anhydromevalonic acid as the acyl substituent on the N^δ atom (Figure 4). The α-amino group of the separate units, which are linked via ester bonds, may be either free or acylated with acetic acid. Siderophores of this genre are produced by *Fusarium roseum, Fusarium cubense,* and other fungi.

Pseudomonas putida is capable of suppressing disease caused by *Fusarium oxysporum,* a wilt pathogen (64). A similar control effect was achieved with Fe·EDDA or EDDA (65). A direct correlation has been observed between siderophore synthesis in fluorescent pseudomonads and their capacity to inhibit germination of chlamydospores of *F. oxysporum* (16b, 16c, 72). The mechanism of wilt control was attributed to competition for iron, the bacteria being assumed to generate a chelator which by some mechanism inhibits the fungi (34). EDDA, which can inhibit fungal germ tube growth, presumably acts in the same manner. An additive effect was observed when both agents were employed. This was assigned to growth stimulation of the *Pseudomonas* spp. and to growth inhibition of the fungus by the synthetic chelator (64, 65). Data on the stability of ferric fusarinine does not seem to be available, although the value can be assumed to be in the range of that for other trihydroxamates, which would be ca 10^{30}. In the case of aerobactin and enterobactin, factors other than binding constant appear to play a prime role in infections of the human caused by *E. coli.*

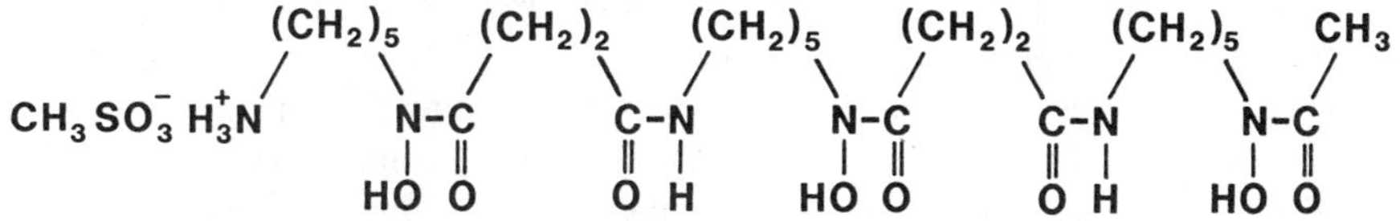

Figure 3   Deferriferrioxamine B mesylate salt, from *Streptomyces* spp., available commercially as Desferal (Ciba-Geigy).

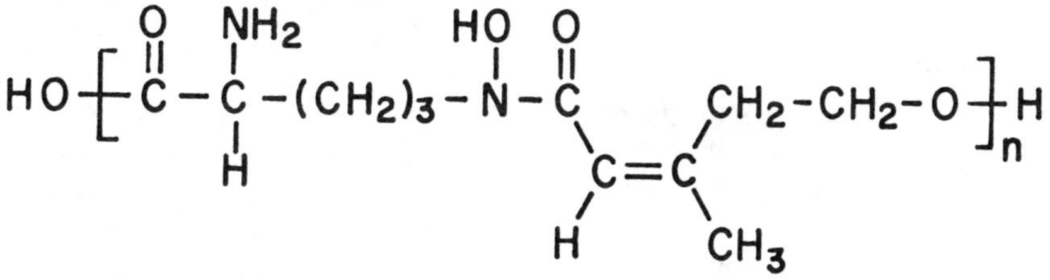

Figure 4   General structure of linear and cyclic (n=3) fusarinines, from *Fusarium* spp. and other fungi. These siderophores may also be acetylated on the α-amino group.

While competition for a limited pool of soluble iron appears to play an important role in suppression of soil-borne diseases, quite different findings have been made for the suppression of disease symptoms in aerial tissues of plants. In the latter case various iron chelators such as 2,3-dihydroxybenzoic acid, EDTA, and a siderophore from a fluorescent pseudomonad were found to stimulate, at high concentration, germination and appressorium formation of *Colletotricum musae* and *Botrytis cinerea* spores both in vitro and in vivo (13, 14, 29). The iron complexes or iron salts inhibited germination and reduced lesion development. Moreover, spores grown in high iron media were found to germinate less readily than those produced in a low iron medium. The inhibitory effect could be reversed by incubation with iron chelators. These data suggest that iron inhibits spore germination in these fungi. The ferrisiderophore ferricrocin has been reported to act at low concentration as a germination factor for conidia of *Neurospora crassa* (32).

Interestingly, iron or iron chelates have also been used to correct symptoms of various viral diseases of plants. Chlorotic lesions are often associated with plant infections caused by viruses, bacteria, and fungi. It is intriguing to speculate that plants, like the mammalian host, attempt to control infection by creating an iron-deficient environment at the site of infection. Of course, pathogen-derived factors such as toxins and siderophores might induce chlorosis of host tissue.

SCHIZOKINEN

This siderophore was so named to recognize its effect in reducing the lag time of growth of *Bacillus megaterium* ATCC19213. As may be seen in Figure 5, it is comprised of a residue of citric acid substituted on the distal carboxyl groups with 1-amino-3-(*N*-hydroxy-*N*-acetyl)aminopropane. Schizokinen is the simplest member of these citrate-containing siderophores which also includes aerobactin and arthrobactin (ferric arthrobactin = "terregens factor"). Miller and coworkers (36, 44) have achieved the chemical synthesis of all three of these siderophores. Like the mycobactins and ferrioxamines, these are examples of hydroxamic acids synthesized by bacteria rather than by fungi. A remarkable feature of schizokinen is the fact that it is produced by such seemingly disparate species as *B. megaterium* and by various strains of cyanobacteria such as *Anabaena* spp. (56, 69).

Schizokinen has the distinction of being the first, and as yet only, siderophore to be identified by chemical and physical methods to be present in soil (1). One kilo (dry weight) of rice paddy soil yielded 0.13 to 0.20 mg of purified siderophore, characterized by paper chromatography and NMR as schizokinen. Blue-green algae, common inhabitants of rice paddies, are suspected to be the source of schizokinen.

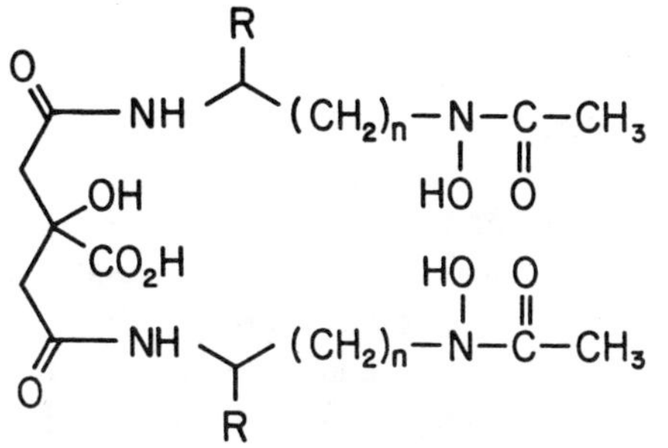

Figure 5 General structure of the citrate-containing siderophores from different bacterial species.

	R	n	Source
Aerobactin	COOH	4	Various genera of enterobacteria
Terregens factor (arthrobactin)	H	4	*Arthrobacter* spp.
Schizokinen	H	2	*Bacillus megaterium, Cyanobacterial* spp.

Aerobactin, a higher homologue of schizokinen, was first isolated from *Aerobacter aerogenes* 62-I and subsequently shown to occur in a number of other enteric bacteria where its genetic determinants may be encoded on large, conjugative plasmids or on the chromosome (10). Aerobactin has been studied extensively from the standpoint of its molecular genetics and has proved to be transcriptionally regulated by iron (11). The biosynthetic and transport genes comprise an operon (16a) contained on plasmid ColV ("V" = "virulence") of clinical isolates of *E. coli* and preceded upstream by a main, exceptionally strong, iron-regulated promoter. A chromosomal locus designated *fur* (*Fe* *u*ptake *r*egulation), first discovered in *Salmonella typhimurium,* has been mapped at 15.7 min on the *E. coli* chromosome (5, 28). The *fur* gene has been cloned and sequenced and shown to code for a 17K polypeptide containing an exceptionally high level of histidine (63). The mechanism of action of the Fur protein is presently unknown, but it has been speculated that it binds iron, as Fe(II), to effect negative control of the multiple iron-regulated operons and genes in *E. coli*. It is pictured as performing this feat through competition for RNA polymerase at the −10 region of the operator located upstream from the regulated genes (51); however, it may well have other functions. Much detail still remains to be worked out regarding the mode of regulation of iron assimilation in prokaryotic species, including the basic mode of action of the Fur protein, and essentially nothing is known about this important process in higher forms of life.

Over 30 years ago Lochhead and his colleagues in Canada isolated from soil species of *Arthrobacter* which both produced and required unique growth factors (41). Thus *Arthrobacter pascens* and other *Arthrobacter* sp. formed a substance, "Terregens Factor," required for growth by *Arthrobacter terregens*. The latter bacterium was found to respond to ferrichrome, to a variety of other hydroxamate-type siderophores, and to extracts of soil. In Berkeley a closely related strain of *Arthrobacter,* isolated via its capacity to degrade puromycin, was similarly found to require siderophores (25). This strain has been deposited in the ATCC as *A. flavescens* 25091. It is commonly used for bioassay of hydroxamate siderophores since it fails to propagate on a basal medium containing yeast extract in the absence of iron chelates (catechols are inactive). The fact that yeast extract can be employed in the assay medium underscores once more the absence of hydroxamate type siderophores in *S. cerevisiae*.

ENTEROBACTIN

In 1970 studies on aromatic biosynthesis in *E. coli* and iron transport in *S. typhimurium* led to the isolation of a novel catechol siderophore named, respectively, enterochelin and enterobactin. The catechol nucleus was first correlated with siderophores in iron-starved cultures of *Bacillus subtilis* where it was found, as 2,3-dihydroxybenzoic acid, conjugated with glycine. Enterobactin (Figure 6), which is the cyclic trimer of 2,3-dihydroxybenzoylserine, forms thermodynamically stable, six-coordinate chelates with Fe(III) at neutral and alkaline pH levels. The high pKa values of the ring hydroxyl groups leads to drastic reduction in affinity for Fe(III) as the pH is lowered below 7, and at higher acidity the iron is held decidedly less firmly than in the trihydroxamates. In addition, at lower pH the complexed ion is reduced by an intramolecular redox reaction which is facilitated in nonaqueous media (31).

Enterobactin is probably produced by all genera of the enteric bacteria. It presently has no special relevance to plant physiology and is listed here mainly because it is regarded as the model catechol-type siderophore. Furthermore, since mutants defective in enterobactin synthesis failed to grow on minimal media containing excess citrate, an iron (III) ligand *S. typhimurium* cannot use, a direct role for siderophores in high-affinity iron assimilation was for the first time established by genetic criteria (57).

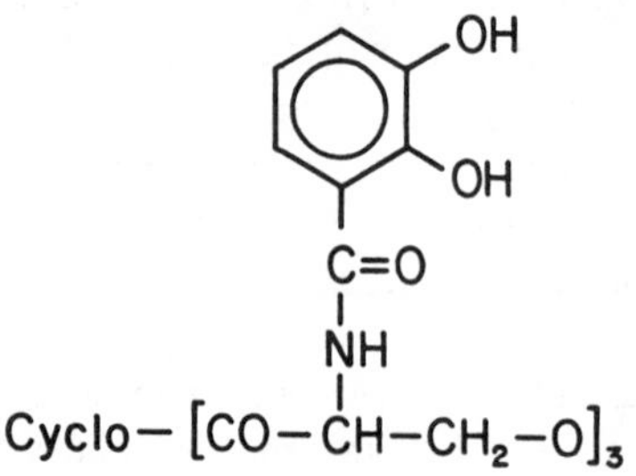

$$\text{Cyclo} - [CO - CH - CH_2 - O]_3$$

Figure 6 Enterobactin, a catechol-type siderophore common to many enteric bacterial species.

Ferric enterobactin is transported by an 81K outer membrane protein in *E. coli*. This receptor also serves as a binding site for colicin B.

Current work with *Erwinia chrysanthemi*, a pathogen of saintpaulia plants, shows that mutants resistant to a bacteriocin and defective in synthesis of one to three low iron-induced outer membrane proteins with molecular weights in the 80 kd range are unable to elicit symptoms of soft rot disease in susceptible plants (18). This is the first, albeit tentative, evidence that siderophore systems may play a role in infections of plant tissue. Unfortunately, the siderophore(s) produced by *E. chrysanthemi* is presently not known. Should enterobactin prove to be responsible for the observed effect, this would be the converse of the findings in clinical isolates of *E. coli* where this siderophore was found to bind tightly to serum albumin and to be inferior to aerobactin as an iron-scavenging agent. There is yet no report in the literature that siderophores are synthesized in vivo in plants, although elboration of these compounds in situ in the case of infections of animals has been demonstrated clearly. Also, in contrasting aerobactin and enterobactin it is entirely possible that the presence of a catechol is better than complete absence of any siderophore.

AGROBACTIN

Agrobactin (Figure 7) may be chemically described as spermidine substituted on its primary amino groups with 2,3-dihydroxybenzoic acid and on its secondary amino group with 2,3-dihydroxybenzoylthreonine, the latter residue cyclized as its oxazoline (52). The first such linear catechol was reported from *Micrococcus* (now *Paracoccus*) *denitrificans* in 1975 by Tait (76), who apparently overlooked the presence of the oxazoline ring. The compound, originally designated Compound III, has been structurally revised and renamed parabactin (55). A third linear catechol-type siderophore is vibriobactin, from *Vibrio cholerae* (26). It is unique in having as its backbone the unusual polyamine nor-spermidine. The two oxazoline rings are unsymmetrically de-

Figure 7   Agrobactin (R=OH) and parabactin (R=H), linear catechol-type siderophores from *Agrobacterium tumefaciens* and *Paracoccus denitrificans*, respectively.

ployed in vibriobactin, a condition which gives rise to duplicated NMR spectra owing to rotation around the tertiary amide bond.

Agrobactin, from *Agrobacterium tumefaciens* B6, was the first siderophore to be examined from the standpoint of its possible role as a virulence factor in plant disease (38). A bacterial siderophore was chosen for this study since, following characterization of the compound, its biological relevance could be ascertained through mutational analysis. The results showed that mutants of *A. tumefaciens* defective in synthesis of agrobactin were still capable of causing crown galls in sunflower. The bacteria were reisolated and shown not to have reverted in planta. In retrospect it seems that this particular system may not have constituted a fair test of siderophores as virulence factors in microorganisms attacking plants. In theory, only a single cell is needed to transform the plant and this does not require much iron-scavenging ability on the part of the pathogen. Secondly, *A. tumefaciens* can transport citrate, and this polycarboxylic acid is present in the plant. Nonetheless, agrobactin may be important to the bacterium for survival outside of plant tissue.

Agrobactin has been found to promote uptake of ^{59}Fe and concomitant synthesis of chlorophyll in both mono- and dicotyledonous plants grown in low iron nutrient solution (6). Other catechol-type siderophores such as parabactin, vibriobactin, and enterobactin have yet to be tested.

RHIZOBACTIN

Siderophore production by the Rhizobia has long been of special interest in view of the prominent role of iron enzymes at several stages in the nitrogen fixation and assimilation process. The iron enzymes and proteins involved include ferredoxin, hydrogenase, nitrogenase, and leghemoglobin. Although some *Rhizobia* are known to excrete bioactive catechols at low iron, none of these compounds have been fully characterized (46a). Rhizobactin, from *Rhizobium meliloti* DM4, represents a novel structure for a siderophore since it is neither catechol nor hydroxamate (Figure 8) (71). It is best described as a natural complexone, a family of synthetic ligands related to EDTA and studied intensively over the years in the inorganic chemistry laboratories of the Eidgenossische Technische Hochschule in Zurich. A ligand of this type can be expected to bind six-coordinate trivalent metal ions to yield mononuclear complexes with a single negative charge. A certain selectivity for Fe(III) is doubtless afforded by the preponderance of oxygen in the coordination sphere and in particular by the presence of the α-hydroxycarboxylate function, the ionization of which will account for the net negative charge on the metal complex.

Little is yet known about the biology of rhizobactin. Only six out of 13 isolates of *R. meliloti* were stimulated in the bioassay, suggesting that the siderophore is strain specific (70).

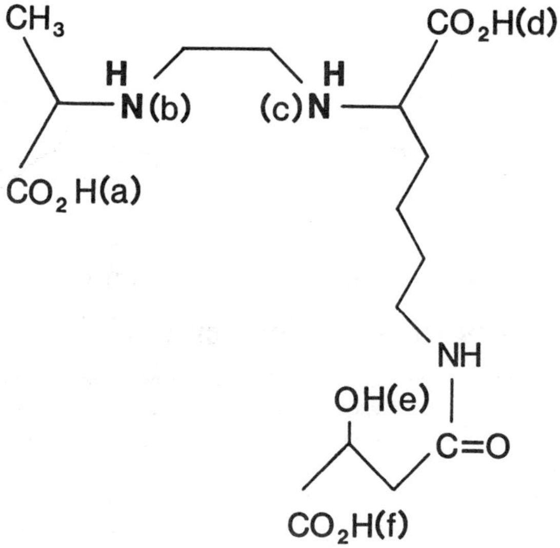

Figure 8   Rhizobactin, from *Rhizobium meliloti*. The letters a, b, c, d, e, and f indicate metal binding centers.

PSEUDOBACTIN

The pseudomonads represent a diversified group of Gram negative bacteria widely distributed in the soil and often found in close association with plant roots. The fluorescent members, which belong to Group 1 on the basis of genetic homology, elaborate yellow-green water-soluble siderophores when cultured at low iron levels. These are variously called pyoverdines and pseudo-bactins. The first member of the series to be fully characterized was pseudobac-tin (77), a linear hexapeptide containing as Fe(III) ligands a yellow-green sub-stituted 2,3-diamino-6,7-dihydroxyquinoline nucleus, an α-hydroxycarboxylate function, and a terminal, cyclic hydroxamic acid (Figure 9). Probably all fluores-cent pseudomonads produce related pseudobactin-like siderophores that differ principally in the number and configuration of the amino acids in the peptide chain. For example, pyoverdine Pa from *Pseudomonas aeruginosa* PAO 1 ATCC 15692 appears to be an octapeptide containing the usual fluorescent chromophore and two hydroxamic acid functions, one of which bears a formyl group as the acyl substituent (82). Some *Azotobacter* spp. form fluorescent yellow-green peptides at low iron growth, but these remain to be characterized (16). The *Azotobacter*, such as *A. vinelandii*, produce in addition 2,3-dihydroxybenzoic acid and N,N'-*bis*-(2,3-dihydroxybenzoyl)-L-lysine (54). These compounds are capable of solubiliz-ing iron from minerals that contain little freely exchangeable iron. Bacteria grown in the presence of iron-laden minerals were depressed for siderophore production. This study highlights the distinction that must be made between total iron and bioavailable iron in any milieu.

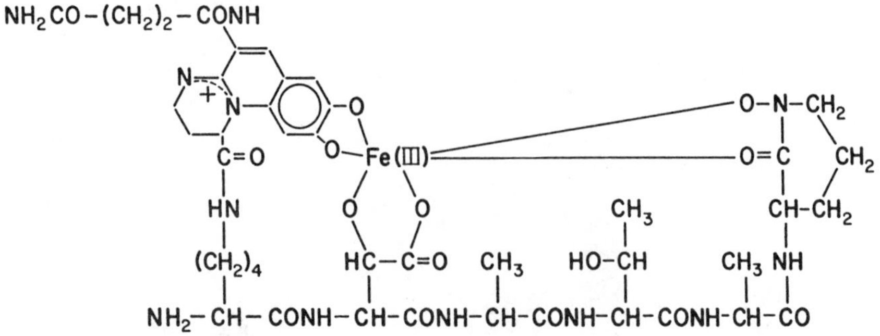

$NH_2CO-(CH_2)_2-CONH$

Figure 9   Pseudobactin from *Pseudomonas* B10, a siderophore characterized by the presence of a yellow-green fluorescent chromophore. Siderophores of the pseudobactin-pyoverdine class are common products of iron-stressed fluorescent *Pseudomonas* spp.

The fluorescent pseudomonads have by now been studied quite intensively for their ability to stimulate plant growth and suppress wilt caused by *Fusarium oxysporum* and take-all root disease caused by *Gaeumannomyces graminis* [for review see Schroth & Hancock (68)]. Nonfluorescent siderophores, closely related to the pyoverdines and pseudobactins, are also produced by the fluorescent pseudomonads. Colorless pseudobactin A slowly converts to pseudobactin in aqueous media and may be a precursor of the latter; both compounds act as siderophores in *Pseudomonas* B10. The usual interpretation is that plant growth promoting rhizobacteria excrete an agent or agents that stimulate or protect the plant by control of nutrient supply, by antibiosis, by hormone action, or by some combination of these mechanisms. The agents most commonly implicated in antibiosis against pathogens are the fluorescent siderophores, a conspicuous product of Group 1 *Pseudomonas* spp. growing on low iron media.

That microorganisms display variable capacity to utilize siderophores of exogenous origin has been known for many years. This principle will be manifest as an iron-reversible antibiosis. An example may be the cyanobacterial blooms in fresh water resulting from elaboration of siderophores not active for competing algal forms (47). Outer membrane proteins (receptors?) appearing under iron deprivation exhibit strain-specific variable electrophoretic mobilities, a phenomenon that could account for the specificity of uptake of ferrisiderophores of the pseudobactin-pyoverdine class (45).

In one study the growth of several test fungi was found to be inhibited by siderophores, but no evidence was adduced for a role of these chelating agents in soil mycostasis (42). However, three *Pseudomonas* isolates, producing mainly siderophores in vitro rather than other inhibitory substances, were found to diminish the narrow rotation yield losses in certain field crops. The mechanism was thought to be via control of the growth or metabolism of deleterious organisms (24). Over half of the isolates in a collection of 112 *Pseudomonas* spp. showed inhibitory action against 5 pathogenic and saprophytic fungi, 5 Gram-positive and 4 Gram-negative bacteria. The results were attributed mainly to production of fluorescent siderophores rather than to other agents that might have been present (23).

Fluorescent pseudomonads isolated from a soil suppressive to *Fusarium* wilt also reduced take-all of wheat caused by *Gaeumannomyces graminis* var. *tritici* and ophiobolus patch caused by *G. graminis* var. *avenae* (87). Since there was no correlation between in vitro antibiosis on agar plates and suppression of disease in potted plants, it was concluded that the mechanism of biocontrol was via competition for iron. The fungal pathogen does not appear to produce siderophores of a type analogous to those of the fluorescent *Pseudomonas* spp., but it is of course conceivable that it could form siderophores of an alternative type.

Strains of *Pseudomonas* spp., selected on the basis of their antibiosis against *G. graminis* var. *tritici,* when applied to seeds significantly suppressed disease in winter wheat grown in the greenhouse or in the field. The effect was believed to result from a direct inhibition of *G. graminis* rather than to an indirect effect on the plant or other soil organisms (81). By mutational analysis siderophores as well as a phenazine antibiotic have been implicated in the control of this fungus (D. M. Weller, personal communication). These compounds may act at different times during infection. Siderophores may be required for growth over the root surface while the antibiotic may be needed during host penetration. Siderophores have also been implicated in suppression of pythium rot of winter wheat (D. M. Weller, personal communication).

Finally, in contrast to the results found with agrobactin, a preparation of pseudobactin from *Pseudomonas* B10 was found to inhibit uptake of iron by higher plants (7). It may thus be that plants, like microorganisms, are selective about the particular siderophore that will serve them as a source of iron.

PHYTOSIDEROPHORES

Plants are known to synthesize a family of linear hydroxy and amino substituted iminocarboxylic acids, several members of which contain the 4-membered azetidine ring (75; Figure 10). Nicotianamine, the initial compound discovered in the series, is characterized by the presence of an amino group alpha to the terminal carboxyl function; in mugineic acid this group is replaced by an hydroxyl, and the further introduction of an hydroxyl function at the 2' position results in a dramatic shift in specificity from softer to harder metal ions. Other compounds here grouped as putative phytosiderophores include avenic acid from oats, 2'-deoxymugineic acid from wheat, 3-hydroxymugineic acid from rye, and distichonic acid from beer barley.

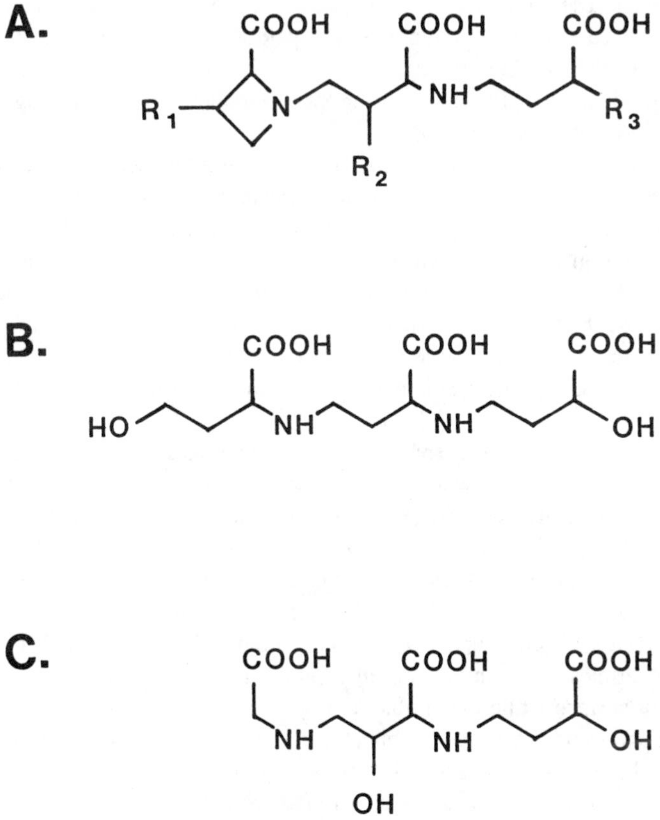

A.

B.

C.

Figure 10 Metal-binding imino-carboxylic acids from higher plants.
A. $R_1 = R_2 = H; R_3 = NH_2$: nicotianamine
    $R_1 = H; R_2 = R_3 = OH$: mugineic acid
    $R_1 = R_2 = H; R_3 = OH$: 2'-deoxymugineic acid
    $R_1 = R_2 = R_3 = OH$: 3-hydroxymugineic acid
B. Avenic acid A.
C. Distichonic acid.

Nicotianamine, formerly known as the "normalizing" factor, is absent in the chloronerva mutant of tomato (67). This is one of the very few known auxotrophic mutants in higher plants. Either grafting or treatment with traces of nicotianamine restores normal growth and development. The structure of nicotianamine suggested a role in iron binding and elicited the appellations cytosiderophore and phytosiderophore. However, these designations may not be appropriate since chloronerva suffers from iron overload rather than deficiency. Furthermore, nicotianamine is not excreted from roots under iron stress, as is mugineic acid. It has recently been speculated that nicotianamine may be an effector or regulator of mineral metabolism (67). Mugineic acid occurs generally in graminaceous plants and appears to have the general attributes of a siderophore (75).

Research on the mechanisms whereby plants acquire iron has shown substantial progress in recent years. Bienfait (8) discovered evidence for the presence of both constitutive ("standard") and induced ("turbo") reducing systems for iron uptake. The standard system, which uses NADH or NADPH as electron source, is apparently present in all cells. The inducible system, on the other hand, is confined to the plasma membrane of the root epidermis and is activated by iron starvation. Ferric EDTA, but not ferrioxamine B or ferric aerobactin, was found to be reduced in roots of iron-deficient *Phaseolus vulgaris* (9).

Romheld & Marschner (62) propose the existence of two separate strategies for uptake of iron when plants are stressed for this nutrient. In the first, which is characteristic of dicots and nongraminaceous monocots, protons are released concomitantly with induction of a reductase in the plasma membrane. In the second, typical of grasses, phytosiderophores (mugineic acid or avenic acid) are released from the roots and form stable Fe(III) chelates believed to be transported by specific uptake systems. In these experiments iron-stressed barley exudate was assumed to be the source of phytosiderophores; ferrioxamine B, ferrichrome, ferric rhodotorulate, and various ethylenediamine carboxylate complexes related to EDTA proved to be relatively poor sources of iron. The mechanism of uptake of chelated iron in grasses evidently differs from the corresponding process in microorganisms in that mugineic acid, the main siderophore, is used by grass species in general and, at least in barley and sorghum, the uptake system is not induced.

CONCLUSION

All recent work confirms and extends the early observation that siderophores are produced generally by Gram-positive and negative bacteria and by fungi. It follows that a finite siderophore level must occur in soil. The actual isolation of a specimen of schizokinen from a field in Texas authenticates this assertion (1). The precise concentration and types of each siderophore will depend on the particular microflora and will represent the balance between mineralization and biosynthesis, both of which have been shown to be induced processes.

There are at least three ways in which siderophores could affect plant life. In the first, the chelates act in the soil to solubilize and transport Fe(III), a very important mineral in plant nutrition (79). Since microorganisms preceded plants in the evolutionary succession, it is reasonable to suppose that the former life forms played, and continue to play, a role in weathering of rocks and formation of soil. It is further reasonable to imagine that plants adapted their iron transport systems so as to exploit those already devised by microorganisms. Although certain siderophores have been found to be effective sources of iron for higher plants, a specific transport system for ferrisiderophores has yet to be demonstrated outside of the microbial world. It is also conceivable that some siderophores could sequester iron in a form not available to some or all plant species.

A second possible effect of siderophores may be via the facilitation of plant disease. A survey (39) of Gram-negative bacterial phytopathogens, including sp. of *Agrobacterium, Erwinia,* and *Pseudomonas,* showed the presence of siderophores or siderophore-like compounds in all cases and, with the exception of the common yeast *S. cerevisiae,* siderophores have been detected in all fungal species examined for their presence. Siderophores are by now firmly established as one very important determinant of virulence for infections of animals. While this conclusion remains to be extended to plants, the very recent work with *E. chrysanthemi* (18) holds the promise that such a demonstration may for the first time be possible. Certain bacterial toxins, notably diphtheria and botulinum, are elaborated only under low iron conditions, and the toxins are solely responsible for manifestation of disease. The situation is less clear cut in plants, but it can be anticipated that the iron status of the host and the relative capacity of the pathogen to acquire iron will similarly influence the course of infection and disease. The fungal pathogen *Stemphylium botryosum,* the causal agent of leaf spot and foliage blight of tomato, elaborates a line of toxins known as stemphyloxins (43). These substituted decalones form colored products with iron (III) and, moreover, production of the toxin is induced by iron added up to ca 2 mg/1; higher levels of iron repress synthesis. The compounds do not appear to act as siderophores, although they may play some role in the low-affinity iron uptake in *S. botryosum.* The fungus does, however, produce three typical iron-repressible hydroxamate-type siderophores. A pathogen of rice, *Pyricularia oryzae,* produces a metal-binding compound, tenuazonic acid, which is toxic toward plants, animals, and bacteria (37). As a member of the tetramic acid family, it contains the 3-acetylpyrrolidine-2,4-dione heterocycle, which confers an affinity for Fe(III) and other transition metal ions. Both free ligand and metal complex are toxic.

The third mechanism is a type of biocontrol in which certain microbial species, such as fluorescent pseudomonads, discourage the growth or metabolic activities of competing microorganisms. There is substantial evidence that siderophores may play some part, albeit not an exclusive one, in this process. Antibiosis, hormone effects, and lytic activity may work in combination with siderophores to afford a microenvironment at the root surface that favors or impedes plant growth. In future efforts designed to disclose the mechanism of plant growth-promoting rhizobacterial action, it is hoped that investigators will use isolated compounds of known structure and purity rather than culture supernatants. This precaution, which has been observed in one case (34), is especially important when working with pyoverdines and pseudobactins since these siderophores readily undergo photolysis and are acid labile. Mutations aimed at elimination of siderophore biosynthesis and transport should be performed with transposable elements, and the nature of the insertional event should be characterized genetically and biochemically. Recourse must be made to isolation of a larger collection of mutants in those situations where chemical mutagenesis is at present the only available option. A careful comparison of the population dynamics of such mutants with their respective prototrophic parents would be prudent. Siderophore synthesis could be required primarily for survival in the soil and/or root colonization by *Pseudomonas* with disease suppression being relegated to other factors, e.g. toxins or hormones formed by an actively growing population of *Pseudomonas* spp.

Finally, it must be stressed that siderophores have in common only the high affinity, and hence selectivity, for iron (III). Apart from this common feature, the complexed ion may differ radically as regards such properties as charge, solubility, exchange rate, and stereochemistry. It is thus clear that no uniform biological effect can be expected of the collection of siderophores possessing the disparate structures recounted in this review.

ACKNOWLEDGMENTS

The authors are indebted to Jeffrey Buyer and Paul Gill for carefully reading the manuscript and to the many colleagues who responded to our plea for reprints or preprints of their work.

Literature Cited

1. Akers, H. A. 1983. Isolation of the siderophore schizokinen from soil of rice fields. *Appl. Environ. Microbiol.* 45: 1704–6
2. Akers, H. A., Magee, K. P. 1985. The siderophore mediated release of iron and magnesium from Mt. St. Helen's ash and silicate rock samples. *Experientia* 41: 522–23
3. Atkin, C. L., Neilands, J. B. 1986. Rhodotorulic acid, a diketopiperazine dihydroxamic acid with growth factor activity. I. Isolation and characterization. *Biochemistry* 7:3734–39
4. Atkin, C. L., Neilands, J. B. 1972. Leaf infections: Siderochromes (natural polyhydroxamates) mimic the "Green Island" effect. *Science* 176:300–2
5. Bagg, A., Neilands, J. B. 1985. Mapping of a mutation affecting regulation of iron uptake systems in *Escherichia coli* K-12. *J. Bacteriol.* 161:450–53

6. Becker, J. O., Messens, E., Hedges, R. W. 1985. The influence of agrobactin on the uptake of ferric iron by plants. *FEMS Microbiol. Ecol.* 31:171–76
7. Becker, J. O., Hedges, R. W., Messens, E. 1985. Inhibitory effect of pseudobactin on the uptake of iron by higher plants. *Appl. Environ. Microbiol.* 49:1090–93
8. Bienfait, H. F. 1985. Regulated redox processes at the plasmalemma of plant root cells and their function in iron uptake. *J. Bioenerg. Biomembr.* 17:73–83
9. Bienfait, H. F., Bino, R. J., van der Bliek, A. M., Duivenvoorden, J. F., Fontaine, J. M. 1983. Characterization of ferric reducing activity in roots of Fe-deficient *Phaseolus vulgaris*. *Physiol. Plant.* 59:196–202
10. Bindereif, A., Neilands, J. B. 1985. Aerobactin genes in clinical isolates of *Escherichia coli*. *J. Bacteriol.* 161:727–35

11. Bindereif, A., Neilands, J. B. 1985. Promoter mapping and transcriptional regulation of the iron assimilation system of plasmid ColV-K30 in *Escherichia coli* K-12. *J. Bacteriol.* 162:1039–46

12. Braun, V. 1985. The unusual features of the iron transport systems of *Escherichia coli*. *Trends Biochem. Res.* 10:75–78

13. Brown, A. E., Swinburne, T. R. 1981. Influence of iron and iron chelators on formation of progressive lesions by *Colletotrichum musae* on banana fruits. *Trans. Br. Mycol. Soc.* 77:119–24

14. Brown, A. E., Swinburne, T. R. 1982. Iron-chelating agents and lesion development by *Botrytis cinerea* on leaves of *Vicia faba*. *Physiol. Plant Pathol.* 21:13–21

15. Castignetti, D., Smarrelli, J. 1984. Siderophore reduction catalyzed by higher plant NADH:nitrate reductase. *Biochem. Biophys. Res. Commun.* 125:52–58

16. Corbin, J. L., Bulen, W. A. 1969. The isolation and identification of 2,3-dihydroxybenzoic acid and 2-N,6-N-di(2,3-dihydroxybenzoyl)-L-lysine formed by iron-deficient *Azotobacter vinelandii*. *Biochemistry* 8:757–62

16a. de Lorenzo, V., Bindereif, A., Paw, B. H., Neilands, J. B. 1986. Expression and mapping of the aerobactin biosynthesis and transport genes of plasmid ColV-K30 in *Escherichia coli* K12. *J. Bacteriol.* 165:570–78

16b. Elad, Y., Baker, R. 1985. Influence of trace amounts of cations and siderophore-producing pseudomonads on chlamydospore germination of *Fusarium oxysporum*. *Ecol. Epidemiol.* 75:1047–52

16c. Elad, Y., Baker, R. 1985. The role of competition for iron and carbon in suppression of chlamydospore germination of *Fusarium* spp. by *Pseudomonas* spp. *Ecol. Epidemiol.* 75:1053–59

17. Emery, T. 1980. Malonichrome, a new iron chelator from *Fusarium roseum*. *Biochim. Biophys. Acta* 629:383–90

18. Expert, D., Toussaint, A. 1985. Bacteriocin-resistant mutants of *Erwinia chrysanthemi*: Possible involvement of iron acquisition in phytopathogenicity. *J. Bacteriol.* 163:221–27

19. Finkelstein, R. A., Sciortino, C. V., McIntosh, M. A. 1983. Role of iron in microbe-host interactions. *Rev. Infect. Dis.* 5:S759–77

20. Garibaldi, J. A., Neilands, J. B. 1955. Isolation and properties of ferrichrome A. *J. Am. Chem. Soc.* 77:2429–30

21. Garibaldi, J. A., Neilands, J. B. 1956. Formation of iron-binding compounds by microorganisms. *Nature* 177:526–27

22. Gause, G. F. 1958. The search for anticancer antibiotics. *Science* 127:506–8

23. Geel, F. P., Schippers, B. 1983. Selection of antagonistic and fluorescent *Pseudomonas* spp. and their root colonization and persistence following treatment of seed potatoes. *Phytopathol. Z.* 108:193–206

24. Geel, F. P., Schippers, B. 1983. Reduction of yield depressions in high frequency potato cropping soil after seed tuber treatments with antagonistic fluorescent *Pseudomonas* spp. *Phytopathol. Z.* 108:207–14

25. Greenberg, J., Barker, H. A. 1962. A ferrichrome-requiring *Arthrobacter* which decomposes puromycin amino nucleoside. *J. Bacteriol.* 83:1163–64

26. Griffiths, G., Sigel, S. P., Payne, S. M., Neilands, J. B. 1984. Vibriobactin, a siderophore from *Vibrio cholerae*. *J. Biol. Chem.* 259:383–85

27. Holzberg, M., Artis, W. M. 1983. Hydroxamate siderophore production by opportunistic and systemic fungal pathogens. *Infect. Immunol.* 40:1134–39

28. Hantke, K. 1984. Cloning of the repressor protein gene of iron-regulated systems in *Escherichia coli* K-12. *Mol. Gen. Genet.* 197:337–41

29. Harper, D. B., Swinburne, T. R., Moore, S. K., Brown, A. E., Graham, H. 1980. A role for iron in germination of conidia of *Colletotrichum musae*. *J. Gen. Microbiol.* 121:169–74

30. Harrington, G. L., Neilands, J. B. 1982. Isolation and characterization of dimerum acid from *Verticillium dahliae*. *J. Plant Nutr.* 5:675–82

31. Hider, R. C. 1984. Siderophore mediated absorption of iron. *Struct. Bonding* 58:25–87

32. Horowitz, N. H., Charlang, G., Horn, G., Williams, N. P. 1976. Isolation and identification of the conidial germination factor of *Neurospora crassa*. *J. Bacteriol.* 127:135–40

33. Jalal, M. A. F., Mocharla, R., Barnes, C. L., Hossain, M. B., Powell, D. R., et al. 1984. Extracellular siderophores from *Aspergillus ochraeceous*. *J. Bacteriol.* 158:683–88

34. Kloepper, J. W., Leong, J., Teintze, M., Schroth, M. N. 1980. *Pseudomonas* siderophores: A mechanism explaining disease-suppressive soils. *Curr. Microbiol.* 4:317–20

35. Lankford, C. L. 1973. Bacterial assimilation of iron. *Crit. Rev. Microbiol.* 2:273–331

36. Lee, B. H., Miller, M. J. 1983. Natural ferric ionophores: Total synthesis of schi-

zokinen, schizokinen A and arthrobactin. *J. Org. Chem.* 48:24–31

37. Lebrun, M.-H., Duvert, P., Gaudemer, F., Deballon, C., Boucly, P. 1985. Complexation of the fungal metabolite tenuazonic acid with copper (II), iron (III), nickel (II) and magnesium (II) ions. *J. Inorg. Biochem.* 24:167–82

38. Leong, S. A., Neilands, J. B. 1981. Relationship of siderophore-mediated iron assimilation to virulence in crown gall disease. *J. Bacteriol.* 147:482–91

39. Leong, S. A., Neilands, J. B. 1982. Siderophore production by phytopathogenic microbial species. *Arch. Biochem. Biophys.* 218:351–59

40. Liu, A., Neilands, J. B. 1984. Mutational analysis of rhodotorulic acid synthesis in *Rhodotorula pilimanae*. *Struct. Bonding* 58:97–106

41. Lochhead, A. G., Burton, M. O., Thexton, R. H. 1952. A bacterial growth factor synthesized by a soil bacterium. *Nature* 170:282–83

42. Lockwood, J. L., Schippers, B. 1984. Evaluation of siderophores as a factor in soil mycostasis. *Trans. Br. Mycol. Soc.* 82:589–94

43. Manulis, S., Kashman, Y., Netzer, D., Barash, I. 1984. Phytotoxins from *Stemphylium botryosum:* Structural determination of stemphyloxin II, production in culture and interaction with iron. *Phytochemistry* 23:2193–98

44. Maurer, P. J., Miller, M. J. 1982. Microbial iron chelators: Total synthesis of aerobactin and its constituent amino acid, N⁶-acetyl-N⁶-hydroxylysine. *J. Am. Chem. Soc.* 104:3096–3101

45. Meyer, J. M., Moci, M., Abdallah, M. A. 1979. Effects of iron on the protein composition of the outer membrane of fluorescent pseudomonads. *FEMS Microbiol. Lett.* 5:395–98

46. Miles, A. A., Khimji, P. L. 1975. Enterobacterial chelators of iron: Their occurrence, detection and relation to pathogenicity. *J. Med. Microbiol.* 8:477–90

46a. Modi, M., Shah, K. S., Modi, V. V. 1985. Isolation and characterization of catechol-like siderophore from cow pea *Rhizobium* RA-1. *Arch. Microbiol.* 141:156–58

47. Murphy, T. P., Lean, D. R. S., Nalewajko, C. 1976. Bluegreen algae: Their excretion of iron-selective chelators enables them to dominate other algae. *Science* 192:900–2

48. Neilands, J. B. 1981. Iron absorption and transport in microorganisms. *Ann. Rev. Nutr.* 1:27–46

49. Neilands, J. B. 1981. Microbial iron compounds. *Ann. Rev. Biochem.* 50:715–31

50. Neilands, J. B. 1982. Microbial envelope proteins related to iron. *Ann. Rev. Microbiol.* 36:285–309

51. Neilands, J. B., Nakamura, K. 1985. Regulation of iron assimilation in microorganisms. *Nutr. Rev.* 43:193–97

52. Ong, S. A., Peterson, T., Neilands, J. B. 1979. Agrobactin, a siderophore from *Agrobacterium tumefaciens*. *J. Biol. Chem.* 254:1860–65

53. Orlando, J. A., Neilands, J. B. 1982. Ferrichrome compounds as a source of iron for higher plants. In *Chemistry and Biology of Hydroxamic Acids,* ed. H. Kehl, pp. 123–29. Basel:Karger. 191 pp.

54. Page, W. J., Huyer, M. J. 1984. Derepression of the *Azotobacter vinelandii* siderophore system using iron-containing minerals to limit iron repletion. *J. Bacteriol.* 158:496–501

55. Peterson, T., Neilands, J. B. 1979. Revised structure of a catecholamide spermidine siderophore. *Tetrahedron Lett.* 50:4805–8

56. Plowman, J. E., Loehr, T. M., Goldman, S. J., Sanders-Loehr, J. 1984. Structure and siderophore activity of ferric schizokinen. *J. Inorg. Biochem.* 20:183–97

57. Pollack, J. R., Ames, B. N., Neilands, J. B. 1970. Iron transport in *Salmonella typhimurium:* Mutants blocked in the biosynthesis of enterobactin. *J. Bacteriol.* 104:635–39

58. Powell, P. E., Szaniszlo, P. J., Reid, C. P. P. 1983. Confirmation of occurrence of hydroxamate siderophores in soil by a novel *Escherichia coli* bioassay. *Appl. Environ. Microbiol.* 46:1080–83

59. Reeves, M. W., Pine, L., Neilands, J. B., Balows, A. 1983. Absence of siderophore activity in *Legionella* sp. grown in iron deficient media. *J. Bacteriol.* 154:324–29

60. Reid, C. P. P., Crowley, D. E., Kim, H. J., Powell, P. E., Szaniszlo, P. J. 1984. Utilization of iron by oat when supplied as ferrated synthetic chelate or as ferrated hydroxamate siderophore. *J. Plant Nutr.* 7:437–47

61. Reid, R. K., Reid, C. P. P., Powell, P. E., Szaniszlo, P. J. 1984. Comparison of siderophore concentrations in aqueous extracts of rhizosphere and adjacent bulk soils. *Pedobiologia* 26:263–66

62. Romheld, V., Marschner, H. 1986. Evidence for the existence of a specific uptake system for iron phytosiderophores in roots of grasses. *Plant Physiol.* In press

63. Schäffer, S., Hantke, K., Braun, V. 1985. Nucleotide sequence of the iron

regulatory gene *fur*. *Mol. Gen. Genet.* 200:110–13

64. Scher, F. M., Baker, R. 1982. Effect of *Pseudomonas putida* and a synthetic chelator on induction of soil suppressiveness to fusarium wilt pathogens. *Phytopathology* 72:1567–73

65. Scher, F. M., Dupler, M., Baker, R. 1984. Effect of synthetic iron chelates on population densities of *Fusarium oxysporum* and the biological control agent *Pseudomonas putida* in soil. *Can. J. Microbiol.* 30:1271–75

66. Schneider, W. 1984. Hydrolysis of iron (III)-chaotic olation versus nucleation. *Comments Inorg. Chem.* 3:205–22

67. Scholz, G., Schlesier, G., Seifert, K. 1985. Effect of nicotianamine on iron uptake by the tomato mutant "chloronerva". *Physiol. Plant.* 63:99–104

68. Schroth, M. N., Hancock, J. G. 1982. Disease-suppressive soil and root-colonizing bacteria. *Science* 216:1376–81

69. Simpson, F. B., Neilands, J. B. 1976. Siderochromes in cyanophyceae. Isolation and characterization of schizokinen from *Anabaena* sp. *J. Phycol.* 12:44–48

70. Smith, M. J., Neilands, J. B. 1984. Rhizobactin, a siderophore from *Rhizobium meliloti*. *J. Plant Nutr.* 7:449–58

71. Smith, M. J., Shoolery, J. N., Schwyn, B., Holden, I., Neilands, J. B. 1985. Rhizobactin, a structurally novel siderophore from *Rhizobium meliloti*. *J. Am. Chem. Soc.* 107:1739–43

72. Sneh, B., Dupler, M., Elad, Y., Baker, R. 1984. Chlamydospore germination of *Fusarium oxysporum* f. sp. *cucumerinum* as affected by fluorescent and lytic bacteria from a fusarium-suppressive soil. *Phytopathology* 74:1115–24

73. Snow, G. N. 1970. Mycobactins: Iron-chelating growth factors from mycobacteria. *Bacteriol. Rev.* 34:99–125

74. Staley, J. T., Jackson, M. J., Palmer, F. E., Adams, J. B., Borns, D. J., et al. 1983. Desert varnish coatings and microcolonial fungi on rocks of the Gibson and Great Victoria deserts, Australia. *BMR J. Aust. Geol. Geophys.* 8:83–87

75. Sugiura, Y., Nomoto, K. 1984. Phytosiderophores. *Struct. Bonding* 58:107–35

76. Tait, G. H. 1975. The identification and biosynthesis of siderochromes formed by *Micrococcus denitrificans*. *Biochem. J.* 146:191–204

77. Teintze, M., Hossain, M. B., Barnes, C. L., Leong, J., Van Der Helm, D. 1981. Structure of ferric pseudobactin, a siderophore from a plant growth promoting *Pseudomonas*. *Biochemistry* 20:6446–57

78. Waksman, S. 1957. Penalty of isolationism. *Science* 125:585

79. Wallace, A. 1982. Historical landmarks in progress relating to iron chlorosis in plants. *J. Plant Nutr.* 5:277–88

80. Warren, R. A. J., Neilands, J. B. 1965. Mechanism of microbial catabolism of ferrichrome A. *J. Biol. Chem.* 240:2055–58

81. Weller, D. M., Cook, R. J. 1981. Control of take-all of wheat with fluorescent pseudomonads. *Phytopathology* 71:1007

82. Wendenbaum, S., Demange, P., Dell, A., Meyer, J. M., Abdallah, M. A. 1983. The structure of pyoverdine Pa, the siderophore of *Pseudomonas aeruginosa*. *Tetrahedron Lett.* 24:4877–80

83. Weinberg, E. D. 1984. Iron withholding: A defense against infection and neoplasia. *Physiol. Rev.* 64:65–102

84. Willson, R. L. 1984. Ill-placed iron, oxygen free radicals and disease; some recent and not so recent radiation studies. In *The Biology and Chemistry of Active Oxygen*, ed. J. V. Bannister, W. H. Bannister, pp. 238–58. New York: Elsevier. 262 pp.

85. Winkelmann, G. 1979. Evidence for steriospecific uptake of iron chelates in fungi. *FEBS Lett.* 97:43–46

86. Winkelmann, G. 1983. Specificity of siderophore iron uptake by fungi. In *The Biological Chemistry of Iron*, ed. B. Dunford, pp. 107–16. Dordrecht: Reidel. 231 pp.

87. Wong, P. T. W., Baker, R. 1984. Suppression of wheat take-all and *Ophiobolus* patch by fluorescent pseudomonads from a *Fusarium*-suppressive soil. *Soil Biol. Biochem.* 16:397–403

Ann. Rev. Plant Physiol. 1986. 37:209–32

PHYSIOLOGY OF ACTINORHIZAL NODULES

J. D. Tjepkema and C. R. Schwintzer

Department of Botany and Plant Pathology, University of Maine, Orono, Maine 04469

D. R. Benson

Microbiology Section, University of Connecticut, Storrs, Connecticut 06268

CONTENTS

INTRODUCTION

Actinorhizal plants rival legumes in the amount of N_2 that they fix on a global basis, yet basic knowledge of many aspects of their biology and physiology is limited or is very recent. A major stimulus to recent studies has been the

0066-4294/86/0601-0209$02.00

isolation of the N_2-fixing endophyte *Frankia* (an actinomycete). The first confirmed isolation was published in 1978 (29), and numerous isolations have followed. These strains have been used in many of the subsequent studies on actinorhizal plants.

Advances in the study of actinorhizal plants have led to recent reviews of several topics, including current reviews of *Frankia* taxonomy (70), the infection process (19, 20), and genetics (89). For these topics we give a brief overview and/or refer the reader to the reviews. In general, our goal has been to introduce the reader to actinorhizal plants and to emphasize areas of current research on *Frankia* and actinorhizal nodules.

Actinorhizal plants are taxonomically diverse, occurring in 8 families and 21 genera (26, 67). Well-known genera include *Alnus* (alder), *Myrica* (e.g. sweet gale), *Ceanothus* (e.g. snow brush), *Elaeagnus* (e.g. autumn olive), and *Casuarina*. In spite of their taxonomic diversity, the actinorhizal plants have some common features. All are perennial dicots (26) and all but *Datisca*, which has herbaceous shoots (49, 163), are woody shrubs or trees (26). Further, they are found primarily in cooler climates, being common in the north and south temperate zones and occurring at high altitudes in the tropics. *Casuarina* and *Myrica* are exceptions and include subtropical and tropical species as well (25, 117).

Ecologically, the majority of actinorhizal plants are pioneers on nitrogen-poor, open sites. In late glacial and postglacial times (117), they were abundant and widely distributed in Europe and North America, where they colonized nitrogen-poor glacial deposits and accelerated soil development by adding nitrogen. They were displaced as increasing soil nitrogen permitted non-nitrogen fixing plants to flourish (116). In modern times actinorhizal plants are much less abundant, being found primarily on sandy and gravelly sites, on exposed raw mineral soil, along streams and in wetlands, and on other nitrogen-poor sites (117). Many of the early successional environments occupied by actinorhizal plants are relatively harsh. For example, water stress and temperature extremes are common at open, sandy sites and anaerobiosis at wet sites. Consequently, most actinorhizal plants adapt to a variety of stresses in addition to their ability to fix nitrogen. Actinorhizal plants are especially important in high latitude countries such as Scandinavia, Canada, and New Zealand, where conditions are not favorable for legumes but actinorhizal plants are abundant and capable of vigorous growth (117).

Economically, actinorhizal plants have potential in forestry, biomass production, land reclamation, and amenity planting. In forestry *Alnus rubra* is used as sawtimber and pulpwood (94) and *Casuarina* as timber and fuel wood (46, 117). In addition, these and many other actinorhizal plants can be interplanted with crop trees or used in rotation with them to improve production on nitrogen-poor sites (48, 80, 127). In land reclamation and amenity planting, actinorhizal plants are especially useful on difficult sites such as acid mine spoils and coastal dunes where their tolerance to extreme conditions is important (26, 38, 117).

FRANKIA ISOLATION TECHNIQUES

Numerous strains of *Frankia* from several actinorhizal plants have been described. Most are from species of *Alnus,* but isolates have also been obtained from 13 other genera (70).

Techniques for isolating frankiae are still being refined. Most isolation protocols involve surface sterilizing the nodule, homogenizing or slicing the nodule aseptically to release the endophyte, and incubating vesicle clusters or nodule pieces in agar or liquid growth medium. Plates or tubes are then examined at intervals of a few days, weeks, or even months, for the outgrowth of colonies (26a, 30, 70). Osmium tetraoxide (63) is probably the most effective surface sterilant because it penetrates the nodule slowly and, with appropriate incubation times, will sterilize only the superficial cellular layers of the nodule. It is, however, highly toxic. Vesicle clusters obtained by homogenizing nodule lobes can be readily purified by collecting and washing them on nylon screens (12); this procedure is simple and allows vesicle clusters to be plated at high density with few contaminants. Although most frankiae can be isolated on relatively simple media some *Alnus glutinosa* strains require inclusion of a lipid fraction from root extracts. The exact nature of the substances involved is not yet known but work on their identification is in progress (26a; A. Quispel, personal communication). Because media requirements for isolating frankiae vary with the strain, the probability of obtaining an isolate can be increased by using more than one medium for each isolation attempt (12, 17). Frankiae have also been isolated from soil (10).

FRANKIA TAXONOMY AND HOST SPECIFICITY

All frankiae isolated to date are gram positive, filamentous, sporulating, branching prokaryotes (68). Therefore, frankiae have been grouped with filamentous soil bacteria in the order Actinomycetales. A new family, Frankiaceae, has been created having *Frankia* as the sole genus (11).

Although an organism can be identified readily as belonging to the genus *Frankia,* considerable morphological and physiological variation occurs within the genus, making species characterization a difficult problem. Thus it has been decided that no species names should be used until a more complete understanding is reached of what constitutes a species of *Frankia.* Rather, strains have been designated by a combination of numbers and letters (70).

Physiologically, frankiae can be separated into at least two major groups. Type "A" strains are genetically heterogeneous, relatively aerobic, use carbohydrates, and are usually not infective on their host plants. Type "B" strains are serologically, genetically, and chemically related, are relatively less tolerant of O_2, use carbohydrates poorly, and are generally infective (able to form nodules) and effective (able to fix nitrogen) on the host plant of origin (70). DNA-DNA hybridization studies have shown that "B" organisms (also known as genogroup 1) have genomic homology levels between 67.4 to 94.1%. Group "A" isolates have less than 39% homology with ArI4, a genogroup 1 strain, and less than 33% homology with EuI1a, a member of the "A" assemblage. The "A" group is therefore genetically as well as physiologically heterogeneous. An et al (5) have identified four possible genogroups using four probe strains in DNA hybridization experiments. Such results suggest that the genetic diversity of frankiae, particularly "A" group frankiae, is likely to be great.

Host-endophyte specificity is a complex topic (89, 96). Moreover, our understanding is limited because infective isolates (and many isolates are not infective!) are presently available for only about half of the actinorhizal genera, and these isolates usually have been tested on only a few host genera. Many of the existing isolates fit into three cross-inoculation groups. The first consists of strains isolated from *Alnus*, *Myrica*, or *Comptonia*, and these usually nodulate all three host genera with the nodules being effective in N_2 fixation. But there are exceptions, such as a number of strains isolated from *Myrica gale*, which form ineffective nodules on *Alnus glutinosa* while forming effective nodules on *Shepherdia*, which is not a member of this cross-inoculation group (123). The second cross-inoculation group consists of strains isolated from the Elaeagnaceae (*Elaeagnus*, *Shepherdia*, and *Hippophae*) and perhaps other members of the Rhamnales such as *Colletia* (47). It also includes a strain of *Frankia* isolated from *Casuarina*, which does not nodulate its original host but does nodulate *Colletia*, *Hippophae*, and *Elaeagnus* (47). The third cross-inoculation group consists of strains isolated from *Casuarina* that do nodulate *Casuarina*. The few such isolates that exist nodulate only *Casuarina* (165), except for one that also nodulates *Myrica gale* (J. G. Torrey, unpublished data).

MORPHOLOGY AND DEVELOPMENT

Frankiae in Culture and Symbiosis

HYPHAE   Frankial hyphae are ultrastructurally similar to those of other actinomycetes (62, 87). They have a double-layered cell wall, with the inner layer participating in cross-wall formation. The outer layer is analogous to "sheaths" that are common on other sporoactinomycetes (54). Unique morphological characteristics of frankiae include their mode of spore formation and their synthesis of specialized structures known as "vesicles" during starvation for nitrogen.

SPORANGIA AND SPORES   Frankiae sporulate in symbiosis and in pure culture. Sporangia vary in appearance from strain to strain, but their general morphology and mode of formation are characteristic of the genus *Frankia*. Typically, sporangia from a single culture have many different ages, sizes, and shapes; they are borne either terminally or in an intercalary position in the hyphae (87). Sporangia may reach sizes of 30 to 60 μm in length, and each may contain from a few to several hundred spores (87).

Frankial sporangia form from thickened hyphae (54, 87). Septa originating from the inner sporangial wall segment the sporangium in several planes. Increases in the thickness of the septa result in the separation of cells which ultimately become spores (54, 69).

Frankial spores probably serve as resistant propagules that are released directly into the medium by free-living frankiae or from nodules by decay. They persist in the soil and are able to infect further generations of host plants (150). Mature spores are nonmotile, irregularly shaped, elliptical, or spherical, and vary in size from 0.8 to 1.5 μm in diameter. Ultrastructurally, mature spores have thick cell walls and dense cytoplasm containing nucleoid regions and conspicuous electron translucent granules that are likely to be food reserves (43, 87).

In addition to sporangia, an isolate from *Casuarina* forms torulose chains of spore-like cells from vegetative hyphae (34). These grow out readily under favorable conditions and may complement sporangia as survival structures.

VESICLES   "Vesicle" is the term applied, for historical reasons, to terminal swellings on the tips of frankial hyphae in symbiosis or in cultures starved for nitrogen. Vesicles take many forms in symbiosis. Frankiae in root nodules from *Alnus, Hippophae, Elaeagnus, Shepherdia, Colletia,* and *Discaria* have spherical vesicles. In *Purshia, Rubus, Ceanothus, Dryas,* and *Chamaebatia* the vesicles are pear-shaped, and those from *Comptonia, Myrica, Cercocarpus, Datisca,* and *Coriaria* are club-shaped to filamentous (50, 88). Frankiae from *Casuarina* apparently do not form vesicles in the nodule (18, 146), although strains isolated from *Casuarina* nodules form spherical vesicles when induced for N_2 fixation in pure culture (45, 85).

Vesicle morphology in the nodule apparently is dictated by the host plant. This conclusion was drawn first from studies using a *Comptonia peregrina* isolate, CpI1, which has club-shaped vesicles in the *Comptonia* nodule. When CpI1 infects *Alnus glutinosa*, spherical vesicles are formed (62). Most frankiae form spherical vesicles when induced in pure culture, suggesting that the spherical shape is the norm and that other shapes in symbiosis are the result of host-endophyte interactions.

When purified from *Frankia* cultures, vesicles are actually lightbulb-shaped with a spherical body and a hyphal stem (88a). Structurally, they are quite similar to vesicles produced in *Alnus*-type root nodules. Young vesicles, termed "provesicles" (42), appear dark in phase contrast microscopy, show few, if any, cross-walls, and are not correlated with nitrogenase activity. Mature vesicles are refractile in phase contrast, may be multilobed in older cultures, form an increasing number of septae as they age, and develop coincident with nitrogenase activity (42, 142).

While moderate levels (1 mM) of nitrogen sources such as NH_4Cl, KNO_3, and glutamine generally suppress vesicle formation (45, 139), abundant vesicle formation has recently been reported in the presence of 5 mM asparagine, arginine, or citrulline (66). In the case of arginine and citrulline, the vesicles had high nitrogenase activities, but there was no nitrogenase activity in the vesicles formed in the presence of asparagine. Similar to the asparagine result, a strain of *Frankia* has been reported where vesicles are formed in the presence of 3.7 mM NH_4Cl, but they lack nitrogenase as indicated by C_2H_2 reduction assays and Western blots analyzed with antisera to nitrogenase from *Rhizobium leguminosarum* (78).

Nodule Initiation

Unlike legume nodules, actinorhizal nodules are modified lateral roots. Each nodule lobe has a central vascular cylinder, and the endophyte is confined to the cortical tissue. As in most legumes, the endophyte may enter the host via deformed root hairs and then spread by penetration through host cell walls within the root (19, 20, 89). However, intercellular infection through the root epidermis, with apoplastic colonization of the root cortex prior to cell penetration, has been reported recently for *Elaeagnus angustifolia* (81). The route of infection is at least partly under host control, because two endophyte strains that infect *Elaeagnus* by an intercellular mechanism are limited to root hair infection in *Myrica gale* (I. M. Miller and D. D. Baker, submitted). Intercellular infection has also been observed in certain legumes.

Very little is known about the control of infection and development of actinorhizal nodules. The concentrations of several hormones in the nodule fluctuate with time of year, with cytokinins being high in the spring, while abscissic acid and gibberellins are high during the fall and winter (153, 157). The role of hormone production by frankiae is unknown, although cultures can synthesize indoleacetic acid at a low rate (155).

CARBON METABOLISM

Hundreds of frankiae are currently in culture, and information on their metabolic versatility is accumulating, even though most grow slowly. Determining if growth occurs on a given substrate can involve weeks and even months of incubation before growth commences (16). Consequently, the absence of growth on a particular compound must be interpreted cautiously (16, 129).

Carbon sources used by frankiae for growth are generally quite diverse. The most effective include short-chain fatty acids like propionate and acetate, fatty acid derivatives such as Tweens, TCA cycle intermediates like succinate or malate, and other organic acids like pyruvate (16, 21, 114, 129). Several strains grow on sugars or sugar alcohols, but growth on these substrates is only rarely superior to growth on organic acids like propionate (28, 71, 129). The inability of many frankiae to use carbohydrates has been attributed to the lack of sugar transport systems (71). Adding Tweens to media containing carbohydrates, or raising the carbohydrate concentration, sometimes improves the growth response (71); Tweens presumably act by increasing the permeability of cell membranes. Because frankiae store sugars in the form of glycogen or trehalose (15, 73) most, if not all, strains probably possess enzymes for carbohydrate catabolism.

Some work has been done on the pathways of carbohydrate and organic acid metabolism by frankiae. Tricarboxylic acid cycle, glyoxylate cycle, and gluconeogenic enzyme activities have been demonstrated in *Frankia* AvcIl growing under the appropriate conditions (4, 22). Recent studies on an *Alnus rubra* isolate, ArI3, indicate that glucose is catabolized through the Embden-Meyerhof-Parnas pathway, like other actinomycetes (32, 72). Propionate is metabolized by conversion to succinate via propionyl-CoA carboxylase (125).

Carbon sources used by frankiae in symbiosis have not been identified. Candidate compounds include short-chain fatty acids, TCA cycle intermediates, or pyruvate (17). Bacterial metabolism in the plant has been studied using vesicle cluster suspensions; such suspensions are purified from crushed nodules. Each vesicle cluster is derived from a single infected plant cell and consists of a central mycelium with vesicles generally formed on the periphery. Vesicle cluster suspensions from *Datisca* nodules oxidize succinate, and succinate-dependent O_2 uptake is enhanced by ADP (4). In addition, vesicle clusters from *Alnus, Datisca,* and *Hippophae* root nodules take up O_2 when supplied with NADH or when supplied with a mixture of glutamate, malate, and NAD^+ (4). This observation has been suggested to indicate that reducing equivalents are transferred between the plant cytosol and microorganism via a malate-aspartate shuttle (2). A similar shuttle involving glutamate has just (1985) been proposed as the major mechanism for the transfer of reducing equivalents and carbon skeletons from the host to the endophyte in the legume-*Rhizobium* symbiosis (58). Cytochromes *a, b, c,* and *o* have been demonstrated in vesicle clusters from *Alnus* (31).

One problem with experiments done on frankiae vesicle clusters derived directly from root nodules is the difficulty of separating plant activities from bacterial activities. Electron micrographs of vesicle clusters released from the nodule show the remains of plant mitochondria and other membranes (4). Thus, it is difficult to draw conclusions from metabolic studies on isolated vesicle clusters (4). Now that frankiae can be grown readily in culture, an improved approach to identifying carbon sources used in symbiosis would be to use mutants defective in specific transport or metabolic functions.

NITROGEN AND HYDROGEN METABOLISM

Frankiae can use several compounds as a nitrogen source including amino acids, urea, nitrate, adenine, uracil, ammonia, and N_2. The most effective nitrogen source for growth of frankiae is ammonia, although some strains grow well on aspartate, asparagine, glutamate, glutamine, urea, or nitrate (66, 115). Unlike free-living rhizobia, frankiae fix N_2 readily in culture (45, 138, 139). Some, if not all, frankiae can grow on N_2, although growth rates are low (42, 114). Some strains apparently prefer N_2 over ammonium (D. D. Baker, personal communication; M. P. Lechevalier, personal communication).

Transport and assimilation of NH_4^+ have been studied most extensively in *Frankia* sp. strain CpI1. Entry of NH_4^+ into the cell occurs by two routes. In media having a high NH_4^+ concentration, NH_4^+ enters cells by passive diffusion in amounts sufficient to sustain cell growth. In nitrogen-free media, an inducible active transport permease mediates the concentrative uptake of NH_4^+ (76). Such active transport systems have been found in many bacteria and fungi, although the frankial transport system has a higher affinity for ammonium than most others tested (76). This frankial NH_4^+ uptake system probably is of little importance in symbiosis, but it may play a role in the survival of frankiae in the soil.

As in other free-living N_2-fixing prokaryotes, glutamine synthetase (GS) is responsible for the primary assimilation of NH_4^+ in frankiae (17, 44, 109). Glutamate dehydrogenase activity has not been detected in frankiae, but alanine dehydrogenase is present at low levels (D. R. Benson, unpublished). In most prokaryotes, GS is regulated in part by covalent attachment of an AMP moiety to each of 12 subunits, leading to attenuation of biosynthetic activity as the enzyme becomes increasingly adenylylated (74). Exceptions include GS from *Bacillus* sp. and heterocystous cyanobacteria which are regulated by cumulative feedback inhibition by glutamine and various other metabolites (145). Glutamine synthetase from the *Casuarina Frankia* sp. strain Dl1 was thought not to be regulated by adenylylation on the basis that snake venom phosphodiesterase (SVP) did not relieve Mg^{2+} inhibition of GS transferase activity (44). Glutamine synthetase from *Frankia* strain CpI1, on the other hand, was shown to be regulated by adenylylation-deadenylylation by 2-dimensional gel analysis of the purified protein and by catalytic changes of the enzyme upon treatment with SVP (D. R. Benson, T. J. Browning and N. A. Noridge, submitted). In many respects, the frankial enzyme is similar to that from *Streptomyces cattleya* (126).

In symbiosis, only low GS activity has been detected in isolated vesicle clusters (3, 23). In isolated vesicles from cultures, GS activity and ammonium transport activity are barely detectable (N. A. Noridge and D. R. Benson, submitted). The possibility therefore exists that vesicles excrete ammonium into the culture medium much as they are proposed to do in symbiosis, although this will be difficult to demonstrate experimentally since N-starved cells take up NH_4^+ so avidly.

The same situation is present in rhizobial and algal symbioses with higher plants (109). Thus, fixed nitrogen may be transferred from the bacterium to the plant as ammonia (4). Once in the plant tissue, ammonium is most likely assimilated via plant GS, which is present in alder root nodules at elevated levels (53). Studies using ^{13}N-labeled ammonium indicate that NH_4^+ is assimilated in alder root nodules by both GS and glutamate dehydrogenase (GDH), although the location of the enzymes in infected or uninfected cells has not been determined (100). The suggestion that GDH plays a role in NH_4^+ assimilation by root nodule cells is consistent with the observation of elevated GDH activity in alder nodules (23); however, it is at odds with the concept that alder GDH is primarily a deaminating enzyme (23). To date, frankiae have not been shown to release ammonium when growing on N_2, or when isolated as vesicle clusters from root nodules. Therefore, the nature of the nitrogen-containing compound transferred in the symbiosis is still speculative.

Purified nitrogenase from the nodules of *Alnus glutinosa* is similar to nitrogenases from other organisms (13). In most instances, nitrogenase activity occurs only after frankiae have formed vesicles, either in culture (42, 138) or during nodule development (79, 106). Recent studies have shown that isolated vesicles have a different protein pattern upon 2-dimensional gel electrophoresis than the vegetative cells (N. A. Noridge and D. R. Benson, submitted). Most telling is the demonstration that nitrogenase is present in vesicles isolated by density gradient centrifugation free of hyphae (88a, 128). Western blots using antisera to nitrogenase from *Rhizobium leguminosarum* also show that there is enrichment of nitrogenase in vesicle fractions of cultures (78). It thus seems likely that vesicles are involved in N_2 fixation, but in the *Casuarina* endophyte nitrogenase activity can occur in the absence of vesicles (85), as will be discussed later.

Aerobic N_2-fixing bacteria often have an H_2-oxidizing hydrogenase (uptake hydrogenase) that is synthesized along with nitrogenase during nitrogen starvation (37). The hydrogenase recaptures electrons lost through the operation of nitrogenase. In a sense, hydrogenase acts as an energy-conserving system within the microorganism (35). For many years, actinorhizal root nodules were known to evolve little, if any, H_2 (83). In fact, nodules were shown to take up

H_2 (101). This observation suggested the presence of uptake hydrogenases in the nodule. Hydrogenase activity was found in cell-free extracts of vesicle clusters from *Alnus glutinosa* nodules (13) and vesicle cluster suspensions from *Alnus glutinosa* (97), *A. rubra,* and *Myrica pensylvanica* (14). All actinorhizal nodules tested to date have H_2-oxidizing activity with the exception of *Alnus incana* inoculated with a strain of *Frankia* from Sweden (110, 112) and nodules collected in autumn which showed decreased or undetectable activity (97). The high frequency of hydrogenase in actinorhizal nodules contrasts with the scarcity of hydrogenase in rhizobial isolates (37). An interesting property of hydrogenase in actinorhizal nodules is that some, but not all, are irreversibly inhibited when nodules are exposed to 10 kPa of C_2H_2 (161).

NITROGEN FIXATION AND OXYGEN

Response of Frankiae to Oxygen

Cultures of frankiae can fix N_2 at atmospheric partial pressure of O_2 (pO_2) (45, 84, 138, 139). Because the nitrogenase of frankiae is as sensitive to O_2 inactivation as other nitrogenases (13), some mechanism must exist that protects the enzyme from O_2. Insight into this mechanism is provided by studies with a strain of *Frankia* isolated from *Casuarina*. When cultures are grown under microaerobic conditions, nitrogenase activity occurs in the absence of vesicles, but is very sensitive to inhibition by O_2 (85). In contrast, cultures that are grown at atmospheric pO_2 form typical vesicles and are not inhibited by atmospheric pO_2. This suggests that vesicles protect nitrogenase from O_2.

A fascinating parallel occurs in cyanobacteria, where cultures lacking heterocysts can fix N_2 only under microaerobic conditions, while aerobic fixation occurs when heterocysts are present (124). Moreover, vesicles both from cultures of frankiae and from nodules have a laminated envelope that is very similar in appearance to the laminated layer that occurs in the walls of heterocysts (64, 65, 142). This laminated layer may impede the diffusion of O_2 into the heterocysts (124) or vesicles (142).

Further evidence for a physical barrier to O_2 diffusion in vesicles comes from studies of O_2 uptake by cultures of frankiae (84, 85). Respiration of cultures that lack vesicles becomes O_2 saturated at low pO_2 values, as would be expected of most microorganisms. But when vesicles are present a much greater pO_2 is required for O_2 saturation, with respiration being O_2 unsaturated even at atmospheric pO_2. The observed kinetics are consistent with diffusion limitation of O_2 uptake into the vesicle.

Factors other than the vesicle wall are also involved in protecting nitrogenase from O_2. One is probably a high rate of O_2 uptake by the vesicle, which keeps the pO_2 low within the diffusion barrier. Another factor is the presence of superoxide dismutase and catalase which may protect nitrogenase and other cell components from oxidation via superoxide and hydroxyl free radicals (122).

The ability of frankia to fix N_2 at atmospheric pO_2 is in marked contrast to free-living rhizobia, which can fix N_2 only at very low pO_2 values. Legume nodules contain diffusion barriers that help to provide the low pO_2 environment needed for N_2 fixation by rhizobia (134). Such diffusion barriers should not be needed in actinorhizal nodules, but in fact are present in some instances, as will be discussed below.

Hemoglobins and Oxygen Diffusion

Hemoglobins occur in all effective legume nodules, in nodules of *Parasponia* (a nonlegume nodulated by rhizobia) (7), and in certain actinorhizal nodules (39, 132). This suggests that hemoglobins may be necessary for effective N_2-fixing symbioses. However, there is little or no hemoglobin in the nodules of *Ceanothus americanus* and *Datisca glomerata* (132, 137; J. D. Tjepkema, unpublished observations), and these nodules apparently fix N_2 at rates comparable to legume and other actinorhizal nodules on a weight basis (50, 137, 140). Thus we are faced with the question of why hemoglobins occur in some actinorhizal nodules and not in others.

Hemoglobins in legume nodules are believed to function by facilitating O_2 transport at the low pO_2 values that occur in the zone of infected tissue (6). The hemoglobin that occurs in relatively high concentrations in the nodules of *Casuarina* and *Myrica* (39, 132, 137) may function in a similar way, because zones of low pO_2 also occur in these nodules. Evidence for low pO_2 comes from studies using an oxygen microelectrode to measure pO_2 in *Myrica gale* (133) and the lack of vesicle formation by the endophyte in *Casuarina* (85). The zones of low pO_2 in *Casuarina* and *Myrica* nodules may result from suberization or lignification of the walls of the infected host cells which could restrict O_2 diffusion into these cells (18, 98). The O_2 affinity of *Casuarina* hemoglobin is high, being similar to that of legume and *Parasponia* hemoglobins (164). This is consistent with a function in O_2 transport at low pO_2.

The absence or low concentration of hemoglobins in nodules of other actinorhizal genera might be related to the relatively free access of O_2 to the endophyte. Under high pO_2 conditions, hemoglobin would have little effect on O_2 transport (6). Free access of O_2 to the endophyte in *Alnus rubra* is suggested by the continuous network of intercellular air spaces from the soil atmosphere to the surface of the infected cells (131, 156). Likewise, the suberized or lignified host walls that may cause low pO_2 in *Casuarina* and *Myrica* are absent from all other genera that have been studied except for *Comptonia*, a close relative of *Myrica* (unpublished observations of R. H. Berg and also D. J. Asa and J. D. Tjepkema).

Nonetheless, the function of hemoglobin when it occurs in low concentrations is of interest. Perhaps it is involved in some function other than O_2 transport or perhaps there are limited regions of low pO_2 where it does function in O_2 transport. Although suberization is absent in the actinorhizal genera that have low hemoglobin, calculations show that the resistance to O_2 diffusion through the cytoplasm of the infected cells could result in zones of low pO_2 (136). This would be especially true if intercellular air spaces are lacking between some of the infected cells (133, 156).

Using Southern blots and cloned leghemoglobin cDNA from soybeans as a probe, cross-hybridizing sequences have been found in DNA from all five genera of actinorhizal plants that have been examined (52, 95). This suggests that actinorhizal hemoglobins are coded for by the host plant rather than the endophyte and are quite similar to legume hemoglobins. Surprisingly, hybridization was also found between leghemoglobin cDNA and DNA from a species of *Betula* (52). No nodulation has ever been observed in this genus, although it is related to *Alnus,* which does form nodules.

The hemoglobin from *Casuarina* has been purified and sequenced to residue 42 from the amino terminal and found to be 56% homologous with soybean and lupin hemoglobin (60). This suggests a common genetic origin with legume hemoglobin. These results combined with those in the preceding paragraph and the occurrence of hemoglobin in *Parasponia* (7) suggest that the genes for hemoglobin may be widespread in the plant kingdom. The possibility exists that hemoglobins may eventually be found in plant tissue other than root nodules.

Response of Nodules to Soil pO_2

The optimal pO_2 for nitrogen fixation by actinorhizal nodules is usually very close to atmospheric levels or 20 kPa (24, 156, 162). As in legume nodules (6), there is probably a balance between the need for O_2 in oxidative phosphorylation and inhibition of nitrogenase by excessive O_2. Mechanisms that prevent or limit inhibition by excessive O_2 were discussed above. Adaptations to the opposite problem, that of low environmental pO_2, have been examined only in *Myrica gale.* This low shrub is common in peatlands where the water table is close to the surface in midsummer (102, 119) and flooding is common during the winter and spring (105).

Myrica gale roots and nodules possess several characteristics that presumably enhance oxygen transport within the plant. The roots can develop extensive internal air spaces (41) and transport oxygen efficiently, as evidenced by radial oxygen diffusion from roots grown in water culture (8). The roots can also produce upward-growing branch roots (105, 107), allowing them to make contact with better aerated surface zones and thus increasing oxygenation of the root system.

Myrica gale nodules also produce upward growing roots, designated nodule roots. These arise singly at the tips of the nodule lobes, are uninfected, and contain extensive intercellular air spaces (141). The extent of nodule root development depends on environmental conditions. Long, thick nodule roots that penetrate the soil surface are produced in very wet soils where pO_2 is low, whereas only short, thin ones are produced in drier, well-aerated soils (107, 120). Nodule roots enhance nitrogenase activity at low ambient pO_2, presumably by increasing the flux of O_2 to the endophyte-containing nodule lobes, but have no effect at atmospheric pO_2 (130). Their ability to increase O_2 flux to the infected tissue is enhanced by their upward growing habit that permits them to contact better aerated surface zones or the free atmosphere itself. Nodule roots also occur in other *Myrica* species, *Comptonia* and *Casuarina* (141), but their function has not been examined.

The adaptations discussed above are not sufficient to permit normal levels of nitrogenase activity when the nodules are flooded. Resumption of nitrogenase activity in spring depends on nodule growth and differentiation of *Frankia* vesicles within newly produced cells in the nodule cortex (see "Seasonal Patterns of Nodule Activity"). Frankiae within flooded *Myrica gale* nodules are able to differentiate vesicles and resume nitrogenase activity, but flooding delays the onset of nitrogenase activity and greatly reduces specific nitrogenase activity and production of new nodule biomass. Interestingly, new nodules can be initiated on adventitious roots growing from the flooded portions of the stems into the surrounding flood waters, thus replacing largely nonfunctional, deeply flooded nodules with new nodules in the aerobic zone (105). As a consequence of its various adaptations to reduced pO_2, *Myrica gale* can survive and make modest growth in water-saturated soils, but it makes much stronger growth in better aerated, moist soils (105, 107).

CONTROL AND ENERGY COST OF NITROGEN FIXATION

N_2 fixation obviously benefits the host plant, but there is a substantial cost because of the use of photosynthate. Under some conditions the rate of N_2 fixation may be limited by lack of sufficient photosynthate. Defoliation can reduce nitrogenase activity in as little as an hour (56), and a decline in available photosynthate may be a factor when lower fixation rates are observed during nightime hours (36). However, increasing the available photosynthate does not necessarily increase nitrogenase activity. When shoot growth, which is a competing sink for photosynthate, is restricted by excision or induced dormancy, there is a decrease rather than an increase in nitrogenase activity (55).

The energy used for N_2 fixation by actinorhizal plants is roughly comparable to that used by legumes, when either nodule respiration (140) or whole plant energy budgets are considered (135). At 52 days after germination, seedlings of *Alnus* used 11.6% of gross photosynthate for nodule respiration and 4.1% for nodule mass (135). Thus energy usage for N_2 fixation is substantial.

Based on theoretical considerations, the energy costs of nitrogen uptake and assimilation are thought to increase in the order of $NH_4^+ < NO_3^- < N_2$ (36, 91,

99). The greater energy cost of N_2 as a nitrogen source is supported by numerous experiments in which both NO_3^- and NH_4^+ are found to inhibit nodule formation and activity (36). Also, as expected from energy cost, *Alnus incana* grew faster with NH_4NO_3 as a nitrogen source than when dependent on N_2 (57). However, the effects of nitrogen source are complex and involve factors other than energy cost, such as the lag in N_2 fixation while nodules form and effects on ion balance and soil/plant acidity (144). The uptake and assimilation of NH_4^+ cause strong acidification, while plants dependent on N_2 generate moderate acidity, and the uptake and metabolism of NO_3^- generates alkalinity. Thus the effect of nitrogen source on growth rate depends partly on pH control of the rooting medium during the experiment (144). Moreover, rooted cuttings of *Alnus incana* fed NH_4^+ excrete nitrogen in an organic form whereas N_2-fixing cuttings do not excrete detectable amounts of nitrogen (111). Thus actinorhizal plants may use N_2 more efficiently than combined nitrogen. In any event, it is clear that N_2 fixation is suppressed when adequate supplies of NO_3^- or NH_4^+ are available from the soil.

Although the subject has received little attention, there is evidence that nodule mass is regulated not only by external supplies of fixed nitrogen but also by the internal demand for nitrogen. In seedlings of *Alnus rubra* and *Myrica gale,* nodule mass was positively correlated with the relative growth rate, being 6–7% of the total dry weight during the period of maximum growth rate and 2–3% as growth rate slowed (135). This suggests that either nodule mass was controlling growth rate or that nodule mass was closely regulated by the plant to meet its demand for nitrogen. Further evidence for control of nodule mass comes from observations of weight per nodule and number of nodules per plant. When only a few nodules are present they each tend to become very large, while when many nodules are present they each remain small (J. D. Tjepkema, unpublished). Thus the total nodule mass per plant tends to be constant in spite of variability in nodule number. A related factor that affects nodule mass may be host-endophyte compatibility. There is evidence that less effective nodules may be compensated for by an increased nodule mass (159).

SEASONAL PATTERNS OF NODULE ACTIVITY

Seasonal patterns of N_2 fixation have been observed in all actinorhizal plants examined to date. These patterns are the product of interactions between the local climate and phenology of the host plant and the *Frankia* endophyte and thus vary with location and host. By far the best understood case is that of *Myrica gale,* a deciduous shrub, in Massachusetts (103, 104, 106, 108, 148). There nitrogenase activity is absent during the winter, appears in mid-May when the leaves unfold, reaches a peak in July or early August when the shoots are fully expanded, and then declines until it is no longer measurable in late

October after all leaves have fallen (103, 106). Excepting extremes in weather, this pattern appears to be stable from year to year (104), and the same general pattern is also found in Scotland (121). Interestingly, temperature has little effect on the general shape of the seasonal curve even though nitrogenase activity is strongly temperature dependent (103).

The seasonal pattern of nitrogenase activity is, however, strongly affected by endophyte and nodule phenology. The endophyte overwinters as colonizing hyphae in partially expanded cortical cells near the nodule lobe apex. Vesicles first appear in newly mature cortical cells coincident with the onset of nitrogenase activity in spring, occupy the bulk of the current season-infected tissue in summer, and disappear accompanied by death of the host cell as nitrogenase activity ceases in autumn (106). Nitrogenase activity also depends on the amount of current season nodule tissue, increasing as nodule biomass increases (105). The annual peak in nitrogenase activity is reached in midsummer as most nodules complete their growth for the season (106). Sporangia begin to differentiate in June in vesicle-containing cells and become more prominant in late summer. The host cell rapidly senesces following the onset of sporangial differentiation and is dead when the sporangia are mature (148).

The seasonal pattern of nodule respiration (CO_2 evolution) in *Myrica gale* is similar to that for nitrogenase activity except that it continues at low rates when nitrogenase activity is absent (108). As might be expected, respiration is greater in spring at the onset of nitrogenase activity when the nodules are actively growing than in autumn at the cessation of nitrogenase activity when there is no growth. When nitrogenase activity is substantial, the bulk of nodule respiration is apparently associated with nitrogenase activity and related processes, resulting in a relatively constant energy cost of nitrogen fixation throughout the growing season ($CO_2:C_2H_4 = 4.0–6.5$ with an annual value of 4.9) (108). Hydrogen evolution is barely detectable throughout the growing season and thus does not represent a significant energy loss (108).

The seasonal patterns of *Myrica gale* nodules are probably typical of winter-deciduous actinorhizal plants in general in temperate climates. Similar patterns of nitrogenase activity in relation to shoot phenology have been observed in several *Alnus* species (1, 40, 82, 92, 143, 159), *Hippophae rhamnoides* (1) and *Purshia tridentata* (33). Periods unfavorable to photosynthesis are likely to alter the shape of the curve, as in *Purshia tridentata* in Oregon where water stress depresses nitrogenase activity during the latter part of the growing season (33). In plants producing heavy seed crops, nitrogenase activity may also be depressed during the period of rapid seed growth in summer (93). The seasonal pattern of respiration in *Alnus glutinosa* is apparently similar to that in *Myrica gale* (1). *Alnus* also has a similar pattern of nodule lobe growth as *Myrica gale*, i.e. formation of new nodule lobe tissues beginning in spring prior to the onset of nitrogenase activity and continuing until midsummer (160). Finally, formation of new vesicle-containing tissues in spring and death of vesicle-containing cells and the enclosed endophyte in autumn has been observed in *Alnus* as well (59, 61, 160).

The seasonal pattern of nitrogenase activity in evergreen species is probably also related to nodule and endophyte phenology although there is only very limited information on them. In Oregon, *Ceanothus velutinus*, an evergreen, has substantially reduced nitrogenase activity in early spring accompanied by nodule lobe growth (77). In the Himalayas, *Alnus nepalensis*, a partially deciduous tree, lacks nitrogenase activity in the majority of its nodules in midwinter and has only low rates in the remainder of the nodules (113). Finally in Egypt, there is one report of the nodules of *Casuarina equisetifolia*, an evergreen, turning brown and losing their nitrogenase activity in late summer (152a).

SPORE (+) AND SPORE (−) STRAINS OF *FRANKIA*

Two types of nodules can be distinguished based on the extent of spore formation *in planta*. These have been designated sp (+) or P (for spore-positive) and sp (−) or N (for spore-negative) (63, 90). Spores and sporangia are abundant in spore (+) and rare in spore (−) nodules (152). Apparently this difference is the result of genetic differences in the endophyte (90, 148, 149). Spore (+) and spore (−) nodules have been observed in several *Alnus* species (90, 152, 159), in *Myrica gale* (106), and in *Comptonia peregrina* (148). They may occur in other genera as well because sporangia have been observed in nodules of *Casuarina* (R. H. Berg, unpublished observations), *Hippophae,* and *Ceanothus* (43).

Spore (+) and spore (−) strains of the endophyte differ in several important respects in addition to the production of spores in nodules. Interestingly, when the endophyte strains are grown in vitro, both form numerous sporangia on most media (9, 27, 90). However, the sporangia formed by spore (+) strains are generally larger and somewhat more numerous than those formed by spore (−) strains (90). In addition, spore (−) strains grow much more rapidly in vitro and form larger, looser colonies than spore (+) strains (90).

A growing body of preliminary evidence indicates that nodules infected by spore (+) and spore (−) *Frankia* strains differ physiologically; spore (−) strains seem to form superior symbioses. Spore (−) nodules are probably more effective than spore (+) nodules in supporting growth of the host plant. Growth of *Alnus glutinosa* and *Alnus rubra* seedlings or plantlets with spore (−) nodules is superior to those with spore (+) nodules (51, 75, 118). *Alnus viridis* spp. *crispa* seedlings inoculated with several isolated strains also grow best with spore (−) strains, but some spore (−) strains are no more effective than spore (+) strains (90). Finally, *Comptonia peregrina* seedlings with spore (−)

nodules make substantially better growth than seedlings bearing spore (+) nodules (147).

Specific nitrogenase activity of spore (−) nodules may also be greater than that of spore (+) nodules, although the number of cases examined to date is small, and there is considerable variation among strains belonging to the same spore type. Specific activity of *Comptonia peregrina* nodules induced with a spore (−) isolate is much greater than in nodules with a spore (+) strain (147). In *Myrica gale,* specific activity of nodules with two spore (−) strains is significantly greater than nodules with a spore (+) strain but a third spore (−) strain produces nodules in which the specific activity does not differ from the spore (+) strain (147). In *Alnus rubra* specific activity of spore (−) nodules is substantially greater than spore (+) nodules (158). In *Alnus glutinosa* mean nitrogenase activities are higher in spore (−) than in spore (+) nodules but the difference is not statistically significant (149).

Finally, the energy cost of nitrogen fixation may be greater in spore (+) than spore (−) nodules. In *Comptonia peregrina* and *Myrica gale* seedlings, nodules induced by two spore (+) strains have substantially larger $CO_2 : C_2H_4$ ratios than spore (−) nodules (147).

The above differences between spore (+) and spore (−) nodules can be expected to have both practical and ecological consequences. Practically, spore (−) strains are more promising for inoculating plants to be used in forestry, land reclamation, and amenity planting. Ecologically, the significance of spore (+) and spore (−) nodules is not clear, although numerous studies of their distribution have been made. In most cases both spore types are found within a larger geographical region. In the Netherlands, spore (+) and spore (−) *Alnus glutinosa* nodules occur within the same stand and even on the same tree (149), but there is some tendency for sp(+) nodules to be associated with local disturbance and sp(−) nodules with high soil moisture content (151). In Finland, spore (+) nodules are predominantly associated with *Alnus incana* and spore (−) nodules with *Alnus glutinosa* (A. Weber, K. van Dijk, and V. Sundman, unpublished observations). In eastern Canada, nodules on *Alnus incana* ssp. *rugosa* and *Alnus viridis* ssp. *crispa* are primarily spore (−) north of latitude 47°N while to the south both spore (+) and spore (−) nodules occur (90). In Massachusetts and Maine, *Comptonia peregrina* nodules are primarily spore (−) (148; C. R. Schwintzer, unpublished observations) and *Myrica gale* nodules both spore (+) and spore (−) (106; C. R. Schwintzer, C. R. Kashanski, and S. A. Lancelle, unpublished observations).

CONCLUDING REMARKS

Actinorhizal plants are very deserving of the attention that they are currently receiving. Not only are they economically important, but they are also promising for use in the creation of new nitrogen-fixing plants by genetic engineering. Frankiae form symbioses with diverse hosts and can fix nitrogen at atmospheric pO_2, thus eliminating the need for a special microaerobic environment in the host tissue. The recent discovery that *Elaeagnus angustifolia* is nodulated by direct intercellular infection (81) further supports the possibility that actinorhizal symbioses can be extended to new host plants.

ACKNOWLEDGMENTS

We thank our colleagues for sending preprints. Our research has been supported by National Science Foundation Grant No. DMB-8315415 (JDT), U.S. Department of Agriculture Grant No. 84-CRCR-1-1433 (CRS), and U.S. Department of Agriculture Grant No. 81-CRCR-1-0651 (DRB).

Literature Cited

1. Akkermans, A. D. L. 1971. *Nitrogen fixation and nodulation of Alnus and Hippophae under natural conditions*. PhD thesis. State Univ., Leiden, Netherlands. 85 pp.
2. Akkermans, A. D. L., Huss-Danell, K., Roelofsen, W. 1981. Enzymes of the tricarboxylic acid cycle and the malate-aspartate shuttle in the N_2-fixing endophyte of *Alnus glutinosa*. *Physiol. Plant.* 53:289–94
3. Akkermans, A. D. L., Roelofsen, W. 1980. Symbiotic nitrogen fixation by actinomycetes in *Alnus*-type root nodules. In *Nitrogen Fixation, Proceedings of the Phytochemical Society of Europe*, ed. W. D. P. Stewart, J. R. Gallon, pp. 279–99. London/New York: Academic. 451 pp.
4. Akkermans, A. D. L., Roelofsen, W., Blom, J., Huss-Danell, K., Harkink, R. 1983. Utilization of carbon and nitrogen compounds by *Frankia* in synthetic media and in root nodules of *Alnus glutinosa, Hippophae rhamnoides*, and *Datisca cannabina*. *Can. J. Bot.* 61:2793–2800
5. An, C. S., Riggsby, W. S., Mullin, B. C. 1985. Relationships of *Frankia* isolates based on deoxyribonucleic acid homology studies. *Int. J. Syst. Bacteriol.* 35:140–46
6. Appleby, C. A. 1984. Leghemoglobin and *Rhizobium* respiration. *Ann. Rev. Plant Physiol.* 35:443–78
7. Appleby, C. A., Tjepkema, J. D., Trinick, M. J. 1983. Hemoglobin in a nonleguminous plant, *Parasponia:* Possible genetic origin and function in nitrogen fixation. *Science* 220:951–53
8. Armstrong, W. 1968. Oxygen diffusion from the roots of woody species. *Physiol. Plant.* 21:539–43
9. Baker, D. D. 1982. A cumulative listing of isolated frankiae, the symbiotic nitrogen-fixing actinomycetes. *Actinomycetales* 17:35–42
10. Baker, D. D., O'Keefe, D. 1984. A modified sucrose fractionation procedure for the isolation of frankiae from actinorhizal root nodules and soil samples. *Plant Soil* 78:23–28
11. Becking, J. H. 1970. *Frankiaceae* fam. nov. *(Actinomycetales)* with one new combination and six new species of the genus *Frankia* Brunchorst 1886, 174. *Int. J. Syst. Bacteriol.* 20:201–20
12. Benson, D. R. 1982. Isolation of *Frankia* strains from alder actinorhizal root nodules. *Appl. Environ. Microbiol.* 44:461–65
13. Benson, D. R., Arp, D. J., Burris, R. H. 1979. Cell-free nitrogenase and hydrogenase from actinorhizal root nodules. *Science* 205:688–89
14. Benson, D. R., Arp, D. J., Burris, R. H. 1980. Hydrogenase in actinorhizal root nodules and root nodule homogenates. *J. Bacteriol.* 142:138–44
15. Benson, D. R., Eveleigh, D. E. 1979. Ultrastructure of the nitrogen-fixing symbiont of *Myrica pensylvanica* L. (bayberry) root nodules. *Bot. Gaz.* 140 (Suppl.):S15–S21
16. Benson, D. R., Hanna, D. 1983. *Frankia* diversity in an alder stand as estimated by sodium dodecyl sulfate—polyacrylamide gel electrophoresis of whole-cell proteins. *Can. J. Bot.* 61:2919–23
17. Benson, D. R., Mazzucco, C. E., Browning, T. J. 1985. Physiological aspects of *Frankia*, In *Nitrogen Fixation*

and *CO₂ Metabolism*, ed. P. W. Ludden, J. E. Burris, pp. 175–82. New York: Elsevier. 445 pp.

18. Berg, R. H. 1983. Preliminary evidence for the involvement of suberization in infection of *Casuarina. Can. J. Bot.* 61: 2910–18

19. Berry, A. M. 1984. The infection process in actinorhizal symbioses: Review of recent research. In *Current Perspectives in Microbial Ecology*, ed. M. J. Klug, C. A. Reddy, pp. 222–29. Washington, DC: Am. Soc. Microbiol. 710 pp.

20. Berry, A. M. 1986. Cellular aspects of root nodule establishment in the *Frankia* symbioses. In *Plant-Microbe Interaction*, Vol. 2, ed. T. Kosuge. New York: Macmillan. In press

21. Blom, J. 1982. Carbon and nitrogen source requirements of *Frankia* strains. *FEMS Microbiol. Lett.* 13:51–55

22. Blom, J., Harkink, R. 1981. Metabolic pathways for gluconeogenesis and energy generation in *Frankia* AvcII. *FEMS Microbiol. Lett.* 11:221–24

23. Blom, J., Roelofsen, W., Akkermans, A. D. L. 1981. Assimilation of nitrogen in root nodules of alder *(Alnus glutinosa)*. *New Phytol.* 89:321–26

24. Bond, G. 1961. The oxygen relation of nitrogen fixation in root nodules. *Z. Allg. Mikrobiol.* 1:93–99

25. Bond, G. 1976. The results of the IBP survey of root-nodule formation in non-leguminous angiosperms. In *Symbiotic Nitrogen Fixation in Plants*, ed. P. S. Nutman, pp. 443–74. Cambridge: Cambridge Univ. Press. 584 pp.

26. Bond, G. 1983. Taxonomy and distribution of non-legume nitrogen-fixing systems. In *Biological Nitrogen Fixation in Forest Ecosystems: Foundations and Applications*, ed. J. C. Gordon, C. T. Wheeler, pp. 55–87. The Hague: Nijhoff. 342 pp.

26a. Burggraaf, A. J. P. 1984. *Isolation, cultivation and characterization of Frankia strains from actinorhizal root nodules*. PhD thesis. State Univ. Leiden, The Netherlands. 179 pp.

27. Burggraaf, A. J. P., Quispel, A., Tak, T., Valstar, J. 1981. Methods of isolation and cultivation of *Frankia* species from actinorhizas. *Plant Soil* 61:157–68

28. Burggraaf, A. J. P., Shipton, W. A. 1983. Studies on the growth of *Frankia* isolates in relation to infectivity and nitrogen fixation (acetylene reduction). *Can. J. Bot.* 61:2774–82

29. Callaham, D., Del Tredici, P., Torrey, J. G. 1978. Isolation and cultivation *in vitro* of the actinomycete causing root nodulation in *Comptonia. Science* 199:899–902

30. Carpenter, C. V., Robertson, L. R. 1983. Isolation and culture of nitrogen-fixing organisms. See Ref. 26, pp. 89–106

31. Ching, T. M., Monaco, P. A., Ching, K. K. 1983. Energy status and cytochromes in isolated endophytic vesicle clusters of red alder root nodules. *Can. J. For. Res.* 13:921–28

32. Cochrane, V. W. 1955. The metabolism of species of streptomyces. VIII. Reactions of the Embden-Meyerhof-Parnas sequence in *Streptomyces coelicolor. J. Bacteriol.* 69:256–63

33. Dalton, D. A., Zobel, D. B. 1977. Ecological aspects of nitrogen fixation by *Purshia tridentata. Plant Soil* 48:57–80

34. Diem, H. G., Dommergues, Y. R. 1985. *In vitro* production of specialized reproductive torulose hyphae by *Frankia* strain ORS 021001 isolated from *Casuarina junghuhniana* root nodules. *Plant Soil.* 87:17–29

35. Dixon, R. O. D. 1972. Hydrogenase in legume root nodule bacteroids: occurrence and properties. *Arch. Microbiol.* 85:193–201

36. Dixon, R. O. D., Wheeler, C. T. 1983. Biochemical, physiological and environmental aspects of symbiotic nitrogen fixation. See Ref. 26, pp. 107–71

37. Eisbrenner, G., Evans, H. J. 1983. Aspects of hydrogen metabolism in nitrogen-fixing legumes and other plant-microbe associations. *Ann. Rev. Plant Physiol.* 34:105–36

38. Fessenden, R. J. 1979. Use of actinorhizal plants for land reclamation and amenity planting in the U.S.A. and Canada. See Ref. 48, pp. 403–19

39. Fleming, A. I., Wittenberg, B. A., Wittenberg, J. B., Appleby, C. A. 1985. Hemoglobin in the root nodules of the actinorhizal genus *Casuarina. Proc. Aust. Biochem. Soc.* 17:24 (Abstr.)

40. Fleschner, M. D., Delwiche, C. C., Goldman, C. R. 1976. Measuring rates of symbiotic nitrogen fixation by *Alnus tenuifolia. Am. J. Bot.* 63:945–50

41. Fletcher, W. W. 1955. The development and structure of the root-nodules of *Myrica gale* L. with special reference to the nature of the endophyte. *Ann. Bot. (NS)* 19:501–13

42. Fontaine, M. S., Lancelle, S. A., Torrey, J. G. 1984. Initiation and ontogeny of vesicles in cultured *Frankia* sp. strain HFPArI3. *J. Bacteriol.* 160:921–27

43. Gardner, I. C. 1976. Ultrastructural studies of non-leguminous root nodules. See Ref. 25, pp. 485–95

44. Gauthier, D. L. 1983. Effect of L-methionine-DL-sulfoximine on acetylene

reduction and vesicle formation in de-repressed cultures of *Frankia* strain D11. *Can. J. Microbiol.* 29:1003–6

45. Gauthier, D. L., Diem, H. G., Dommergues, Y. 1981. *In vitro* nitrogen fixation by two actinomycete strains isolated from *Casuarina* nodules. *Appl. Environ. Microbiol.* 41:306–8

46. Gauthier, D. L., Diem, H. G., Dommergues, Y. R., Ganry, F. 1985. Assessment of N_2 fixation by *Casuarina equisetifolia* inoculated with *Frankia* ORS021001 using ^{15}N methods. *Soil Biol. Biochem.* 17:375–79

47. Gauthier, D. L., Frioni, L., Diem, H. G., Dommergues, Y. 1984. The *Colletia spinosissima-Frankia* symbiosis. *Acta Oecol./Oecol. Plant.* 5(19):231–39

48. Gordon, J. C., Wheeler, C. T., Perry, D. A., eds. 1979. *Symbiotic Nitrogen Fixation in the Management of Temperate Forests.* Corvallis: Forest Res. Lab., Oregon State Univ. 501 pp.

49. Hafeez, F., Akkermans, A. D. L., Chaudhary, A. H. 1984. Observations on the ultrastructure of *Frankia* sp. in root nodules of *Datisca cannabina* L. *Plant Soil* 79:383–402

50. Hafeez, F., Chaudhary, A. H., Akkermans, A. D. L. 1984. Physiological studies on N_2-fixing root nodules of *Datisca cannabina* L. and *Alnus nitida* Endl. from Himalaya region in Pakistan. *Plant Soil* 78:129–46

51. Hall, R. B., McNabb, H. S. Jr., Maynard, C. A., Green, T. L. 1979. Toward development of optimal *Alnus glutinosa* symbioses. *Bot. Gaz.* 140(Suppl.):S120–26

52. Hattori, J., Johnson, D. A. 1985. The detection of leghemoglobin-like sequences in legumes and non-legumes. *Plant Mol. Biol.* 4:285–92

53. Hirel, B., Perrot-Rechenmann, C., Maudinas, B., Gadal, P. 1982. Glutamine synthetase in alder *(Alnus glutinosa)* root nodules. Purification, properties and cytoimmunochemical localization. *Physiol. Plant.* 55:197–203

54. Horriere, F., Lechevalier, M. P., Lechevalier, H. A. 1983. *In vitro* morphogenesis and ultrastructure of a *Frankia* sp. ArI3 (Actinomycetales) from *Alnus rubra* and a morphologically similar isolate (AirI2) from *Alnus incana* subsp. *rugosa*. *Can. J. Bot.* 61:2843–54

55. Huss-Danell, K., Sellstedt, A. 1983. Nitrogenase activity in response to restricted shoot growth in *Alnus incana*. *Can. J. Bot.* 61:2949–55

56. Huss-Danell, K., Sellstedt, A. 1985. Nitrogenase activity in response to darkening and defoliation of *Alnus incana*. *J. Exp. Bot.* 36:1352–58

57. Ingestad, T. 1980. Growth, nutrition, and nitrogen fixation in grey alder at varied rates of nitrogen addition. *Phsiol. Plant.* 50:353–64

58. Kahn, M. L. 1985. A model of nutrient exchange in the *Rhizobium*-legume symbiosis. In *Nitrogen Fixation Research Progress*, ed. H. J. Evans, P. J. Bottomley, W. E. Newton, pp. 193–99. Dordrecht: Nijhoff. 731 pp.

59. Käppel, M., Wartenberg, H. 1958. Der Formenwechsel des *Actinomyces alni* Peklo in den Wurzeln von *Alnus glutinosa* Gaertner. *Arch. Mikrobiol.* 30:46–63

60. Kortt, A. A., Burns, J. E., Inglis, A. S., Appleby, C. A., Trinick, M. J., et al. 1985. Hemoglobins from the nitrogen-fixing root nodules of non-legumes and their relationship to the legume hemoglobins. *Proc. Aust. Biochem. Soc.* 17:43 (Abstr.)

61. Krebber, O. 1932. Untersuchungen über die Wurzelknöllchen der Erle. *Arch. Mikrobiol.* 3:588–608

62. Lalonde, M. 1979. Immunological and ultrastructural demonstration of nodulation of the European *Alnus glutinosa* (L.) Gaertn. host plant by an actinomycetal isolate from the North American *Comptonia peregrina* (L.) Coult. root nodule. *Bot. Gaz.* 140(Suppl.):S35–S43

63. Lalonde, M., Calvert, H. E., Pine, S. 1981. Isolation and use of *Frankia* strains in actinorhizae formation. In *Current Perspectives in Nitrogen Fixation*, ed. A. H. Gibson, W. E. Newton, pp. 296–99. Canberra: Austr. Acad. Sci. 534 pp.

64. Lalonde, M., DeVoe, I. W. 1976. Origin of the membrane envelope enclosing the *Alnus crispa* var. *mollis*. Fern. root nodule endophyte as revealed by freeze-etching microscopy. *Physiol. Plant Pathol.* 8:123–29

65. Lalonde, M., Knowles, R., DeVoe, I. W. 1976. Absence of "void area" in freeze-etched vesicles of the *Alnus crispa* var. *mollis* Fern. root nodule endophyte. *Arch. Microbiol.* 107:263–67

66. Lamont, H., Torrey, J. G., Young, P. 1985. Asparagine and glutamine as nitrogen sources controlling development and nitrogenous activity of Frankia sp. strain HFP CpI1. See Ref. 58, p. 365

67. Lechevalier, M. P. 1983. Cataloging *Frankia* strains. *Can. J. Bot.* 61:2964–67

68. Lechevalier, M. P. 1984. The taxonomy of the genus *Frankia*. *Plant Soil* 78:1–6

69. Lechevalier, M. P., Lechevalier, H. A. 1979. The taxonomic position of the actinomycetic endophytes. See Ref. 38, pp. 111–23

70. Lechevalier, M. P., Lechevalier, H. A. 1986. Genus *Frankia* Brunchorst 1886,

174[AL]. In *Bergey's Manual of Systematic Bacteriology*, Vol. 4, ed. S. T. Williams. Baltimore: Williams & Wilkins. In press

71. Lechevalier, M. P., Ruan, J. 1984. Physiology and chemical diversity of *Frankia* spp. isolated from nodules of *Comptonia peregrina* (L.) Coult. and *Ceanothus americanus* L. *Plant Soil* 778:15–22

72. Lopez, M. F., Torrey, J. G. 1985 Enzymes of glucose metabolism in *Frankia* sp. *J. Bacteriol.* 162:110–16

73. Lopez, M. F., Whaling, C. S., Torrey, J. G. 1983. The polar lipids and free sugars of *Frankia* in culture. *Can. J. Bot.* 61:2834–42

74. Magasanik, B. 1982. Genetic control of nitrogen assimilation in bacteria. *Ann. Rev. Genet.* 16:135–68

75. Malcolm, D. C., Hooker, J. E., Wheeler, C. T. 1985. *Frankia* symbiosis as a source of nitrogen in forestry: A case study of symbiotic nitrogen-fixation in a mixed *Alnus-Picea* plantation in Scotland. *Proc. R. Soc. Edinburgh* 85B:263–82

76. Mazzucco, C. E., Benson, D. R. 1984. (14C)Methylammonium transport by *Frankia* sp. strain CpI1. *J. Bacteriol.* 160:636–41

77. McNabb, D. H., Cromack, K. Jr. 1983. Dinitrogen fixation by a mature *Ceanothus velutinus* (Dougl.) stand in the Western Oregon Cascades. *Can. J. Microbiol.* 29:1014–21

78. Meesters, T. M., van Genesen, S. T., Akkermans, A. D. L. 1986. Growth, acetylene reduction activity and localization of nitrogenase in relation to vesicle formation in *Frankia* strains Cc1.17 and Cp 1.2. *Arch. Microbiol.* In press

79. Mian, S., Bond, G. 1978. The onset of nitrogen fixation in young alder plants and its relation to differentiation in the nodular endophyte. *New Phytol.* 80:187–92

80. Mikola, P., Uomala, P., Mälkönen, E. 1983. Application of biological nitrogen fixation in European silviculture. See Ref. 26, pp. 279–94

81. Miller, I. M., Baker, D. D. 1986. The initiation, development and structure of root nodules in *Elaeagnus angustifolia* L. (Elaeagnaceae). *Protoplasma.* In press

82. Moiroud, A., Capellano, A. 1979. Etude de la dynamique de l'azote à haute altitude. I. Fixation d'azote (réduction de l'acétylène) par *Alnus viridis. Can. J. Bot.* 57:1979–85

83. Moore, A. W. 1964. Note on nonleguminous nitrogen-fixing plants in Alberta. *Can. J. Bot.* 42:952–55

84. Murry, M. A., Fontaine, M. S., Tjepke-ma, J. D. 1984. Oxygen protection of nitrogenase in *Frankia* sp. HFPArI3. *Arch. Microbiol.* 139:162–66

85. Murry, M. A., Zhang, Z., Torrey, J. G. 1985. Effect of O₂ on vesicle formation, acetylene reduction and O₂-uptake kinetics in *Frankia* sp. HFPCcI3 isolated from *Casuarina cunninghaniana. Can. J. Microbiol.* In press

86. Deleted in proof

87. Newcomb, W., Callaham, D., Torrey, J. G., Peterson, R. L. 1979. Morphogenesis and fine structure of the actinomycetous endophyte of nitrogen-fixing root nodules of *Comptonia peregrina. Bot. Gaz.* 140(Suppl.):S22–S34

88. Newcomb, W., Heisey, R. M. 1984. Ultrastructure of actinorhizal root nodules of *Chamaebatia foliolosa* (Rosaceae). *Can. J. Bot.* 62:1697–1707

88a. Noridge, N. A., Benson, D. R. 1986. Isolation and nitrogen-fixing activity of *Frankia* sp. strain CpI1 vesicles. *J. Bacteriol.* In press

89. Normand, P., Lalonde, M. 1985. The genetics of actinorhizal *Frankia:* a review. *Plant Soil* In press

90. Normand, P., Lalonde, M. 1982. Evaluation of *Frankia* strains isolated from provenances of two *Alnus* species. *Can. J. Microbiol.* 28:1133–42

91. Pate, J. S., Atkins, C. A., Rainbird, R. M. 1981. Theoretical and experimental costing of nitrogen fixation and related processes in nodules of legumes. See Ref. 63, pp. 105–16

92. Pizelle, G. 1975. Variations saisonnierès de l'activité nitrogénasique des nodules d'*Alnus incana* (L.) Moench et d'*Alnus cordata* (Lois.) Desf. *C.R. Acad. Sci. Paris D 281* (23):1829–32

93. Pizelle, G. 1984. Seasonal variations of the sexual reproductive growth and nitrogenase activity (C₂H₂) in mature *Alnus glutinosa. Plant Soil* 78:181–88

94. Resch, H. 1979. Industrial uses and utilization potential for red alder. See Ref. 48, pp. 444–50

95. Roberts, M. P., Jafar, S., Mullin, B. C. 1986. Leghemoglobin-like sequences in the DNA of four actinorhizal plants. *Plant Mol. Biol.* In press

96. Rodriguez-Barrueco, C., Subramaniam, P. 1986. Host-endophyte specificity in *Frankia* symbiosis. In *New Trends in Biological Nitrogen Fixation*, ed. N. S. Subba Rao. New Delhi: Oxford and IBH Publ. Co. In press

97. Roelofsen, W., Akkermans, A. D. L. 1979. Uptake and evolution of H₂ and reduction of C₂H₂ by root nodules and nodule homogenates of *Alnus glutinosa. Plant Soil* 52:571–78

98. Schaede. R. 1939. Die Actinomyceten-Symbiose von *Myrica gale*. *Planta* 29:32–46
99. Schubert, K. R. 1982. *The Energetics of Biological Nitrogen Fixation*. Rockville, MD: Am. Soc. Plant Physiol. 30 pp.
100. Schubert, K. R., Coker, G. T. III. 1981. Ammonia assimilation in *Alnus glutinosa* and *Glycine max*. *Plant Physiol*. 67:662–65
101. Schubert, K. R., Evans, H. J. 1976. Hydrogen evolution: A major factor affecting the efficiency of nitrogen fixation in nodulated symbionts. *Proc. Natl. Acad. Sci. USA* 73:1207–11
102. Schwintzer, C. R. 1978. Vegetation and nutrient status of northern Michigan fens. *Can. J. Bot.* 56:3044–51
103. Schwintzer, C. R. 1979. Nitrogen fixation by *Myrica gale* root nodules in a Massachusetts wetland. *Oecologia* 43:283–94
104. Schwintzer, C. R. 1983. Nonsymbiotic and symbiotic nitrogen fixation in a weakly minerotrophic peatland. *Am. J. Bot.* 70:1071–78
105. Schwintzer, C. R. 1985. Effect of spring flooding on endophyte differentiation, nitrogenase activity, root growth and shoot growth in *Myrica gale*. *Plant Soil* 87:109–24
106. Schwintzer, C. R., Berry, A. M., Disney, L. D. 1982. Seasonal patterns of root nodule growth, endophyte morphology, nitrogenase activity, and shoot development in *Myrica gale*. *Can. J. Bot.* 60:746–57
107. Schwintzer, C. R., Lancelle, S. A. 1983. Effect of water-table depth on shoot growth, root growth, and nodulation of *Myrica gale* seedlings. *J. Ecol.* 71:489–501
108. Schwintzer, C. R., Tjepkema, J. D. 1983. Seasonal pattern of energy use, respiration, and nitrogenase activity in root nodules of *Myrica gale*. *Can. J. Bot.* 61:2937–42
109. Scott, D. B. 1978. Ammonia assimilation in N$_2$-fixing systems. In *Limitations and Potentials for Biological Nitrogen Fixation in the Tropics*, ed. J. Dobereiner, R. H. Burris, A. Hollaender, pp. 223–35. New York/London: Plenum. 398 pp.
110. Sellstedt, A., Huss-Danell, K. 1984. Growth, nitrogen fixation and relative efficiency of nitrogenase in *Alnus incana* grown in different cultivation systems. *Plant Soil* 78:147–58
111. Sellstedt, A., Huss-Danell, K. 1985. Biomass production and nitrogen utilization by *Alnus incana* when grown on N$_2$ or NH$_4^+$. See Ref. 58, p. 369
112. Sellstedt, A., Huss-Danell, K., Ahlquist,

A.-S. 1986. Nitrogen fixation and biomass production in symbiosis between *Alnus incana* and *Frankia* strains with different hydrogen metabolism. *Physiol. Plant*. In press
113. Sharma, E., Ambasht, R. S. 1984. Seasonal variation in nitrogen fixation by different ages of root nodules of *Alnus nepalensis* plantations, in the eastern Himalayas. *J. Appl. Ecol.* 21:265–70
114. Shipton, W. A., Burggraaf, A. J. P. 1982. A comparison of the requirements for various carbon and nitrogen sources and vitamins in some *Frankia* isolates. *Plant Soil* 69:149–61
115. Shipton, W. A., Burggraaf, A. J. P. 1983. Aspects of the cultural behaviour of *Frankia* and possible cultural implications. *Can. J. Bot.* 61:2783–92
116. Silvester, W. B. 1976. Ecological and economic significance of the non-legume symbioses. *Proc. 1st Int. Symp. Nitrogen Fixation, Pullman, 1974*, ed. W. E. Newton, C. J. Nyman, pp. 489–506. Pullman: Wash. State Univ. Press. 717 pp.
117. Silvester, W. B. 1977. Dinitrogen fixation by plant associations excluding legumes. In *A Treatise on Dinitrogen Fixation, Sect. 4*, ed. R. W. F. Hardy, A. H. Gibson, pp. 141–90. New York: Wiley. 527 pp.
118. Simon, L., Stein, A., Cote, S., Lalonde, M. 1985. Performance of *in vitro* propagated *Alnus glutinosa* (L.) Gaertn. clones inoculated with *Frankiae*. *Plant Soil* 87:125–33
119. Spence, D. H. N. 1964. The macrophytic vegetation of freshwater lochs, swamps and associated fens. In *The Vegetation of Scotland*, ed. J. H. Burnett, pp. 306–425. Edinburgh: Oliver & Boyd. 613 pp.
120. Sprent, J. I., Scott, R. 1979. The nitrogen economy of *Myrica gale* and its possible significance for the aforestation of peat soils. See Ref. 48, pp. 234–42
121. Sprent, J. I., Scott, R., Perry, K. M. 1978. The nitrogen economy of *Myrica gale* in the field. *J. Ecol.* 66:657–68
122. Steele, D. B., Stowers, M. D. 1985. Enzymatic mechanisms for the protection of nitrogenase from oxygen in *Frankia*. See Ref. 58, p. 439
123. St-Laurent, L., Lalonde, M. 1984. *Isolation and characterization of Frankia strains isolated from Myrica gale* L. Presented at Current Research on *Frankia* and Actinorhizal Plants Int. Symp., Montmorency Forest, Laval Univ, Quebec
124. Stewart, W. D. P. 1980. Some aspects of structure and function in N$_2$-fixing cya-

nobacteria. *Ann. Rev. Microbiol.* 34:497–536

125. Stowers, M. D., Kulkarni, R. K., Steele, D. B. 1986. Intermediary carbon metabolism in *Frankia*. *Arch. Microbiol.* In press

126. Streicher, S. L., Tyler, B. 1981. Regulation of glutamine synthetase activity by adenylylation in the Gram-positive bacterium *Streptomyces cattleya*. *Proc. Natl. Acad. Sci. USA* 78:229–33

127. Tarrant, R. F. 1983. Nitrogen fixation in North American forestry: Research and application. See Ref. 26, pp. 261–77

128. Tisa, L. S., Ensign, J. C. 1985. Characterization of vesicles isolated from *Frankia* isolate EAN1pec. See Ref. 58, p. 440

129. Tisa, L., McBride, M., Ensign, J. C. 1983. Studies of growth and morphology of *Frankia* strains EAN1pec, EuI1c, CpI1, and ACN1AC. *Can. J. Bot.* 61:2919–23

130. Tjepkema, J. D. 1978. The role of oxygen diffusion from the shoots and nodule roots in nitrogen fixation by root nodules of *Myrica gale*. *Can. J. Bot.* 56:1365–71

131. Tjepkema, J. D. 1979. Oxygen relations in leguminous and actinorhizal nodules. See Ref. 48, pp. 175–86

132. Tjepkema, J. D. 1983. Hemoglobins in the nitrogen-fixing root nodules of actinorhizal plants. *Can. J. Bot.* 61:2924–29

133. Tjepkema, J. D. 1983. Oxygen concentration within the nitrogen-fixing root nodules of *Myrica gale* L. *Am. J. Bot.* 70(1):59–63

134. Tjepkema, J. D. 1984. Oxygen, hemoglobins, and energy usage in actinorhizal nodules. In *Advances in Nitrogen Fixation Research*, ed. C. Veeger, W. E. Newton, pp. 467–73. The Hague: Nijhoff. 760 pp.

135. Tjepkema, J. D. 1985. Utilization of photosynthate for nitrogen fixation in seedlings of *Myrica gale* and *Alnus rubra*. See Ref. 17, pp. 183–92

136. Tjepkema, J. D. 1986. Oxygen regulation and energy usage in actinorhizal symbiosis. See Ref. 96. In press

137. Tjepkema, J. D., Murry, M. A. 1985. Heme content and diffusion limitation of respiration and nitrogenase activity in nodules of *Casuarina cunninghamiana*. See Ref. 58, p. 370

138. Tjepkema, J. D., Ormerod, W., Torrey, J. G. 1980. Vesicle formation and acetylene reduction activity in *Frankia* sp CpI1 cultured in defined nutrient media. *Nature* 287:633–35

139. Tjepkema, J. D., Ormerod, W., Torrey, J. G. 1981. Factors affecting vesicle formation and acetylene reduction (ni-

trogenase activity) in *Frankia* sp. CpI1. *Can. J. Microbiol.* 27:815–23

140. Tjepkema, J. D., Winship, L. J. 1980. Energy requirements for nitrogen fixation in actinorhizal and legume root nodules. *Science* 209:279–81

141. Torrey, J. G., Callaham, D. 1978. Determinate development of nodule roots in actinomycete-induced root nodules of *Myrica gale*. *Can. J. Bot.* 56:1357–64

142. Torrey, J. G., Callaham, D. 1982. Structural features of the vesicle of *Frankia* sp. CpI1 in culture. *Can. J. Microbiol.* 28:749–57

143. Tripp, L. N., Bezdicek, D. F., Heilman, P. E. 1979. Seasonal and diurnal patterns and rates of nitrogen fixation by young red alder. *For. Sci.* 25:371–80

144. Troelstra, S. R., Van Dijk, K., Blacquiere, T. 1985. Effects of N source on proton excretion, ionic balance and growth of *Alnus glutinosa* (L.) Gaertner: Comparison of N_2 fixation with single and mixed sources of NO_3 and NH_4. *Plant Soil* 84:361–85

145. Tumer, N. E., Robinson, S. J., Haselkorn, R. 1983. Different promoters for the *Anabaena* glutamine synthetase gene during growth using molecular or fixed nitrogen. *Nature* 306:337–42

146. Tyson, J. H., Silver, W. S. 1979. Relationship of ultrastructure to acetylene reduction (N_2 fixation) in root nodules of *Casuarina*. *Bot. Gaz.* 140 (Suppl.):S44–S48

147. VandenBosch, K. A., Torrey, J. G. 1984. Consequences of sporangial development for nodule function in root nodules of *Comptonia peregrina* and *Myrica gale*. *Plant Physiol.* 76:556–60

148. VandenBosch, K. A., Torrey, J. G. 1985. Development of endophytic *Frankia* sporangia in field- and laboratory-grown nodules of *Comptonia peregrina* and *Myrica gale*. *Am. J. Bot.* 72:99–108

149. van Dijk, C. 1978. Spore formation and endophyte diversity in root nodules of *Alnus glutinosa* (L.) Vill. *New Phytol.* 81:601–15

150. van Dijk, C. 1979. Endophyte distribution in the soil. See Ref. 38, pp. 84–94

151. van Dijk, C. 1984. *Ecological aspects of spore formation in the Frankia-Alnus symbiosis*. PhD thesis. State Univ., Leiden, The Netherlands. 154 pp.

152. van Dijk, C., Merkus, E. 1976. A microscopical study of the development of a spore-like stage in the life cycle of the root-nodule endophyte of *Alnus glutinosa* (L.) Gaertn. *New Phytol.* 77:73–91

152a. Wahab, A. M. A. 1980. Nitrogen-fixing nonlegumes in Egypt: I. Nodulation and N_2 (C_2H_2) fixation by *Casuarina equisetifolia*. *Z. Allg. Mikrobiol.* 20:3–12

153. Watts, S. H., Wheeler, C. T., Hillman, J. R., Berrie, A. M. M., Crozier, A., Math, V. B. 1983. Abscisic acid in the nodulated root system of *Alnus glutinosa*. *New Phytol.* 95:203–8

154. Deleted in proof

155. Wheeler, C. T., Crozier, A., Sandberg, F. 1984. The biosynthesis of indole-3-acetic acid by *Frankia*. *Plant Soil* 78:99–104

156. Wheeler, C. T., Gordon, J. C., Ching, T. M. 1979. Oxygen relations of the root nodules of *Alnus rubra* Bong. *New Phytol.* 82:449–57

157. Wheeler, C. T., Henson, I. E., McLaughlin, M. E. 1979. Hormones in plants bearing actinomycete nodules. *Bot. Gaz.* 140 (Suppl.):S52–S57

158. Wheeler, C. T., Hooker, J. E., Crowe, A., Berrie, A. M. M. 1986. The improvement and utilization in forestry of nitrogen fixation by actinorhizal plants with special reference to *Alnus* in Scotland. *Plant Soil*. In press

159. Wheeler, C. T., McLaughlin, M. E., Steele, P. 1981. A comparison of symbiotic nitrogen fixation in Scotland in *Alnus glutinosa* and *Alnus rubra*. *Plant Soil* 61:169–88

160. Wheeler, C. T., Watts, S. H., Hillman, J. R. 1983. Changes in carbohydrates and nitrogenous compounds in the root nodules of *Alnus glutinosa* in relation to dormancy. *New Phytol.* 95:209–18

161. Winship, L. J., Martin, K. J. 1985. Inhibition of uptake hydrogenase by acetylene in actinorhizal nodules and free-living *Frankia* in culture. See Ref. 58, p. 371

162. Winship, L. J., Tjepkema, J. D. 1985. Nitrogen fixation and respiration by attached root nodules of *Alnus rubra* Bong.: Effects of temperature and oxygen concentration. *Plant Soil* 87:91–107

163. Winship, L. J., Chaudhary, A. H. 1979. Nitrogen fixation by *Datisca glomerata*: a new addition to the list of actinorhizal diazotrophic plants. See Ref. 48, p. 485

164. Wittenberg, J. B., Wittenberg, B. A., Trinick, M. J., Gibson, O. H., Fleming, A. I., et al. 1985. Hemoglobins which supply oxygen to intracellular prokaryotic symbionts. See Ref. 58, p. 354

165. Zhang, Z., Lopez, M. F., Torrey, J. G. 1984. A comparison of cultural characteristics and infectivity of *Frankia* isolates from root nodules of *Casuarina* species. *Plant Soil* 78:79–90

Ann. Rev. Plant Physiol. 1986. 37:233–46

FRUCTOSE 2,6-BISPHOSPHATE AS A REGULATORY METABOLITE IN PLANTS[1]

Steven C. Huber

U.S. Department of Agriculture, Agricultural Research Service, North Carolina State University, Raleigh, North Carolina 27695–7631

CONTENTS

Hexose phosphates play a central role in plant carbohydrate metabolism. The degradation of hexose phosphates via glycolysis and the oxidative pentose phosphate pathway is a common characteristic of plant cells, whereas hexose phosphate production, via photosynthesis (reductive pentose phosphate pathway) or gluconeogenesis, occurs in only certain tissues. Some cells, then, have the ability to both degrade and form hexose sugars, and certain enzymes must be highly regulated to prevent futile cycling. Within the past five years, two major discoveries have been made that modify our understanding of plant carbohydrate metabolism: (*a*) the discovery of fructose 2,6-bisphosphate

[1]The US Government has the right to retain a nonexclusive royalty-free license in and to any copyright covering this paper.

(F26BP),[2] a new regulatory metabolite, and (*b*) the discovery of a pyrophosphate-dependent phosphofructokinase in plants.

The aim of this chapter is to review the role of F26BP in the regulation of metabolism in higher plants. Detailed treatment of the overall pathways of hexose metabolism is beyond the scope of this paper. Where necessary or beneficial, reference to animal systems will be made.

OCCURRENCE IN PLANT TISSUES

The occurrence of F26BP in plants was established first in mung bean seedlings by Sabularse & Anderson in 1981 (28). Since that time, F26BP has been identified in a range of tissues, including isolated guard cells (14), leaves (7), and storage tissues (44). With the development of an extremely sensitive bioassay (44, 45), the concentration of F26BP in tissues can be measured accurately, and changes in the metabolite under different conditions can be monitored. In general, the concentration of F26BP in tissues is in the range of 0.1 to 1 nmol/g fresh weight (17, 21, 37, 44). Assuming that F26BP is restricted to the cytosol (5), and that the latter is 10% of the tissue volume, the concentration of F26BP would range from 1 to 10 μM. Approximately tenfold higher concentrations of F26BP have been measured in Jerusalem artichoke tubers (44). This does not appear to be a general characteristic of storage tissues because potato tubers contain less than 0.5 nmol F26BP/g fresh wt (44). Earlier estimates of extremely high concentrations of F26BP in leaf tissue (up to 30 nmol/g fresh wt) (7) have not been confirmed. Plant tissues, in general, can

[2]Abbreviations used:

DHAP,	dihydroxyacetone phosphate
FBP,	fructose 1,6-bisphosphate
FBPase,	fructose 1,6-bisphosphatase (EC 3.1.3.11)
F26BP,	fructose 2,6-bisphosphate
F26BPase,	fructose 2,6-bisphosphatase (EC 3.1.3)
F6P,	fructose 6-phosphate
F6P,2K,	fructose 6-phosphate,2-kinase (EC 2.7.1)
G1P,	glucose 1-phosphate
G1,6P,	glucose 1,6-bisphosphate
G6P,	glucose 6-phosphate
Hexose-P,	hexose phosphates
P-esters,	phosphate esters
PFK,	phosphofructokinase
PFP,	pyrophosphate: fructose 6-phosphate phosphotransferase (EC 2.7.1.90)
PGA,	3-phosphoglycerate
Ru5P,	ribulose 5-phosphate
SPS,	sucrose phosphate synthase
Triose-P,	triose phosphates
UDPGlc,	UDP-glucose
6PGlu,	6-phosphogluconate.

rapidly adjust the concentration of F26BP severalfold, indicating the capacity for F26BP metabolism (synthesis/degradation). This aspect will be discussed in more detail later.

ENZYMES REGULATED BY F26BP

To date, seven enzymes have been identified that are regulated by F26BP; four of these are activated by F26BP and three are inhibited (Table 1). In general, enzymes that may be involved in glycolysis are activated by F26BP, whereas at least one enzyme of gluconeogenesis is inhibited. Of central importance are the opposing effects of F26BP on enzymes involved in the interconversion of F6P and FBP. Soon after the discovery of a PPi-linked PFK (termed PFP) in plant tissues (4), it was recognized that the enzyme from most tissues was activated strongly by F26BP (28, 29). Although the enzyme is freely reversible, it is the only enzyme in addition to ATP-dependent PFK that is capable of converting F6P to FBP. It is also important to note that ATP-PFK from plant tissues is unaffected by F26BP (7). Animal tissues do not contain PPi-PFK, but the ATP-dependent PFK is strongly activated by F26BP (reviewed in 15, 43). It is probable that PFP functions, at least under some conditions, in the glycolytic direction, and this will be discussed in more detail below. Formation of F6P from FBP during gluconeogenesis is catalyzed by a specific FBPase; the cytosolic, but not the plastid enzyme is strongly inhibited by F26BP (7). The differential effect of F26BP on enzymes involved in F6P/FBP interconversion constitutes the central core of a regulatory scheme whereby the flux of carbon through major pathways may be regulated.

In addition, two novel enzymes that may be involved in sucrose utilization have been reported to be activated by F26BP. In tissues that import sucrose and are active in cell wall biosynthesis and/or starch storage, sucrose synthase is thought to play a key role in sucrose breakdown: sucrose + UDP $\rightleftharpoons$ UDPGlc + fructose. The UDPGlc formed may be converted to G1P via UDPGlc pyrophosphorylase or by UDPGlc phosphorylase, the latter of which has been reported to be activated by F26BP and involves Pi rather than PPi as substrate (13). This work needs to be confirmed, as do studies conducted to determine how widely distributed UDPGlc phosphorylase activity is. A source of PPi (for either UDPGlc pyrophosphorylase or the forward direction of PFP) could be the pyrophosphorolytic cleavage of ATP, which is activated by F26BP (10). The occurrence of ATP pyrophosphorylase in tissues other than corn scutellum has not been established. Phosphoglucomutase is common to pathways of hexose breakdown and utilization, and it has been determined that F26BP can replace G1,6P as obligate cofactor (11). The existence of G1,6P in plants has not been firmly established, and thus F26BP may assume this role in plants. It is interesting that 6PGlu dehydrogenase is inhibited strongly by F26BP (27). Hence, F26BP may activate glycolysis but inhibit the oxidative pentose phos-

Table 1 Enzymes regulated by F26BP

Enzyme	Reaction catalyzed	Tissue	Effect[a]	Reference
UDPGlc phosphorylase	UDPGlc + Pi $\rightleftharpoons$ UDP + G1P	Potato tuber	A	13
ATP pyrophosphorylase	ATP $\longrightarrow$ AMP + PPi	Corn scutellum	A	10
PFP	F6P + PPi $\rightleftharpoons$ FBP + Pi	Many tissues	A	3, 4, 19a, 26, 29, 32
Phosphoglucomutase	G1P $\rightleftharpoons$ G6P	Mung bean seedlings, animal sources	A	11
Cytosolic FBPase	FBP $\longrightarrow$ F6P + Pi	Spinach leaf, castor bean	I	16, 20, 38
6PGlu DH	6PGlu + NADP $\longrightarrow$ Ru5P + NADPH + CO_2	Cytosolic and plastid isozymes from castor bean endosperm	I	27
Trehalose phosphorylase	Trehalose + Pi $\rightleftharpoons$ G1P + glucose	*Euglena gracilis*	I	27a

[a]A, activation; I, inhibition.

phate pathway. Not only the rate, but also the pathway involved in hexose degradation may be controlled by F26BP. In some algae, trehalose is a major reserve carbohydrate, and the metabolism of trehalose may be regulated by F26BP as a result of inhibition of the reversible enzyme trehalose phosphorylase (27a).

REGULATION OF F26BP CONCENTRATION

Central to an understanding of the role of F26BP in metabolism is an appreciation of the biochemical control mechanisms that regulate F26BP concentration. F26BP metabolism involves a specific kinase and phosphatase that catalyze the reactions:

$$
\begin{array}{ccc}
\text{ATP\quad ADP} & & \text{Pi} \\
& & \\
\text{F6P} \longleftrightarrow & \text{F26BP} \longrightarrow & \text{F6P} \\
& & \\
\text{F6P,2K} & & \text{F26BPase}
\end{array}
$$

In animal tissues, both activities are catalyzed by the same protein (9). Whether this is also true in plants has not been established, but the two activities do co-purify (5, 6, 35).

Metabolic Control of F6P,2K/F26BPase

The concentration of F26BP in vivo will be a function of the relative activities of F6P,2K and F26BPase, both of which are sensitive to metabolic regulation (fine control). The kinase and phosphatase have been characterized from several tissues, including spinach leaf (5, 6, 35), maize leaf (33) and root (32, 36), castor bean endosperm (21), and several storage or sink tissues (36). In general, similar properties have been identified. F6P,2K exhibits sigmoidal saturation kinetics with respect to F6P; Pi is a strong activator that increases $V_{\max}$ and induces hyperbolic saturation kinetics for F6P. Even in the presence of Pi, the K_m for F6P is sufficiently high (0.4 to 0.6 mM) so that changes in F6P concentration in situ would be expected to influence kinase activity. DHAP and PGA are inhibitors, and each interacts differently with the activator Pi. PGA inhibits F6P,2K even in the absence of Pi, and the two are antagonistic. DHAP, in contrast, has no effect on kinase activity in the absence of Pi, and Pi cannot relieve the inhibition (35). This explains why DHAP was not reported as an effector of F6P,2K in the original study (6). The partially purified F6P,2K preparations studied also catalyze the release of phosphate from the 2-position of F26BP to yield F6P. The activity is relatively specific for F26BP and does not require a divalent cation. The K_m for F26BP is about 1 μM (38), and the activity is inhibited by Pi and F6P (6, 21, 38). There are no known activators of the phosphatase. Sensitivity to effectors varied greatly when enzymes from different sink tissues were compared (36), suggesting that differences in the regulation of metabolism in various tissues may need to be investigated.

Differences also appear to exist in regulatory properties of the enzyme from maize (a C_4 plant) and spinach (a C_3 plant). With both enzymes, Pi is a strong activator of the kinase and an inhibitor of the phosphatase. However, with the maize enzyme, DHAP and PGA inhibit F26BPase as well as F6P,2K (33). Furthermore, phosphoenolpyruvate also strongly inhibits both activities, and several metabolites that have no effect on the spinach enzyme selectively inhibit either F6P,2K or F26BPase: oxaloacetate (F6P,2K), pyruvate, and UDPGlc (F26BPase) (33). The metabolic and physiological significance of these differences in properties remains to be established.

Possible Coarse Control

As described above, metabolic or "fine" control of F6P,2K/F26BPase has been firmly established. The enzyme from animal tissues is also highly regulated by metabolic intermediates and is also subject to covalent modification, which constitutes an overriding control mechanism. In liver, a cAMP-dependent protein kinase phosphorylates the protein, which results in inhibition of the kinase and activation of the phosphatase activity (9, 15). The specific protein kinase is subject to hormonal control, and thus F26BP metabolism in animal tissues is regulated at several levels. Preliminary attempts to phosphorylate the partially purified spinach leaf enzyme with cAMP-dependent protein kinase have been unsuccessful (5). However, the possibility that a specific protein kinase is involved, or that some other covalent modification mechanism exists, has not been investigated. It seems likely that a coarse level of control is involved, and this will be an important area for future study.

ROLE IN GLYCOLYSIS

The capacity for hexose degradation via glycolysis and the oxidative pentose phosphate pathway is regarded as a universal characteristic of plant cells (1), including mature photosynthetic tissues (34). One of the central questions regarding the role of F26BP in glycolysis concerns whether PFK and/or PFP function in the formation of FBP from F6P. Because PFK is irreversible, it can only function in glycolysis, whereas PFP, being freely reversible, could function in either the glycolytic or gluconeogenic direction. If PFP is to function in the glycolytic direction, an adequate pool of PPi must be present. Recent studies have verified that a substantial PPi pool (5 to 39 nmol/g fresh wt) exists in various tissues of pea and corn seedlings (8, 30). In addition, Smyth et al (31) have recently demonstrated a PPi-dependent glycolytic pathway (F6P to pyruvate) in extracts from pea seeds that was responsive to added F26BP. Experiments such as these establish the potential for PPi-linked glycolysis, but the actual flux of carbon through either pathway remains to be established. Two fundamentally different mechanisms have been proposed to explain how the freely reversible PPi-PFK may actually function in glycolysis; these are now considered separately.

Metabolite-Mediated Interconversion of Enzymes

Recently, Buchanan and coworkers (2, 46, 47) postulated that PFK and PFP can be reversibly interconverted in vitro by incubation with certain metabolites. Specifically, F26BP catalyzed the conversion of PFP to PFK, which being irreversible can only function in the glycolytic direction. The principal evidence for the F26BP-mediated conversion of PFP to PFK is measurement of an initial rapid reaction velocity (hyperactive phase) that is not sustained beyond a few minutes. The initial burst of ATP-dependent activity in the presence of F26BP is not catalyzed by PFK, but rather by metabolism of PPi (present as a contaminant in ATP) by PFP (a contaminant of the PFK preparation). Consequently, PFK from castor bean endosperm that is prepared absolutely free of contaminating PFP activity does not yield hyperactive kinetics with F26BP (24). In addition, apparent metabolite conversion of PFK to PFP does not result in a decrease in actual PFK activity (12), as might be expected if one form was converted to the other.

Similarly, the apparent conversion of PFK to PFP in the presence of UDPGlc plus PPi can be explained by utilization of UTP by PFK (22). The UTP is formed from the UDPGlc plus PPi by action of UDPGlc pyrophosphorylase, a contaminant in the auxiliary enzymes used in the PFK assay. In the absence of auxiliary enzymes, PFK from rabbit muscle could not use PPi as substrate (22). To date, there is no convincing evidence to support the notion of "metabolite-mediated catalyst conversion."

These results also emphasize the need for extreme caution about enzyme and reagent purity when studying F26BP effects. Because F26BP is an effective regulator of enzyme activities at low concentrations (μM), but enzyme assays generally involve substrate amounts (mM), trace contaminants can give erroneous results. New workers who come into this area need to be acutely aware of the need for caution.

Molecular Forms of PFP

A working model to explain how F26BP and PFP may be involved in glycolysis has been forwarded by Black and coworkers (2a). It was originally observed that the activity of PFP and its sensitivity to F26BP changed markedly during germination and development of pea seedlings (48). Gel filtration experiments resolved two peaks of PFP activity in pea (48) and wheat (50) seedlings: a large form that was relatively insensitive to F26BP and a small form that was activated to a much greater extent. Subsequent work with pea seedlings demonstrated the existence of two interconvertible forms of PFP, with sedimentation coefficients of 6.3 and 12.7S in sucrose density gradient ultracentrifugation (49). The large form has a high ratio of glycolytic to gluconeogenic activity, whereas the reverse is true for the small form. Importantly, F26BP promotes the association of the small form to yield the large form. Thus, the presence of

F26BP results in a molecular form of PPi-PFK that favors glycolysis (2a). The wheat enzyme contains two subunits, designated α and β; the large form is $\alpha_2\beta_2$ and the small form is probably β_2 (50).

PFP has been partially purified and kinetically characterized from castor bean endosperm (18, 19, 23, 26), mung bean hypocotyl (28, 29), spinach leaf (7), potato tuber (45), and pineapple leaves (3). With the exception of the enzyme from pineapple leaves (3), enzyme activity is virtually dependent upon F26BP. Differences in relative F26BP activation may reflect variable F26BP contamination of some commercial sources of F6P (25); however, species differences noted within the same study cannot be attributed to this (4). Contamination of F6P with F26BP may also explain sigmoidal kinetics; when this problem is eliminated, hyperbolic kinetics for F6P are observed with the enzyme from castor bean endosperm (25), which confirms the earlier results obtained with the mung bean enzyme (28).

The enzyme is confined to the cytosol and has similar activities and responses to effectors in both the forward and reverse directions; F26BP activates and Pi inhibits the reactions. However, stimulation of the reverse reaction (F6P production) by F26BP is substantially less than in the forward reaction (FBP production). Activation of PFP by F26BP can be modulated by the concentrations of F6P and Pi; F6P decreases the K_a for F26BP, whereas Pi increases the K_a and reduces V_{max} (26). Thus, an increase in the F6P/Pi ratio would increase the activation of PFP at a fixed, limiting F26BP concentration. As described above, an increase in the F6P/Pi ratio might also result in increased F26BP concentration by activation of F6P,2K. Thus, activity of PFP in vivo may be a function of the concentrations of F6P, Pi, and F26BP.

ROLE IN GLUCONEOGENESIS

It is generally accepted that F26BP plays a role in the regulation of gluconeogenesis. The basis for this involvement is inhibition of cytoplasmic FBPase, which catalyzes the first irreversible step in the conversion of triose-P to sucrose: Fru 1,6-P$_2$ → F6P + Pi.

Effects on FBPase

Cytosolic FBPase has been purified from spinach leaves (16, 38, 41) and castor bean endosperm (20) and characterized with respect to F26BP regulation. In general, the properties of the enzyme from these two tissues are very similar. In the absence of effectors, the K_m for FBP is quite low (about 2 μM). Micromolar concentrations of F26BP cause marked inhibition of enzyme activity and convert hyperbolic saturation kinetics to sigmoidal (20, 38). Thus, it appears that F26BP is not a simple competitive inhibitor, but it may be an allosteric competitive inhibitor that binds to a site other than the catalytic site. Plant FBPase is also inhibited by AMP and Pi, but the inhibition is weak and positive cooperativity is not induced. However, a striking synergism between AMP and F26BP has been observed with both spinach leaf and castor bean endosperm FBPase (20, 38, 41). Furthermore, the combined effect of AMP and F26BP is further amplified by Pi. The complex regulation of cytosolic FBPase by F26BP and several other effectors appears to explain to a large extent how sucrose formation may be regulated in leaves under different conditions.

Photosynthetic Tissues

In the light, chloroplasts export triose-P to the cytosol where they are converted to sucrose. The rate of triose-P conversion to sucrose must be regulated with respect to the rate of CO_2 assimilation (which determines triose-P availability); without such regulation, stromal metabolites could be depleted rapidly and photosynthesis strongly inhibited under suboptimal conditions. The fact that this does not occur is taken to indicate that a feedforward (39) control of photosynthetic sucrose formation exists. When photosynthetic rates are high, and the rate of sucrose production exceeds the rate of sucrose export, an increased proportion of photosynthate is diverted from sucrose toward starch. This constitutes a feedback (40) type of control that affects partitioning of fixed carbon between starch and sucrose. It is clear that F26BP plays an important role in both feedforward and feedback control mechanisms.

FEEDFORWARD CONTROL   The pioneering experiments of Stitt et al (39) established that an inverse relationship exists between photosynthetic rate and leaf F26BP concentration. The increases in F26BP concentration would inhibit cytosolic FBPase activity in situ, thereby slowing the rate of sucrose formation under limiting conditions. In this situation, the rise in F26BP concentration can be attributed to lowered concentrations of DHAP, which would increase F6P,2K activity. The concentration of DHAP changes with the photosynthetic rate to a much greater extent than other metabolites; consequently, Stitt et al (38) have assessed the physiological status of leaf tissue under different conditions on the basis of DHAP content of the leaf material. Also, it is likely that as the concentration of P-esters declines as photosynthesis is reduced, the concentration of Pi will be increased. Thus, Pi and DHAP will fluctuate inversely, and for a small change in their absolute concentrations, the Pi/DHAP ratio can change to a greater extent. Under limiting conditions for photosynthesis, the large increase in the Pi/DHAP concentration ratio would activate F6P,2K activity, and the increase in Pi would inhibit F26BPase activity. Another factor contributing to the reduction in FBPase activity would be lower concentrations of the substrate FBP.

FEEDBACK CONTROL   When sucrose production exceeds the rate of assimilate export, sucrose will accumulate to a higher level than normal and photosynthate will then be diverted to starch. This shift in partitioning between starch and sucrose is associated with increased concentrations of F26BP (17, 40), which would inhibit cytosolic FBPase and thereby slow carbon flux into sucrose. This would then restrict export of triose-P out of the chloroplast and thereby stimulate starch synthesis. It is clear that F26BP plays a role in this feedback control mechanism, but all of the details are not understood at present.

Under conditions of excess sucrose accumulation, there is an accumulation of DHAP and hexose-P (17, 40), and this is presumably associated with a reciprocal drop in Pi concentration. The observed increase in F6P concentration is greater than that of DHAP. Recall that these P-esters affect F6P,2K in opposite directions; F6P can be considered an activator (as substrate in a reaction with a relatively high K_m), whereas DHAP is an inhibitor of the kinase. In addition, F6P directly inhibits F26BPase activity. Collectively, the net result is an increase in F26BP concentration that along with other shifts in metabolites would substantially inhibit FBPase activity in situ (38).

Even though F26BP plays a key role in the feedback mechanism, it may not be the primary or initial factor. As sucrose accumulates and the rate of sucrose formation declines, there is an increase in the concentrations of UDPGlc and F6P, the substrates of SPS (40). This indicates that SPS activity is regulated and has been inhibited in situ by some mechanism. Thus, in feedback control, F26BP concentration may be adjusted in response to the regulation of SPS activity (17a). This is an important area for future study.

Compartmentation in C_4 Plants

A new concept in the area of C_4 photosynthesis is that sucrose formation may be strictly compartmented within the mesophyll cells of maize. Usuda & Edwards (42) initially concluded that at least some of the enzymes involved in sucrose biosynthesis are enriched in mesophyll cells, and recently Soll et al (33) established that F6P,2K and F26BPase were primarily, if not exclusively, localized in the mesophyll tissue. Hence, it appears that the enzymatic capacity and associated regulatory systems of sucrose and starch metabolism are strictly compartmented between the mesophyll and bundle sheath cells. Future studies will undoubtedly examine the intercellular distribution of other key enzymes, namely, PFP and PFK. It should be stressed that the localization of sucrose synthesis in the mesophyll cell has only been suggested for maize; earlier work with other C_4 species conclusively established that SPS activity is distributed between the two cell types (for review see 7a). More extensive studies with a range of species from the three C_4 subgroups will be required to resolve this issue.

Nonphotosynthetic Tissues

Leaves are not the only tissues active in gluconeogenesis. During germination and early growth of certain plants such as castor bean, there is a high rate of conversion of fat to sucrose in the endosperm. A cytoplasmic FBPase that is displaced from equilibrium in vivo (1) has been long recognized as an important regulated step in the gluconeogenic pathway. Castor bean endosperm contains F26BP, the enzymes involved in its metabolism (21), and an F26BP-inhibited cytosolic FBPase (20). Various treatments that are known to inhibit gluconeogenesis result in a rapid increase in F26BP concentration in this tissue (21). Hence, F26BP appears to play an important role in the regulation of gluconeogenesis in nonphotosynthetic as well as photosynthetic tissues.

SUMMARY

F26BP plays a central role in the regulation of carbon metabolism in plants. It appears to function as an activator of glycolysis and inhibitor of gluconeogenesis. F26BP and the enzymes involved in its synthesis and degradation have been identified in a range of tissues, and it is likely the occurrence of this regulatory system will be universal in plant tissues. Metabolic fine control of the activity of F6P,2K/F26BPase is well established, but the significance of differences in properties of the enzyme from various sources needs to be established. Another question concerns whether some coarse control mechanism(s) exists along with the metabolic control of F26BP metabolism.

A number of enzymes have been identified that are regulated by F26BP. The cytosolic FBPase and PFP occur widely in plants, but the distribution of some of the other regulated enzymes has not been explored. Even though PFP occurs widely, can it operate in both directions in situ? Many specific questions remain to be answered, but a general scheme is now available that can be used as a working model to study the regulation of carbon metabolism. The next few years should reveal new insights in this area.

ACKNOWLEDGMENTS

I am grateful to the numerous colleagues who shared their preprints and unpublished data with me. I also thank Drs. C. C. Black and P. S. Kerr, and Ms. W. Kalt-Torres for reading the manuscript and making helpful suggestions, and Ms. Adina Horner for her expert assistance in preparing the manuscript. This has been part of the cooperative investigations of the USDA-ARS and NCARS. This is paper No. *10136* of the Journal Series of the NCARS, Raleigh, North Carolina.

244 HUBER

Literature Cited

1. Ap Rees, T. 1980. Integration of pathways of synthesis and degradation of hexose phosphates. In *The Biochemistry of Plants: A Comprehensive Treatise. Carbohydrates: Structure and Function*, ed. J. Preiss, 3:1–41. New York: Academic

2. Balogh, A., Wong, J. H., Wotzel, C., Soll, J., Cséke, C., Buchanan, B. B. 1984. Metabolite-mediated catalyst conversion of PFK and PFP: A mechanism of enzyme regulation in green plants. *FEBS Lett.* 169:287–92

2a. Black, C. C., Smyth, D. A., Wu, M-X. 1985. Pyrophosphate-dependent glycolysis and regulation by fructose 2,6-bisphosphate in plants. In *Nitrogen Fixation and CO_2 Metabolism*, ed. P. W. Ludden, J. E. Burris, pp. 361–70. New York: Elsevier. 445 pp.

3. Carnal, N. W., Black, C. C. Jr. 1979. Pyrophosphate-dependent 6-phosphofructokinase, a new glycolytic enzyme in pineapple leaves. *Biochem. Biophys. Res. Commun.* 86:20–26

4. Carnal, N. W., Black, C. C. Jr. 1983. Phosphofructokinase activities in photosynthetic organisms. The occurrence of pyrophosphate-dependent 6-phosphofructokinase in plants. *Plant Physiol.* 71:150–55

5. Cséke, C., Buchanan, B. B. 1983. An enzyme synthesizing fructose 2,6-bisphosphate occurs in leaves and is regulated by metabolite effectors. *FEBS Lett.* 155:139–42

6. Cséke, C., Stitt, M., Balogh, A., Buchanan, B. B. 1983. A product-regulated fructose 2,6-bisphosphatase occurs in green leaves. *FEBS Lett.* 162:103–6

7. Cséke, C., Weeden, N. F., Buchanan, B. B., Uyeda, K. 1982. A special fructose bisphosphate functions as a cytoplasmic regulatory metabolite in green leaves. *Proc. Natl. Acad. Sci. USA* 79:4322–26

7a. Edwards, G. E., Huber, S. C. 1981. The C_4 pathway. In *The Biochemistry of Plants: Photosynthesis*, ed. M. D. Hatch, N. K. Boardman, 8:238–81. New York: Academic. 521 pp.

8. Edwards, J., Ap Rees, T., Wilson, P. M., Morrell, S. 1984. Measurement of the inorganic pyrophosphate in tissues of *Pisum sativum* L. *Planta* 162:188–91

9. El-Maghrabi, M. R., Pilkis, S. J. 1984. Rat liver 6-phosphofructo 2-kinase/fructose 2,6-bisphosphatase: A review of relationships between the two activities of the enzyme. *J. Cell. Biochem.* 26:1–17

10. Eschevarria, E., Humphreys, T. 1984. Involvement of sucrose synthase in sucrose catabolism. *Phytochemistry* 23: 2173–78

11. Galloway, C. M., Dugger, W. M., Black, C. C. 1985. The activation of phosphoglucomutase by fructose 2,6-bisphosphate. *Plant Physiol.* 77S:110

12. Gancedo, J. M. 1984. Metabolite-mediated catalyst conversion of PFK and PFP: Can PFK really be converted to PFP? *FEBS Lett.* 175:369–70

13. Gibson, D. M., Shine, W. E. 1983. Uridine diphosphate glucose breakdown is mediated by a unique enzyme activated by fructose 2,6-bisphosphate in *Solanum tuberosum*. *Proc. Natl. Acad. Sci. USA* 80:2491–94

14. Hedrich, R., Raschke, K., Stitt, M. 1985. A role for fructose 2,6-bisphosphate in regulating carbohydrate metabolism in guard cells. *Plant Physiol.* In press

15. Hers, H. G., Van Schaftingen, E. 1982. Fructose 2,6-bisphosphate 2 years after its discovery. *Biochem. J.* 206:1–12

16. Herzog, B., Stitt, M., Heldt, H. W. 1984. Control of photosynthetic sucrose synthesis by fructose 2,6-bisphosphate. III. Properties of the cytosolic fructose 1,6-bisphosphatase. *Plant Physiol.* 75: 561–65

17. Huber, S. C., Bickett, D. M. 1984. Evidence for control of carbon partitioning by fructose 2,6-bisphosphate in spinach leaves. *Plant Physiol.* 74:445–47

17a. Huber, S. C., Kerr, P. S., Kalt-Torres, W. 1985. Regulation of sucrose formation and movement. In *Regulation of Carbohydrate Partitioning in Photosynthetic Tissues*, ed. R. C. Heath, J. Preiss, pp. 199–215. Baltimore: Williams & Wilkins

18. Kombrink, E., Kruger, N. J. 1984. Inhibition by metabolic intermediates of pyrophosphate: fructose 6-phosphate phosphotransferase from germinating castor bean endosperm. *Z. Pflanzenphysiol.* 114:443–53

19. Kombrink, E., Kruger, N. J., Beevers, H. 1984. Kinetic properties of pyrophosphate: Fructose 6-phosphate phosphotransferase from germinating castor bean endosperm. *Plant Physiol.* 74:395–401

19a. Kowalczyk, S., Januszewska, B., Cymerska, E., Maslowski, P. 1984. The occurrence of inorganic pyrophosphate: D-fructose-6-phosphate 1-phosphotransferase in higher plants. I. Initial characterization of partially purified enzyme from *Sanseviera trifasciata* leaves. *Physiol. Plant.* 60:31–37

20. Kruger, N. J., Beevers, H. 1984. Effect of fructose 2,6-bisphosphate on the kinetic properties of cytoplasmic fructose 1,6-bisphosphatase from germinating castor bean endosperm. *Plant Physiol.* 76:49–54

21. Kruger, N. J., Beevers, H. 1985. Synthesis and degradation of fructose 2,6-bisphosphate in endosperm of castor bean seedlings. *Plant Physiol.* 77:358–64

22. Kruger, N. J., Dennis, D. T. 1985. A source of apparent pyrophosphate: Fructose 6-phosphate phosphotransferase activity in rabbit muscle phosphofructokinase. *Biochem. Biophys. Res. Commun.* 126:320–26

23. Kruger, N. J., Dennis, D. T. 1985. Molecular properties of pyrophosphate: Fructose 6-phosphate phosphotransferase from potato tubers. *Plant Physiol.* 77S:114

24. Kruger, N. J., Dennis, D. T. 1985. Reassessment of an apparent hyperactive form of phosphofructokinase from plants. *Plant Physiol.* 78:645–48

25. Kruger, N. J., Kombrink, E., Beevers, H. 1983. Fructose 2,6-bisphosphate as a contaminant of commercially obtained fructose 6-phosphate: Effect on PPi: fructose 6-phosphate phosphotransferase. *Biochem. Biophys. Res. Commun.* 117: 37–42

26. Kruger, N. J., Kombrink, E., Beevers, H. 1983. Pyrophosphate: fructose 6-phosphate phosphotransferase in germinating castor bean seedlings. *FEBS Lett.* 153:409–12

27. Miernyk, J. A., MacDougall, P. S., Dennis, D. T. 1984. In vitro inhibition of the plastid and cytosolic isozymes of 6-phosphogluconate dehydrogenase from developing endosperm of *Ricinus communis* by fructose 2,6-bisphosphate. *Plant Physiol.* 76:1093–94

27a. Miyatake, K., Kuramoto, Y., Kitaoka, S. 1984. Fructose 2,6-bisphosphate, a potent regulator of carbohydrate metabolism, inhibits trehalose phosphorylase from protist *Euglena gracilis*. *Biochem. Biophys. Res. Commun.* 122:906–11

28. Sabularse, D. C., Anderson, R. L. 1981. D-Fructose 2,6-bisphosphate: A naturally occurring activator for inorganic pyrophosphate: D-fructose 6-phosphate 1-phosphotransferase in plants. *Biochem. Biophys. Res. Commun.* 103: 848–55

29. Sabularse, D. C., Anderson, R. L. 1981. Inorganic pyrophosphate: D-fructose 6-phosphate 1-phosphotransferase in mung

beans and its activation by D-fructose 1,6-bisphosphate and D-glucose 1,6-bisphosphate. *Biochem. Biophys. Res. Commun.* 100:1423–29

30. Smyth, D. A., Black, C. C. Jr. 1984. Measurement of the pyrophosphate content of plant tissues. *Plant Physiol.* 75:862–64

31. Smyth, D. A., Wu, M.-X. 1984. Pyrophosphate and fructose 2,6-bisphosphate effects on glycolysis in pea seed extracts. *Plant Physiol.* 76:316–20

32. Smyth, D. A., Wu, M.-X., Black, C. C. Jr. 1984. Phosphofructokinase and fructose 2,6-bisphosphatase activities in developing corn seedlings (*Zea mays* L.). *Plant Sci. Lett.* 33:61–70

33. Soll, J., Wotzel, C., Buchanan, B. B. 1985. Enzyme regulation in C_4 photosynthesis. Identification and localization of activities catalyzing the synthesis and hydrolysis of fructose 2,6-bisphosphate in corn leaves. *Plant Physiol.* 77:999–1003

34. Stitt, M., Ap Rees, T. 1978. Pathways of carbohydrate oxidation in leaves of *Pisum sativum* and *Triticum aestivum*. *Phytochemistry* 17:1251–56

35. Stitt, M., Cséke, C., Buchanan, B. B. 1984. Regulation of fructose 2,6-bisphosphate concentration in spinach leaves. *Eur. J. Biochem.* 143:89–93

36. Stitt, M., Cséke, C., Buchanan, B. B. 1986. Occurrence of a metabolite-regulated enzyme synthesizing fructose 2,6-bisphosphate in plant sink tissues. *Physiol. Veg.* In press

37. Stitt, M., Gerhardt, R., Kürzel, B., Heldt, H. W. 1983. A role for fructose 2,6-bisphosphate in the regulation of sucrose synthesis in spinach leaves. *Plant Physiol.* 72:1139–41

38. Stitt, M., Heldt, H. W. 1985. Control of photosynthetic sucrose synthesis by fructose 2,6-bisphosphate. VI. Regulation of the cytosolic fructose 1,6-bisphosphatase in spinach leaves in an interaction between metabolic intermediates and fructose 2,6-bisphosphate. *Plant Physiol.* 79:599–608

39. Stitt, M., Herzog, B., Heldt, H. W. 1984. Control of photosynthetic sucrose synthesis by fructose 2,6-bisphosphate. I. Coordination of CO_2 fixation and sucrose synthesis. *Plant Physiol.* 75:548–53

40. Stitt, M., Kürzel, B., Heldt, H. W. 1984. Control of photosynthetic sucrose synthesis by fructose 2,6-bisphosphate. II. Partitioning between sucrose and starch. *Plant Physiol.* 75:554–60

41. Stitt, M., Mieskes, G., Soling, H. D., Heldt, H. W. 1982. On a possible role of

fructose 2,6-bisphosphate in regulating photosynthetic metabolism in leaves. *FEBS Lett.* 145:217–22

42. Usuda, H., Edwards, G. E. 1980. Localization of glycerate kinase and some enzymes for sucrose synthesis in C_3 and C_4 plants. *Plant Physiol.* 65:1017–22

43. Uyeda, K., Furuya, E., Richards, C. S., Yokoyama, M. 1982. Fructose-2,6-P_2, chemistry and biological function. *Mol. Cell. Biochem.* 48:97–120

44. Van Schaftingen, E., Hers, H. G. 1983. Fructose 2,6-bisphosphate in relation with the resumption of metabolic activity in slices of Jerusalem artichoke tubers. *FEBS Lett.* 164:195–200

45. Van Schaftingen, E., Lederer, B., Bartrons, R., Hers, H. G. 1982. A kinetic study of pyrophosphate: fructose 6-phosphate phosphotransferase from potato tubers. Application to a microassay of fructose 2,6-bisphosphate. *Eur. J. Biochem.* 129:191–95

46. Wong, J. H., Balogh, A., Buchanan, B. B. 1984. Pyrophosphate functions as phosphoryl donor with UDP-glucose-treated mammalian phosphofructokinase. *Biochem. Biophys. Res. Commun.* 121:842–47

47. Wong, J. H., Balogh, A., Wotzel, C., Soll, J., Buchanan, B. B. 1984. Metabolite-mediated catalyst conversion in C_3 and C_4 plants. *Plant Physiol.* 75S:53

48. Wu, M.-X., Smyth, D. A., Black, C. C. Jr. 1983. Fructose 2,6-bisphosphate and the regulation of pyrophosphate-dependent phosphofructokinase activity in germinating pea seeds. *Plant Physiol.* 73: 188–91

49. Wu, M.-X., Smyth, D. A., Black, C. C. Jr. 1984. Regulation of pea seed pyrophosphate-dependent phosphofructokinase: Evidence for interconversion of two molecular forms as a glycolytic regulatory mechanism. *Proc. Natl. Acad. Sci. USA* 81:5051–55

50. Yan, T. F. J., Tao, M. 1984. Multiple forms of pyrophosphate: D-fructose-6-phosphate 1-phosphotransferase from wheat seedlings. Regulation by fructose 2,6-bisphosphate. *J. Biol. Chem.* 259: 5087–92

Ann. Rev. Plant Physiol. 1986. 37:247–74
Copyright © 1986 by Annual Reviews Inc. All rights reserved

CARBON DIOXIDE AND WATER VAPOR EXCHANGE IN RESPONSE TO DROUGHT IN THE ATMOSPHERE AND IN THE SOIL[1]

E.-D. Schulze

Lehrstuhl Pflanzenökologie, Universität Bayreuth, D-858 Bayreuth, Federal Republic of Germany

CONTENTS

INTRODUCTION

The exchange of carbon dioxide (CO_2) and water vapor between plants and the atmosphere is regulated in the long term (days to weeks) by changes in leaf area

---

[1] Abbreviations used: A, CO_2-assimilation; c_i, intercellular CO_2 concentration; E, transpiration; g, leaf conductance; Δw, difference in mole fraction of vapor between leaf and air; Δ, change of a parameter; ψ, water potential. Subscripts m, r, s, x refer to mesophyll, root, soil, and xylem respectively.

0066-4294/86/0601-0247$02.00

and by the development of a photosynthetic apparatus in the leaf mesophyll, and in the short term (hours to days) by adjustment of photosynthetic capacity and changes in stomatal aperture. Regulation of gas exchange is important for plant performance, to maintain growth without dessication in an atmosphere in which the gradient for CO_2 uptake between the ambient air and the intercellular air spaces is about 0.1 $mPaPa^{-1}$, whereas the gradient in water vapor concentration between the hydrated living plant tissue and the atmosphere is for most situations greater than 10 $mPaPa^{-1}$. Since nature has not evolved a membrane that is permeable to CO_2 but nonpermeable to water vapor (23, 104), CO_2 uptake for photosynthesis and water vapor loss of transpiration take place through the same pores in the epidermis, the stomata, which inevitably will result in a proportionally greater water loss than CO_2 uptake. In general, the flux of water vapor from the leaf to the atmosphere is often about 2 orders of magnitude greater than the opposite flux of CO_2 (36), and in some crop species the amount of water transported per day through a leaf is 10 times the leaf weight (12). Therefore, water is a predominant factor in determining the geographic distribution of vegetation on a global scale; also, in agriculture the occurrence of drought appears to be the most important yield factor (11).

Stomatal and photosynthetic responses to stimuli such as light, CO_2, and temperature are understood well enough (94, 141) to allow useful predictions of rates of photosynthesis (32). In contrast, our understanding of stomatal and photosynthetic responses to water stress and models of their function under drought are still nonmechanistic (33, 61).

Two major control loops have been proposed for stomatal responses to plant water status (94): (a) a feedback response via leaf water status which leads to stomatal closure when zero turgor in the bulk leaf tissue is reached (129), and (b) a direct (feedforward) response to the difference in mole fraction of water vapor between leaf and air (63, 77, 92). These concepts lead to conflicting interpretations if plant responses are observed under field conditions and if both transpiration and photosynthesis are considered simultaneously. During the course of a day stomata may close before zero turgor is reached (107) or—if air humidity is high—may remain open at zero turgor (109, 130, 134). Air humidity may also affect CO_2 assimilation directly, which would in turn influence leaf conductance (3, 37, 98, 138). Additionally, independent effects of soil water status on conductance have been reported recently (9, 39, 40).

This review describes the current hypotheses of how humidity and plant and soil water status may interact and regulate stomatal conductance and photosynthesis. As a number of reviews have recently covered related fields, this chapter will not discuss in detail stomatal metabolic physiology (94, 141) or ecophysiology (107), nor will it discuss stomatal responses to light, CO_2, and temperature (94, 141). For general plant responses to water stress, the reader is referred to the reviews by Hsiao (46), Begg & Turner (5), and Bradford & Hsiao

(14). For water flow in plants the reader is referred to the detailed review by Boyer (12). Recent evaluations of the equations of water and CO_2 diffusion through stomata are found in von Caemmerer & Farquhar (19) and Leuning (67).

This review will focus on the effects of 1. humidity, 2. leaf water potential and leaf turgor, and of 3. soil water status on leaf conductance, transpiration, and CO_2 assimilation.

POSSIBLE INTERRELATIONSHIPS BETWEEN LEAF CONDUCTANCE, AIR HUMIDITY, AND SOIL WATER STATUS

Stomata respond to several stimuli which are coupled directly or indirectly to evaporative conditions in the atmosphere or to soil water status. It is quite clear that "flow charts" of possible control processes are often oversimplifications, but they serve the important purpose of emphasizing that numerous interactions of control processes exist, even if only a single evironmental parameter, say humidity, is varied. A flow chart is used here (Figure 1) to identify various control steps and to show how water vapor and soil water status may interact in regulating stomatal conductance, neglecting all the other environmental parameters.

A change in the difference of water vapor concentration (mole fraction) between a leaf and its surrounding air, $\Delta(\Delta w)$[1], might affect the leaf simultaneously in three different ways: 1. via a change in stomatal and cuticular transpiration $[\Delta E/\Delta(\Delta w)]$; 2. via a direct effect on leaf conductance $[\Delta g/\Delta(\Delta w)]$; and, at least in some cases, 3. via effects of humidity on the photosynthetic apparatus, as expressed by changes in the A/c_i curve $[\Delta A/\Delta(\Delta w)]$. Each of these effects will cause a number of important secondary perturbations in the leaf. The change in transpiration will affect the water potential in mesophyll cells where the water evaporates $(\Delta\psi_m/\Delta E)$. This in turn will have three different consequences: (a) it may change leaf conductance $(\Delta g/\Delta\psi_m)$, either as a hydraulic response through turgor changes in the epidermis or as a metabolic response; (b) it could affect the photosynthetic apparatus $(\Delta A/\Delta\psi_m$ as expressed by a change in the A/c_i curve) in response to local water stress or other yet unknown processes; or (c) it will influence the water potential of the xylem $[(\Delta\psi_x/\Delta\psi_m)$, depending on the hydraulic conductance within the leaf and on the ratio of the area of a single leaf to the total leaf area]. A change in leaf diffusive conductance will close the "feedback loop" through its effect on transpiration $(\Delta E/\Delta g)$. At the same time, Δg will change the intercellular CO_2 concentration $(\Delta c_i/\Delta g)$, which may alter CO_2 assimilation $(\Delta A/\Delta c_i)$, depending on the CO_2 assimilation response to CO_2.

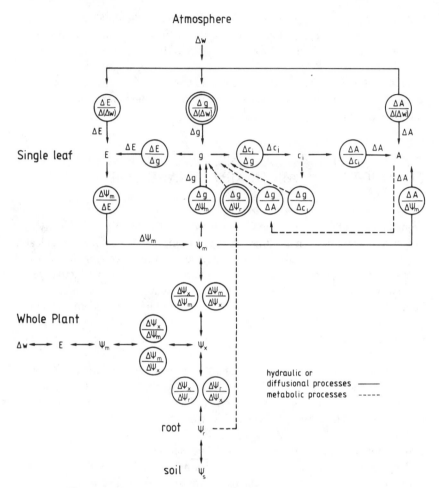

Figure 1 Schematic presentation of possible effects of a change in the leaf/air vapor concentration difference Δw, above a single leaf or a whole plant and of the soil water potential ψ_s, on transpiration E, leaf conductance g, intercellular CO_2 concentration c_i, CO_2 assimilation A, and plant water potential ψ. The symbol Δ represents change of a parameter, the subscripts m, x, r, and s refer to mesophyll, xylem, root, and soil respectively.

Photosynthesis as well as c_i may also modulate leaf conductance through metabolic signals ($\Delta g/\Delta c_i$; $\Delta g/\Delta A$).

Soil water status (ψ_s) will also affect leaf water status. Depending on the hydraulic conductance in the flow path through the plant, there will be a change in the root and in the leaf xylem water potential ($\Delta\psi_x/\Delta\psi_r$) with progressive soil drought. However, ψ_x of a single leaf can be affected, independent of ψ_s, by changing transpiration and hence water potential of the whole plant [$\Delta\psi_x$

$_{leaf}/\Delta\psi_{x\ whole\ plant}$ (109)]. A change in xylem water status will affect the leaf mesophyll and will act through the same feedback loops that were affected by humidity. In addition, as the soil water potential changes, there may be a direct metabolic signal from the roots to the shoot which affects stomata ($\Delta g/\Delta\psi_s$), independent of the hydraulic changes operating via the water potential.

The model in Figure 1 is tentative and future research may add or delete different components. It is the aim of this review to discuss how effective some of these control loops are, since the interrelationships of atmospheric and soil water status are quite complicated, consisting not only of "fluxes" of water or CO_2, but also of metabolic interactions. It should become quite clear that single-parameter response curves are inadequate to describe stomatal responses unless all other parameters are either held constant or the concurrent changes in all other factors are measured.

GAS EXCHANGE RESPONSES TO CHANGES IN AIR HUMIDITY

Responses of Stomata

If leaf conductance (including stomata, cuticle, and boundary layer) was always constant, transpiration would rise linearly with increasing Δw. However, the latter is generally not the case. In an earlier study (110) on *Prunus armeniaca*, transpiration did not increase linearly with rising Δw, and even declined once Δw exceeded 15 mPaPa^{-1}. Since then, this phenomenon has been observed in many species (107), not only under natural field conditions, but also in controlled environment chambers (65, 125) and even in excised leaves supplied with water through the petiole (O. L. Lange, personal communication). Also, the phenomenon can be simulated for natural conditions (61, 111) and has a large effect on plant water use efficiency (113).

By using the isolated epidermis of the fern *Polypodium vulgare* and of lamb's lettuce, *Valerianella locusta,* it has been demonstrated clearly that stomata respond directly to changes in air humidity (63). In both species the epidermis is naturally attached only in small areas to the mesophyll and can easily be stripped off without any injury to the epidermal cells and without any adhering mesophyll cell fragments. By changing air humidity, Lösch (68) was able to induce more than 100 opening and closing reactions in the same pair of guard cells in an isolated epidermis over a period of 4 days, without any sign of a change in response. Thus, the humidity response is a property of the epidermis independent of the mesophyll in these species. It is also independent of any signals linked to photosynthesis in epidermal cells surrounding the guard cells, since *Valerianella*, in contrast to *Polypodium*, has no epidermal chloroplasts.

Possible mechanisms involved in the stomatal response to changes in air humidity have been reviewed by several authors (71, 77, 120), but as Boyer

(12) pointed out, until we know more about the pathways of water within the leaf, the exact mechanism remains unclear. It is difficult to establish un-equivocally that stomata respond to atmospheric humidity simply from measurements of transpiration alone since any change in plant water loss will cause a simultaneous change in leaf water potential. This makes it ambiguous whether stomata respond directly to changes in air humidity or indirectly via changes in the mesophyll water status (107, 133). Unless transpiration declines with increasing Δw above a critical Δw and leaf water status increases (109, 110, 112) one cannot conclude that there is a dominant direct effect of humidity (31).

A number of possible interactions between stomatal response, transpiration, and leaf water potential during changes of Δw are distinguished in Figure 2 (left): 1. At constant leaf conductance, transpiration is proportional to Δw and leaf water potential change is inversely related to Δw (97). 2. If stomata respond to changes in leaf transpiration in such a way as to minimize further decreases in leaf water potential (or turgor), it is expected that transpiration would initially increase but then level off at a constant maximum level in dry air (assuming there is only a proportional reaction and not an overreaction of stomata). In this case, leaf conductance would show a curvilinear response approaching a minimum value, i.e. a decreasing slope of the $g/\Delta w$ relationship, and leaf water potential would decrease in a curvilinear fashion to reach a corresponding constant minimum value in dry air (23, 50). This type of control is referred to as feedback control because changes in leaf conductance would be explained by stomatal responses to changes in leaf water potential which depend on leaf water loss (23, 31). It will be shown later that this response curve could also be explained by a porous cuticular membrane. 3. If epidermal water loss and turgor of the epidermis determine stomatal opening independent of stomatal transpiration and mesophyll water status, then it is expected with increasing evaporative demand that leaf transpiration would reach a maximum and then decrease, whereas leaf water potential would reach a corresponding minimum and then increase again (109, 110). This type of response is referred to as feedforward control (23). Mathematical treatment of this phenomenon (31) shows that a response which leads to decreased transpiration in dry air could be explained by loss of water that bypasses the stomatal pore (72, 73). However, flow patterns of water at the cellular level in the leaf necessary to distinguish between feedforward and feedback mechanisms have not been established, and it should be clear that feedback and feedforward refer to the end result of a control operation and not to the mechanisms underlaying the control.

Because of current technical difficulties in measuring cuticular conductance in the presence of stomata, we do not even know its exact magnitude. General-ly, cuticular conductances are fairly small: for fruit cuticles (which have no stomata) the cuticular conductance ranges between 0.01 and 10 mmol $m^{-2}s^{-1}$,

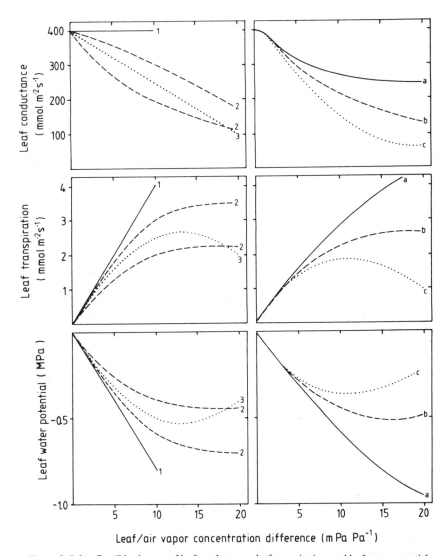

Figure 2 (left) Possible changes of leaf conductance, leaf transpiration, and leaf water potential with variations in leaf/air vapor concentration difference. 1. no response of conductance to changing humidity; 2. feedback control of leaf water potential on leaf conductance for different minimum water potentials; 3. direct (proportional) response of leaf conductance to changing humidity. Assumption: a proportional effect of transpiration on water potential.

(right) Possible changes of leaf conductance to variations in leaf/air vapor concentration difference, assuming that the stomatal response is a function of a humidity-dependent permeability change of a porous membrane as measured by Schönherr (104: Figure 6.1 and 6.5) for *Solanum melongena* (a), *Citrus aurantium* (b) and *Polymethacrylate* (c). Assumption: 20 C, 10 mmol m^{-2}s^{-1} maximum cuticular conductance, 400 mmol m^{-2}s^{-1} maximum leaf conductance, proportional effect of cuticular transpiration on leaf conductance, and stomatal closure at 150 μmol m^{-2}s^{-1} (line a and b) or 100 μmol m^{-2}s^{-1} (line c) cuticular transpiration.

for leaves estimates of cuticular conductance range between 1 and 20 mmol $m^{-2}s^{-1}$ (104).

I will first try to explain mechanisms that lead to the direct (feedforward) response of the stomata to changing Δw; the indirect (feedback) response will be discussed in the following section. An indication of how stomata respond to Δw changes can be derived from the experiments of Lange et al (63) and Lösch (68). The isolated epidermis floating on water required a minute air bubble underneath the stomata in order to show a humidity response. If the mesophyllic surface of the epidermis was totally flooded with water, stomata did not respond to changes in humidity, but stomata of the isolated epidermis responded also to a decrease in water potential of the floating solution (70). This implies that a gradient in water relations parameters between epidermis and mesophyll or within the epidermis is required for the humidity response to function. This gradient would be caused by a liquid flow resistance in the pathway of the water flow, and this resistance is overcome when the epidermis is flooded. The exact location of this flow resistance may be species specific. In intact leaves no consistent gradient in turgor was detected in *Tradescantia* between epidermal cells and cells subsidiary to guard cells (144) or between epidermal cells in relation to their distance from veins or stomatal cells (17). Therefore, the epidermis of Tradescantia is well equilibrated in its water relations, and a gradient in water relations parameters should develop between the epidermis and the mesophyll. Additionally, water losses from the inner side of the guard cells to the substomatal chamber seem generally not to be important for closing stomata in dry air, because Sheriff (119) observed condensation of liquid water on the inner side of the guard cells of intact leaves of *Tradescantia* and other species in high light when stomata were closed by exposure to dry air. In some species water may not readily be lost from the inner walls of the guard cells. This may be explained by the observation of Boyer (12), who showed for *Pyrus* that the cuticle can extend far into the substomatal cavity. Internal suberization of leaf tissues was also demonstrated by Scott (115). An internal leaf cuticle suggests that the water lost by stomatal transpiration would evaporate from deeper layers in the mesophyll (12), where a very low cell wall resistance has been postulated (51). But it is obvious that the extension of leaf internal cuticle needs further attention and may be species specific. In gymnosperms the cuticle extends only to the central pore (142).

It becomes clear from the above discussion that the water relations of the epidermis need to be studied for further understanding of the stomatal humidity response. Since the pioneer work of Meidner & Edwards (79), our knowledge is confined to studies with only one plant species, *Tradescantia virginiana*. Its large epidermal cells have a high volumetric elastic modulus (125 bar) and a hydraulic conductivity of 10^{-6} to 10^{-7} cm $s^{-1}bar^{-1}$ (127). Thus its epidermis equilibrates very quickly to changes in water potential of single epidermal cells.

Figure 3 shows simultaneous measurements of leaf transpiration, leaf conductance, epidermal turgor, and xylem water potential in *T. virginiana* (116). Following a 0.8 mPa Pa^{-1} drop in Δw (first dotted line in Figure 3), transpiration decreased 0.5 mmol m^{-2}s^{-1} and epidermal turgor rose 0.13 MPa without significant time lag. It appears significant that a steady state turgor exists in the epidermis which changes to a new steady state with humidity. But only after the epidermal cells had reached a higher turgor level (second dotted line in Figure 3) did the stomata begin to open, and it then took about 1 h for the stomatal aperture to reach a new steady state. During the transition of stomatal adjustment, transpiration rose without a consistent effect on epidermal turgor. This experiment shows no hydropassive effect of epidermal turgor on leaf conductance. Since the hydraulic tissue conductivity of epidermal and subsidiary cells is large enough to allow for rapid water equilibration (17), the stomatal response can only be explained by a turgor-dependent metabolic response. This agrees with Lösch (69), who reported that potassium is exchanged by guard cells during their response to humidity although not as a primary event.

The experiment described in Figure 3 suggests that a hydroactivated process is a necessary step in the control of a direct response of stomata to changes in air humidity. But the relationship between the extent of the hydroactive process and the turgor of epidermal cells is not known, and also the processes which determine the epidermal turgor in relation to air humidity are unclear. A new approach toward understanding the epidermal water relations may result from the work of Schönherr (104). He distinguished cuticles as being either homogeneous "solubility membranes" or inhomogeneous "porous membranes." For the homogeneous cuticular membrane, the permeability (or diffusive conductance) of the cuticle is independent of the relative humidity and the water flux is proportional to the vapor pressure difference between leaf and air. Water moves by diffusion of individual molecules and clusters of water molecules do not form. In this model there is only a water/cuticular polymer interaction and no water/water interaction within the cuticle. The permeability is the same when measured in a vapor/membrane/vapor-system or in a vapor/membrane/water-system. This permeability characteristic would account for the direct response of stomata to changes in humidity in a strict sense, but water permeability of homogeneous cuticular membranes is known to be extremely low (even lower than that of Teflon) and may not allow for sufficient cuticular water loss to permit a humidity response to take place. Only in very small epidermal cells could the relative volume change become large enough at low cuticular transpiration to cause a turgor change.

In contrast to the homogeneous nature of solubility membranes, Schönherr (104) pointed out that the porous membrane contains traces (or small amounts) of polar substances. The hydration of these polar groups is a function of the

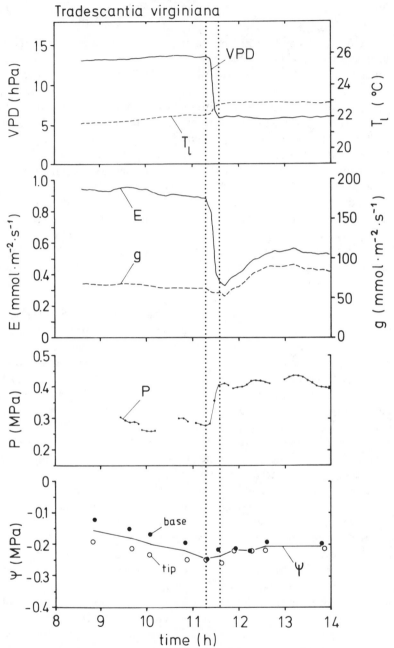

Figure 3 The effect of a step decrease in vapor pressure deficit VPD, on leaf transpiration E, leaf conductance g, turgor of epidermal cells P, and xylem water potential ψ, below (base) and above (tip) the point of turgor measurement. T_l: leaf temperature. The two dotted lines indicate the start (left) and the end (right) of the VPD change. Adapted from Shackel & Brinckmann (116).

atmospheric humidity. With increasing relative humidity the number of water molecules surrounding the polar groups increases. Membrane permeability to water increases in relation to the water content of the porous membrane. Schönherr postulated that water sorption to polar polymers leads to the formation of water clusters or to continuous water-filled channels. Liquid water continuity exists across a porous membrane that is exposed to water on one side and vapor on the other. The liquid/vapor transition takes place on the side of the membrane that is exposed to the air. Cuticular conductance cannot be calculated from the vapor pressure difference across the membrane since the hydration of polar polymers changes with the water potential of the air, which is proportional to the relative air humidity. Therefore, cuticular conductance is related to the relative air humidity rather than to Δw (104). Evidence by electron microscopy supporting the existance of these channels in the cuticle is given by Sack & Paolillo (102). They find electron-dense fibrils that reach to the surface in the cuticle of *Fumaria* guard cells. Maercker (73) used diffusion of tritiated water and local staining with hygroscopic coatings to demonstrate regions of increased cuticular water loss above the anticlinal walls of the stomatal and subsidary cells (75, 76). Most plant cuticles investigated so far have been found to be heterogeneous, and the conductance can reach values of the cuticular matrix (J. Schönherr, personal communication).

As a hypothesis one could propose that a porous membrane changes its conductance (g_c) with the relative humidity of the air [$g_c = f(\%rh)$], and cuticular transpiration (E_c) depends on the vapor concentration in the air ($E_c = \Delta w\, g_c$). Cuticular transpiration would then regulate the turgor of the epidermis, which in turn would govern in a hydroactivated process the size of the stomatal aperture. The hydroactivated stomatal response is proportional to epidermal turgor (37a). Fischer (35) showed that the aperture is proportional to the guard cell turgor. Figure 2 (right) shows the response of stomatal conductance to changes in Δw on the basis of measured changes in permeability of porous membranes with humidity as published by Schönherr (104), but assuming a proportionality between E_c and g_s. A porous membrane would explain transpiration-vapor pressure response curves which were previously understood as a feedforward regulation (decreasing E at high Δw), as well as transpiration/vapor pressure responses that were previously interpreted as feedback response via mesophyll water stress (increasing or constant E at high Δw). The concept of a porous membrane would account for the observations that most humidity response curves are not linear but curvilinear in the range of high leaf conductances and that many response curves do not reach low conductance values even when the air is very dry, e.g. *Vigna unguiculata* (107). It could explain the observation that a minimum vapor pressure deficit is required before stomata start to close (58, 59); this is related to the fact that the permeability of the cuticle may change in a sigmoid fashion with relative humidity (104). A porous membrane could be the basis of the observation that the response of stomata to changes in humidity is to a certain extent under environmental control (80). Porous membranes may also explain the observed movement of ions through cuticles during foliar fertilization or leaching (30).

In order to permit epidermal turgor changes that are independent of the transpiration stream, a water potential gradient should exist between the epidermis and the xylem. Indeed, Shackel & Brinckmann (116) found in *Tradescantia* a gradient between epidermal cells and xylem that is 0.3 MPa at 1.0 mmol $m^{-2}s^{-1}$ transpiration, the epidermis always having a lower water potential than the xylem. In *Tradescantia* (144) the water potential gradient is related to the somewhat lower hydraulic conductivity and smaller area of water flow in the mesophyll cells when compared to the epidermal cells. Regulation of water flow is very likely to be under symplastic control. The finding of a steady state water potential and turgor gradient between mesophyll and epidermis still does not explain why the metabolic response of the guard cells is proportional to the epidermal turgor change.

Another mechanism of stomatal response to humidity changes—mediated directly by turgor changes in the guard cells—has been documented (2). In this case guard cells are lacking a proper cuticle at a location in the throat of the stomatal aperture. They probably lose water at a greater rate from this spot and this leads to turgor changes in the guard cells. However, the spot of higher water loss is at least partially under guard cell control since with closure the two guard cells touch each other at that spot of a deficient cuticle. Therefore, this mechanism acts somewhat like a feedback system: i.e. transpiration should not decrease at high evaporative demand as long as stomata respond proportionally to the water loss from the spot of deficient cuticle.

Responses of the Photosynthetic Apparatus

Midday stomatal closure from low atmospheric humidity was previously thought to be the main mechanism besides temperature causing the midday depression of CO_2 uptake in *Prunus* and other desert species under field conditions (107, 111). The midday depression has been simulated recently in controlled environment chambers (125). If nothing but stomata causes this reduction in CO_2 exchange at noon, for example in direct response to dry air, then the supply of the mesophyll with CO_2 must be reduced when the demand in the mesophyll remains unchanged. However, in steady state experiments, the observed decline in c_i was less than expected (110), indicating that besides a stomatal response, the biochemical properties of the photosynthetic apparatus in the mesophyll did change. In natural environments with changing humidity conditions, c_i often was remarkably constant (112, 137).

A number of investigators conducted experiments to assess the effect of air humidity on CO_2 assimilation. They found that the rate of photosynthesis at CO_2 saturation decreased at low air humidity (3, 37, 98, 137), and that there was an aftereffect of dry air that continued to depress the rate of photosynthesis (117). In *Xanthium* the CO_2 saturated rate of photosynthesis was inhibited by dry air (117) whereas in *Arbutus* the initial slope and the saturated rate of the CO_2 response curve both decreased (98, 126).

The mechanism of the changes in the CO_2 response of photosynthesis induced by dry air and high transpiration is not understood at all. Photoinhibition (88, 89), induced by low CO_2, is probably not a factor, since a low supply of external CO_2 during the determination of CO_2 response curves does not have the same effect, except at mesophyll internal CO_2 concentrations of less than 100 ppm (74, 124) or after very prolonged stomatal closure (M. Küppers, personal communication). Sharkey (117) explains the reduction of the CO_2-saturated photosynthesis at high transpiration by water potential gradients that develop locally within the leaf tissue between the xylem and the site of evaporation. But when water potential of a test leaf was changed by changing whole plant transpiration under conditions of constant transpiration of the test leaf, there was no effect on the photosynthetic capacity of that test leaf in a climate-controlled test cuvette (59, 109, 134). However, when only a small section of a test leaf was subjected to increases in transpiration, decreases in the CO_2 assimilation capacity were found. Also, it was observed that CO_2 assimilation increased again when transpiration decreased in very dry air. This points to a mesophyll internal mechanism that is related to the leaf water loss (117). Interestingly, recent experiments by Heber et al (43) demonstrate rapid water movement at the thylakoid level with changes in the water supply, but these changes of the thylakoid water status had no effect on photosynthesis unless additional and as yet unknown solute effects occured.

Of fundamental importance to this discussion is which water relations parameter affects photosynthesis. The chemical potential of water does not affect photosynthesis except with very extreme dehydration (10, 62, 87). There is more and more evidence that plant metabolism does not respond directly to plant water potential per se but rather to factors varying in concert with water potential such as turgor (5, 14). However, it is difficult to imagine how cell organelles detect turgor changes. Photosynthesis has been shown to be sensitive to changes in osmotic pressure and cell volume. It is known that osmotic stress reduces photosynthesis in chloroplasts, protoplasts, and leaf discs (56), and this may result from stromal acidification (6).

Photosynthesis is also sensitive to changes in cell volume rather than to changes in water potential (27, 56, 57). The relative change in volume ($\Delta V/V$) is approximately proportional to the change in water potential ($\Delta \psi$) and inversely proportional to the sum of elastic modulus (ϵ) and osmotic potential (Π), i.e. $\Delta V/V = \Delta \psi / (\epsilon + \Pi)$ (123). Therefore, because diurnal changes of water potential may well be in the range of 1 to 2 MPa, it is very likely that significant changes in the volume of mesophyll cells occur that could reduce photosynthesis (27, 54). This becomes even more obvious if we consider that the elastic modulus of mesophyll cells is about 1/10 that of epidermal cells (144), and therefore the volume changes of cells in the mesophyll are expected to be larger than of cells in the epidermis. The parameter "relative water content" may be a better parameter (136) than "water potential" to explain physiological changes in the leaf.

In addition to these osmotic and volume effects, plant hormones may also act to modulate photosynthesis and stomatal response during the response to dry air. Burschka et al (18) found a three-to-fourfold increase in abscisic acid (ABA) content in *Arbutus* leaves during midday. It seems possible that ABA released into the apoplast during turgor or volume changes acts on photosynthesis and stomata (18) and causes the aftereffects of dry air exposure (110). According to Raschke & Hedrich (96), the changes in CO_2 assimilation and stomatal conductance could also be explained by effects of ABA (21). The effects of ABA on photosynthesis and stomatal action will be explained in more detail in the following section.

GAS EXCHANGE RESPONSES TO CHANGES IN MESOPHYLL WATER STATUS

It was shown in the previous section that stomatal responses to air humidity can result from turgor changes in the epidermis that can occur independent of mesophyll water status. The specific role of mesophyll water status on stomatal function is now evaluated.

Transient Responses

Abrupt changes in water supply to the leaf xylem will cause turgor changes in the mesophyll as well as in the epidermis which open or close stomata passively, but in a direction opposite what is normally associated with changes in water status. A sudden reduction in water supply might lead to transient passive stomatal opening, and a sudden increase in leaf water supply might cause transient stomatal closure. This hydropassive effect (122) was described as early as 1898 by Darwin (25), who explained it as the consequence of a time lag between pressure changes in the epidermal cells and in the guard cells around the stomatal pore. The effect is commonly observed after cutting a leaf (48). Raschke (91) studied this response in more detail by applying vacuum or hydrostatic pressure to the cut end of excised leaves of *Zea mays*.

The equilibration of leaf water status following an abrupt change in the water supply frequently results in stomatal oscillations, but ultimately the stomata adjust to apertures that are dependent on the pretreatment of the leaf and the actual leaf environment (92, 121). Therefore, the transient stomatal response indicates the existence of a hydraulic continuity in the leaf, but it does not explain steady state gas exchange responses. Therefore, recent measurements of light scattering as a leaf was excised indicate large water fluxes at the thylakoid membranes which inhibit photosynthesis if excision takes place under water (43). Therefore, passive stomatal responses that are caused by excising a leaf under water may cause a reaction at the photosynthetic level in the mesophyll that does not occur when the leaf is excised in air.

Steady State Responses to Changes in Leaf Water Status

Stomata close when the plant or the leaf "runs out of water." This has been demonstrated with plants in the field (128), or in pots (53), with detached leaves (132), with excised trees (100), and with twigs in which the water flow had been blocked by freezing with dry ice (38). These observations led Hsiao (46) and Turner (129) to the concept of a "threshold response" of stomata whereby stomata close at zero turgor of the leaf (not epidermis), i.e. the closure may occur at different levels of leaf water potential, depending on the osmotic pressure and adjustment of the leaf (52, 131, 132). However, since stomata have an independent ion and osmotic regulation of turgor (94), they may not follow mesophyll turgor caused simply by hydraulic effects. There is a need to distinguish between bulk leaf turgor and guard cell turgor.

Stomatal closure at leaf wilting (zero bulk leaf turgor) is very likely mediated by ABA (93). To understand the effects of ABA, one must distinguish between the total ABA content of a plant—which is to the largest extent stored in the symplast—and the ABA released to the apoplast. Pierce & Raschke (85, 86) showed that in detached, slowly drying leaves, increased ABA synthesis occurred at a threshold close to zero mesophyll turgor. The threshold did not depend on the leaf water potential, but on leaf turgor, because the water potential at zero turgor could be varied by 0.6 MPa as a result of osmotic adjustment without changing the threshold. However, biosynthesis of ABA is too slow to explain stomatal closure (44). Moreover, the ABA content of leaves does not correlate with stomatal functioning (95). Burschka et al (18) found the same concentration of ABA in irrigated (ψ −0.1 to −2.5 MPa) and dry (ψ −5 MPa) Arbutus trees with open and closed stomata respectively. By contrast, injection of a minute amount of ABA into the xylem (pmol cm^{-2} leaf area^{-1}) closed stomata and reduced photosynthesis such that the leaf internal partial pressure of CO_2 remained constant in Arbutus (18a). The addition of only 1 fmol mm^{-2} leaf area $^{-1}$ caused a 5% reduction of leaf conductance in Xanthium, and this represents less than 2% of the total ABA content of the leaf (95). The reason for the potent effect of apoplastic ABA on the functioning of stomata may be the result of its action at the outer plasmalemma, where it interferes with ion transport (41, 45). Thus it is apoplastic ABA rather than total

ABA that affects stomatal function (95). ABA synthesis may be regarded as a form of chemical stress integrator. During repeated episodes of low turgor, ABA accumulates in the chloroplasts where it is available for release to the apoplast in response to stress events even at stress levels that may be too mild to cause ABA synthesis. The redistribution of ABA between symplast and apoplast appears to be a diffusional process following pH gradients (24, 55), or a turgor- or volume-regulated process (42) that may occur not only at wilting but also at positive turgor. If stress is released, the ABA is metabolized (139).

It is intriguing to explain water stress-induced stomatal closure as an effect of ABA, but many of the observations were made by using detached leaves or by artificially adding ABA to the xylem water. Also, when leaves are cut under water additional solute effects may occur (43). In the intact plant, the effect of ABA in the positive turgor range may depend upon additional internal factors or upon pretreatments. Raschke & Hedrich (96) found that stomata of plants became sensitive to CO_2 changes when ABA was added to the xylem stream. This effect was enhanced by repeated application of ABA, and eventually stomata closed. However, guard cells were able to escape from the effect of ABA over time. The mechanism of this escape is unknown, but it may be linked to the metabolism (139) or it may be mediated by other plant hormones, because, for example, cytokinins can modify the response of stomata to ABA (7, 8). This shows that in the intact plant with its range of hormonal responses, stomata may not regulate in concert with the water status of the leaf. In fact, 1. stomata may remain open even in wilted leaves (44, 109, 130, 134); 2. during the day stomata may reach a maximum aperture when leaf water potential (and turgor) reaches a minimum (low) (59, 112, 114); and 3. stomata may close in dry soil at positive turgor and in a progressive rather than a threshold manner (40, 70, 134). In *Gossypium* stomata closed when turgor was still 0.5 MPa. Furthermore, ABA was not found in the apoplast (xylem pressure exudate) until leaf turgor dropped to 0.1 MPa (1, 90). The range of the response of leaf conductance to leaf water potential during soil drying may be modified by atmospheric conditions (134), leaf morphology (66), and other adaptive mechanisms, all of which may interfere with the anticipated threshold response.

The contradictory evidence that exists for the relation between leaf conductance and leaf water potential (or leaf turgor) in intact plants arises to a large extent from the fact that leaf water potential is related to whole plant transpiration through the hydraulic conductance in the plant (Figure 1). The difficulty in identifying causal relationships becomes most obvious from an experiment by Turner et al (133) in which a variety of morphologically different plant species were compared during changes in humidity. In all cases stomata closed in dry air, but closure was not sufficient to compensate for the effect imposed on transpiration and leaf water potential by increasing Δw. In dry air leaf water potentials dropped in proportion to plant water loss and to the hydraulic resistance in the plant. Since conductance, leaf water potential, and humidity change in concert, it is impossible to be sure which change is primary and which secondary. *Vigna* maintained a constant and high turgor over a broad range of Δw, and most other species showed a linear correlation between water potential and leaf conductance changes. This illustrates the experimental difficulty in analyzing feedback systems and the danger of interpreting a correlation between two parameters as a causal relationship. The following analysis will show that it may not be the leaf water status that controlled stomata in the intact plant.

In order to understand the effect of leaf water potential (or turgor) on stomatal conductance in an intact plant system, leaf water potential must be perturbed by a mechanism other than soil drought or transpiration of the test leaf. One possible perturbation is to keep a test leaf in an environment of constant humidity and light while the whole plant transpiration (and thus the whole plant as well as the test leaf water potential) is changed by altering humidity (40, 109, 134). Figure 4 (left) shows the difference produced in leaf conductance and leaf water potential (as measured in a test leaf under constant humidity and light conditions) by changing the humidity about the whole plant under conditions of soil drying. The curve expressing leaf conductance as a function of leaf water potential of an isolated leaf could be shifted about 0.8 MPa just by changing whole plant transpiration. It will be shown in the following section that in this experiment stomatal closure was actually related to soil water status rather than to leaf water status (Figure 4 right).

In summary, leaf water potential and turgor can be manipulated through transpiration changes, and stomata close with drying soil but at different levels of leaf water potential (or turgor), depending on the transpiration rate of the test leaf. When water potential of an individual leaf is manipulated via changes in transpiration of whole plants, the lowest potentials occur when transpiration is high, generally when stomata are open, and this effect is reversible. By contrast, when leaf water potential is manipulated by soil dryness, the response is no longer easily reversible (i.e. rewatering will not cause immediate complete stomatal opening) and stomata close progressively, independent of the humidity and leaf water potential. Differences in the aerial and soil environment will to a large extent explain the variation in reported correlations (23, 44, 46) between leaf water potential and water flux or leaf conductance.

GAS EXCHANGE RESPONSES TO CHANGES IN SOIL WATER STATUS

Experimental Evidence

There is a wealth of data in the agricultural literature that relates plant water loss to soil water status. In most of these studies it is assumed that control is exerted

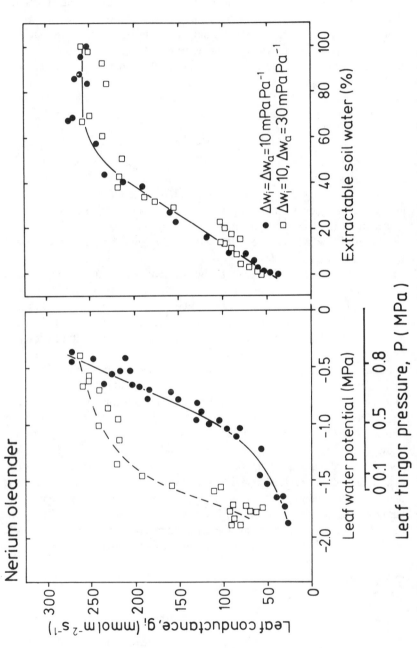

Figure 4 Leaf conductance (g_l) of a test leaf in a cuvette at constant vapor pressure deficit $\Delta w_i = 10 \text{ mPaPa}^{-1}$ and drying soil as related to leaf water potential and turgor (left) or extractable soil water (right). The whole plant was exposed to different vapor pressure deficits (Δw_a) of either to 10 mPaPa^{-1} (closed symbols)

via leaf water status. Yet there is also evidence from field studies that stomatal responses correlate with soil water supply but not with leaf water status (4). Evidence favoring an independent control of leaf conductance by changes in soil water status is obtained from experiments of Gollan et al (40) and Turner et al (134) in which leaf water potential of *Nerium* and *Helianthus* could be substantially perturbed (Figure 4: 0.8 MPa) by changes of leaf transpiration without altering leaf conductance. However, stomata did close irreversibly (at least in the short term) when soil water was depleted by 30% (Figure 4 *right*). This is still indirect evidence for an effect of soil water status on leaf conductance. Since leaf and soil water status changed in parallel, an effect of leaf water status that is secondary to changes in soil water status cannot be ruled out, although it would need to operate over a wide range of leaf water potential and turgor.

A new experimental approach is required to demonstrate a direct signal from the root to the stomata. One very elegant method was developed by Passioura (83) to maintain full turgor in a leaf while the soil dries. His technique of pressurizing the root allows maintenance of leaf water potential close to 0 MPa without affecting the change in water potential gradient between soil and root in a drying soil. When plants were subjected to progressively drying soil and when leaf water potential was thus maintained near 0 MPa, stomata closed in dry soil although the leaves of these plants had never experienced any water deficit (39).

Possible Mechanisms

The mechanism by which the root water status controls leaf conductance is not known, although root effects on stomata have been reported for changing soil temperatures (60) or for conditions of flooding (15, 82). It is easier to rule out certain factors than to prove clearly the role of a factor.

There is evidence against an effect of CO_2 assimilation on conductance. Stomatal closure with dry soil is not consistently correlated with changes in CO_2 assimilation, although the photosynthetic capacity may also change with dry soil (117). Schulze & Hall (107) showed that with drying soil stomata of C_3 plants generally decrease their aperture prior to changes in the photosynthetic capacity at ambient CO_2 concentrations. This effect may depend on leaf morphology. In the mesomorphic leaf of *Helianthus*, under drying soil conditions CO_2 assimilation changed with leaf conductance in such a manner that c_i was virtually unchanged except in very dry soil (134). But in the sclerophyllous leaves of *Nerium*, photosynthetic capacity remained virtually unchanged and c_i decreased steadily because of stomatal closure in dry soil (40).

Stomata could respond to the changing water supply in drying soil. Küppers (59) showed a general positive correlation between stomatal conductance and the soil/root/leaf hydraulic conductivity. Woody species with a high hydraulic conductivity had a higher stomatal conductance than woody species with low hydraulic conductivity. Similar observations were made with plants growing under desert conditions (99, 103). Therefore, it appears as if changes in hydraulic conductivity from the soil and through the plant in drying soil affect stomatal aperture.

In the plant stem, flow resistance changes mainly because of cavitation within xylem vessels (28, 135, 143). Yet the correlation Küppers (59) found between stomatal and hydraulic conductivity was independent of the leaf water potential. Therefore, the control mechanism is not located at the stem but at the root level.

There are changes in the hydraulic pathway at the root/soil interface, where the water flow through the soil to the root changes with soil drought (83, 84). Progressive pruning experiments (16) with *Helianthus* roots showed a proportional effect on leaf conductance up to stomatal closure, but very little effect on the hydraulic conductance.

There appears to be a signal coming from the root which is independent of its hydraulic conductance. Such a signal could be produced by a change in water flow through the soil that will result in a change in the water status of the growing region of the root (84). Visible changes in root diameter occur with transpiration during the day as an indication that considerable turgor changes occur (47). Davies et al (26) described a split root experiment in which one part of the root was dried while the other part was kept wet. The dry side of the root tip had a water potential .25 MPa lower and a turgor 0.3 MPa lower in the root tip than those in the wet root. This shows that the root tip can experience water stress quite independent of the whole plant.

A change in root water status will affect active ion uptake (12, 101) and ion concentration in the xylem (105, 108). While this might alter the pH that could influence the distribution of ABA, pH changes are likely to be small since the conducting system of the plant is probably well buffered, but this is not well studied. It is not at all clear to what extent changes in ion uptake and xylem fluid affect stomata (43). A correlation between leaf conductance and nitrogen supply was found in mistletoe (29, 106).

Probably the most important effect of drying soil is a change in root growth (13, 20, 26, 34). The total root biomass was found to increase with mild water stress (118) but decreases with progressive drought (81). A change in the age and the activity of the root system will have an effect on the synthesis and release of plant hormones, in particular cytokinins, which are biosynthesized in active meristems. There are numerous observations that cytokinins increase transpiration, open stomata (49), and act on photosynthesis (78). Zeatin and kinetin caused some reversal of ABA-mediated closure in *Zea mays,* but not in *Commelina* (7). The most convincing data on effects of root hormone relations on stomata of water-stressed plants come from split pot (22) or split root experiments (9). Drying the soil around one part of the split root system of *Zea mays* and keeping the other parts well watered resulted in a reduction of stomatal aperture, even though leaf water potential, turgor, and ABA content remained unaffected. The observation on ABA is surprising since it is known that ABA is transported not only in the xylem but also in the phloem (140). Kinetin and zeatin applied to leaf pieces from these plants reversed the effect of drying soil on stomata. Blackman & Davies (9) suggest that a continuous supply of cytokinin from the roots may be necessary in *Zea* to sustain stomatal opening, and interruption of this supply triggered by soil drying may signal the shoot to restrict stomatal opening and thereby water use.

CONCLUDING REMARKS

The effects of drought have fascinated plant scientists for over a century. But because of the difficulty of actually measuring plant water status at the cellular level, because of the sensitive transformations of hydraulic into metabolic signals, and because of the great analytical difficulties in measuring substances in the apoplast, our understanding of the mechanisms of atmospheric and soil drought on plant functioning has remained schematic for a long time. There are advanced models describing the end results of controls and knowledge of changes in some aspects of cellular metabolism in response to drought, but the understanding of plant functioning at the tissue and organ level under conditions of drought has emerged only recently. The following facts and hypotheses may stand out as particularly important:

1. Air humidity and soil water status act on transpiration and CO_2 assimilation via effects on the stomatal aperture and via independent effects on the photosynthetic metabolism. There is no indication of a direct effect of CO_2 assimilation on leaf conductance under conditions of drought.

2. The steady state responses of stomata are all linked to metabolic processes in the guard cells. Models based on hydraulic feedback and feedforward effects refer to the end result of a control process but not to the mechanism at the cellular level.

3. Three responses of stomata appear to be of major importance for explaining the control of gas exchange under conditions of drought: (*a*) a direct response to humidity, (*b*) a direct response to root functioning, (*c*) a response to mesophyll internal conditions. These responses are now discussed in more detail:

3a. The humidity response of the stomata is a property of the epidermis and related to its steady state turgor. The epidermal turgor results from a balance between the rate of water supply from the mesophyll and the rate of epidermal water loss. The latter is determined by the vapor pressure deficit in the air, leaf temperature, and cuticular conductance. It appears that the cuticle is a porous membrane, the permeability of which is related to the relative humidity and not to the vapor pressure deficit. The epidermal turgor triggers a yet unknown metabolic process in the guard cells to respond proportionally.

3b. Stomata respond directly to a signal from the roots under conditions of soil drought. It appears that the signal is related to the metabolic activity of the root, and it is probably related to the metabolism of cytokinins, which tend to open stomata and counteract the effects of ABA in the leaves.

3c. Stomata respond to ABA in the leaf apoplast, but the effect is dependent on pretreatments with ABA, and the guard cells are able to escape from the effect of ABA over time when ABA is supplied artificially. The metabolism of ABA and the content of ABA in the bulk leaf are not related to its action on stomata. Apparently, only a very small amount of the total ABA content needs to be released into the apoplast adjacent to the guard cells for a significant response in the stomata, but the process of ABA release into the apoplast is unknown (pH, volume, or solute effects). It cannot be ruled out that the compartmentation of ABA is involved also during the direct response of stomata to humidity and root functioning. So far the ABA effect is considered as an "emergency" response that takes place under conditions of desiccation, but this view may have to be corrected with finer methods of analysis.

4. Humidity may act on the photosynthetic metabolism (change of the A/c_i curve) via a process as yet unknown. Water stress in the mesophyll seems not to be a cause, since shrinking and swelling of the thylakoids by water uptake and water loss appears to have no effect on photosynthesis, but photosynthesis changes if the solute environment in the apoplast changes.

5. Intrinsic photosynthesis (A/c_i curve) decreases also under conditions of soil drought, but the effect is species and treatment specific. Generally, stomata are more sensitive to changes in the root environment and function than photosynthesis.

This review suggests that the direct responses of stomata to humidity and root functioning are of primary importance for plant performance under conditions of drought, especially under natural conditions (Figure 1:double circled responses). There is virtually no effect of bulk leaf water potential on stomatal reaction, but pH, volume, or solute effects in the apoplast may perturb the plant hormone balance that acts on stomata. One may generalize that plants do not care about the plant water status as long as they can maintain the water fluxes in the epidermis and in the roots. Although it has been suggested that drought has independent effects on assimilation and conductance, it is interesting to note that with changing humidity the photosynthetic capacity is adjusted in some cases such that c_i remains constant. This may be caused by the fact that the same metabolic factor acts on stomata and photosynthesis. This review emphasizes the importance of plant hormones, ABA, and cytokinin in controlling assimilation and stomatal conductance under conditions of drought.

There are many apsects of drought that remain to be explored and that are not sufficiently covered in this review. At the whole plant level the linkage from the single leaf to the whole plant and canopy level and to growth needs further attention. At the cellular level we need more understanding of the transport, compartmentation, and action of plant hormones and of the conditions in the leaf cell wall. I have tried to summarize possible mechanisms for humidity and soil drought actions on transpiration and CO_2 assimilation. The reader should be aware, however, that these hypotheses are based on observations that were made on very few plant species. Considering the diversity of habitats and specialization of plant life forms, one should expect additional mechanisms and response types to emerge in the future.

ACKNOWLEDGMENTS

My sincere thanks to my friends Howard and Juli Graves, UC Berkeley, for editing my English. I would also like to thank Dr. J. S. Boyer, Dr. T. C. Hsiao, Dr. K. Raschke, and Dr. N. C. Turner for helpful comments and discussions during preparation of this article. My special thanks to Mrs. Evelin Amberg for typing this manuscript. The research performed by the author was supported by the Deutsche Forschungsgemeinschaft (SFB 137).

Literature Cited

1. Ackerson, R. C. 1982. Synthesis and movement of abscisic acid in water-stressed cotton leaves. *Plant Physiol.* 69:609–13
2. Appleby, R. F., Davies, W. J. 1983. A possible evaporation site in the guard cell wall and the influence of leaf structure on the humidity response by stomata of woody plants. *Oecologia* 56:30–40
3. Ball, M. C., Farquhar, G. D. 1984. Photosynthetic and stomatal responses of two mangrove species, *Aegiceras corniculatum* and *Avicennia marina*, to long term salinity and humidity conditions. *Plant Physiol.* 74:1–6
4. Bates, L. M., Hall, A. E. 1982. Diurnal and seasonal responses of stomatal conductance for cowpea plants subjected to different levels of environmental drought. *Oecologia* 54:304–8
5. Begg, J. E., Turner, N. C. 1976. Crop water deficits. *Adv. Agron.* 28:161–217
6. Berkowitz, G. A., Gibbs, M. 1983. Reduced osmotic potential effects on photosynthesis. Site-specific effects of osmotically induced stromal acidification. *Plant Physiol.* 72:1100–9
7. Blackman, P. G., Davies, W. J. 1983. The effects of cytokinins and ABA on stomatal behaviour of maize and *Commelina*. *J. Exp. Bot.* 34:1619–26
8. Blackman, P. G., Davies, W. J. 1984. Age-related changes in stomatal response to cytokinins and abscisic acid. *Ann. Bot.* 54:121–25
9. Blackman, P. G., Davies, W. J. 1985. Root to shoot communication in maize plants of the effects of soil drying. *J. Exp. Bot.* 36:39–48
10. Boyer, J. S. 1976. Photosynthesis at low water potentials. *Philos. Trans. R. Soc. London Ser. B* 273:501–12
11. Boyer, J. S. 1982. Plant productivity and environment. *Science* 218:443–48
12. Boyer, J. S. 1985. Water transport. *Ann. Rev. Plant Physiol.* 36:473–516
13. Boyer, J. S., Cavalieri, A. J., Schulze, E.-D. 1985. Control of the rate of cell enlargement: Excision, wall relaxation, and growth-induced water potentials. *Planta* 163:527–43
14. Bradford, K. J., Hsiao, T. C. 1982. Physiological responses to moderate water stress. See Ref. 64, pp. 264–324
15. Bradford, K. J., Hsiao, T. C. 1982. Stomatal behavior and water relations of water logged tomato plants. *Plant Physiol.* 70:1508–13
16. Briggs, G. M., Wiebe, H. H. 1982. The effects of root pruning on the water relations of *Helianthus annuus* L. *J. Exp. Bot.* 33:966–76

17. Brinckmann, E., Tyerman, S. D., Steudle, E., Schulze, E.-D. 1984. The effect of different growing conditions on water relations parameters of leaf epidermal cells of *Tradescantia virginiana* L. *Oecologia* 62:110–17

18. Burschka, C., Tenhunen, J. D., Hartung, W. 1983. Diurnal variations in abscisic acid content and stomatal response to applied abscisic acid in leaves of irrigated and nonirrigated *Arbutus unedo* plants under naturally fluctuating environmental conditions. *Oecologia* 58:128–31

18a. Burschka, C., Lange, O. L., Hartung, W. 1985. Effects of abscisic acid on stomatal conductance and photosynthesis in leaves of intact *Arbutus unedo* plants under natural conditions. *Oecologia* 67: 593–95

19. Caemmerer, S. von, Farquhar, G. D. 1981. Some relationships between the biochemistry of photosynthesis and the gas exchange of leaves. *Planta* 153:376–87

20. Caldwell, M. M., Camp, L. B. 1974. Belowground productivity of two cool desert communities. *Oecologia* 17:123–30

21. Cornic, G., Miginiac, E. 1983. Nonstomatal inhibition of net CO_2 uptake by (±) abscisic acid in *Pharbitis nil*. *Plant Physiol.* 73:529–33

22. Coutts, M. P. 1981. Leaf water potential and control of water loss in droughted sitka spruce seedlings. *J. Exp. Bot.* 131: 1193–1201

23. Cowan, I. R. 1977. Stomatal behaviour and environment. *Adv. Bot. Res.* 4:117–228

24. Cowan, I. R., Raven, J. A., Hartung, W., Farquhar, G. D. 1982. A possible role for abscisic acid in coupling stomatal conductance and photosynthetic carbon metabolism in leaves. *Aust. J. Plant Physiol.* 9:489–98

25. Darwin, F. 1898. Observations on stomata. *Philos. Trans. R. Soc. London Ser. B (Biol. Sci.)* 190:531–621

26. Davies, W. J., Metcalfe, J., Lodge, T. A., da Costa, H. R. 1986. Plant growth substances and the regulation of growth under drought. *Aust. J. Plant Physiol.* In press

27. Dietz, K.-J., Heber, U. 1983. Carbon dioxide gas exchange and the energy status of leaves of *Primula palinuri* under water stress. *Planta* 158:349–56

28. Dixon, M. A., Grace, J., Tyree, M. T. 1984. Concurrent measurements of stem density, leaf and stem water potential, stomatal conductance and cavitation on a sapling of *Thuia occidentalis* L. *Plant Cell Environ.* 7:615–18

29. Ehleringer, J. R., Schulze, E.-D., Ziegler, H., Lange, O. L., Farquhar, G. D., Cowan, I. R. 1985. Xylem-tapping mistletoes: water or nutrient parasites. *Science* 227:1479–81

30. Epstein, E. 1972. *Mineral Nutrition of Plants: Principles and Perspectives.* New York / London / Sydney / Toronto: Wiley. 412 pp.

31. Farquhar, G. D. 1978. Feedforward responses of stomata to humidity. *Aust. J. Plant Physiol.* 5:787–800

32. Farquhar, G. D., Caemmerer, S. von. 1982. Modelling of photosynthesic response to environmental conditions. See Ref. 64, pp. 549–88

33. Farquhar, G. D., Wong, S. C. 1984. An empirical model of stomatal conductance. *Aust. J. Plant Physiol.* 11: 191–210

34. Fernandez, O. A., Caldwell, M. M. 1975. Phenology and dynamics of root growth of three cool semi-desert shrubs under field conditions. *J. Ecol.* 63:703–714

35. Fischer, R. A. 1973. The relationship of stomatal aperture and guard-cell turgor pressure in *Vicia faba*. *J. Exp. Bot.* 24:387–99

36. Fischer, R. A., Turner, N. C. 1978. Plant productivity in the arid and semiarid zones. *Ann. Rev. Plant Physiol.* 29:277–317

37. Forseth, I. N., Ehleringer, J. R. 1983. Ecophysiology of two solar tracking desert winter annuals. III. Gas exchange responses to light, CO_2 and VPD in relation to long-term drought. *Oecologia* 57:344–51

37a. Frensch, J. 1986. Der Wasserhaushalt von Epidermiszellen in *Tradescantia* bei Änderungen der Luftfeuchtigkeit. Thesis Bayreuth

38. Glatzel, G. 1983. Mineral nutrition and water relations of hemiparasitic mistletoes: A question of partitioning. Experiments with *Loranthus europaeus* on *Quercus petraea* and *Quercus robur*. *Oecologia* 56:193–201

39. Gollan, T., Passioura, J. B., Munns, R. 1985. Soil water status affects the stomatal conductance of fully turgid wheat and sunflower leaves. *Aust. J. Plant Physiol.* 13: In press

40. Gollan, T., Turner, N. C., Schulze, E.-D. 1985. The responses of stomata and leaf gas exchange to vapour pressure deficits and soil water content. III. In the sclerophyllous woody species *Nerium oleander*. *Oecologia* 65:356–62

41. Hartung, W. 1983. The site of action of abscisic acid at the guard cell plasmalemma of *Valerianella locusta*. *Plant Cell Environ.* 6:427–28

42. Hartung, W., Kaiser, W. M., Burschka, C. 1983. Release of abscisic acid from leaf strips under osmotic stress. *Z. Pflanzenphysiol.* 112:131–38
43. Heber, U., Neimanis, S., Lange, O. L. 1986. Stomatal aperture, photosynthesis and water fluxes in mesophyll cells as affected by the abscission of leaves. Simultaneous measurements of gas exchange, light scattering and chlorophyll fluorescence. *Planta.* In press
44. Henson, I. E., Alagarswamy, G., Bidinger, F. R., Mahalakshmi, V. 1982. Stomatal responses of pearl millet (*Pennisetum americanum* (L.) Leeke) to leaf water status and environmental factors in the field. *Plant Cell Environ.* 5:65–74
45. Hornberg, C., Weiler, E. W. 1984. High-affinity binding sites for abscisic acid on the plasmalemma of *Vicia faba* guard cells. *Nature* 310:321–24
46. Hsiao, T. C. 1973. Plant responses to water stress. *Ann. Rev. Plant Physiol.* 24:519–70
47. Huck, M. G., Klepper, B., Taylor, H. M. 1970. Diurnal variations in root diameter. *Plant Physiol.* 45:529–30
48. Ivanow, L. 1928. Zur Methodik der Transpirationsbestimmung am Standort. *Ber. Dtsch. Bot. Ges.* 46:306–10
49. Jewer, P. C., Incoll, L. D. 1980. Promotion of stomatal opening in the grass *Anthephora pubescens* Nees by a range of natural and synthetic cytokinins. *Planta* 150:218–21
50. Jones, H. G. 1983. *Plants and Microclimate.* Cambridge: Univ. Press. 323 pp.
51. Jones, H. G., Higgs, K. H. 1980. Resistance to water loss from the mesophyll cell surface in plant leaves. *J. Exp. Bot.* 31:545–53
52. Jones, M. M., Turner, N. C. 1978. Osmotic adjustment in leaves of sorghum in response to water deficits. *Plant Physiol.* 61:122–26
53. Jordan, W. R., Ritchie, J. T. 1971. Influence of soil water stress on evaporation, root absorption, and internal water status of cotton. *Plant Physiol.* 48:783–88
54. Kaiser, W. M. 1982. Correlation between changes in photosynthetic activity and changes in total protoplast volume in leaf tissue from hygro-, meso- and xerophytes under osmotic stress. *Planta* 154:538–45
55. Kaiser, W. M., Hartung, W. 1981. Uptake and release of abscisic acid by isolated photoautotrophic mesophyll cells, depending on pH gradients. *Plant Physiol.* 68:202–6
56. Kaiser, W. M., Kaiser, G., Prachuab, P. K., Wildman, S. G., Heber, U. 1981. Photosynthesis under osmotic stress. Inhibition of photosynthesis of intact chloroplasts, protoplasts, and leaf slices at high osmotic potentials. *Planta* 153:416–22
57. Kaiser, W. M., Stepper, W., Urbach, W. 1981. Photosynthesis of isolated chloroplasts and protoplasts under osmotic stress. *Planta* 151:375–80
58. Körner, C. L., Cochrane, P. M. 1985. Stomatal responses and water relations of *Eucalyptus pauciflora* in summer along an elevational gradient. *Oecologia* 66:443–55
59. Küppers, M. 1984. Carbon relations and competition between woody species in a Central European hedgerow. II. Stomatal responses, water use, and hydraulic conductivity in the root/leaf pathway. *Oecologia* 64:344–54
60. Küppers, M., Hall, A. E., Schulze, E.-D. 1982. Effects of day-to-day changes in root temperature on leaf conductance to water vapour and CO_2 assimilation rates of *Vigna unguiculata* L. Walp. *Oecologia* 52:116–20
61. Küppers, M., Schulze, E.-D. 1986. An empirical model of net photosynthesis and leaf conductance for the simulation of diurnal courses of CO_2 and H_2O exchange. *Aust. J. Plant Physiol.* In press
62. Lange, O. L. 1969. Experimentell-ökologische Untersuchungen an Flechten der Negev-Wüste. I. CO_2-Gaswechsel von *Ramalina maciformis* (DEL.) BORY unter kontrollierten Bedingungen im Laboratorium. *Flora Abt. B* 158:324–59
63. Lange, O. L., Lösch, R., Schulze, E.-D., Kappen, L. 1971. Responses of stomata to changes in humidity. *Planta* 100:76–86
64. Lange, O. L., Nobel, P. S., Osmond, C. B., Ziegler, H., eds. 1982. *Physiological Plant Ecology II. Water relations and carbon assimilation.* Berlin/Heidelberg/New York: Springer Verlag. *Encycl. Plant Physiol.*, Vol. 12B. 747 pp.
65. Lange, O. L., Tenhunen, J. D., Braun, M. 1982. Midday stomatal closure in Mediterranean type sclerophylls under simulated habitat conditions in an environmental chamber. I. Comparison of the behaviour of various European Mediterranean species. *Flora* 172:563–57
66. Larcher, W., De Moraes, J. A. V. P., Bauer, H. 1981. Adaptive responses of leaf water potential, CO_2-exchange and water use efficiency of *Olea europaea* during drying and rewatering. In *Components of Productivity of Mediterranean-Climate Regions—Basic and Applied Aspects*, ed. N. S. Margaris, H. A.

Mooney, pp. 77–84. The Hague/Boston/London: Dr. W. Junk

67. Leuning, R. 1983. Transport of gases into leaves. *Plant Cell Environ.* 6:181–94

68. Lösch, R. 1977. Responses of stomata to environmental factors. Experiments with isolated epidermal strips of *Polypodium vulgare*. 1. Temperature and humidity. *Oecologia* 29:85–97

69. Lösch, R. 1978. Veränderungen im stomatären Kaliumgehalt bei Änderungen von Luftfeuchte und Umgebungstemperatur. *Ber. Dtsch. Bot. Ges.* 91:645–56

70. Lösch, R. 1979. Responses of stomata to environmental factors in experiments with isolated epidermal strips of *Polypodium vulgare*. II. Leaf bulk water potential, air humidity, and temperature. *Oecologia* 39:229–38

71. Lösch, R., Tenhunen, J. D. 1981. Stomatal responses to humidity—phenomenon and mechanism. In *Stomatal Physiology*, ed. P. G. Jarvis, T. A. Mansfield, pp. 137–61. Cambridge: Univ. Press

72. Maercker, U. 1965. Zur Kenntnis der Transpiration der Schließzellen. *Protoplasma* 60:61–78

73. Maercker, U. 1965. Mikroautoradiographischer Nachweis tritiumhaltigen Transpirationswassers. *Naturwissenschaften* 52:15

74. Mahall, B. E., Schlesinger, W. H. 1982. Effects of irradiance on growth, photosynthesis, and water use efficiency of seedlings of the chaparral shrub, *Ceanothus megacarpus*. *Oecologia* 54:291–99

75. Maier-Maercker, U. 1979. Peristomatal transpiration and stomatal movement: A controversial view. I. Additional proof of peristomatal transpiration by hygrophotography and a comprehensive discussion in the light of recent results. *Z. Pflanzenphysiol.* 91:25–43

76. Maier-Maercker, U. 1979. II. Observation of stomatal movements under different conditions of water supply and demand. *Z. Pflanzenphysiol.* 91:157–72

77. Maier-Maercker, U. 1983. The role of peristomatal transpiration in the mechanism of stomatal movement. *Plant Cell Environ.* 6:369–80

78. Meidner, H. 1967. The effect of kinetin on stomatal opening and the rate of intake of carbon dioxide in mature primary leaves of barley. *J. Exp. Bot.* 18:556–61

79. Meidner, H., Edwards, M. 1975. Direct measurement of turgor pressure potentials of guard cells, I. *J. Exp. Bot.* 26:319–30

80. Mooney, H. A., Chu, C. 1983. Stomatal responses to humidity of coastal and interior populations of a California shrub. *Oecologia* 57:148–50

81. Nagarajah, S., Schulze, E.-D. 1983. Responses of *Vigna unguiculata* (L.) Walp. to atmospheric and soil drought. *Aust. J. Plant Physiol.* 10:385–94

82. Osonubi, O., Fasehun, F. E., Fasidi, I. O. 1985. The influence of soil drought and partial water logging on water relations of *Gmelina arborea* seedlings. *Oecologia* 66:126–31

83. Passioura, J. B. 1980. The transport of water from soil to shoot in wheat seedlings. *J. Exp. Bot.* 31:333–45

84. Passioura, J. B. 1984. Hydraulic resistance of plants. I. Constant or variable? *Aust. J. Plant Physiol.* II:335–40

85. Pierce, M., Raschke, K. 1980. Correlation between loss of turgor and accumulation of abscisic acid in detached leaves. *Planta* 148:174–82

86. Pierce, M., Raschke, K. 1981. Synthesis and metabolism of abscisic acid in detached leaves of *Phaseolus vulgaris* L. after loss and recovery of turgor. *Planta* 153:156–65

87. Potter, J. R., Boyer, J. S. 1973. Chloroplast response to low leaf water potentials. II. Role of osmotic potential. *Plant Physiol.* 51:993–97

88. Powles, S. B., Björkman, O. 1982. Photoinhibition of photosynthesis: Effect on chlorophyll fluorescence at 77 K in intact leaves and in choroplast membranes of *Nerium oleander*. *Planta* 156:97–107

89. Powles, S. B., Chapman, K. S. R., Osmond, C. B. 1980. Photoinhibition of intact attached leaves of C_4 plants: Dependence on CO_2 and O_2 partial pressure. *Aust. J. Plant Physiol.* 7:737–47

90. Radin, J. W., Ackerson, R. C. 1981. Water relations of cotton plants under nitrogen deficiency. III. Stomatal conductance, photosynthesis, and abscisic acid accumulation during drought. *Plant Physiol.* 67:115–19

91. Raschke, K. 1970. Stomatal response to pressure changes and interruptions in the water supply of detached leaves of *Zea mays*. *Plant Physiol.* 45:415–23

92. Raschke, K. 1970. Leaf hydraulic system: Rapid epidermal and stomatal responses to changes in water supply. *Science* 167:189–91

93. Raschke, K. 1975. Simultaneous requirement of carbon dioxide and abscisic acid for stomatal closing in *Xanthium strumarium* L. *Planta* 125:243–59

94. Raschke, K. 1979. Movements of stomata. *Encyclopedia of Plant Physiology*

(NS) Vol. 7. *Physiology of Movements,* ed. W. Haupt, M. E. Feinleib, pp. 383–441. Berlin/Heidelberg/New York: Springer-Verlag

95. Raschke, K. 1982. Involvement of abscisic acid in the regulation of gas exchange: Evidence and inconsistancies. In *Plant Growth Substances,* ed. P. F. Wareing, pp. 581–90. London: Academic

96. Raschke, K., Hedrich, R. 1985. Simultaneous and independent effects of abscisic acid on stomata and the photosynthetic apparatus in whole leaves. *Planta* 163: 105–18

97. Rawson, H. M., Begg, J. E., Woodward, R. G. 1977. The effects of atmospheric humidity on photosynthesis, transpiration and water use efficiency of leaves of several plant species. *Planta* 134:5–10

98. Resemann, A., Raschke, K. 1984. Midday depression in stomatal and photosynthesis activity of leaves of *Arbutus unedo* are caused by large water vapour pressure differences between leaf and air. *Plant Physiol.* (Suppl.)75:66

99. Roy, J., Mooney, H. A. 1982. Physiological adaptation and plasticity of water stress of coastal and desert populations of *Heliotropium curassavicum* L. *Oecologia* 52:370–75

100. Running, S. W. 1980. Field estimates of root and xylem resistances in *Pinus contorta* using root excision. *J. Exp. Bot.* 31:555–69

101. Russell, R. S., Barber, D. A. 1960. The relationships between salt uptake and the absorption of water by intact plants. *Ann. Rev. Plant Physiol.* 11:127–40

102. Sack, F. D., Paolillo, D. J. 1983. Stomatal pore and cuticle formation in *Funaria. Protoplasma* 116:1–13

103. Sánchez-Diáz, M. F., Mooney, H. A. 1979. Resistance to water transfer in desert shrubs native to Death Valley, California. *Physiol. Plant.* 46:139–46

104. Schönherr, J. 1982. Resistance of plant surfaces to water loss: Transport properties of cutin, suberin, and associated lipids. See Ref. 64, pp. 154–79

105. Schulze, E.-D., Bloom, A. J. 1984. Relationship between mineral nitrogen influx and transpiration in radish and tomato. *Plant Physiol.* 76:827–28

106. Schulze, E.-D., Ehleringer, J. R. 1984. The effect of nitrogen supply on growth and water-use efficiency of xylemtapping mistletoes. *Planta* 162:268–75

107. Schulze, E.-D., Hall, A. E. 1982. Stomatal responses, water loss and CO_2 assimilation rates of plants in contrasting environments. See Ref. 64, pp. 181–230

108. Schulze, E.-D., Koch, G., Percival, F., Mooney, H. A., Chu, C. 1985. The nitrogen balance of *Raphanus sativus* x *raphanistrum* plants. I. Daily nitrogen use under high nitrate supply. *Plant Cell Environ.* 8:713–20

109. Schulze, E.-D., Küppers, M. 1979. Short-term and long-term effects of plant water deficits on stomatal responses to humidity in *Corylus avellana* L. *Planta* 146:319–26

110. Schulze, E.-D., Lange, O. L., Buschbom, U., Kappen, L., Evenari, M. 1972. Stomatal responses to changes in humidity in plants growing in the desert. *Planta* 108:259–70

111. Schulze, E.-D., Lange, O. L., Evenari, M., Kappen, L., Buschbom, U. 1974. The role of air humidity and leaf temperature in controlling stomatal resistance of *Prunus armeniaca* L. under desert conditions. I. A simulation of the daily course of stomatal resistance. *Oecologia* 17:159–70

112. Schulze, E.-D., Lange, O. L., Evenari, M., Kappen, L., Buschbom, U. 1975. See Ref. 111. II. The significance of leaf water status and internal carbon dioxide concentration. *Oecologia* 18:219–33

113. Schulze, E.-D., Lange, O. L., Evenari, M., Kappen, L., Buschbom, U. 1975. See Ref. 111. III. The effect on water use efficiency. *Oecologia* 19:303–14

114. Schulze, E.-D., Turner, N. C., Glatzel, G. 1984. Carbon, water and nutrient relations of two mistletoes and their hosts: A hypothesis. *Plant Cell Environ.* 7:293–99

115. Scott, F. M. 1948. Internal suberization of plant tissues. *Science* 108:654–55

116. Shackel, K. A., Brinckmann, E. 1985. *In-situ* measurement of epidermal cell turgor, leaf water potential and gas exchange in *Tradescantia virginiana* L. *Plant Physiol.* 78:66–70

117. Sharkey, T. D. 1984. Transpirationinduced changes in the photosynthetic capacity of leaves. *Planta* 160:143–50

118. Sharp, R. E., Davies, W. J. 1985. Root growth and water uptake by maize plants in drying soil. *J. Exp. Bot.* 36:In press

119. Sheriff, D. W. 1977. Evaporation sites and distillation in leaves. *Ann. Bot.* 41:1081–82

120. Sheriff, D. W. 1984. Epidermal transpiration and stomatal responses to humidity: Some hypotheses explored. *Plant Cell Environ.* 7:669–77

121. Sheriff, D. W., McGruddy, E. 1976. Changes in leaf viscous flow resistance following excision, measured with a new porometer. *J. Exp. Bot.* 27:1371–75

122. Stålfelt, M. G. 1929. Die Abhängigkeit

der Spaltöffnungsreaktionen von der Wasserbilanz. *Planta* 8:287–340

123. Steudle, E., Tyerman, S. D., Wendler, S. 1983. Water relations of plant cells. In *Effects of Stress on Photosynthesis*, ed. R. Marcelle, H. Clijsters, M. van Poucke, pp. 95–109. The Hague/Boston/London: Dr. W. Junk

124. Taylor, S. E., Terry, N. 1984. Limiting factors in photosynthesis. V. Photochemical energy supply colimits photosynthesis at low values of intercellular CO_2 concentration. *Plant Physiol.* 75:82–86

125. Tenhunen, J. D., Lange, O. L., Braun, M. 1981. Midday stomatal closure in Mediterranean type sclerophylls under simulated habitat conditions in an environmental chamber. II. Effect of the complex of leaf temperature and air humidity on gas exchange of *Arbutus unedo* and *Quercus ilex*. *Oecologia* 50:5–11

126. Tenhunen, J. D., Lange, O. L., Gebel, J., Beyschlag, W., Weber, J. A. 1984. Changes in photosynthetic capacity, carboxylation efficiency, and CO_2-compensation point associated with midday stomatal closure and midday depression of net CO_2 exchange of leaves of *Quercus suber*. *Planta* 162:193–203

127. Tomos, A. D., Steudle, E., Zimmermann, U., Schulze, E.-D. 1981. Water relations of leaf epidermal cells of *Tradescantia virginiana*. *Plant Physiol.* 68:1135–43

128. Turner, N. C. 1974. Stomatal behaviour and water status of maize, sorghum and tobacco under field conditions. II. At low soil water potential. *Plant Physiol.* 53:360–65

129. Turner, N. C. 1974. Stomatal response to light and water under field conditions. *R. Soc. NZ Bull.* 12:423–32

130. Turner, N. C. 1975. Concurrent comparisons of stomatal behaviour, water status, and evaporation of maize in soil at high or low water potential. *Plant Physiol.* 55:932–36

131. Turner, N. C., Begg, J. E., Rawson, H. M., English, S. D., Hearn, A. B. 1978. Agronomic and physiological responses of soybean and sorghum crops to water deficits. III. Components of leaf water potential, leaf conductance, $^{14}CO_2$ photosynthesis and adaptation to water deficits. *Aust. J. Plant Physiol.* 5:179–94

132. Turner, N. C., Begg, J. E., Tonnet, M. L. 1978. Osmotic adjustment of sorghum and sunflower crops in response to water deficits and its influence on the water potential at which stomata close. *Aust. J. Plant Physiol.* 5:597–608

133. Turner, N. C., Schulze, E.-D., Gollan, T. 1984. The responses of stomata and leaf gas exchange to vapour pressure deficits and soil water content. I. Species comparisons at high soil water contents. *Oecologia* 63:338–42

134. Turner, N. C., Schulze, E.-D., Gollan, T. 1985. The responses of stomata and leaf gas exchange to vapour pressure deficits and soil water content. II. In the mesophyllic herbaceous species *Helianthus annuus*. *Oecologia* 65:348–55

135. Tyree, M. T., Dixon, M. A., Tyree, E. L., Johnson, R. 1984. Ultrasonic acoustic emissions for the sap-wood of cedar and hemlock. *Plant Physiol.* 75:988–92

136. Walter, A., Kreeb, K. 1970. Die Hydratation und Hydratur des Protoplasmas der Pflanzen und ihre ökophysiologische Bedeutung. *Protoplasmatologia* Vol. 2 C6. Wien/New York: Springer-Verlag. 306 pp.

137. Weber, J. A., Tenhunen, J. D., Lange, O. L. 1986. Effects of temperature at constant air dew point on leaf carboxylation efficiency and CO_2 compensation point of different leaf types. *Planta* 166:81–88

138. Wong, S. C., Cowan, I. R., Farquhar, G. D. 1979. Stomatal conductance correlates with photosynthetic capacity. *Nature* 282:424–26

139. Zeevart, J. A. D. 1983. Metabolism of abscisic acid and its regulation in *Xanthium* leaves during and after water stress. *Plant Physiol.* 71:477–81

140. Zeevart, J. A. D., Boyer, G. L. 1984. Accumulation and transport of abscisic acid and its metabolites in *Ricinus* and *Xanthium*. *Plant Physiol.* 74:934–39

141. Zeiger, E. 1983. The biology of stomatal guard cells. *Ann. Rev. Plant Physiol.* 34:441–75

142. Ziegler, H. 1986. Stomatal evolution. In *Stomatal Function*, ed. E. Zeiger, G. D. Farquhar, I. R. Cowan. Stanford: Stanford Univ. Press. In press

143. Zimmermann, M. H., Milburn, J. A. 1982. Transport and storage of water. See Ref. 64, pp. 135–52

144. Zimmermann, U., Hüsken, D., Schulze, E.-D. 1980. Direct turgor pressure measurements in individual leaf cells of *Tradescantia virginiana*. *Planta* 149:445–53

Ann. Rev. Plant Physiol. 1986. 37:275–308
Copyright © 1986 by Annual Reviews Inc. All rights reserved

STEROL BIOSYNTHESIS

P. Benveniste

Laboratoire de Biochimie Végétale, Institut de Botanique, 67083, Strasbourg, France

CONTENTS

INTRODUCTION

This chapter deals mainly with sterol biosynthesis in vascular plants, a subject treated in detail in recent reviews (53, 54, 95). An encyclopedic coverage of the topic is not possible in the available space, so only the most recent work reported in the literature, and especially that which in my opinion opens new frontiers in the field, will be presented. This review is organized as follows: after a general overview of plant sterol biosynthesis, I discuss more thoroughly some particular steps, and finally explain why rationally designed sterol biosynthesis inhibitors (SBIs) are powerful tools to reveal hidden aspects of sterol biosynthesis.

Some aspects of sterol biosynthesis in vascular plants cannot be separated from corresponding ones in animals and in lower eukaryotes; therefore, when necessary I take a comparative approach. This is based on excellent material

275

0066-4294/86/0601-0275$02.00

available on sterol biosynthesis in fungi (84, 100) and in animals (46). Sterols from marine organisms will not be considered here because this exciting area has been reviewed recently (37). Most prokaryotes (bacteria, cyanobacteria, etc) do not synthesize sterols but often produce instead sterol-like molecules such as the hopanoids (see 115).

Although vascular plants generally contain a complex mixture of sterols, a remarkable unity emerges from careful analysis of the constituents of this mixture. In most cases (95) three 4-desmethyl sterols—24-methyl cholesterol (*11*A or *11*B), sitosterol (*11*C), and stigmasterol (*11*G)—are encountered in large amounts. *11*C and *11*G, which often constitute more than 70% of total sterols, have the α-chirality at C-24[1]; 24-methyl cholesterol consists of a mixture of the α- and β-epimers (24-R and 24-S respectively) at C-24[1] (53, 54, 95). These stereochemical aspects will be discussed in detail because they are important biosynthetically and phylogenetically. In addition to the three major sterols, plants contain a great variety of minor 4-desmethyl-, 4α-methyl-, and 4,4-dimethyl sterols that can be precursors of *11*A, B, C, and G.

KEY STEPS OF STEROL BIOSYNTHESIS

The Lanosterol-Cycloartenol Bifurcation

PHYLOGENETIC ASPECTS   A biosynthetic scheme leading to sterols of vascular plants is tentatively proposed in Scheme 2, which represents the steps behind squalene 2,3-oxide. The steps before squalene 2,3-oxide have not been considered in this present article. Because many of the enzymes involved do not have absolute substrate specificity, we have proposed a matrix of alternative routes rather than a single linear route (84, 99, 100). It is generally accepted that the biosynthetic pathway from acetate to squalene 2,3-oxide is common in animals, vascular plants, and fungi. Downstream from squalene 2,3-oxide there is a dichotomy between nonphotosynthetic eukaryotic phyla (animals, fungi, etc) where squalene oxide is cyclized to lanosterol (*4L*) and photosynthetic eukaryotic phyla (algae, bryophytes, tracheophytes, etc) where it is cyclized to cycloartenol (*1L*) (41, 53, 95). At first glance there seem to be exceptions to this rule. The cycloartanol pathway exists in the parasitic plants *Cuscuta europaea* and *Orobanche lutea* (117) and in *Astasia longa,* a white nonphotosynthetic naturally occurring mutant of *Euglena gracilis* (116), but these organisms, although nonphotosynthetic, evidently belong to a photosynthetic phylum, and therefore the presence of the cycloartenol pathway is not surprising.

Recent studies have shown that the lanosterol pathway is present in the fungi *Achlya radiosa* (A. Kerkenaar et al, unpublished observations) and *Sap-*

---

[1]α- and β-Chirality at C-24 are defined in Scheme 1. See (95) for a full discussion. The nomenclature R and S has been used in some special cases.

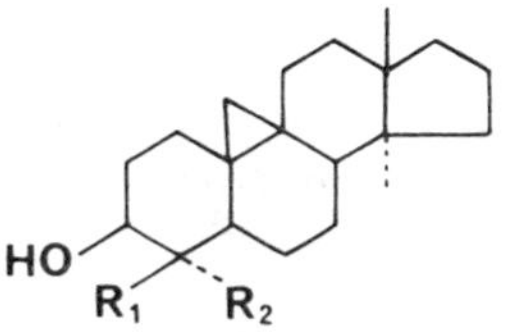

$R_1 = R_2 = CH_3$ 1
$R_1 = H$ $R_2 = CH_3$ 2
$R_1 = R_2 = H$ 3

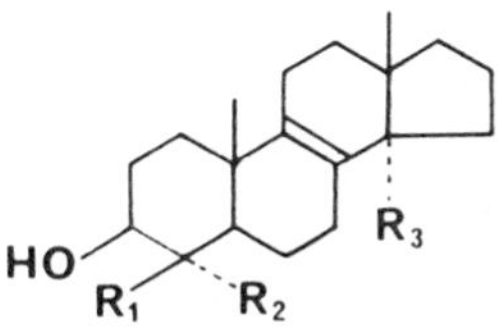

$R_1 = R_2 = R_3 = CH_3$ 4
$R_1 = H$ $R_2 = R_3 = CH_3$ 5
$R_1 = R_2 = H$ $R_3 = CH_3$ 6
$R_1 = R_3 = H$ $R_2 = CH_3$ 7
$R_1 = R_2 = R_3 = H$ 8

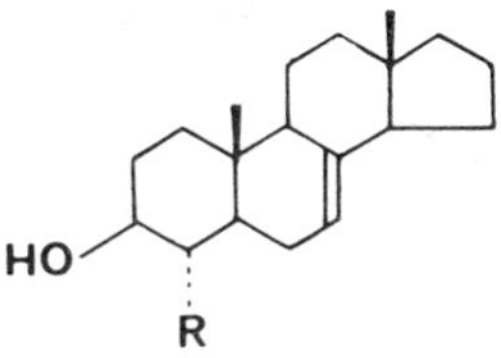

$R = CH_3$ 9
$R = H$ 10

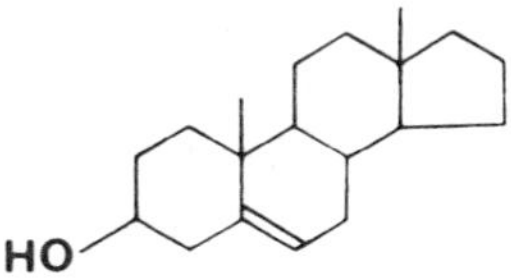

11

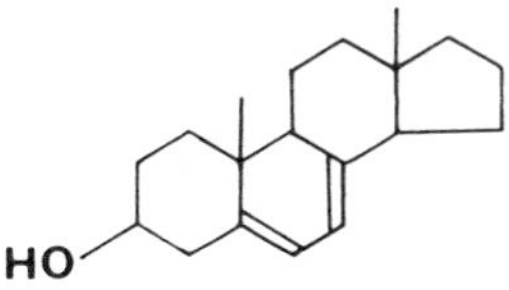

12

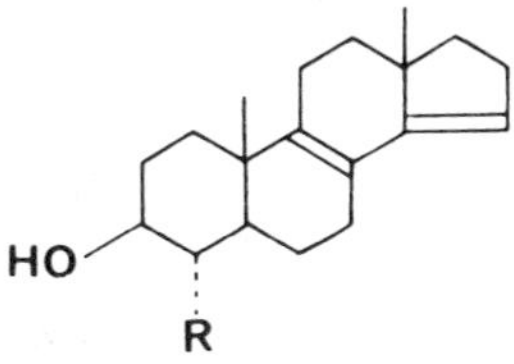

$R = CH_3$ 13
$R = H$ 14

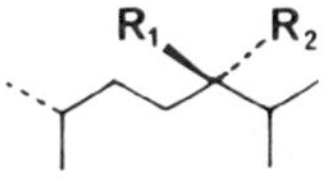

$R_1 = CH_3 , R_2 = H$ A
$R_1 = H , \quad R_2 = CH_3$ B
$R_1 = C_2H_5 , R_2 = H$ C
$R_1 = H , \quad R_2 = C_2H_5$ D
$R_1 = R_2 = H$ E

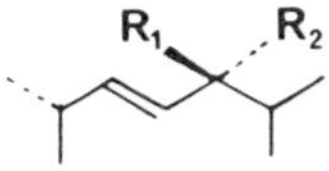

$R_1 = H, R_2 = CH_3$ F
$R_1 = C_2H_5 , R_2 = H$ G
$R_1 = H , R_2 = C_2H_5$ H

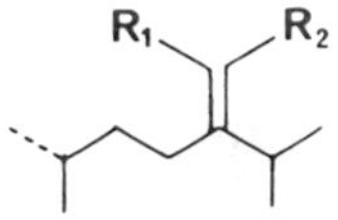

$R_1 = R_2 = H$ I
$R_1 = CH_3 , R_2 = H$ J
$R_1 = H , \quad R_2 = CH_3$ K

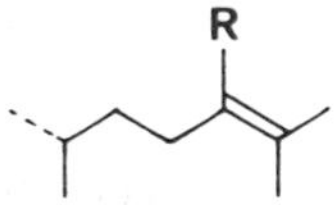

$R = H$ L
$R = CH_3$ M
$R = C_2H_5$ N

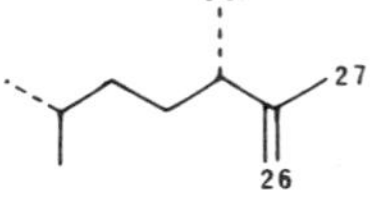

$R = CH_3$ Q
$R = C_2H_5$ P

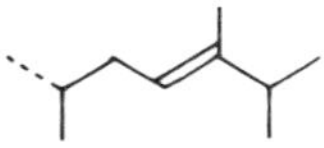

R

Scheme 1   *(top)* The various sterol nuclei and *(bottom)* side chains present in vascular plants. As defined in (95), side chain A has the configuration (24-R); side chain B has the configuration (24-S). Reproduced with permission from T. W. Goodwin (53).

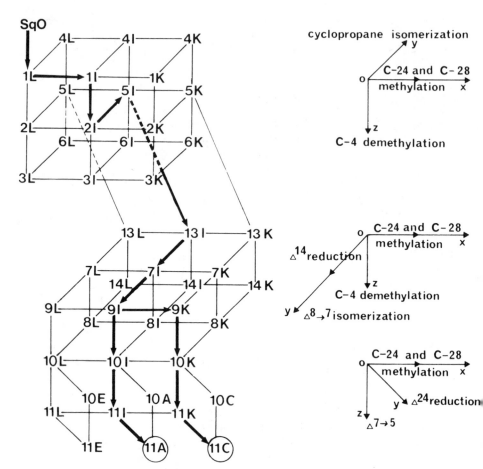

Scheme 2 Biosynthesis of 24-methyl-(*11*A) and 24-ethyl (*11*C) cholesterol in vascular plants. The structure of sterols is indicated by a number and a letter as defined in Scheme 1. The direction of the biosynthetic flow is indicated by the coordinate axes. The reactions are generally irreversible. The preferred biosynthetic pathway operating in most vascular plants is underlined by heavy arrows. SqO: squalene-2(3)-oxide.

rolegnia ferax (13). This at first seems reasonable because the lanosterol pathway has been found in fungi, but further study is needed, because it has been suggested that the oomycetes, of which *Achlya* and *Saprolegnia* are members, would derive from algae. Investigations of the ability of one other oomycete *(Phytophthora cactorum)*, previously believed to convert 9β,19-cyclosteroids to various sterols not possessing a cyclopropane ring (140), to cyclize |2-¹⁴C|squalene oxide or to metabolize |2-³H|cycloartenol have shown that this fungus, according to classical views, was unable to utilize both substrates. All these results support earlier affirmations that fungi do not produce 9β,19-cyclopropyl sterols and do not contain any vestigial cyclosteroid isomerase (48, 94).

Until now one of the most intriguing results obtained in the same context is the discovery that the cycloartenol pathway is without any doubt present in *Acanthamoeba polyphaga* (107), an organism that is classified in the protozoa. Such a result shows that *Acanthamoeba* belongs to a different phylum from that of *Dictyostelium* which synthesizes its sterols from lanosterol. All these results stress the need to find other biochemical criteria to clear up protist phylogeny.

MECHANISM OF THE CYCLIZATION OF SQUALENE-2(3)-OXIDE   This enzymatic reaction (Scheme 3), which involves the chair-boat-chair-boat folding of the molecule (41), has been discussed in previous articles and reviews (31, 36, 41, 137, 138). Although this is a fascinating enzyme, not much work has been done on it. Because of subtle changes in its catalytic pathway, this enzyme catalyzes the cyclization of (3-S)-2,3-oxidosqualene into lanosterol or cycloartenol. The mechanistic differences underlying the catalytic activity of this enzyme constitute a challenge for everybody interested in the fundamental aspects of sterol biosynthesis.

Enzymatic cyclization of the all-*trans* 2,3 oxidosqualene is believed to be triggered by a general acid-catalyzed epoxide-ring opening assisted by the neighboring π-bond (138). The concertedness of the ensuing overall annulation and backbone rearrangements is a matter of debate (36). However, for entropic reasons and from experimental evidence, the reaction is more likely to proceed through a series of discrete conformationally rigid carbocationic intermediates (137) that result in a tetracyclic protosteryl intermediate (31) (a in Scheme 3). After a series of hydride and methyl transpositions, the following step in the pathway leading to lanosterol (4L) or cycloartenol (1L) would involve a transient carbocationic species at C-9 (b in Scheme 3; 53, 95). The way by which this carbocation is stabilized by the enzyme (covalent bond, ion pair, or simply interaction with a negatively charged surface of the enzymatic active site) is still a matter of debate, but is probably of major importance for the last step of the reaction. This consists of the removal of a 19H to give 1L or of a 8β H to give 4L. An alternative route is possible to reach 4L that would involve a direct 9β H elimination from a in Scheme 3. Whereas the 9β H abstraction leads to the thermodynamic product and can be performed chemically (31), the 19H abstraction must obey a very strict kinetic control because the cyclopropane ring formation is not favored thermodynamically.

It is possible that the C-9 carbocation stabilization by the active site of the enzyme may play an important role in this process, possibly in destabilizing the structure of the carbocationic intermediate in a way that the 4β- and the 10-methyl groups become closer each other. The result is an unfavorable

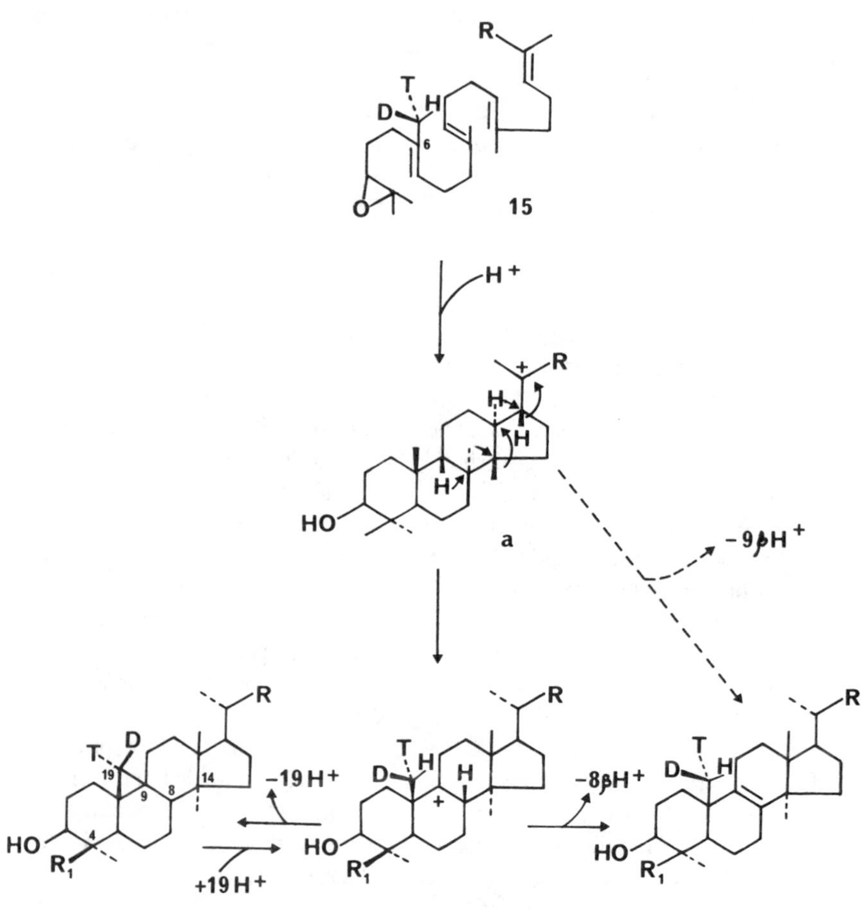

1L $R_1 = CH_3$ | b $R_1 = CH_3$ | 4L $R_1 = CH_3$

2I $R_1 = H$ | e $R_1 = H$ | 5I $R_1 = H$

Scheme 3 Biosynthesis of cycloartenol (*1*L) in photosynthetic eukaryotic phyla: *15*→ *a*→ *b* →*1*L. Biosynthesis of lanosterol (*4*L) in nonphotosynthetic eukaryotic phyla: *15* →*a*→ *b*→ *4*L or *15*→ *a*→ *4*L. Cyclopropane ring opening catalyzed by the cycloeucalenol(*2*I)-obtusifoliol (*5*I) isomerase: *2*I →*e* →*5*I.

4β-methyl-10-methyl interaction at the active site of the enzyme, which can be relaxed by the cyclopropane formation through the 19H abstraction (F. Schuber, personal communication). To check the likelihood of these considerations, it would be necessary to trap the carbocationic species at C-8 or C-9. The way to reach this goal will be discussed in the second part of this chapter, devoted to the use of sterol biosynthesis inhibitors.

The mechanism of the formation of the cyclopropane ring has been studied with 2,3-oxidosqualene carrying a chiral methyl group at C-6. The cycloartenol (1L) formed from this substrate by a microsomal preparation from *Ochromonas malhamensis* was examined by tritium NMR, which showed that one predominant species contained tritium in the pro-19R position coupled to deuterium in the pro-19S position (2). Therefore, ring closure takes place with retention of configuration (Scheme 3). The same conclusion resulted from experiments with the cyclase enzyme from *Zea mays* seedlings (16, 107a). Moreover, those results suggest that the subsite of the cyclase involved in the C-19 hydrogen abstraction is situated above the C-cycle of the C-9 carbocationic intermediate (16, 107a).

The Alkylation Reactions

The alkylation reactions lead to a second important specific character of tracheophyte sterols, i.e. the presence of methyl or ethyl groups at C-24 (Scheme 1; 53, 95). The alkylation reaction occurs at the level of cycloartenol in tracheophytes; therefore, it is logical to consider it just after *1L* formation. Scheme 2 shows all possibilities of methylation occurring in vascular plants. The two C-24 and C-28-methylations are represented by horizontal lines. They catalyze the following reactions: Δ^{24}-sterol + S-adenosylmethionine (AdoMet) → 24-methylene sterol (or a 24-methyl-25-methylene sterol) + S-adenosylhomocysteine (AdoHCy) and 24-methylene sterol + AdoMet → 24-ethylidene sterol (or a 24-ethyl-25-methylene sterol) + AdoHCy (23, 95). The occurrence of these reactions in nature is justified by the identification of sterols possessing the side chains I and K with various nuclei (1 to 14).

Whereas 24-ethylidene sterols with a $\Delta^{Z-24(28)}$ are widely found [e.g. isofucosterol (*11*K) or 24-ethylidene lophenol (*9*K)] and appear to be intermediates of (24R)-24-ethyl sterol biosynthesis (52, 72, 78), 24-ethylidene sterols with a $\Delta^{E-24(28)}$ [e.g. fucosterol (*11*J)] are rarely found and do not seem to be involved in 24-ethyl sterol biosynthesis (72). They seem to be end products in some organisms (e.g. Phaeophyceae) (53, 95). Although 24-methyl- and 24-ethyl-25(27)-methylene sterols are not as widely found in nature as 24-methylene- and 24-ethylidene sterols, they may also play an important role in the biosynthesis of 24-methyl and 24-ethyl sterols as shown below. For more clarity they have not been considered in Scheme 2. The identification of 24-methylene and 24-ethylidene sterols with most of the possible sterol nuclei (1 to 14): cyclofontumienol (2K) in *Fontumia latifolia* (90), 4α,14α-dimethyl-5α-stigmasta-8,Z-24(28)-dien-3β-ol (5K) in suspension cultures of *Rubus fruticosus* cells grown in the presence of Fenarimol (124), 4α-methyl-5α-stigmasta-8,14,Z-24(28)-trien-3β-ol (*13*K) in *Vernonia anthelmintica* (44), 4,4,14α-trimethyl-5α-stigmasta-8,24(28)-dien-3β-ol (4K) in *Mucor rouxii* (119) etc suggest that the two alkylation reactions are not specific of a precise sterol nucleus. Recent experiments have been conducted in our laboratory to ascertain this point (112): microsomes from suspension cultures of *Rubus fruticosus* cells have been incubated in the presence of |methyl-^{14}C AdoMet| and various $\Delta^{24(25)}$- and $\Delta^{24(28)}$ sterols. When the AdoMet-Δ^{24}-sterol-C-24-methyltransfer-ase was considered, it was shown that cycloartenol (*1L*), the best substrate (control = 100%) of the series, was much better than lanosterol (*4L*) (9% of the control) in agreement with previous studies (42, 143). By contrast, 31-nor cycloartenol (*2L*) (35%) was as efficiently methylated as 31-nor lanosterol (*5L*) (34%), suggesting that the differential capacities of 9β,19-cyclopropyl- and Δ^8-sterols to be methylated are fully expressed when the 4α-methyl is present such as in *1L* and *4L*. When this methyl group is absent such as in *2L* and *5L*, selectivity is no longer observed. Desmosterol (*11L*) is still methylated (25%), although less efficiently than cycloartenol (*1L*). The acetate of *1L* is not at all methylated (0%), showing that a free 3β-OH group is absolutely necessary for the enzymatic methylation. When $\Delta^{24(28)}$ sterols are used as substrates of the enzyme catalyzing the second methylation, 24-methylene lophenol (*9I*) (100%) was shown to be the best substrate of the series; however, 4α-methyl fecosterol (*7I*) (80%) was almost as efficient as *9I*. Episterol (*10I*) (58%), which lacks the 4α-methyl group, gave only 50% of the maximal activity. Fecosterol (*8I*) was still less efficient, while 24-methylene cholesterol (*11I*) (5%) gave very little activity. 24-Methylene cycloartanol (*1I*), cycloeucalenol (*2I*), and obtusifoliol (*5I*) were not significantly methylated. Therefore, the molecular features in-volved in the capability of $\Delta^{24(28)}$ sterols to be substrates of the $\Delta^{24(28)}$-sterol methyltransferase are as follows: (*a*) the 4α-methyl group and the Δ^7 (or Δ^8) double bond appear to be important; (*b*) the presence of a methyl group at C-14 (and possibly also of the cyclopropane ring) abolishes the substrate capacity; therefore, whereas $\Delta^{24(25)}$- 9β,19-cyclopropyl sterols (e.g. *1L*) are the best substrates for the Δ^{24}-sterols methyltransferase, $\Delta^{24(28)}$-9β,19-cyclopropyl sterols (e.g. *2I*) are not substrates for $\Delta^{24(28)}$-sterol methyltransferase. In other words, the 9β,19-cyclopropane ring must be opened and the 14α-methyl group removed before a sterol could be methylated by the $\Delta^{24(28)}$-sterol methylase (112), as shown in the biosynthetic pathway proposed previously (53, 95) and outlined in Scheme 2.

    The mechanism of the first methylation has been thoroughly studied in *Trebouxia sp* (86), and the major results have been summarized in Scheme 4:1. A single enzyme converts cycloartenol (*1L*) into 24-methylene cycloartanol (*1I*) and cyclolaudenol (*1Q*). 2. At the active site of this enzyme the reaction is initiated by addition of the methyl group from AdoMet to C-24 of the substrate; the methylation step involves the Si-face of the double bond and proceeds by an inversion mechanism at the methyl center. 3. Following formation of the carbocation, a hydride ion migrates from C-24 to C-25 (*c→d*) in a reversible reaction. The migration occurs in such a way that the Z-methyl group of the starting material (*1*L) is transformed specifically into the proR-methyl group of the product (*1*I). In other words, the methyl insertion and the hydride migration are occurring on opposite faces of the original double bond. 4. Starting from *c*, a proton loss from what was originally Z-methyl group leads to *1*Q; this reaction displays an isotope effect $K_H/K_D = 3.5$. 5. Proton loss from *d* to 1I is also observed. Since loss from an (R)-methyl group generates a compound with a Z-double bond, this last step is obviously mediated by a basic subsite of the enzyme, the protonated form of which is located specifically on the Re face of the double bond of the final product. Thus it is now established that the alkylation step and the proton loss are occurring on opposite faces of the nodal plane of the substrate double bond. Similar results concerning proton loss and the stereochemical course of the hydride migration had been observed previously for ergosterol biosynthesis in both yeast and *Claviceps paspali* (5).

Such a study brings new and important data concerning the mechanism of the sterol side chain biosynthesis in one alga. It is of major importance to know whether the pathway of Scheme 4 would apply to the case of vascular plants.

High-resolution NMR studies (97, 118) have shown that 24-ethyl sterols from most higher plants have the configuration 24α except for 24-ethyl sterols from *Clerodendrum* spp. (103), *Kalenchoë diagremontiana* (97), and the seeds of *Cucurbita pepo* (24). The situation is considerably different when one considers 24-methyl sterols where NMR reveals the occurrence of a mixture of 24α- and 24β-isomers (83, 118). In *Zea mays* 24β-isomers even predominate (147). The pathway for the synthesis of 24α-alkyl sterols in higher plants is indicated in pathways 1 and 3 (Scheme 5). Experiments with $|2\text{-}^{14}C, (4R)\text{-}4\text{-}^3H_1|$ mevalonic acid in *Nicotiana tabacum, Dioscorea tokoro* (135), *Larix decidua* (53), *Medicago sativa,* and *Spinacea oleracea* (6) demonstrated that the hydrogen at C-24 in cycloartenol is not present in stigmasterol (*11*G) or in spinasterol (*10*G). However, it is retained at C-25 in isofucosterol (*11*K), which is synthesized by intact pine (106) and barley (77) seedlings. Furthermore, barley embryos cultured in the presence of $|C\text{-}^2H_3|$ methionine synthesized methyl and ethyl sterols that contained only two and four deuteriums, respectively (77), suggesting the involvement of 24-methylene- and 24-ethylidene intermediates; these latter (e.g. *9*I and *9*K) have been isolated in the same material (77) and other sources (61).

The stereochemistry of hydrogen migration from C-24 to C-25 during the alkylation in *Pinus pinea* is such that the C-25 pro-E-methyl group of the Δ^{24}-sterol precursor (possibly *1*L) becomes the pro-R isopropyl methyl group of isofucosterol (*11*K) (98). This is the opposite of that found in fungi and in *Trebouxia spp.* (see Scheme 4) and may reflect that the enzymatic mechanism that leads to 24α-alkyl sterols in vascular plants is different from that leading to 24β-alkyl sterols in cryptogams. The proposed $\Delta^{24(25)}$ intermediates (*11*M and

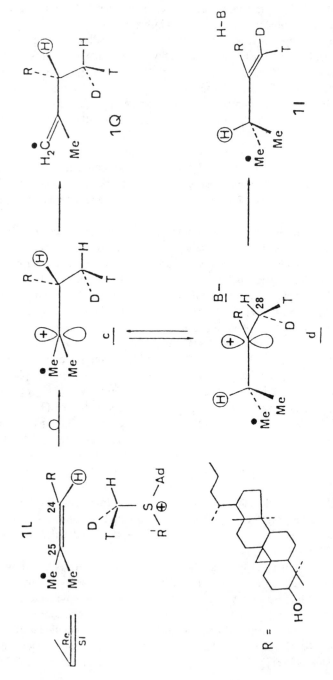

Scheme 4   Steric course of the C-24 alkylation step in *Trebouxia spp.* Reproduced with permission from D. Arigoni (5, 86).

11N) are based on (a) the existence of $\Delta^{24(25)}$-24-methyl sterols (such as 11M) in *Withania somnifera* (80) and seeds of some Solanaceae (61) and the presence of $\Delta^{24(25)}$-24-ethyl sterols (e.g. 11N and 10N) in *Zea mays* (87) and *Helianthus annuus* (53, 65) respectively; (b) the elimination of the hydrogen at C-24 as discussed above (6, 135); (c) the observation that 31-nor cyclobranol (2M) accumulate dramatically in *Zea mays* (15) and in *Triticum sativum* (M. F. Costet et al, unpublished results) treated with Fenpropimorph, a systemic fungicide and an inhibitor of the cycloeucalenol-obtusifoliol isomerase (111) (Scheme 5).

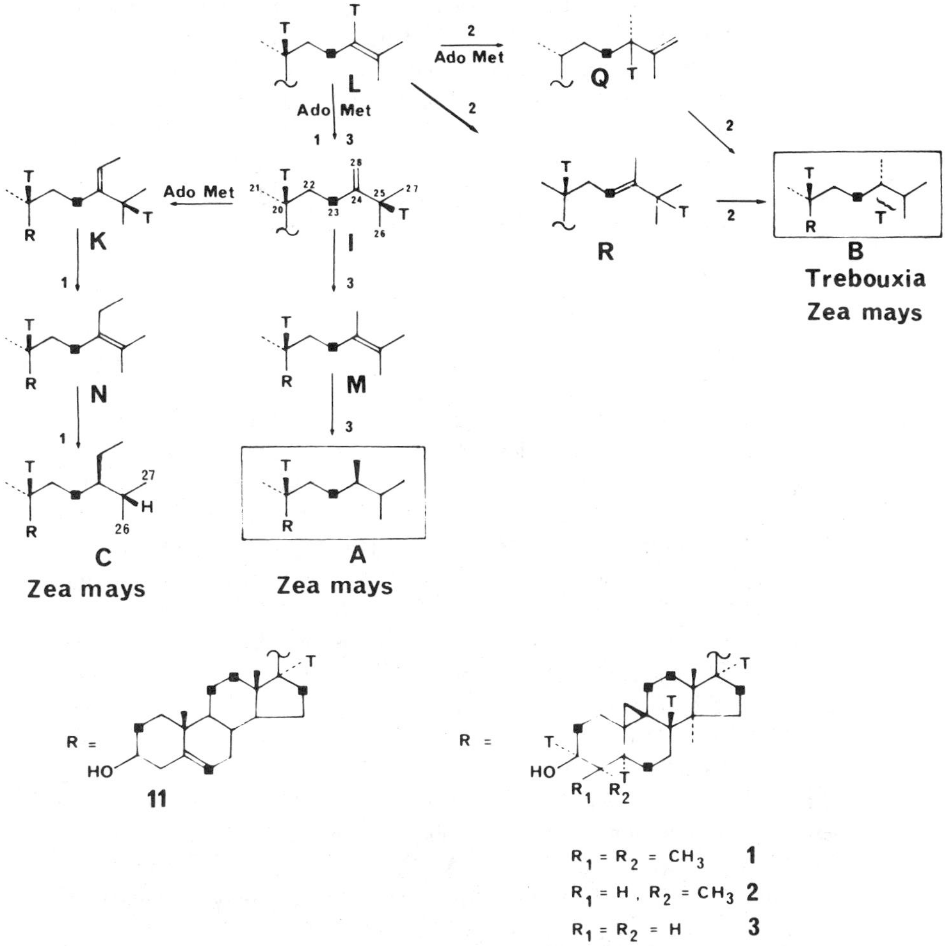

Scheme 5 Biosynthetic routes leading to 24α-ethyl-(route 1), 24β-methyl-(route 2) and 24α-methyl-(route 3) sterols in plants treated with |(4-R)-4-³H₁,5-¹⁴C| mevalonic acid. ■¹⁴C-atom originating from C-5 of mevalonic acid. Adapted from Zakelj & Goad (147).

If the route for the synthesis of 24β-methyl sterols in higher plants is the same as in green algae (e.g. *Trebouxia* spp) (53, 95), then in experiments with |2-^{14}C, (4R)-4-^{3}H$_1$| mevalonic acid, the ^{14}C/^{3}H ratio in the mixture of epimeric 24-methyl sterols should fall between 5:3 (100% β-epimers, with ^{3}H retained at C-24) and 5:2 (100% α-epimers with no ^{3}H retained at C-24). Recent experiments with *Zea mays* seedlings have confirmed this (147). The observed ^{14}C/^{3}H ratio of the mixture of 24-methyl sterols was 5:2.82, which fitted well with the ratio of the known relative amounts of the α- and β-epimers as determined by NMR (147). Furthermore, if the 24β-methyl sterol synthesis follows that in *Trebouxia* spp, cyclolaudenol (*1*Q) should be formed together with 24-methylene cycloartanol (*1*I) in higher plants. The presence of *1*Q has been demonstrated recently along with *1*I in *Zea mays* (147), giving additional support to the operation of pathway 2 in this material and possibly in other higher plants. Recently in etiolated maize coleoptiles two Δ^{23} sterols: ergosta-5,23-dien-3β-ol (*11*R) and 5α-ergosta-7,23-dien-3β-ol (*10*R) have been found along with their precursors (*1*R, *2*R, *5*R, *9*R) (64, 121, 122). Moreover, it has been demonstrated that cyclosadol (*1*R) was formed directly by C-24 methylation of cycloartenol and did not originate from isomerization of 24-methylene cycloartanol (Scheme 5) (88). It has been suggested that *11*R after hydrogenation could give the 24β-methyl sterols contained in etiolated maize coleoptiles (147). Therefore, this cyclosadol (*1*R) pathway would add to the already identified cyclolaudenol (*1*Q) and 24-methylene cycloartanol (*1*I) pathways.

The use of Fenpropimorph, a fungicide that inhibits the cycloeucalenol-obtusifoliol isomerase (COI), leads to a dramatic accumulation of 9β,19-cyclopropyl sterols such as 24-methyl pollinastanol (*3*A and *3*B) in various plants (12, 15, 125). By contrast to the situation in control plants, where 24-ethyl sterols predominate, 24-methyl sterols were more than 98% of total cyclopropyl sterols. In addition, 24-methyl cyclopropyl sterols were a mixture of 24α and 24β-24-methyl epimers and are similar in that respect to the 24-methyl cholesterol (*11*A and *11*B) of control plants. As shown above, the presence of *11*A and *11*B has been explained by the operation of route 3, which leads to the 24α-24-methyl epimer (*11*A) via Δ$^{24(28)}$- and Δ$^{24(25)}$-sterols, and route 2, which leads to the 24β-24-methyl sterol via hydrogenation of a Δ^{23}- or a Δ^{24}-sterol intermediate (Scheme 5).

Such intermediates have been sought in excised *Zea mays* axes grown aseptically in the presence of Fenpropimorph and either |5-^{14}C| mevalonic acid or |CH$_3$-^{2}H$_3$|-L-methionine. Whereas Δ$^{24(28)}$- (e.g. *1*I) and Δ$^{24(25)}$- (e.g. 2M) cyclopropyl sterols were found in relatively large amounts, only low radioactivity was associated to Δ^{25}-sterols (*1*Q and 2Q). GC/MS Analysis of the sterols from axes grown in the presence of |CH$_3$-^{2}H$_3$|-L-methionine showed that Δ$^{24(28)}$-cyclopropyl sterols contained only two ^{2}H atoms at C-28 as expected and that the 24-methyl pollinastanol (*3*A and *3*B) contained species with two ^{2}H atoms and no species with three ^{2}H atoms. These results gave evidence that both (24α)- and (24β)-epimers originate from a common $\Delta^{24(28)}$ precursor. After incubation of the axes with $|5$-^{14}C, (4R)-4-^{3}H$_1|$ mevalonic acid, the 24-methyl pollinastanol had a ^{3}H $:$ ^{14}C atomic ratio of 4 : 6, which is consistent with the intermediacy of a $\Delta^{24(25)}$-sterol. All these data are in accordance with a pathway where $\Delta^{24(28)}$-cyclopropyl sterols are isomerized to give $\Delta^{24(25)}$-cyclopropyl sterols, which in turn would give pollinastanol (*3*A and *3*B). Stereochemical aspects of the C-24 alkylation in higher plants have been studied recently using $|1,2$-^{13}C$_2|$ acetate (129–129a). During sitosterol (*11*c) biosynthesis in *Physalis peruviana*, it has been shown that the C-26 (pro-R methyl group at C-25) originated from C-2 of mevalonic acid and C-27 (pro-S-methyl group) from C-6 (129). In the same plant, the ^{13}C labeling pattern of the two methyl groups at C-25 of dihydrobrassicasterol (*11*B) was shown to differ from that of campesterol (*11*A), giving evidence that the configuration at C-25 in *11*A and *11*B is opposite (129a).

Opening of the Cyclopropane Ring

The cycloeucalenol (*2*I)-obtusifoliol (*5*I) isomerase (COI) which opens the 9β,19-cyclopropane ring of *2*I to give *5*I has no significant action on cycloartenol (*1*L) and 24-methylene cycloartanol (*1*I) (25, 57). This explains the absence of lanosterol (*4*L) and 24-methylene lanosterol (*4*I) from most plant tissues examined except the seeds of several solanaceae (62, 66), and the latex of *Euphorbia helioscopia* (104) where the conversion of *1*L to *4*L has been demonstrated (105). Therefore, in most plants the first intermediate without a cyclopropane ring is obtusifoliol (*5*I), so alkylation at C-24 and removal of a methyl group from C-4 must occur prior to the opening of the ring as shown in Scheme 2 (preferred pathway). The COI has been detected in cell-free preparations from tissue cultures of *Rubus fruticosus*, *Nicotiana tabacum*, and in *Zea mays*, *Phaseolus aureus*, and *Pisum sativum* seedlings (25, 57; M. Taton, unpublished results). It was not detectable in rat liver (50, 57) and yeast homogenates (3, 100), suggesting strongly that, as for the 2(3)-oxidosqualene-cycloartenol cyclase, the COI would be a biochemical marker of tracheophytes and possibly of photosynthetic eukaryotic phyla.

When the reaction catalyzed by the COI is carried out in ^{2}H$_2$O with cycloeucalenol (*2*I) as substrate, the incorporation of one deuterium into the C-19-methyl group of obtusifoliol (*5*I) is observed (108). This result is consistent with a general acid catalyzed cyclopropane ring opening that would lead to a carbocationic intermediate *e* and that would be followed by a regiospecific 8Hβ proton elimination (Scheme 3). The failure of *1*L and *1*I to be effective substrates has been ascribed to the presence of a 4β-methyl group that hinders the binding of *1*L or *1*I on the active site of the enzyme (107a). Experiments with (19S)-$|19$D, 19T$|$ 31-norcycloartenol (*2*L) and the maize enzyme that opens the cyclopropane ring demonstrated that ring opening occurs with retention of configuration (16). This result suggests that as for the 2(3)-oxidosqualene-cycloartenol cyclase (see above) the subsite of the enzyme that is responsible for the general acid catalyzed cyclopropane ring opening, would be located above the C cycle of the substrate (16, 107a).

Later Stages

REMOVAL OF METHYL GROUPS   The ubiquitous presence of cycloeucalenol in higher plants and the previous discussion on the opening of the cyclopropane ring give evidence that the first methyl group removed in the conversion of cycloartenol into alkylated sterols is one of the two located at C-4. This is different from the reactions in animals where the C-14 methyl group is first removed in the biosynthesis of cholesterol (46).

Experiments with *Polypodium vulgare* showed that the 4α-methyl group of cycloartenol arose specifically from C-2 of mevalonic acid and that it is this group that is removed in its conversion into 31-nor cycloartenol (2L). From this work and other experiments (47, 74), it is possible that the removal of the first C-4 methyl group in plants involves the same mechanism as that functioning in animals. The 14α-methyl group is usually lost next, probably at the obtusifoliol (5I) level, and the next intermediate is the $\Delta^{8,14}$-diene (13I). This compound as well as several other $\Delta^{8,14}$-sterols have been shown to accumulate dramatically in suspension cultures of *Rubus fruticosus* cells grown in the presence of 15-azasterol which inhibits the sterol Δ^{14}-reductase (127; see below), and this result suggests strongly that as in animals and yeast, $\Delta^{8,14}$ sterols are the products of the 14α-methyl demethylase reaction. The mechanism of this reaction has been studied mostly in yeast cell-free systems and has been discussed thoroughly elsewhere (100). The 14α-methyl demethylase from yeast has been purified extensively and has been shown to be a cytochrome P-450-containing enzyme having several monooxygenase activities involved in the oxidation of the 14α-methyl group into formic acid (146).

The isolation and the study of this enzyme in higher plants remains a very stimulating challenge since it may have different properties than the rat liver and yeast enzymes. As in rat liver (49), it has been shown that the removal of the 14α-methyl group in *Calendula officinale* involves the formation and reduction of a Δ^{14} double bond during which the 15α-hydrogen is exchanged and the 15β-hydrogen epimerized (22). The involvement of cytochrome P-450 is also highly probable in plants since the 14α-methyl demethylase reaction is strongly inhibited in suspension cultures of *Rubus fruticosus* by Fenarimol (124), a fungicide capable of binding with cytochrome P-450 in yeast and other fungi (59).

The removal of the second 4α-methyl group has been studied in *Ochromonas* by following the conversion of |2,24-³H₃| obtusifoliol (5I) into poriferasterol

(*11*H). It was revealed that the hydrogen in the 4β (axial) position is inverted to the 4α(equatorial) position during the demethylation (73). Other studies concerning the two C-4 demethylations do not seem to exist, although these steps have great interest in higher plants. First of all, the order of demethylation is different in higher plants (4,14,4) from that (14,4,4) in animals and fungi. As the two 4-demethylations operate in very different substrates—24-methylene cycloartanol (*1*I) and 24-methylene lophenol (*9*I) for the first and second demethylases respectively—it is possible that these operations are catalyzed by two different enzymes whereas in animals the two 4-demethylations, operating on two similar substrates—4,4-dimethyl-zymosterol and 4α-methyl zymosterol (*7*L)—could be catalyzed by a single enzyme.

CONVERSION OF Δ^8 STEROLS INTO Δ^5 STEROLS   Since the second demethylation at C-4 should probably operate on Δ^7-sterols such as *9*I or *9*K, we expect a reaction that occurs before this demethylation, namely the $\Delta^8 \rightarrow \Delta^7$-sterol isomerization. This step involves the stereospecific loss of the 7β-hydrogen in *Ochromonas malhamensis* (131), *Camellia sinensis* and *Clerodendron campbellii* (51), *Larix decidua* (50), *Calendula officinalis* (130), *Taxus baccata,* and *Polypodium vulgare* (53). This is the same stereospecificity as found in rats (49) but opposite to that in yeast (1, 21) and *Aspergillus niger* (14). The $\Delta^8 \rightarrow \Delta^7$-sterol isomerase in suspension cultures of *Rubus fruticosus* cells is inhibited by AY 9944, a hypocholesterolemic drug, resulting in a dramatic accumulation of Δ^8-sterols (especially *7*I, *7*K, *8*C and *8*K) (123). 24-Ethyl-Δ^8-sterols (such as *7*K, *8*K, and *8*C) have been encountered very rarely in tracheophytes. Therefore, the use of AY 9944 has shown that Δ^8-sterols are also intermediates of sterol biosynthesis in higher plants. Recently Δ^8-sterols have been isolated in seed oil from various cucurbitaceae plants (63).

The conversion of Δ^7-sterols [e.g. Δ^7-avenasterol (*10*K) or episterol (*10*I)] into Δ^5-sterols in higher plants has been demonstrated in vivo (106); this step probably involves the intermediacy of $\Delta^{5,7}$-sterols since these latter have been detected in a few plants (30). Moreover, an enzyme converting Δ^7-cholestenol (*10*E) into $\Delta^{5,7}$-cholestenol (*12*E) in rat liver has been described recently (69, 114). Formation of the Δ^5 bond involves stereochemical removal of the 5α- and 6α-hydrogens as demonstrated in *Ochromonas malhamensis* (132), *Larix decidua* (50), and *Calendula officinalis* (130). This is the same stereospecificity as observed in the rat (49). This enzymatic activity would not be present in all plants since in some cases, Δ^7-sterols (e.g. spinasterol (*9*G) or *9*P) predominate and Δ^5-sterols are absent. This is especially true in *Spinacia oleracea* (95), *Phytolacca esculenta* (145), *Lophocereus schottii* (19), *Saponaria officinalis* (58), and several plants mentioned in (95). This characteristic seems to be associated with some families of plants (cucurbitaceae, chenopodiaceae, cactaceae, etc) and could be related to some specific functions, possibly at the membrane level.

Very little is known of the pathway involved in the reduction of the Δ^7 bond in plants, but as discussed previously (142), it should involve *trans* addition of hydrogens at 7α and 8β, with the former arising from NADPH and the latter from the medium. The occurrence in several plants of both Δ^7- and Δ^5-avenasterol (9K and 11K respectively; P. Schmitt, unpublished experiments) suggests that the 5-dehydrogenation and the Δ^7-reduction steps would principally occur on a Z,24-ethylidene sterol as shown in Scheme 2, but this hypothesis does not exclude other minor pathways.

Finally, the insertion of the Δ^{22} bond remains a mysterious reaction in higher plants. There is some evidence that the Δ^{22}-desaturation would be a very late reaction, as suggested by results showing the conversion of sitosterol (11C) to stigmasterol (11G) (54, 92) and data showing that after $|2\text{-}^{14}C|$ mevalonic incorporation in etiolated coleoptiles of *Zea mays*, the specific radioactivity of the acetate of 11G is much lower than that of the acetate of 11C (54). This situation contrasts with that in *Cucurbita maxima*, where it has been suggested that spinasterol, a $\Delta^{7,22}$-sterol (10G), would be a precursor of dihydrospinasterol, a Δ^7-sterol (10D) (20). The Δ^{22} desaturase has been shown to be a microsomal cytochrome P-450 monooxygenase in *Saccharomyces cerevisiae* (55).

USE OF ENZYME INHIBITORS TO STUDY STEROL BIOSYNTHESIS

General Concepts

Enzyme inhibitors are very useful in studying metabolic pathways. When acting in vivo on their target enzyme they often lead to the accumulation of the substrate of the enzyme and reveal intermediates which, because of their high turnover, are present in tissues in amounts too small to be detected. In order to play its expected role, the inhibitor should bind with its target enzyme and in turn hinder the binding of the physiological substrate. Among the various classes of enzyme inhibitors, one can distinguish reversible inhibitors (e.g. ground state analog inhibitors and transition state analog inhibitors) on the one hand and irreversible inhibitors (e.g. affinity labels and suicide inhibitors) on the other. Of course, the clues used to design enzyme inhibitors come from an understanding of the mechanistic details of enzyme-catalyzed reactions. The reader interested in increasing his knowledge in that area can refer to a number of excellent articles (67, 133, 139).

REVERSIBLE INHIBITORS   This group can be subdivided in a first approximation into ground state analog inhibitors and transition state (TS) analog inhibitors. The former mimics a substrate in the way it binds to the enzyme whereas the latter takes advantage of additional binding interactions available only to the TS by incorporating key structural elements of the unstable, TS form of the substrate in the stable structure of the inhibitor (11, 38, 102) (Equation 1). The result is that TS analog

$$E + S \rightleftharpoons ES \rightarrow |ES|\ddagger \rightarrow ES' \rightarrow EP \rightleftharpoons E + P \qquad 1.$$

S, P: substrate, product of the enzymatic reaction in their ground state; E: enzyme; [ES]‡: activated enzyme-substrate complex (transition state); ES': high-energy intermediate stabilized by the enzyme.

inhibitors may have a much higher affinity for the active site of the enzyme than traditional ground state analog inhibitors (11, 144). In many cases, the mechanisms of action of enzymes involve enzyme-bound high-energy intermediates (HEI) that are often better defined than TS. In such cases it is likely that the energy of the appropriate TS approximates that of the HEI. Therefore, the same strategy as that developed above for TS analog inhibitors would apply for HEI analog inhibitors (38).

From the considerations discussed in the first part of this article it is evident that several enzymes of sterol biosynthesis catalyze reactions involving carbocationic intermediates (Scheme 6). Some of them, possibly being stabilized transiently by the active site of the enzyme as a part of the enzyme mechanism, may be considered as typical HEIs. This is especially the case for enzyme-catalyzed isomerizations, alkylations, hydrogenations, cyclizations, etc. As shown also in Scheme 6, a general strategy to mimic carbocationic HEIs could be to replace the carbon atom bearing the positive charge by a nitrogen atom; the resulting tertiary amine, having a pKa close to 10, is protonated at physiological pH (7.4) and the resulting stable ammonium derivative displays a positive charge at a position identical to that occupied by the unstable HEI and could be considered a potential HEI analog inhibitor.

Such a strategy had been already applied with success to design new inhibitors of glycosylases (75) and of α mannosidases (76). The inhibition of the phosphoenol pyruvate-shikimate-3-phosphate synthase by the potent herbicide glyphosate probably also follows the same principles (4). The examples quoted below give evidence that the TS (HEI) analog concept applies quite well when enzymes involved in sterol biosynthesis are concerned and permits researchers to design potent sterol biosynthesis inhibitors (SBIs).

IRREVERSIBLE INHIBITORS   Irreversible inhibitors generally bind covalently to the enzyme, usually through electrophilic attack of the inhibitor on an enzyme-bound nucleophile. To gain specificity for reactive electrophiles, two

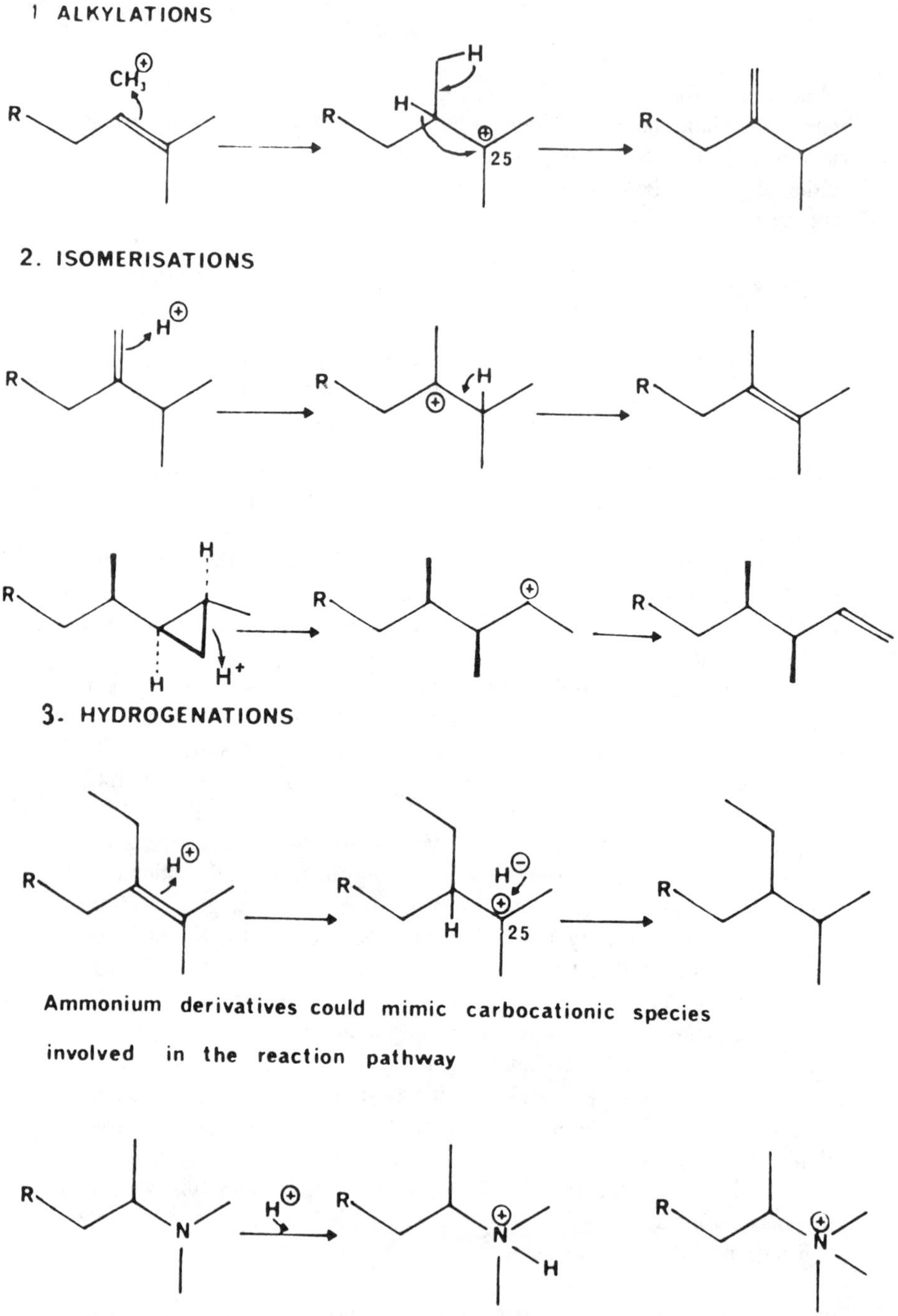

1 ALKYLATIONS

2. ISOMERISATIONS

3. HYDROGENATIONS

Ammonium derivatives could mimic carbocationic species involved in the reaction pathway

Scheme 6 Enzymic reactions of sterol biosynthesis involving demonstrated or postulated carbocationic intermediates. R = tetracyclic sterolic nucleus. The cyclopropane isomerisation example was adapted from Zimmerman et al (148).

general strategies have evolved. The first is the concept of "active site-directed" irreversible inhibitors of "affinity labels" (133); the second is that of "suicide" inhibitors (133, 139). Until now there have been few examples of their use in the case of sterol biosynthesis enzymes.

Inhibition of the 2(3)-Oxidosqualene Cyclase

The 2(3)-oxidosqualene cyclases represent a group of enzymes which convert 2(3)-oxidosqualene (*15*) into polycyclic triterpenoids such as cycloartenol (*1L*), lanosterol (*4L*), α- and β-amyrins, etc (33; see scheme 3). Enzymic cyclization of the all-*trans* (*15*) implies the polarization of the C-2 oxygen bond leading to a charge deficiency at C-2 of *15* (Scheme 7).

Taking into account the strategy developed above, the possibility of designing compounds that would be selective and potent inhibitors of the oxidosqualene cyclase has been investigated. Accordingly, 2-aza-2,3-dihydrosqualene (*16*), 2-aza-2,3-dihydrosqualene-N-oxide (*17*), as well as several derivatives of *16* and *17* (*18* to *25*) (27, 34, 40) have been synthesized

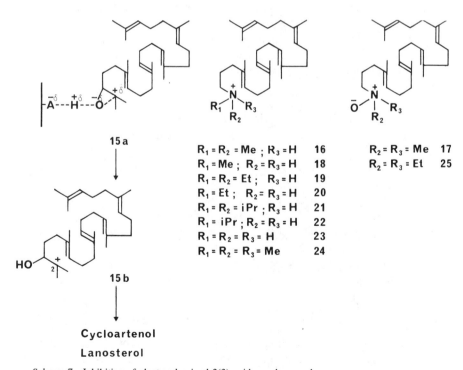

15 a

$R_1 = R_2 = Me$; $R_3 = H$	16
$R_1 = Me$; $R_2 = R_3 = H$	18
$R_1 = R_2 = Et$; $R_3 = H$	19
$R_1 = Et$; $R_2 = R_3 = H$	20
$R_1 = R_2 = iPr$; $R_3 = H$	21
$R_1 = iPr$; $R_2 = R_3 = H$	22
$R_1 = R_2 = R_3 = H$	23
$R_1 = R_2 = R_3 = Me$	24

$R_2 = R_3 = Me$ 17
$R_2 = R_3 = Et$ 25

15 b

Cycloartenol

Lanosterol

Scheme 7   Inhibition of plant and animal 2(3)-oxidosqualene cyclases.

and shown to be powerful inhibitors of both 15-lanosterol cyclase from rat liver (27, 40) and 15-cycloartenol (β-amyrin) cyclase from plant tissues (27, 34, 40) ($K_i/K_m = 10^{-3}$ for 25, the most potent inhibitor of the series). To explain these results, it has been suggested that tertiary amines such as 16, being protonated at physiological pH, could show some similarities with the transient HEI C-2 carbocation (15b) that results from the protonation of the oxiran ring and that the N-oxide (17), because of its strong dipolar moment, presents structural and electronic similarities with one possible TS (15a) involved in the first step of the enzymatic cyclization of 15. The quaternary derivative (24) was shown to be as strongly inhibitory as the parent tertiary amine (16), suggesting that the protonated form of the amine was the inhibitory species. A chemical structure-inhibitory activity relationships study showed that the diethyl derivatives (19 and 25) were the most potent inhibitors of the series (40).

Among other molecules tested possessing a positively charged nitrogen atom at the pH of the incubation, 2(3)-epiminosqualene, which possess an aziridine in place of the oxiran ring, was shown by Corey et al (32) to be a potent inhibitor of liver or pea seedling 15-cyclases and found to be a reversible, noncompetitive inhibitor of both cyclases (40). Similarly, U 18666 A (3β-(β-diethylaminoethoxy)-androst-5-en-17-one), a known hypocholesterolemic drug in mammalian cells (26), was shown to be a very potent inhibitor of both animal and plant cyclases (40). In contrast, AMO 1618, a plant growth regulator which, besides its known effect on the biosynthesis of gibberellins (44), has been reported to affect the biosynthesis of sterols by interfering with the cyclization of 15 in plant systems and in animals (39, 93), failed to inhibit the cyclization of 15 in microsomes from plant and animals (40). N-Dodecyl imidazole (NDI) has been shown to inhibit the 15-cyclase in a rat brain extract (35) and proved to be an excellent inhibitor of both the rat liver and pea seedling cyclases (40).

A recent structure-activity relationships study performed recently on rat liver and yeast 15-cyclases with various N-(1-n-dodecyl)-heterocycles have given evidence that none inhibited the yeast enzyme whereas the rat liver enzyme was efficiently inhibited (85). 4,4,10β-Trimethyl-$trans$-decal-3β-ol, which has been described as inhibitor of the 15-cyclase of Chinese hamster ovary cells (28), was found to be a noncompetitive inhibitor of the rat liver microsomal enzyme and presented no activity toward the higher plants cyclases (40). These last results underline subtle differences between the cyclases.

Inhibition of the S-Adenosyl-L-Methionine-Sterol-C-24-Methyltransferase

The sterol-C-24-methyltransferase catalyzes the insertion of a methyl group into a Δ^{24} acceptor sterol. It has been demonstrated that the Δ^{24} sterol substrate for this enzyme in plants is cycloartenol (1L) (42, 143), whereas zymosterol

(8L) is the best substrate in yeast (89). When incubated in the presence of AdoMet and the enzyme, 1L is converted essentially into 24-methylene cycloartanol (11). The mechanism of this reaction has been studied in *Trebouxia spp* (Scheme 4). It is still not known whether the pathway drawn in Scheme 4 applies strictly to vascular plants; nevertheless, it is likely that the basic principles involved in its elaboration remain valid, as for instance the involvment of the carbocationic species (*c* and *d*), possibly interacting transiently with the active site of the enzyme and the stereospecificity of the different steps of the reaction (5, 86). The HEI (*c*) has been tentatively mimicked by replacing the C-25 with a positively charged nitrogen atom (e.g. protonated amine). The resulting 25-azacycloartanol (26) has been tested in a cell-free system from maize seedlings that can alkylate exogenously added 1L in the presence of AdoMet and was shown to inhibit strongly the AdoMet-cycloartenol-C-24-methyltransferase (AdoMet CMT) activity (K_i = 30 nM) (91, 110). This K_i value was three orders of magnitude lower than the K_m values for the two substrates (K_M^{1L} = 30 μM; K_M^{AdoMet} = 40 μM). The inhibition was shown to be noncompetitive with respect to both 1L and AdoMet (109).

In order to determine the molecular parameters of the inhibition and to gain information about its mechanism, various azasteroids and analogues have been synthesized and assayed. The following results have been obtained: 1. The presence of a positive charge at position 25 was found to be the major cause of the inhibition since electrostatically neutral isosteric compounds (e.g. 27, 28) possessing a carbon in place of the nitrogen atom were not inhibitory. The positive charge leading to inhibition may be conferred by a protonated amine, a quaternary ammonium group, as well as by a sulfonium or an arsonium group (compounds 26, 29, 30, 31 respectively). 2. The most potent inhibitor of the series contained a tertiary N-oxide function (compound 32). 3. A steroid-like structure of the inhibitor was also important. 4. The presence of a free 3β-hydroxy group and the bent conformation of cycloartenol, which are essential molecular features of the substrate for the methylation reaction, were no longer required to observe inhibition. These results are consistent with the idea that C-25 heteroatoms (N, As, and S), substituted triterpenoid derivatives possessing a positive charge at position 25, are analogs of a carbocationic HEI involved during the reaction catalyzed by the AdoMet-CMT (109).

25-Azacycloartanol (26) has also been shown to inhibit the second methylation reaction catalyzed by the AdoMet-24-methylene lophenol (10 I)-28-methyltransferase but less efficiently than the first methyl transferase (110). When the culture medium of suspension cultures of *Rubus fruticosus* cells was supplemented with 26, the growth was severely inhibited at concentrations of 26 close to 1 μM. The sterol profile of cells treated with 26 was strongly modified. The major result consisted of a strong decrease of the content in 24-ethyl sterols and particularly in sitosterol (11C), and a striking accumulation of sterols nonalkylated at C-24 such as cycloartenol (*1*L), cholesterol (*11*E), and desmosterol (*11*L). The strong accumulation of *1*L and a more modest amount of 9I is consistent with the preferred biosynthetic pathway of Scheme 2, where the central positions occupied by *1*L and 9I in the two methylation reactions have been thrown into relief (126).

It had been observed previously that 25-aza-24-dihydrozymosterol and other various azasteroids blocked in vivo and in vitro the C-24 methylation of zymosterol in yeast (7, 99, 100), resulting in a strong accumulation of zymosterol (*8*L), cholesta-5,7,24-trienol (*12*L), and cholesta-5,7,22,24-tetraen-3β-ol. Kinetic inhibition studies with partially purified 24-sterol methyltransferase and several azasterols suggest that azasterols act noncompetitively with respect to zymosterol and are competitive inhibitors with respect to AdoMet (99). To explain the very strong inhibitions obtained, the authors have suggested that the electronic doublet carried on the nitrogen atom could mimic the nucleophilic Δ^{24} bond. A similar explanation has been proposed to explain the inhibition of an AdoMet-Δ^{24}-sterol-C-24-methyltransferase by triparanol in a cell-free extract from pea seedlings (81).

Another target of 25-azasterol is the Δ^{24}-sterol reductase. The inhibition of this enzyme had been recognized much earlier (17) in rat liver treated with 25-azacholesterol and has been discussed (81), taking into account the resemblances between the Δ^{24}-sterol hydrogenation reaction with the C-methylation reaction. The fact that 25-azacholesterol and several other azasteroids lead to an accumulation of desmosterol in insects is also consistent with the inhibition of the Δ^{24}-sterol reductase involved in the dealkylation of 24-ethyl sterols (134).

Inhibition of Cycloeucalenol-Obtusifoliol Isomerase, of $\Delta^8 \rightarrow \Delta^7$-Sterol Isomerase, and of Δ^{14}-Reductase

Treatment of suspension cultures of *Rubus fruticosus* and maize seedlings with tridemorph (*33*) or fenpropimorph (*34*), which are systemic fungicides (128), leads to a dramatic accumulation of 9β,19-cyclopropyl sterols. Among them, 24-methyl pollinastanol (*3*A, *3*B) (40% of total sterols), cycloeucalenol (2I), and 24-dihydrocycloeucalenol (2A, 2B) predominate (12, 15, 125). In addition to 9β,19-cyclopropyl sterols, various amounts of Δ^8- and of $\Delta^{8,14}$-sterols are detectable (125). Similar results have been obtained with suspension cultures of carrot, tobacco, and soybean treated with tridemorph (60). The inhibition of ergosterol biosynthesis in *Saccharomyces cerevisiae, Botrytis cinerea,* and *Ustilago maydis* by tridemorph, fenpropimorph, and fenpropidin (*35*) lead to an accumulation of Δ^8- and $\Delta^{8,14}$-sterols (9, 68, 71, 79).

All these results suggest strongly that the COI (in higher plants only), the $\Delta^8 \rightarrow \Delta^7$-sterol isomerase, and the Δ^{14}-reductase in both plants and fungi are target enzymes for N-substituted morpholines. This has been confirmed by direct assay of the chemicals on the isolated enzymes. As shown recently (111), fenpropimorph, tridemorph, and various other N-substituted morpholines strongly inhibit the COI ($K^{\text{fenpropimorph}}_i$ = 0.4 μM) and the $\Delta^8 \rightarrow \Delta^7$-sterol isomerase in a cell-free system of maize seedlings; in addition, the same fungicides inhibit the Δ^{14}-reductase in microsomes of yeast (70). By contrast, these molecules are only slightly active on the AdoMet-CMT and the 2(3)-oxidosqualene cyclase (M. Taton, unpublished results).

To explain the properties of N-substituted morpholines it has been suggested that at physiological pH (7.4) the morpholines are easily protonated and exhibit some similarities with carbocationic HEIs (e,f,g) involved in the reaction catalyzed by the COI, the $\Delta^8 \rightarrow \Delta^7$-sterol isomerase, and the Δ^{14}-reductase (12) (Scheme 8). These HEIs share in common the fact that their positive charges (located at C-9, C-8, and C-14 in e, f, g respectively) are very close to each other. As seen in Scheme 8, various conformers ($33i$, ii, iii) of the flexible N-substituted morpholins can be written in such a way that the positively charged nitrogen coincides with the C-9, C-8, and C-14 carbocations present in the HEIs. Another mechanism has been proposed to explain the differential action of 33, 34, and 35 on the three enzymes in yeast (9).

N-Benzyl-8-aza-4α,10-dimethyl-$trans$-decal-3β-ol (36) was designed to mimic the C-8 and the C-9 carbocationic HEI occurring during the reactions catalyzed by the $\Delta^8 \rightarrow \Delta^7$-sterol isomerase and the COI respectively (Scheme 8). In accordance with the "transition state analogues" theory, this analogue of an HEI was found to be a very potent and specific inhibitor (K_i/K_m 10^{-3}) of the two enzymatic reactions both in vitro and in vivo (113). 8-Aza-Decalins such as 36 were considered as leader molecules of a new class of rationally designed SBIs which could be of value for agronomical applications as potential fungicides.

The natural occurrence of $\Delta^{8,14}$-sterols in the cactus Lophocereus schottii has been reported (19). This cactus contains also large amounts of pilocereine, an isoquinolein alkaloid which could be responsible for the inhibition of the Δ^{14}-reductase and of the presence in this plant of an interrupted pathway (19). In a routine search for fungicides, scientists at Eli Lilly Company observed that the common fungus Geotrichum flavo-brunneum produced antifungal compounds. The principal active ingredient has been identified as 15-aza-24-methylene-D-homo-cholesta-8,14-dien-3β-ol (37) (18, 100) and has been shown to inhibit strongly the yeast Δ^{14}-reductase (56, 100) and later to lead to an accumulation of $\Delta^{8,14}$-sterols in suspension cultures of Rubus fruticosus cells (127). The fact that the protonated form of 37 looks somewhat like the carbocationic intermediate (g) is worthy of consideration in the preceding context (Scheme 8).

Inhibition of Other Enzymes

A great deal of interest has been concentrated on the inhibition of 14α-methyl-demethylase. It has been recognized that several chemicals having fungicidal

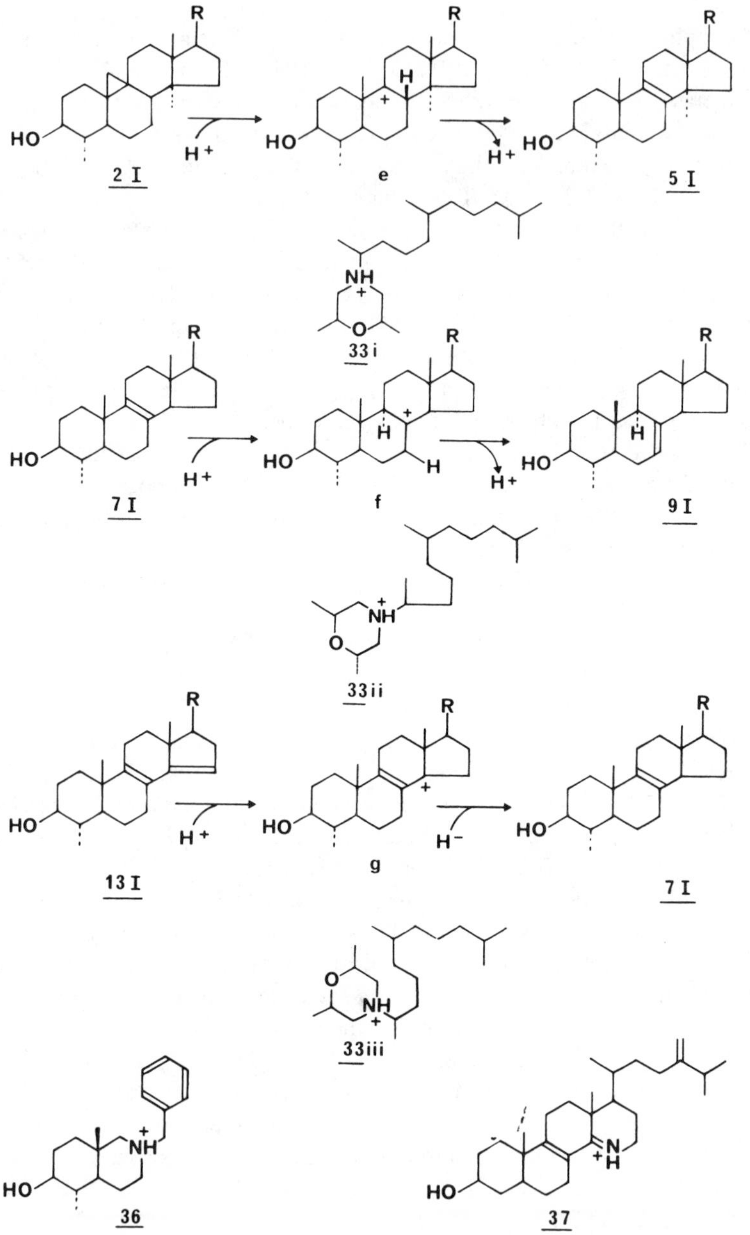

Scheme 8   Inhibition of the cycloeucalenol (2I)-obtusifoliol (5I) isomerase, the Δ^8-(7I) → Δ^7-(9I) sterol isomerase and the $\Delta^{8,14}$-sterol (13I) reductase. The protonated tridemorph has been represented in three different conformations (33i,ii,iii) mimicking the three carbocationic high-energy intermediates (e,f,g respectively).

activities are also inhibitors of ergosterol biosynthesis (128 and references cited therein). This is true of a huge family of molecules containing pyridine, pyrimidine, imidazole, and triazole groups such as propiconazole, triadimefon, fenarimol, dichlobutrazol, etc (120, 136; Scheme 9). These sterol biosynthesis inhibitors (SBIs) have been shown to inhibit the 14α-methyl sterol demethylase in various fungi (59, 141). The potency of these molecules appears to rest on their ability to bind with the 14α-methyl sterol binding site of the enzyme by their hydrophobic moiety and to be strong ligands of the heme iron of cytochrome P-450 (the prosthetic group of the enzyme) by their heterocyclic function (141) (Scheme 9). 14α-Methyl sterols accumulate also in three species of *Chlorella* grown in a culture medium containing triarimol |α-(2,4-dichlorophenyl-α-phenyl-5 pyrimidine methanol|. The introduction of the Δ^{22}-bond is also inhibited. It has been suggested that triarimol may be a specific inhibitor of cytochrome P-450 which participates in one or more reactions, leading to the oxidative elimination of the 14α-methyl group (43).

The effect of these chemicals on vascular plants has been studied in a few cases. Suspension cultures of *Rubus fruticosus* cells treated with fenarimol accumulate high amounts of 14α-methyl sterols [especially obtusifoliol (5I) and 14α-methyl fecosterol (6I)], giving evidence that the 14α-methyl demethylase from higher plants is also strongly inhibited by this chemical (124). Treatment of *Hordeum vulgare* seedlings with propiconazole and triadimefon leads also to an accumulation of 14α-methyl sterols (P. Ullmann, unpublished results). The results described above have given useful clues for the rational design of new SBIs that are of great interest as plant growth regulators (8) and fungicides for agricultural and medical uses.

The 4-methyl-demethylases are not inhibited by the SBIs quoted above; 4α-methyl cholesterol demethylase from rat liver has been shown to be irreversibly inactivated by 4α-(cyanomethyl)-5α-cholestan-3β-ol (10), a novel suicide inhibitor. The formation of an electrophilic acylcyanhydrine in the active site of the enzyme has been suggested. No inhibitors have been reported until now for the last steps of vascular plant biosynthesis. AY 9944, which has been shown to inhibit the $\Delta^{5,7}$-sterol-Δ^7-reductase in rat liver (29), does not affect this enzyme in suspension cultures of *Rubus fruticosus* cells but leads to a strong accumulation of various Δ^8-sterols (123). The effect of AY 9944 on algae is complex, as shown by the pioneer work of Patterson et al (101). In *Chlorella emersonii* there was inhibition of removal of the C-14 methyl group, of the C-28 transmethylation, and of desaturation at C-22, while in *C. ellipsoidea* the inhibition was mainly on the Δ^{14} reductase and to a lesser extent on the $\Delta^8 \rightarrow \Delta^7$-sterol isomerase (101).

As specified at the beginning, we have considered only the enzymes situated downstream to squalene 2,3-oxide (*15*). However, interesting inhibitors have been reported to act on enzymes occurring before *15*. This is true of compactin,

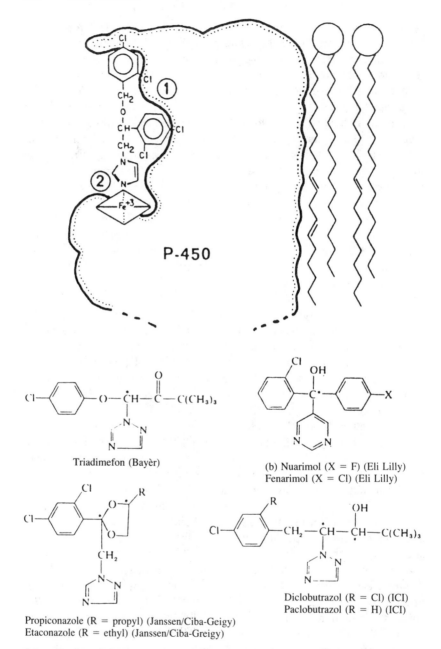

Scheme 9 Speculative representation of miconazole binding to the sterol 14α-demethylation complex. 1. Substrate-binding site; 2. heme iron-binding site. Structure of some inhibitors of the 14α-methyl demethylase. Adapted from (59) with permission of the authors.

an extremely potent inhibitor of the 3-hydroxy-3-methylglutaryl coenzyme A reductase of animal and plant origin (8a), and of an ammonium analog of a carbocationic intermediate involved in the reaction pathway catalyzed by the presqualene pyrophosphate-squalene synthetase (120a).

Also of interest is the fact that 2-aza-2,3-dihydrosqualene (*16*), 25-azacycloartanol (*26*), fenpropimorph (*34*), and other compounds mentioned above were consistently found to be noncompetitive inhibitors (40, 109, 111). At first sight this finding does not seem to favor the hypothesis that at least *16* and *26* have been designed as TS or HEI analog inhibitors. However, remember that we are dealing with complex anisotropic systems and with substrates and inhibitors that are strongly hydrophobic; therefore, classical enzyme kinetics may not apply.

CONCLUSION

As shown above, the available SBIs constitute a panoply of molecular tools that permit control of several key enzymes involved in sterol biosynthesis. There are far too many applications to describe here, but SBIs when used in vivo are very useful in revealing hidden metabolic pathways (Scheme 2). When an enzyme such as the COI is blocked in vivo, the preferred pathway is interrupted and lateral pathways are expressed. This observation results from the fact that some of the enzymes situated downstream to the aimed target are not totally specific. They can, for instance, catalyze the 4-demethylation of cycloeucalenol (*2I*) to give 24-methylene pollinastanol (*3I*); *2I* and *3I* can also be methylated at C-28 to give cyclofontumienol (*2K*) and 24-ethylidene pollinastanol (*3K*), which are generally not found naturally in plants. Therefore, plant cells treated with COI inhibitors accumulate products (*1, 2, 3*) situated at the front of the cube at the top of Scheme 2. Likewise, treatment of plant cells with 14α-methyl de-methylase inhibitors leads to the accumulation of products (14α-methyl sterols) (*4, 5, 6*) situated at the rear of the same cube. Inhibitors of Δ^{14}-reductase, $\Delta^8 \rightarrow \Delta^7$-sterol isomerase lead to $\Delta^{8,14}$-sterol and Δ^8-sterol accumulation situated at the rear (*13,14*) and the middle (*7,8*) of the parallelepipede. A general partial biosynthetic sequence underlined with heavy arrows in Scheme 2 was proposed earlier (53) on the basis of the information quoted above; it is clear, however, that the number of possible permutations resulting from the several alkylation mechanisms in conjunction with the changes required to the sterol nuclei allows a multitude of possible routes from cycloartenol (*1L*) to 24-methyl cholesterol (*11*A and *11*B), sitosterol (*11*C), or stigmasterol (*11*G) (53). Only a part of these possibilities is indicated in the matrix of alternative routes depicted in Scheme 2. Studies with yeast mutants have given crucial information for determining the order of the main steps (C-24 methylation, 14-methyl de-methylation, $\Delta^8 \rightarrow \Delta^7$-isomerization, 22-dehydrogenation, 24-reduction, etc) occurring during ergosterol biosynthesis (9a, 100). Until now, biochemical mutants from somatic plant cells affected in sterol biosynthesis have never been characterized. However, some trials in that area appear to be promising (29a), and further work should be strongly encouraged.

ACKNOWLEDGMENTS

I warmly thank Bernadette Bastian for her help in preparing this manuscript. My special thanks also to the authors cited for having kindly informed me of their studies. The original work described herein was supported by grants from the Centre National de la Recherche Scientifique and the Ministere de la Recherche et de la Technologie.

Literature Cited

1. Akhtar, M., Rahimtula, A. D., Wilton, D. C. 1970. The stereochemistry of hydrogen elimination from C-7 in cholesterol and ergosterol biosynthesis. *Biochem. J.* 117:539–42

2. Altman, L. J., Han, C. Y., Bertolino, A., Haudy, G., Laugani, D., et al. 1978. Stereochemistry of the 1,3-proton loss from a chiral methyl group in the biosynthesis of cycloartenol as determined by tritium NMR spectroscopy. *J. Am. Chem. Soc.* 100:3235–37

3. Anding, C., Parks, L. W., Ourisson, G. 1974. Enzymic modifications of cyclopropane sterols in yeast cell-free system. *Eur. J. Biochem.* 43:459–63

4. Anton, D. L., Hedstrom, L., Fish, S. M., Abeles, R. H. 1983. Mechanism of enolpyruvyl shikimate-3-phosphate synthase exchange of phosphoenolpyruvate with solvent protons. *Biochemistry* 22:5903–8

5. Arigoni, D. 1978. Stereochemical studies of enzymic C-methylations. *Ciba Found. Symp.* 60:243–61

6. Armarego, W. L. F., Goad, L. J., Goodwin, T. W. 1973. Biosynthesis of α-spinasterol from $(2\text{-}^{14}C,(4R)\text{-}4\text{-}^{3}H_1)$ mevalonic acid by *Spinacea oleracea* and *Medicago sativa*. *Phytochemistry* 12:2181–87

7. Avruch, L., Fischer, S., Pierce, H. Jr., Oehlschlager, A. C. 1976. The induced biosynthesis of 7-dehydrocholesterols in yeast: Potential sources of new provitamin D_3 analogs. *Can. J. Biochem.* 54:657–65

8. Bach, T. J. 1985. Selected natural and synthetic enzyme inhibitors of sterol biosynthesis as molecular probes for *in vivo* studies concerning the regulation of plant growth. *Plant Science* 39:183–87

8a. Bach, T. J., Lichtenthaler, H. K. 1983. Mechanisms of inhibition by mevinolin

(MK 803) or microsome-bound radish and of partially purified yeast HMG-CoA reductase (EC.1.1.1.34). *Z. Naturforsch. Teil C* 38:212–19

9. Baloch, R. I., Mercer, E. I., Wiggins, T. E., Baldwin, B. C. 1984. Inhibition of ergosterol biosynthesis in *Saccharomyces cerevisiae* and *Ustilago maydis* by tridemorph, fenpropimorph and fenpropidin. *Phytochemistry* 23:2219–26

9a. Bard, M., Woods, R. A., Barton, D. H. R., Corrie, J. E. T., Widdowson, D. A. 1977. Sterol mutants of *Saccharomyces cerevisiae:* Chromatographic analyses. *Lipids* 12:645–54

10. Bartlett, D. L., Robinson, C. H. 1982. Mechanism-based inactivation of 4-methylsterol oxidase by 4α-(cyanomethyl)-5α-cholestan-3β-ol. *J. Am. Chem. Soc.* 104:4729–30

11. Bartlett, P. A., Marlowe, C. K. 1983. Phosphonamidates as transition-state analogue inhibitors of thermolysin. *Biochemistry* 22:4618–24

12. Benveniste, P., Bladocha, M., Costet, M. F., Ehrhardt, A. 1984. Use of inhibitors of sterol biosynthesis to study plasmalemma structure and function. In *Journal of the Annual Proceedings of the Phytochemical Society of Europe*, ed. A. M. Boudet, G. Allibert, G. Marigo, P. J. Lea, 24:283–300. Oxford: Clarendon. 334 pp.

13. Berg, L. R., Patterson, G. W., Lusby, W. R. 1983. Effects of triarimol and tridemorph on sterol biosynthesis in *Saprolegnia ferax. Lipids* 18:448–52

14. Bimpson, T., Goad, L. J., Goodwin, T. W. 1969. The stereochemistry of hydrogen elimination at C-6, C-22, and C-23 during ergosterol biosynthesis by *Aspergillus fumigatus* Fres. *Chem. Commun.* 1969:297–98

15. Bladocha, M., Benveniste, P. 1983.

Manipulation by Tridemorph, a systemic fungicide, of the sterol composition of maize leaves and roots. *Plant Physiol.* 71:756–62

16. Blättler, W. A. 1978. *Sterischer Verlauf der Bildung und Öffnung des Cyclopropanringes in der Biosynthese von Phytosterinen.* PhD thesis. Eidgenöss. Tech. Hochsh, Zürich

17. Blohm, T. R., Stevens, V. L., Kariya, T., Alig, H. N. 1970. Effects of clomiphene *cis* and *trans* isomers on sterol metabolism in the rat. *Biochem. Pharmacol.* 19:2231

18. Boeck, L. D., Hoehn, M. M., Westhead, J. E., Wolter, R. K., Thomas, D. N. 1975. New azasteroidal antifungal antibiotics from *Geotrichum flavobrunneum.* Discovery and fermentation studies. *J. Antibiot.* 28:95–101

19. Campbell, C. E., Kircher, H. W. 1980. Senita cactus:A plant with interrupted sterol biosynthesis pathways. *Phytochemistry* 19:2777–79

20. Caputo, O., Delprino, L., Viola, F., Caramiello, R., Balliano, G. 1983. Biosynthesis of sterols and triterpenoids in tissue cultures of *Cucurbita* maxima. *Planta Med.* 47:176–80

21. Caspi, E., Ramm, P. J. 1969. Stereochemical differences in the biosynthesis of C_{27} Δ^7-steroidal intermediates. *Tetrahedron Lett.* 3:181–85

22. Caspi, E., Sliwowski, J., Robichaud, C. S. 1975. Biosynthesis of sitosterol from (2R)- and (2S)-|2-^{3}H| mevalonic acid in the pea. The incorporation of a 15α-tritium atom derived from (3RS, 2R)-[2-^{14}C, 2-^{3}H] mevalonic acid. *J. Am. Chem. Soc.* 97:3820–22

23. Castle, M., Blondin, G. A., Nes, W. R. 1963. Evidence for the origin of the ethyl group of β-sitosterol. *J. Am. Chem. Soc.* 85:3306–7

24. Cattel, L., Balliano, G., Caputo, O. 1979. Sterols and triterpenes from *Cucurbita maxima* Duch. *Planta Med.* 37:264–67

25. Cattel, L., Delprino, L., Benveniste, P., Rahier, A. 1979. Effect of the configuration of the methyl group at C-4 on the capacity of 4-methyl-9,19-cyclosteroids to be substrates of a cyclopropane cleavage enzyme from maize. *J. Am. Oil Chem. Soc.* 56:6–11

26. Cenedella, R. J. 1980. Concentration-dependent effects of AY-9944 and U 18666A on sterol synthesis in brain. *Biochem. Pharmacol.* 29:2751–54

27. Cerutti, M., Delprino, L., Cattel, L., Bouvier-Navé, P., et al. 1985. N-oxide as potential function in the design of enzyme inhibitors. Application to 2,3-oxidosqualene-sterol cyclases. *Chem.*

Commun. 1985:1054–55

28. Chang, T. Y., Schiavoni, E. S., McCrae, K. R., Nelson, J. A., Spencer, T. A. 1979. Inhibition of cholesterol biosynthesis in Chinese hamster ovary cells by 4,4,10β-trimethyl-trans-decal-3β-ol. *J. Biol. Chem.* 254:11258–63

29. Chappel, C. I., Dvornik, D., Hill, P., Kraml, M., Gaudry, R. 1963. Trans-1,4-bis-(2-chlorobenzylaminomethyl) cyclohexane dihydrochloride (AY 9944): a novel inhibitor of cholesterol biosynthesis. *Circulation* 28:651

29a. Chiu, P.-L., Bottino, P. J., Patterson, G. W. 1980. Sterol composition of nystatin and amphotericin B resistant tobacco calluses. *Lipids* 15:50–54

30. Chiu, P.-L., Patterson, G. W., Fenner, G. P. 1985. Sterols of bryophytes. *Phytochemistry* 24:263–66

31. Corey, E. J., Lin, K., Yamamoto, H. 1969. Separation of the cyclisation and rearrangement processes of sterol biosynthesis. Enzymic formation of a protosterol derivation. *J. Am. Chem. Soc.* 91:2132–34

32. Corey, E. J., Ortiz de Montellano, P. R., Lin, K., Dean, P. D. G. 1967. 2,3-Iminosqualene, a potent inhibitor of enzymic cyclization of 2,3-oxidosqualene to sterols. *J. Am. Chem. Soc.* 89:2797–98

33. Dean, P. D. G. 1972. The cyclase of triterpene and sterol biosynthesis. *Steroidologia* 2:143–57

34. Delprino, L., Balliano, G., Cattel, L., Benveniste, P., Bouvier, P. 1983. Inhibition of higher plant 2,3-oxidosqualene cyclase by 2-aza-2,3-dihydrosqualene and its derivatives. *Chem. Commun.* 1983:381–82

35. Dennick, R. G., Dean, P. D. G. 1974. Squalene-2(3)-epoxide-lanosterol cyclase in developing rat brain. *J. Neurochem.* 23:261–66

36. Dewar, M. J. C., Reynolds, C. H. 1984. Ground-states of molecules. Complexes as intermediates in reactions. Biomimetic cyclization. *J. Am. Chem. Soc.* 106:1744–50

37. Djerassi, C. 1981. Recent studies in the marine sterol field. *Pure Appl. Chem.* 53:873–90

38. Douglas, K. T. 1983. Transition-state analogs in drug design. *Chem. Ind.* 1983:311–15

39. Douglas, T. J., Paleg, L. G. 1972. Inhibition of sterol biosynthesis by 2-isopropyl-4-dimethylamino-5-methylphenyl-1-piperidine carboxylate methyl chloride in tobacco and rat-liver preparations. *Plant Physiol.* 49:417–20

40. Duriatti, A., Bouvier-Navé, P., Benveniste, P., Schuber, F., Delprino, L.,

et al. 1985. In vitro inhibition of animal and higher plants 2,3-oxidosqualene-sterol cyclases by 2-aza-2,3-dihydro-squalene and derivatives, and by other ammonium containing molecules. *Biochem. Pharmacol.* 34:2765–77

41. Eschenmoser, A., Ruzicka, L., Jeger, O., Arigoni, D. 1955. Zur Kenntnis der Triterpene. Eine stereochemische Interpretation der biogenetischen Isoprenregel bei den Triterpenen. *Helv. Chim. Acta* 38:1890–1904

42. Fonteneau, P., Hartmann-Bouillon, M. A., Benveniste, P. 1977. A 24-methylene lophenol C-28 methyltransferase from suspension cultures of bramble cells. *Plant Sci. Lett.* 10:147–55

43. Frasinel, C., Patterson, G. W., Dutky, S. R. 1978. Effect of triarimol on sterol and fatty-acid composition of three species of *Chlorella*. *Phytochemistry* 17:1567–70

44. Frost, D. J., Ward, J. P. 1970. Sterols in *Vernonia-anthelmintica* seed 8,14,(Z)-24(28)stigmastatrienol, a new phytosterol. *Rec. Trav. Chem.* 89:1054–56

45. Frost, R. G., West, C. A. 1977. Properties of kaurene synthetase from *Marah macrocarpus*. *Plant Physiol.* 59:22–29

46. Gaylor, J. L. 1981. Formation of sterols in animals. In *Biosynthesis of Isoprenoid Compounds*, ed. J. W. Porter, S. L. Spurgeon, 1:482–543. New York: Wiley. 558 pp.

47. Ghisalberti, E. L., de Souza, N. J., Rees, H. H., Goad, L. J., Goodwin, T. W. 1969. Biological removal of the 4α-methyl group during the conversion of cycloartenol into 31-norcycloartenol in *Polypodium vulgare* Lin. *Chem. Commun.* 1969:1403–8

48. Gibbons, G. F., Goad, L. J., Goodwin, T. W., Nes, W. R. 1971. Concerning the role of lanosterol and cycloartenol in steroid biosynthesis. *J. Biol. Chem.* 246:3967–76

49. Goad, L. J. 1970. Sterol biosynthesis. *Biochem. Soc. Symp.* 29:45–77

50. Goad, L. J., Gibbons, G. F., Bolger, L. M., Rees, H. H., Goodwin, T. W. 1969. Incorporation of [2-^{14}C,(5-R)-5-^{3}H$_1$] mevalonic acid into cholesterol by a rat liver homogenate and into β-sitosterol and 28-isofucosterol by *Larix decidua* leaves. *Biochem. J.* 114:885–92

51. Goad, L. J., Goodwin, T. W. 1972. The biosynthesis of plant sterols. *Prog. Phytochem.* 3:113–98

52. Goad, L. J., Lenton, J. R., Knapp, F. F., Goodwin, T. W. 1974. Phytosterol side-chain biosynthesis. *Lipids* 9:582–95

53. Goodwin, T. W. 1979. Biosynthesis of terpenoids. *Ann. Rev. Plant Physiol.* 30:369–404

54. Grunwald, C. 1975. Plant sterols. *Ann. Rev. Plant Physiol.* 26:209–36

55. Hata, S., Nishino, T., Katsuki, H., Aoyama, Y., Yoshida, Y. 1983. Two species of cytochrome P-450 involved in ergosterol biosynthesis of yeast. *Biochem. Biophys. Res. Commun.* 116:162–66

56. Hays, P. R., Parks, L. W., Pierce, H. D. Jr., Oehlschlager, A. C. 1977. Accumulation of ergosta-8,14-dien-3β-ol by *Saccharomyces cerevisiae* cultured with an azasterol antimycotic agent. *Lipids* 12:666–68

57. Heintz, R., Benveniste, P. 1974. Plant sterol metabolism. Enzymatic cleavage of the 9β,19β-cyclopropane ring of cyclopropyl sterols in bramble tissue cultures. *J. Biol. Chem.* 249:4267–74

58. Henry, M., Chantalat-Dublanche, I. 1985. Isolation of spinasterol and its glucoside from cell suspension of *Saponaria officinalis* L. *Planta Med.* In press

59. Henry, M. J., Sisler, H. D. 1984. Effects of sterol biosynthesis-inhibiting fungicides on cytochrome-P-450 oxygenations in fungi. *Pestic. Biochem. Physiol.* 22:262–75

60. Hosokawa, G., Patterson, G. W., Lusby, W. R. 1984. Effect of triarimol, tridemorph and triparanol on sterol biosynthesis in carrot, tobacco and soybean suspension cultures. *Lipids* 19:449–56

61. Itoh, T., Ishii, T., Tamura, T., Matsumoto, T. 1978. Four new and other 4α-methylsterols in the seeds of solanaceae. *Phytochemistry* 17:971–77

62. Itoh, T., Jeong, T. M., Hirano, Y., Tamura, T., Matsumoto, T. 1977. Occurrence of lanosterol and lanosterol in seeds of red pepper. *Steroids* 29:569–77

63. Itoh, T., Matsumoto, T. 1985. Sterols and triterpene alcohols of cucurbitaceae plants. *Lipids*. In press

64. Itoh, T., Shimizu, N., Tamura, T., Matsumoto, T. 1981. 24-methyl-E-23-dehydrolophenol, a new sterol and two other 24-methyl-5-Δ^{23}-sterols in *Zea mays* germ oil. *Phytochemistry* 20:1353–56

65. Itoh, T., Tamura, T., Matsumoto, T. 1977. 4-Desmethylsterols in the seeds of solanaceae. *Steroids* 30:425–33

66. Itoh, T., Tamura, T., Matsumoto, T. 1977. Triterpene alcohols in the seeds of solanaceae. *Phytochemistry* 16:1723–26

67. Jencks, W. P. 1975. Binding-energy, specificity, and enzymic catalysis:the Circe effect. *Adv. Enzymol.* 43:219–410

68. Kato, T., Shoami, M., Kawase, Y.

1980. Comparison of tridemorph with buthiobate in antifungal mode of action. *J. Pestic. Sci.* 5:69–79

69. Kawata, S., Trzaskos, J. M., Gaylor, J. L. 1985. Microsomal enzymes of cholesterol biosynthesis from lanosterol. Purification and characterization of Δ^7-sterol 5-desaturase of rat liver microsomes. *J. Biol. Chem.* 260:6609–17

70. Kerkenaar, A. 1986. Mechanism of action of morpholins. In *Modern, Selective Fungicides*, ed. H. Lyr. Jena: Fischer Verlag. In press

71. Kerkenaar, A., Uchiyama, M., Versluis, G. G. 1981. Specific effects of tridemorph on sterol biosynthesis in *Ustilago maydis*. *Pestic. Biochem. Physiol.* 16:97–104

72. Knapp, F. F., Goad, L. J., Goodwin, T. W. 1977. The conversion of 24-ethylidene sterols into poriferasterol by the alga *Ochromonas malhamensis*. *Phytochemistry* 16:1677–81

73. Knapp, F. F., Goad, L. J., Goodwin, T. W. 1977. Stereochemistry of C-4 demethylation during conversion of obtusifoliol into poriferasterol by the alga *Ochromonas malhamensis*. *Phytochemistry* 16:1683:88

74. Knapp, F. F., Nicholas, H. J. 1971. The biosynthesis of 31-nor cyclolaudenone in *Musa sapientum*. *Phytochemistry* 10:97–102

75. Lai, H.-Y. L., Axelrod, B. 1973. 1-Aminoglycosides, a new class of specific inhibitors of glycosidases. *Biochem. Biophys. Res. Commun.* 54:463–68

76. Lalégerie, P., Legler, G., Yon, J. M. 1982. The use of inhibitors in the study of glycosidases. *Biochimie* 64:977–1000

77. Lenton, J. R., Goad, L. J., Goodwin, T. W. 1975. Sitosterol biosynthesis in *Hordeum vulgare*. *Phytochemistry* 14:1523–28

78. Lenton, J. R., Hall, J. L., Smith, A. R. H., Ghisalberti, E. L., Rees, H. H., et al. 1971. The utilization of potential phytosterol precursors by *Ochromonas malhamensis*. *Arch. Biochem. Biophys.* 143:664–74

79. Leroux, P., Gredt, M. 1984. Etudes sur les inhibiteurs de la biosynthèse des stérols fongiques:I. Fongicides provoquant l'accumulation de 4-desmethyl sterols. *Agronomie* 3:123–30

80. Lockley, W. J. S., Roberts, D. P., Rees, H. H., Goodwin, T. W. 1974. 24-methyl-cholesta-5,24(25)-dien-3β-ol:a new sterol from *Withania somnifera*. *Tetrahedron Lett.* 43:3773–76

81. Malhotra, H. C., Nes, W. R. 1971. The mechanism of introduction of alkyl groups at C-24 of sterols. IV. Inhibition

by Triparanol. *J. Biol. Chem.* 246:4934–37

82. Deleted in proof

83. McKean, M. L., Nes, W. R. 1977. Evidence for separate intermediates in the biosynthesis of 24α- and 24β-sterols in tracheophytes. *Phytochemistry* 16:683–88

84. Mercer, E. I. 1984. The biosynthesis of ergosterol. *Pestic. Sci.* 15:133–55

85. Mercer, E. I., Morris, P. K., Baldwin, B. C. 1985. Differences in the inhibitory effects of N-(1-n-dodecyl)-heterocycles on the 2,3-oxidosqualene lanosterol-cyclase of rat liver and yeast. *Comp. Biochem. Physiol.* 80:341–46

86. Mihailovic, M. M. 1984. *Biosynthesis of phytosterols in Trebouxia spp:Steric course of the C-alkylation step*. PhD thesis. Eidgenöss. Tech. Hochsh., Zürich

87. Misso, N. L. A., Goad, L. J. 1984. Investigations on the Δ^{23}-, $\Delta^{24(28)}$ and Δ^{25}-sterols of *Zea mays*. *Phytochemistry* 23:73–82

88. Misso, N. L. A., Goad, L. J. 1983. The synthesis of 24-methylene cycloartanol, cyclosadol and cyclolaudenol by a cell-free preparation from *Zea mays* shoots. *Phytochemistry* 22:2473–79

89. Moore, J. T., Gaylor, J. L. 1969. Isolation and purification of an S-adenosyl-methionine:Δ^{24}-sterol methyltransferase from yeast. *J. Biol. Chem.* 244:6334–40

90. Mukam, L., Charles, G., Hentchoya, J., Njimi, T., Ourisson, G. 1973. Le cyclofuntumienol, 4α-methylsterol en C_{13} de *Funtumia elastica*. *Tetrahedron Lett.* 29:2779–82

91. Narula, A. S., Rahier, A., Benveniste, P., Schuber, F. 1981. 24-Methyl-25-azacycloartanol, an analog of a carbonium ion high energy intermediate, is a potent inhibitor of (S)-adenosyl-L-methionine:sterol C-24-methyl-transferase in higher plant cells. *J. Am. Chem. Soc.* 103:2408–9

92. Navari-Izzo, F., Izzo, R. 1984. In vivo conversion of 4-^{14}C sitosterol and 22,23-^{3}H sitosterol to stigmasterol in barley seedlings. *Phytochemistry* 23:769–72

93. Nes, W. D., Douglas, T. J., Lin, J.-T., Heftmann, E., Paleg, L. G. 1982. Regulation of isopentenoid biosynthesis by plant growth retardants in *Nicotiana tabacum*. *Phytochemistry* 21:575–79

94. Nes, W. D., Heupel, R. C., Le, P. H. 1984. A comparison of sterol biosynthesis in fungi and tracheophytes and its phylogenetic and functional implications. In *Structure, Function and Metabolism of Plant Lipids*, ed. P. A. Siegenthaler, W. Eichenberger, pp. 207–

16. Amsterdam/New York/Oxford: Elsevier

95. Nes, W. R. 1977. The biochemistry of plant sterols. *Adv. Lipid Res.* 15:233–324

96. Nes, W. R., Krevitz, K., Behzadan, S. 1976. Configuration at C-24 of 24-methyl and 24-ethyl cholesterol in tracheophytes. *Lipids* 11:118–26

97. Nes, W. R., Krevitz, K., Joseph, J., Nes, W. D., Harris, B., et al. 1977. The phylogenetic distribution of sterols in tracheophytes. *Lipids* 12:511–27

98. Nicotra, F., Ronchetti, F., Russo, G., Lugaro, G., Casellato, M. 1981. Stereochemical fate of the isopropylidene methyl groups of lanosterol during the biosynthesis of isofucosterol in *Pinus pinea*. *J. Chem. Soc. Perkin Trans. 1* 1981:498–502

99. Oehlschlager, A. C., Angus, R. H., Pierce, A. M., Pierce, H. D. Jr., Srinivasan, R. 1984. Azasterol inhibition of Δ^{24}-sterol methyltransferase in *Saccharomyces cerevisiae*. *Biochemistry* 23:3582–89

100. Parks, L. W. 1978. Metabolism of sterols in yeast. *CRC Crit. Rev. Microbiol.* 6:301–41

101. Patterson, G. W., Doyle, P. J., Dickson, L. G., Chan, J. T. 1974. Novel sterols from inhibited *Chlorella spp. Lipids* 9:567–74

102. Pauling, L. 1946. Molecular architecture and biological reactions. *Chem. Eng. News* 24:1375–77

103. Pinto, W. J., Nes, W. R. 1985. 24β-Ethylsterols, n-alkanes and n-alkanols of *Clerodendrum splendens*. *Phytochemistry* 24:1095–97

104. Ponsinet, G., Ourisson, G. 1967. Biosynthèse in vitro des triterpènes dans le latex d'*Euphorbia*. *Phytochemistry* 6:1235–43

105. Ponsinet, G., Ourisson, G. 1968. Aspects particuliers de la biosynthèse des triterpènes dans le latex d'*Euphorbia*. *Phytochemistry* 7:757–64

106. Raab, K. H., de Souza, N. J., Nes, W. R. 1968. The H-migration in the alkylation of sterols at C-24. *Biochim. Biophys. Acta* 152:742–48

107. Raederstorff, D., Rohmer, M. 1984. Le cycloarténol, précurseur des stérols chez le protozoaire *Acanthamoeba polyphaga*. *C. R. Acad. Sci. Paris* 299:57–60

107a. Rahier, A. 1980. *Biosynthèse des stérols chez les plantes. Etude du mode d'action de deux enzymes membranaires: la cycloeucalenol-obtusifoliol isomérase et la SAM cycloartenol C-24 méthyltransférase.* PhD thesis. Univ. Strasbourg, France

108. Rahier, A., Cattel, L., Benveniste, P.

1977. Mechanism of the enzymatic cleavage of the 9β,19-cyclopropane ring of cycloeucalenol. *Phytochemistry* 16:1187–92

109. Rahier, A., Génot, J. C., Schuber, F., Benveniste, P., Narula, A. S. 1984. Inhibition of S-adenosyl-L-methionine sterol C-24-methyltransferase by analogues of a carbocationic ion high energy intermediate. *J. Biol. Chem.* 259:15215–23

110. Rahier, A., Narula, A. S., Benveniste, P., Schmitt, P. 1980. 25-azacycloartanol, a potent inhibitor of S-adenosyl-L-methionine-sterol-C-24 and C-28-methyltransferase in higher plant cells. *Biochem. Biophys. Res. Commun.* 92:20–25

111. Rahier, A., Schmitt, P., Huss, B., Benveniste, P., Pommer, E. H. 1985. Chemical structure-activity relationships of the inhibition of sterol biosynthesis by N-substituted morpholines in high plant cells. *Pestic. Biochem. Physiol.* 25:112–24

112. Rahier, A., Taton, M., Bouvier-Navé, P., Schmitt, P., Benveniste, P., et al. 1985. Design of high energy intermediate analogue to study sterol biosynthesis in higher plants. *Lipids.* In press

113. Rahier, A., Taton, M., Schmitt, P., Benveniste, P., Place, P., Anding, C. 1985. Inhibition of $\Delta^8 \rightarrow \Delta^7$-sterol isomerase and of cycloeucalenol-obtusifoliol isomerase by N-benzyl-8-aza-4α,10-dimethyl-*trans*-decal-3β-ol, an analogue of a carbocationic high energy intermediate. *Phytochemistry* 24:1223–32

114. Reddy, V. V. R., Kupfer, D., Caspi, E. 1978. Mechanism of C-5 double bond introduction in the biosynthesis of cholesterol by rat liver microsomes. *J. Biol. Chem.* 252:2797–801

115. Rohmer, M., Bouvier-Navé, P., Ourisson, G. 1984. Distribution of hopanoid triterpenes in prokaryotes. *J. Gen. Microbiol.* 130:1137–50

116. Rohmer, M., Brandt, R. D. 1973. Les stérols et leurs précurseurs chez *Astasia longa* Pringsheim. *Eur. J. Biochem.* 36:446–54

117. Rohmer, M., Ourisson, G., Benveniste, P., Bimpson, T. 1975. Sterol biosynthesis in heterotrophic plant parasites. *Phytochemistry* 14:727–30

118. Rubinstein, I., Goad, L. J., Clague, A. D. H., Mulheirn, L. J. 1976. The 220 MHz NMR spectra of phytosterols. *Phytochemistry* 15:195–200

119. Safe, S., Safe, L. M. 1975. Isolation and synthesis of E-24-ethylidene-5α-lanost-8-en-3β-yl acetate. *Can. J. Chem.* 53:3247–49

120. Sancholle, M., Weete, J. D., Montant,

C. 1983. Effects of triazoles on fungi. I. Growth and cellular permeability. *Pestic. Biochem. Physiol.* 21:31–44

120a. Sandifer, R. M., Thompson, M. D., Gaughan, R. G., Poulter, C. D. 1982. Squalene synthetase. Inhibition by an ammonium analog of a carbocationic intermediate in the conversion of presqualene pyrophosphate to squalene. *J. Am. Chem. Soc.* 104:7376–78

121. Scheid, F., Benveniste, P. 1979. Ergosta-5,23-dien-3β-ol and ergosta-7,23-dien-3β-ol, two new sterols from *Zea mays* etiolated coleoptiles. *Phytochemistry* 18:1207–9

122. Scheid, F., Rohmer, M., Benveniste, P. 1982. Biosynthesis of $\Delta^{5,23}$-sterols in etiolated coleoptiles from *Zea mays*. *Phytochemistry* 21:1959–67

123. Schmitt, P., Benveniste, P. 1979. Effect of AY 9944 on sterol biosynthesis in suspension cultures of bramble cells. *Phytochemistry* 18:445–50

124. Schmitt, P., Benveniste, P. 1979. Effect of fenarimol on sterol biosynthesis in suspension cultures of bramble cells. *Phytochemistry* 18:1659–65

125. Schmitt, P., Benveniste, P., Leroux, P. 1981. Accumulation of 9β,19-cyclopropyl sterols in suspension cultures of bramble cells cultured with tridemorph. *Phytochemistry* 20:2153–58

126. Schmitt, P., Narula, A. S., Benveniste, P., Rahier, A. 1981. Manipulation by 25-azacycloartanol of the relative percentage of C_{10}, C_9 and C-8 side chain sterols in suspension cultures of bramble cells. *Phytochemistry* 20:197–201

127. Schmitt, P., Scheid, F., Benveniste, P. 1980. Accumulation of $\Delta^{8,14}$-sterols in suspension cultures of bramble cells cultured with an azasterol antimycotic agent (A25 822B). *Phytochemistry* 19:525–30

128. Schwinn, F. J. 1984. Ergosterol biosynthesis inhibitors. An overview of their history and contribution to medicine and agriculture. *Pestic. Sci.* 15:40–47

129. Seo, S., Uomori, A., Yoshimura, Y., Takeda, K. 1983. Stereospecificity in the biosynthesis of phytosterol side chains:^{13}C NMR assignments of C-26 and C-27. *J. Am. Chem. Soc.* 105:6343–44

129a. Seo, S., Uomori, A., Yoshimura, Y., Takeda, K. 1984. Biosynthesis of 24-methylsterols from [1,2-^{13}C$_2$] acetate: Dihydrobrassicasterol and campesterol in tissue cultures of *Physalis peruviana* and ergosterol in yeast. *Chem. Commun.* 1984:1174–76

130. Sliwowski, J., Kasprzyk, Z. 1974. Stereospecificity of stereo biosynthesis in

Calendula officinalis flowers. *Phytochemistry* 13:1451–57

131. Smith, A. R. H., Goad, L. J., Goodwin, T. W. 1968. The stereochemistry of hydrogen elimination at C-7 and C-22 in phytosterol biosynthesis by *Ochromonas malhamensis*. *Chem. Commun.* 1968: 926–27

132. Smith, A. R. H., Goad, L. J., Goodwin, T. W. 1968. The stereochemistry of hydrogen elimination at C-6 and C-23 in phytosterol biosynthesis by *Ochromonas malhamensis*. *Chem. Commun.* 1968: 1259–60

133. Stark, G. R., Bartlett, P. A. 1983. Design and use of potent, specific enzyme inhibitors. *Pharmacol. Ther.* 23:45–78

134. Svoboda, J. A., Thompson, M. J., Robbins, W. E., Kaplanis, J. N. 1978. Insect steroid metabolism. *Lipids* 13:742–53

135. Tomita, Y., Uomori, A. 1970. Mechanism of the biosynthesis of the ethyl side chain at C-24 of stigmasterol in tissue cultures of *Nicotiana tabacum* and *Dioscorea tokoro*. *Chem. Commun.* 1970: 1416–17

136. Van den Bossche, H., Lauwers, W., Willemsens, G., Marichal, P., Cornelissen, F., Cools, W. 1984. Molecular basis of the antimycotic and antibacterial activity of N-substituted imidazoles and triazoles; the inhibition of isoprenoid biosynthesis. *Pestic. Sci.* 15:188–98

137. Van Tamelen, E. E. 1982. Bioorganic characterization and mechanism of the 2,3-oxidosqualene-lanosterol conversion. *J. Am. Chem. Soc.* 104:6480–81

138. Van Tamelen, E. E., James, D. R. 1977. Overall mechanism of terpenoid terminal epoxide polycyclizations. *J. Am. Chem. Soc.* 99:950–52

139. Walsh, C. 1982. Suicide substrates: mechanism-based enzyme inactivators. *Tetrahedron* 38:871–909

140. Warner, S. A., Eierman, D. F., Sovocool, G. W., Domnas, A. J. 1982. Cycloartenol-derived sterol biosynthesis in the peronosporales. *Proc. Natl. Acad. Sci. USA* 79:3769–72

141. Wiggins, T. E., Baldwin, B. C. 1984. Binding of azole fungicides related to diclobutrazol to cytochrome P-450. *Pestic. Sci.* 15:206–09

142. Wilton, D. C., Munday, K. A., Skinner, S. J. M., Akhtar, M. 1968. Biological conversion of 7-dehydrocholesterol into cholesterol and comments on reduction of double bonds. *Biochem. J.* 106:803

143. Wojciechowski, L. J., Goad, L. J., Goodwin, T. W. 1973. S-Adenosyl-L-methionine cycloartenol methyltransferase activity in cell-free system from *Tre-*

bouxia and *Scenedesmus obliquus. Biochem. J.* 136:405–12

144. Wolfenden, R. 1976. Transition state analog inhibitors and enzyme catalysis. *Ann. Rev. Biophys. Bioeng.* 5:271–306

145. Woo, W. S., Kang, S. S. 1985. Triterpenoids and sterols from seeds of *Phytolacca esculenta. Phytochemistry* 24:1116–67

146. Yoshida, Y., Aoyama, Y. 1984. Yeast cytochrome P-450 catalyzing lanosterol 14α-demethylation. I. Purification and spectral properties. *J. Biol. Chem.* 259:1655–60

147. Zakelj, M., Goad, L. J. 1983. Observations on the biosynthesis of 24-methylcholesterol and 24-ethyl cholesterol by *Zea mays. Phytochemistry* 22:1931–36

148. Zimmerman, M. P., Li, H-T., Duax, W. L., Weeks, C. M., Djerassi, C. 1984. Stereochemical effects in cyclopropane ring openings. Biomimetic ring openings of all isomers of 22,23-methylenecholesterol acetate. *J. Am. Chem. Soc.* 106:5602–12

Ann. Rev. Plant Physiol. 1986. 37:309–34

MEMBRANE-BOUND NAD(P)H DEHYDROGENASES IN HIGHER PLANT CELLS

I. M. Møller

Department of Plant Physiology, University of Lund, Box 7007, S-220 07, Lund, Sweden

W. Lin

Central Research and Development Department, Experimental Station, E. I. du Pont de Nemours & Company, Wilmington, Delaware 19898, USA

CONTENTS

INTRODUCTION

NAD(P)H dehydrogenases are enzymes that catalyze the oxidation of NADH or NADPH by another redox catalyst such as cytochrome (EC 1.6.2), quinone

(EC 1.6.5), iron-sulphur protein (EC 1.6.7), or other acceptor (EC 1.6.99). They do not use O_2 as the acceptor (45, 136). The enzymes are often flavoproteins containing FMN[1] or FAD as the prosthetic group.

Some important and well-characterized membrane-bound NAD(P)H dehydrogenases are found in most or all eukaryotic cells—the mitochondrial NADH dehydrogenase and the cyt P-450 system in the ER. However, higher plant cells contain other membrane-bound NAD(P)H dehydrogenases in the mitochondrial inner membrane, in the glyoxysomal membrane, in the tonoplast, and in the plasmalemma. Thus, most major plant membranes contain an NAD(P)H dehydrogenase.

In this chapter, we review the properties and possible physiological significance of these membrane-bound NAD(P)H dehydrogenases with an emphasis on the bioenergetic aspects. In several cases, we do not know the nature of the natural acceptor and we will, therefore, use the term NAD(P)H dehydrogenase about any enzyme that can catalyze transfer of electrons from NADH (NADPH) to a synthetic acceptor like ferricyanide or duroquinone or to added cyt c.

MITOCHONDRIA

Mammalian mitochondria contain two NADH dehydrogenases. One is located in the outer membrane, while the other—the so-called Complex I—is located on the inner surface of the inner membrane. In addition to these two dehydrogenases, plant mitochondria appear to contain two more NADH dehydrogenases, one on either side of the inner membrane (for a schematic drawing, see Figure 1). Our knowledge about the NADH dehydrogenase on the outer membrane is scant, and little new information has become available since past reviews (72, 99, 101). In contrast, our knowledge about the three dehydrogenases located on the inner membrane has recently expanded rapidly, and a number of previously unexplained results can now be understood. It will, therefore, be possible to discuss the properties and possible physiological significance of these dehydrogenases in some detail (for a brief review, see 102; and for an encyclopedic, historical review, see 104).

Outer Membrane

The outer membrane of mammalian mitochondria contains a cyt b, an NADH dehydrogenase, and other enzymes that are not relevant to this review (72). Isolated outer membranes of plant mitochondria also contain a rotenone- and antimycin-insensitive NADH-cyt c reductase like their mammalian counter-

---

[1]Abbreviations used: CAM, crassulacean acid metabolism; cyt, cytochrome; DCCD, N,N'-dicyclohexylcarbodiimide; EDTA, ethylene diaminetetraacetic acid; EGTA, ethylene glycol-bis(β-aminoethylether)N,N',N'-tetraacetic acid; ER, endoplasmic reticulum; FMN, flavine mononucleotide; MDH, malate dehydrogenase; SHAM, salicylhydroxamic acid.

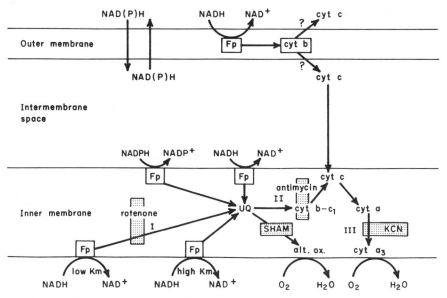

Figure 1   The localization of NAD(P)H dehydrogenases in plant mitochondria. Fp, flavoprotein; UQ, ubiquinone; I, II, III indicate the sites of energy-conservation, i.e. H^+ pumping.

parts (24, 93). It consists of a flavin dehydrogenase and a cyt b (29, 90, 93). The cyt b is different from the P-450 found in the microsomal membranes (90). The NADH dehydrogenase is specific for the 4α-proton of NADH, and this underlines the similarity with the mammalian outer membrane dehydrogenase which is also 4α-specific (29, 72).

The dehydrogenase shows high affinity for NADH (K_m 2.4 μM), high activity with NADH and different electron acceptors (ferricyanide, dichlorophenolindophenol, cyt c) and much lower activity (10% or less) with NADPH (24, 90, 93). The NADH-cyt c reductase is particularly specific for NADH, whereas the reductase in microsomes shows significant NADPH-cyt c activity (10–30% of the activity with NADH) (24, 29, 90, 93).

The outer membrane NADH dehydrogenase is insensitive to EGTA, rotenone, amytal, SHAM, antimycin, and Ca^{2+}. It is strongly inhibited by p-chloromercuribenzoate and mersalyl, indicating that SH-groups are involved (24, 90, 93), and by 2,4-di- and 2,4,5-trichlorophenoxyacetic acid (73). Since the latter two compounds do not affect microsomal NADH-cyt c reductase, this is a further indication that the outer membrane dehydrogenase is distinct from the microsomal enzyme.

The physiological role of the dehydrogenase-cyt b system in the outer mitochondrial membrane is unknown. It is induced in potato slices during prolonged washing (27). In both mammalian and plant mitochondria, it has been possible to cause antimycin-insensitive KCN-sensitive oxygen consumption via the outer membrane by the addition of NADH and oxidized cyt c (72). This is observed in mixtures of outer and inner membranes (or in mitochondria with damaged outer membranes) and is interpreted as the result of reduction of soluble cyt c on the outer membrane by the antimycin A-insensitive NADH-cyt c reductase. The reduced cyt c then passes the electrons on to the cyt c oxidase by interacting at the cyt c binding site on the outer surface of the inner membrane (Figure 1; 27, 89). Such a scheme could conceivably function in vivo, provided sufficient soluble cyt c is present in the intermembrane space, and provided the cyt c-reducing sites are on the inner surface of the outer membrane and not on the outer surface (72; Figure 1). However, it is difficult to see what advantage it has over the direct oxidation of NADH on the outer surface of the inner membrane, except perhaps to be active under conditions where the direct oxidation of NADH by the external NADH dehydrogenase is inactivated—at very low concentrations of free Ca^{2+} or when other substrates are oxidized simultaneously (see below).

The External NAD(P)H Dehydrogenase of the Inner Membrane

GENERAL PROPERTIES   Mammalian mitochondria are generally assumed to oxidize exogenous NADH only if the inner membrane is damaged (61; but see also 114). In contrast, all plant mitochondria with one known exception can oxidize exogenous NADH. Mitochondria from fresh red beet (*Beta vulgaris*) roots do not oxidize NADH. The capacity of the isolated mitochondria to oxidize NADH develops during aging of root slices, but not in the presence of cycloheximide. This indicates that cytoplasmic protein synthesis is necessary for the induction of NADH oxidation in red beet slices and that at least one polypeptide essential for exogenous NADH oxidation is coded for in the nucleus (116). It is the only report to date of genetic control over an NADH dehydrogenase in the inner membrane of plant mitochondria.

NADH oxidation shows ADP/O ratios like those obtained with succinate and about two-thirds of those with malate, and it is insensitive to rotenone. Both observations indicate that the first coupling site is bypassed (Figure 1; 43, 131). In mitochondria without an alternative oxidase, the oxidation of NADH is totally inhibited by antimycin A or KCN, just like the oxidation of other substrates. In mitochondria from nonthermogenic tissues which contain an alternative oxidase (e.g. cassava, sweet potatoes, washed potato slices, green leaves, sycamore cells, wheat shoots), NADH appears to have selective access to the cytochrome oxidase pathway, since it is more inhibited by antimycin A or KCN than electron flow from other substrates. In mitochondria from thermogenic tissues, like the spadices of *Arum*, this selectivity is not found (see 57).

With oxygen as the terminal acceptor, exogenous NADH oxidation has the following characteristics:

1. NADH is usually oxidized faster than other substrates and especially faster than the NAD$^+$-linked substrates (11, 131; see also 141 and references therein). It is oxidized particularly rapidly in thermogenic spadices where it can reach 4000 nmol O$_2$ min^{-1}mg^{-1} (56).

2. The pH optimum of NADH oxidation is 6.8–7.2 in mitochondria from *Arum maculatum* spadices, Jerusalem artichoke tubers (82, 83, 86), potato tubers (3), and spinach leaves (32). The K_m that depends on the cation composition of the assay medium and on the tissue from which the mitochondria are isolated is between 10 and 100 μM (2, 32, 43, 54, 81).

3. NADH oxidation is strongly inhibited by sulfhydryl group reagents in mitochondria from Jerusalem artichoke tubers, white turnips *(Brassica rapa)*, and *Arum* spadices (24, 82). The degree of inhibition is pH dependent (82). It may also be species-dependent, as Arron & Edwards (3) could not find any inhibition of NADH oxidation in potato mitochondria by sulfhydryl group reagents. Dicumarol is a very potent inhibitor of exogenous NADH oxidation and inhibits completely at 50 μM with a K_i of less than 5 μM (24). A similar high degree of inhibition was found for platanetin, a flavonoid derivative (129). The latter K_i was 2 μM for NADH oxidation, whereas it was 180 and 200 μM when succinate or malate were the substrates.

SPECIFIC REQUIREMENT FOR Ca^{2+} Exogenous NADH oxidation specifically requires Ca^{2+} for activity (15, 18, 32, 77, 79, 87). Plant mitochondria are isolated with sufficient membrane-bound Ca^{2+} to ensure full NADH oxidation activity in the absence of added Ca^{2+} (79), and only when this membrane-bound Ca^{2+} is removed with chelators such as EDTA, EGTA, and citrate is an inhibition of NADH oxidation seen (15, 18, 79, 87). From the experiments of Cowley & Palmer (18), it was calculated that exogenous NADH oxidation requires ca 1 μM free Ca^{2+} for activity (102), and this was confirmed by direct measurements using a Ca^{2+}-specific electrode (87).

The dependence on Ca^{2+} is strongly pH-dependent and disappears at low pH (82, 86). There is at present no satisfactory explanation for this observation. The requirement for Ca^{2+} does not appear to involve calmodulin (121), and it disappears when the enzyme is solubilized (16, 17).

The role that Ca^{2+} plays in NADH oxidation is still unresolved, but Ca^{2+} could facilitate the interaction of the enzyme with the respiratory chain (Complex III). The ease with which chelators extract Ca^{2+} is very dependent on the metabolic state of the mitochondria. If EGTA (or EDTA) is added before NADH, a strong inhibition is observed while the addition of chelator after NADH results initially in a very weak inhibition, especially if the surface potential is highly negative (see below). It appears that mitochondria actively oxidizing NADH bind Ca^{2+} more tightly than resting mitochondria (79).

Phytochrome in the Pfr form causes a small stimulation of external NADH oxidation of oat mitochondria in vitro which is reversed by red light. It has been suggested that this results from Ca^{2+} moving out of the mitochondrial matrix and activating the enzyme (118). Thus, the external NADH dehydrogenase may be involved in the photomorphogenic responses of plants.

STIMULATION BY CATIONS   In a low-salt medium and in the presence of sufficient Ca^{2+} to ensure maximal activity, the addition of salts will stimulate NADH oxidation. Salts containing monovalent, divalent, and trivalent cations give maximal stimulation at 50–100 mM, 1–10 mM, and 0.1–0.2 mM, respectively (31, 32, 39, 48, 80, 81). The effect is correlated with the reduction in the size of the negative surface potential of the membrane caused by electrostatic screening of fixed membrane charges (48). Subsequent studies have shown that changes in the apparent K_m(NADH) under different ionic conditions can be explained by the effect of the surface potential on the effective concentration of NADH (negatively charged at neutral pH) near the active site of the enzyme (32, 80, 81). However, electrostatic screening also increases the V_{max} of NADH oxidation with O_2 as electron acceptor, and one or more steps in the cytochrome oxidase pathway beyond ubiquinone appear to be involved (80, 81). It is possible that the lateral distribution of enzyme complexes in the inner membrane is affected by changes in the surface potential.

OXIDATION OF EXOGENOUS NADPH   Plant mitochondria also oxidize exogenous NADPH. Rates of NADPH oxidation comparable to NADH oxidation have been found in mitochondria from leaves, seedlings, tubers, and roots from a variety of species including CAM, C3, and C4 plants (2, 3, 32, 54, 81–83, 94, 105). The oxidation of NADPH and NADH show similar ADP/O ratios, respiratory control, and sensitivity to antimycin A, KCN, and rotenone. However, there are several differences: (a) NADPH oxidation is more sensitive to sulfhydryl group inhibitors than NADH oxidation (3, 82). (b) The K_m for NADPH is always higher than for NADH, in some cases by as much as a factor of ten (3, 32, 54, 81, 82). (c) The optimum pH is 5.5–6.8 for NADPH oxidation, depending on the tissue, which is lower than for NADH oxidation (32, 82, 83). (d) NADPH oxidation appears to require higher concentrations of free Ca^{2+} to be active than does NADH oxidation (2, 3, 94, 105). NADPH oxidation is induced by Ca^{2+} at neutral pH in mitochondria from pea leaf (94) and maize leaf (105), but not in mitochondria from spinach leaves (32; D. A. Day, M. Neuburger and R. Douce, personal communication). This may point to an interesting difference in the function of mitochondria in the leaves of different species.

NADPH oxidation is not caused by a phosphatase splitting off phosphate to form NADH from NADPH (54, 83), and there is no evidence for the presence of transhydrogenase activity in intact mitochondria that could catalyze the reduction of NAD^+ by NADPH (83). Based on the differences between the properties of NADH and NADPH oxidation, it appears most likely that two separate dehydrogenases exist, one for each coenzyme, as shown in Figure 1.

ISOLATION OF EXTERNAL NADH DEHYDROGENASE   Douce et al (29) found a 4β-specific NADH dehydrogenase in the supernatant after osmotic rupture of the outer membrane of mung bean mitochondria. The same fraction also contained an NADH-reducible flavoprotein that was thought to be part of the dehydrogenase. Three recent reports have described the partial solubilization of the inner mitochondrial membrane with detergents and the subsequent separation and partial purification of one or more NAD(P)H dehydrogenases (16, 17, 52). In two of these studies, a rotenone-insensitive NADH dehydrogenase was isolated from *Arum* mitochondria. This dehydrogenase contained a flavin and showed pH optimum and K_m for NADH similar to the external NADH oxidation of the intact mitochondria (16, 17). However, the polypeptide composition of these two preparations differed so much that it is not possible to draw any firm conclusions concerning the molecular weight or polypeptide composition of the external NADH dehydrogenase. Nevertheless, this is a research area with the potential for rapid development in the next few years.

NADH Dehydrogenases on the Inside of the Inner Membrane

In mammalian mitochondria, rotenone or piericidin causes a complete inhibition of electron transport to oxygen via the internal NADH dehydrogenase. In plant mitochondria, only a partial inhibition is observed, which is thought to be caused by inhibition of a similar enzyme. The residual respiration of plant mitochondria in the presence of inhibitors results from another NADH dehydrogenase.

MAMMALIAN MITOCHONDRIA   This NADH dehydrogenase, which is also called Complex I, forms one of the points of entry into the respiratory chain and catalyzes the transfer of electrons from NADH to ubiquinone. The purified dehydrogenase is a very complex protein containing FMN, 6 iron-sulfur centers, and at least 24 different polypeptides, with a total molecular mass of 700,000 or more. It catalyzes rotenone-sensitive electron transfer between NADH and ubiquinone with concomitant proton translocation (Site I of energy conservation) (110).

PLANT MITOCHONDRIA
The effect of Complex I inhibitors   As mentioned above, rotenone does not fully inhibit the oxidation of NAD^+-linked substrates in plant mitochondria. The degree of inhibition depends on the conditions under which it is measured. There is little or no inhibition of malate oxidation in State 4 (44, 74), while a partial inhibition by rotenone in State 3 has commonly been observed (44, 74,

130). The efficiency of rotenone in bringing about an inhibition is much lower—by a factor of 500—in plant than in animal mitochondria (44).

Amytal, another inhibitor of Complex I in mammalian mitochondria, also inhibits only partially the oxidation of malate by plant mitochondria. Its K_I is in the millimolar range and three orders of magnitude higher than that of rotenone (44).

Finally, piericidin A is a very potent inhibitor of the internal NADH dehydrogenase in plant mitochondria, having a K_I of 1 nmol (mg protein)$^{-1}$ or less. The degree of inhibition varied with the NAD$^+$-linked substrate used, but was not complete (11).

Molecular properties of the rotenone-sensitive dehydrogenase   In the presence of rotenone, proton extrusion in State 3 with NAD$^+$-linked substrates is reduced by 2–3 H$^+$/O or by one-third to one-half of the total (88, 91), and the subsequent ADP/O ratio is reduced to about two-thirds the value measured in the absence of rotenone (11, 74, 103). Thus, rotenone-sensitive respiration is linked to proton extrusion just as in mammalian mitochondria. Furthermore, three out of the six iron-sulfur centers found in mammalian Complex I have been identified in plant mitochondria (88, 104, 126). It appears reasonable to conclude that the inner membrane of plant mitochondria contains an NADH dehydrogenase that strongly resembles the mammalian Complex I. No description of the isolation and characterization of this dehydrogenase has been published.

Rotenone-insensitive oxidation of endogenous NADH   As shown above, this oxidation is not linked to proton extrusion or ATP synthesis at the first site of energy conservation, and it thus resembles the oxidation of external NADH in several ways. However, several lines of evidence show the presence of a distinct rotenone-insensitive dehydrogenase on the inner surface of the inner membrane: (*a*) Rotenone (piericidin)-insensitive, but antimycin-sensitive reduction of ferricyanide can be observed with NAD$^+$-linked substrates, indicating that ferricyanide is accepting the electrons at cytochrome *c* after the antimycin block (Figure 1; 11, 74). (*b*) On the basis of detailed studies on the temperature dependence of the rate of substrate oxidation in broad bean (*Vicia faba*) mitochondria, Marx & Brinkmann (75) could distinguish between rotenone-insensitive malate oxidation and external NADH oxidation. (*c*) The stimulation of rotenone-insensitive oxidation of, e.g., α-ketoglutarate by NAD$^+$ is prevented when entry of NAD$^+$ into the matrix is blocked by N-4-azido-2-nitrophenyl-4-aminobutyryl-3'-NAD$^+$ (23). (*d*) Rotenone-insensitive oxidation of NAD$^+$-linked substrates can be observed in intact mitochondria, even when the external NADH dehydrogenase is completely inhibited by EGTA (103). (*e*) Møller & Palmer (84) isolated about 50%

inside-out submitochondrial particles from Jerusalem artichoke mitochondria and inhibited the external NADH dehydrogenase facing the assay medium in right-side-out vesicles with EGTA. Under these conditions, two NADH dehydrogenase activities were observed, and it was concluded that they were both associated with the inner surface of the inner membrane. One was rotenone-sensitive and had a K_m(NADH) of 7 μM, the other was rotenone-insensitive and had a K_m(NADH) of 80 μM (84). The V_{max} for the two enzymes were similar. The former high-affinity dehydrogenase is assumed to be the classic proton-pumping Complex I, whereas the latter low-affinity dehydrogenase is the dehydrogenase unique to plant mitochondria. It should be noted that we do not know whether the low-affinity dehydrogenase results from a separate enzyme or from a bypass within Complex I (84). The large difference between the two internal NADH dehydrogenases in the affinity for NADH is of central importance for the interpretation of results obtained with NAD$^+$-linked substrates, especially malate, and for an understanding of the potential regulation of the Krebs cycle, as discussed below.

Malate oxidation by isolated plant mitochondria   Malate has a central role in plant metabolism (59). It is involved in such diverse physiological processes as carbohydrate and lipid breakdown, CAM and C_4 metabolism, maintenance of pH and electrical balance of the cytosol, stomatal movements, and others (35, 59, 85, 139, 141). The oxidation of malate by isolated plant mitochondria shows complex responses to substrate concentration, concentration of coenzymes, pH, respiratory state, and the presence of inhibitors. These responses can be understood on the basis of (*a*) the presence of the two internal NADH dehydrogenases with different kinetic properties (see above), and (*b*) the special properties of the matrix enzymes oxidizing malate (see below). We will demonstrate how the interactions between these enzymes determine the reduction level of the pyridine nucleotides and thus the concentration of NADH in the matrix under all conditions. This regulates the route and inhibitor sensitivity of electron transport. There is at present almost complete agreement on this interpretation (11a, 23, 84, 85, 95, 97, 102, 103, 134, 141). We will first describe the properties of the enzymes responsible for malate oxidation and then give some examples of the complex behavior of malate oxidation and how they can be rationalized.

Enzymes oxidizing malate   Malate is oxidized by two enzymes in the matrix of plant mitochondria, NAD$^+$-malic enzyme (EC 1.1.1.39) and NAD$^+$-malate dehydrogenase (EC 1.1.1.37). The equilibrium of malate dehydrogenase strongly disfavors product formation, e.g. at pH 7.2 it requires about 45,000 times more malate than oxaloacetate in the matrix to achieve a 50% reduction of NAD$^+$ [(NADH)/ (NAD$^+$) = 1.0]. Thus, the presence of even low concentrations of oxaloacetate causes the concentration of NADH to be negligible. The efficient removal of oxaloacetate is, therefore, necessary if malate dehydrogenase is to contribute NADH to the respiratory chain.

Malate dehydrogenase is by far the most active enzyme in the Krebs cycle (10, 85). Therefore, the reaction has been suggested to be near or at equilibrium in the matrix and to return rapidly to equilibrium after this has been disturbed, e.g. by addition of malate (9, 84, 102, 103).

Malic enzyme that catalyzes the decarboxylation of malate to pyruvate and the concomitant reduction of NAD^+ to NADH is much less active than malate dehydrogenase (85). Isolated malic enzyme has pH optimum at pH 6.8–6.9 (71). The enzyme is reversible in the presence of high substrate concentrations, but the rate is much lower than in the forward reaction (71). Depending on the ionic strength and the concentration of malate and citrate, the enzyme exists in three different oligomeric forms with different V_{max} and K_m (38). This may explain the observed optimum and subsequent decrease in pyruvate production when leaf mitochondria are incubated with increasing concentrations of malate (97, 134).

Malate oxidation: Observations and explanations   Observation: Oxaloacetate inhibits the oxidation of all NAD^+-linked substrates strongly (11, 28, 140). Explanation: Oxaloacetate enters the matrix rapidly on a special carrier (25) and forces malate dehydrogenase backwards, causing a complete oxidation of matrix pyridine nucleotides (95). Only when almost all the oxaloacetate has been reduced by the NADH produced by the active dehydrogenase (e.g. isocitrate dehydrogenase in citrate oxidation or malic enzyme in malate oxidation) and the concentration of NADH starts rising does the oxygen consumption resume. Oxaloacetate clearly has access to all NADH in the matrix.

Observation: External NAD^+ stimulates rotenone-insensitive oxygen uptake with all NAD^+-linked substrates even when the external NADH dehydrogenase has been inactivated (11, 23, 103). Explanation: NAD^+ enters the matrix and shifts the equilibrium of malate dehydrogenase so that the concentration of NADH rises and engages the low-affinity, rotenone-insensitive dehydrogenase. Thus, there is no evidence for the presence of a transmembrane transhydrogenase (23, 103) which has previously been postulated to transfer reducing equivalents from the matrix to external NAD^+. The explanation for the stimulation of respiration by NAD^+ is based on the very important, recent observation that NAD^+ (as well as other coenzymes) is accumulated by intact plant mitochondria in an energy-dependent manner (96, 130). The control of mitochondrial metabolism through a regulation of coenzyme levels by influx and efflux has been proposed (96). This promises to be an area where significant progress in our understanding of the role of mitochondria in the metabolism of plant cells will be made in coming years. It is especially exciting as coenzyme exchange across the inner mitochondrial membrane is unknown in yeast (132) or mammals. Either these mitochondria synthesize their own coenzymes, or coenzyme transport is so slow that is has escaped detection. The importance of maintaining the NAD$^+$ level in the mitochondrial matrix is accentuated by the observation that *Helminthosporium maydis* race T toxin may have its toxic effect by causing a release of NAD$^+$ and possibly coenzyme A from mitochondria of T-line maize (6).

Observation: Malate oxidation shows complex State 3/State 4 cycles. State 3 becomes progressively slower and State 4 is biphasic; first a very slow phase, followed by a faster phase (58, 95, 103). Explanation: During State 3 oxaloacetate accumulates, the concentration of NADH decreases as a consequence, the low-affinity dehydrogenase becomes less engaged, and the rate decreases. At the transition from State 3 to State 4, the high-affinity dehydrogenase is inactivated by the high ATP/ADP ratio; as the low-affinity dehydrogenase is still restricted by lack of NADH, the total rate is very slow. Malic enzyme, however, produces NADH continuously and forces malate dehydrogenase backwards until all the oxaloacetate has disappeared. At the point the NADH concentration starts rising in the matrix, the low-affinity dehydrogenase becomes engaged, and a faster State 4 is observed (95, 103).

The response of malate oxidation to pH (95, 97, 130, 140), rotenone or piericidin (11, 11a, 44, 95, 103, 140), and to different concentrations of malate (11, 97, 134) can be rationalized using the same principles as in the previous examples.

Lance and coworkers have proposed a selective interaction between malic enzyme and the rotenone-insensitive dehydrogenase on the one hand and malate dehydrogenase and the rotenone-sensitive dehydrogenase on the other (57, 120). This interaction is probably not structural, but results from the fact that the equilibrium of malate dehydrogenase does not normally permit its forward reaction when the NADH concentration is so high as to engage the low-affinity, rotenone-insensitive NADH dehydrogenase (see 141 for a thorough discussion of this point).

Inhibition of malate oxidation by cytokinins has been reported (13, 76). However, since it requires as much as 0.5 mM benzylaminopurine, the most active cytokinin, to be effective, and since the naturally occurring cytokinin zeatin has little or no effect (13, 76), this observation is, in our opinion, unlikely to be of physiological relevance.

Simultaneous Oxidation of Several Substrates

In vivo it is likely that the mitochondria are presented with several substrates at the same time. NADH, NADPH, pyruvate, and malate are the most likely; but in specialized tissues, succinate (lipid degradation in castor bean) or glycine (photorespiration in leaves) may contribute significantly (35, 141). The rates of oxidation of individual substrates are not additive, and the question arises of which substrate takes precedence. In green tissues, glycine takes precedence over the oxidation of all other substrates (5, 22, 30, 135), which can be seen as a mechanism for ensuring the continuous operation of the photorespiratory cycle. Glycine decarboxylase apparently competes successfully with other NAD^+-linked enzymes for NAD^+ in the matrix (22). Even when there is no competition for coenzymes, interactions between substrates are observed. Thus, electrons from internal succinate and NADH dehydrogenases take precedence over electrons from the external dehydrogenase (5, 19, 22). The mechanism by which this occurs is unknown, but Moore & Rich (88) suggest that it can be explained on the basis of simple kinetic assumptions.

Connection of NADH Dehydrogenases with the Respiratory Chain

All of the inner membrane dehydrogenases are thought to donate their electrons to ubiquinone from where the electrons will either pass down the cytochrome pathway to cytochrome oxidase and oxygen with concomitant proton extrusion and ATP synthesis, or down the alternative pathway to the alternative oxidase and oxygen without any energy coupling (Figure 1; 57, 88). It has already been mentioned that electrons from the external NADH dehydrogenase appear to have preference for the cytochrome pathway. This observation does not agree with the concept of one mobile pool of ubiquinone equally accessible to all donors and acceptors, and it may point to a nonrandom distribution of complexes in the inner mitochondrial membrane (88). An extreme example of nonrandom distribution of electron transport components is found in the photosynthetic membranes (1). Submitochondrial particles of different polarity and possibly deriving from different regions of the inner mitochondrial membrane have recently been isolated (49). Use of these particles should make it possible to determine whether similar mechanisms for regulating the lateral distribution of membrane proteins operate in thylakoid (1) and mitochondrial membranes.

Substrate oxidation in State 4 is less sensitive to rotenone, antimycin, or cyanide than in State 3 (44). Rotenone-insensitivity and cyanide-insensitivity of mitochondrial respiration develop in parallel during aging of potato tuber slices (99, 100). Finally, addition of NAD^+ stimulates cyanide-insensitive, as well as rotenone-insensitive, malate oxidation (57, 120). These observations point at the simultaneous engagement of a rotenone- and cyanide-insensitive pathway under conditions where the concentration of NADH is high. This is one reason why Palmer (100) proposed two independent parallel chains from NADH to oxygen, either sensitive to all inhibitors or completely insensitive. The observations are also consistent with the "overflow" hypothesis of Lambers (see 55 and references therein), which says that the alternative pathway is used to burn up excess carbohydrates.

TONOPLAST

The latex from the rubber tree, *Hevea brasiliensis,* contains a vacuolar compartment, the so-called lutoids. The membrane of the vesicles contained an NADH-cyt c reductase and two b-type cytochromes, neither of which was of the cyt P-450 type. The NADH-cyt c reductase was not active with NADPH, and the enzyme did not react directly with O_2 (92). In contrast, Chrestin et al (14) reported that NADH, as well as NADPH, gave rise to O_2 consumption when added to lutoid vesicles and that superoxide, peroxide, and other active oxygen species were formed as the products. It was suggested that this would result in the degradation of the lutoid membrane and the release of factors that would cause the coagulation of the latex in vivo.

NADH-cyt c reductase activity caused a decrease in the proton gradient across the lutoid membrane (inside acid) as measured by [^{14}C]-methylamine distribution. This was interpreted to result from proton pumping out of the lumen into the medium (21). An alternative interpretation is that release of protons in the reductase reaction (see section on glyoxysomes) caused an acidification of the medium, thereby decreasing the pH difference between medium and lumen. More experiments are required to determine the extent and direction of proton pumping in these vesicles [see (70a) for a detailed discussion of different models for H^+ pumping in the tonoplast and the plasmalemma].

We are not aware of any reports on NADH dehydrogenase activity in ordinary tonoplasts. The increasing research interest in the properties of isolated intact vacuoles and tonoplast vesicles makes it likely, however, that a redox system similar to that present in other plant membranes will be found also in ordinary tonoplasts in the near future.

ENDOPLASMIC RETICULUM

Mammalian microsomal membranes contain a flavoprotein NADPH-cyt c reductase linked to a cyt P-450, as well as a flavoprotein NADH-cyt c reductase linked to a cyt b_5. These systems may interact. The cyt P-450 is a so-called mixed-function oxidase that catalyzes the insertion of an oxygen atom into the substrate molecule ($X + O_2 + DH_2 \rightarrow XO + D + H_2O$, where X is the substrate and DH_2 normally is NADPH). The cyt P-450 in liver microsomes has an extremely wide specificity, whereas cyt P-450 from other mammalian tissues has a more narrow specificity. The enzyme functions in the production of secondary metabolites, as well as having a control role in the detoxification of xenobiotics (for a review, see 137).

Microsomes/ER of plant cells contain similar components. The NADPH-cyt c reductase catalyzes the hydroxylation of many hydrophobic compounds. It is implicated in the biosynthetic pathways leading to, e.g., lignins, flavonoids, cutins, suberin, gibberellins, alkaloids, and phytoalexins (see 137 and references therein). The NADH-cyt b_5 reductase is thought to be involved in the desaturation of fatty acids (124b) and may also participate in the reduction of cyt P-450 in mono-oxygenase reactions (137). Since many of the studies on the enzymology of cyt P-450 were carried out on crude microsomal preparations, some of the above reactions may turn out to be catalyzed by cyt P-450 in other membrane systems such as the plasmalemma (see below).

We are not aware of any reports on proton pumping or ion transport caused by the ER redox systems. However, crude microsomal preparations have frequently been reported to take up ions, and it is possible that some of this activity is associated with ER vesicles. The possible effect of NAD(P)H oxidation on proton pumping and ion transport should be resolved through the use of purified preparations of ER vesicles with a well-defined polarity.

GLYOXYSOMAL MEMBRANE

Glyoxysomal membranes isolated from castor beans contained large amounts of flavin, NADH-cyt c reductase, NADH-ferricyanide reductase, and cyt b_5 and P-420/450. However, the NADPH-cyt c reductase activity was very low (41 and references therein). The reductases may have binding sites for NADH on both sides of the membrane. Recently, it was reported that the acidification of the medium in isolated castor bean glyoxysomes caused by glyoxysomal NADH-cyt c reductase activity (NADH and cyt c added to the medium) was unaffected by uncoupler and was probably the result of the proton released in the reaction: NADH + 2 cyt $c(Fe^{3+}) \rightarrow NAD^+ + 2$ cyt $c(Fe^{2+}) + H^+$ (70). This does not exclude the possibility that protons are pumped into the lumen of the glyoxysomes during the reaction. Luster & Donaldson (70) suggested that NADH generated inside the glyoxysomes as a result of turnover of the glyoxylate cycle may be transferred across the glyoxysomal membrane to external acceptors.

PLASMALEMMA

The plasma membrane of mammalian, bacterial, and yeast cells is known to contain a redox system (for reviews see 20a and 68). NADH is oxidized by an NADH dehydrogenase with concomitant pumping of protons out into the medium, thus generating an electrochemical proton gradient. This gradient is then used to drive solute uptake (e.g. amino acids, organic acids, or inorganic ions) (34, 50, 113). Recent evidence suggests that a similar redox system including an NADH dehydrogenase is found in the plasmalemma of higher plant cells (for a recent brief review see 70a).

Traditionally, proton pumping ATPases have been thought to be solely responsible for the creation of a trans-plasmalemma proton gradient and thus indirectly for ion uptake into plant cells (for reviews see 42, 106, 125, 127). The existence of an alternative system for energizing the plasmalemma will, therefore, have a profound influence on our interpretation/understanding of ion uptake. For this reason, we will review the evidence for and properties of this plasmalemma redox system in detail, although our present knowledge is quite limited, as will become apparent.

Evidence for the Redox System

Evidence for the existence of a redox system in the plasmalemma has come from three sources here listed in descending order of complexity: (a) intact tissues such as intact roots or root segments, (b) single cells or protoplasts, and (c) purified plasmalemma vesicles.

WHOLE TISSUE AND PROTOPLASTS   One of the first demonstrations of a redox system in the plasmalemma was by Novak & Ivankina (46, 98), using *Elodea* and *Vallisneria* leaves. The evidence included histochemical staining and the participation of NADH in acidification of the medium and in the formation of an electrochemical transmembrane potential. It was suggested that the terminal acceptor of electrons from NADH was O_2 and that protons were pumped out as a result of plasmalemma electron transport (46, 98).

Oxidation of exogenous NADH   A similar redox system has been described for root protoplasts (see 67 for a brief review) and intact roots. The addition of exogenous NADH to a suspension of maize root protoplasts resulted in an increase in O_2 consumption, hyperpolarization of the membrane potential, increase of inorganic ion influx into the protoplasts, and increase in proton efflux from the protoplasts (64–66). Sugarcane protoplasts isolated from suspension cell culture oxidized exogenous NADH and NADPH at equal rates with a concomitant increase in O_2 uptake. In this case, however, alkalinization of the medium, acidification of the cytoplasm, as well as inhibition of K^+ and sugar uptake were reported (128). The reason for this difference between two types of protoplasts is not known, but it may be species-related.

NADH addition to corn root segments also resulted in increased O_2 uptake (65), although extensive washing of the segments may be necessary to elicit this response (53). Addition of NADH to carrot protoplasts leads to increased O_2 consumption and K^+ uptake (78). NADH was oxidized by intact oat roots and maize root segments and this oxidation was stimulated by ferricyanide (108, 119).

Reduction of ferricyanide   Carrot protoplasts and cells reduced ferricyanide, which cannot cross the plasmalemma (20), with a concomitant acidification of the medium (12). This reduction appeared to be dependent on cytoplasmic NAD(P)H as glycolytic inhibitors inhibited ferricyanide reduction (4, 20). Ferricyanide was also reduced by intact roots of oat (119), maize (33), or beans *(Phaseolus vulgaris)* (8, 122). Concomitant with ferricyanide reduction by bean roots, the transmembrane potential was depolarized (123) and the endogenous NADPH concentration decreased (124). The endogenous concentration of NADPH also decreased in corn roots during ferricyanide reduction, whereas the concentration of NADH increased (108). The latter observation, as well as other correlations between changes in rate of ferricyanide reduction and NADPH content (108), indicate that NADPH may be the natural electron donor on the cytoplasmic side of the plasmalemma.

The above observations led Lin (66) and Rubinstein et al (119) to propose the presence of two NAD(P)H dehydrogenases in the plasmalemma of root cells possibly linked to the same redox components: one system should work with exogenous NADH, the other with endogenous NAD(P)H.

PURIFIED PLASMALEMMA   Purified plasmalemma fractions are characterized by glucan synthetase II activity, binding of silicotungstic acid or N-naphthyl-phthalamic acid, and vanadate-sensitive activity of a K^+-stimulated, Mg^{2+}-dependent ATPase (47, 60, 109, 127, 138). Membrane vesicles characterized by these criteria always contain appreciable NAD(P)H-cyt c reductase activity (47, 60, 138). Pupillo & De Luca (107) reported the presence of a quinone-dependent NAD(P)H oxidase activity in a plasmalemma fraction from *Cucurbita* hypocotyls. Purified right-side-out plasmalemma vesicles consume exogenous NADH and NADPH with concomitant O_2 uptake (78a). Thus, results obtained on isolated plasmalemma are consistent with those obtained on intact tissues and protoplasts—all show the presence of an NAD(P)H oxidation system(s). We will first consider the molecular properties of this (these) system(s) and then discuss its (their) possible physiological significance.

Components of the Redox System

In mammalian cells, it is well established that electron transfer systems of diverse complexity are associated with the plasmalemma; and it seems that such redox systems may be involved in energizing metabolite transport and in regulating membrane-bound enzymatic and transport activities (20a, 36, 111). Redox constituents have also been detected in plasma membrane-enriched preparations from other eukaryotes (20a).

Evidence collected so far suggests that purified plasmalemma preparations contain NAD(P)H-cyt c reductase activity (see above), a flavin NAD(P)H dehydrogenase (47, 112), and a b-type cyt with an α-band at 556–560 nm that is reducible by blue light or dithionite (47, 63, 112, 138). The b-type cyt is likely to be of the P-420/450 type, as the CO-spectrum shows peaks at 420 and 450 nm (51). The participation of quinones in the redox system has also been suggested (66), and quinones do act as electron mediators in some systems (107).

Cytoplasmic NAD(P)H is a likely candidate for the natural electron donor to the redox system (20, 108, 124). However, NADH can also donate electrons to the outer surface of the plasmalemma of roots (53, 64, 108, 119), protoplasts (65), and purified right-side-out plasmalemma vesicles (78a). It is possible that NADH is produced by malate dehydrogenase in the cell wall and that the oxidation of this NADH by the plasmalemma redox system produces H_2O_2 or oxygen radicals used in cell wall synthesis (37).

A range of inhibitors of plasmalemma ATPase activity were applied to corn root protoplasts. Neither of the inhibitors affected basal respiration or extra O_2 consumption due to exogenous NADH (66). However, NADH-stimulated net H^+ efflux and net K^+ influx were inhibited selectively by sulfhydryl group reagents and by DCCD and ethidium bromide (66). Misra et al (78) found that DCCD inhibited NADH-stimulated O_2 consumption, as well as K^+ and sucrose uptake. These results indicate that a sulfhydryl group, as well as a proton channel, are involved in the functioning of the redox system.

Although we now know some of the components of the plasmalemma redox system, there are still many unanswered questions: Are there more components involved such as, e.g., quinone as suggested by Lin (66)? What is the pathway of electron transfer? How is it coupled to H^+ extrusion? How is the coupling to ion and sugar uptake affected? What is (are) the natural electron acceptor(s) for the redox system? How are the relative activities of the ATPase and the redox system regulated?

Physiological Significance

The ability of living plant cells to accumulate solutes greatly in excess of their concentration in the environment not only allows substrates and essential nutrients to move to particular areas of the tissue, but also maintains osmotic pressure and, hence, the volume of plant cells. The regulation of transport at the cell surface and across the plasmalemma is, therefore, vital to the understanding of cell growth and metabolism. The existence of active transport across plant membranes is well established, but the mechanism(s) responsible is(are) poorly understood. One major stumbling block is the identity of the energy source utilized in active transport.

The primary active transport process in higher plants is thought to be electrogenic proton transport (106, 125). The proton-motive force, or the electrochemical proton gradient generated by proton pumping, provides the driving force for transport of different solutes. Measurements of the electrical potential and pH in intact plant cells and isolated intact vacuoles suggest that electrogenic proton pumps are localized in the plasmalemma (a proton extrusion process) (125) and in the tonoplast (proton transport from cytoplasm to vacuole) (127, 133). Proton pumps are important not only for their role in mediating active transport of nutrients, but they are also directly involved in various physiological processes (124a). Two mechanisms appear to drive H^+ transport across the plasmalemma: (a) the hydrolysis of ATP catalyzed by membrane-bound ATPases (106, 125, 127), and (b) electron transport through a membrane-bound redox system energized by light or by the oxidation of NADH or NADPH (see above and 70a). The in vivo function of the latter system is not well understood. The proton gradient generated may drive the uptake of inorganic ions (64, 66, 67) and sugars (78). Transmembrane electron transport across the plasmalemma of intact bean roots causes a reduction of ferricyanide and other ferric complexes to ferro-complexes which are then taken up by the plants. This system is inducible by iron deficiency in dicotyledonous plants and has been suggested to be the primary mechanism for iron uptake (see 7). Many years ago, Lundegårdh (69) and Robertson (117) proposed a direct link between respiration-linked electron transport and ion uptake into roots. It seems likely that this theory will experience a revival in the coming years.

Guard cells respond to light. The swelling of guard cell protoplasts in response to light demonstrates that photoreception is within the guard cell proper (142, 143). Light stimulates proton extrusion in guard cells, induces photophosphorylation in guard cell chloroplasts, and, in the blue region, sustains wide stomatal apertures at very low quantum fluxes. Stomatal responses to light quality depend on the sensitivity of two photoreceptor systems of the guard cells, the antenna chlorophylls in the chloroplasts, and a blue light photoreceptor, presumably in the plasmalemma (142). The latter is probably part of the redox system described above and may be part of or identical to the NAD(P)H dehydrogenase (51, 62, 138).

The respiration (= O_2 consumption) of plant roots, especially the relative contribution of the alternative oxidase under different growth conditions, has attracted much attention. Cyanide, inhibitor of the cytochrome oxidase pathway, and SHAM, inhibitor of the alternative pathway (see Figure 1), have been used in these experiments. It is assumed that only mitochondrial O_2 consumption is sensitive to the inhibitors (for reviews see 55, 57). The presence of a cyanide-sensitive, O_2-consuming redox system in roots (119) or in purified plasmalemma vesicles from roots (78a) is contrary to this assumption. Furthermore, SHAM stimulates O_2 consumption by plasmalemma vesicles from wheat roots by ten- to twentyfold in the presence of NADH, and this oxidation is also

Table 1 Summary of the properties of NAD(P)H dehydrogenases and associated redox components in plant membranes[a]

Membrane	Property			
	e^- −donor	e^- −acceptor	Redox components	Suggested function
Mitochondrial				
Outer	NADH	cyt c, FeCN, DCPIP	flavin cyt b	?
Inner				
Outer surface	NAD(P)H	natural acceptor: UQ	linked to respiratory chain (see Fig. 1 and Ref. 88)	ATP synthesis; transport
Inner surface				
rotenone-sensitive	NADH	as above	as above	as above
rotenone-insensitive	NADH	as above	as above	ATP synthesis? transport? overflow mechanism
Tonoplast (lutoid)	NADH NADPH?	cyt c, O_2?	flavin? 2 cyt b	H$^+$ pumping? production of active oxygen species
ER	NAD(P)H	FeCN? cyt c	flavin, cyt b, cyt P-450	H$^+$ pumping? hydroxylation
Glyoxysomal	NADH NADPH?	FeCN, cyt c	flavin, cyt b, cyt P-420/450	H$^+$ pumping? export of reducing equivalents
Plasmalemma	NAD(P)H	FeCN, other Fe complexes, cyt c, O_2	flavin, cyt b, cyt P-420/450, quinone?	H$^+$ pumping? blue light receptor, cell wall synthesis

[a]FeCN, ferricyanide; DCPIP, dichlorophenolindophenol; UQ, ubiquinone.

cyanide sensitive (78a). SHAM-stimulated O_2 consumption by the plasmalemma may have been the reason for the increase in respiration observed when SHAM was added to pea roots (26). Irrespective of the mechanism of the SHAM-induced respiration, its presence may make it necessary to reevaluate results on root respiration obtained through the use of SHAM and cyanide.

SUMMARY AND PERSPECTIVES

The properties and possible function of the known membrane-bound NAD(P)H dehydrogenases and associated redox components are summarized in Table 1. Most major plant membranes have been reported to contain an NAD(P)H dehydrogenase system. The exceptions are the chloroplast and nuclear membranes, but we expect that even in these membranes similar redox systems will be found in the near future. The presence of a similar redox system in the cytoplasmic organelles and the plasmalemma is consistent with the endosymbiont theory of organelle evolution. However, in spite of the similarity in the composition of the redox systems in different membranes, their physiological role appears to vary widely (Table 1).

One of the factors that will affect the in vivo activity of NAD(P)H dehydrogenases is the concentration of free NAD(P)H in the compartment which the enzyme faces. For example, the activity of the plasmalemma redox system is correlated with the total concentration of NADPH in the whole tissue as discussed above. However, we know only a little about the distribution (40) and movement (see section regarding inner mitochondrial membrane) of pyridine nucleotides between the compartments of plant cells, and next to nothing about the state of the pyridine nucleotides (bound/free) within the compartments. This is an area where more research is sorely needed.

Further advances in our understanding of the properties and function of the membrane-bound NAD(P)H dehydrogenases will initially come from studies on purified membrane systems. The membrane vesicles should preferably be of a well-defined polarity, such as is obtained through phase partitioning (49, 60). However, an integration of studies at the membrane, cellular (protoplast), and whole plant (or plant organ) level will be necessary to further understand their role in plants. In particular, the results should help us understand better the intracellular as well as intercellular transport phenomena.

ACKNOWLEDGMENTS

We wish to thank numerous colleagues for help in the preparation of this manuscript. One of us (IMM) gratefully acknowledges support from the Swedish Natural Sciences Research Council. We also thank Lena Strandh (Lund) and Lynne K. Baker (Du Pont) for typing the manuscript.

Literature Cited

1. Anderson, J. M. 1986. Photoregulation of the composition, function, and structure of thylakoid membranes. *Ann. Rev. Plant Physiol.* 37:93–136
2. Arron, G. P., Edwards, G. E. 1979. Oxidation of reduced nicotinamide adenine dinucleotide phosphate by plant mitochondria. *Can. J. Biochem.* 57:1392–99
3. Arron, G. P., Edwards, G. E. 1980. Oxidation of reduced nicotinamide adenine dinucleotide phosphate by potato mitochondria. *Plant Physiol.* 65:591–94
4. Barr, R., Craig, T. A., Crane, F. L. 1985. Transmembrane ferricyanide reduction in carrot cells. *Biochim. Biophys. Acta* 812:49–54
5. Bergman, A., Ericson, I. 1983. Effects of NADH, succinate and malate on the oxidation of glycine in spinach leaf mitochondria. *Physiol. Plant.* 59:421–27
6. Bervillé, A., Ghazi, A., Charbonnier, M., Bonavent, J.-F. 1984. Effects of methomyl and *Helminthosporium maydis* toxin on matrix volume, proton motive force, and NAD accumulation in maize (*Zea mays* L.) mitochondria. *Plant Physiol.* 76:508–17
7. Bienfait, H. F. 1985. Regulated redox processes at the plasmalemma of plant root cells and their function in iron uptake. *J. Bioenerg. Biomembr.* 17:73–83
8. Bienfait, H. F., Bino, R. J., van der Bliek, A. M., Duivenvoorden, J. F., Fontaine, J. M. 1983. Characterization of ferric-reducing activity in roots of Fe-deficient *Phaseolus vulgaris*. *Physiol. Plant.* 59:196–202
9. Bowman, E. J., Ikuma, H. 1976. Regulation of malate oxidation in isolated mung bean mitochondria. II. Role of adenylates. *Plant Physiol.* 58:438–46
10. Bowman, E. J., Ikuma, H., Stein, H. J. 1976. Citric acid cycle activity in mitochondria isolated from mung bean hypocotyls. *Plant Physiol.* 58:426–32
11. Brunton, C. J., Palmer, J. M. 1973. Pathways for the oxidation of malate and reduced pyridine nucleotide by wheat mitochondria. *Eur. J. Biochem.* 39:283–91
11a. Bryce, J. H., Wiskich, J. T. 1985. Effect of NAD and rotenone on the partitioning of malate oxidation between malate dehydrogenase and malic enzyme in isolated plant mitochondria. *Aust. J. Plant Physiol.* 12:229–39
12. Chalmers, J. D. C., Coleman, J. O. D., Walton, N. J. 1984. Use of an electrochemical technique to study plasmamembrane redox reactions in cultured cells of *Daucus carota* L. *Plant Cell Rep.* 3:243–46
13. Chauveau, M., Dizengremel, P., Roussaux, J. 1983. Interaction of benzylaminopurine with electron transport in plant mitochondria during malate oxidation. *Plant Physiol.* 73:945–48
14. Chrestin, H., Bangratz, J., d'Auzac, J., Jacob, J. L. 1984. Role of the lutoidic tonoplast in the senescence and degeneration of the laticifiers of *Hevea brasiliensis*. *Z. Pflanzenphysiol.* 114:261–68
15. Coleman, J. O. D., Palmer, J. M. 1971. Role of Ca^{2+} in the oxidation of exogenous NADH by plant mitochondria. *FEBS Lett.* 17:203–8
16. Cook, N. D., Cammack, R. 1984. Purification and characterization of the rotenone-insensitive NADH dehydrogenase of mitochondria from *Arum maculatum*. *Eur. J. Biochem.* 141:573–77
17. Cottingham, I. R., Moore, A. L. 1984. Partial purification and properties of the external NADH dehydrogenase from cuckoo-pint (*Arum maculatum*) mitochondria. *Biochem. J.* 224:171–79
18. Cowley, R. C., Palmer, J. M. 1978. The interaction of citrate and calcium in regulating the oxidation of exogenous NADH in plant mitochondria. *Plant Sci. Lett.* 11:345–50
19. Cowley, R. C., Palmer, J. M. 1980. The interaction between exogenous NADH oxidase and succinate oxidase in Jerusalem artichoke (*Helianthus tuberosus*) mitochondria. *J. Exp. Bot.* 31:199–207
20. Craig, T. A., Crane, F. L. 1981. Evidence for a transplasma membrane electron transport system in plant cells. *Proc. Indiana Acad. Sci.* 90:150–55
20a. Crane, F. L., Sun, I. L., Clark, M. G., Grebing, C., Löw, H. 1985. Transplasma-membrane redox systems in growth and development. *Biochim. Biophys. Acta* 811:233–64
21. Cretin, H. 1983. Efflux transtonoplastique de protons lors due fonctionnement d'un systéme transporteur d'electrons (la NADH-cytochrome *c* reductase) membranaire des vacuolysosomes du latex d'*Hevea brasiliensis*. *C. R. Acad. Sci., Ser. III* 296:137–42
22. Day, D. A., Neuburger, M., Douce, R. 1985. Interactions between glycine decarboxylase, the tricarboxylic acid cycle and the respiratory chain in pea leaf mitochondria. *Aust. J. Plant Physiol.* 12:119–30

23. Day, D. A., Neuburger, M., Douce, R., Wiskich, J. T. 1983. Exogenous NAD⁺ effects on plant mitochondria. A reinvestigation of the transhydrogenase hypothesis. *Plant Physiol.* 73:1024–27

24. Day, D. A., Wiskich, J. T. 1975. Isolation and properties of the outer membrane of plant mitochondria. *Arch. Biochem. Biophys.* 171:117–23

25. Day, D. A., Wiskich, J. T. 1984. Transport processes of isolated plant mitochondria. *Physiol. Vég.* 22:241–61

26. de Visser, R., Blacquière, T. 1984. Inhibition and stimulation of root respiration in *Pisum* and *Plantago* by hydroxamate. *Plant Physiol.* 75:813–17

27. Dizengremel, P. 1977. Increased participation of the outer mitochondrial membrane in the oxidation of exogenous NADH during aging of potato slices. *Plant Sci. Lett.* 8:283–89

28. Douce, R., Bonner, W. D. Jr. 1972. Oxalacetate control of Krebs cycle oxidations in purified plant mitochondria. *Biochem. Biophys. Res. Commun.* 47:619–24

29. Douce, R., Mannella, C. A., Bonner, W. D. Jr. 1973. The external NADH dehydrogenases of intact plant mitochondria. *Biochim. Biophys. Acta* 292:105–16

30. Dry, I. B., Day, D. A., Wiskich, J. T. 1983. Preferential oxidation of glycine by the respiratory chain of pea leaf mitochondria. *FEBS Lett.* 158:154–58

31. Earnshaw, M. J. 1975. The mechanism of K⁺-stimulated exogenous NADH oxidation in plant mitochondria. *FEBS Lett.* 59:109–12

32. Edman, K., Ericson, I., Møller, I. M. 1985. The regulation of exogenous NAD(P)H oxidation in spinach leaf mitochondria by pH and cations. *Biochem. J.* 232:471–77

33. Federico, R., Giartosio, C. E. 1983. A transplasma membrane electron transport system in maize root. *Plant Physiol.* 73:182–84

34. Garcia, M. L., Benavides, J., Gimenez-Gallego, G., Valdivieso, F. 1980. Coupling between NADH and metabolite transport in renal brush border membrane vesicles. *Biochemistry* 19:4840–43

35. Gardeström, P., Edwards, G. E. 1985. Leaf mitochondria (C3 + C4 + CAM). In *Encyclopedia of Plant Physiology: Higher Plant Cell Respiration* (NS), ed. R. Douce, D. A. Day, 18:314–46. Berlin/New York: Springer-Verlag. 522 pp.

36. Goldenberg, H. 1982. Plasma membrane redox activities. *Biochim. Biophys. Acta* 694:203–23

37. Gross, G. G. 1977. Cell wall-bound malate dehydrogenase from horseradish. *Phytochemistry* 16:319–21

38. Grover, S. D., Wedding, R. T. 1982. Kinetic ramifications of the association-dissociation behavior of NAD malic enzyme. A possible regulatory mechanism. *Plant Physiol.* 70:1169–72

39. Hackett, D. P. 1961. Effects of salts on DPNH oxidase activity and structure of sweet potato mitochondria. *Plant Physiol.* 36:445–52

40. Hampp, R., Goller, M., Füllgraf, H., Eberle, I. 1985. Pyridine and adenine nucleotide status, and pool sizes of a range of metabolites in chloroplasts, mitochondria and the cytosol/vacuole of *Avena* mesophyll protoplasts during dark/light transition: Effect of pyridoxal phosphate. *Plant Cell Physiol.* 26:99–108

41. Hicks, D. B., Donaldson, R. P. 1982. Electron transport in glyoxysomal membranes. *Arch. Biochem. Biophys.* 215:280–88

42. Hodges, T. K. 1976. ATPases associated with membranes of plant cells. In *Encyclopedia of Plant Physiology*, (NS) ed. A. Pirson, M. H. Zimmermann, 2A:260–83. Berlin/New York: Springer-Verlag. 400 pp.

43. Ikuma, H., Bonner, W. D. Jr. 1967. Properties of higher plant mitochondria. I. Isolation and some characteristics of tightly coupled mitochondria from dark-grown mung bean hypocotyls. *Plant Physiol.* 42:67–75

44. Ikuma, H., Bonner, W. D. Jr. 1967. Properties of higher plant mitochondria. III. Effects of respiratory inhibitors. *Plant Physiol.* 42:1535–44

45. IUB 1978. Recommendations of the Nomenclature Committee of the International Union of Biochemistry on the Nomenclature and Classification of Enzymes. *Enzyme Nomenclature 1978.* New York: Academic

46. Ivankina, N. G., Novak, V. A. 1980. H⁺-transport across plasmalemma. H⁺-ATPase or redox-chain? In *Plant Membrane Transport: Current Conceptual Issues,* ed. R. M. Spanswick, W. J. Lucas, J. Dainty, pp. 503–4. Amsterdam: Elsevier/North-Holland Biomed. Press

47. Jesaitis, A. J., Heners, P. R., Hertel, R., Briggs, W. R. 1977. Characterization of a membrane fraction containing a *b*-type cytochrome. *Plant Physiol.* 59:941–47

48. Johnston, S. P., Møller, I. M., Palmer, J. M. 1979. The stimulation of exogenous NADH oxidation in Jerusalem artichoke mitochondria by screening of charges on the membranes. *FEBS Lett.* 108:28–32

49. Kay, C. J., Ericson, I., Gardeström, P., Palmer, J. M., Møller, I. M. 1985. Generation and purification of submitochondrial particles of different polarities from plant mitochondria. *FEBS Lett.* 193:169–74
50. Kilberg, M. S., Christensen, H. N. 1979. Electron-transferring enzyme in the plasma membrane of the Ehrlich ascites tumor cell. *Biochemistry* 18:1525–30
51. Kjellbom, P., Larsson, C., Askerlund, P., Schelin, C., Widell, S. 1985. Cytochrome P-450/420 in plant plasma membranes: A possible component of the blue-light-reducible flavoprotein-cytochrome complex. *Photochem. Photobiol.* 42:779–83
52. Klein, R. R., Burke, J. J. 1984. Separation procedure and partial characterization of two NAD(P)H dehydrogenases from cauliflower mitochondria. *Plant Physiol.* 76:436–41
53. Kochian, L. V., Lucas, W. J. 1985. Potassium transport in corn roots. III. Perturbation by exogenous NADH and ferricyanide. *Plant Physiol.* 77:429–36
54. Koeppe, D. E., Miller, R. J. 1972. Oxidation of reduced nicotinamide adenine dinucleotide phosphate by isolated corn mitochrondria. *Plant Physiol.* 49:353–57
55. Lambers, H. 1985. Respiration in intact plants and tissues: Its regulation and dependence on environmental factors, metabolism and invaded organisms. See Ref. 35, pp. 418–73
56. Lance, C., Chauveau, M. 1975. Evolution des activités oxydatives et phosphorylantes des mitochondries de l'*Arum maculatum* L. au cours du développement de l'inflorescence. *Physiol. Vég.* 13:83–94
57. Lance, C., Chauveau, M., Dizengremel, P. 1985. The cyanide-resistant pathway of plant mitochondria. See Ref. 35, pp. 202–57
58. Lance, C., Hobson, G. E., Young, R. E., Biale, J. B. 1967. Metabolic processes in cytoplasmic particles of the avocado fruit. IX. The oxidation of pyruvate and malate during the climacteric cycle. *Plant Physiol.* 42:471–78
59. Lance, C., Rustin, P. 1984. The central role of malate in plant metabolism. *Physiol. Vég.* 22:625–41
60. Larsson, C. 1985. Plasma membranes. In *Methods of Plant Analysis (NS)*, ed. H. F. Linskens, J. F. Jackson, pp. 85–104. Berlin: Springer-Verlag.
61. Lehninger, A. L. 1955. Oxidative phosphorylation. *Harvey Lect.* 1953–1954:176–215

62. Leong, T.-Y., Briggs, W. R. 1981. Partial purification and characterization of a blue light-sensitive cytochrome-flavin complex from corn membranes. *Plant Physiol.* 67:1042–46
63. Leong, T.-Y., Vierstra, R. D., Briggs, W. R. 1981. A blue light-sensitive cytochrome-flavin complex from corn coleoptiles. Further characterization. *Photochem. Photobiol.* 34:697–703
64. Lin, W. 1982. Isolation of NADH oxidation system from the plasmalemma of corn root protoplasts. *Plant Physiol.* 70:326–28
65. Lin, W. 1982. Responses of corn root protoplasts to exogenous NADH: Oxygen consumption, ion uptake and membrane potential. *Proc. Natl. Acad. Sci. USA* 79:3773–76
66. Lin, W. 1984. Further characterization of the transport property of plasmalemma NADH oxidation system in isolated corn root protoplasts. *Plant Physiol.* 74:219–22
67. Lin, W. 1985. Energetics of membrane transport in protoplasts. *Physiol. Plant.* 65:102–8
68. Löw, H., Crane, F. L. 1978. Redox function in plasma membranes. *Biochim. Biophys. Acta* 515:141–51
69. Lundegårdh, H. 1955. Mechanisms of absorption, transport, accumulation, and secretion of ions. *Ann. Rev. Plant Physiol.* 6:1–24
70. Luster, D. G., Donaldson, R. P. 1985. Characterization of electron transport activities of the glyoxysomal membranes. *Plant Physiol.* 73:147(Suppl.)
70a. Lüttge, U., Clarkson, D. T. 1985. Mineral nutrition: Plasmalemma and tonoplast redox activities. *Prog. Bot.* 47:73–86
71. Macrae, A. R. 1971. Isolation and properties of a 'malic' enzyme from cauliflower bud mitochondria. *Biochem. J.* 122:495–501
72. Mannella, C. A. 1985. The outer membrane of plant mitochondria. See Ref. 35, pp. 106–33
73. Mannella, C. A., Bonner, W. D. Jr. 1978. 2,4-Dichlorophenoxyacetic acid inhibits the outer membrane NADH dehydrogenase of plant mitochondria. *Plant Physiol.* 62:468–69
74. Marx, R., Brinkmann, K. 1978. Characteristics of rotenone-insensitive oxidation of matrix-NADH by broad bean mitochondria. *Planta* 142:83–90
75. Marx, R., Brinkmann, K. 1979. Effect of temperature on the pathways of NADH-oxidation in broad-bean mitochondria. *Planta* 144:359–65
76. Miller, C. O. 1982. Cytokinin modifica-

tion of mitochondrial function. *Plant Physiol.* 69:1274–77

77. Miller, R. J., Dumford, S. W., Koeppe, D. E., Hanson, J. B. 1970. Divalent cation stimulation of substrate oxidation by corn mitochondria. *Plant Physiol.* 45:649–53

78. Misra, P. C., Craig, T. A., Crane, F. L. 1984. A link between transport and plasma membrane redox system(s) in carrot cells. *J. Bioenerg. Biomembr.* 18:143–52

78a. Møller, I. M., Bérczi, A. 1985. Oxygen consumption by purified plasmalemma vesicles from wheat roots: Stimulation by NADH and salicylhydroxamic acid (SHAM). *FEBS Lett.* 193:180–84

79. Møller, I. M., Johnston, S. P., Palmer, J. M. 1981. A specific role for Ca^{2+} in the oxidation of exogenous NADH by Jerusalem-artichoke *(Helianthus tuberosus)* mitochondria. *Biochem. J.* 194:487–95

80. Møller, I. M., Kay, C. J., Palmer, J. M. 1984. Electrostatic screening stimulates rate-limiting steps in mitochondrial electron transport. *Biochem. J.* 223:761–67

81. Møller, I. M., Palmer, J. M. 1981. Charge screening by cations affects the conformation of the mitochondrial inner membrane. *Biochem. J.* 195:583–88

82. Møller, I. M., Palmer, J. M. 1981. The inhibition of exogenous NAD(P)H oxidation in plant mitochondria by chelators and mersalyl as a function of pH. *Physiol. Plant.* 53:413–20

83. Møller, I. M., Palmer, J. M. 1981. Properties of the oxidation of exogenous NADH and NADPH by plant mitochondria. Evidence against a phosphatase or a nicotinamide nucleotide transhydrogenase being responsible for NADPH oxidation. *Biochim. Biophys. Acta* 638:225–33

84. Møller, I. M., Palmer, J. M. 1982. Direct evidence for the presence of a rotenone-resistant NADH dehydrogenase on the inner surface of the inner membrane of plant mitochondria. *Physiol. Plant.* 54:267–74

85. Møller, I. M., Palmer, J. M. 1984. Regulation of the tricarboxylic acid cycle and organic acid metabolism. In *The Physiology and Biochemistry of Plant Respiration,* ed. J. M. Palmer, pp. 105–22. Cambridge: Cambridge Univ. Press. 195 pp.

86. Møller, I. M., Palmer, J. M., Johnston, S. P. 1983. Inhibition of exogenous NADH oxidation in plant mitochondria by chlorotetracycline in the presence of calcium ions. *Biochim. Biophys. Acta* 725:289–97

87. Moore, A. L., Åkerman, K. E. O. 1982. Ca^{2+} stimulation of the external NADH dehydrogenase in Jerusalem artichoke *(Helianthus tuberosus)* mitochondria. *Biochem. Biophys. Res. Commun.* 109:513–17

88. Moore, A. L., Rich, P. R. 1985. Organization of the respiratory chain and oxidative phosphorylation. See Ref. 35, pp. 134–72

89. Moreau, F. 1976. Electron transfer between outer and inner membranes in plant mitochondria. *Plant Sci. Lett.* 6:215–21

90. Moreau, F. 1978. The electron transport system of outer membranes of plant mitochondria. In *Plant Mitochondria,* ed. G. Ducet, C. Lance, pp. 77–84. Amsterdam: Elsevier/North Holland Biomed. Press. 454 pp.

91. Moreau, F., Davy de Virville, J. 1985. Stoichiometry of proton translocation coupled to substrate oxidation in plant mitochondria. *Plant Physiol.* 77:118–23

92. Moreau, F., Jacob, J.-L., Dupont, J., Lance, C. 1975. Electron transport in the membrane of lutoids from the latex of *Hevea brasiliensis. Biochim. Biophys. Acta* 396:116–24

93. Moreau, F., Lance, C. 1972. Isolement et propriétés des membranes externes et internes de mitochondries végétales. *Biochimie* 54:1335–48

94. Nash, D., Wiskich, J. T. 1983. Properties of substantially chlorophyll-free pea leaf mitochondria prepared by sucrose density gradient separation. *Plant Physiol.* 71:627–34

95. Neuburger, M., Day, D. A., Douce, R. 1984. The regulation of malate oxidation in plant mitochondria by the redox state of endogenous pyridine nucleotides. *Physiol. Vég.* 22:571–80

96. Neuburger, M., Day, D. A., Douce, R. 1985. Transport of NAD^+ in Percoll purified potato tuber mitochondria. Inhibition of NAD^+ influx and efflux by N-4-azido-2-nitrophenyl-4-aminobutyryl-3'-NAD^+. *Plant Physiol.* 78:405–10

97. Neuburger, M., Douce, R. 1980. Effect of bicarbonate and oxaloacetate on malate oxidation by spinach leaf mitochondria. *Biochim. Biophys. Acta* 589:176–89

98. Novak, V. A., Ivankina, N. G. 1978. Nature of electrogenesis and ion transport in plant cells. *Dokl. Biophys.* 242:1229–32 (Engl. transl.)

99. Palmer, J. M. 1976. The organization and regulation of electron transport in plant mitochondria. *Ann. Rev. Plant Physiol.* 27:133–57

100. Palmer, J. M. 1979. The 'uniqueness' of plant mitochondria. *Biochem. Soc. Trans.* 7:246–52

101. Palmer, J. M., Coleman, J. O. D. 1974. Multiple pathways of NADH oxidation in the mitochondrion. *Horizons Biochem. Biophys.* 1:220–60
102. Palmer, J. M., Møller, I. M. 1982. Regulation of NAD(P)H dehydrogenases in plant mitochondria. *Trends Biochem. Sci.* 7:258–61
103. Palmer, J. M., Schwitzguébel, J.-P., Møller, I. M. 1982. Regulation of malate oxidation in plant mitochondria. *Biochem. J.* 208:703–11
104. Palmer, J. M., Ward, J. A. 1985. The oxidation of NADH by plant mitochondria. See Ref. 35, pp. 173–201
105. Petit, P. 1984. Mitochondria from the mesophyll cells of *Zea mays* leaves. *J. Plant Physiol.* 116:351–64
106. Poole, R. J. 1978. Energy coupling for membrane transport. *Ann. Rev. Plant Physiol.* 29:437–60
107. Pupillo, P., De Luca, L. 1982. Pyridine nucleotide-linked dehydrogenases (quinone-dependent) in plasma membrane and endoplasmic reticulum of plant cells. In *Plasmalemma and Tonoplast: Their Function in the Plant Cell*, ed. D. Marmé, E. Marrè, R. Hertel, pp. 321–26. Amsterdam: Elsevier Biomed. Press
108. Qiu, Z.-S., Rubinstein, B., Stern, A. I. 1985. Evidence for electron transport across the plasma membrane of *Zea mays* root cells. *Planta* 165:383–91
109. Quail, P. H. 1979. Plant cell fractionation. *Ann. Rev. Plant Physiol.* 30:425–84
110. Ragan, C. I. 1980. The molecular organization of NADH dehydrogenase. *Sub-Cell. Biochem.* 7:267–307
111. Ramasarma, T., Crane, F. L. 1981. Does vanadium play a role in cellular regulation? *Curr. Top. Cell Regul.* 20:247–301
112. Ramirez, J. M., Gallego, G. G., Serrano, R. 1984. Electron transfer constituents in plasma membrane fractions of *Avena sativa* and *Saccharomyces cerevisiae*. *Plant Sci. Lett.* 34:103–10
113. Ramos, S., Kaback, H. R. 1977. The electrochemical proton gradient in *Escherichia coli* membrane vesicles. *Biochemistry* 16:848–54
114. Rasmussen, U. F., Rasmussen, H. N. 1985. The NADH oxidase system (external) of muscle mitochondria and its role in the oxidation of cytoplasmic NADH. *Biochem. J.* 229:631–41
115. Deleted in proof
116. Rayner, J. R., Wiskich, J. T. 1983. Development of NADH oxidation by red beet mitochondria on slicing and aging of the tissues. *Aust. J. Plant Physiol.* 10:55–63

117. Robertson, R. N. 1960. Ion transport and respiration. *Biol. Rev.* 35:231–64
118. Roux, S. J. 1984. Ca^{2+} and phytochrome action in plants. *BioScience* 34:25–29
119. Rubinstein, B., Stern, A. I., Stout, R. G. 1984. Redox activity at the surface of oat root cells. *Plant Physiol.* 76:386–91
120. Rustin, P., Moreau, F., Lance, C. 1980. Malate oxidation in plant mitochondria via malic enzyme and the cyanide-insensitive electron transport pathway. *Plant Physiol.* 66:457–62
121. Schwitzguébel, J.-P., Nguyen, T. D., Siegenthaler, P.-A. 1985. Calmodulin is not involved in the regulation of exogenous NADH oxidation by plant mitochondria. *Physiol. Plant.* 63:187–91
122. Sijmons, P. C., Bienfait, H. F. 1983. Source of electrons for extracellular Fe(III) reduction in iron-deficient bean roots. *Physiol. Plant.* 59:409–15
123. Sijmons, P. C., Lanfermeijer, F. C., de Boer, A. H., Prins, H. B. A., Bienfait, H. F. 1984. Depolarization of cell membrane potential during transplasma membrane electron transfer to extracellular electron acceptors in iron-deficient roots of *Phaseolus vulgaris* L. *Plant Physiol.* 76:943–46
124. Sijmons, P. C., van den Briel, W., Bienfait, H. F. 1984. Cytosolic NADPH electron donor for extracellular Fe^{3+} reduction in iron-deficient bean roots. *Plant Physiol.* 75:219–21
124a. Smith, F. A., Raven, J. A. 1979. Intracellular pH and its regulation. *Ann. Rev. Plant Physiol.* 30:289–311
124b. Soliday, C. L., Kolattukudy, P. E. 1977. Biosynthesis of cutin. ω-hydroxylation of fatty acids by the endoplasmic reticulum fraction from germinating *Vicia faba*. *Plant Physiol.* 59:1116–21
125. Spanswick, R. M. 1981. Electrogenic ion pumps. *Ann. Rev. Plant Physiol.* 32:267–89
126. Storey, B. T. 1980. Electron transport and energy coupling in plant mitochondria. In *The Biochemistry of Plants*, ed. D. D. Davies, 2:125–95. New York: Academic
127. Sze, H. 1984. H^+-translocating ATPases of the plasma membrane and tonoplast of plant cells. *Physiol. Plant.* 61:683–91
128. Thom, M., Maretzki, A. 1985. Evidence for a plasmalemma redox system in sugarcane. *Plant Physiol.* 77:873–76
129. Tissut, M., Ravanel, P., Macherel, D. 1984. Study of some natural and synthetic compounds inhibiting the electron transfer in plant mitochondria. *Physiol. Vég.* 22:607–14
130. Tobin, A., Djerdjour, B., Journet, E.,

Neuburger, M., Douce, R. 1980. Effect of NAD$^+$ on malate oxidation in intact plant mitochondria. *Plant Physiol.* 66: 225–29

131. Tomlinson, P. F. Jr., Moreland, D. E. 1975. Cyanide-resistant respiration of sweet potato mitochondria. *Plant Physiol.* 55:365–69

132. von Jagow, G., Klingenberg, M. 1970. Pathways of hydrogen in mitochondria of *Saccharomyces carlsbergensis*. *Eur. J. Biochem.* 12:583–92

133. Wagner, G. J., Lin, W. 1982. An active proton pump of intact vacuole isolated from tulip petals. *Biochim. Biophys. Acta* 689:261–66

134. Walker, G. H., Oliver, D. J. 1983. Changes in the electron transport chain of pea leaf mitochondria metabolizing malate. *Arch. Biochem. Biophys.* 225:847–53

135. Walker, G. H., Oliver, D. J., Sarojini, G. 1982. Simultaneous oxidation of glycine and malate by pea leaf mitochondria. *Plant Physiol.* 70:1465–69

136. Walsh, C. 1978. Chemical approaches to the study of enzymes catalyzing redox transformations. *Ann. Rev. Biochem.* 47:881–931

137. West, C. A. 1980. Hydroxylases, monooxygenases, and cytochrome P-450. See Ref. 126, pp. 317–64

138. Widell, S., Larsson, C. 1984. Blue light effects and the role of membranes. In *Blue Light Effects in Biological Systems,* ed. H. Senger, pp. 177–84. Berlin: Springer-Verlag

139. Wiskich, J. T. 1980. Control of the Krebs cycle. See Ref. 126, pp. 243–78

140. Wiskich, J. T., Day, D. A. 1982. Malate oxidation, rotenone-resistance, and alternative path activity in plant mitochondria. *Plant Physiol.* 70:959–64

141. Wiskich, J. T., Dry, I. B. 1985. The tricarboxylic acid cycle in plant mitochondria: Its operation and regulation. See Ref. 35, pp. 281–313

142. Zeiger, E. 1983. The biology of stomatal guard cells. *Ann. Rev. Plant Physiol.* 34:441–75

143. Zeiger, E., Hepler, P. K. 1977. Light and stomatal function: Blue light stimulates swelling of guard cell protoplasts. *Science* 196:887–89

Ann. Rev. Plant Physiol. 1986. 37:335–61

THE CONTROL BY STATE TRANSITIONS OF THE DISTRIBUTION OF EXCITATION ENERGY IN PHOTOSYNTHESIS[1,2]

David C. Fork

Department of Plant Biology, Carnegie Institution of Washington, Stanford, California 94305-1297

Kazuhiko Satoh

Department of Pure and Applied Sciences, University of Tokyo, 3-8-1, Komaba Meguro-ku, Tokyo, 153 Japan

CONTENTS

[1]CIW-DPB Publication No. 885

[2]Abbreviations used: CCCP, carbonyl cyanide *m*-chlorophenyl hydrazone; chl, chlorophyll; DAD, diaminodurene; DBMIB, 2,5-dibromo-3-methyl-6-isopropyl-*p*-benzoquinone; DCCD, dicyclohexyl carbodiimide; DCMU, 3-(3,4-dichlorophenyl)-1, 1-dimethylurea; DCPIP, 2,6-dichlorophenol indophenol; HES, high energy state; LHC, light-harvesting chl protein; MV, methylviologen; PSI, II, photosystem I, II; Q_A, the primary stable quinone acceptor of photosystem II; Q_B, the secondary quinone acceptor of photosystem II; TPTC, triphenyl tin chloride.

INTRODUCTION

In almost all algae and higher plants the absorption of light at any particular wavelength results in unbalanced excitation of the two photosystems of photosynthesis. At wavelengths below about 670 nm, PSII absorbs more light than does PSI in higher plants and green algae. At wavelengths longer than this, PSI absorption becomes predominant. This situation of unbalanced absorption is exaggerated in the red and blue-green algae where PSII absorbs about 80–90% of the wavelengths in the region from about 470 to 660 nm. However, above 660 as well as below about 450 nm PSI becomes the dominant absorbing photosystem. Since electron flow from water to carbon dioxide requires the operation in series of both PSI and PSII, it is necessary, in order to achieve maximum efficiency, that both photosystems operate at appropriate rates. State transitions provide a mechanism whereby more balanced excitation of the two photosystems can be achieved. The phenomenon of state transitions was discovered independently by Murata (104, 106) in the red algae *Porphyridium cruentum* and *Porphyra yezoensis* and by Bonaventura & Myers (26) in the green alga, *Chlorella pyrenoidosa* (see also 144, 145). These changes of state usually take place reversibly over a time scale of minutes and are induced by excess excitation absorbed by a particular photosystem.

STATE I AND STATE II

State I results from overexcitation of PSI by blue or red light absorbed predominantly by PSI in the case of the red or blue-green algae or by far-red light in green algae and higher plants. State II results from excess excitation of PSII by the wavelengths absorbed predominantly by PSII (light II).

In State I (the high fluorescent state) the absorption cross-section of PSII for light II is increased and/or energy transfer from PSII to PSI is retarded, allowing more excitation energy to reach or remain longer in PSII and thus increasing the probability of its being reemitted as fluorescence. In State II (the low fluorescent state) the absorption cross-section of PSI for light II is increased and/or energy transfer from PSII to PSI is enhanced, giving rise to a decreased probability of fluorescence being emitted by PSII and thus a decreased level of fluorescence.

In State I the quantum yield in low light intensities would be expected to increase for reactions associated with PSII. Conversely, the quantum yield of reactions associated with PSI such as oxidation of P700, cytochrome f, and of Q_A would be expected to decrease. The reverse would be true for State II.

Spillover vs Absorption Cross-Section

The distribution of light energy between the two photosystems can be altered in two different ways: by an increase in energy transfer from PSII to PSI ("spillover"; 111) or by changes in the fraction of light energy absorbed by a particular photosystem (change in cross-section). Ley (87) found that the cross-section for PSII (measured as relative flash yield for O_2 evolution) in the red alga *P. cruentum* was 50% larger in State I than in State II.

In spillover, energy transfer would proceed in the thermodynamically favored direction from PSII to PSI and not vice versa, and would presumably necessitate a close physical proximity between the antennae of the two photosystems. The transfer would be affected by the redox state of the reaction centers of PSII and be greater when the traps were reduced (closed) and less when the traps were open. By contrast, a change in the cross-section of a particular photosystem implies that the antennae pigments themselves move, thus changing their relationship and the amount of excitation energy that they deliver to a particular photosystem. Changes in amount of light energy delivered to PSII as a result of such a movement would not be affected by the redox state of the reaction centers of PSII. A quantitative treatment of these parameters has been provided by Butler's bipartite (or tripartite) model of energy distribution in the photosynthetic apparatus (30, 31, 33, 88, 148, 149).

Measurements of chl fluorescence which, for example, follow the redox state of Q_A (one of the primary quinone stable acceptors of PSII), via the kinetics of fluorescence in the presence of DCMU at room temperature, can distinguish between a mechanism for energy redistribution involving an alteration in absorption cross-section from one involving changes in spillover. A general decrease of fluorescence intensity such that F_m and F_o decrease in proportion to each other would suggest a mechanism involving a decrease in cross-section of PSII. For a complete review of fluorescence parameters in plants see (51). A decrease of fluorescence intensity such that F_m decreased relatively more than F_o (a preferential decrease of F_v) would indicate a mechanism involving increased spillover from PSII to PSI (provided that it could also be demonstrated that the energy lost to PSII resulted in enhanced PSI activity).

It is still not clear which of the two phenomena (changes in spillover or absorption cross-section) underlies state transitions. At first Murata (104) suggested changes in spillover, but after phosphorylation of LHC-II, the cross-section also seems to be affected (see section on Phosphorylation of LHC-II in vitro). Recently, Weis (147) has reported an interesting phenomenon. He measured the State I to State II transition in spinach leaves at various temperatures and found that at 15°C only the absorption cross-section was affected whereas at 35° spillover was also seen (see section on Temperature Effects on State Transitions).

MEASUREMENT OF STATE CHANGES

Fluorescence Emission Spectra

The change in the distribution of light energy between PSI and PSII resulting from state changes is manifested as changes in emission spectra of fluorescence. Murata (104) originally observed that the emission spectra of P. cruentum at 77K changed upon illumination of the sample before cooling. Preillumination with green light (567 nm) absorbed by phycobilins decreased the F685 and F695 fluorescence bands associated with PSII and increased the PSI band at 715 nm. Fluorescence bands at 648 and 664 nm from phycobilins were unchanged by green preillumination. Exposure to far-red light (694 nm) absorbed predominantly by chl a caused little change in fluorescence yield of any of the three bands. Since there was no change in the intensity of the fluorescence of the phycobilin bands, Murata postulated that a conformational change of the lamellae induced by preillumination altered the distance between, and the mutual orientations of, chl a molecules, giving rise to alterations in the rate of excitation energy transfer between them, and further suggested that the process was controlled in some way by electron transport since the changes induced by phycobilin preillumination were inhibited by DCMU and did not occur at liquid nitrogen temperature.

Preillumination at the growth temperature of the thermophilic blue-green alga (cyanobacterium) Synechococcus lividus with far-red light absorbed predominantly by PSI (light I) led to the high fluorescent (State I) condition. Cells kept in darkness or illuminated with light absorbed by PSII (light II) at 632.8 nm both produced the identical low fluorescent state, State II. The difference spectrum between the State I and the State II condition showed increased fluorescence at 685 nm and near 755 nm with a main maximum around 695 nm (47).

In an elegant series of measurements, Bruce et al (29) demonstrated with fluorescence spectroscopy having picosecond time resolution that chl a associated with PSII was uncoupled from energy transfer to PSI with their demonstration that the decay lifetime ($t_{1/e}$) of fluorescence at 695 nm was 1.6 and 2.8 times longer for P. cruentum and A. nidulans cells, respectively, in State I compared to State II. The time-resolved fluorescence emission spectra obtained at 77K by Bruce et al (29) showed similarities to decay rates measured for these algae at physiological temperatures (155). The spectral measurements made at 77K for cells in State I and II showed differences in energy transfer characteristics as influenced by state transitions without the complication of other fluorescence intensity changes that are seen at physiological temperatures.

Kinetics of Fluorescence

A transient increase of chl fluorescence that is seen after switching from light I to light II (or turning off background light I) has been used as a monitor of State I to State II transitions. However, a difficulty arises using these changes because, besides state transitions, changes of fluorescence intensity can be produced by reactions such as formation of high-energy states (HES) in thylakoid membranes (78, 79a), photoinhibition (119; for a review see 118), and the so-called Kautsky transients seen upon exposure of a dark-adapted sample to light. Photoinhibition is not reversible over the same time scale as are state transitions, however.

The characteristic features of the Kautsky transient have been labeled O, I, D, P, S, M, T, where I, P, and M designate transitory periods of high fluorescence while D, S, and T denote periods of minimum fluorescence intensity. Explanations for the origin of each of the features of the Kautsky transient have been put forward (52, 115, 126, 127). Therefore, studies employing fluorescence as a monitor of state transitions must be able to distinguish the contribution, if any, of each of these mechanisms and to separate them so that each can be observed separately. Some of the features of the Kautsky transient are seen only in intact leaves and intact chloroplasts or algal cells. Thus, the use of thylakoid membranes can eliminate some of these transients and may simplify interpretation of results.

Reports (132, 133) of rapid state I to II transitions occurring in the order of a few milliseconds in leaves based on measurements of the I-D dip of the Kautsky fluorescence transient measured simultaneously at wavelengths above 715 nm and below 690 nm seem better accounted for on the basis of reabsorption effects since intact leaves have high optical densities and strongly scatter the incident actinic light. The actinic beam is thus attenuated as it passes through the leaf. Cells lying deeper within the leaf receive weaker excitation. The resulting fluorescence is reabsorbed as it passes out of the leaf, 685 nm fluorescence being more strongly reabsorbed than long wavelength fluorescence. These types of effects may explain the different kinetics seen (as well as the differences in fluorescence between the upper and lower sides of the leaf). In fact, a clear I-D dip (as well as a P-S decline) was observed in a chl b-less barley mutant in which state changes were not observed (90). For a further discussion of these problems see (92).

In many plants the Kautsky transients are prominent and overlap fluorescence changes produced by the state transitions. One way to avoid this difficulty is to choose a material such as the blue-green algae, e.g. *Synechococcus,* where Kautsky changes constitute only about 10% of the total fluorescence signal and fluorescence transients produced by state changes are large.

Addition of DCMU also seems to be a good way to isolate the state changes. DCMU does not prevent the State II to I transition (43, 44, 88, 125, 128, 150), and in the presence of this inhibitor the Kautsky transients are eliminated and the fluorescence time course is less complex, rising from an initial low level (F_o) to a maximum level (F_m) as the traps of PSII become closed. However, there is a further slow and large increase of fluorescence in the presence of DCMU that occurs over a time scale of about a minute (100, 101, 116, 117). The slow rise seen in the presence of DCMU cannot be related to redox changes of Q_A since the traps of PSII are already closed (reduced). It has been suggested that this slow fluorescence increase in fact reflects a transition from State II to State I (40, 101, 128, 150). The slow fluorescence rise in DCMU is affected by uncouplers of phosphorylation such as CCCP and methylamine but not by inhibitors of phosphorylation such as phlorizin or DCCD (101, 128). However, Catt et al (41) and Sane et al (125) have seen marked effects of DCCD on state changes. It is likely that the State II-I transition is connected, at least in some plants, with proton movements mediated by cyclic electron flow in PSI (see section on Cyclic Electron Transport and Proton Gradients).

It is usually noted that DCMU inhibits the State I to II transition (44). This inhibition would not be expected if the return to State II is a dark process (47, 125). However, even though noncyclic electron flow is abolished by DCMU, a cyclic electron flow can still be maintained in light II that would inhibit the formation of State II by driving the reverse State II to I transition. Under very low light intensities (or in the dark; 125) DCMU did not inhibit the State I to II transition (128).

Physiological Measurements

In addition to measurement of O_2 by means of polarographic techniques, as was done originally by Bonaventura & Myers (26), it is possible to measure oxygen evolution in leaves and lichens by photoacoustic techniques. In an important series of experiments, Canaani & Malkin (38, 39) and Canaani et al (36) used fluorescence and photoacoustic techniques to measure state transitions in intact leaves. Emerson enhancement (112) was found to be a sensitive and quantitative indicator of the ratio of excitation energy distribution between PSI and PSII. Enhancement (determined as the ratio of the yields of oxygen evolution in the presence and absence of far-red background light) in State II was found to be as low as 1.06 while enhancement in State I ranged from 1.4 to 1.8. A transition from State II to State I gave rise to an increase in the yield of oxygen evolution by intermittent PSII light measured photoacoustically in the presence of background light I. In the absence of background light, however, the yield of oxygen evolution was smaller in State I than in State II. Canaani & Malkin (38) found that state changes in higher plant leaves resulted in changes in the cross-sections of light absorbed by each of the two photosystems. Thus, in State I the fraction of light absorbed by PSII (β) was 0.64 while that absorbed by PSI (α) was 0.36. In State II β was 0.46 and α was 0.43. No evidence for a change in the transfer of energy from PSII to PSI as a result of state transitions was seen in leaves (see also 94).

Interestingly, mutant barley lacking the LHC-II complex failed to show state transitions, suggesting that this pigment complex, at least in higher plants and green algae, is an essential requirement for state transitions (39, 90). Canaani & Malkin (38) suggested that their findings support the idea of a migrating light-harvesting pigment complex which associates itself more with PSII (LHC-II) in State I and is more evenly distributed between PSI and II in State II (see section on Phosphorylation of LHC-II in vivo).

MECHANISMS UNDERLYING STATE CHANGES

Control of Energy Distribution by Cations

Murata (105) and Homann (63) noted that Mg^{2+} addition to spinach chloroplasts increased the fluorescence of the PSII bands at 684 and 695 nm and decreased the PSI bands at 735 nm at 77K. Mg^{2+} addition also increased F_v, the variable component of fluorescence. Since Mg^{2+} addition also accelerated the rate of the DCPIP Hill reaction mediated by PSII and decreased the rate of $NADP^+$ reduction in the presence of reduced DCPIP and DCMU that was sensitized by PSI, Murata postulated that Mg^{2+} controlled the distribution of excitation energy between the two pigment systems. An increase of Mg^{2+} ions (or other cations) was suggested to retard spillover of excitation absorbed by chl of PSII to PSI and a decrease of Mg^{2+} to enhance spillover. The decrease of the PSII fluorescence bands seen originally in *P. cruentum* (104) during the State I-II transition was thus postulated to be controlled, as in spinach chloroplasts, by a decrease in concentration of Mg^{2+} ions in the thylakoids that occurred upon illumination of this alga with green light. Wong & Govindjee (153) measured polarization of chl fluorescence at room temperature in pea thylakoids and noted that low concentrations of mono- and divalent cations had antagonistic effects on excitation energy migration. Na^+ decreased energy migration among PSII units and increased transfer to PSI. Both effects were reversed by Mg^{2+}. Divalent cations increased energy transfer from chl *b* to chl *a* of PSII. The cation-induced changes in relative fluorescence intensities at 77K were greater than changes in lifetimes (quantum efficiencies). These differences were explained by Wong et al (154) on the basis of changes in excitation transfer from PSII to PSI and of changes in the population of fluorescent species.

Since the rate of the PSI reaction $DCPIP_{red} \rightarrow$ methylviologen (MV) is decreased by Mg^{2+} addition (27, 105), even at saturating light intensities (27, 28, 120), it is unlikely that Mg^{2+} acts by altering the distribution of excitation energy in favor of PSII (7, 26, 73, 74, 107). Bose & Ramanujam (28) suggested a heterogeneity for PSI where one fraction of PSI becomes unavailable for electron transport from DCPIP $\rightarrow$ MV in the presence of Mg^{2+}, which may hinder the approach of PSI donors by electrostatic screening (see also 72).

Krause (78, 79, 79a) and Barber and co-workers (11, 12) have seen that chloroplasts isolated so as to retain their outer membranes showed slow fluorescence changes comparable with those observed with intact leaves. A light-induced fluorescence decrease (quenching) was explained as being produced by the formation of the ΔpH component of the HES upon efflux of Mg^{2+} from the grana to the stroma that was accompanied by an inward pumping of H^+.

Cation translocation thus seemed in some respects to be controlling state transitions. However, light I and II have antagonistic effects on state changes, but they both produce electric field changes in the same direction (48), inducing an increased positive charge accumulation inside the thylakoid and a concomitant efflux of Mg^{2+}. Cations seem to affect spillover of energy from PSII to PSI but, as mentioned previously, state transitions, at least in leaves, seem to operate by a mechanism involving changes in cross-section (38, 94). Contrary to the idea proposed by Bennoun & Jupin (18), Krupinska et al (80) suggested, on the basis of results obtained with synchronously cultured *Scenedesmus obliquus,* that light-induced state changes in vivo and Mg^{2+}-induced changes of fluorescence properties of chloroplast particles show an inverse behavior indicating that a change in the ionic environment is not the main mechanism of state changes. Hodges & Barber (62) confirmed the findings of Telfer et al (140) and Horton & Black (70) that cation and phosphorylation effects are two independent processes.

Although the cation migration hypothesis appears to offer a satisfying explanation for state transitions, it nevertheless has some serious drawbacks. Some aspects of control of state transitions may, however, be linked to cation movements (see section on Dark State, for example).

Phosphorylation of LHC-II in vitro

Phosphorylation of LHC-II has recently been proposed to be the mechanism underlying the State I-II transition in higher plants and green algae (1, 9, 14, 16, 58, 66, 137). This phosphorylation is thought to be mediated by a Mg^{2+}-dependent, membrane-bound kinase (13, 15) whose activity is controlled by the redox state of plastoquinone (1, 66, 69, 139). Thus, plastoquinone is the sensor that detects unbalanced excitation of the two photosystems. Reduction of plastoquinone upon absorption of light II leads to activation of this kinase which, in the presence of ATP and Mg^{2+}, phosphorylates LHC-II. Addition of negatively charged phosphate groups to the LHC-II is thought to induce mutual repulsion, causing the proteins to move laterally and approach more closely the PSI centers in the unstacked stromal regions (86). This would allow PSI reaction centers to receive more of the energy absorbed by LHC-II (increase their cross-section). The State I–State II transition is thus the light-driven step requiring absorption of quanta by PSII. Only a small fraction of the LHC-II can be involved in these movements because, according to the model of Anderson (2), the PSI units are located almost entirely in the stroma thylakoids and in the exposed, end membranes and margins of grana stacks. These regions have been estimated to contain about 10–20% of the total LHC-II (5). Barber's model excludes the LHC-II from the nonappressed membranes and the PSI complexes from the grana partitions (9). Thus PSI and PSII units are largely separated from each other in different membrane areas. It has been reported (9) that only about 20% of the LHC-II is phosphorylated, which agrees with Andersson and Anderson's estimate of the amount of exposed PSII units mentioned above.

A pigment-protein complex containing chlorophylls a and b instead of phycobilins that was obtained from the prokaryotic blue-green alga *Prochloron* was found by Schuster et al (134) to become phosphorylated in vitro in the absence of light. Presumably this dark phosphorylation would tend to keep cells always in State II, an advantage for an organism living in deep waters where far-red light absorbed by PSI is in short supply.

Conditions that bring about the oxidation of plastoquinone (darkness in chloroplasts or excitation of photosystem I) cause the reverse State II–State I transition whereby a membrane-bound phosphatase dephosphorylates the previously phosphorylated LHC-II. Loss of the negatively charged phosphate groups would permit the LHC-II to migrate away from the PSI centers, thereby decreasing the possibility of energy input from LHC-II to PSI and allowing more energy to reach PSII centers (State I). Although several of the polypeptides that are associated with PSII are phosphorylated with the same kinetics as the LHC-II, only the kinetics of the dephosphorylation of LHC-II follow the kinetics of the decrease of the ratio of fluorescence emission at 735/685 nm at 77K (138), suggesting that phosphorylation of the LHC-II polypeptide is what controls energy distribution changes.

Haworth et al (56, 57) analyzed phosphorylated and nonphosphorylated pea and barley thylakoid membranes using the Butler tripartite model (31). In phosphorylated membranes, the proportion of absorbed quanta transferred directly to PSI (α, the absorption cross-section of PSI), was increased significantly as was $k_{T(II \to I)}$, the rate constant for energy transfer (spillover) from PSII to PSI. Energy transfer between LHC-II and PSII was diminished. These changes were reversible upon dephosphorylation. Tsala & Strasser (142), however, found in pea chloroplasts that phosphorylation changed only the absorption cross-section for PSI and not spillover from PSII to I (cf 91).

Studies of fluorescence kinetics with picosecond time resolution using spinach chloroplast fragments showed that phosphorylation produced a quenching of the slow (1.1–2.0 ns) decay phase (equivalent to quenching of variable fluorescence) and an increase of the middle (300–750 ps) phase decay component (55). The increased middle phase yield of fluorescence may arise from less efficient energy transfer between LHC-II and chl a associated with PSII (113) or from increased energy transfer to less rapidly functioning β centers of PSII (53). Haworth et al (55) interpreted their results as indicating a decreased antenna size of PSII as a result of phosphorylation [for a review see (77)].

Analysis of pea chloroplasts by electron microscopy showed that proteins of about 8 nm in diameter fracturing on the P face of the thylakoid membrane moved from the stacked regions of the thylakoids to the stroma lamellae upon phosphorylation and returned again upon dephosphorylation. Kyle et al (82) found two populations of LHC-II in isolated pea thylakoid membranes—one tightly associated with PSII and unable to move following phosphorylation and another which, upon phosphorylation, moves laterally from appressed regions of the grana to the stroma lamellae and serves as an antenna to PSI. The principal phosphorylated polypeptides migrating to the stroma were found to be in the 26–27 kDa range. Subsequent to migration the chl a/b ratio decreased and F681 at 77K increased in stroma lamellae. These data and those of Gerola et al (50) and Black et al (25) support the so-called mobile antenna hypothesis (84), where only the LHC-II antenna moves to the stroma, rather than the idea proposed by Barber (9, 10), where an entire complex of PSII, along with its antenna, moves following phosphorylation.

Incorporation of cholesterol hemisuccinate into thylakoid membranes of pea, which reduced their fluidity (135), suppressed state transitions presumably as a result of inhibition of this lateral migration of phosphorylated LHC-II in the more rigid membranes (54).

Biggins (20) investigated the effect of membrane phosphorylation on linear dichroism at room temperature using pea thylakoids oriented in a magnetic field. No changes were seen in linear dichroism which would indicate a reorientation of specific pigments. However, small changes in selective scattering of polarized light were seen which suggest the membrane phosphorylation had resulted in a small decrease in thylakoid stacking; this is in agreement with the findings of Kyle et al (84) of a 23% decrease in the amount of stacked membranes after phosphorylation.

The alteration of energy distribution favoring system I upon phosphorylation of LHC-II as a result of a state I-II transition were significant enough so that increased activity of PSI would be expected after such a transition. Such was indeed the case as Farchaus et al (45) and Horton & Black (69) found an increased quantum yield for partial reactions of PSI (DCPIP$_{red}$→MV and DAD$_{red}$→MV) after phosphorylation of chloroplast fragments of pea or spinach (see also 136). Interestingly, it was necessary to adjust the experimental protocol to reduce the amount of photoinhibition that occurred upon the illumination treatment used to activate the chloroplast kinase. When this was done phosphorylation brought about a decrease of the PSII reaction H$_2$O→DAD$_{ox}$ of about 15–20% (45). Perhaps the larger decrease of PSII activity (33%) seen by Steinback et al (138) resulted from a contribution of photoinhibition to the decrease in activity of PSII in addition to that produced by phosphorylation of the LHC-II.

In *Synechococcus lividus* the photooxidation of cyt c_{553} by PSI under light-limiting conditions and in the presence of DCMU and reduced DBMIB to inhibit reduction of the cytochrome by reduced plastoquinone was about 30% lower in State I compared to State II (47). By contrast, Haworth & Melis (59) were unable to see any effect on the kinetics of P700 oxidation of phosphorylation of spinach thylakoids and suggested that a State I to II transition did not increase the absorption cross-section of PSI. Markwell et al (95) found that zinc ions stimulated phosphorylation of LHC-II, but there was no accompanying decrease of PSII fluorescence.

Energy transfer between adjacent units of PSII are known to occur (32, 34, 75). This effect results in the curve of the fluorescence increase with time in the light as PSII traps become reduced to assume a sigmoidal rather than an exponential shape. This "fluorescence rise curve" was further analyzed into two components, a fast nonexponential component (termed the α component) and a slower, exponential β component (97–99). The α component was suggested to reflect PSII units that can transfer energy among themselves and the β component thought to represent disconnected PSII units that cannot transfer energy among themselves but probably only to PSI units (separate packages). Phosphorylation has been found to lead to a reversible decrease of the α component and an increase of the β component (64, 68, 81), suggesting that in State II there is a decrease in interunit energy transfer in PSII and, by inference, an increased possibility for energy transfer from PSII to PSI.

Hodges & Barber (61) have argued that the two components seen in the fluorescence rise curve may simply be a manifestation of the existence of two populations of PSII centers some more easily inhibited by DCMU than others. Also, Telfer et al (140) and Horton & Black (70) concluded that changes in α and β levels induced either by phosphorylation of LHC-II or changes in cation levels did not support the concept of two different structural forms of PSII, as suggested by Anderson & Melis (4).

Phosphorylation of LHC-II in vivo

Most of the conclusions concerning the role of phosphorylation of LHC-II in controlling state transitions have been reached on the basis of studies done in vitro utilizing chloroplasts or thylakoid membrane preparations. However, Owens & Ohad (114) supplied intact cells of *Chlamydomonas reinhardtii* with ^{32}P orthophosphate and separated the phosphorylated polypeptides. Phosphorylation of thylakoid membrane polypeptides was completely independent of illumination. Similar incorporation of phosphate was seen in cells illuminated with red or far-red light, in cells given white light or kept in darkness, or in cells treated with DCMU or in controls. At least 80% of the membrane phosphoproteins were in the continuously phosphorylated state.

Owens & Ohad (114) postulated that there may be two populations of LHC-II, one whose phosphorylation affects mainly the interaction such as seen by Horton & Black (67) between LHC-II and PSII in the grana, and another population in the stroma lamellae where phosphorylation affects its interaction with PSI units. Another interpretation of these results is that phosphorylation of LHC-II is not the mechanism controlling state changes in *Chlamydomonas*. Wollman & Delepelaire (152) concluded in their study of *C. reinhardtii* that state changes both in vitro and in vivo are associated with phosphate incorporation into certain thylakoid polypeptides and that this incorporation did not depend upon the presence of PSII centers but was mediated by the redox state electron transport carriers.

Parallel measurements of ATP-induced changes in fluorescence behavior of thylakoid membranes isolated from *Chlorella* and the State I to State II transition of intact cells suggested to Saito et al (123) that phosphorylation of LHC-II is involved in this transition in vivo. By contrast, Sane et al (125) found that light II induced a State I to II transition in *Chlorella* when photophosphorylation was blocked either by uncouplers or energy transfer inhibitors, suggesting that photophosphorylation is not related to the development of State II. Baker et al (8) failed to find a correlation between the light-induced State I to II transition and protein phosphorylation in wheat thylakoids. Moreover, addition of zinc ions inhibited the transition but stimulated LHCP phosphorylation. Sane et al (124) saw in chloroplasts a heat-induced development of State II as identified by fluorescence characteristics and electron transport activities in samples not supplied with ATP or substrates for phosphorylation, suggesting that this state can develop independently of LHC-II phosphorylation (see section on Temperature Effects on State Changes).

State I-II transitions were studied in vivo in leaves of tobacco by Canaani et al (35, 37) by using both photoacoustic and fluorimetric techniques. Immersion of leaves in a solution containing Na fluoride abolished the State II to State I transition presumably by blocking the dephosphorylation of LHC-II. This conclusion, however, requires the acceptance that Na fluoride treatment of leaves specifically affected only the phosphatase catalyzing the LHC-II de-phosphorylation. However, Na fluoride was specific in its effect in that NaCl did not produce a similar result. It is clear from all of the above-discussed results that more work is needed in order to elucidate the role that polypeptide phosphorylation plays in state transitions in vivo.

Cyclic Electron Transport and Proton Gradients

The question of whether a reversible protein phosphorylation event is function-ing to control state transitions in a red alga was addressed by Biggins et al (23). Although some 12 polypeptides were found to be phosphorylated in the light, there was no specific difference between cells in State I or II, suggesting that a mechanism involving a phosphorylation of a light-harvesting pigment complex like that proposed for higher plants (1, 13, 15, 69) is not functional in blue-green and red algae. Moreover, in the blue-green alga *Synechococcus lividus* the redox state of plastoquinone (see section on Phosphorylation of LHC-II in vitro) had no influence on the State I to II transition nor on the State II to I transition (128).

A State II to I transition in the presence of light absorbed preferentially by PSI is inhibited by DBMIB (21, 128), which prevents oxidation of reduced PQ, and by MV which draws electrons from PSI (128). Thus a cyclic electron transport pathway mediated by PSI via the cytochrome b_6-f complex must be required for the transition to State I, and this electron transport must be energetically coupled since proton ionophores such as CCCP prevent the transition (24, 128).

Biggins et al (23) proposed for *P. cruentum* that changes in the localized charge distribution, perhaps induced by localized gradients of protons gener-ated by PSI or by changes in redox levels of electron transport components, lead to changes in the physical proximity between the two photosystems in such a way as to alter the probability of energy transfer between them. Evidence that such movements are involved in state transitions comes from experiments done with cells of *P. cruentum* chemically cross-linked with glutaraldehyde, which fixes them in a particular state and prevents further changes (21). Low tempera-ture in thermophilic algae also eliminates state transitions, presumably by hindering movement of pigment complexes in the rigid gel phase of the thylakoid membrane lipids (see section on Temperature Effects on State Changes).

Sane et al (125) found the uncoupler of photophosphorylation, CCCP, to block completely the development of State I in *Chlorella* cells. In addition, both DCCD and TPTC, inhibitors of ATP synthesis, also prevented adaptation to State I, which suggests that membrane energization is necessary for the de-velopment of State I. As mentioned previously, in the presence of all three of these inhibitors dark incubation did not return the cells to State II, but illumina-tion with light II did bring about State II. It was suggested by Sane et al (125) that certain redox changes of electron transport carriers are necessary to permit State II formation.

Biggins & Bruce (22) saw, as did Ried & Reinhardt (122), that the State I to II transition in red algae showed a photochemical dose dependence under relatively low excitation intensities and was biphasic. In *P. cruentum* the slow phase was correlated with the induction of photosynthesis and could be eliminated if the preceding time spent in State I was very short. With high intensity light the fast phase dominated and transition to State II was completed within 1 s. The photochemical turnover time of the fast phase was determined with single turnover flashes to be less than 30 ms. The State II to I transition was not dose dependent when induced by single-turnover flashes. Conversion to State I required only 15 turnovers of PSI when the dark interval between flashes was of the order of 400 ms. This optimum frequency of 2.5 Hz corresponds to the turnover time of cyclic electron transport in *P. cruentum* (19, 96) and is consistent with the mechanism proposed for red and blue-green algae mentioned previously, where transition to State I is driven by cyclic electron flow around PSI that generates localized electrochemical potential differences of H^+.

OTHER STATES

Dark State

An unsettled question relating to State I-II transitions is the state that exists upon dark adaptation. Murata (104) originally proposed that the dark state was the high fluorescent State I in *P. cruentum,* and Bonaventura & Myers (26) suggested that *Chlorella* was in the same state in the dark. However, Bennoun (17) and Ley & Butler (89) suggested that the dark state was a mixed condition of States I and II in *Chlamydomonas* and *P. cruentum,* respectively. Chow et al (42) suggested that State I existed in the dark in higher plants. By contrast, Ried & Reinhardt (121) found a number of species of red algae to be in State II in the dark, and Williams & Salamon (151) and Williams et al (150) found *Chlorella* to be in State II in the dark, as did Sane et al (125). Satoh & Fork (129) reported the dark state to be State II in the red alga *Porphyra perforata,* the blue-green alga *Synechococcus lividus* (47), and the green alga *Scenedesmus obliquus* (130).

As mentioned previously, difficulty arises when fluorescence measurements are used to study state transitions in that Kautsky transients and state transition have to be distinguished from each other with regard to their effects on fluorescence. When this confusion was avoided by using materials where the rates of dark recovery of the Kautsky effect and state transitions were different (121), or the Kautsky effect was small compared to state transitions (47), then the dark state was found to be State II. Measurement of O_2 evolution either by photoacoustic techniques (38, 39) or polarographically is a more direct and perhaps safer way to assess state transitions, and Williams & Salamon (151) used the latter technique with *Chlorella* to show it existed in State II in the dark. According to Weis (146), the dark state in spinach varies with the temperature; below 15°C leaves attain State I and above 25°C, State II. It will be very interesting to know whether a similar phenomenon can be observed in algae. Sane et al (125) found that the dark state was not always identical to State II because in the presence of CCCP, DCCD, or TPTC dark incubation did not induce State II, while light II was still able to induce State II.

Catt et al (40, 41) have proposed that State I and II and the dark state represent three distinct states. Although the dark state and State II both showed the expected lower level of F_m compared to State I in DCMU-poisoned *Chlorella*, it was observed that the F_m level was lower in the dark-adapted cells compared to cells in State II. Thus, it was suggested that State II and the dark state were different in origin. Two light-driven processes were suggested to be involved: a low light intensity, wavelength-independent change and a higher intensity, wavelength-dependent change reflecting a State I to II transition. The low intensity effect was thought to be associated with changes in the local ionic environment within the chloroplast and the State I to II transition with phosphorylation of LHC-II of the type discussed already (1, 65).

Thus, in the dark the fluorescence is at a level slightly lower than State II. The low intensity wavelength-independent reaction would lead to Mg^{2+} efflux from the thylakoids and produce a fluorescence increase to a level near State I. Phosphorylation of LHC-II would bring about a fluorescence decrease as the State I to II transition occurred. Catt et al (41) suggested that light preferentially absorbed by PSII had little or no effect on dark-adapted algae (cf 47) because the fluorescence decrease produced by LHC-II phosphorylation during a State I-II transition was compensated for by a fluorescence increase produced by the low light intensity driven Mg^{2+} efflux. The fluorescence changes attributed to a State II to I transition reported for a *Scenedesmus* mutant lacking LHC-II (130) were thus explained as being produced instead by a low light intensity Mg^{2+} efflux reaction (41) and not by state changes.

As explained in the section on cation-induced control of state changes, an explanation for state transitions based on a mechanism involving cation movement contains some major shortcomings, although it does retain attractive features. The proposals of Catt et al (41) described above suggest that H^+ influx and Mg^{2+} efflux caused a fluorescence increase, an idea that is counter to the widely observed high energy-induced fluorescence quenching upon influx of H^+ and efflux of Mg^{2+} ions. State II and the dark state were thought to be two separate states because the dark state had a lower fluorescence level than did State II. The State I to II transition occurred with a half time of about a minute while the State I to dark transition required about 7 min (41). This is not particularly strong evidence that the dark state and State II are different states because light II might hasten a process that would occur slowly in the dark.

The dark state was proposed to be the nonphosphorylated state. If so, we must, under certain conditions, be able to find another dark phosphorylated state. The result that the return in the dark to this nonphosphorylated state from either State I or State II proceeded with the same half time can be explained only by assuming that the rate of dephosphorylation and that of dark recovery from the low light state are the same, an idea that has yet to be demonstrated.

Low Light Intensity State

Canaani & Malkin (39) described a new "low light-intensity state" in leaves of higher plants in which short wavelength excitation (640 nm) did not reach the PSI reaction centers and in which the yield of O_2 evolution dropped nearly to zero. This state was induced by low light intensities, typically around 1 W/m^2. Surprisingly, in the low light state the fluorescence yield remained virtually constant and did not increase dramatically, as would be expected, as 640 nm light, not transferred to PSI, would reduce all the PSII reaction centers. This low fluorescent state was explained as being maintained by a cyclic flow of electrons around PSII which would lead to no net oxygen evolution but would keep the steady-state level of fluorescence low. Note that their low light intensity state is different from that of Catt et al (41) mentioned above.

Canaani & Malkin (39) proposed that the low light state resulted from a detachment (or change in orientation) of the major part of the antennae of PSI with respect to the reaction centers of PSI. However, even in this state the reaction centers of PSI remain photochemically active because they retain a certain amount of "core" far-red absorbing chl pigments (less than 30 molecules) closely associated with them (60).

Reassociation of the major antennae to the reaction centers of PSI is induced by exposure of leaves to "normal" light levels or by a short period of high light intensity. This recovery, which lasts some minutes, is preceded by a photochemical reaction triggered by the absorption of about 1 photon per 20 chl molecules. Since the PQ pool also has about 20 chl molecules associated with it (93), Canaani & Malkin suggested that its reduction triggers the low light state similar to its proposed triggering of the State I-II transition. A chl-b-less barley mutant, lacking both LHC-II (141) and the newly discovered light-harvesting chl a/b protein associated with PSI, designated LHC-I (60), did not show a transition to the low light state, suggesting that LHC-I and perhaps LHC-II are necessary for the adaptation to low light intensities. However, care should be used in drawing conclusions based on experiments using mutants because, for example, 650 nm light is no longer light II in chl b-less mutants.

It is possible, however, to explain the near zero levels of O_2 evolution in low light on the basis of a mechanism other than that postulated by Canaani & Malkin (39) where low light intensities induce a detachment of CP-I from the

PSI reaction center. More evidence will be needed to support the concept of a detachable PSI antenna since it is known that LHC-I is tightly bound to the reaction centers of PSI and that these centers have more than 60 chl molecules associated with each P700 (103).

Even though we accept the concept of a detachable LHC-II, it is still difficult to explain simply by this mechanism the absence of O_2 evolution at light levels below about 0.2 Wm^{-2} such as seen by Canaani & Malkin (39) in leaves in the low light intensity state because some electrons can still flow through PSI and II. Rather, this suggests that some other reactions are involved in the absence of O_2 evolution such as inactivation of the CO_2-fixing enzymes, ferredoxin-$NADP^+$ reductase (126), or ATPase (102).

State III

A new light state was found to exist in the intertidal red alga *Porphyra perforata* (129, 131) whereby preillumination with light absorbed by PSII decreased PSII activity without having an effect on PSI activity. A transition to State III was marked by parallel decreases of the two PSII fluorescence bands seen at 77K at 685 and 695 nm without an increase in the PSI fluorescence band at 735 nm. The time course of fluorescence in *Porphyra* treated with DCMU showed an initial fluorescence level upon illumination (F_o) that was markedly reduced in State III, as was the rate of rise of variable fluorescence (F_v). These observations suggest that light II treatment did not change the transfer of energy from PSII to I but rather altered the arrangements of the pigments in PSII in some way so that less energy arrived at the reaction centers of PSII.

Since this alga is often exposed in its natural habitat to potentially photoinhibiting high irradiance levels, often in combination with severe dehydration and high salinity, it is possible that a transformation into State III represents a protective mechanism whereby excess quanta are not delivered to PSII but are diverted to some other pathway allowing harmless deexcitation.

The State II to III transition has both a light and dark process (131). Light-induced proton transport across the thylakoid membranes seemed to initiate the transition. The dark process of the transition continued for several minutes afterwards. A prolonged dark period returned the algae to State II, but light I accelerated this transition from State III to II, perhaps as a result of a membrane potential generated across the thylakoid membranes by cyclic electron flow around PSI. More work is needed to understand the functional significance of State III.

STATE TRANSITIONS AND PHOTOPROTECTION

It is reasonable to assume that state transitions, in addition to their role in optimizing photosynthesis, may serve to protect plants against damage induced as a result of absorption of excess quanta. State III, as discussed above, may be one of the mechanisms allowing *Porphyra perforata* to thrive in an environment producing recurring periods of severe stress. Horton & Lee (71) found that phosphorylation of chloroplast thylakoid membranes stimulated what appeared to be a cyclic flow of electrons around PSII. A similar conclusion was reached by Jursinic & Kyle (76) on the basis of studies on pea thylakoids. A cyclic flow of electrons around PSII could serve as a means of dissipating excess light energy absorbed by the reaction centers of PSII and thereby prevent photoinhibition. Kyle et al (83) proposed that plants have developed a coordinated defense against photodamage which involves the rapid repair of photo-damaged Q_B protein along with State II induction from LHC-II phosphorylation whereby quanta may be redirected from PSII to PSI (or perhaps cycled around PSII).

TEMPERATURE EFFECTS ON STATE TRANSITIONS

In an interesting set of experiments, Sane et al (124) showed that mild high temperature treatment (45–48°C for 2 min) induced State II in *Chlorella* cells, *Canna* leaves, and spinach chloroplasts and that cooling returned the samples to State I. Heated cells retained the ability to undergo state changes since light I irradiation at elevated temperatures returned the cells to State I. Sane et al (125) had previously seen that the dark-adapted ("relaxed") cells of *Chlorella* acquired the characteristics of State II (see section on Dark State). Since elevated temperatures also induced State II, it was proposed that the increased fluidity of the membranes at higher temperatures facilitated this relaxation to State II. It was further suggested that the heat-induced transition to State II might represent a type of protective mechanism against photooxidative damage since in State II at elevated temperatures more of the quanta absorbed by PSII would be transferred to PSI which is more resistant to photooxidative damage.

Weis (146–148) measured fluorescence kinetics in intact spinach leaves at 77K as well as fluorescence emission spectra in "diluted" leaves that were prepared by grinding the frozen leaf and mixing the resulting powder with ice and quartz crystals, a procedure that minimized reabsorption effects (148). A State I to II transition produced the expected increase in fluorescence of PSI at 735 nm and decrease of PSII fluorescence at 695 nm. The kinetics of the transition to State II were the same in leaves at 15° and 32°C, but at 32°C the extent of the change was increased. The ratio of F_I/F_{II} increased from 1.4 to 2.0 in passing from 5 to 35°C. Using Butler & Kitajima's model (33), an analysis was made of the effect of a transition to State II in leaves at 15° and 35°C on the absorption cross-section of PSI (α) and on the transfer of energy (spillover) from PSII to PSI [$k_{T(II \rightarrow I)}$]. At 15°C a transition to State II increased α by about 15% and had no affect on $k_{T(II \rightarrow I)}$. At 35°C α increased by 20% but $k_{T(II \rightarrow I)}$ increased by 60%. The increased spillover seen at the higher temperature was attributed to an increased lateral mobility of the pigment complexes in the thylakoid membranes (147).

While high temperatures may facilitate state transitions, low temperatures impede them. Low temperature effects are most readily observed in the thermophilic blue-green algae where it is possible to have thylakoid membrane lipids passing from the liquid crystalline into the gel phase at temperatures above 0°C. Measurements made with *Anacystis nidulans* grown at 38°C showed that state changes were drastically reduced below about 22°C, and for cells grown at 28°C this critical temperature was near 13°C. These temperatures correspond to the onset of gel phase formation as the temperature is lowered in this alga (6, 49, 108–110, 143). In the thermophilic blue-green alga *Synechococcus lividus* grown at 55°C, both the State II to I and the I to II transitions declined at temperatures below 44°C, the phase transition temperature for this alga (46, 47, 128). In some cases a second critical temperature was seen near 38°C. It was possible to "freeze" these algae in one state or the other by suddenly lowering the temperature to below the phase transition temperature (~20°C for example), thus allowing other measurements to be made without changing the state of the cells.

These critical temperatures for state transitions in the blue-green algae suggest that movement of molecules, perhaps the chl proteins, in the bulk lipids of the thylakoids is necessary in order that state transitions take place. The second transition temperature near 38°C may correspond to a phase transition of minor lipids more closely associated with the chl proteins (128). The need for mobility of pigment assemblies was also demonstrated in the glutaraldehyde fixation experiments of Biggins (21; see section on Cyclic Electron Transport and Proton Gradients).

CONCLUDING COMMENTS

Many investigations of state transitions have revealed that the pigment assemblies of photosynthetic plants are dynamic and rearrange themselves to best absorb and distribute available light to maximize photosynthesis. Much of the present effort in this area is now concentrated on elucidating the mechanism(s) by which this process takes place. Reversible phosphorylation of LHC-II is an attractive mechanism to explain energy redistribution that can clearly be demonstrated in vitro. The demonstration of the operation of this mechanism in vivo is a much more difficult task, and evidence seems about equally divided between positive and negative results.

As suggested by Larkum & Barrett (85), perhaps the formation of stacked thylakoid membranes in green plants and most algae is a mechanism for absorbing more effectively the green and yellow wavelengths, as a result of increased scattering and pigment concentration effects, without the expense of synthesizing N-rich phycobilins as has been done in the blue-green and red algae. Anderson (2, 3) has pointed out that thylakoid stacking which is particularly well developed in shade plants reduces the amount of energy transferred from PSII to PSI since LHC-II is located only on the exposed surfaces and the amount of spillover is roughly proportional to the ratio of exposed to appressed membranes. Red and blue-green algae have approached the problem of adapting to environments ranging from high light to extreme shade by employing nonstacked thylakoids having phycobilisomes to absorb green and yellow wavelengths and, as has been pointed out in this review, these algae have prominent state transitions. It is therefore not surprising that these plants may use a mechanism different from that of green plants and other algae to control light energy distribution.

State transitions may also play a role in protecting plants from photooxidative damage by limiting the amount of light energy reaching PSII and redirecting it to PSI for dissipation. Since relatively little work has been focused in this direction, it is apparent that more studies on the role played by state transitions in photoadaptation and photoprotection will further advance our understanding of how plants have accomodated to widely different habitats.

ACKNOWLEDGMENTS

Many thanks to J. Biggins, S. Bose, D. Bruce, O. Canaani, Govindjee, P. Haworth, S. Malkin, E. Weis, and W. P. Williams, who provided preprints of their manuscripts or who gave suggestions during the preparation of this manuscript.

Literature Cited

1. Allen, J. F., Steinback, K. E., Arntzen, C. J. 1981. Chloroplast protein phosphorylation couples plastoquinone redox state to distribution of excitation energy between photosystems. *Nature* 291:25–29
2. Anderson, J. M. 1981. Consequences of spatial separation of photosystem 1 and 2 in thylakoid membranes of higher plant chloroplasts. *FEBS Lett.* 124:1–10
3. Anderson, J. M. 1982. The significance of grana stacking in chlorophyll *b*-containing chloroplasts. *Photobiochem. Photobiophys.* 3:225–41
4. Anderson, J. M., Melis, A. 1983. Localization of different photosystems in separate regions of chloroplast membranes. *Proc. Natl. Acad. Sci. USA* 80:745–49
5. Andersson, B., Anderson, J. M. 1980. Lateral heterogeneity in the distribution of chlorophyll-protein complexes of the thylakoid membranes of spinach chloroplasts. *Biochim. Biophys. Acta* 593:427–40
6. Armond, P. A., Staehelin, L. A. 1979. Lateral and vertical displacements of integral membrane proteins during lipid phase transition in *Anacystis nidulans*. *Proc. Natl. Acad. Sci. USA* 76:1901–5
7. Arntzen, C. J., Briantais, J. M. 1975. Chloroplast structure and function. In *Bioenergetics of Photosynthesis*, ed. Govindjee, pp. 52–113. New York: Academic. 698 pp.
8. Baker, N. R., Markwell, J. P., Webber, A. N., Thornber, J. P. 1984. The role of light-harvesting complex phosphorylation in mediating the state 1–state 2 transition: a re-examination. In *Advances in Photosynthesis Research*, ed. C. Sybesma, 3:319–22. The Hague/Boston/Lancaster: Martinus Nijhoff/Dr. W. Junk. 927 pp.
9. Barber, J. 1982. Influences of surface charges on thylakoid structure and func-

tion. *Ann. Rev. Plant Physiol.* 33:261–95
10. Barber, J. 1983. Membrane conformational changes due to phosphorylation and the control of energy transfer in photosynthesis. *Photobiochem. Photobiophys.* 5:181–90
11. Barber, J., Telfer, A. 1974. Ionic regulation in chloroplasts as monitored by prompt and delayed chlorophyll fluorescence. In *Membrane Transport in Plants,* ed. U. Zimmerman, J. Dainty, pp. 281–88. New York/Heidelberg/Berlin: Springer-Verlag. 473 pp.
12. Barber, J., Telfer, A., Mills, J., Nicolson, J. 1974. Slow chlorophyll fluorescence changes in isolated intact chloroplasts: Evidence for cation control. *Proc. 3rd Int. Congr. Photosynthesis,* ed. M. Avron, 1:53–63. Amsterdam/Oxford/New York: Elsevier. 806 pp.
13. Bennett, J. 1977. Phosphorylation of chloroplast membrane polypeptides. *Nature* 269:344–46
14. Bennett, J. 1979. Chloroplast phosphoproteins phosphorylation of polypeptides of the light-harvesting chlorophyll protein complex. *Eur. J. Biochem.* 99:133–37
15. Bennett, J. 1983. Regulation of photosynthesis by reversible phosphorylation of the light-harvesting chlorophyll a/b protein. *Biochem. J.* 212:1–13
16. Bennett, J., Steinback, K. E., Arntzen, C. J. 1980. Chloroplast phosphoproteins: Regulation of excitation energy transfer by phosphorylation of thylakoid membrane proteins. *Proc. Natl. Acad. Sci. USA* 77:5253–57
17. Bennoun, P. 1974. Correlation between states I and II in algae and the effect of magnesium on chloroplasts. *Biochim. Biophys. Acta* 368:141–47
18. Bennoun, P., Jupin, H. 1974. The relationship between thylakoid stacking, state I and state II phenomena in whole cells and the cation effects in chloroplasts of *Chlamydomonas reinhardi.* See Ref. 12, 1:163–69
19. Biggins, J. 1973. The kinetic behavior of cytochrome *f* in cyclic and non-cyclic electron transport in *Porphyridium cruentum. Biochemistry* 12:1165–70
20. Biggins, J. 1982. Thylakoid conformational changes accompanying membrane protein phosphorylation. *Biochim. Biophys. Acta* 679:479–82
21. Biggins, J. 1983. Mechanism of the light state transition in photosynthesis. I. Analysis of the kinetics of cytochrome *f* oxidation in State 1 and State 2 in the red alga, *Porphyridium cruentum. Biochim. Biophys. Acta* 724:111–17

22. Biggins, J., Bruce, D. 1985. Mechanism of the state transition in photosynthesis. III. Kinetics of the state transition in *Porphyridium cruentum. Biochim. Biophys. Acta* 806:230–36
23. Biggins, J., Campbell, C. L., Bruce, D. 1984. Mechanism of the light state transition in photosynthesis. II. Analysis of phosphorylated polypeptides in the red alga *Porphyridium cruentum. Biochim. Biophys. Acta* 767:138–44
24. Biggins, J., Campbell, C. L., Creswell, L. L., Wood/Brown, E. A. 1984. Mechanism of the light state transition in *Porphyridium cruentum.* See Ref. 8, 3:303–6
25. Black, M., Horton, P., Foyer, C. 1984. Effects of protein phosphorylation on the properties of thylakoid membranes. See Ref. 8, 3:315–18
26. Bonaventura, C., Myers, J. 1969. Fluorescence and oxygen evolution from *Chlorella pyrenoidosa. Biochim. Biophys. Acta* 189:366–83
27. Bose, S., Mullet, J. E., Hoch, G. E., Arntzen, C. J. 1981. Effects of cations on photosystem I electron transport. *Photobiochem. Photobiophys.* 2:45–52
28. Bose, S., Ramanujam, P. 1984. Heterogeneity in photosystem I. Evidence from cation-induced decrease in photosystem 1 electron transport. *Biochim. Biophys. Acta* 764:40–45
29. Bruce, D., Biggins, J., Steiner, T., Thewalt, M. 1985. Mechanism of the light state transition in photosynthesis. IV. Picosecond fluorescence spectroscopy of *Anacystis nidulans* and *Porphyridium cruentum* in state I and State II at 77 K. *Biochim. Biophys. Acta* 806:237–46
30. Butler, W. L. 1977. Chlorophyll fluorescence: A probe for electron transport and energy transfer. In *Encyclopedia of Plant Physiology* (NS), ed. A. Trebst, M. Avron, 5:149–67. Berlin: Springer-Verlag. 730 pp.
31. Butler, W. L. 1978. Energy distribution in the photochemical apparatus of photosynthesis. *Ann. Rev. Plant Physiol.* 29:345–78
32. Butler, W. L. 1980. Energy transfer between photosystem II units in a connected package model of the photochemical apparatus of photosynthesis. *Proc. Natl. Acad. Sci. USA* 77:4697–4701
33. Butler, W. L., Kitajima, M. 1975. Energy transfer between photosystem II and photosystem I in chloroplasts. *Biochim. Biophys. Acta* 396:72–85
34. Butler, W. L., Magde, D., Berens, S. J. 1983. Fluorescence lifetimes in the bipartite model of the photosynthetic apparatus with α, β heterogeneity in photosys-

tem II. *Proc. Natl. Acad. Sci. USA* 80:7510–14

35. Canaani, O., Barber, J., Malkin, S. 1984. Evidence that phosphorylation and dephosphorylation regulate the distribution of excitation energy between the two pigment photosystems of photosynthesis *in vivo*: Photoacoustic and fluorimetric study of an intact leaf. *Proc. Natl. Acad. Sci. USA* 81:1614–18

36. Canaani, O., Cahen, D., Malkin, S. 1982. Photosynthetic chromatic transitions and Emerson enhancement effects in intact leaves studied by photoacoustics. *FEBS Lett.* 150:142–46

37. Canaani, O., Cahen, D., Malkin, S. 1984. Photoacoustics as a probe for photosynthetic O_2 evolution and energy storage in an intact leaf—Distribution of excitation energy between PS II and PS I. See Ref. 8, 3:331–34

38. Canaani, O., Malkin, S. 1984. Distribution of light excitation in an intact leaf between the two photosystems of photosynthesis. Changes in the absorption cross-sections following State 1-State 2 transitions. *Biochim. Biophys. Acta* 766: 513–24

39. Canaani, O., Malkin, S. 1984. Physiological adaptation to a newly observed low light intensity state in intact leaves, resulting in extreme imbalance in excitation energy distribution between the two photosystems. *Biochim. Biophys. Acta* 766:525–32

40. Catt, M., Saito, K., Williams, W. P. 1984. State I/State II and dark adaptation in green and blue-green algae. See Ref. 8, 3:295–98

41. Catt, M., Saito, K., Williams, W. P. 1984. Excitation energy distribution in green algae: The existence of two independent light-driven control mechanisms. *Biochim. Biophys. Acta* 767:39–47

42. Chow, W. S., Telfer, A., Chapman, D. J., Barber, J. 1981. State 1–state 2 transition in leaves and its association with ATP-induced chlorophyll fluorescence quenching. *Biochim. Biophys. Acta* 638:60–68

43. Duysens, L. N. M. 1972. Oxygen evolution in light flashes; changes in energy transfer to reaction center 2 upon pigment state transitions and inhibition by DCMU and FCCP. *Proc. 2nd Int. Congr. Photosynthesis Res.*, ed. G. Forti, M. Avron, A. Melandri, 1:19–25. The Hague; Dr. Junk. 860 pp.

44. Duysens, L. N. M., Talens, A. 1969. Reactivation of reaction center II by a product of photoreaction I. In *Progress in Photosynthesis Research*, ed. H. Metz-ner, 2:1073–81. Tübingen: Int. Union Biol. Sci. 589 pp.

45. Farchaus, J. W., Widger, W. R., Cramer, W. A., Dilley, R. A. 1982. Kinase-induced changes in electron transport rates of spinach chloroplasts. *Arch. Biochem. Biophys.* 217:362–67

46. Fork, D. C., Murata, N., Sato, N. 1979. Effect of growth temperature on the lipid and fatty acid composition and the dependence on temperature of light-induced redox reactions of cytochrome f and of light energy redistribution in the thermophilic blue-green alga *Synechococcus lividus*. *Plant Physiol.* 63:524–30

47. Fork, D. C., Satoh, K. 1983. State I-State II transitions in the thermophilic blue-green alga (cyanobacterium) *Synechococcus lividus*. *Photochem. Photobiol.* 37:421–27

48. Fowler, C. F., Kok, B. 1974. Direct observation of a light-induced electric field in chloroplasts. *Biochim. Biophys. Acta* 357:308–18

49. Furtado, D., Williams, W. P., Brain, A. P. R., Quinn, P. J. 1979. Phase separations in membranes of *Anacystis nidulans* grown at different temperatures. *Biochim. Biophys. Acta* 555:352–57

50. Gerola, P., Torti, F., Jennings, R. C. 1984. Energy coupling between protein-chlorophyll complexes in chloroplasts. The effect of membrane phosphorylation. See Ref. 8, 3:307–10

51. Govindjee, Amesz, J., Fork, D. C., eds. 1986. *Light Emission by Plants and Bacteria*. New York: Academic. In press

52. Govindjee, Papageorgiou, G. 1971. Chlorophyll fluorescence and photosynthesis: Fluorescence transients. *Photophysiology* 6:1–50

53. Gulotty, R. J., Fleming, G. R., Alberte, R. S. 1982. Low-intensity picosecond fluorescence kinetics and excitation dynamics in barley chloroplasts. *Biochim. Biophys. Acta* 682:322–31

54. Haworth, P. 1983. Protein phosphorylation-induced State I–State II transitions are dependent on thylakoid membrane microviscosity. *Arch. Biochem. Biophys.* 226:145–54

55. Haworth, P., Karukstis, K. K., Sauer, K. 1983. Picosecond fluorescence kinetics in spinach chloroplasts at room temperature. Effects of phosphorylation. *Biochim. Biophys. Acta* 725:261–71

56. Haworth, P., Kyle, D. J., Arntzen, C. J. 1982. A demonstration of the physiological role of membrane phosphorylation in chloroplasts, using the bipartite and tripartite models of photosynthesis. *Biochim. Biophys. Acta* 680:343–51

57. Haworth, P., Kyle, D. J., Arntzen, C. J.

1982. Protein phosphorylation and excitation energy distribution in normal, intermittent-light-grown, and a chlorophyll *b*-less mutant of barley. *Arch. Biochem. Biophys.* 218:199–206

58. Haworth, P., Kyle, D. J., Horton, P., Arntzen, C. J. 1982. Chloroplast membrane protein phosphorylation. *Photochem. Photobiol.* 36:743–48

59. Haworth, P., Melis, A. 1983. Phosphorylation of chloroplast thylakoid membrane proteins does not increase the absorption cross-section of photosystem 1. *FEBS Lett.* 160:277–80

60. Haworth, P., Watson, J. L., Arntzen, C. J. 1983. The determination, isolation and characterization of a light-harvesting complex which is specifically associated with photosystem I. *Biochim. Biophys. Acta* 724:151–58

61. Hodges, M., Barber, J. 1983. The significance of the kinetic analysis of fluorescence induction in DCMU-inhibited chloroplasts in terms of photosystem 2 connectivity and heterogeneity. *FEBS Lett.* 160:177–81

62. Hodges, M., Barber, J. 1984. Analysis of chlorophyll fluorescence quenching by DBMIB as a means of investigating the consequences of thylakoid membrane phosphorylation. *Biochim. Biophys. Acta* 767:102–7

63. Homann, P. 1969. Cation effects on the fluorescence of isolated chloroplasts. *Plant Physiol.* 44:932–36

64. Horton, P. 1982. Heterogeneity in photosystem II. *Biochem. Soc. Trans.* 10:338–40

65. Horton, P., Allen, J. F., Black, M. T., Bennett, J. 1981. Regulation of phosphorylation of chloroplast membrane polypeptides by the redox state of plastoquinone. *FEBS Lett.* 125:193–96

66. Horton, P., Black, M. T. 1980. Activation of adenosine 5′ triphosphate-induced quenching of chlorophyll fluorescence by reduced plastoquinone. *FEBS Lett.* 119:141–44

67. Horton, P., Black, M. T. 1981. Light-induced redox changes in chloroplast cytochrome *f* after phosphorylation of membrane proteins. *FEBS Lett.* 132:75–77

68. Horton, P., Black, M. T. 1981. Light-dependent quenching of chlorophyll fluorescence in pea chloroplasts induced by adenosine 5′-triphosphate. *Biochim. Biophys. Acta* 635:53–62

69. Horton, P., Black, M. T. 1982. On the nature of the fluorescence decrease due to phosphorylation of chloroplast membrane proteins. *Biochim. Biophys. Acta* 680:22–27

70. Horton, P., Black, M. T. 1983. A comparison between cation and protein phosphorylation effects on the fluorescence induction curve in chloroplasts treated with 3-(3,4-dichlorophenyl)-1 1-dimethylurea. *Biochim. Biophys. Acta* 722:214–18

71. Horton, P., Lee, P. 1983. Stimulation of a cyclic electron-transfer pathway around photosystem II by phosphorylation of chloroplast thylakoid proteins. *FEBS Lett.* 162:81–84

72. Itoh, S. 1978. Electrostatic state of the membrane surface and the reactivity between ferricyanide and electron transport components inside chloroplast membranes. *Plant Cell Physiol.* 19:149–66

73. Jennings, R. C., Garlaschi, F. M., Gerola, P. D., Etzion-Katz, R., Forti, G. 1981. Proton-induced grana formation in chloroplasts. Distribution of chlorophyll-protein complexes and photosystem II photochemistry. *Biochim. Biophys. Acta* 638:100–7

74. Jennings, R. C., Gerola, P. D., Forti, G., Garlaschi, F. M. 1979. The influence of proton-induced grana formation on partial electron-transport reactions in chloroplasts. *FEBS Lett.* 106:247–50

75. Joliot, A., Joliot, P. 1964. Étude cinétique de la réaction photochemique libérant l'oxygene au cours de la photosynthese. *C.R. Acad. Sci.* 258:4622–25

76. Jursinic, P. A., Kyle, D. J. 1983. Changes in the redox state of the secondary acceptor of photosystem II associated with light-induced thylakoid phosphorylation. *Biochim. Biophys. Acta* 723:37–44

77. Karukstis, K. K., Sauer, K. 1983. Fluorescence decay kinetics of chlorophyll in photosynthetic membranes. *J. Cell Biochem.* 23:131–58

78. Krause, G. H. 1973. The high-energy state of the thylakoid system as indicated by chlorophyll fluorescence and chloroplast shrinkage. *Biochim. Biophys. Acta* 292:715–28

79. Krause, G. H. 1974. Changes in chlorophyll fluorescence in relation to light-dependent cation transfer across thylakoid membranes. *Biochim. Biophys. Acta* 333:301–13

79a. Krause, G. H., Weis, E. 1984. Chlorophyll fluorescence as a tool in plant physiology. II. Interpretation of fluorescence signals. *Photosynth. Res.* 5:139–57

80. Krupinska, K., Akoyunoglou, G., Senger, H. 1985. Relationship between light-induced stateI/II-transitions and Mg²⁺-effect on fluorescence, tested in

synchronous cultures of *Scenedesmus obliquus*. *Photochem. Photobiol.* 41:159–64

81. Kyle, D. J., Haworth, P., Arntzen, C. J. 1982. Thylakoid membrane protein phosphorylation leads to a decrease in connectivity between photosystem II reaction centers. *Biochim. Biophys. Acta* 680:336–42

82. Kyle, D. J., Kuang, T-Y., Watson, J. L., Arntzen, C. J. 1984. Movement of a subpopulation of the light-harvesting complex (LHC-II) from grana to stroma lamellae as a consequence of its phosphorylation. *Biochim. Biophys. Acta* 765:89–96

83. Kyle, D. J., Ohad, I., Arntzen, C. J. 1985. Molecular mechanisms of compensation to light stress in chloroplast membranes. *UCLA Symp.* (NS)22:51–69

84. Kyle, D. J., Staehelin, L. A., Arntzen, C. J. 1983. Lateral mobility of the light-harvesting complex in chloroplast membranes controls excitation energy distribution in higher plants. *Arch. Biochem. Biophys.* 222:527–41

85. Larkum, A. W. D., Barrett, J. 1983. Light-harvesting processes in algae. *Adv. Bot. Res.* 10:1–219

86. Larsson, U. K., Jergil, B., Andersson, B. 1983. Changes in the lateral distribution of the light-harvesting chlorophyll-*a/b*-protein complex induced by its phosphorylation. *Eur. J. Biochem.* 136:25–29

87. Ley, A. C. 1984. Effective cross-sections in *Porphyridium cruentum*. Implications for energy transfer between phycobilisomes and photosystem II. *Plant Physiol.* 74:451–54

88. Ley, A. C., Butler, W. L. 1977. The distribution of excitation energy between Photosystem I and Photosystem II in *Porphyridium cruentum*. In *Photosynthetic Organelles*, Special issue of *Plant Cell Physiol.*, pp. 33–46

89. Ley, A. C., Butler, W. L. 1980. Energy distribution in the photochemical apparatus of *Porphyridium cruentum* in state I and state II. *Biochim. Biophys. Acta* 592:349–63

90. Lieberman, J. R., Bose, S., Arntzen, C. J. 1978. Requirement of the light-harvesting pigment protein complex for magnesium ion regulation of excitation energy distribution in chloroplasts. *Biochim. Biophys. Acta* 502:417–29

91. Lombard, F., Strasser, R. J. 1984. Evidences for spillover changes during State 1 to State 2 transition in green leaves. See Ref. 8, 3:271–74

92. Malkin, S., Armond, P. A., Mooney, H. A., Fork, D. C. 1981. Photosystem II

photosynthetic unit sizes from fluorescence induction in leaves. Correlation to photosynthetic capacity. *Plant Physiol.* 67:570–79

93. Malkin, S., Kok, B. 1966. Fluorescence induction studies in isolated chloroplasts. I. Number of components involved in the reaction and quantum yields. *Biochim. Biophys. Acta* 126:413–32

94. Malkin, S., Telfer, A., Barber, J. 1985. Quantitative analysis of state 1–state 2 transitions in intact leaves using modulated fluorimetry—evidence for changes in absorption cross-section of the two photosystems during state transitions. *Biochim. Biophys. Acta* In press

95. Markwell, J. P., Baker, N. R., Bradbury, M., Thornber, J. P. 1984. Use of zinc ions to study thylakoid protein phosphorylation and the state 1–state 2 transition *in vitro*. *Plant Physiol.* 74:348–54

96. Maxwell, P. C., Biggins, J. 1976. Role of cyclic electron transport in photosynthesis as measured by the photoinduced turnover of P_{700} in vivo. *Biochemistry* 15:3975–81

97. Melis, A., Duysens, L. N. M. 1979. Biphasic energy conversion kinetics and absorbance difference spectra of photosystem II of chloroplasts. Evidence for two different photosystem II reaction centers. *Photochem. Photobiol.* 29:373–82

98. Melis, A., Homann, P. H. 1975. Kinetic analysis of the fluorescence induction in 3-(3,4-dichlorophenyl)-1, 1-dimethylurea poisoned chloroplasts. *Photochem. Photobiol.* 21:431–37

99. Melis, A., Homann, P. H. 1978. A selective effect of Mg^{2+} on the photochemistry of one type of reaction center in photosystem II of chloroplasts. *Arch. Biochem. Biophys.* 190:523–30

100. Mohanty, P., Govindjee. 1973. Light-induced changes in the fluorescence yield of chlorophyll *a* in *Anacystis nidulans* II. The fast changes and the effect of photosynthetic inhibitors on both the fast and slow fluorescence induction. *Plant Cell Physiol.* 14:611–29

101. Mohanty, P., Govindjee. 1973. Light-induced changes in the fluorescence yield of chlorophyll *a* in *Anacystis nidulans* I. Relationship of slow fluorescence changes with structural changes. *Biochim. Biophys. Acta* 305:95–104

102. Morita, S., Itoh, S., Nishimura, M. 1982. Correlation between the activity of membrane-bound ATPase and the rate of flash-induced 515-nm absorbance change in chloroplasts in intact leaves, assayed by means of rapid isolation of

chloroplasts. *Biochim. Biophys. Acta* 679:125–30

103. Mullet, J. E., Burke, J. J., Arntzen, C. J. 1980. Chlorophyll proteins of photosystem I. *Plant Physiol.* 65:814–22

104. Murata, N. 1969. Control of excitation transfer in photosynthesis. I. Light-induced change of chlorophyll *a* fluorescence in *Porphyridium cruentum*. *Biochim. Biophys. Acta* 172:242–51

105. Murata, N. 1969. Control of excitation transfer in photosynthesis II. Magnesium ion-dependent distribution of excitation energy between two pigment systems in spinach chloroplasts. *Biochim. Biophys. Acta* 189:171–81

106. Murata, N. 1970. Control of excitation transfer in photosynthesis. IV. Kinetics of chlorophyll *a* fluorescence in *Porphyra yezoensis*. *Biochim. Biophys. Acta* 205:379–89

107. Murata, N. 1971. Effects of monovalent cations on light energy distribution between two pigment systems of photosynthesis in isolated spinach chloroplasts. *Biochim. Biophys. Acta* 226:422–32

108. Murata N., Fork, D. C. 1975. Temperature dependence of chlorophyll *a* fluorescence in relation to the physical phase of membrane lipids in algae and higher plants. *Plant Physiol.* 56:791–96

109. Murata, N., Ono, T. 1981. The lipid phase of membranes and its influence on the photosynthetic activities in the blue-green alga, *Anacystis nidulans. Proc. 5th Int. Congr. Photosynthesis,* ed. G. Akoyunoglou, 6:473–81. Philadelphia: Int. Sci. Serv. 769 pp.

110. Murata, N., Troughton, J. H., Fork, D. C. 1975. Relationship between the transition of the physical phase of membrane lipids and photosynthetic parameters in *Anacystis nidulans* and lettuce and spinach chloroplasts. *Plant Physiol.* 56:508–17

111. Myers, J. 1963. Enhancement. In *Photosynthetic Mechanisms of Green Plants,* ed. B. Kok, A. Jagendorf, pp. 301–17. Washington DC: Natl. Acad. Sci.-Natl. Res. Council Publ. No. 1145. 766 pp.

112. Myers, J. 1971. Enhancement studies in photosynthesis. *Ann. Rev. Plant Physiol.* 22:289–312

113. Nairn, J. A., Haehnel, W., Reisberg, P., Sauer, K. 1982. Picosecond fluorescence kinetics in spinach chloroplasts at room temperature. Effects of Mg^{2+}. *Biochim. Biophys. Acta* 682:420–29

114. Owens, G. C., Ohad, I. 1982. Phosphorylation of *Chlamydomonas reinhardti* chloroplast membrane proteins *in vivo* and *in vitro. J. Cell Biol.* 93:712–18

115. Papageorgiou, G. 1975. Chlorophyll fluorescence: An intrinsic probe of photosynthesis. In *Bioenergetics of Photosynthesis,* ed. Govindjee, pp. 319–71. New York: Academic. 698 pp.

116. Papageorgiou, G., Govindjee. 1967. Changes in intensity and spectral distribution of fluorescence. Effect of light treatment on normal and DCMU-poisoned *Anacystis nidulans. Biophys. J.* 7:375–89

117. Papageorgiou, G., Govindjee. 1968. Light-induced changes in the fluorescence yield of chlorophyll *a in vivo.* I. *Anacystis nidulans. Biophys. J.* 8:1299–1315

118. Powles, S. B. 1984. Photoinhibition of photosynthesis. *Ann. Rev. Plant Physiol.* 35:15–44

119. Powles, S. B., Osmond, C. B. 1979. Photoinhibition of intact attached leaves of C_3 plants illuminated in the absence of both carbon dioxide and of photorespiration. *Plant Physiol.* 64:982–88

120. Ramanujam, P., Bose, S. 1983. Correlation between thylakoid stacking and cation-induced decrease in photosystem I electron transport at saturating light intensity. *Arch. Biochem. Biophys.* 221:238–42

121. Ried, A., Reinhardt, B. 1977. Distribution of excitation energy between photosystem I and photosystem II in red algae. II. Kinetics of the transition between State 1 and State 2. *Biochim. Biophys. Acta* 460:25–35

122. Ried, A., Reinhardt, B. 1980. Distribution of excitation energy between Photosystem I and Photosystem II in red algae. III. Quantum requirements of the induction of a State 2-State 1 transition. *Biochim. Biophys. Acta* 592:76–86

123. Saito, K., Williams, W. P., Allen, J. F., Bennett, J. 1983. Comparison of ATP-induced and State 1/State 2-related changes in excitation energy distribution in *Chlorella vulgaris. Biochim. Biophys. Acta* 724:94–103

124. Sane, P. V., Desai, T. S., Tatake, V. G., Govindjee. 1984. Heat induced reversible increase in photosystem 1 emission in algae, leaves and chloroplasts: Spectra, activities, and relation to state changes. *Photosynthetica* 18:439–44

125. Sane, P. V., Furtado, D., Desai, T. S., Tatake, V. G. 1982. A study of state changes in *Chlorella:* The effect of uncoupler and energy transfer inhibitors. *Z. Naturforsch. Teil C* 37:458–63

126. Satoh, K. 1981. Fluorescence induction and activity of ferredoxin-$NADP^+$ reductase in *Bryopsis* chloroplasts. *Biochim. Biophys. Acta* 638:327–33

127. Satoh, K. 1982. Mechanism of photoactivation of electron transport in intact *Bryopsis* chloroplasts. *Plant Physiol.* 70:1413–16

128. Satoh, K., Fork, D. C. 1983. The relationship between state II to state I transitions and cyclic electron flow around photosystem I. *Photosynth. Res.* 4:245–56

129. Satoh, K., Fork, D. C. 1983. A new mechanism for adaptation to changes in light intensity and quality in the red alga, *Porphyra perforata*. I. Relation to State 1-State 2 transitions. *Biochim. Biophys. Acta* 722:190–96

130. Satoh, K., Fork, D. C. 1983. State 1-State 2 transitions in the green alga *Scenedesmus obliquus*. *Photochem. Photobiol.* 37:429–34

131. Satoh, K., Fork, D. C. 1983. A new mechanism for adaptation to changes in light intensity and quality in the red alga, *Porphyra perforata*. II. Characteristics of state II-state III transitions. *Photosynth. Res.* 4:61–70

132. Schreiber, U., Vidaver, W. 1976. The I-D fluorescence transient. An indicator of rapid energy distribution changes in photosynthesis. *Biochim. Biophys. Acta* 440:205–14

133. Schreiber, U., Vidaver, W. 1976. Rapid light induced changes of energy distribution between photosystems I and II. *FEBS Lett.* 62:194–97

134. Schuster, G., Owens, G. C., Cohen, Y., Ohad, I. 1984. Light independent phosphorylation of the chlorophyll *a, b*-protein complex in thylakoids of the prokaryote *Prochloron*. See Ref. 8, 3:283–86

135. Shinitzky, M., Skornick, Y., Haran-Ghera, N. 1979. Effective tumor immunization induced by cells of elevated membrane-lipid microviscosity. *Proc. Natl. Acad. Sci. USA* 76:5313–16

136. Solis, B., Strasser, R. J. 1984. Changes in the photochemical activities of thylakoids during phosphorylation of the light-harvesting complex. See Ref. 8, 3:275–78

137. Staehelin, L. A., Arntzen, C. J. 1983. Regulation of chloroplast membrane function: Protein phosphorylation changes the spatial organization of membrane components. *J. Cell Biol.* 97:1327–37

138. Steinback, K. E., Bose, S., Kyle, D. J. 1982. Phosphorylation of the light-harvesting chlorophyll-protein regulates excitation energy distribution between photosystem II and photosystem I. *Arch. Biochem. Biophys.* 216:356–61

139. Telfer, A., Barber, J. 1981. ATP-dependent State I-State II changes in isolated pea thylakoids. *FEBS Lett.* 129:161–65

140. Telfer, A., Hodges, M., Barber, J. 1983. Analysis of chlorophyll fluorescence induction curves in the presence of 3-(3,4-dichlorophenyl)-1,1-dimethylurea as a function of magnesium concentration and NADP-activated light-harvesting chlorophyll *a/b*-protein phosphorylation. *Biochim. Biophys. Acta* 724:167–75

141. Thornber, J. P., Highkin, H. R. 1974. Composition of the photosynthetic apparatus of normal barley leaves and a mutant lacking chlorophyll *b*. *Eur. J. Biochem.* 41:109–16

142. Tsala, G., Strasser, R. J. 1984. Energy distribution changes during phosphorylation of the light-harvesting complex in thylakoids. See Ref. 8, 3:279–82

143. Tsukamoto, Y., Ueki, T., Mitsui, T., Ono, T., Murata, N. 1980. Relationship between growth temperature of *Anacystis nidulans* and phase transition temperature of its thylakoid membranes. *Biochim. Biophys. Acta* 602:673–75

144. Wang, R. T., Myers, J. 1974. On the state 1-State 2 phenomenon in photosynthesis. *Biochim. Biophys. Acta* 347:134–40

145. Wang, R. T., Myers, J. 1976. On the distribution of excitation energy to two photoreactions of photosynthesis. *Photochem. Photobiol.* 23:405–10

146. Weis, E. 1984. Short term acclimation of spinach to high temperatures. *Plant Physiol.* 74:402–7

147. Weis, E. 1985. Light- and temperature-induced changes in the distribution of excitation energy between Photosystem I and Photosystem II in spinach leaves. *Biochim. Biophys. Acta* 807:118–26

148. Weis, E. 1985. Chlorophyll fluorescence at 77K in intact leaves: Characterization of a technique to eliminate artifacts related to self-absorption. *Photosynth. Res.* 6:73–86

149. Williams, W. P. 1977. The two photosystems and their interactions. In *Primary Processes of Photosynthesis*, ed. J. Barber, pp. 99–146. Amsterdam/Elsevier: North-Holland Biomed. Press. 516 pp.

150. Williams, W. P., Furtado, D., Nutbeam, A. R. 1980. Comparison between State I/State II adaptation in a unicellular green alga and high energy state quenching in isolated spinach chloroplasts. *Photochem. Photobiophys.* 1:91–102

151. Williams, W. P., Salamon, Z. 1976. Enhancement studies on algae and iso-

lated chloroplasts Part I. Variability of photosynthetic enhancement in *Chlorella pyrenoidosa*. *Biochim. Biophys. Acta* 430:282–99

152. Wollman, F.-A., Delepelaire, P. 1984. Correlation between changes in light energy distribution and changes in thylakoid membrane polypeptide phosphorylation in *Chlamydomonas reinhardtii*. *J. Cell. Biol.* 98:1–7

153. Wong, D., Govindjee. 1979. Antagonistic effects of mono- and divalent cations on polarization of chlorophyll fluorescence in thylakoids and changes in excitation energy transfer. *FEBS Lett.* 97:373–77

154. Wong, D., Merkelo, H., Govindjee. 1979. Regulation of excitation transfer by cations: Wavelength-resolved fluorescence lifetimes and intensities at 77K in thylakoid membranes of pea chloroplasts. *FEBS Lett.* 104:223–26

155. Yamazaki, I., Mimuro, M., Murao, T., Yamazaki, T., Yoshihara, K., Fujita, Y. 1984. Excitation energy transfer in the light-harvesting antenna system of the red alga *Porphyridium cruentum* and the blue-green alga *Anacystis nidulans:* Analysis of time-resolved fluorescence spectra. *Photochem. Photobiol.* 39:233–40

Ann. Rev. Plant Physiol. 1986. 37:363–76

ALTERATION OF GENE EXPRESSION DURING ENVIRONMENTAL STRESS IN PLANTS

Martin M. Sachs and Tuan-Hua David Ho

Department of Biology, Washington University, St. Louis, Missouri 63130

CONTENTS

INTRODUCTION

Higher plants are subjected to a large number of environmental and biological stresses. These adverse conditions may interfere with normal growth and development and in crop plants will lower food quality and yield. During the last few years several groups have initiated studies on the response of plants to stress with emphasis on the analysis of gene expression.

Animals appear to respond to a wide variety of stress situations by synthesizing "heat-shock" proteins. Plants also synthesize heat-shock proteins in response to high temperature stress. However, plants synthesize different sets of proteins during other stress conditions such as flooding, high salt con-

363

0066-4294/86/0601-0363$02.00

centrations, or intense UV light exposure. This area of research is of interest not only because it will expand our understanding of coordinate gene expression in higher organisms, but studies on the stress-induced proteins and the genes encoding them may lead to the engineering of crop plants more resistant to normally encountered stress conditions.

Anaerobiosis and heat shock are two stress conditions that have received a great amount of attention in recent years. The response of plants to these stress conditions is to alter gene expression by synthesizing novel polypeptides. The new protein synthesis is actually biphasic, a synthesis of "early proteins" and then "late proteins." This review details advances made in these two areas and highlights what is known to occur with respect to gene expression changes brought about by other types of environmental stress.

ANAEROBIC STRESS

The physiological and ultrastructural aspects of the response to anaerobiosis in plants have been reviewed extensively (5, 36, 37, 46). In this section we review the effect of anaerobiosis on plant gene expression. Anaerobic treatment results in a drastic alteration in the pattern of protein synthesis in maize seedlings (64). Preexisting (aerobic) protein synthesis is repressed while selective synthesis of new polypeptides is initiated (64). This is most likely a plant's natural response to flooding.

Studies on the maize anaerobic response stemmed from the extensive analysis of the maize alcohol dehydrogenase (ADH) system by Drew Schwartz and coworkers (reviewed in 22, 23, 25). Hageman & Flesher (28) were the first to show that ADH activity increases as a result of flooding maize seedlings. Freeling (21) later reported that ADH activity increased at a zero order rate between 5 and 72 hours of anaerobic treatment, reflecting a simultaneous expression of two unlinked genes *Adh1* and *Adh2*. Schwartz (70) showed that ADH activity is required to allow maize seeds and seedlings to survive flooding. ADH is the major terminal enzyme of fermentation in plants and is responsible for recycling NAD^+ during anoxia. It has been suggested that ethanolic fermentation permits tight cytoplasmic pH regulation; thus preventing acidosis from competing lactic or malic fermentation (58–60).

The maize anaerobic response is analogous to the heat-shock response observed in many organisms including plants (see below). There is a repression of preexisting protein synthesis and de novo synthesis of a new set of proteins. Except for one possible overlap, anaerobiosis induces a different set of proteins in maize than heat shock (13, 40, 64). In an early study (50), anaerobic treatment was shown to cause a near complete dissociation of polysomes and a rapid repression of protein synthesis in soybean roots. Maize seedlings respond in the same way (E. S. Dennis and A. J. Pryor, unpublished), but this was shown to result from a redirection of protein synthesis (64). As is the case of many heat-shock systems (cf 66), in maize seedlings subjected to anoxia (e.g. an argon atmosphere), there is an immediate repression of preexisting (aerobic) protein synthesis along with the induction of a new set of proteins (64). In the first 5 hours of anaerobic treatment there is a transition period during which there is a rapid increase in the synthesis of a class of polypeptides with an approximate molecular weight of 33 kd. These have been referred to as the transition polypeptides (TPs). After approximately 90 min of anoxia, the synthesis of an additional group of ~20 polypeptides is induced. This group of 20 anaerobic polypeptides (ANPs) represents greater than 70% of the total labeled amino acid incorporation after 5 h of anaerobiosis. By this time the synthesis of the TPs is at a minimal level; however, these polypeptides accumulate to a high level during early anaerobiosis and have shown by pulse-chase experiments to be very stable. The synthesis of the ANPs continues at a constant rate for up to 72 h of anaerobic treatment, at which time protein synthesis decreases concurrently with the start of cell death (64).

The identities of some of the ANPs are now known. The isozymes of alcohol dehydrogenase, encoded by the *Adh1* and *Adh2* genes, have been identified as ANPs through the use of genetic variants (20, 63). More recently, glucose-6-phosphate isomerase (41), fructose-1,6-diphosphate aldolase (42), and sucrose synthetase (P. Starlinger and W. Werr, personal communication) have been identified as ANPs. Pyruvate decarboxylase activity has also been shown to be induced by anaerobiosis (48, 75) and therefore may also be an ANP. The identities and functions of the remaining ANPs or of the TPs are as yet unknown. However, five of the ANPs so far identified are glycolytic enzymes, and sucrose synthetase is also involved in glucose metabolism. In light of the inability of maize seedlings to survive even 5 h of flooding in the absence of an active *Adh1* gene (70), it appears that at least one function of the anaerobic response is to enable the plant to produce as much ATP as possible when there is an oxygen deficit, as would occur in water-logged soils.

In the presence of air, each maize organ examined, including the roots, coleoptile, mesocotyl, endosperm, scutellum, and anther wall synthesizes a tissue-specific spectrum of polypeptides. Under anaerobic conditions all of the above organs synthesize only the ANPs. Moreover, except for a few characteristic qualitative and quantitative differences, the patterns of anaerobically induced protein synthesis in these diverse organs are remarkably similar (55). On the other hand, maize leaves, which have emerged from the coleoptile, do not incorporate labeled amino acids under anaerobic conditions and do not survive even a brief exposure to anaerobiosis (55).

The shift in pattern of protein synthesis during anaerobiosis has been observed in many other plant species including rice (6, 57), sorghum, barley, pea, and carrot (Freeling laboratory, unpublished). In anaerobically treated barley aleurone cells, lactate dehydrogenase (LDH) activity increases (31) as does ADH activity and mRNA levels (32). In the barley system the gibberellic acid (GA_3) induction of α-amylase synthesis and its mRNA accumulation is suppressed by anaerobiosis. The "anaerobic" pattern of protein synthesis is perturbed only slightly by GA_3 when aleurone cells incorporate labeled amino acids in a nitrogen atmosphere (31).

The rapid repression of preexisting protein synthesis caused by anaerobic treatment is correlated with a near complete dissociation of polysomes in soybeans (50) and maize (E. S. Dennis and A. J. Pryor, personal communication). This does not result from degradation of "aerobic" mRNA, because the mRNAs encoding the preexisting proteins remain translatable in an in vitro system at least 5 h after anaerobic treatment is initiated (64). This is in agreement with the observation that the polysomes dissociated by anaerobiosis rapidly reform up to 80–90% of their pretreatment levels, even in the absence of new RNA synthesis, when soybean seedlings are returned to air (50).

The induction of anaerobic polypeptide synthesis occurs in two phases: the immediate synthesis of the TPs and then the appearance, after a 90 min lag, of the approximately 20 ANPs. In heat shock, the appearance of the heat-shock polypeptides (hsp) is immediate with kinetics of induction very similar to the rapid appearance of the TPs. Both the hsps and the TPs exhibit a decreased rate of synthesis after 5 h of stress treatment (13, 64). It is interesting that the only potential overlap in protein synthesis observed in maize in response to heat shock or anaerobiosis is that of a 33 kd polypeptide TP and a similar sized hsp (40). In soybeans there appears to be a ~27–28 kd hsp that is induced by a number of different environmental insults, including both heat shock and anaerobiosis (15).

The molecular analysis of the maize anaerobic response was initiated by synthesizing cDNA clones from high molecular weight poly(A) RNA of anaerobically treated maize seedling roots (24). Cloned anaerobic-specific cDNAs were identified by colony hybridization analysis, by differential hybridization to labeled cDNAs of mRNA from anaerobic and aerobic roots. Colonies were selected that hybridized specifically with "anaerobic" cDNAs. The anaerobic-specific cDNA clones were grouped into families on the basis of cross-hybridization to each other, and several of these families were analyzed by hybrid-selected translation and by RNA gel blot (northern) hybridization. The *Adh1* and *Adh2* cloned cDNA families were subsequently identified from this anaerobic-specific cDNA clone library, and these cDNA clone families and the genes encoding them were analyzed extensively (17, 18).

The anaerobic-specific cDNA clones were used as probes to measure gene expression. The levels of mRNA hybridizable to these cDNAs increase during anaerobic treatment. This has been quantified rigorously in the case of *Adh1* and *Adh2*. The kinetics of mRNA increase are the same for *Adh1* and *Adh2*. In both cases, the mRNA level first appears to increase at 90 min of anaerobic treatment. The mRNA level continues to increase until it plateaus at fiftyfold above the aerobic level at 5 h of anaerobiosis. This level is maintained until after 48 h in the case of *Adh1* but starts declining after 10 h in the case of *Adh2* (18). This pattern of mRNA level increase and decrease is reflected in the previously described rates of in vivo anaerobic protein synthesis for ADH1 and ADH2 (64). Sucrose synthetase (ANP87) also appears to have the same kinetics of mRNA level increase (30). In vitro run-off transcription experiments (62; L. Beach, personal communication) show an increase in transcription rate of the *Adh1* gene during anoxia, strongly indicating that the increase in the levels of anaerobic-specific mRNAs resulting from anaerobiosis is the result of induced transcription of the *Anp* genes.

A comparison of the regions of the *Adh1* and *Adh2* genes upstream from the site of transcription initiation reveals only a few islands of homology. These include an 11 bp homologous region that includes the "TATA box" (27) and three additional 8 bp regions of homology. One or more of these sequences might account for the anaerobic control of these genes (18), much as the 10 bp consensus sequence determined in the heat-shock system (56) appears to be responsible for the high-temperature induction of *hsp* gene transcription. In vitro mutagenesis coupled with transformation and gene expression studies as well as the analysis of the 5' regions of the other anaerobic genes will be necessary to determine which if any of these homologous regions might be important in regulating the induced expression of the anaerobic response genes.

HIGH-TEMPERATURE STRESS

Alteration of Gene Expression During Heat Stress

The most readily observable response to heat stress in many organisms is the induction of heat-shock proteins (hsp) (for review, see 66). This phenomenon was first investigated in the fruit fly, *Drosophila melanogaster*. When *Drosophila* cells are shifted from their normal growth temperature (25°C) to an increased temperature (37°C), normal protein synthesis ceases with the concomitant synthesis of a novel set of hsp. This alteration of gene expression is accompanied by the regression of preexisting polytene chromosome puffs and the generation of new ones (for review see 3). The hsp are not detectable at 25°C except in embryonic tissues, and their induction at 37°C is rapid; they appear within a few minutes after heat shock begins (3). The mRNAs encoding the hsp result from de novo transcription and are selectively translated during heat shock (3).

Three different size groups of hsp are present in higher plants that have been studied, including soybean (45), pea (4), tobacco (1, 4), tomato (65), and maize (8, 9, 13): 1. the large hsp group ranging in size from 68 to 104 kd. This group is ubiquitous among all the organisms including bacteria, animals, and plants; 2. the intermediate size group between 20 and 33 kd; and 3. the small hsp group, about 15–18 kd in size, is unique to higher plants. All of these proteins appear to be coordinately expressed when the tissue is under heat stress. The optimal condition for hsp induction in higher plants is a drastic temperature upshift to 39–41°C. However, hsp also can be induced if there is a gradual temperature increase such as a 2.5°C increase per hour, a condition closer to what occurs in the field (1, 12). Hsp synthesis can be detected within 20 min of heat shock, and the increase in transcript levels of some *hsp* genes in 3–5 min (44, 67). However, the induction of hsp appears to be transient; it lasts only for a few hours despite the continuous presence of heat-shock temperatures (13, 67). In maize roots, the hsp are synthesized up to 4 h, followed by a quick decline upon prolonged heat shock (13). While the synthesis of hsp decreases after 4 h of heat shock, two additional "late" proteins, 62 and 50 kd in size, begin to be synthesized (13). The "late" proteins first appear after 4 h of heat shock and continue to be synthesized for at least 20 h. This biphasic induction of new proteins during heat shock appears to be similar to that observed during anaerobiosis described above. It is not yet known whether the synthesis of the "early" hsp is a prerequisite for the induction of the late proteins. So far, most of the research effort is still centered around the traditional or "early" set of hsp.

The induction of hsp follows closely with the increase of the levels of their transcripts. Liquid hybridization studies conducted by Key's group using cloned cDNA probes have shown that about 20 different species of hsp18 mRNA accumulate to 20,000 copies per cell within 2 h of heat-shock treatment in soybean (67). When a heat-shocked tissue is returned to the normal temperature (e.g. 28°C) the synthesis of hsp decreases over the next few hours with a concomitant decline in the levels of gene transcripts. Several plant *hsp* genes have been cloned and sequenced including the genes for maize hsp70 (71) and several encoding soybean hsp18 (16, 43, 68). The maize *hsp70* gene appears to be very similar to the *Drosophila* counterpart with 75% sequence homology in the coding region (71). The three soybean *hsp18* genes share greater than 90% homology in their deduced amino acid sequences. They also share similarities with the *Drosophila* small hsp (22–27 kd) in the hydropathy profiles (52a). Besides the TATA box (27), sequences related to the *Drosophila* heat-shock consensus regulatory element (CT-GAA--TTC-AG) (56) are found −48 to −62 base pairs 5' to the start of transcription (69). In addition, there are secondary heat-shock consensus elements (decameric palindromic sequences: AGAAATTTTCT;CTnGAAnnTTCnAG) located further upstream (52a). The DNA sequence analysis of these plant *hsp* genes supports the view that the molecular mechanism involved in the induction of *hsp* genes is highly conserved among eukaryotes. To further analyze the promoter regions, Gurley et al (unpublished) have introduced one of the soybean small *hsp* genes into primary sunflower tumors via Ti plasmid mediated transformation. The soybean *hsp* gene containing 3.25 kb of 5' flanking sequences is strongly transcribed in a thermoinducible manner in sunflower tumors. The 5' flanking region can be deleted to -192 base with no effect on the levels of thermoinduction, but it results in a large increase in basal transcription. Additional deletion to position -95 reduces both the thermoinducible and basal transcription. Further analysis could pinpoint more precisely the sequences that are important for the induction of *hsp* genes.

Besides heat shock, many other factors including osmotic stress, high salt, DNP, arsenite, anaerobiosis, and high concentrations of ABA, ethylene, or auxins, can also induce the synthesis of certain hsp (15, 44). In soybean, arsenite and cadmium induce a normal spectrum of hsp, yet the other factors seem to induce hsp27 specifically. In maize ABA also induces hsp70 (33). Although the physiological significance of this type of induction is not known, these factors can certainly be used as additional tools to investigate the molecular mechanisms underlying hsp expression.

The induction of hsp is accompanied by the alteration of expression of other genes. In *Drosophila* hsp are essentially the only proteins that are synthesized during heat shock. The expression of normal proteins is suppressed, and many preexisting mRNAs are sequestered rather than being degraded (3, 72). When returned to normal temperature the synthesis of normal protein resumes in the absence of new transcription (52). Thus, while the induction of hsp in *Drosophila* is likely at the transcriptional level, the repression of synthesis of normal proteins is at the translational level. In yeast, the preexisting mRNA are not sequestered but undergo normal turnover (52). Hence, in the absence of continuous synthesis of these preexisting mRNAs, their levels decrease gradually during heat shock. In soybean, hsp are also the only proteins synthesized during heat shock (45). Vierling & Key (73) have examined the effect of heat shock on the synthesis of ribulose 1,5-bisphosphate carboxlase in a soybean cell culture line. The synthesis of both the large and small subunits of this enzyme is suppressed by 80% during heat shock. The levels of mRNA for the small subunit (encoded in the nuclear genome) also decreases during heat shock. In contrast, changes in the synthesis of large subunits (encoded in the chloroplast genome) show little relationship to the corresponding mRNA levels; large subunit mRNA levels remain relatively unchanged by heat shock. Since the expression of small subunit gene is likely to be suppressed during heat shock, it is not yet clear whether the decrease in the levels of this mRNA is the consequence of normal turnover or heat-shock enhanced degradation. Unlike the case in soybean, the synthesis of many normal proteins continues in cereal grains during heat shock (12, 13). However, because the rate of total protein synthesis is unchanged, there appears to be suppression of the synthesis of some of the normal proteins to compensate for the increase in hsp synthesis. Belanger et al (4a) have used the aleurone layers of barley seeds to study the effect of heat shock on the synthesis of normal proteins. This tissue synthesizes α-amylase and protease in response to gibberellic acid (GA$_3$) treatment at normal temperature (25°C). The mRNA coding for α-amylase has been shown to be extremely stable, and its level reaches the maximum after 12 h of GA$_3$ treatment when the synthesis of α-amylase continues even in the absence of new transcription (35). When aleurone layers are subjected to heat shock treatment after α-amylase is fully induced, hsp are synthesized, whereas the synthesis of α-amylase is quickly suppressed to less than 15% of the original level with a corresponding decrease in the levels of α-amylase mRNA. Upon recovery from heat shock, the synthesis of α-amylase resumes but only in the presence of new transcription. Similar observations have been made with the GA$_3$-induced protease as well as all the other secretory proteins. The same authors have also examined the ultrastructural changes in the heat-shocked aleurone cells and found that heat shock causes a fast and substantial disruption of rough endoplasmic reticulum (RER) (4a). Since all the secretory proteins are synthesized on RER, the disruption of this structure could be related to the decrease in synthesis of secretory proteins and the enhanced degradation of their mRNA during heat shock. It is clear from the above work that heat shock caused an *enhanced* degradation of some of the preexisting mRNA coding for normal proteins, a phenomenon that has not been observed in other systems.

In contrast to *Drosophila* and soybean, both maize and barley appear to have proteins whose synthesis is neither suppressed nor enhanced by heat shock. It is not yet known why the synthesis of certain proteins is preserved while the synthesis of other proteins is quickly suppressed.

Potential Physiological Roles of Hsp

Although ubiquitin (7), an ATP-dependent protease (26), and enolase (36a) have been shown to be induced by heat shock in animal and microbial systems, the exact function and identity of the remaining hsp remain unclear. It has been shown that the induction of hsp allows the cells to establish thermotolerance in many organisms (66), i.e. to survive at a temperature that is normally lethal. Lin et al (51) have shown that briefly subjecting soybean seedlings to 45°C followed by incubation at 28°C results in the induction of hsp and a concomitant establishment of thermotolerance. In maize roots the temperature regime and time course of hsp induction also correlate closely with those for the establishment of thermotolerance (12). The thermotolerant maize plants not only survive at the lethal temperature but slightly outgrow the control plants when they are returned to normal temperatures.

All the tissues in maize plants, with the exception of germinating pollen, synthesize hsp during heat shock (2, 14). It is known that germinating pollen is more sensitive to high temperature than other tissues (34), a phenomenon that also suggests that the ability to synthesize hsp is related to thermotolerance. However, this view is questioned by the observation that it is possible to establish thermotolerance in germinating *Tradescantia* pollen without the induction of hsp (2).

One approach to study the potential role of hsp is to study the cellular location of these proteins, which could be related to the function of specific organelles. Lin et al (51) have found that in soybean hsp15–18 and hsp69–70 are associated with nuclei, mitochondria, and ribosomes, hsp22–24 with mitochondrial fractions, and hsp84 and 92 are in the postribosomal supernatant. They have also observed that the hsp induced by other factors such as arsenite are not associated with the organelles mentioned above, indicating that the association of hsp with specific organelles is heat mediated. More recently, E. Vierling et al (unpublished) have convincingly demonstrated that in soybean, pea, and maize, the precursor of hsp 22 (28 kd in size) and 27 can be taken up by isolated chloroplasts in vitro. In contrast, Nover et al (53) have found practically no association of hsp with mitochondria or proplastids in heat-shocked tomato cultures. They also find no association of hsp18 with any of the organelles mentioned above in tomato. Instead, over 50% of hsp18 becomes associated with cytoplasmic "heat-shock granules" whose function is unknown (54). In earlier work the same research group has reported that heat shock causes the phosphorylation of a ribosomal protein (65). Although this ribosomal protein is not a hsp, this observation does imply that heat shock causes alterations of ribosomal functions. Cooper (12) has done an investigation of the cellular location of maize hsp using sucrose gradient centrifugation. Hsp29 (possibly equivalent to the soybean hsp27) is found to be associated with mitochondria, hsp18 and 70 with the plasma membranes, hsp25 and 72 with the ER, and the higher molecular weight hsp such as hsp79–83 are in the soluble fractions. The association of hsp with membrane fractions suggests a role of hsp in the altered membrane functions that usually occur at elevated temperatures.

It has been shown that hsp accumulate in field grown, heat-stressed plants (10, 44), indicating these proteins could be essential for the plants to cope with the natural stress conditions. However, it is apparent that a well-defined role for any of the plant hsp has not yet been elucidated. To this end a clear-cut investigation on the cellular location of hsp using immunohistochemical methods with monospecific antibodies could provide useful information.

OTHER ENVIRONMENTAL STRESS RESPONSES

Proteins Induced in Salt-Adapted Cells

In recent years there has been increased interest in the breeding of salt-resistant crop plants to be used in coastal areas and lands that require heavy irrigation. Many culture cell lines adapted to high salt conditions have been isolated. As an initial attempt to study the biochemical mechanism of salt adaptation, Ericson & Alfinito (19) and Singh et al (71a) have examined the protein profiles in a NaCl-adapted (salt-adapted) line of *Nicotiana tabacum* and a non-NaCl-adapted line. Several proteins are found to be more abundant in salt-adapted cells, and another protein of 26 kd is unique in salt-adapted cells. When the salt-adapted cells are transferred to the low salt medium, the levels of 20 and 32 kd proteins return to those in the nonadapted cells. However, the 26 kd protein is present for at least two passages. The 20 kd protein has 4% hydroxyproline and 11.3% proline, which is unusually high, yet less than the hydoxyproline-rich glycoproteins in cell walls. R. A. Bressan's group (unpublished) has purified the 26 kd protein and found that this protein is also induced by ABA which accumulates in the salt-adapted cells. The physiological role of these proteins is not yet known.

Response to Heavy Metals

When plants are grown in soils contaminated with high concentrations of heavy metals, several metabolic activities could be adversly affected. However, plants can be selected that can grow on normally toxic concentrations of cadmium or copper. The molecular mechanisms behind this heavy metal tolerance are not yet understood; however, low molecular weight, cysteine-rich, soluble, metal-binding proteins, similar to the metallothioneins studied extensively in animals (74), have been found in resistant plant cells (38). In *Datura innoxia* cells that are resistant to cadmium, such a metal-binding protein has been found and de novo synthesis of this protein is induced by cadmium (39). A similar copper-binding protein has been found in the roots of *Mimulus guttatus* (61). The amount of this Cu-binding, metallothionein-like protein increases after 5 h of exposure to a solution containing Cu ions.

Response to UV Light Exposure

Plants can be damaged by the ultraviolet rays of intense sunlight. As a defense, plants produce flavonoid pigments that absorb the detrimental UV radiation and can minimize damage to the plant. Cell suspension cultures of parsley produce and accumulate flavonoids when treated with ultraviolet light (29). It was found that UV treatment causes the coordinate induction of phenylalanine ammonia-lyase, 4-coumarate:CoA ligase, chalcone synthase, and UDP-apiose synthase, all enzymes required for flavonoid biosynthesis. It was shown, using cDNA probes, that the levels of mRNA encoding these enzymes increase during UV treatment (47) and that this is the result of increased transcription rates of the genes (11).

Water Stress-Induced Proteins

Despite the intensive research effort in the field of water stress, very few water stress-induced proteins have been detected. It has been shown that two of the hsp, hsp70 in maize and hsp27 in soybean, can be also induced by water stress (15, 33).

CONCLUDING REMARKS

All of the environmental stress conditions studied cause alterations of gene expression resulting in the induction of new proteins and the repression of at least some normally expressed proteins. It is believed that these stress-induced proteins allow plants to make biochemical and structural adjustments that enable them to cope with the stress conditions. In the case of anaerobic stress, the function of the induced ADH and some of the other ANPs has been well elucidated. However, the role(s) of TPs remains to be resolved. The physiological role of hsp is not yet clear, although the synthesis of these proteins appears to be related to the acquisition of thermotolerance. The UV induction of enzymes involved in the synthesis of flavonoid pigments should certainly protect the plants from being damaged by the irradiation. The induction of metal-binding proteins appears to be a mechanism to sequester excessive amounts of heavy metals. To date little is known concerning the roles of the salt and water stress-induced proteins.

The DNA sequences encoding of many of these stress proteins, and those regulating their expression, have been isolated and are presently being characterized. With the advancement in transformation techniques it is conceivable that the genes coding for stress proteins from a tolerant plant could be introduced to sensitive plants to enhance their resistance to stress conditions. This will have an enormous impact in allowing the growth of crop plants in areas where they presently cannot be grown.

ACKNOWLEDGMENTS

We would like to thank our colleagues who have provided us with their latest research information. We also extend our appreciation to Peter Brown, Gerg Heck, Liang-Shiou Lin, Jih-Jing Lin, Becky Peck, and Joseph Varner for their critical reading of this manuscript.

Literature Cited

1. Altschuler, M., Mascarenhas, J. P. 1982. Heat shock proteins and effects of heat shock in plants. *Plant Mol. Biol.* 1:103–15
2. Altschuler, M., Mascarenhas, J. P. 1982. The synthesis of heat-shock and normal proteins at high temperatures in plants and their possible roles in survival under heat stress. See Ref. 66, pp. 321–27
3. Ashburner, M., Bonner, J. J. 1979. The induction of gene activity in *Drosophila* by heat shock. *Cell* 17:241–54

4. Barnett, T., Altschuler, M., McDaniel, C. N., Mascarenhas, J. P. 1980. Heat-shock induced proteins in plant cells. *Dev. Genet.* 1:331–40
4a. Belanger, F. C., Brodl, M. R., Ho, T.-H. D. 1986. Heat shock causes destabilization of specific mRNAs and destruction of endoplasmic reticulum in barley aleurone cells. *Proc. Natl. Acad. Sci. USA.* In press
5. Bennett, D. C., Freeling, M. 1986. Flooding and the anaerobic stress response. In *Model Building in Plant*

Physiology/Biochemistry, ed. D. W. Newman, K. G. Wilson. Boca Raton, FL: CRC Press. In press

6. Bertani, A., Menegus, F., Bollini, R. 1981. Some effects of anaerobiosis on protein metabolism in rice roots. *Z. Pflanzenphysiol.* 103:37–43

7. Bond, U., Schlesinger, M. J. 1985. Ubiquitin is a heat shock protein in chicken embryo fibroblasts. *Mol. Cell. Biol.* 5:949–56

8. Baszczynski, C. L., Walden, D. B., Atkinson, B. G. 1982. Studies on the heat shock response in different tissues and genotypes of maize. *Maize Genet. Coop. Newsl.* 56:112–13

9. Baszczynski, C. L., Walden, D. B., Atkinson, B. G. 1982. Regulation of gene expression in corn *(Zea mays, L.)* by heat shock. *Can. J. Biochem.* 60:569–79

10. Burke, J. J., Hatfield, J. L., Klein, R. R., Mullet, J. E. 1985. Accumulation of heat shock proteins in field-grown cotton. *Plant Physiol.* 78:394–98

11. Chappell, J., Hahlbrock, K. 1984. Transcription of plant defense genes in response to UV-light or fungal elicitor. *Nature* 311:76–78

12. Cooper, P. 1985. *The heat shock proteins of maize: Their induction, cellular location and potential function.* PhD thesis. Univ. Ill., Urbana

13. Cooper, P., Ho, T.-H. D. 1983. Heat shock proteins in maize. *Plant Physiol.* 71:215–22

14. Cooper, P., Ho, T.-H. D., Hauptmann, R. M. 1984. Tissue specificity of the heat-shock response in maize. *Plant Physiol.* 75:431–41

15. Czarnecka, E., Edelman, L., Schöffl, F., Key, J. L. 1984. Comparative analysis of physical stress responses in soybean seedlings using cloned heat shock cDNAs. *Plant Mol. Biol.* 3:45–58

16. Czarnecka, E., Gurley, W. B., Nagao, R. T., Mosquera, L., Key, J. L. 1985. DNA sequence and transcript mapping of a soybean gene encoding a small heat shock protein. *Proc. Natl. Acad. Sci. USA* 82:3726–30

17. Dennis, E. S., Gerlach, W. L., Pryor, A. J., Bennetzen, J. L., Inglis, A., et al. 1984. Molecular analysis of the alcohol dehydrogenase *(Adh1)* gene of maize. *Nucleic Acids Res.* 12:3983–4000

18. Dennis, E. S., Sachs, M. M., Gerlach, W. L., Finnegan, E. J., Peacock, W. J. 1985. Molecular analysis of the alcohol dehydrogenase 2 *(Adh2)* gene of Maize. *Nucleic Acids Res.* 13:727–43

19. Ericson, M. C., Alfinito, S. H. 1984. Proteins produced during salt stress in tobacco cell culture. *Plant Physiol.* 74:506–9

20. Ferl, R. J., Dlouhy, S. R., Schwartz, D. 1979. Analysis of maize alcohol dehydrogenase by native-SDS two dimensional electrophoresis and autoradiography. *Mol. Gen. Genet.* 169:7–12

21. Freeling, M. 1973. Simultaneous induction by anaerobiosis or 2,4-D of multiple enzymes specified by two unlinked genes: Differential *Adh1–Adh2* expression in maize. *Mol. Gen. Genet.* 127:215–27

22. Freeling, M., Bennett, D. C. 1985. Maize *Adh1. Ann. Rev. Genet.* 19:297–324

23. Freeling, M., Birchler, J. A. 1981. Mutants and variants of the *Alcohol dehydrogenase-1* gene in maize. In *Genetic Engineering: Principles and Methods,* ed. J. K. Setlow, A. Hollaender, 3:223–64. New York: Plenum

24. Gerlach, W. L., Pryor, A. J., Dennis, E. S., Ferl, R. J., Sachs, M. M., Peacock, W. J. 1982. cDNA cloning and induction of the alcohol dehydrogenase gene *(Adh1)* of maize. *Proc. Natl. Acad. Sci. USA* 79:2981–85

25. Gerlach, W. L., Sachs, M. M., Llewellyn, D., Finnegan, E. J., Dennis, E. S. 1986. Maize alcohol dehydrogenase; A molecular perspective. In *Plant Gene Research, Vol. 3. "A Genetic Approach to Plant Biochemistry,"* ed. A. D. Blonstein, P. J. King. New York: Springer-Verlag. In press

26. Goff, S. A., Goldberg, A. L. 1985. Production of abnormal proteins in *E. coli* stimulates transcription of lon and other heat shock genes. *Cell* 41:587–95

27. Goldberg, M. L. 1979. *Sequence analysis of Drosophila histone genes.* PhD thesis. Stanford Univ., Stanford, CA.

28. Hageman, R. H., Flesher, D. 1960. The effect of anaerobic environment on the activity of alcohol dehydrogenase and other enzymes of corn seedlings. *Arch. Biochem. Biophys.* 87:203–9

29. Hahlbrock, K. 1981. Flavonoids. In *The Biochemistry of Plants,* ed. E. E. Conn, 7:425–29. New York: Academic

30. Hake, S., Kelley, P. M., Taylor, W. C., Freeling, M. 1985. Coordinate induction of alcohol dehydrogenase 1, aldolase, and other anaerobic RNAs in maize. *J. Biol. Chem.* 260:5050–54

31. Hanson, A. D., Jacobsen, J. V. 1984. Control of lactate dehydrogenase, lactate glycolysis, and α-amylase by O_2 deficit in barley aleurone layers. *Plant Physiol.* 75:566–72

32. Hanson, A. D., Jacobsen, J. V., Zwar, J. A. 1984. Regulated expression of three

alcohol dehydrogenase genes in barley aleurone layers. *Plant Physiol.* 75:573–81

33. Heikkila, J. J., Papp, J. E. T., Schultz, G. A., Bewley, J. D. 1984. Induction of heat shock protein messenger RNA in maize mesocotyls by water stress, abscisic acid, and wounding. *Plant Physiol.* 76:270–74

34. Herrero, M. P., Johnson, R. R. 1980. High temperature stress and pollen viability of maize. *Crop Sci.* 20:796–800

35. Ho, T.-H. D., Varner, J. E. 1974. Hormonal regulation of mRNA metabolism in barley aleurone layers. *Proc. Natl. Acad. Sci. USA* 71:4783–86

36. Hook, D. D., Crawford, R. M. M. 1978. *Plant Life in Anaerobic Environments.* Ann Arbor: Ann Arbor Sci. Publ. 564 pp.

36a. Iida, H., Yahara, I. 1985. Yeast heat-shock protein of M_r 48,000 is an isoprotein of enolase. *Nature* 315:688–70

37. Jackson, M. B. 1985. Ethylene and responses of plants to soil waterlogging and submergence. *Ann. Rev. Plant Physiol.* 36:145–74

38. Jackson, P. J., Naranjo, C. M., McClure, P. R., Roth, E. J. 1985. The molecular response of cadmium resistant *Datura innoxia* cells to heavy metal stress. In *Cellular and Molecular Biology of Plant Stress,* ed. J. L. Key, T. Kosuge, pp. 145–60. New York: Liss

39. Jackson, P. J., Roth, E. J., McClure, P. R., Naranjo, C. M. 1984. Selection, isolation, and characterization of cadmium-resistant *Datura innoxia* suspension cultures. *Plant Physiol.* 75:914–18

40. Kelley, P. M., Freeling, M. 1982. A preliminary comparison of maize anaerobic and heat-shock proteins. See Ref. 66, pp. 315–19

41. Kelley, P. M., Freeling, M. 1984. Anaerobic expression of maize glucose phosphate isomerase I. *J. Biol. Chem.* 259:673–77

42. Kelley, P. M., Freeling, M. 1984. Anaerobic expression of maize fructose-1,6-diphosphate aldolase. *J. Biol. Chem.* 259:14180–83

43. Key, J. L., Gurley, W. B., Nagao, R. T., Czarnecka, E., Mansfield, M. A. 1985. Multigene families of soybean heat shock proteins. In *Molecular Form and Function of the Plant Genome,* ed. L. van Vloten-Doting, G. Groot, T. Hall. New York: Plenum

44. Key, J. L., Kimpel, J. A., Lin, C. Y., Nagao, R. T., Vierling, E., et al. 1985. The heat shock response in soybean. See Ref. 38, pp. 161–79

45. Key, J. L., Lin, C. Y., Chen, Y. M. 1981. Heat shock proteins of higher plants. *Proc. Natl. Acad. Sci. USA* 78:3526–30

46. Kozlowski, T. T. 1984. *Flooding and Plant Growth.* Orlando: Academic. 356 pp.

47. Kuhn, D. N., Chappell, J., Boudet, A., Hahlbrock, K. 1984. Induction of phenylalanine ammonia-lyase and 4-coumarate:CoA ligase mRNAs in cultured plant cells by UV light or fungal elicitor. *Proc. Natl. Acad. Sci. USA* 81:1102–6

48. Laszlo, A., St. Lawrence, P. 1983. Parallel induction and synthesis of PDC and ADH in anoxic maize roots. *Mol. Gen. Genet.* 192:110–17

49. Deleted in proof

50. Lin, C. Y., Key, J. L. 1967. Dissociation and reassembly of polyribosomes in relation to protein synthesis in the soybean root. *J. Mol. Biol.* 26:237–47

51. Lin, C. Y., Roberts, J. K., Key, J. L. 1984. Acquisition of thermotolerance in soybean seedlings: Synthesis and accumulation of heat shock proteins and their cellular localization. *Plant Physiol.* 74:152–60

52. Lindquist, S. 1981. Regulation of protein synthesis during heat shock. *Nature* 294:311–14

52a. Nagao, N. T., Czarnecka, E., Gurley, W. B., Schöffl, F., Key, J. L. 1985. Genes for low-molecular-weight heat shock proteins of soybeans: Sequence analysis of a multigene family. *Mol. Cell. Biol.* 5:3417–28

53. Nover, L., Scharf, K. D. 1984. Synthesis, modification, and structural binding of heat shock proteins in tomato cell cultures. *Eur. J. Biochem.* 139:303–13

54. Nover, L., Scharf, K. D., Neumann, D. 1983. Formation of cytoplasmic heat shock granules in tomato cell cultures and leaves. *Mol. Cell Biol.* 3:1648–55

55. Okimoto, R., Sachs, M. M., Porter, E. K., Freeling, M. 1980. Patterns of polypeptide synthesis in various maize organs under anaerobiosis. *Planta* 150:89–94

56. Pelham, H. R. B., Bienz, M. 1982. A synthetic heat-shock promoter element confers heat-inducibility on the herpes simplex virus thymidine kinase gene. *EMBO J.* 1:1473–77

57. Pradet, A., Mocquot, B., Raymond, P., Morisset, C., Aspart, L., Delseny, M. 1985. Energy metabolism and synthesis of nucleic acids and proteins under anoxic stress. See Ref. 38, pp. 227–45

58. Roberts, J. K. M., Andrade, F. H., Anderson, I. C. 1985. Further evidence that cytoplasmic acidosis is a determinant of

flooding intolerance in plants. *Plant Physiol.* 77:492–94
59. Roberts, J. K. M., Callis, J., Jardetzky, O., Walbot, V., Freeling, M. 1984. Cytoplasmic acidosis as a determinant of flooding intolerance in plants. *Proc. Natl. Acad. Sci. USA* 81:6029–33
60. Roberts, J. K. M., Callis, J., Wemmer, D., Walbot, V., Jardetzky, O. 1984. Mechanism of cytoplasmic pH regulation in hypoxic maize root tips and its role in survival under hypoxia. *Proc. Natl. Acad. Sci. USA* 81:3379–83
61. Robinson, N. J., Thurman, D. A. 1986. Purification of a metallothinein-like Cu-complex from *Mimulus guttatus*. Submitted for publication
62. Rowland, L. J., Strommer, J. N. 1985. Insertion of an unstable element in an intervening sequence of maize *Adh1* affects transcription but not processing. *Proc. Natl. Acad. Sci. USA* 82:2875–79
63. Sachs, M. M., Freeling, M. 1978. Selective synthesis of alcohol dehydrogenase during anaerobic treatment of maize. *Mol. Gen. Genet.* 161:111–15
64. Sachs, M. M., Freeling, M., Okimoto, R. 1980. The anaerobic proteins of maize. *Cell* 20:761–67
65. Scharf, K. D., Nover, L. 1982. Heat shock induced alterations of ribosomal protein phosphorylation in plant cell cultures. *Cell* 30:427–37
66. Schlesinger, M. J., Ashburner, M., Tissieres, A. 1982. *Heat Shock From Bacteria to Man.* New York: Cold Spring Harbor Lab. 440 pp.
67. Schöffl, F., Key, J. L. 1982. An analysis of mRNA's for a group of heat shock proteins of soybean using cloned cDNA's. *J. Mol. Appl. Genet.* 1:301–14

68. Schöffl, F., Key, J. L. 1983. Identification of a multigene family for small heat shock proteins in soybean and physical characterization of one individual gene coding region. *Plant Mol. Biol.* 2:269–78
69. Schöffl, F., Raschke, E., Nagao, R. T. 1984. The DNA sequence analysis of soybean heat-shock genes and identification of possible regulatory promoter elements. *EMBO J.* 3:2491–97
70. Schwartz, D. 1969. An example of gene fixation resulting from selective advantage in suboptimal conditions. *Am. Nat.* 103:479–81
71. Shah, D. M., Rochester, D. E., Krivi, G. G., Hironaka, C. M., Mozer, T. J., et al. 1985. Structure and expression of maize HSP 70 gene. See Ref. 38, pp. 181–200
71a. Singh, N. K., Handa, A. K., Hasegawa, P. M., Bressan, R. A. 1985. Proteins associated with adaptation of cultured tobacco cells to NaCl. *Plant Physiol.* 79:126–37
72. Storti, R. V., Scott, M. P., Rich, A., Pardue, M. L. 1980. Translational control of protein synthesis in response to heat shock in *D. melanogaster* cells. *Cell* 22:825–34
73. Vierling, E., Key, J. L. 1985. Ribulose 1,5-bisphosphate carboxylase synthesis during heat shock. *Plant Physiol.* 78:155–62
74. Webb, M. 1979. The metallothioneins. In *The Chemistry, Biochemistry, and Biology of Cadmium,* ed. M. Webb, pp. 195–266. Amsterdam: Elsevier/North Holland
75. Wignarajah, K., Greenway, H. 1976. Effect of anaerobiosis on activities of alcohol dehydrogenase and pyruvate decarboxylase in roots in *Zea mays. New Phytol.* 77:575–84

Ann. Rev. Plant Physiol. 1986. 37:377–405

BIOPHYSICAL CONTROL OF PLANT CELL GROWTH

Daniel Cosgrove

Department of Biology, Pennsylvania State University, University Park, Pennsylvania 16802

CONTENTS

INTRODUCTION

The childhood tale of Jack and the Beanstalk includes an episode in which bean seedlings grow up through the clouds in a single night. Although we may smile at such a notion, the event does raise questions. Why don't plants grow as fast in our world? What limits the growth rate of real plants?

This review examines the prevailing concept that plant cell growth results from turgor-driven yielding of the cell wall. This is a physical description of how plant cells increase in size during growth and morphogenesis, thereby

0066-4294/86/0601-0377$02.00

enabling plants to create a full leaf canopy, a tall stem structure, and an extensive root system. For purposes of discussion, the theme of turgor-driven cell growth will be dissected into two broad questions: what characteristics of the cell wall enable it to yield (extend irreversibly) in a controlled fashion for prolonged periods of time, and how is water transported to and absorbed by growing cells to maintain turgor pressure in the face of prolonged yielding of the cell wall? Various facets of these questions will be examined in turn, with an emphasis on the most recent work and on efforts to define the physical constraints limiting plant growth.

The limited nature of this review precludes discussion of many aspects pertinent to growth. Readers are referred to recent reviews in this series, especially those by Taiz on cell wall mechanical properties (91), Silk on the kinematics of plant growth (82), Boyer on water transport (6), Morgan on osmoregulation (66), Eisinger on ethylene-induced expansion (27), and Labavitch on cell wall turnover (48). Other recent reviews deal with leaf growth (40, 94), hormone action on cell walls (16), and wall biochemistry (56).

THE PHYSICS OF IRREVERSIBLE CELL ENLARGEMENT

Irreversible enlargement of plant cells results from two interdependent physical processes: water absorption and cell wall yielding. Water absorption increases the volume of growing cells (which typically consist of 85%–95% water), whereas wall yielding generates the driving force for water uptake.

This physical conception of growth may be summarized as follows (14, 18, 19, 40, 76). Under normal conditions, the wall of a turgid cell is elastically (reversibly) stretched by the protoplast, simultaneously giving rise to a stress in the load-bearing parts of the wall and a compression of the cell contents (i.e. a hydrostatic pressure or turgor pressure). Growth is initiated when load-bearing elements in the wall yield, thereby inducing a contraction (or "relaxation") of elastically distended wall elements and a reduction in wall stress. Since wall stress and turgor pressure constitute equal and opposite forces, turgor pressure and wall stress decrease simultaneously. Such relaxation reduces the cell water potential and gives rise to a passive water influx, which in turn increases cell volume and extends the cell wall.

The crux of this model is that cell growth starts with a reduction in wall stress because of irreversible yielding of the wall, and that water absorption and physical expansion are spontaneous consequences of this initial modification of the wall. In the past some authors have suggested that growth can be initiated by increases in cell osmotic pressure or by decreases in the elastic modulus of the cell wall. While such changes by themselves may indeed induce a water influx, such water flow will be transient and reversible and should not be imagined as the cause of growth. Because water transport is passive and reversible, the irreversible aspect of growth must arise from the yielding of the cell wall, not from water transport (19, 76).

This summary presents a physical description of short-term growth and does not address many questions of interest for long-term growth. For example, how do assimilation of CO_2, absorption of nutrients, and synthesis of membranes, proteins, cell wall components, and other cell constituents tie into this model? In most cases, these biochemical processes influence physical enlargement only indirectly. To cite one instance, water absorption during cell expansion tends to dilute the solutes within the cells. If this proceeded unabated, growth would soon cease as cell osmotic pressure and turgor pressure fell toward zero. Synthesis, transport, and assimilation of new solutes counteract this tendency and thus support continued growth. More encompassing (and complicated) models of plant growth include the influence of these secondary processes on the primary processes of wall expansion and water absorption and also include feedback between the various processes that support plant growth (e.g. 9, 40). Such models are useful conceptually, but we are a long way from being able to evaluate them quantitatively.

Growth Equations

A corollary of the preceding model of plant growth is that cell turgor pressure is in dynamic balance between wall yielding which tends to dissipate turgor pressure and water uptake which tends to restore turgor pressure. Lockhart (50) formulated a quantitative analysis of steady-state growth of single cells. Although other models for growth have been proposed (33, 36), Lockhart's analysis has served for the past two decades as perhaps the most useful framework for discussing the physics of growth.

Using a slightly different formulation from that employed by Lockhart, water uptake by a cell in solution can be defined as a relative rate of volume increase $[(1/V)dV/dt]$ and is given by:

$$(1/V)\ dV/dt = A\ Lp\ (\sigma\Delta\pi - P)/V = L\ (\sigma\Delta\pi - P) \qquad 1.$$

Table 1 lists the symbols used in this and following equations. According to Equation 1, the conductance L is a transport coefficient which determines the rate of water entry into a cell for any given driving force $(\sigma\Delta\pi - P)$. Equation 1 has strong theoretical and empirical justification and validation (25, 61, 104).

To arrive at an analogous expression for the rate of wall yielding, Lockhart (50) adopted the idea that irreversible wall extension depended on the wall stress in excess of a minimum stress or yield threshold. Because wall stress

Table 1   Definition of major symbols used in the text

Lp	membrane hydraulic conductivity (cm s^{-1} MPa^{-1})
σ	solute reflection coefficient (dimensionless)
$\Delta\pi$	osmotic pressure difference between the cell and its surroundings (MPa)
P	hydrostatic pressure (MPa)
V	cell volume (cm^3)
A	cell surface area (cm^2)
L	volumetric hydraulic conductance (s^{-1} MPa^{-1}); for single cells this is equal to $A \cdot Lp/V$;
ψ	water potential (MPa)
ϕ	cell wall yielding coefficient [extensibility] (s^{-1} MPa^{-1})
Y	minimum turgor necessary for growth [yield threshold] (MPa)
ϵ	volumetric elastic modulus of the cell (MPa)
D_c	free energy diffusivity for water flow through a tissue via the cell-to-cell pathway alone (cm^2 s^{-1})
D_t	free energy diffusivity for water flow through a tissue (cm^2 s^{-1})
π_{cw}	osmotic pressure of the cell wall solution (MPa)
r	cell radius (cm)

normally arises from cell turgor, he proposed the rather simple function (again, in modified form):

$$(1/V)\ dV/dt = \phi\ (P - Y) \qquad\qquad 2.$$

which states that the relative growth rate $(1/V)dV/dt$ depends in a linear fashion on turgor pressure in excess of a critical turgor (Y, the "yield threshold") and on a wall yielding coefficient (ϕ). This coefficient was called wall extensibility by Lockhart. Unfortunately, this term has also been used by other workers to refer to many other quantities that are quite different from ϕ, e.g. elastic extensibility, Instron extensibility, plastic extensibility, and so on. This has led to some confusion in the literature between the coefficient (ϕ) governing wall expansion during plant growth and some of these other mechanical parameters. For this reason, the term extensibility will generally be avoided in this review. When it is used, it will refer to ϕ in the above equation. Equation 2 should be viewed as an approximate relation to describe the apparent dependence of wall expansion on turgor pressure. The utility and empirical validity of this relation will be examined in a later section.

In this model of growth, turgor pressure acts as the link that coordinates water uptake with cell wall yielding. Because of cell wall yielding, turgor pressure and water potential of a growing cell are at least slightly displaced from equilibrium. The effective water potential disequilibrium ($\Delta\Psi_{eff} = \sigma\Delta\pi - P$) under conditions of steady-state growth depends on the value of ϕ relative to L and is given by:

$$\Delta\Psi_{eff} = \frac{\phi}{\phi + L} \quad (\sigma\Delta\pi - Y) \qquad\qquad\qquad 3.$$

This relation is obtained by setting Equations 1 and 2 equal to each other, substituting $P = (L\pi + \phi Y)/(\phi + L)$ and solving for $\Delta\Psi_{eff}$. The ratio $\phi/(\phi + L)$ is a useful measure of the limitation of growth by water transport. When L is much larger than ϕ, this ratio approaches zero. Therefore $\Delta\Psi_{eff}$ approximates zero ($P \approx \sigma\Delta\pi$) and growth is limited by ϕ. At the other extreme when L is much smaller than ϕ, the ratio approaches unity, P drops nearly to Y and growth is limited by L. When the ϕ equals L, the ratio equals 1/2 and growth is equally limited by water transport and wall yielding.

Lockhart's analysis strictly applies only to single, isolated cells in which the cell hydraulic conductance (L) is determined by Lp and the geometry of the cell. In multicellular tissues, the pathway for water flow is more complicated. Before considering multicellular growth, however, it is instructive to examine whether water transport (low L) may limit growth of single isolated cells, using a pea cortical cell as an example. The volumetric hydraulic conductance (L) for a single cell is given by $Lp \cdot A/V$. Using $Lp = 10^{-4}$ cm s^{-1} MPa^{-1}, $V = 0.25 \cdot 10^{-6}$ cm^3, and $A = 0.27 \cdot 10^{-3}$ cm^2 for a typical pea stem cortical cell (22), L may be calculated to be 0.1 s^{-1} MPa^{-1}. For a cell growing at the rapid rate of 10% per h, the driving force for water absorption would be a mere 0.3 kPa. In other words, cell wall yielding reduces turgor pressure by less than one thousandth of the turgor pressure typical of such cells (19, 22). This calculation shows that for single isolated cells, water transport does not represent a rate-limiting process for growth. Calculations made for the giant-celled alga *Nitella* lead to similar conclusions (18, 67). For such cells, water potential equilibrium is nearly perfect, and growth is controlled entirely by the wall-yielding process (e.g. Equation 2).

GROWTH OF MULTICELLULAR TISSUES

Organ growth differs from single cell growth in at least two important respects. First, organs such as stems, leaves, or roots are composed of a variety of tissue and cell types that may differ in their growth properties. Second, to reach a growing cell embedded in the middle of a stem or leaf, water must traverse a much more circuitous path than is the case for an isolated cell. In the case of tissues, the hydraulic conductance relevant for growth is not simply that of the growing cell ($Lp \cdot A/V$), but must also include the effect of this longer pathway on the value of L.

Variation in Growth Properties

CELL LAYERS   Plant organs consist of a variety of cell types, and it is likely that these cells have different physical properties. For instance, there is a long-standing hypothesis that the epidermis controls the growth of coleoptiles, stems, and leaves (53, 93). Because the various cell layers are constrained by mechanical linkages between cell walls to grow at a uniform rate, the only practical approach to date has been to measure average or bulked parameters (2, 7, 8, 19, 40, 81, 99, 100). When all the cells are growing at the same rate, this seems justified. However, no physical analysis of growth has yet been offered that rigorously deals with such inhomogeneities. As our methods for measuring the growth properties of cell and tissues improve, we will need a theoretical analysis and evaluation of this practice.

GROWTH GRADIENTS   A related problem—one that is inherent in all growing axes such as roots and stems—is that cells at various positions along an axis grow at different rates. Silk (82) has summarized some of the mathematical and physiological properties of such gradients and makes a strong case for the necessity of coming to terms with such spatial and temporal variation in growth in order to understand developmental patterns of plants. Presumably such spatial gradients in growth rate depend upon gradients in one or more of the biophysical parameters governing growth. Unfortunately, there is little direct evidence on this matter.

In recent years there have been a number of biophysical studies of the growth of stems (7, 19), leaves (40, 99, 100), and other organs (101) that contain spatial gradients in growth. For the most part, these studies have ignored spatial variation in growth rate and have treated the tissues as if they had uniform properties. For instance, single values for water potential, turgor pressure, etc were measured or calculated. When these measurements are restricted to small portions of the growing region, the values obtained should represent averages for that part of the growing axis. Hsiao et al (40) emphasize that these values are useful approximations, but that ultimately we need full characterization of the gradients in physical properties of growing axes. Such characterization will require methods with very fine spacial resolution. In this respect, the use of the pressure probe for growth studies (19, 22, 23, 87) holds great promise because it has resolution at the cell level.

Geometry of Water Flow

PATHWAYS OF WATER FLOW   Because of the geometry of most organs, the pathway for water uptake by growing cells is longer and more complicated than that for an isolated cell. In the case of an epidermal leaf cell, water must travel from the soil, through the root cortex and endodermis, through the long xylem path, and finally across a layer of several to many cells before it can be absorbed and thus contribute to cell expansion. Some studies (4, 65) have suggested that the major resistance to water flow along this pathway is located along the last leg of the journey, i.e. the pathway from the xylem to the growing cell. Other studies have found that resistances in the root (3, 74, 75) or along the xylem pathway (90) may also restrict water flow.

Water transport through nonvascularized tissues can occur via two parallel pathways: the cell wall (or apoplast) pathway and the cell-to-cell pathway (63, 64, 87, 102). The cell wall pathway includes water movement through wall layers between cells and along surface films lining intercellular air spaces. The cell-to-cell pathway includes two conceptual pathways. Water may move from vacuole to vacuole with intermediate steps in the cytoplasm and the cell wall. Water may also move symplastically without crossing membranes, i.e. through plasmodesmata. Evaluation of water flow via the symplast has proved difficult because of the extreme sensitivity of calculated volume flux to the pore radius (34, 80). However, recent dye-injection experiments by Goodwin (29) indicate that the effective plasmodesmatal pore diameter may be only 3–5 nm in *Elodea*. This figure is smaller than that suggested by ultrastructural studies (34, 80). Assuming that dye injection did not cause restriction of the pore by callose formation or by some other type of plugging, one may calculate that bulk water movement via the symplast would constitute only a negligible fraction of the total water flux involved in growth. Similar dye-injection studies are needed to ascertain whether growing tissues likewise have symplastic connections with such small effective diameters.

Despite several recent investigations (42, 85, 87, 102), the relative contributions of the two pathways to total water flux through tissues remains uncertain. Because of lack of membranes, the cell wall should offer a low-resistance path for water flow; however, the cross-sectional area of this pathway is much less than that for the cell-to-cell pathway, so total volume flux via the cell wall will be correspondingly reduced. It is clear from many recent pressure-probe experiments that the cell wall pathway does not dominate water transport in tissues (22, 85, 86, 95, 102). As emphasized recently by Boyer (6), the fact that cell half-times for water potential equilibration are much faster than tissue half-times for such equilibration suggests that water moves to some extent via the cell-to-cell path. Otherwise, the half-times would be the same.

Recent efforts to assess the relative contributions of the two pathways have produced different and contradictory conclusions. Jones et al (42) and Steudle & Jeschke (85) measured the Lp of the various cell layers in wheat and barley roots. They modeled the overall hydraulic conductance of the root as a series of cell conductances (ignoring the cell wall pathway) and then compared this value with experimental measures of whole-root hydraulic conductance. Both studies concluded that the major pathway for water flow was via the cell-to-cell pathway. However, in a recent update, Jones et al (43) reversed their conclusion when they found whole-root hydraulic conductances that were tenfold larger than originally reported. In the original experiments roots were excised and allowed to exude for 15 h prior to the experiment. Jones et al (43) found that this pretreatment induced a considerable reduction in root conductance, although the basis for this reduction was not investigated.

Other attempts to estimate water flux through the cell-to-cell pathway have been carried out by Steudle and coworkers (22, 85, 86, 102), who have used the pressure probe to measure the half-time ($T_{1/2}$) for water potential equilibration of individual cells. Using the theory developed by Philip (73), they then computed a diffusivity (D_c) for water movement via the cell-to-cell pathway alone with the approximation:

$$D_c = \alpha \cdot Lp \cdot r (\epsilon + \pi) \approx \frac{\ln(2)}{2} \cdot \frac{r^2}{T_{1/2}(\text{cell})} \qquad 4.$$

where r is the cell radius, ϵ is the volumetric elastic modulus, and α is a shape factor (~ 1). This was compared with empirical estimates of the diffusivity (D_t) for water movement based upon the kinetics of tissue swelling or shrinkage. When water moves exclusively by the cell-to-cell pathway, D_t equals D_c. In *Kalanchoë* leaves (86), the ratio D_c/D_t was 0.75, in growing pea epicotyls (22) it was 0.50, in corn midrib tissue (102) it varied from 0.36 in the region farthest from the xylem to about 6 in the regions closer to the xylem, and in soybean hypocotyls (87) it was 4. Although the authors did not comment upon it (87, 102), ratios greater than 1 indicate that something is wrong with the analysis, because D_c should not be larger than D_t. Such discrepancies may be the result of inhomogeneities in cellular characteristics, leading to overestimation of D_c. Also D_t may be underestimated because of artifacts produced by the large dehydrations to which some tissues (87, 102) were subjected.

Despite such problems, these studies suggest that water movement through tissues may be transported in large part via the cell-to-cell pathway. An important caveat to note, however, is that the estimates of D_c are very approximate. I would hazard the guess that such D_c values are reliable only within a factor of 2 or 3. One major source of error lies in variation of cell size, which can substantially influence calculated values for D_c (see Equation 4). Such uncertainties underscore the need for better and more direct methods to estimate water flow via the cell wall pathway.

GROWTH-SUSTAINING WATER POTENTIAL GRADIENTS   Because of the anatomical arrangement of leaves, stems, and roots, water must travel a greater distance to reach an epidermal cell than to reach a cell bordering the xylem. Because hydraulic conductance depends in part on path length, cells at different positions in an organ will experience different hydraulic conductances. These cells grow at similar rates, however, which means that the driving force for water uptake varies with cell position.

In a seminal analysis of this situation, Molz & Boyer (65) modeled the *shape* of internal water potential gradients which are needed to sustain the growth of a uniformly elongating cylinder of tissue. As water moves from the internal xylem toward the epidermis, water potential decreases at first relatively steeply and then becomes nearly constant near the epidermis (this analysis assumes nontranspiring conditions). The gradient is not linear because cells near the xylem must transport not only water necessary for their own growth, but also water for the growth of all the cells beyond them. Therefore, the water flux is high near the xylem but reduces to zero at the epidermis.

In a further extension of this model of steady-state growth, Silk & Wagner (83) modeled water potential gradients in a growing corn root and included spacial gradients in growth rate along the root axis in their analysis. Their results predict concentric, egg-shaped isopotential lines around the growing zone of the root, with the most negative water potentials in the region of maximal growth.

These studies show that, in principle, the water potential of growing tissues is not uniform but varies with position in the tissue. The next section examines efforts to determine the magnitude of these gradients. The question of the size of these internal water potential gradients is closely tied up with the question of what limits plant growth.

DOES WATER TRANSPORT LIMIT GROWTH?

Water Potentials of Growing Tissues

For tissues with low hydraulic conductance, the above analyses predict that substantial internal gradients in water potential will arise because wall yielding dissipates turgor pressure and the low conductance of the water pathway prevents full equilibration. Such gradients are essential for sustaining water flux into the growing tissue through a pathway of low conductance. However, they also reduce the average turgor pressure of the tissue and thereby restrict wall yielding and growth. (Large transpirational fluxes also reduce tissue water potential and growth. This is a distinct effect.)

There is general agreement that growing tissues frequently have water potentials lower than that expected of well-hydrated tissues, even when transpiration is stopped. In an early study of the water relations of growing tissue, the water potential of rapidly elongating oat coleoptiles was estimated from the rate of osmotic exchange with mannitol solutions to be between -0.08 and -0.25 MPa (77). Subsequent studies of growing leaves, stems, and other organs measured water potentials between -0.1 and -0.4 MPa using thermocouple psychrometry (2, 65, 101). More recently the pressure probe (41) was used to make direct measurements of turgor pressure in growing pea epicotyls (19, 22, 23). By combining measurements of P and π, water potentials of about -0.25 MPa were calculated. It should be noted that in all these studies measurements were made with nontranspiring tissue. Thus water movement was solely the result of cell growth. These data indicate that the turgor pressure of growing tissue is reduced by 10 to 50% from that expected in well-hydrated tissue.

Although there seems to be widespread agreement about the magnitude of the water potentials of growing tissues, consensus is lacking regarding the interpretation of these reduced water potentials. One view was described above: that the water flux for cell growth requires a water potential difference of about -0.3 MPa to overcome the resistance encountered by water moving to the growing tissue (6, 65, 101). The main evidence for this view is that growing tissues appear to have lower water potentials than nongrowing tissues. This interpretation implies that the hydraulic conductance of growing tissue is low enough that water transport partially limits the rate of growth. Some studies with soybean hypocotyls have suggested that L restricts growth by about 50%, i.e. $L \approx \phi$ (6, 7). This conclusion, if true, is important because it implies that the control of growth may reside in the water transport properties of the tissue, not in the cell wall properties.

Recent studies, however, have obtained data inconsistent with this interpretation (19–21). Cosgrove & Cleland (21) reasoned that if a water potential of -0.25 MPa was necessary to sustain the water influx for rapid growth of pea stems, then a reduced growth rate (= water flux) should be associated with a higher water potential and turgor pressure (provided L and π stayed constant). They measured turgor pressure in intact pea seedlings using the pressure probe, and altered the growth rate by decapitation, cold temperature, cyanide, and auxin application. These treatments changed the growth rate by twentyfold, yet turgor pressure and water potential of the tissue remained nearly unaffected. The largest change in turgor pressure was 0.05 MPa. Parallel experiments showed that the hydraulic conductance was the same at high and at low growth rates. Moreover, direct pressure probe measurements of the radial gradient in turgor pressure in the cortex found only small (0.05 MPa) differences in turgor pressure across the cortex of the pea internode (21). These results suggested that the water potential gradient that sustains growth in pea epicotyls was -0.05 MPa at most. Curiously, the low water potential of the growing tissue appeared to be uncoupled to the water flux for growth.

In a second test of the assumption that low water potentials resulted from low hydraulic conductance, internodes were infiltrated with water (21). Infiltration short-circuits the normal water pathway for growth, so water potential should closely approach equilibrium. Contrary to this prediction, water potential increased by only 0.05 MPa (from -0.24 MPa to -0.19 MPa after infiltration). Thus the low water potentials did not seem to be traceable to low hydraulic conductance; rather, Cosgrove & Cleland (21) suggested that solutes in the free space might account for the low water potentials.

To test this hypothesis, cell wall free-space solution was extracted using three different techniques (20). For all three methods, π of the wall solution (π_{cw}) extracted from apical pea stems was about 0.18 MPa. Solutions obtained from basal (nongrowing) stem tissue contained half this concentration. Stem tissues of young cucumber and soybean seedlings were also examined and showed a similar pattern to that found in pea seedlings: high π_{cw} in apical (growing) regions and lower π_{cw} in basal regions.

It should be noted that these high values of π_{cw} are not corrected for dilution by infiltration water. When this is done, values 50% to 100% higher are obtained, depending on the tissue. However, Cosgrove & Cleland (20) showed by perfusion studies that cell wall solutes are in dynamic equilibrium with some (presumably intracellular) source of solutes. Thus it appears that the undisturbed value of π_{cw} lies somewhere between the extracted values and the values corrected for dilution, probably closer to the extracted values.

These studies (20, 21) concluded that the water potential of growing tissues is held low by the osmotic pressure of the apoplast solution. Hydraulic conductance is large enough that only small water potential gradients are necessary to sustain water influx for growth. This interpretation accounts for both the low water potentials frequently measured for growing tissue and the fact that the apparent water potential gradients do not dissipate when growth is inhibited (2, 8, 21, 23).

Boyer (6) has recently raised three objections to this idea. First, he said that infiltration of the tissue should upset equilibria between free solutes and sorbed or compartmented solutes. Thus, he argued, the extracted apoplast solution will contain some solutes released only after infiltration. This argument, however, refutes itself. In the equilibrium established after infiltration, the apoplast solution will contain the same or a somewhat lower concentration of solutes as compared with the undisturbed wall solution. Without any correction for dilution, the osmotic pressure of the extracted solutions ranged from 0.1 to 0.2 MPa for growing stem tissues of cucumber, soybean, and pea—that is, they were large enough to account for most of the low water potentials measured in growing tissues. In a second objection, Boyer pointed out the possibility of contamination by solutes released from the phloem of cut segments. Cosgrove & Cleland (20) specifically tested for this and related artifacts of excision by collecting apoplast solution using mild perfusion of *intact* pea seedlings. In these experiments there were no cut surfaces for contamination, yet the extracted wall solutions contained the same osmolality as that obtained from excised and washed segments. Third, Boyer argued that in the experiment in which growth was reduced by low temperature, turgor pressure was measured before new water potential gradients had sufficient time to become established and stable. He claimed that tissues require 20–200 min for such stabilization. In contrast to this claim, pea internodes were shown to have half-times for such equilibration on the order of 1 min (19, 21, 22). In the low-temperature experiment cited by Boyer, turgor pressure was measured in the interval from 5 to 15 min after transfer to the low temperature. A low and stable growth rate was established in these experiments within 1–2 min. Thus, sufficient time was alloted for the establishment of a new steady turgor pressure. Moreover, in another experiment in the same series, turgor pressure was measured 90 to 120 min after seedlings were decapitated to reduce their growth rate. Growth rate was reduced by 88%, yet the water potential increased by only 0.05 MPa. In these experiments plenty of time elasped to permit turgor pressures and water potentials to stabilize.

It is worth noting that pressure probe studies of growing pea and soybean stem cells (21, 22, 87) have found remarkably high membrane hydraulic conductivities ($Lp \approx 10^{-4}$ cm s^{-1} MPa^{-1}). These values are comparable to those measured for *Nitella* and represent the highest Lps reported for vascular plants (104). In calculating the theoretical gradients in water potential needed to sustain growth, Molz & Boyer (65) used an estimate of Lp an order of magnitude lower than these actual values and therefore calculated overly large gradients.

From the foregoing it appears that the water potential gradients sustaining growth are small and that the reduced water potentials of some growing tissues are largely the result of apoplastic solutes. A third explanation for reduced water potentials has been proposed: that they arise after excision of the tissue because of continued wall relaxation (1, 18, 40). This possibility is relevant when tissue is excised and isolated from a water supply, as commonly practiced in thermocouple psychrometry (1, 8, 11, 101). Such wall relaxation artifacts will be considered below in the section on stress relaxation.

Driving Forces for Water Uptake

The results discussed above raise important and fundamental questions about the mechanism of water movement through plant tissues. Frequently it is stated that water moves through plant tissues in response to a water potential gradient (6, 63). This is true only for an ideal osmotic membrane—one that selectively blocks solute transport yet permits water movement, i.e. the reflection coefficient (σ) is 1. In the more general case (25, 104) the driving force for water movement is given by ($\sigma\Delta\pi - \Delta P$). Plant tissues are quite different from ideal membranes in that they consist of cells embedded in a porous cell wall matrix. Although the reflection coefficient of cell membranes is close to 1 for most physiological solutes (97, 104), cell walls are certainly much less selective, so the overall reflection coefficient of many tissues is probably much less than 1; however, there seems to be surprisingly little quantitative data on this point.

Because σ does not always equal 1, some authors (70, 104) have made a strong case against the use of ψ gradients to describe water transport. Moreover, Nobel [(67), pp. 503–4] illustrates an example in which water may flow *against* a ψ gradient, e.g. within the phloem water moves in response to a pressure gradient that may oppose a gradient in ψ.

Thus apoplastic solutes will reduce tissue water potential but will be relatively ineffective in generating water movement for growth (20). Biochemists meet a similar situation when they make sucrose gradients to separate membranes and cell organelles by centrifugation. Despite the fact that water may move freely in these tubes, such water potential gradients persist for many hours or days and dissipate only as quickly as the sucrose can diffuse to equilibrium, which is to say, very slowly.

Recognition of apoplastic solutes is particularly important when tissue hydraulic conductance is calculated in the conventional way by relating water flux to $\Delta\psi$. Large errors in estimates of L can arise because π_{cw} decreases tissue ψ but may not generate water movement. For instance, in growing pea internodes twentyfold changes in L under various growing conditions could be calculated [see Table 3 in (21)], but these were not physically real. Similar errors may account for some reports in the literature of apparent changes in hydraulic conductance (4, 8). High concentrations of apoplast solutes have also been reported for other tissues (49) and may be more widespread than is commonly recognized.

These arguments point out that the driving force for water uptake during growth must ultimately be traced to a reduction in P (however slight) because of stress relaxation of the loosened wall. Since water movement across membranes is very rapid, $\Delta\psi$ across the plasmalemma is negligible under normal conditions (21, 63, 102). Any reduction in cell ψ because of wall relaxation will induce transfer of water from the apoplast to the cell and thereby increase the tension of the apoplast solution (i.e. a more negative pressure in the apoplast will develop). Consequently, water will flow into the growing tissue in response to the reduction in P of both the expanding cells and the surrounding apoplast.

Propagation of Water Across Tissues

The question of how rapidly alterations in water potential and turgor pressure propagate across growing tissues is an important one. Such information may be used to assess whether water transport limits growth. For single nongrowing cells, the half-time for equilibration of ψ is given by:

$$T_{1/2} = \frac{V \ln(2)}{A \, Lp \, (\epsilon + \pi)} = \frac{\ln(2)}{L \, (\epsilon + \pi)} \qquad 5.$$

where ϵ is the volumetric elastic modulus of the cell (25, 72, 104). Equation 5 provides a convenient way to estimate the Lp of cell membranes and is commonly employed in pressure probe studies (21, 22, 85, 87, 95, 104).

As first pointed out by Cosgrove (18) and more recently by Ortega (69), the half-time for growing tissues may also be influenced by the yielding properties of the cell wall. The relevant half-time is given by:

$$T_{1/2} = \frac{\ln(2)}{L(\epsilon+\pi) + \phi\epsilon}$$

6.

It should be noted that Equation 6 is a slightly different expression from that developed in previous analyses, where simpler (69) or more complicated (18) expressions were obtained. Ortega (69) correctly pointed out that the additional (and erroneous) terms in Cosgrove's analysis arose because of differentiation of a linearized form of the definition of ϵ. Ortega's derivation, on the other hand, was oversimplified by the assumption that cell π does not change as the cell shrinks or swells. The correct form shown in Equation 6 can be readily derived from Ortega's Equation 18, using the method of Dainty (25) to express $\Delta\pi$ as a function of P. Cell π is then given by $\pi_o - \pi_o P/\epsilon$, where π_o is the osmotic pressure of the cell at zero turgor pressure.

Equation 6 reduces to Equation 5 if ϕ is much smaller than L (18, 19). Recently it was verified by numerical analysis that Equation 5, which strictly applies to single cells, may also be used in an approximate fashion to estimate the effective hydraulic conductance of whole tissues (19), provided $L \gg \phi$. Thus the rate of swelling or shrinkage of whole tissues can provide an alternative method to estimate whole tissue L, one that avoids major errors caused by apoplastic solutes.

Table 2 shows the half-time for shrinking or swelling of various plant tissues. In these experiments, the tissue was induced to shrink or swell by use of osmotica, by evaporation followed by immersion in water, or by changes in the hydrostatic pressure of the xylem or of the ambient atmosphere. In most cases the half-times are approximately one to two min, which indicates rapid water movement, but some tissues exhibit much slower rehydrations.

These kinetic data may be used to evaluate the water potential disequilibrium caused by growth (19, 28, 58). Assuming $(\epsilon + \pi)$ to be 10 MPa (19, 65), tissues with a 1 min half-time will have an L of 4.2 MPa^{-1} h^{-1}. For such tissue the water potential gradient required to support growth at the rapid rate of 10% h^{-1} would be about 0.024 MPa. This value is only 5% of the osmotic pressure of many growing tissues and so represents a small gradient. This confirms the validity of using Equation 5 rather than the more general Equation 6 (that is, ϕ is small relative to L). On the other hand, for tissues with half-times of 30 min, water potential gradients will be so large that turgor pressure will be clamped at the yield threshold by wall expansion and L will dominate the rate of growth. Then Equation 5 may not validly be used since ϕ is as large or larger than L. These calculations illustrate the key importance of these half-times.

In attempting to sort out the validity of the half-times represented in Table 2,

Table 2 Half-times for establishment of steady state after change in external water potential

Tissue	$T_{1/2}$ (min)	Method	Reference
oat coleoptile[a]	1.5–3	Δlength w/mannitol & NaCl	(77)
oat coleoptile	1.7	Δlength w/evapor. + immersion	(77)
rye coleoptile[b]	1.2	Δlength w/osmoticum	(31)
pea epicotyl[b]	1	Δlength w/mannitol	(22)
pea epicotyl	1	Δlength w/root pressure	(21)
pea epicotyl	2	Δlength w/rehydration after stress relaxation	(19)
sunflower leaf	1.4	pressure bomb exudation	(55)
sunflower leaf	2.4	pressure bomb exudation	(5)
wheat leaf	0.4	pressure bomb exudation	(57)
castorbean leaf[c]	~3	water uptake w/evapor. + immersion	(62)
corn leaf	~1	ΔP w/xylem pressure	(102)
corn leaf	10	Δweight w/evap. to 85% + immersion	(102)
corn leaf[d]	1–4	ΔΨ w/root pressure	(68)
soybean hypocotyl	19	Δweight w/evap. to 85% + immersion	(87)
soybean hypocotyl	30	ΔΨ w/evap. to Ψ = −0.6 MPa & rehydration	(7; Fig. 7D)
barley root[a]	1.7–12.9	ΔP w/mannitol	(85)
wheat coleoptile[a]	30	Δlength w/mannitol	(28)

[a]cuticle not abraded.
[b]cuticle abraded or removed.
[c]Half-time computed as the average of the first 87.5% (~8 min) of rehydration.
[d]Estimated from graphs in paper.

two points should be kept in mind regarding the experiments in which an external osmoticum is used to cause swelling or shrinkage. First, the half-time may be artificially lengthened by slow penetration of osmoticum across unstirred layers and into the tissue (22, 77). This problem can be particularly acute when the cuticle is left intact or when the solution flows by the tissue at a slow rate. This probably explains the very slow (30 min) half-time reported for wheat coleoptiles (28). By way of contrast, the half-time was 1.2 min for rye coleoptiles in which the cuticle was abraded and high flow rates for the osmoticum were used (31). Thus the long half-time for wheat may be misleading. A second problem with these experiments has to do with the geometry of water flow. When an external osmoticum is used, water is exchanged across the outer surface of the tissue, whereas during growth of shoots (but not roots) and for most of the other types of experiments reported in Table 2 the exchange surface is internal (i.e. at the xylem). For an organ such as a pea stem, this difference in geometry can lead to a shortening of the half-time by a factor of about 2 (D. J. Cosgrove, unpublished results). The faster rate of water movement comes about because the surface area for water exchange is larger and the mean length of the water pathway is shorter when water is transferred across the outer surface as compared with exchange at the xylem.

Some of the particularly long half-times in Table 2 come from "sorption experiments" in which tissue was allowed to dehydrate by evaporation until its weight was reduced to 85% of its initial value (7, 8, 65, 87, 102). Then the tissue was recut, submerged under water, and periodically reweighed. Two recent studies have examined the basis for such slow rehydration kinetics (87, 102). Westgate & Steudle (102) investigated water flow in the midrib of corn leaves. They found that when water flow was induced by pressure changes in the xylem, changes in water potential propagated very quickly through the tissue (half-time about 1 min). In contrast, water moved at about 1/30th of the rate in sorption experiments. Very similar results were reported for soybean hypocotyls by Steudle & Boyer (87). To explain the discrepancy in water flow rates, they proposed the following scheme (87, 102). Under conditions of the sorption experiments, the reduction in ψ in the apoplast was mainly caused by apoplastic solutes. Since the wall has a low reflection coefficient, π_{cw} was ineffective in driving water absorption. Therefore, water moved primarily by the cell-to-cell pathway in the sorption experiments. On the other hand, when hydrostatic pressure in the xylem was altered, water moved very rapidly through the cell wall pathway because the wall is very open to pressure-driven flow.

This hypothesis, however, leaves several questions unanswered. Normally evaporative dehydration is thought to reduce the hydrostatic pressure in both the cell and cell wall phases (67, p. 94). However, Steudle & Boyer (87) suggest that π_{cw} increases greatly as dehydration proceeds. How or why this might occur or be coordinated with dehydration is not made clear. Furthermore, if the reflection coefficient of the wall is nearly 0 as postulated (87), cells in the middle of the tissue should be incapable of absorbing water from the outside until π_{cw} decreases. This is so because cells are in close equilibrium with ψ of the surrounding apoplast (86, 87, 95, 102).

In published discussions of sorption experiments (63–65, 87, 102) there does not seem to be much consideration of what effect dehydration to 85% has on the kinetics of water movement. Two effects are likely. First, when tissues are dehydrated by 15%, both turgor pressure and the volumetric elastic modulus (ϵ) will be greatly reduced. In wheat leaves, a 15% dehydration reduced ϵ from 25 MPa to nearly 0 MPa (24). Because the half-time for transient water flow is strongly dependent on ϵ (see Equations 4 to 6), large dehydrations will slow the kinetics of rehydration via the cell-to-cell pathway and lengthen $T_{1/2}$, perhaps by more than a factor of 10. Second, the porosity of the cell wall pathway is probably sensitive to the hydration state of the tissue. As water is removed from the tissue, wall capillaries will shrink, thereby increasing their resistance to water flow. Since volume flux is sensitive to the fourth power of the radius of such pores, this effect could be quite large. Moreover, there may be hysteresis effects, as occurs in soils when they dry out and rehydrate. These considerations emphasize the need to study the kinetics of swelling or shrinking without imposing large changes in turgor pressure on the tissue. Further work is required to evaluate the influence of dehydration on rehydration kinetics and to decide whether some tissues indeed have such slow rates of water movement.

YIELDING OF CELL WALLS

Stress Relaxation

For more than 15 years researchers have examined the mechanical properties of isolated cell walls by stress relaxation methods. These and related methods for measuring the mechanical properties of walls have recently been discussed by Taiz (91). He and others (71, 94) point out that the parameters obtained by these in vitro tests with dead wall preparations are not equivalent to the in vivo wall yield coefficient ϕ or to the yield threshold of Equation 2. Although these in vitro methods do not quantitatively measure ϕ or Y, nevertheless they have been useful in studying how hormones and environmental conditions alter wall properties and in some cases give good correlations with growth rate (16, 17, 46, 51, 91).

Recently a new in vivo stress relaxation technique has been devised which does permit quantitative measures of Y and ϕ of growing tissue (19, 23). The method evolved from the realization that when growing tissue is removed from its water supply, loosening of the wall will continue, with a consequent reduction in turgor pressure and water potential. Various earlier studies suggested or showed that wall relaxation gives rise to errors in measurements of water potential of growing tissue (1, 18, 61). Cosgrove et al (23) showed that this process could be used to measure the yield threshold for growth of pea internodes. They used the pressure probe and isopiestic psychrometry to follow the reduction in turgor pressure or water potential of excised growing segments that were isolated from a water supply. Turgor pressure of rapidly growing tissue relaxed during the first 60–90 min to a turgor pressure of 0.3 MPa and remained nearly constant thereafter, whereas slowly growing tissue reached the same value only after several hours (see Figure 1). Stress relaxation did not occur in nongrowing basal stem segments or in growing tissue excised and supplied with water. Subsequent reports by Hsiao et al (40) and by Boyer et al (7) have also documented wall relaxation in corn leaves and soybean hypocotyls.

In a recent further development, Cosgrove (19) analyzed the time course for

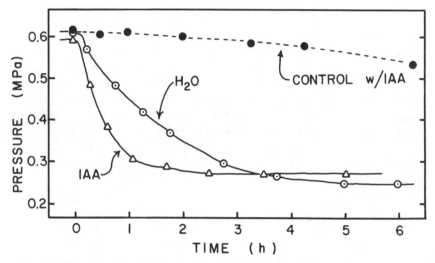

Figure 1 Stress relaxation of etiolated pea internode segments. Growing segments were excised and pretreated with water or 10^{-5} M indole-3-acetic acid (IAA); after a stable growth rate was established, the solutions were removed and the time course for stress relaxation was followed by measuring turgor pressure of 8–12 cortical cells. In the plot labeled CONTROL w/IAA, the tissue was kept in contact with the solution to prevent stress relaxation. The slow drop in turgor pressure in this case is the result of solute dilution during growth (19).

stress relaxation of pea stems. Wall relaxation caused an exponential decay in turgor pressure with a time course given by:

$$P_t = Y + (P_o - Y) \exp(-\phi \epsilon t) \qquad\qquad 7.$$

where P_t is turgor pressure at time t, and P_o is the initial turgor pressure. The rate of relaxation was increased by auxin treatment (see Figure 1), which was shown to stimulate growth solely by increasing ϕ. Moreover, values of ϕ obtained for pea stems by this stress relaxation analysis agreed reasonably well with steady-state estimates of ϕ (obtained using Equation 2). For rapidly growing pea internodes, ϕ was about 0.24 MPa^{-1} h^{-1}.

Boyer et al (7) recently devised a new "guillotine psychrometer" to investigate the effects of excision on the water potential of growing soybean hypocotyls. They reported that upon excision ψ dropped very rapidly (within 5 min) by about 0.1 MPa. Subsequently, ψ declined more slowly and variably. They interpreted the immediate drop in ψ to be relaxation to the yield threshold, with subsequent changes in ψ resulting from reduction in the yield threshold. However, this interpretation is inconsistent with their other estimates of ϕ and L based upon steady-state measures of growth rate and ψ. They estimated L to be 0.28 MPa^{-1}h^{-1} and ϕ to be 0.34 MPa^{-1}h^{-1}. Using 3 MPa as an estimate of ϵ

for soybean hypocotyls (87), we may use this value of ϕ in Equation 7 to compute the time constant for stress relaxation to be $1.0 \, h^{-1}$. Thus for stress relaxation to reach equilibrium ($\sim$5 half-times) ought to require 3.5 h (not 5 min), if their estimate of ϕ is valid. On the other hand, if the tissue really relaxed to the yield threshold within 5 min, ϕ would have to be about 14 $MPa^{-1}h^{-1}$. Since this is more than 40 times larger than their estimate of L, no relaxation would be seen as P would already be nearly at the yield threshold.

In view of the theory presented above and also the stress relaxation results obtained with peas (19, 23), it seems more likely that the slow reduction in ψ noted by Boyer et al (7) resulted from wall relaxation. Unfortunately, control experiments in which excised tissues were kept in contact with a water supply to prevent stress relaxation were not reported in their study. If wall relaxation is relatively slow, what then accounts for the initial rapid drop in ψ? Recent experiments suggest that it may be caused by release of hydrostatic pressure ("root pressure") in the xylem (D. J. Cosgrove, unpublished data). When hypocotyls of soybean or cucumber were cut, the root stump continued to exude for one to several hours, indicating a positive xylem pressure. Measurements with displacement transducers showed that the stem shrinks slightly upon excision, and pressure probe measurements verified a drop in stem turgor pressure upon excision, as reported for the guillotine experiments (7). However, the rapid reduction in turgor occurred even when the excision was made under water, showing that the immediate drop in P was not caused by stress relaxation. In contrast, water availability did prevent the slower reduction in P, as would be expected if this slower phase were the result of stress relaxation. One of the implications of these results is that ψ propagates rather quickly through the soybean tissue, because the changes in ψ noted by Boyer et al (7) were complete within 5 min.

It is possible to estimate the ratio ϕ/L by taking the ratio of the half-time for swelling or shrinkage to the half-time for stress relaxation (19). This relationship is given by:

$$\frac{T_{1/2} \, \text{swelling}}{T_{1/2} \, \text{stress relax.}} = \frac{\phi\epsilon}{L(\epsilon+\pi) + \phi\epsilon} \approx \frac{\phi}{L} \quad (\text{for } \epsilon >> \pi \text{ and } \phi >> L) \qquad 8.$$

For pea internodes, this ratio was found to be about 1/9 for rapidly growing tissue and 1/25 for slowly growing tissue (19). These results reinforce earlier conclusions (20–22) that L in pea stems is large and has only a minor effect on growth. If the interpretation given above for relaxation in soybean hypocotyls turns out to be correct, then the ratio ϕ/L for soybeans will be of similar magnitude to that for pea stems. This contrasts with the estimated ratio of 1/1 given by Boyer et al (7).

Turgor Pressure and Wall Yielding

One of the tenets of Lockhart's analysis of plant growth (50) is that the rate of wall expansion depends on turgor pressure. A number of studies (2, 10, 14, 28, 31, 40, 45) have supported this notion, although in many cases changes in turgor pressure were only inferred from changes in external osmotica and not measured.

There have also been some studies with results at variance with this idea. Kuzmanoff & Evans (47) found that the growth rate of lentil roots remained constant over a wide range of external osmotic pressures. They speculated that ϕ or Y may have changed in response to external osmoticum such that the growth rate remained constant. However, these workers did not measure turgor pressure, nor did they measure ϕ or Y. Other recent studies on water stress reported no correlation between growth rate and turgor pressure (54, 60, 98). These results have led some to question the importance of turgor pressure in wall extension (6, 28). However, there may be other reasons for the anomalous results.

Wall relaxation of the growing tissue may be a culprit in these studies. Expanding tissues are liable to undergo stress relaxation after excision, with the more rapidly expanding tissue relaxing at a faster rate than the slowly growing tissue (see Figure 1). In the study by Matsuda & Riazi (54), leaf water potential and turgor pressure were estimated by the Shardakov procedure with a 2-h equilibration in sorbitol solutions. For the studies of corn leaves (60, 98), the water potential of excised leaf tissue was measured by isopiestic thermocouple psychrometry with equilibration times of many hours. It seems likely that these long equilibration times would have allowed stress relaxation to reduce turgor pressure to the yield threshold for growth. Such an interpretation is in accord with the stress relaxation behavior of pea internodes (19, 23), soybean hypocotyls (7), bean leaves (100), soybean leaves (1), and castorbean leaves (61).

The failure of some recent studies to detect reduction in ψ following excision of the growing tissue of maize leaves (60, 98, 101) and soybean hypocotyls (8) is paradoxical and remains to be fully explained. Cavalieri & Boyer (11) found evidence of reduction in the water potential of mature tissue attached to expanding tissue, but no reduction in the water potential of the growing tissue itself was found. The more recent experiments of Boyer et al (7), however, do show stress relaxation in soybean hypocotyls, suggesting that earlier calculations of turgor pressure in growing tissues by psychrometric methods were in error. Hsiao et al (40) found that the growing region of maize leaves undergoes reduction in ψ subsequent to excision because of wall relaxation. They concluded that the lack of dependence of expansion on turgor for maize leaves (60) and probably for barley leaves (54) was an excision artifact. When they carried out ψ measurements at low temperature to inhibit cell expansion, they found a linear dependence of growth rate on turgor pressure.

Goering et al (28) found that when wheat coleoptiles were incubated in a graded series of mannitol solutions, the growth rate began to decline only with concentrations greater than 80 mM ($\pi = 0.2$ MPa). Similar observations have been made for pea internodes (23) and for lentil roots (47). Goering et al suggested that growth was independent of P at high turgor pressures. However, no measurements of turgor pressure were made, and it is not at all certain that changing the external osmotic pressure from 0 to 0.1 MPa or 0.2 MPa induces an equivalent change in turgor pressure in a tissue with an intact cuticle such as they used. For instance, if the apoplast contained solutes equivalent to 0.2 MPa, as has been found for some stem tissues (20), then one might expect little change in turgor pressure until external π exceeded π_{cw}.

KINETICS OF STRESS RELAXATION   Several observations of the stress relaxation behavior of pea stems reinforce the idea that the rate of wall yielding depends on turgor pressure. The exponential decline in turgor pressure (Figure 1) closely approximates the predicted behavior, which was calculated assuming the rate of wall yielding depends in a linear fashion on the quantity $(P-Y)$. If wall yielding were independent of turgor pressure, then one might predict a constant rate of wall relaxation. This did not occur in pea stems (19, 23) nor apparently in other tissues (1, 7, 61, 100). That turgor pressure, and not water potential, is the relevant physical quantity governing wall relaxation is indicated by the following observation. When pea stem segments were excised and immediately allowed to undergo stress relaxation, they relaxed to a turgor pressure of 0.3 MPa, corresponding to a water potential of about -0.5 MPa (23). When the segments were first preincubated on water and then isolated from water to begin stress relaxation, they again relaxed to a turgor pressure of 0.3 MPa. However, this P corresponded to a water potential of -0.3 MPa because during preincubation on water the osmotic pressure of the pea segments was reduced by 0.2 MPa (23). This reduction in osmotic pressure resulted in part from dilution of solutes as the segments elongated and in part from leakage of solutes to the medium (D. J. Cosgrove, unpublished data). These results further support the hypothesis that wall yielding depends on turgor pressure in excess of the yield threshold.

OSMOTIC PRESSURE MAINTENANCE   Maintaining a suitable intracellular osmotic pressure is essential for growth, because only by this means can cells maintain the positive turgor pressure necessary for wall expansion. In pea epicotyls, cell osmotic pressure decreases about only 0.1 MPa from the top to the bottom of the growing region (19), and in maize roots it remains constant (39). In these cases it appears that solute accumulation by the growing cells is sufficient to prevent the large dilution in osmotic pressure that would occur if the growing cells failed to accumulate solutes. Studies of the growth of excised oat coleoptiles indicate that solute uptake from the medium is controlled by a complex interaction between growth rate, solute availability, and a maximum internal solute concentration (88, 89). Katsumi et al (44) suggest that gibberellic acid may stimulate cucumber stem elongation by increasing the osmotic pressure of the growing cells. Although there are other suggestions that solute fluxes into expanding tissue may control cell growth in the long term, there does not seem to be an adequate theoretical analysis to test these ideas. Research needs to be directed toward filling the large gap in our understanding of how growing cells maintain cell osmotic pressure during long periods of enlargement.

When plants undergo mild water stress, cell expansion is inhibited but cell solutes begin to accumulate. This is viewed as an adaptive response by plants to maintain turgor pressure and is discussed in detail in recent reviews (9, 40, 84, 96). In some cases solute accumulation during water stress was reported to maintain a high turgor pressure in spite of reduced water potentials (40, 60, 98). Problems with these data because of stress relaxation were noted above. Despite such problems, there appears to be growing evidence that water stress may modify wall properties such that the wall expands at a lower rate for any given turgor pressure (26, 40, 55, 92, 98). Hsiao et al (40) noted such a change within 2 h of imposition of water stress. How such changes in wall properties come about and in response to what stimuli remain to be worked out. A recent suggestion of Van Volkenburgh & Boyer (98) is that wall acidification may be involved (98).

The Yield Threshold

As numerous studies have shown, when turgor pressure of growing tissue is reduced, growth likewise is reduced and finally ceases at turgor pressures above zero (10, 12, 40, 45). Thus turgor evidently must exceed a minimum or threshold value before growth can occur. The pioneering studies by Green with *Nitella* showed that Y was apparently adjustable (30, 32). The yield threshold increased or decreased within certain limits to compensate for changes in turgor pressure, with the result that growth rate tended to stabilize. However, the lowest attainable value of Y was about 0.2 MPa. At turgor below this value, growth would not resume.

In higher plants the value of the yield threshold varies somewhat according to tissue and conditions. In pea stems, Y was found to be 0.29 MPa and was not influenced by decapitation, auxin, or fusicoccin treatment (19, 23). In maize leaves, Y was about 0.37 MPa (40) and apparently increased during water stress (38). In bean leaves, Y varied between 0.4 and 0.2 MPa, depending upon the age of the leaf (100). Thus, although the number of studies is admittedly limited, it appears that Y for plant tissues lies in the range of 0.2 to 0.4 MPa, and seems to be adjustable under some conditions.

One topic that has attracted only meager attention is the nature of the yield threshold. At least two different interpretations of the yield threshold are possible. First, the existence of a nonzero Y may indicate the requirement for a minimum hydrostatic pressure within the cell for wall loosening to persist. There is ample evidence, for example, that membrane transport is affected by turgor pressure (104). It would seem that cellular control of wall loosening must occur via the plasmalemma, although we do not yet understand the nature of this control; therefore, alterations in membrane properties as a function of turgor might account for a yield threshold. Relevant to the acid-growth hypothesis, Cleland (15) reported that auxin-stimulated acidification of the medium by oat coleoptiles was reduced by water stress (incubation on 0.15 to 0.45 M mannitol). Masuda et al (52) reported similar findings. Thus it is possible that the biochemical basis for wall loosening is sensitive to turgor pressure.

The second, and more common, interpretation of the yield threshold is that the load-bearing elements in the wall require a minimum elastic strain for wall yielding to occur. This idea is supported by experiments in which turgor pressure was reduced by osmoticum or by other means, but elongation continued when wall tension was maintained by external force (35, 78). Moreover, studies of dead cell wall preparations by creep or stress relaxation analysis have found some indication of a yield point (13, 59, 103). It is not yet clear how these in vitro measures of yield points compare with in vivo measures of yield thresholds for higher plant tissues. Métraux & Taiz (59) estimated the yield stress of isolated wall preparations of *Nitella* under unilateral stress, but as Taiz (91) points out, quantitative comparison of such values with in vivo yield thresholds is difficult because of the differences in force vectors (turgor exerts a multiaxial stress on walls).

Although the bulk of the evidence supports the idea that a minimum stress in the wall is necessary for growth, the molecular mechanism for this behavior is still open to speculation. It is possible that a minimum wall stress is required to enable cross-links in the wall to be broken, or perhaps to enable entangled wall polymers to slip past one another. Further work is needed to elucidate the molecular basis of the yield threshold and its alteration under various growing conditions.

If we accept that the yield threshold represents a requirement for a minimum wall stress in the load-bearing regions of the wall, then cell size may have a substantial influence on the minimum turgor required for growth. The following calculation illustrates this principle. For a cylindrical cell, the longitudinal stress is given by $P \cdot r / 2d$, where d is wall thickness. Hence, if a cell doubled in radius during growth, turgor pressure could decrease by 1/2 and still maintain the same wall stress (assuming constant wall thickness). Using these relationships it should be possible to calculate the minimum wall stress necessary for growth from the values given above for the yield threshold. Unfortunately, the relevant wall thickness (d) to use for these calculations is the thickness of the load-bearing region of the wall. The results of Richmond et al (79) suggest that this may be substantially less than the total wall thickness. In *Nitella* only the inner 25% of the wall bears any stress, and similar conclusions may apply to the wall of higher plants as well (see 91). Until this point is better resolved, these calculations of in vivo yield stresses are on shaky ground.

Because wall expansion is driven by turgor pressure in excess of Y, the value of $(P-Y)$ is important. Green found that P exceeded Y by only 0.02 MPa in *Nitella*. For this reason, small changes in P could elicit large changes in growth rate. However, the growth disturbances were transient because Y appeared to change, bringing $(P-Y)$ back toward the value of 0.02 MPa. In growing pea internodes, $(P-Y)$ for nontranspiring tissue was found to be much larger, about 0.3 MPa (19). In maize leaves, $(P-Y)$ for well-hydrated tissue appeared to be about 0.25 MPa (40), whereas in bean leaves it was only about 0.1 MPa. Milburn (61) described preliminary experiments with castorbean leaves which imply that $(P-Y)$ in this tissue might be as large as 0.9 MPa. These values suggest that the effective driving force for growth of well-hydrated plant tissues $(P-Y)$ is about one-half the value of turgor pressure. In other words, a 10% increase in P ought to translate into a 20% increase in growth rate. Such amplification is presumably the basis for the high sensitivity of cell expansion to water stress (37). Under field conditions the value of P will depend additionally on the rate of transpiration, water availability, and other factors, and so growth may be even more sensitive to changes in turgor than indicated above.

SUMMARY

Turgor pressure plays a central role in coordinating cell wall extension with water uptake by growing cells. Physical analysis of these processes, as represented by the growth equations, provides a useful framework for evaluating the physical constraints that limit the growth rate of plants. Three types of factors emerge from these analyses: wall-yielding parameters (ϕ, Y), a water transport coefficient (L), and the osmotic pressure of the growing cells. Recent studies have measured these parameters, but to date we have valid data for very few growing organs and under restricted growing conditions.

There remain several gaps in our understanding of water transport for growth. The role of apoplastic solutes in growing tissues is yet to be addressed. Our models of water transport have not properly considered how apoplastic solutes may influence water movement; this is a case where experimental results point to a gap in theory. Recent demonstrations of wall relaxation highlight a fundamental flaw with many of the early measures of the water potential of growing tissue; these studies need to be reevaluated with our new knowledge.

A number of studies have noted rapid swelling or shrinking kinetics for growing tissues. Such results imply that only small gradients in water potential are needed to sustain cell enlargement. As a corollary, growth of such tissues is controlled predominantly by the yielding properties of the cell wall. However, in tissues where the kinetics are slow (half-times greater than 5 or 10 min), large disequilibria will develop even at moderate growth rates. In such cases, the growth rate may be restricted by the water transport characteristics of the tissue. The basis for these slow rehydration half-times needs further research.

The ability of growing tissue to maintain a high internal osmotic pressure is essential for growth, yet we have little insight into the mechanisms controlling cell osmotic pressure. Because cell osmotic pressure determines the maximum turgor pressure a cell can generate, we need to know more about how solute fluxes into growing tissues are controlled. This is particularly important for our understanding of plant adaptations to water stress. Regarding the growth equations, future theoretical analysis needs to be directed toward identifying and measuring the transport coefficients that govern buildup and maintenance of cell osmotic pressure. Until this is done it will be difficult to evaluate solute transport as a limiting factor for growth.

Finally, the physical and biochemical natures of stress relaxation and yielding of the cell wall remain to be discovered. The technique of in vivo stress relaxation offers a new tool to examine wall yielding without complications from water transport. Although wall yielding ultimately gives rise to cell growth, we must integrate our studies of cell wall properties with studies of water transport and solute accumulation to obtain a complete understanding of plant growth and its control.

ACKNOWLEDGMENTS

I wish to thank numerous colleagues who provided me with preprints and reprints of their work, in particular T. C. Hsiao, E. Steudle, M. E. Westgate, and E. Van Volkenburgh. Special thanks to R. E. Cleland and L. Taiz for stimulating discussions and to R. H. Hamilton for helpful criticism of the manuscript. Unpublished results reported here were supported by a grant from the Department of Energy.

Literature Cited

1. Baughn, J. W., Tanner, C. B. 1976. Excision effects of leaf water potential of five herbaceous species. *Crop Sci.* 16: 184–90
2. Boyer, J. S. 1968. Relationship of water potential to growth of leaves. *Plant Physiol.* 43:1056–62
3. Boyer, J. S. 1971. Resistances to water transport in soybean, bean, and sunflower. *Crop Sci.* 11:403–7
4. Boyer, J. S. 1974. Water transport in plants: Mechanism of apparent changes in resistance during absorption. *Planta* 117:187–207
5. Boyer, J. S. 1977. Regulation of water movement in whole plants. *Symp. Soc. Exp. Biol.* 31:455–70
6. Boyer, J. S. 1985. Water transport. *Ann. Rev. Plant Physiol.* 36:473–516
7. Boyer, J. S., Cavalieri, A. J., Schulze,

E.-D. 1985. Control of the rate of cell enlargement: Excision, wall relaxation, and growth-induced water potentials. *Planta* 163:527–43

8. Boyer, J. S., Wu, G. 1978. Auxin increases the hydraulic conductivity of auxin-sensitive hypocotyl tissue. *Planta* 139:227–37

9. Bradford, K. J., Hsiao, T. C. 1982. Physiological responses to moderate water stress. In *Encyclopedia of Plant Physiology: Physiological Plant Ecology II, Water Relations and Carbon Assimilation* (NS), ed. O. L. Lange, P. S. Nobel, C. B. Osmond, H. Ziegler, 12B:263–323. Berlin/New York: Springer-Verlag

10. Bunce, J. A. 1977. Leaf elongation in relation to leaf water potential in soybean. *J. Exp. Bot.* 28:156–61

11. Cavalieri, A. J., Boyer, J. S. 1982. Water potentials induced by growth in soybeans. *Plant Physiol.* 69:492–96

12. Cleland, R. E. 1959. Effect of osmotic concentration on auxin action and on irreversible expansion of the *Avena* coleoptile. *Physiol. Plant.* 12:809–25

13. Cleland, R. E. 1971. The mechanical behavior of isolated *Avena* coleoptile walls subjected to constant stress. *Plant Physiol.* 47:805–11

14. Cleland, R. E. 1976. The control of cell enlargement. *Symp. Soc. Exp. Biol.* 31:101–15

15. Cleland, R. E. 1976. Fusicoccin-induced growth and hydrogen ion excretion of *Avena* coleoptiles: Relation to auxin responses. *Planta* 128:201–6

16. Cleland, R. E. 1981. Wall extensibility: Hormones and wall extension. In *Encyclopedia of Plant Physiology: Plant Carbohydrates II. Extracellular Carbohydrates* (NS), ed. W. Tanner, F. A. Loewus, 13B:225–76. Berlin/Heidelberg/New York: Springer-Verlag

17. Cleland, R. E. 1984. The Instron technique as a measure of immediate-past wall extensibility. *Planta* 160:514–20

18. Cosgrove, D. J. 1981. Analysis of the dynamic and steady-state responses of growth rate and turgor pressure to changes in cell parameters. *Plant Physiol.* 68:1439–46

19. Cosgrove, D. J. 1985. Cell wall yield properties of growing tissues. Evaluation by in vivo stress relaxation. *Plant Physiol.* 78:347–56

20. Cosgrove, D. J., Cleland, R. E. 1983. Solutes in the free space of growing stem tissues. *Plant Physiol.* 72:326–31

21. Cosgrove, D. J., Cleland, R. E. 1983. Osmotic properties of pea stem in-

ternodes in relation to growth and auxin action. *Plant Physiol.* 72:332–38

22. Cosgrove, D. J., Steudle, E. 1981. Water relations of growing pea epicotyl segments. *Planta* 153:343–50

23. Cosgrove, D. J., Van Volkenburgh, E., Cleland, R. E. 1984. Stress relaxation of cell walls and the yield threshold for growth: Demonstration and measurement by micro-pressure probe and psychrometer techniques. *Planta* 162:46–52

24. Cutler, J. M., Shahan, K. W., Steponkus, P. L. 1979. Characterization of internal water relations of rice by a pressure-volume method. *Crop Sci.* 19:681–85

25. Dainty, J. 1976. Water relations in plant cells. In *Encyclopedia of Plant Physiology* (NS), ed. U. Luettge, M. Pitman, 2:12–35. Berlin: Springer-Verlag

26. Davies, W. J., Van Volkenburgh, E. 1983. The influence of water deficit on the factors controlling the daily pattern of growth of *Phaseolus* trifoliates. *J. Exp. Bot.* 34:987–99

27. Eisinger, W. 1983. Regulation of pea internode expansion by ethylene. *Ann. Rev. Plant Physiol.* 34:225–40

28. Goering, H., Ehwald, R., Bleiss, W. 1984. Osmotic relations of growing coleoptile segments. In *Membrane Transport in Plants*, ed. W. J. Dram, K. Janáček, R. Rybová, K. Sigler, pp. 83–89. Prague: Academia. 592 pp.

29. Goodwin, P. B. 1983. Molecular size limit for movement in the symplast of the *Elodea* leaf. *Planta* 157:124–30

30. Green, P. B. 1968. Growth physics in *Nitella*: A method for continuous in vivo analysis of extensibility based on a micromanometer technique for turgor pressure. *Plant Physiol.* 43:1169–84

31. Green, P. B., Cummins, W. R. 1974. Growth rate and turgor pressure: Auxin effect studied with an automated apparatus for single coleoptiles. *Plant Physiol.* 54:863–69

32. Green, P. B., Erickson, R. O., Buggy, J. 1971. Metabolic and physical control of cell elongation rate. In vivo studies in *Nitella*. *Plant Physiol.* 47:423–30

33. Grenetz, P. S., List, A. Jr. 1973. A model for predicting growth responses in plants to changes in external water potential: *Zea mays* primary roots. *J. Theor. Biol.* 39:29–45

34. Gunning, B. E. S. 1976. Introduction to plasmodesmata. In *Intercellular Communication in Plants: Studies on Plasmodesmata*, ed. B. E. S. Gunning, A. W. Robards, pp. 1–12. Berlin: Springer-Verlag

35. Hager, A., Menzel, H., Krauss, A. 1971. Versuche und Hypothese zur Primaerwirkung des Auxins beim Streckungswachstum. *Planta* 100:47–75

36. Hettiaratchi, D. R. P., O'Callaghan, J. R. 1974. A membrane model of plant cell extension. *J. Theor. Biol.* 45:459–65

37. Hsiao, T. C., Acevedo, E. 1974. Plant responses to water deficits, water-use efficiency, and drought resistance. *Agric. Meteorol.* 14:59–84

38. Hsiao, T. C., Jing, J. 1984. Biophysical parameters underlying slower expansion of water-stressed maize leaves: Shifts in yield threshold counter to osmotic adjustment. *Plant Physiol.* 75(1):S174 (Abstr.)

39. Hsiao, T. C., Silk, W. K., Diedenhofen, U., Matson, C. 1985. Spacial distribution of osmoticum and potassium and their deposition rates in the primary root of corn, *Zea mays*. *Plant Physiol.* 77(1):35 (Abstr.)

40. Hsiao, T. C., Silk, W. K., Jing, J. 1985. Leaf growth and water deficits: biophysical effects. In *Control of Leaf Growth*, ed. N. R. Baker, W. D. Davies, C. Ong. *Soc. Exp. Biol. Seminar Ser.* No. 27. Cambridge: Cambridge Univ. Press. In press

41. Huesken, D., Steudle, E., Zimmermann, U. 1978. Pressure probe technique for measuring water relations of cells in higher plants. *Plant Physiol.* 61:158–63

42. Jones, H., Tomos, A. D., Leigh, R. A., Wyn Jones, R. G. 1983. Water-relation parameters of epidermal and cortical cells in the primary root of *Triticum aestivum* L. *Planta* 158:230–36

43. Jones, H., Tomos, A. D., Leigh, R. A., Wyn Jones, R. G. 1984. The integration of cellular and whole-root hydraulic conductivity in wheat and maize. See Ref. 28, pp. 113–14

44. Katsumi, M., Kazama, H., Kawamura, N. 1980. Osmotic potential of the epidermal cells of cucumber hypocotyls as affected by gibberellin and cotyledons. *Plant Cell Physiol.* 21:933–37

45. Kirkham, M. B., Gardner, W. R., Gerloff, G. C. 1972. Regulation of cell division and cell enlargement by turgor pressure. *Plant Physiol.* 49:961–62

46. Kutschera, U., Schopfer, P. 1986. Effect of auxin and abscisic acid on cell wall extensibility in maize coleoptiles. *Planta*. In press

47. Kuzmanoff, K. M., Evans, M. L. 1981. Kinetics of adaptation to osmotic stress in lentil (*Lens culinaris* Med.) roots. *Plant Physiol.* 68:224–27

48. Labavitch, J. M. 1981. Cell wall turn-

over in plant development. *Ann. Rev. Plant Physiol.* 32:385–406

49. Leigh, R. A., Tomos, A. D. 1983. An attempt to use isolated vacuoles to determine the distribution of sodium and potassium in cells of storage roots of red beets (*Beta vulgaris* L. *Planta* 159:469–75

50. Lockhart, J. A. 1965. An analysis of irreversible plant cell elongation. *J. Theor. Biol.* 8:264–75

51. Masuda, Y. 1978. Auxin-induced cell wall loosening. *Bot. Mag. Tokyo Spec. Issue* 1:103–23

52. Masuda, Y., Sakurai, N., Tazawa, M., Teruo, S. 1978. Effect of osmotic shock on auxin-induced cell extension, cell wall changes and acidification in *Avena* segments. *Plant Cell Physiol.* 19:857–67

53. Masuda, Y., Yamamoto, Y. 1972. The control of auxin-induced stem elongation by the epidermis. *Physiol. Plant.* 27:109–15

54. Matsuda, K., Riazi, A. 1981. Stress-induced osmotic adjustment in growing regions of barley leaves. *Plant Physiol.* 68:571–76

55. Matthews, M. A., Van Volkenburgh, E., Boyer, J. S. 1984. Acclimation of leaf growth to low water potentials in sunflower. *Plant Cell Environ.* 7:199–206

56. McNeil, M., Darvill, A. G., Fry, S. C., Albersheim, P. 1984. Structure and function of the primary cell walls of plants. *Ann. Rev. Biochem.* 53:625–63

57. Melkonian, J. J., Wolfe, J., Steponkus, P. L. 1982. Determination of the volumetric modulus of elasticity of wheat leaves by pressure-volume relations and the effect of drought conditioning. *Crop Sci.* 22:116–23

58. Métraux, J.-P., Richmond, P. A., Taiz, L. 1980. Control of cell elongation in *Nitella* by endogenous cell wall pH gradients. *Plant Physiol.* 65:204–10

59. Métraux, J.-P., Taiz, L. 1978. Transverse viscoelastic extension in *Nitella* I. Relation to growth rate. *Plant Physiol.* 61:135–38

60. Michelena, V. A., Boyer, J. S. 1982. Complete turgor maintenance at low water potentials in the elongating region of maize leaves. *Plant Physiol.* 69:1145–49

61. Milburn, J. A. 1979. *Water Flow in Plants*, pp. 150–51. London/New York: Longman. 225 pp.

62. Milburn, J. A., Weatherley, P. E. 1971. The influence of temperature on the process of water uptake by detached leaves and leaf discs. *New Phytol.* 70:929–38

63. Molz, F. J., Ferrier, J. M. 1982. Mathe-

matical treatment of water movement in plant cells and tissue: A review. *Plant Cell Environ.* 5:191–206

64. Molz, F. J., Ikenberry, E. 1974. Water transport through plant cells and cell walls: Theoretical development. *Soil Sci. Soc. Am. Proc.* 38:699–704

65. Molz, F. J., Boyer, J. S. 1978. Growth-induced water potentials in plant cells and tissues. *Plant Physiol.* 62:423–29

66. Morgan, J. M. 1984. Osmoregulation and water stress in higher plants. *Ann. Rev. Plant Physiol.* 35:299–319

67. Nobel, P. 1983. *Biophysical Plant Physiology and Ecology*. San Francisco: Freeman. 608 pp.

68. Nulsen, R. A., Thurtell, G. W., Stevenson, K. R. 1977. Response of leaf water potential to pressure changes at the root surface of corn plants. *Agron. J.* 69:951–54

69. Ortega, J. K. E. 1985. Augmented growth equation for cell wall expansion. *Plant Physiol.* 78:318–20

70. Passioura, J. B. 1984. Hydraulic resistance of plants. I. Constant or variable? *Aust. J. Plant Physiol.* 11:333–39

71. Penny, P., Penny, D. 1978. Rapid responses to phytohormones. In *Phytohormones and Related Compounds: A Comprehensive Treatise*, ed. D. S. Letham, P. B. Goodwin, T. J. V. Higgins, 2:537–97. Berlin: Springer-Verlag

72. Philip, J. R. 1958. The osmotic cell, solute diffusibility, and the plant water economy. *Plant Physiol.* 33:264–71

73. Philip, J. R. 1958. Propagation of turgor and other properties through cell aggregations. *Plant Physiol.* 33:271–74

74. Radin, J. W., Boyer, J. S. 1982. Control of leaf expansion by nitrogen nutrition in sunflower plants. Role of hydraulic conductivity and turgor. *Plant Physiol.* 69:771–75

75. Radin, J. W., Eidenbock, M. P. 1984. Hydraulic conductance as a factor limiting leaf expansion of phosphorus-deficient cotton plants. *Plant Physiol.* 75:372–77

76. Ray, P. M., Green, P. B., Cleland, R. E. 1972. Role of turgor in plant cell growth. *Nature* 239:163–64

77. Ray, P. M., Ruesink, A. W. 1963. Osmotic behavior of oat coleoptile tissue in relation to growth. *J. Gen. Physiol.* 47:83–101

78. Rayle, D. L., Cleland, R. E. 1972. The in-vitro acid-growth response: Relation to in-vivo growth responses and auxin action. *Planta* 104:282–96

79. Richmond, P. A., Métraux, J.-P., Taiz, L. 1980. Cell expansion patterns and directionality of wall mechanical properties in *Nitella*. *Plant Physiol.* 65:211–17

80. Robards, A. W., Clarkson, D. T. 1976. The role of plasmodesmata in the transport of water and nutrients across roots. In *Intercellular Communication in Plants: Studies on Plasmodesmata*, ed. B. E. S. Gunning, A. W. Robards, pp. 181–99. Berlin: Springer-Verlag

81. Schopfer, P., Plachy, C. 1984. Control of seed germination by abscisic acid. II. Effect on embryo water uptake in *Brassica napus* L. *Plant Physiol.* 76:155–60

82. Silk, W. K. 1984. Quantitative descriptions of development. *Ann. Rev. Plant Physiol.* 35:479–518

83. Silk, W. K., Wagner, K. K. 1980. Growth-sustaining water potential distributions in the primary corn root. A noncompartmented continuum model. *Plant Physiol.* 66:859–63

84. Steponkus, P. L., Cutler, J. M., O'Toole, J. C. 1980. Adaptation to water stress in rice. In *Adaptation of Plants to Water and High Temperature Stress*, ed. N. C. Turner, P. J. Kramer, pp. 401–17. New York: Academic

85. Steudle, E., Jeschke, W. D. 1983. Water transport in barley roots: Measurement of root pressure and hydraulic conductivity of roots in parallel with turgor and hydraulic conductivity of root cells. *Planta* 158:237–48

86. Steudle, E., Smith, J. A. C., Luettge, U. 1980. Water-relation parameters of individual mesophyll cells of the Crassulacean acid metabolism plant *Kalanchoë daigremontiana*. *Plant Physiol.* 66:1155–63

87. Steudle, J. S., Boyer, J. S. 1985. Hydraulic resistance to radial water flow in growing hypocotyl of soybean measured by a new pressure-perfusion technique. *Planta* 164:189–200

88. Stevenson, T. T., Cleland, R. E. 1981. Osmoregulation in the *Avena* coleoptile in relation to auxin and growth. *Plant Physiol.* 67:749–53

89. Stevenson, T. T., Cleland, R. E. 1982. Osmoregulation in the *Avena* coleoptile: Control of solute uptake in peeled sections. *Plant Physiol.* 69:292–95

90. Stroshine, R. L., Rand, R. H., Cooke, J. R., Cutler, J. M., Chabot, J. F. 1985. An analysis of resistance to water flow through wheat and tall fescue leaves during pressure chamber efflux experiments. *Plant Cell Environ.* 8:7–18

91. Taiz, L. 1984. Plant cell expansion: Regulation of cell wall mechanical properties. *Ann. Rev. Plant Physiol.* 35:585–657

92. Termaat, A., Passioura, J. B., Munns, R. 1985. Shoot turgor does not limit shoot growth of NaCl-affected wheat and barley. *Plant Physiol.* 77:869–72
93. Thimann, K. V., Schneider, C. L. 1938. Differential growth in plant tissues. *Am. J. Bot.* 25:627–41
94. Tomos, A. D. 1985. The physical limitations of leaf cell expansion. See Ref. 40
95. Tomos, A. D., Steudle, E., Zimmermann, U., Schulze, E.-D. 1981. Water relations of leaf epidermal cells of *Tradescantia virginiana*. *Plant Physiol.* 68:1135–43
96. Turner, N. C., Jones, M. M. 1980. Turgor maintenance by osmotic adjustment: A review and evaluation. See Ref. 84, pp. 87–103
97. Tyerman, S. D., Steudle, E. 1982. Comparison between osmotic and hydrostatic water flows in a higher plant cell: Determination of hydraulic conductivities and reflection coefficients in isolated epidermis of *Tradescantia virginiana*. *Aust. J. Plant Physiol.* 9:461–79
98. Van Volkenburgh, E., Boyer, J. S. 1985. Inhibitory effects of water deficit on maize leaf elongation. *Plant Physiol.* 77:190–94
99. Van Volkenburgh, E., Cleland, R. E. 1981. Control of light-induced bean leaf expansion: Role of osmotic pressure, wall yield stress, and hydraulic conductivity. *Planta* 153:572–77
100. Van Volkenburgh, E., Cleland, R. E. 1985. Wall yield threshold and effective turgor in growing bean leaves. *Planta*. In press
101. Westgate, M. E., Boyer, J. S. 1984. Transpiration- and growth-induced water potentials in maize. *Plant Physiol.* 74:882–9
102. Westgate, M. E., Steudle, E. 1985. Water transport in the midrib tissue of maize leaves. *Plant Physiol.* 78:183–91
103. Yamamoto, R., Kawamura, H., Masuda, Y. 1974. Stress relaxation properties of the cell wall of growing intact plants. *Plant Cell Physiol.* 15:1073–82
104. Zimmermann, U., Steudle, E. 1978. Physical aspects of water relations of plant cells. *Adv. Bot. Res.* 6:45–117

Ann. Rev. Plant Physiol. 1986. 37:407–38
Copyright © 1986 by Annual Reviews Inc. All rights reserved

RAPID GENE REGULATION
BY AUXIN

Athanasios Theologis

Department of Biological Chemistry, Washington University School of Medicine, St. Louis, Missouri 63110

CONTENTS

0066-4294/86/0601-0407$02.00

INTRODUCTION

The primary mechanism of action of plant hormones in general, and of auxins in particular, typified by indoleacetic acid (IAA),[1] continues to elude us even after 55 years of intense investigation since the discovery of IAA (104). Elucidation of the primary action of auxin has been hampered by our inability to detect specific, rapid responses causative of known auxin-regulated phenomena. Auxin effects are generally divided into two main categories: rapid responses such as cell elongation (13, 69), one of the fastest hormonal responses known with a latent period of 10–25 min (22, 33, 95), and long-term responses such as cell division, differentiation, and morphogenesis with a latent period of hours or days (87). The rapidity of the initiation of cell elongation has made the phenomenon most attractive for investigating the primary mechanism of action of the hormone.

Two major theories have been proposed to explain auxin-induced cell elongation. In the 1960s the gene activation hypothesis proposed that auxin regulates the synthesis of specific mRNAs coding for polypeptides necessary for the growth process (42). Supporting this view were observations of enhanced incorporation of radioactive precursors into RNA and proteins (6, 26, 35, 60, 61, 66, 89) after prolonged treatment with auxin (several hours). Auxin's effect on mRNAs or specific proteins, rapid enough to cause cell elongation, however, could not be detected by techniques available at that time even though auxin fully stimulates growth within 10 to 25 min (22, 33, 90, 95). Consequently, during the 1970s attention was focused on the cell membrane and diverted away from gene expression (16, 17, 33, 71), and a second theory, the acid growth theory, was proposed which suggested that cell enlargement is initiated by auxin-induced proton secretion (29, 70). No experimental evidence convincingly demonstrated the direct action of the hormone on a transport system such as proton pumping (39, 100), although the inhibition of auxin-induced proton secretion by protein synthesis inhibitors (14, 16, 17, 71) indicated that auxin may act indirectly by stimulating a proton pump through

[1]*Abbreviations:*

bp, base-pair(s);	Kd, kilodaltons; M_r, molecular weight
ACC,	1-aminocyclopropane-1-carboxylic acid
cDNA,	complementary DNA
pCIB,	p-chlorophenoxyisobutyric acid
2,4-D,	2,4-dichlorophenoxyacetic acid
IAA,	indoleacetic acid
IPA,	N^6-(Δ^2–isopentenyl)adenine
NAA,	naphthalene-1-acetic acid
PAA,	phenylacetic acid
SDS-PAGE,	sodium dodecyl sulfate-polyacrylamide gel electrophoresis
2,4,5-T,	2,4,5-trichlorophenoxyacetic acid

newly synthesized polypeptides which could mediate the growth process in any of several ways. More recently, Vanderhoef & Dute (92) merged the two theories, suggesting that auxin-induced growth involves two phases: an early response resulting simply from auxin-induced proton secretion and a second, later phase thought to be mediated by a separate primary action of auxin on gene expression.

The development of in vitro translation systems for mRNAs and of two-dimensional electrophoresis of translated proteins has made possible the identification of selective mRNA changes in peas, soybean, and corn within 10–20 min after exposure to the hormone (85, 86, 110–112). Recombinant DNA technology has led to the isolation of complementary DNA sequences for some of these early auxin-responsive mRNAs (28, 84, 102), and the isolation of cDNA clones has allowed a detailed characterization of the hormonal response and an examination of the possible relationship between auxin-induced proton secretion and auxin-regulated mRNA induction.

This chapter reviews the experimental evidence demonstrating early mRNA changes (within 10–20 min) by IAA in pea stem segments (84–86), a classical system for studying auxin action on growth. These results are compared with those obtained by other laboratories using other systems, principally soybean (28, 102, 111, 112). "Late mRNA" responses (greater than 2 hours) are also discussed. The accumulated experimental evidence indicates that auxin has the capacity to act at the transcriptional or post-transcriptional level very rapidly, concomitantly with or earlier than the initiation of cell elongation and proton secretion. Whether the proteins coded by the early auxin-regulated mRNAs are responsible for the initiation of cell elongation has not yet been established; however, the mRNA induction is the fastest known for any plant growth regulator and may represent a primary response to auxin. A model uniting auxin-induced proton secretion, mRNA induction, and cell elongation is proposed.

AUXIN-INDUCED CHANGES OF PROTEINS IN PLANT TISSUES

Soybean Hypocotyl

The search for specific auxin-induced changes of proteins in plant tissues has been thwarted primarily by the unavailability of sensitive analytical techniques for resolving complex mixtures of proteins. Patterson & Trewavas (66) failed to demonstrate conclusively specific changes in the proteins synthesized in pea epicotyl tissue treated with 20 μM IAA for 2 hr using the double-label technique. More recently, Bates & Cleland (4, 5), using the same experimental approach, also were unable to detect protein changes in oat coleoptiles after treatment with 10 μM IAA for 40 min to 6 hr. However, Zurfluh & Guilfoyle

(108, 109) were able to detect specific changes in proteins synthesized in elongating and mature sections of soybean hypocotyl after treatment with 50 and 500 μM 2,4-D, respectively. The hypocotyl sections were labeled with high specific activity [^{35}S]methionine in the presence and absence of the hormone up to 3 hrs, and the labeled polypeptides were analyzed by one- and two-dimensional SDS-PAGE. The most significant differences observed in elongating segments are the appearance of a polypeptide of M_r 100 Kd (pI 6.1) and two polypeptides of M_r 40 Kd (pIs 6.1 and 5.9), and the disappearance of a polypeptide of M_r 40 Kd (pI 6.3) in auxin-treated tissue. The polypeptides synthesized after 1 hr of exposure to auxin are similar but not identical to those observed after 3 hr. The 100 Kd protein observed at 1 hr persists up to 3 hr. Zurfluh & Guilfoyle (108, 109) found that the detection of auxin-induced polypeptide synthesis in elongating soybean hypocotyl was masked by wound-induced protein synthesis that occurs at the ends of the incubated sections, and they found it necessary to remove the wounded ends prior to analysis of the polypeptides.

Treatment of the elongating soybean segments with 550 μM isopentenyl-adenosine, a cytokinin, inhibits the auxin-induced cell elongation by 70% after 5 hrs of incubation and also inhibits [^{35}S]methionine incorporation into protein by 60% (108). Furthermore, cytokinin treatment alters the pattern of proteins synthesized in untreated and auxin-treated soybean hypocotyls. This experimental evidence suggests, but does not conclusively prove, that the inhibition of cell elongation by cytokinins results from inhibition of the synthesis of the auxin-induced polypeptides.

In contrast to elongating hypocotyl sections, mature (basal) sections fail to elongate when exposed to auxin. A 3 hr treatment with 2,4-D induces and represses a large number of polypeptides that are not the same as those induced by the hormone in elongating sections, indicating that auxin-induced polypeptide synthesis is tissue specific (108, 109).

Tobacco Mesophyll Protoplast

Auxin is necessary for cell wall regeneration as well as for the induction of mitosis in protoplasts isolated from various plant tissues. In the absence of the hormone, protoplasts initially blocked in G_0 phase do not progress through the phases of the cell cycle (54). Meyer et al (52, 53) were able to detect rapid and specific changes in the proteins synthesized in tobacco mesophyll protoplasts exposed to 4.5 μM 2,4-D. Among 250 [^{35}S]-labeled polypeptides resolved by SDS-PAGE, 9 proline-rich polypeptides (M_r 30–70 Kd) were decreased in amount, while 2 polypeptides (M_r 30 Kd) were enhanced. Time-course experiments (53) showed that the auxin-inducible polypeptides are detected within 30 min of hormone application and reach a constant level after 2 to 4 hr. On the other hand, the two proteins whose levels are reduced by auxin are affected after 6 hr. It has been postulated that the latter are involved in cell wall formation, based on the observation that dichlorobenzonitril (a weed killer) prevents cell wall formation and also the reduction of synthesis of the proline-rich polypeptides by auxin (52). Meyer et al (52) have suggested that the auxin-induced polypeptides are involved in initiating mitosis; however, this has not been experimentally substantiated.

Strawberry Fruit

Nitsch showed that the achenes on the surface of the strawberry receptacle control strawberry fruit development and that auxin substitutes for these structures (59). Veluthambi & Poovaiah (97) studied the effect of auxin on polypeptide synthesis in developing strawberry fruit and found that an 81 Kd polypeptide appears between 5 and 10 days after pollination, and two other polypeptides (M_r 76 and 37 Kd) appear after 10 days. Removal of the achenes (auxin source) from the 5 to 10-day-old fruits results in inhibition of fruit growth and no synthesis of the 81, 76, and 37 Kd polypeptides. Application of NAA to fruits deprived of auxin results in the resumption of growth and also in the appearance of these polypeptides. Furthermore, removal of the achenes results in the appearance of two other polypeptides, M_r 52 and 57 Kd, respectively. Exogenous NAA prevents the appearance of these two polypeptides, indicating a possible role in inhibiting strawberry fruit growth.

AUXIN-INDUCED CHANGES OF mRNAs IN PLANT TISSUES ASSESSED BY IN VITRO TRANSLATION

Late Changes in mRNAs

The auxins clearly have the capacity to alter the biosynthesis of particular proteins in various plant systems. If proteins have a role in initiating the biological phenomena known to be mediated by the hormone and constitute the primary hormonal response, then hormonally induced changes in the concentration of a particular protein should be detectable prior to the onset of the phenomenon and shortly after introduction of the hormone into the experimental system. The protein changes so far detected and described in the previous section do not qualify as primary responses of the hormone in a given experimental system and accordingly are characterized as "late responses." It may be our inability to label in vivo proteins fast enough with high specific activity that prevents detection of rapid hormonal effects by two-dimensional SDS-PAGE and subsequent fluorography. In addition, studies thus far on hormonally induced protein changes do not reveal the mechanism by which they are mediated. For example, do the protein changes result from differences in mRNA levels affected by the hormone at the transcriptional level, or are they an effect of the hormone at the translational or post-translational level or on the stability of the proteins or mRNAs?

In an effort to determine experimentally whether auxin rapidly alters the abundance of specific mRNA sequences, several investigators isolated and translated mRNAs in an in vitro biosynthetic system in the presence of radiolabeled amino acids and analyzed the translated products by two-dimensional SDS-PAGE. (This approach allows analysis of the most abundant mRNA sequences.) Baulcombe et al (8) were the first to show differences among the in vitro translation products of polyadenylylated mRNAs isolated from 2,4-D untreated and treated elongating soybean hypocotyls. Among 220 polypeptides resolved, 20 are enhanced by auxin and another 20 reduced. Unfortunately, neither the auxin concentration nor the rapidity of the induction was reported (the earliest change recorded is after 5 hr of hormone treatment).

A second experimental approach for detecting auxin-induced changes in mRNAs is the technique of cDNA/RNA hybridization. By this technique, which is less sensitive than in vitro translation, Baulcombe et al (8) were able to detect an auxin-induced reduction of two highly abundant sequences. There is no detectable effect on the low abundance class of polyadenylylated mRNAs.

It is well documented that auxin enhances rRNA synthesis in soybean hypocotyls because of an increase in the number of RNA polymerase I molecules as well as its specific activity (26). Recently, Gantt & Key (24) were able to demonstrate by in vitro translation and two-dimensional SDS-PAGE that 2,4-D enhances the level of translatable mRNAs coding for 19 and 10 polypeptides associated with the large (60S) and small (40S) ribosomal subunit, respectively, in soybean hypocotyls. An eightfold increase in translational activity is noticed after 24 hr of treatment with 2.5 mM 2,4-D. Using cDNA clones for nine ribosomal mRNA sequences, the same investigators found that the enhancement in translational activity results from the net increase in the levels of their corresponding mRNAs. Time-course experiments show no detectable change in the mRNA levels after 2 hr of hormonal treatment (25). These auxin-induced changes do not constitute the primary mechanism of auxin action because of their timing; they simply indicate that the hormone has the capacity to alter the levels of the ribosomal protein mRNAs as a consequence of the primary response.

Early Changes in mRNAs

PEA EPICOTYL VERSUS SOYBEAN HYPOCOTYL: A COMPARISON   The auxin-induced mRNA changes described above are not rapid enough to implicate them as the primary response to the hormone. More rapid mRNA responses (within 10–20 min) have recently been detected in pea epicotyl and soybean hypocotyls by two independent laboratories (85, 86, 111, 112). The experiments with pea epicotyl were carried out with 8 mm long segments excised from elongating third internodes of dark-grown 7-day-old Alaska pea seedlings. Segments were incubated for 2 hrs without auxin, and then with or without 20 μM IAA for up to 12 hr. Polyadenylylated mRNA isolated after various periods of hormone exposure was translated in vitro and [^{35}S]methionine-labeled translation products analyzed by two-dimensional SDS-PAGE with subsequent autoradiography. The experiments with soybean hypocotyls were done with segments 12 mm long from the elongating region of the hypocotyl. Segments were incubated for 2 hr without auxin and then incubated with or without 50 μM 2,4-D. Similar methods were used for in vitro translation assays of soybean mRNAs (111, 112), except that mRNA enhancement was judged from the visual appearance of autoradiographic spots and not quantitatively.

Figure 1 shows portions of autoradiograms of two-dimensional separations of [^{35}S]methionine-labeled polypeptides obtained by in vitro translation of pea stem poly(A)$^+$-RNA in the wheat germ system. Incubation for 6 hr with IAA causes a substantial increase in translational activity of mRNAs coding for polypeptides 1 through 6 (compared with the untreated control (Figure 1A). Detailed time-course experiments (85) have shown that translational products #1 and #2 increase after 2 hr and thus have been termed "late IAA-regulated mRNAs." On the other hand, translation product #6 (doublet) increases after 30 min, and #3, 4, and 5 increase within 20 min of exposure to IAA and are thus "early IAA-regulated mRNAs."

Zurfluh & Guilfoyle (111) similarly observed early mRNA increases in soybean hypocotyl segments treated with 2,4-D. After 15 min, they detected one translational product similar to #5 with a M_r of 25 Kd and a position on the pH gradient at about pH 6.5. A second sequence is also enhanced by 2,4-D at 15 min in the same M_r region but located at about pH 5.8; it corresponds closely with #3 of the pea system. After exposing soybean hypocotyl segments for 30 min to 2,4-D, eight enhanced mRNAs were detected, nine after 45 min, and ten sequences after 60 min. These observations qualitatively agree with the results obtained with pea epicotyl, indicating a temporal cascade of auxin-affected sequences. As in the pea system, auxin-induced decreases in mRNAs within 2 hr of treatment were not detected. Most of the auxin-enhanced soybean sequences (111) do not correspond to those recognized in peas (except for #3 and #5). Nothing comparable to the series of 2,4-D enhanced soybean translation products at about 27 Kd, between pH 5.5 and 5.8, is recognized in pea autoradiograms. In addition, the prominent IAA-enhanced sequences #1, 2, and 4 in peas (Figure 1) do not correspond with anything on the soybean autoradiograms. Bevan & Northcote (9), on the other hand, found enhanced soybean translation products of M_r 35 and 45 Kd, similar to the pea polypeptides #1 and 2 (Figure 1), 2 hr after exposure to auxin of soybean tissue in culture. They detected no changes, however, in mRNAs corresponding to the early enhanced pea sequences. These results indicate only a partial similarity

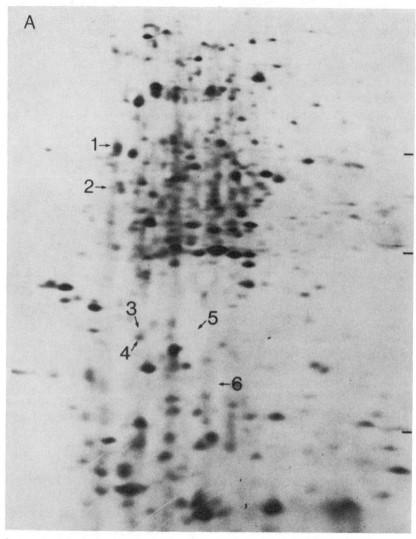

between the auxin response systems in the closely related plants, pea and soybean. The response is specific to auxins, as is indicated by the identical pattern of translation products enhanced by 2,4-D or IAA after 2 hr of exposure (111).

Many induced mRNA sequences are detected from the basal part of intact soybean hypocotyls sprayed with a high concentration of 2,4-D (2.5 mM) or induced by ethylene but not with low concentrations of 2,4-D (200 μM) (112). It is believed that the mRNAs detected with high concentrations of auxin are

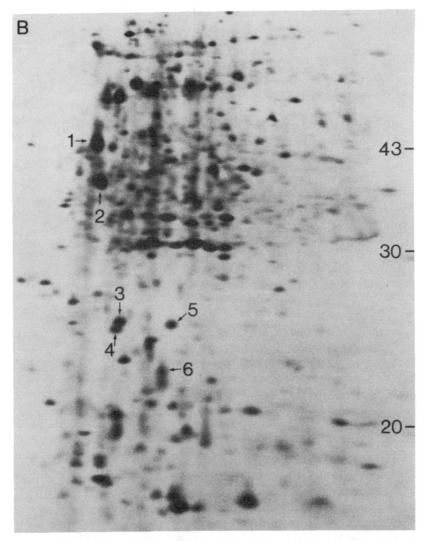

Figure 1   Enhancement of specific translational products in pea epicotyl tissue by 20 μM IAA. A. Control-6 hr. B. IAA-6 hr. For experimental details see (86).

induced by ethylene whose production is stimulated by high auxin concentrations and that the sequences enhanced by lower auxin concentrations are independent of ethylene.

CORN COLEOPTILE   Zurfluh & Guilfoyle compared the in vitro translation products from IAA-treated and untreated corn coleoptiles and found rapid auxin-induced changes with 50 μM IAA (110). At least four mRNA sequences coding for four polypeptides, ranging in M_r from 25 to 30 Kd and having pIs 5.7 to 6.4, are enhanced by IAA. Two of the four are enhanced after 10 min of hormone treatment, and a consistent increase in all four is observed after 20, 30, and 60 min. Whether these mRNAs are the same or related to those detected in pea and soybean remains to be determined.

OTHER PLANT SYSTEMS   In vitro translation experiments carried out by Meyer et al (52) have shown that the auxin-induced increase in specific protein levels in tobacco protoplasts coincides with an increase in their corresponding mRNAs (53).

CLONING THE AUXIN-REGULATED mRNA SEQUENCES

It is obvious from the in vitro translation studies that the auxins rapidly and specifically potentiate certain mRNA sequences in various plant tissues. Highly sensitive techniques are necessary to detect the early mRNA changes that are beyond the limits of detection of techniques available in the past. The specific potentiation in translational activity of polyadenylylated mRNA sequences in pea and soybean tissue has been attributed to an increase in the amount of translatable mRNA brought about by activation of transcription, post-transcriptional processing, or mRNA stabilization (86, 102, 111). However, the possibility also exists that the hormone alters the translatability of preexisting mRNAs by events such as polyadenylylation or capping (78). Whether auxins cause an actual increase in the amounts of specific mRNA sequences can best be answered by obtaining DNA sequences complementary to auxin-regulated mRNAs using recombinant DNA technology.

Soybean Hypocotyl

Baulcombe & Key (7) were the first to isolate cDNA sequences in soybean hypocotyl for mRNAs that are decreased in response to 2,4-D. Kinetic hybridization analysis of cDNA probes previously revealed that a minor abundant class of RNA sequences is markedly reduced in the poly(A)$^+$-RNA isolated from 2,4-D treated tissue (8). A cDNA library was constructed with poly(A)$^+$-RNA isolated from untreated hypocotyls of 5-day-old soybean seedlings into the Pst I site of pBR322 using the G-C tailing procedure (7), and 4000 ampicillin-sensitive, tetracycline-resistant colonies were obtained. Differential colony hybridization using [^{32}P]-labeled cDNAs synthesized from poly(A)$^+$-RNA isolated from untreated and auxin-treated soybean hypocotyls led to the isolation of 12 recombinant clones to mRNAs decreased in the presence of auxin. These clones fall into three groups. The first group is represented by clone p3 which has a cDNA insert of 0.43 Kb and hybridizes to an mRNA of

1000 nucleotides long (Table I). Hybrid selected translation reveals that this mRNA codes for a polypeptide of M_r 33 Kd (8). The second group represented by clone p9 (4 clones) has a cDNA insert of 0.33 Kb and hybridizes to an mRNA of 350 nucleotides long (8). The primary translation product encoded by this mRNA is a 10 Kd polypeptide. The third group is represented by clone p11 with a cDNA insert of 0.42 Kb and hybridizes to an mRNA of 750 nucleotides long. The polypeptide encoded by this mRNA is approximately 25 Kd (8). Time-course experiments indicate that the mRNAs hybridized to clones p3, p9 and p11 decrease 66-, 9-, and 20-fold, respectively, after 4 hr and almost no hybridization was detected after 20 hr of auxin treatment. The 2,4-D concentration was not reported, and it is not possible to know if the observed mRNA changes are "early" or "late" since these were not examined after short time exposures (less than 1 hr) to 2,4-D.

Walker & Key (102), utilizing the same cloning procedure, constructed a cDNA library using poly(A)$^+$-RNA from soybean hypocotyl segments treated with low auxin concentration (24 μM 2,4-D). From a library of 14,000 recombinants they isolated two clones, pJCW1 and pJCW2 (Table I), that hybridize to two mRNAs of similar size which are induced five- to eightfold after 6 hr of auxin treatment. Both auxin-regulated mRNA sequences are induced rapidly (after 30 min for pJCW1 and 15 min for pJCW2).

More recently, Hagen et al (28) isolated cDNA clones to four rapidly induced auxin-regulated mRNAs (latent period 30 min) by differential colony hybridization of a cDNA library constructed from poly(A)$^+$-RNA from intact soybean hypocotyls sprayed with 2.5 mM 2,4-D. The clones hybridize to four different mRNAs, pGH1-4 (Table I), varying in size from 1000 to 2400 nucleotides long. Presently, we do not know the size of the primary translational products coded by these sequences nor whether these sequences correspond to those previously identified by two-dimensional SDS-PAGE.

Pea Epicotyl

The isolation of cDNA clones for the IAA-inducible mRNAs in pea epicotyl (Figure 1) was facilitated by fractionating poly(A)$^+$-RNA from 6 hr control and IAA-treated pea tissue on sucrose gradients (84). The fractionation partitioned the sequences into two groups: a "light" fraction containing sequences #4 and #6 and a "heavy" fraction containing the other four (1, 2, 3, and 4). These fractions provided the mRNA for the synthesis of double-stranded cDNA and subsequent cloning into the unique EcoRI site of the cloning vector, λgt10 (32). The efficiency of the cDNA cloning was 5×10^5 to 8×10^5 recombinant phages per 2 μg poly(A)$^+$-RNA. Differential plaque filter hybridization (77) permitted the isolation of recombinant phages to the early IAA-regulated mRNAs. 2800 phages were screened from the hybrid pool, and the lighter fraction with cDNA probes synthesized by reverse transcription of RNA from the same fraction of

Table 1 Characteristics of the auxin-regulated cDNA clones isolated from pea and soybean tissue

cDNA clone	Tissue	cDNA insert size (Kb)	mRNA size (nucleotides)	Source of poly A$^+$-RNA used for construction of the cDNA library	Level of expression of the mRNAs in the intact tissue (% total poly A$^+$-RNA)	Fold increase or decrease of the mRNA after treatment of tissue segments with auxin	Earliest time of mRNA changes determined experimentally	References
p3	S[a]	0.43	1000	Intact soybean hypocotyls[c] from 5-day-old untreated seedlings	1	66↓	4 hr	(8)
p9	"	0.33	350		0.9	9↓	"	"
p11	"	0.42	750		0.2	20↓	"	"
pJCW1	"	0.74	1100	Excised soybean hypocotyls incubated with 23 μM 2,4-D for 3 hr	0.01	5↑ in 6 hr	30 min	(102)
pJCW2	"	0.72	1000		<0.01	8↑ "	15 "	"
pGH1	"	1.20	1700	Intact soybean hypocotyls sprayed with 2.5 mM 2,4-D for 2 hr	<0.01	2↑ in 4 hr	30 min	(28)
pGH2	"	0.20	1000		<0.001	7↑ "	40 "	"
pGH3	"	1.30	2400		0.001	30↑ "	30 "	"
pGH4	"	0.30	1100		0.01	3↑ "	30 "	"
pIAA4/5	P[b]	0.70	950	Excised pea epicotyls treated with 20 μM IAA for 6 hr	0.01	50↑ in 2 hr	10–15 min	(84)
pIAA6	"	0.72	850		0.01	50↑ "	10–15 "	"

P: Pea epicotyl.
S: Soybean hypocotyl.
[a] The soybean cDNA clones were isolated by differential colony hybridization of cDNA libraries constructed in the vector pBR322.
[b] The pea cDNA clones were isolated by differential plaque filter hybridization of cDNA libraries constructed in the vector λgt10.
[c] Intact hypocotyls were treated with an unknown 2,4-D concentration.

the control and IAA gradients. Table I shows the characteristics of the IAA-inducible cDNA clones isolated from pea epicotyl. Clone pIAA 4/5 with an insert of 750 bp hybridizes to an mRNA of 950 nucleotides. Clone pIAA6 with an insert of 720 bp hybridizes to an mRNA of 840 nucleotides. No cross-hybridization between the clones is detected. Isolated clones correspond to mRNAs coding for the translational products 4, 5, and 6, as evidenced by the identical electrophoretic mobility of their hybrid-selected translation products (84).

Clone pIAA 4/5 selects two mRNAs, one corresponding to polypeptide #4 and the other to #5 (84). The second translational product may be a consequence of a post-translational modification such as acetylation or processing; however, these activities were not detected in the translation system used (84). It is possible that the clone selects the two mRNAs because it contains two fused cDNAs with an insert lacking an internal EcoRI site, although this seems unlikely because it was isolated from the cDNA library constructed from an mRNA fraction containing a small amount of mRNA coding for polypeptide 5. An alternative explanation is that sequence homology between mRNAs 4 and 5 may permit selection of the two mRNAs by clone pIAA4/5. [^{32}P]-labeled pIAA 4/5 DNA hybridizes to a single 2400 bp EcoRI fragment of pea DNA at an approximately single copy level of hybridization (A. Theologis, unpublished). The two mRNAs could be transcribed from the same gene and processed differentially (18, 44, 56, 74, 107), or transcribed from separate but partially homologous genes. Isolation and structural analysis of the gene to distinguish between the two possibilities is required. A third possibility is that polypeptides 4 and 5 may be encoded by allelic variants of one gene, with the clone pIAA 4/5 carrying one of these genes but selecting mRNAs encoded by both alleles.

The characteristics of the cDNA clones isolated from soybean and pea (7, 28, 84, 102) are compared in Table I. The induced mRNAs represent 0.001 to 0.01% and the decreased sequences represent 0.2 to 1% of the total poly(A)$^+$-RNA in the intact tissues. The former increase rapidly (10–30 min after hormone treatment) by 2- to 50-fold in excised segments, whereas the latter decrease 9- to 66-fold over 4 hr of hormone treatment. The mRNA potentiation in tissue segments by auxin cannot be attributed to tissue cutting because application of 2,4-D on intact soybean seedlings (28) or IAA on pea seedlings (A. Theologis, unpublished) further enhances the expression of the mRNAs. The induced mRNAs increase rapidly (10 to 30 min) from 2- to 50-fold in excised tissues after auxin treatment, whereas the auxin-regulated reduced mRNAs are decreased 9- to 66-fold over 4 hr of hormone treatment. The auxin-induced mRNAs are further enhanced in intact soybean seedlings exposed to 2,4-D (28) and intact pea seedlings exposed to IAA (A. Theologis, unpublished). The auxin-responsive cDNA clones described by Walker & Key (102) do not share any sequence homology with those described by Hagen et al

(28), and it is not known whether the pea cDNA clones isolated by Theologis et al (84) share any homology with the soybean cDNA clones (28, 102).

DEVELOPMENTAL REGULATION OF THE AUXIN-RESPONSIVE GENES

Is the level of expression of the auxin-regulated genes correlated with endogenous levels of auxin? The data accumulated thus far, using soybean or pea cDNAs for hybridization analysis of mRNAs isolated from various parts of the growing soybean or pea seedling, do not give a clear answer. For example, the decreased auxin-regulated mRNAs identified by Baulcombe & Key (7) are found at higher concentrations in the basal part (zone C) of the soybean seedling where auxin concentration is low, and lowest in zone A (hook) where auxin concentration is high. This finding indicates a positive correlation between levels of expression and auxin concentration. On the other hand, the endogenous levels of the RNA sequences homologous to pJCW1 and pJCW2 (Table I) are higher in actively elongating tissue (zone B) than in apical (zone A) or mature tissue (zone C), indicating a negative correlation between endogenous auxin levels and the degree of expression of these genes (102). The levels of the RNA sequences homologous to the clones isolated by Hagen et al (28) are approximately the same in the various regions of the soybean hypocotyl. Finally, RNA hybridization analysis with the IAA-regulated cDNA from pea tissue (pIAA4/5 and pIAA6) shows that they are present in the third and second internode (84). The hook, a region of high auxin concentration in the etiolated pea seedling, expresses the auxin-regulated genes at very low levels (84). Thus the sensitivity of the tissue to endogenous levels of the hormone for expressing the auxin-regulated genes is variable and independent of the endogenous auxin concentration.

CHARACTERIZATION OF THE HORMONAL RESPONSE

A detailed characterization of the hormone response has been possible with the isolation of cDNA clones to auxin-regulated mRNAs from pea and soybean tissue (27, 28, 84, 102, 103). Three criteria are useful in establishing that the response to a hormone in general represents the direct or primary action of the hormone; the induction of mRNA should occur rapidly, it should be specific, and it should be unaffected by protein synthesis inhibitors (72).

Rapidity of the Response

The kinetics of accumulation of the IAA-inducible mRNAs in pea epicotyl in response to 20 μM IAA is shown in Figure 2. After 10 min, an increase of pIAA4/5 mRNA is detected and continues over 2 hr. In the absence of IAA this mRNA is present at low levels. However, pIAA6 mRNA starts to accumulate after 20 min with very low levels expressed in the absence of IAA. Similar induction kinetics were obtained with 2,4-D-induced mRNAs in elongating soybean hypocotyls. Walker & Key (102) detected an increase in the pJCW1 and pJCW2 mRNAs after 30 and 15 min, respectively, of 2,4-D administration, and similar kinetics have been obtained with the four cDNA clones isolated by Hagen et al (28), who found that the earliest time of mRNA change is 30 min after hormone administration. Nuclear runoff experiments in isolated nuclei from 2,4-D treated soybean plumules, however, show that the mRNAs hybridizing to clones pGH1-4 are enhanced within 5 min after application of 100 μM 2,4-D (27).

In pea epicotyl, IAA-induced enhancement of specific mRNAs (within 10–15 min) is earlier than the initiation of cell elongation which has a latent period of 21–23 min (33). For the soybean hypocotyl system, the latent period of the auxin-regulated mRNA is almost the same or a little longer than the latent period of cell elongation (11.8 min), as determined by Vanderhoef & Stahl (93). Furthermore, nuclear runoff experiments (27) indicate a 5 min latent period for the mRNA. These responses, considerably faster than gibberellin induction of α-amylase mRNA in aleurone cells (30, 36) or phytochrome-mediated photoinduction of selected mRNAs (88), are comparable to the most rapid mRNA inductions by mammalian (3, 50, 72, 81, 101) and insect hormones (2). In pea epicotyl the originally observed enhancement in translational activity of pIAA4/5 and pIAA6 mRNA clearly is the result of an increase in the amount of the corresponding mRNAs (Figure 2) and not polyadenylylation of preexisting RNA, because the enhancement is seen in hybridization analysis of total nucleic acids (84). Whether the auxin-induced enhancement of specific mRNAs in soybean tissue results from polyadenylylation of preexisting RNA has not been determined.

Activation of transcription, post-transcriptional processing, or selective stabilization of mRNA by auxin can all account for the hormonally induced mRNA accumulation (28, 84, 102). α-Amanitin completely inhibits the accumulation of pIAA4/5 and pIAA6 mRNA, as well as auxin-induced cell elongation and H^+ secretion after 2 hr of treatment (84). This observation, while suggesting an absolute requirement of transcription for the induced cell elongation and H^+ secretion, does not conclusively implicate IAA acting at the transcriptional level, because hormonally regulated mRNA synthesis, processing, or stabilization would all give similar results. An increase in the rate of transcription and stabilization of the mRNA (50, 65) accounts for the accumulation of ovalbumin and conalbumin by steroid hormones in chick oviduct, and similar conclusions have been reached for the accumulation of vitellogenin mRNA by estrogen in *Xenopus laevis* (78). Recently, nuclear runoff experiments by Vannice et al (96) have suggested that the glucocorticoid-

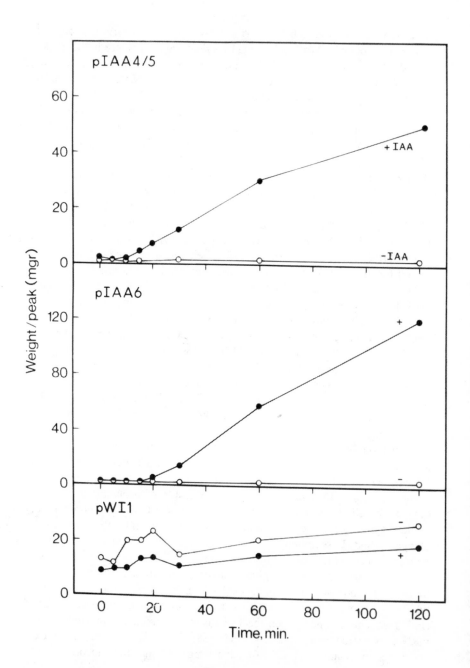

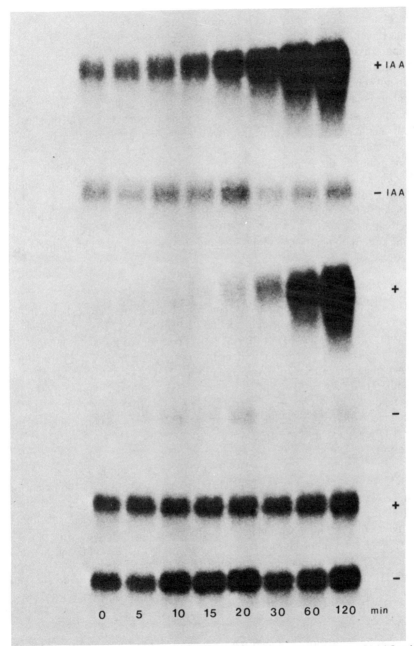

Figure 2 Induction kinetics of the IAA-inducible mRNAs in pea epicotyl tissue. pIAA4/5 and pIAA6 are two IAA-responsive cDNAs, and pW11 is a nondifferential cDNA used as internal control for the RNA hybridization analyses. See (84) for experimental details. Reproduced with permission from the *Journal of Molecular Biology*.

mediated induction of α_1-acid glycoprotein is regulated at the processing level, and Hagen & Guilfoyle (27), using similar techniques, have proposed that the rapid auxin-induced mRNA accumulation is under transcriptional control. However, nuclear runoff experiments with isolated nuclei from treated tissue do not conclusively implicate auxin acting at the transcriptional level, because enhancement of runoff transcripts will be observed in the assay whether the hormone acts at the transcriptional or post-transcriptional level. The rapidity of the auxin response is compatible with transcriptional activation based on estimated maximum transcriptional rates of 12 to 18 transcripts per minute (38). For example, the synthesis of a 6500 nucleotide mRNA such as vitellogenin requires about 7 min (3). The time required to synthesize a 25,000 M_r protein (corresponding to a transcript of roughly 1000 nucleotides) would be approximately 60 seconds, suggesting that there is ample time for transcriptional activation by auxin (IAA or 2,4-D). The development of an in vitro transcription system for plant genes will eventually allow these various possibilities to be tested.

Specificity of the Response

The specificity of the hormonal response has been clearly demonstrated in the pea epicotyl and the soybean hypocotyl systems (27, 84, 103). Incubation of tissue with 20 μM 2,4-D or 20 μM NAA, compounds known to induce cell elongation and H^+ secretion, as does IAA (84), results in the expression of the IAA-inducible mRNAs pIAA4/5 and pIAA6. However, when tissue is treated with 20 μM PAA or 20 μM pCIB, two IAA analogs without biological activity, the induction is not observed (84). Walker & Key (103) similarly have shown that incubation of soybean tissue with 50 μM IAA or 25 μM NAA induces the 2,4-D regulated mRNAs, pJCW1 and pJCW2 mRNAs. Nuclear runoff experiments have shown that, at concentrations of 100 μM, the auxins IAA, 2,4-D, 2,4,5-T, and NAA are all effective in increasing the accumulation of pGH1–4 mRNA sequences (27). There is some variability in the extent of mRNA accumulation with the various auxins, with IAA being less effective than the synthetic auxins. Two naturally occurring structural analogs without auxin activity, tryptophan and indolealdehyde, are unable to induce the pGH1–4 mRNAs. Cyclohexylacetic acid, 1-cyclohexenyl acetic acid, and benzoic acid (27) elicit variable responses with the 2,4-D regulated sequences. With pGH1 and pGH3, these synthetic nonauxin analogs are largely ineffective in inducing a transcriptional response, but with pGH2 and pGH4, benzoic acid is as effective as IAA, and the other two nonauxin analogs are 25 to 50% as effective as the synthetic auxins (27).

The accumulation of the pIAA4/5 and pIAA6 mRNAs in pea tissue is unaffected by the plant hormones kinetin, gibberellic acid, abscissic acid, or ethylene (84). Similarly, the accumulation of the 2,4-D-regulated mRNAs, pJCW1 and pJCW2 in soybean tissue, is unaffected by the cytokinin IPA, ethylene, or its precursor ACC (103). That cytokinins do not prevent auxin-mediated mRNA accumulation, although they inhibit auxin-induced cell elongation (93), suggests that these mRNAs are not involved in the latter process and that the cytokinins act at a step (e.g. translational) subsequent to the hormone-stimulated accumulation of these mRNAs. The effect of other plant hormones on the accumulation of pGH1–4 mRNA in soybean tissue has not yet been investigated. Finally, stress conditions such as anaerobiosis (N_2 for 2 hr), heat shock (40°C for 2 hr), and cold shock (4°C for 2 hr) are ineffective in inducing pIAA4/5 and pIAA6 mRNAs in pea tissue (84). Similar conclusions have been reached for the 2,4-D regulated cDNAs pGH1–4 in soybean tissue (27).

Effect of Protein Synthesis Inhibitors

The effect of protein synthesis inhibitors on the induction of hormonally mediated mRNAs can provide insight into the primary mechanism of action of a hormone. The effect of five such inhibitors on the expression of the pIAA4/5 and pIAA6 mRNAs in pea epicotyl has been investigated in detail by Theologis et al (84). Incubation with 20 μM cycloheximide in the absence of IAA results in the induction of the pIAA4/5 and pIAA6 mRNAs. Administration of two other protein synthesis inhibitors, emetine (300 μM) or anisomycin (300 μM), also causes the accumulation of these mRNAs. However, puromycin (500 μM) and 2-S-aminoethyl cysteine are without effect (84). None of the inhibitors interferes with the IAA induction of the mRNAs. In all cases where the inhibitors are effective in inducing the mRNAs in the absence of IAA, protein synthesis inhibition is higher than 80%; however, when induction is not affected, protein synthesis inhibition is incomplete (84). The ineffectiveness of puromycin to inhibit protein synthesis may be the result of its inability to penetrate the pea tissue. The same may be true for the inhibitor 2-S-aminoethyl cysteine, a lysine analog, although the possibility also exists that either the endogenous pool of lysine is high or that the protein(s) regulating the expression of the gene do not contain lysine.

In sharp contrast to these findings, treatment of soybean hypocotyls with the same cycloheximide concentration (20 μM) does not induce the accumulation of the pGH1–4 mRNAs (27). The inhibitor prevents the 2,4-D mediated mRNA accumulation for clones pGH1, pGH2, and pGH4 by 80% after 2 hr of treatment, but does not affect the 2,4-D mediated induction of pGH3 mRNA. These findings have been interpreted by Hagen & Guilfoyle (27) to suggest that the induction of pGH1, pGH2, and pGH4 mRNAs requires proteins with a high turnover rate, whereas the induction of pGH3 mRNA requires a protein with a longer half life. However, the results also suggest that the accumulation of the pGH3 mRNA constitutes a primary hormonal response, whereas the induction of pGH1, pGH2, and pGH4 mRNA is a secondary hormonal response because of its dependence to concomitant protein synthesis. The effect of protein synthesis inhibition on the accumulation of the 2,4-D-regulated mRNAs pJCW1 and pJCW2 in soybean has not yet been investigated. The above data support the conclusion that the two pea cDNA clones pIAA4/5 and pIAA6 and the soybean clone pGH3 fulfill the requirements of being the primary response to auxin.

MODEL FOR REGULATION OF THE AUXIN GENES

The finding that three structurally and mechanistically dissimilar protein synthesis inhibitors—cycloheximide, emetine, and anisomycin (67)—mimic the IAA-induced mRNA accumulation in pea epicotyl tissue (84) suggests that the corresponding genes are under the control of a rapidly turning-over protein. Figure 3 shows two possible control mechanisms for the transcriptional regulation of the auxin genes. According to the negative control model, a short-lived protein repressor (R) interacts with the promoter region of the auxin-regulated gene and prevents its transcription. Protein synthesis inhibition by cycloheximide results in a rapid decrease in the level of the repressor, releasing its

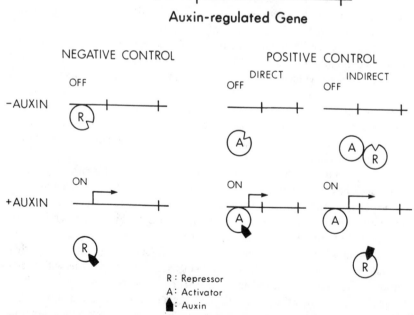

Figure 3   Models for the regulation of the auxin-inducible genes.

control over gene transcription. Auxin, on the other hand, binds to the repressor and inactivates it, thus altering the functionality of the repressor (whereas cycloheximide alters its concentration). The kinetics of the mRNA accumulation by cycloheximide are slower than those by IAA (A. Theologis, unpublished) and support the view that the mechanisms of mRNA induction by IAA and cycloheximide are different.

The second model presented in Figure 3 predicts that the auxin genes are under positive control. Direct positive control requires an activator molecule (A), a protein with a short half-life which is inactive in the absence of auxin and active in its presence. The data obtained with the protein synthesis inhibitors eliminate the direct positive control mechanism as a possible regulatory mechanism for the auxin genes in pea tissue because in the presence of cycloheximide the genes are expressed. The indirect positive control mechanism requires an activator molecule (A), which in the absence of hormone interacts with a protein repressor molecule (R), resulting in the inactivity of the former. Auxin binds to the repressor and releases the activator (A), which then interacts positively with RNA polymerase II, resulting in transcription of the gene. According to this view, cycloheximide lowers the levels of the rapidly turning-over repressor molecule. The negative control mechanism presented in Figure 3 is a simplification of the regulatory mechanism of the lactose operon of *E. coli* (55). The positive control model (indirect) appears to be a main regulatory mechanism found in simple eukaryotic organisms such as yeast and *Neurospora* (51, 63). Recently, a case of negative control has been demonstrated in yeast (37).

All three models in Figure 3 can also explain the behavior of the pGH3 mRNA induction in soybean tissue (27) as long as the repressor (R) or activator (A) molecule is not turning over as rapidly as in the pea tissue. Whether the pGH3 gene is under positive direct or indirect control cannot be distinguished. For the pGH1,2,4 mRNAs, these models cannot prevail because the auxin-mediated induction of these mRNAs is inhibited by cycloheximide. If the mRNAs whose synthesis is decreased in the presence of auxin (7) prove to be the primary response for 2,4-D, the models in Figure 3 can also explain their regulation if modified. The modification would require that the upper panel represent the events occurring in the presence of the hormone (auxin binding to the repressor, activator, or repressor-activator complex would result in cessation of gene expression) and that the lower panel represent events occurring in the absence of the hormone.

The models shown in Figure 3 require that the hormonally regulated genes are not constitutively expressed. It is possible, however, that the genes are transcribed at the same rate in the presence or absence of the hormone, and in the latter case the transcripts are not detected because of their rapid degradation. Auxin could then stabilize the mRNAs post-transcriptionally, e.g. by altering the K_m of an enzymatic component involved in the stability of the mRNAs. Similarly, the effect of cycloheximide on the expression of the IAA-regulated mRNAs in pea tissue can be attributed to the stabilization of the constituitively expressed transcripts, by preventing the synthesis of a labile enzyme, e.g. a nuclease. If auxin indeed acts at the post-transcriptional level, then the models in Figure 3 are untenable. Auxin-regulated genes are not alone in having their transcripts enhanced by protein synthesis inhibitors, and they share this characteristic with other regulated mammalian genes such as those for tyrosine aminotransferase (21), metallothionein-I (49), thionein (20), phosphoenoyl-pyruvate carboxyl kinase (12, 57), b-interferon (73), c-myc (40, 46), and actin (19).

DOSE-RESPONSE CURVES

It has been axiomatic that the dose responses of plants and plant sections to hormones are linear functions of the logarithm of the applied hormone concentration (91). The dose-response curves for the 2,4-D-regulated mRNAs pGH1–4 in soybean tissue (27) are reminiscent of those obtained with olfactory receptors but not with steroid hormone receptors (41). The pIAA4/5 and pIAA6 mRNAs in pea tissue (84) are exceptions and show a dose-dependent increase in their expression up to 10^{-4} M IAA, near maximal induction at a concentration of 10^{-5} M, and a plateau at 10^{-4} M IAA. These response curves are reminiscent of that obtained for prolactin synthesis in response to estradiol (45) except that four orders of magnitude less hormone are required for prolactin induction. This may be attributed to differences in effective hormone concentration intracellularly or hormone receptor affinities.

ARE AUXIN-REGULATED GENES UBIQUITOUS?

Hagen & Guilfoyle (27) were the first to detect 2,4-D directed mRNA accumulation in mung beans and green beans using the pGH1–4 cDNAs. They were unable, however, to determine whether 2,4-D induces homologous mRNAs in pea tissue (27). It still remains to be determined whether the IAA-regulated mRNAs in pea tissue and also the 2,4-D-inducible mRNAs pJCW1 and pJVW2 in soybean tissue have homologous counterparts in other plants (monocots or dicots).

IS AUXIN-INDUCED H$^+$ SECRETION THE DRIVING FORCE FOR RAPID mRNA INDUCTION?

The discovery that auxin rapidly induces specific gene expression raises the question whether the observed mRNA potentiation is mediated by auxin-induced H^+ secretion, currently considered the primary response to the hormone (29, 71). Theologis et al (84) examined the effect of inhibition of the auxin-induced H^+ secretion on the accumulation of the hormonally regulated mRNAs (84). Cerulenin, an inhibitor of fatty acid biosynthesis (84), vanadate, an inhibitor of H^+-ATPases (79, 82), and mannitol all prevent the H^+ secretion and cell elongation but are ineffective in preventing the auxin-induced mRNA accumulation. Cycloheximide, an inhibitor of protein synthesis (67), behaves similarly, except that it induces the hormonally regulated mRNA in the absence of auxin (84). Stimulation of H^+ secretion with fusicoccin (15, 47), however, does not cause induction of auxin-regulated mRNAs in pea (84) or soybean (103) tissue. Recent studies with [^{31}P]NMR have shown that treatment of pea epicotyl with IAA and fusicoccin does not alter the cytoplasmic or vacuolar pH (83). Thus the experimental evidence overwhelmingly rejects any role of auxin-induced H^+ secretion in the rapid mRNA induction, and it can be confidently stated that pH changes cannot be the driving force for the mRNA induction in pea or soybean tissue, as has been postulated for other systems (62).

RELATIONSHIP BETWEEN EARLY AUXIN-INDUCED mRNA ACCUMULATION, H^+ SECRETION, AND CELL ELONGATION: A HYPOTHESIS

Hormonal induction of specific mRNAs leads in vivo to changes in the synthesis of their corresponding polypeptides (65). The question arises whether there is a role for the proteins coded by the early auxin-regulated mRNAs in IAA-induced H^+ secretion and/or cell elongation. The inability of H^+ secretion to mediate the rapid mRNA accumulation offers two possible explanations for the mechanism of auxin action. First, the hormone may have two independent primary mechanisms of action (parallel model), one whereby it activates H^+ secretion and the other whereby it causes the accumulation of specific mRNAs. Second, the rapid mRNA accumulation may be the primary response to the hormone, with H^+ secretion a secondary consequence (series model).

Prior to the discovery of rapid gene regulation by auxin, Vanderhoef & Dute (92) proposed the parallel model to explain auxin's action in initiating cell growth. To accommodate the widely accepted view that auxins do not act rapidly at the gene level (22), the two responses were proposed to have different latent periods. Cell wall acidification that occurs with a lag period of approx·imately 15 min in soybean was thought to initiate cell growth (phase I) without the participation of biosynthetic events. Furthermore, without experimental verification, auxin-regulated cell wall material deposition was proposed to have a lag period of 50 min and to involve activation of transcriptional and translational machinery (92), leading to phase II of cell growth. In view of the induction kinetics of the early mRNAs, it is possible that they are somehow involved in phase II of cell growth; however, the parallel model is untenable on two accounts. First, in peas, cell enlargement and H^+ secretion are abolished by inhibitors of protein and RNA synthesis (84) soon after the application of the hormone, indicating an absolute dependence of the early cell enlargement and initiation of H^+ secretion on these two processes. [The inability of protein synthesis inhibitors to inhibit phase I of cell growth in soybean may reflect a permeability problem (94).] Second, a direct effect of auxin on a H^+-ATPase has yet to be demonstrated (and the proposed cell wall "loosening enzymes" also have yet to be purified).

In view of the above observations, the acid growth theory of auxin action needs to be reexamined. The induced mRNA accumulation is detected 10 min after auxin application in pea tissue (84, Figure 2), 5 to 10 min earlier than the induced H^+ secretion. Furthermore, the mRNA accumulation occurs 11 to 14 min earlier than the initiation of cell elongation (33). The accumulated experimental evidence is compatible with the series model, represented by the sequence of events:

mRNA induction $\rightarrow$ H^+ secretion $\rightarrow$ cell elongation,

and represented pictorially at the cellular level in Figure 4. H^+ secretion, a consequence of the rapid mRNA induction that constitutes the primary response to auxin, is proposed to be an expression of the growth process rather than its cause. Plant cell growth is a biosynthetic event, with the synthesis of cell wall materials required during cell expansion to avoid cell breakage. Growth proceeds by the intussusception of cell wall materials (transported in secretory vesicles) into the preexisting cell wall (68). The continuous flow of materials into the wall results in a decrease of the cell's turgor pressure with concomitant stimulation of water uptake leading to cell expansion. It is proposed that auxin regulates the secretory process responsible for cell wall formation by regulating the availability of specific mRNAs coding for polypeptides that facilitate the exocytosis (fusion-fission) of secretory vesicles filled with cell wall materials. Such proteins, e.g. synexin, have been identified in mammalian systems (31) and are known to facilitate fusion of secretory vesicles.

The model in Figure 4 may be considered a complex metabolic pathway with auxin regulation at the head of the pathway (gene transcription) and the secretory process at the tail. Auxin-induced protein secretion depicted in Figure 4 is a consequence of the enhanced biosynthetic activity induced by the hormone. The model predicts the presence of two H^+ extrusion systems: one, activated indirectly by auxin, is linked to the secretory process, so that when the latter is activated by auxin so is the H^+ secretion system. The second system is activated directly by fusicoccin, is localized on the cell membrane (1, 47, 80), and is independent of the secretory process. The first type of H^+ extrusion

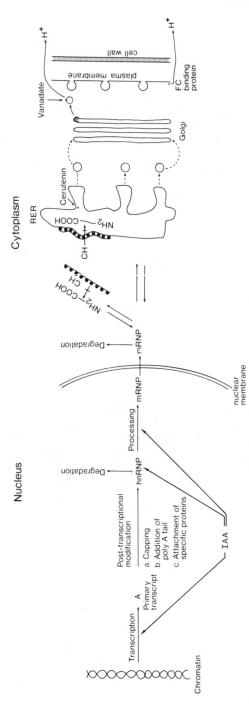

Figure 4   A model relating auxin-induced mRNA accumulation, H⁺ secretion, and cell elongation.

system explains why auxin-induced H^+ secretion is abolished by cycloheximide (15, 71, 84) and the fatty acid biosynthesis inhibitor cerulenin (84). With inhibition of synthesis of the components of the secretory vesicles, the secretory process is itself inhibited, and consequently so are auxin-induced H^+ secretion and cell elongation.

H^+-ATPases are widespread in plant tissues (79, 82). Since vanadate inhibits auxin-induced H^+ secretion (34), it is proposed that a H^+-ATPase is associated with the secretory vesicles, or may be more extensively associated with the endomembrane system from the Golgi to the plasma membrane. Vanadate-sensitive H^+-ATPases associated with the plasma membrane have been widely studied (79, 82), and it is noteworthy that a vanadate-insensitive H^+-ATPase has recently been found in Golgi-like material in corn tissue (11). It remains to be seen if a vanadate-sensitive enzyme is indeed present in secretory vesicles. Since fusicoccin activates a vanadate-sensitive H^+-ATPase and since the fusicoccin H^+ secretion is insensitive to cycloheximide (15, 47, 75) and cerulenin (A. Theologis, unpublished), it is proposed that the fungal toxin-induced H^+ secretion is independent of the secretory process. Fusicoccin-induced cell elongation, therefore, is viewed to be biochemically distinct from that induced by IAA, and its transient nature (15, 43) and independence from protein and fatty acid synthesis suggest that it is mediated by exocytosis of preexisting secretory vesicles.

The role of the secretory vesicle associated with H^+-ATPase indirectly activated by auxin is shown in Figure 5. The intussusception of a secretory vesicle (sv) filled with cell wall material into the cell wall is thermodynamically an uphill process because the chemical potential μ_j of species j inside the vesicle μ_j^{sv} is much smaller than that in the cell wall μ_j^{cw}. This is because the activity (a_j) and the electrochemical potential (E) of species j in the secretory vesicle are smaller than they are in the cell membrane-cell wall complex. According to the model in Figure 5, the $\Delta\mu_j$ is satisfied by an electrochemical gradient generated by the electrogenic H^+-ATPase localized in the secretory route (Figure 4). Proton electrochemical gradients generated by H^+-ATPases serve as the driving force for active transport of inorganic and organic cations, anions, and sugars in various tissues (32). The auxin-induced H^+ secretion shown in Figure 4 is proposed to be the expression of a H^+-ATPase whose electrochemical potential is the driving force for the transport of secretory vesicles responsible for the IAA-induced cell growth. It is interesting that in mammalian systems newly endocytosed materials are exposed to low pH soon after internalization (23), and acidification of intracellular compartments are mechanistically involved in processes that normally occur soon after adsorptive or receptor-mediated endocytosis (48, 76). Similarly, it may be that acidification of the cell wall is a prerequisite of exocytosis in plants.

It has been stated by Palade (64) that "Transport operations are dominated by

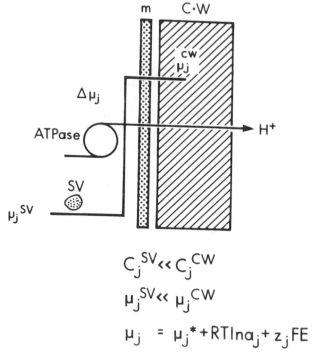

$$C_j^{SV} \ll C_j^{CW}$$

$$\mu_j^{SV} \ll \mu_j^{CW}$$

$$\mu_j = \mu_j^* + RT\ln a_j + z_j FE$$

Figure 5 An electrogenic H^+-ATPase involved in the transport of secretory vesicles to the cell wall. sv = secretory vesicle; m = cell membrane; cw = cell wall; μ_j = chemical potential of species j; μ_j^* = standard chemical potential; a_j = activity of species j which is equal to $\gamma_j 8 c_j$; where γ_j = activity coefficient and c_j = concentration; R = gas constant; T = temperature; z_j = charge number of species j; F = Faraday constant; E = electrical potential.

specific membrane interactions which lead to membrane fusion-fission and eventually to continuity established between the interacting compartments. Somewhere in this chain of reaction energy is required . . ." Auxin-induced H^+ secretion in plants may be the expression of the energy requirement to which Professor Palade has alluded.

CONCLUSIONS AND FUTURE DIRECTIONS

It is now evident that auxin induces specific mRNAs in pea and soybean tissue. The rapidly induced mRNAs are believed to mediate H^+ secretion and cell elongation, while late mRNAs, indirectly regulated by auxin, may code for proteins associated with the aspartyl-conjugating system for auxin (98), ACC synthetase (106), cellulases (99) or other hydrolases (105), or polypeptides necessary for cell division, differentiation, or adventitious meristem initiation (87).

On the horizon are numerous fronts to pursue that will increase our understanding of the mechanism of action of auxin in particular and hormones in general. Subcellular localization of the polypeptides coded by the early and late mRNAs is of paramount importance in elucidating the role of these mRNAs. Expression of full length DNAs with prokaryotic expression vectors will permit the isolation of large amounts of proteins coded by these mRNAs, facilitating their subcellular localization with homologous antibodies. In addition, precise determination of the half lives of the mRNAs (in the presence or absence of auxins) is necessary to determine whether the hormone alters their stability. Also, examination of latent periods of mRNA induction, H^+ secretion, and cell elongation in various systems will contribute to information on the kinetics of induction of the three phenomena.

At the structural level, characterization of the auxin genes and identification of DNA sequences specifically involved in the auxin response is a task of the future, as is the search for homologous genes in a variety of plants to determine if the genes are ubiquitous. Since auxin regulates the establishment of tumors in plants (58), the question arises whether the auxin-regulated genes are related to oncogenes (10). Finally, the early regulated genes and their associated regulatory regions will be valuable tools in developing an in vitro transcription system in plants to elucidate the biochemical machinery responsible for transcriptional activation by auxin.

ACKNOWLEDGMENTS

The support of the National Institutes of Health (GM 35447) and National Science Foundation (DCB 84-21157) is gratefully acknowledged. Many thanks to my friend Thanh Huynh.

Literature Cited

1. Aducci, P., Ballio, A., Federico, R., Montesano, L. 1982. Studies on fusicoccin-binding sites. In *Plant Growth Substances*, ed. P. F. Wareing, pp. 395–404. Proc. 11th Int. Conf. Plant Growth Substances. London: Academic
2. Ashburner, M., Chihara, C., Meltzer, P., Richards, G. 1973. Temporal control of puffing activity in polytene chromosomes. *Cold Spring Harbor Symp. Quant. Biol.* 38:655
3. Baker, H. J., Shapiro, D. J. 1978. Rapid accumulation of vitellogenin messenger RNA during secondary estrogen stimulation of *Xenopus laevis*. *J. Biol. Chem.* 253:4521–24
4. Bates, G. W., Cleland, R. E. 1979. Protein synthesis and auxin-induced growth: Inhibitor studies. *Planta* 145:437–42
5. Bates, G. W., Cleland, R. E. 1980. Protein patterns in the oat coleoptile as influenced by auxin and by protein turnover. *Planta* 148:429–36
6. Baulcombe, D. C., Giorgini, J., Key, J. L. 1980. The effect of auxin on the polyadenylated RNA of soybean hypocotyls. In *Genome Organization and Expression in Plants*, ed. C. J. Leaver, pp. 175–85. New York: Plenum
7. Baulcombe, D. C., Key, J. L. 1980. Polyadenylated RNA sequences which are reduced in concentration following auxin treatment of soybean hypocotyls. *J. Biol. Chem.* 255:8907–13
8. Baulcombe, D. C., Kroner, P. A., Key, J. L. 1981. Auxin and gene regulation. In *Levels of Genetic Control in Development*, ed. S. Subtelny, U. K. Abbott, pp. 83–97. 39th Symp. Soc. Dev. Biol. New York: Liss

9. Bevan, M., Northcote, D. H. 1981. Some rapid effects of synthetic auxins on mRNA levels in cultured plant cells. *Planta* 152:32–35
10. Bishop, J. M. 1985. Viral oncogens. *Cell* 42:23–38
11. Chanson, A., Taiz, L. 1985. Evidence for an ATP-dependent proton pump on the golgi of corn coleoptiles. *Plant Physiol.* 78:232–40
12. Cimbala, M. A., Lamers, W. H., Nelson, K., Monahan, J. E., Warren, H. Y., Hanson, R. W. 1982. Rapid changes in the concentration of phosphoenolpyruvate carboxykinase mRNA in rat liver and kidney. *J. Biol. Chem.* 257:7629–36
13. Cleland, R. 1971. Cell wall extension. *Ann. Rev. Plant Physiol.* 22:197–222
14. Cleland, R. 1973. Auxin-induced hydrogen ion excretion from *avena* coleoptiles. *Proc. Natl. Acad. Sci. USA* 70:3092–93
15. Cleland, R. E. 1976. Fusicoccin-induced growth and hydrogen ion excretion of *avena* coleoptiles: Relation to auxin responses. *Planta* 128:201–6
16. Cleland, R. E. 1976. Kinetics of hormone-induced H^+ excretion. *Plant Physiol.* 58:210–13
17. Cleland, R. E., Rayle, D. L. 1978. Auxin, H^+-excretion and cell elongation. In *The Controlling Factors in Plant Development*, ed. H. Shibaoka, M. Furuya, M. Katsumi, A. Takimoto, pp. 125–39. Tokyo: Botanical Mag.
18. Early, P., Rogers, J., Davis, M., Calame, K., Bond, M., et al. 1980. Two mRNAs can be produced from a single immunoglobulin μ gene by alternative RNA processing pathways. *Cell* 20:313–19
19. Elder, P. K., Schmidt, L. J., Ono, T., Getz, M. J. 1984. Specific stimulation of actin gene transcription by epidermal growth factor and cycloheximide. *Proc. Natl. Acad. Sci. USA* 81:7476–80
20. Enger, M. D., Rall, L. B., Walters, R. A., Hildebrand, C. E. 1980. Regulation of induced thionein gene expression in cultured mammalian cells: Effects of protein synthesis inhibition on translatable thionein mRNA levels in regulatory variants of the CHO cells. *Biochem. Biophys. Res. Commun.* 93:343–48
21. Ernest, M. J., Delap, L., Feigelson, P. 1978. Induction of hepatic tyrosine amino transferase mRNA by protein synthesis inhibitors. *J. Biol. Chem.* 253:2895–97
22. Evans, M., Ray, P. M. 1969. Timing of the auxin response in coleoptiles and its implications regarding auxin action. *J. Gen. Physiol.* 53:1–20
23. Forgac, M., Cantley, L., Wiedenmann,

B., Altstiel, L., Branton, D. 1983. Clathrin-coated vesicles contain an ATP-dependent proton pump. *Proc. Natl. Acad. Sci. USA* 80:1300–3
24. Gantt, J. S., Key, J. L. 1983. Auxin-induced changes in the level of translatable ribosomal protein messenger ribonucleic acids in soybean hypocotyl. *Biochemistry* 22:4131–39
25. Gantt, J. S., Key, J. L. 1985. Coordinate expression of ribosomal protein mRNAs following auxin treatment of soybean hypocotyls. *J. Biol. Chem.* 260:6175–81
26. Guilfoyle, T. J., Lin, C. Y., Chen, Y. M., Nagao, R. T., Key, J. L. 1975. Enhancement of soybean RNA polymerase I by auxin. *Proc. Natl. Acad. Sci. USA* 72:69–72
27. Hagen, G., Guilfoyle, T. J. 1985. Rapid induction of selective transcription by auxins. *Mol. Cell. Biol.* 5:1197–1203
28. Hagen, G., Kleinschmidt, A., Guilfoyle, T. J. 1984. Auxin-regulated gene expression in intact soybean hypocotyl and excised hypocotyl sections. *Planta* 162: 147–53
29. Hager, A., Merzel, H., Krauss, A. 1971. Versuche und Hypothese zur Primärwirkung des Auxins beim Streckungswachstrum. *Planta* 100:47–75
30. Higgins, T. J. V., Zwar, J. A., Jacobsen, J. V. 1976. Gibberellic acid enhances the level of translatable mRNA for α-amylase in barley aleurone layers. *Nature* 260:166–69
31. Hong, K., Duzgunes, N., Papahadjopoulos, D. 1981. Role of synexin in membrane fusion. *J. Biol. Chem.* 256: 3641–44
32. Huynh, T. V., Young, R. A., Davis, R. W. 1985. Constructing and screening cDNA libraries in λgt10 and λgt11. In *The DNA Cloning Techniques: A Practical Approach*, ed. D. M. Glover, pp. 49–78. London: IRL Press
33. Jacobs, M., Ray, P. M. 1976. Rapid auxin-induced decrease in free space pH and its relationship to auxin-induced growth in maize and pea. *Plant Physiol.* 58:203–9
34. Jacobs, M., Taiz, L. 1980. Vanadate inhibition of auxin-enhanced H^+ secretion and elongation in pea epicotyls and oat coleoptiles. *Proc. Natl. Acad. Sci. USA* 77:7242–46
35. Jacobsen, J. V. 1977. Regulation of ribonucleic acid metabolism by plant hormones. *Ann. Rev. Plant Physiol.* 28:537–64
36. Jacobsen, J. V., Beach, L. R. 1985. Control of transcription of α-amylase and rRNA genes in barley aleurone pro-

toplasts by gibberellin and abscisic acid. *Nature* 316:275–77

37. Johnson, A. D., Herskowitz, I. 1985. A repressor (MATa2 product) and its operator control expression of a set of cell type specific genes in yeast. *Cell* 42:237–47

38. Kafatos, F. C. 1972. The cocoonase zymogen cells of silk moths: A model of terminal cell differentiation for specific protein synthesis. *Curr. Top. Dev. Biol.* 7:125–92

39. Kasamo, K. 1979. Characterization of membrane-bound Mg^{++}-activated ATPase isolated from the lower epidermi of tobacco leaves. *Plant Cell Physiol.* 20:281–92

40. Kelly, K., Cochran, B. H., Stiles, C. D., Leder, P. 1983. Cell-specific regulation of the c-myc gene by lymphocyte mitogens and platelet-derived growth factor. *Cell* 35:603–10

41. Kende, H., Gardner, G. 1976. Hormone binding in plants. *Ann. Rev. Plant Physiol.* 27:267–90

42. Key, J. L. 1969. Hormones and nucleic acid metabolism. *Ann. Rev. Plant Physiol.* 20:449–74

43. Kutschera, U., Schopfer, P. 1985. Evidence against the acid-growth theory of auxin action. *Planta* 163:483–93

44. Liaw, C. W., Towle, H. C. 1984. Characterization of a thyroid hormone-responsive gene from rat. *J. Biol. Chem.* 259:7253–60

45. Lieberman, M. E., Maurer, R. A., Gorski, J. 1978. Estrogen control of prolactin synthesis *in vitro*. *Proc. Natl. Acad. Sci. USA* 75:5946–49

46. Makino, R., Hayashi, K., Sugimura, T. 1984. c-myc Transcript is induced in rat liver at a very early stage of regeneration or by cycloheximide treatment. *Nature* 310:697–98

47. Marrè, E. 1979. Fusicoccin: A tool in plant physiology. *Ann. Rev. Plant Physiol.* 30:273–88

48. Matlin, K. S., Reggio, H., Helenius, A., Simons, K. 1982. Pathway of vesicular stomatitis virus entry leading to infection. *J. Mol. Biol.* 156:609–31

49. Mayo, K. E., Palmiter, R. D. 1981. Glucocorticoid regulation of metallothionein-I mRNA synthesis in cultured mouse cells. *J. Biol. Chem.* 256:2621–24

50. McKnight, G. S., Palmiter, R. D. 1979. Transcriptional regulation of the ovalbumin and conalbumin genes by steroid hormones in chick oviduct. *J. Biol. Chem.* 254:9050–58

51. Metzenberg, R. L. 1979. Implications of some genetic control mechanisms in *neurospora*. *Microbiol. Rev.* 43:361–83

52. Meyer, Y., Aspart, L., Chartier, Y. 1984. Auxin-induced regulation of protein synthesis in tobacco mesophyll protoplasts cultivated *in vitro*. *Plant Physiol.* 75:1027–33

53. Meyer, Y., Aspart, L., Chartier, Y. 1984. Auxin-induced regulation of protein synthesis in tobacco mesophyll protoplasts cultivated *in vitro*. *Plant Physiol.* 75:1034–39

54. Meyer, Y., Chartier, Y. 1981. Hormonal control of mitotic development in tobacco protoplasts. *Plant Physiol.* 68:1273–78

55. Miller, J. H., Reznikoff, W. S., eds. 1980. *The Operon.* New York: Cold Spring Harbor Lab.

56. Nabeshima, Y., Kuriyama, Y. F., Muramatsu, M., Ogata, K. 1984. Alternative transcription and two modes of splicing result in two myosin light chains from one gene. *Nature* 308:333–38

57. Nelson, K., Cimbala, M. A., Hanson, R. W. 1980. Regulation of phosphoenol pyruvate carboxykinase (GTP) mRNA turnover in rat liver. *J. Biol. Chem.* 255:8509–15

58. Nester, E. W., Gordon, M. P., Amasino, R. M., Yanofsky, M. F. 1984. Crown gall: A molecular and physiological analysis. *Ann. Rev. Plant Physiol.* 35:387–413

59. Nitsch, J. P. 1950. Growth and morphogenesis of the strawberry as related to auxin. *Am. J. Bot.* 37:211–15

60. Noodén, L. D., Thimann, K. V. 1963. Evidence for a requirement for protein synthesis for auxin-induced cell enlargement. *Proc. Natl. Acad. Sci. USA* 50:194–200

61. Noodén, L. D., Thimann, K. V. 1966. Action of inhibitors of RNA and protein synthesis on cell enlargement. *Plant Physiol.* 41:157–64

62. Nuccitelli, R., Heiple, J. M. 1982. Summary of the evidence and discussion concerning the involvement of pH_1 in the control of cellular functions. In *Intracellular pH: Its Measurement, Regulation and Utilization in Cellular Functions*, ed. R. Nuccitelli, D. W. Deamer, pp. 567–86. New York: Liss

63. Oshima, Y. 1982. Regulatory circuits for gene expression: The metabolism of galactose and phosphate. In *The Molecular Biology of the Yeast Saccharomyces. Metabolism and Gene Expression*, ed. J. N. Strathern, E. W. Jones, J. R. Broach, pp. 159–80. New York: Cold Spring Harbor Lab.

64. Palade, G. 1977. Concluding remarks. In *The International Cell Biology 1976–1977*, ed. B. R. Brinkley, K. R. Porter,

pp. 337–40. New York: Rockefeller Univ. Press

65. Palmiter, R. D. 1975. Quantitation of parameters that determine the rate of ovalbumin synthesis. *Cell* 4:189–97

66. Patterson, B. D., Trewavas, A. J. 1967. Changes in the pattern of protein synthesis induced by 3-indolylacetic acid. *Plant Physiol.* 42:1081–86

67. Pestka, S. 1971. Inhibitors of ribosome functions. *Ann. Rev. Microbiol.* 25:487–562

68. Ray, P. M. 1967. Radioautographic study of cell wall deposition in growing plant cells. *J. Cell Biol.* 35:659–74

69. Ray, P. M. 1974. The biochemistry of the action of indoleacetic acid on plant growth. *Recent Adv. Phytochem.* 7:93–122

70. Rayle, D. L., Cleland, R. 1970. Enhancement of wall loosening and elongation by acid solutions. *Plant Physiol.* 46:250–53

71. Rayle, D. L., Cleland, R. E. 1977. Control of plant cell enlargement by hydrogen ions. *Curr. Top. Dev. Biol.* 11:187–214

72. Ringold, G. M. 1979. Glucocorticoid regulation of mouse mammary tumor virus gene expression. *Biochim. Biophys. Acta* 560:487–508

73. Ringold, G. M., Dieckmann, B., Vannice, J. L., Trahey, M., McCormick, F. 1984. Inhibition of protein synthesis stimulates the transcription of human β-interferon genes in Chinese hamster ovary cells. *Proc. Natl. Acad. Sci. USA* 81:3964–68

74. Rozek, C. E., Davidson, N. 1983. *Drosophila* has one myosin heavy-chain gene with three developmentally regulated transcripts. *Cell* 32:23–34

75. Rubinstein, B., Cleland, R. E. 1981. Responses of *avena* coleoptiles to suboptimal fusicoccin: Kinetics and comparisons with indoleacetic acid. *Plant Physiol.* 68:543–47

76. Russell, J. T. 1984. ΔpH, H^+ diffusion potentials, and Mg^{2+} ATPases in neurosecretory vesicles isolated from bovine neurohypophyses. *J. Biol. Chem.* 259:9496–9507

77. St. John, T. P., Davis, R. W. 1979. Isolation of galactose-inducible DNA sequences from *Saccharomyces cerevisiae* by differential plaque filter hybridization. *Cell* 16:443–52

78. Shapiro, D. 1982. Steroid hormone regulation of vitellogenin gene expression. *CRC Crit. Rev. Biochem.* 12:187–203

79. Spanswick, R. M. 1981. Electrogenic ion pumps. *Ann. Rev. Plant Physiol.* 32:267–89

80. Stout, R. G., Cleland, R. E. 1980. Partial characterization of fusicoccin binding to receptor sites on oat root membranes. *Plant Physiol.* 66:353–59

81. Swaneck, G. E., Kreuzaler, F., Tsai, M. J., O'Malley, B. W. 1979. Absence of an obligatory lag period in the induction of ovalbumin mRNA by estrogen. *Biochem. Biophys. Res. Commun.* 88:1412–18

82. Sze, H. 1985. H^+-translocating ATPases: Advances using membrane vesicles. *Ann. Rev. Plant Physiol.* 36:175–208

83. Talbott, L. D., Roberts, J. K. M., Ray, P. M. 1984. Effect of IAA- and fusicoccin-stimulated proton extrusion on internal pH of pea cells. *Plant Physiol.* 75: Abstr. 232

84. Theologis, A., Huynh, T. V., Davis, R. W. 1985. Rapid induction of specific mRNAs by auxin in pea epicotyl tissue. *J. Mol. Biol.* 183:53–68

85. Theologis, A., Ray, P. M. 1982. Changes in messenger RNAs under the influence of auxins. See Ref. 1, pp. 43–57

86. Theologis, A., Ray, P. M. 1982. Early auxin-regulated polyadenylylated mRNA sequences in pea stem tissue. *Proc. Natl. Acad. Sci. USA* 79:418–21

87. Thimann, K. V. 1969. The auxins. In *The Physiology of Plant Growth and Development,* ed. M. B. Wilkins, pp. 2–45. New York: McGraw-Hill

88. Tobin, E. M., Silverthorne, J. 1985. Light regulation of gene expression in higher plants. *Ann. Rev. Plant Physiol.* 36:569–93

89. Trewavas, A. J. 1968. Effect of IAA on RNA and protein synthesis. *Arch. Biochem. Biophys.* 123:324–35

90. Trewavas, A. J. 1976. Plant growth substances. In *The Molecular Aspects of Gene Expression in Plants,* ed. J. A. Bryant, pp. 249–326. London: Academic

91. Trewavas, A. J. 1981. How do plant growth substances work? *Plant Cell Environ.* 4:203–28

92. Vanderhoef, L. N., Dute, R. R. 1981. Auxin-regulated wall loosening and sustained growth in elongation. *Plant Physiol.* 67:146–49

93. Vanderhoef, L. N., Stahl, C. A. 1975. Separation of two responses to auxin by means of cytokinin inhibition. *Proc. Natl. Acad. Sci. USA* 72:1822–25

94. Vanderhoef, L. N., Stahl, C. A., Lu, T. Y. S. 1976. Two elongation responses to auxin respond differently to protein synthesis inhibition. *Plant Physiol.* 58:402–4

95. Vanderhoef, L. N., Stahl, C. A., Wil-

liams, C. A., Brinkmann, K. A., Green-field, J. C. 1976. Additional evidence for separable responses to auxin in soybean hypocotyl. *Plant Physiol.* 57:817–19

96. Vannice, J. L., Taylor, J. M., Ringold, G. M. 1984. Glucocorticoid-mediated induction of a_1-acid glycoprotein: Evidence for hormone-regulated RNA processing. *Proc. Natl. Acad. Sci. USA* 81:4241–45

97. Veluthambi, K., Poovaiah, B. W. 1984. Auxin-regulated polypeptides change at different stages of strawberry fruit development. *Plant Physiol.* 75:349–53

98. Venis, M. A. 1964. Induction of enzymatic activity by indolyl-3-acetic acid and its dependence on synthesis of ribonucleic acid. *Nature* 202:900–1

99. Verma, D. P. S., MacLachlan, G. A., Byrne, H., Ewings, D. 1975. Regulation and *in vitro* translation of messenger ribonucleic acid for cellulase from auxin-treated pea epicotyls. *J. Biol. Chem.* 250:1019–26

100. Vreugdenhil, D., Burgers, A., Libbenga, K. R. 1979. A particle-bound auxin receptor from tobacco pith callus. *Plant Sci. Lett.* 16:115–21

101. Walker, J. C., Kaye, A. M. 1981. mRNA for the rat uterine estrogen-induced protein. *J. Biol. Chem.* 256:23–26

102. Walker, J. C., Key, J. L. 1982. Isolation of cloned cDNAs to auxin-responsive poly(A)$^+$ RNAs of elongating soybean hypocotyl. *Proc. Natl. Acad. Sci. USA* 79:7185–89

103. Walker, J. C., Legocka, J., Edelman, L., Key, J. L. 1985. An analysis of growth regulator interactions and gene expres-sion during auxin-induced cell elongation using cloned complementary DNAs to auxin-responsive messenger RNAs. *Plant Physiol.* 77:847–50

104. Went, T. W. 1928. Wuchstoff and Wachstum. *Rec. Trav. Bot. Neerl.* 29:379–96

105. Wong, Y. S., MacLachlan, G. A. 1980. 1,3-β-D-glucanases from *pisum sativum* seedlings. *Plant Physiol.* 65:222–28

106. Yang, S. F., Hoffman, N. E. 1984. Ethylene biosynthesis and its regulation in higher plants. *Ann. Rev. Plant Physiol.* 35:155–89

107. Young, R. A., Hagenbuchle, O., Schibler, U. 1981. A single mouse α-amylase gene specifies two different tissue-specific mRNAs. *Cell* 23:451–58

108. Zurfluh, L. L., Guilfoyle, T. J. 1980. Auxin-induced changes in the patterns of protein synthesis in soybean hypocotyl. *Proc. Natl. Acad. Sci. USA* 77:357–61

109. Zurfluh, L. L., Guilfoyle, T. J. 1981. Auxin-induced nucleic acid and protein synthesis in the soybean hypocotyl. See Ref. 8, pp. 99–118

110. Zurfluh, L. L., Guilfoyle, T. J. 1982. Auxin-induced changes in the population of translatable messenger RNA in elongating maize coleoptile sections. *Planta* 156:525–27

111. Zurfluh, L. L., Guilfoyle, T. J. 1982. Auxin-induced changes in the population of translatable messenger RNA in elongating sections of soybean hypocotyl. *Plant Physiol.* 69:332–37

112. Zurfluh, L. L., Guilfoyle, T. J. 1982. Auxin- and ethylene-induced changes in the population of translatable messenger RNA in basal sections and intact soybean hypocotyl. *Plant Physiol.* 69:338–40

Ann. Rev. Plant Physiol. 1986. 37:439–66

STRUCTURAL ANALYSIS OF PLANT GENES[1]

G. Heidecker

National Cancer Institute, National Institutes of Health, Frederick, Maryland 21701

J. Messing

Waksman Institute, Rutgers, The State University, Piscataway, New Jersey 08854

CONTENTS

GENE EXPRESSION DURING PLANT DEVELOPMENT

Phenotypic Variation of Plant Development

The reproductive phase is one of the most important stages of development in cereal plants with respect to their usefulness to man. Consequently, there has been a long-standing effort to understand and control this stage at the physiolog-

---

[1]The US Government has the right to retain a nonexclusive royalty-free license in and to any copyright covering this paper.

ical and genetic level. Particularly, the genetic control of phenotypic variations in storage protein, starch, and anthocyanin synthesis has been studied biochemically and by clonal analysis of somatic mutations, and the latter analysis was instrumental in the study of cell differentiation and tissue development. In contrast to animal cells, plant cells are fixed in their location, so the position of each cell and the subsequent somatically induced phenotypic modifications reveal distinctive patterns that can be used to analyze control of gene action during development (72, 107).

Simple techniques as, for instance, iodine staining for starch or analysis of kernel morphology and transparency have been used to study variants in maize that are affected in carbohydrate and protein synthesis. This has led to the genetic characterization of loci that control the synthesis of storage proteins such as the zeins, enzymes involved in starch biosynthesis such as glucosyltransferase or sucrose synthetase, and enzymes responsible for anthocyanin synthesis. Many of these functions are induced during endosperm development, a specialized tissue differentiated for seed nutrition.

Seed Development

After the pollen has landed on the silk of an ear of corn whose ovules are ready for fertilization, a tube extends from the pollen grain, penetrates the silk, and migrates through the long silk to the ovule. There it deposits the two sperm cells of the pollen into the embryo sac. One cell fuses with the egg cell to give rise to the embryo, and the other combines with the diploid central cell in the embryo sac to form the triploid primary endosperm nucleus (83).

Therefore, different organs of the kernel develop from cells of different genetic makeup. The outer layer or pericarp derives from maternal diploid cells surrounding the embryo sac. The endosperm tissue under the pericarp, including the aleurone layer or the outer layer of cells of the endosperm tissue, is derived from the triploid cell of the second fertilization event and carries two sets of chromosomes of maternal and one set of chromosomes of paternal genetic information. Finally, the embryo carries an equal set of chromosomes from both parents; it is the only seed component that develops into the progeny plant, while the other two are seed-limited. It is important to understand these events in order to appreciate the genetic consequences and the agricultural importance of the endosperm tissue.

During kernel maturation, at the height of photosynthesis of the adult plant, the enormous biosynthetic activity is channeled into the endosperm in the form of sugars and amino acids before dormancy of the seed commences. During germination of the seed, carbohydrates and proteins are broken down into sugars and amino acids to support the development of the young plant. It is this energy transfer that has been selected for by man, and its genetic control will occupy our curiosity far into the future.

ZEIN PROTEINS AND GENES

The Prolamin Fraction of Maize Storage Proteins

About 10 days after pollination, DNA and protein synthesis dramatically increase and continue at increased levels during the further development of the endosperm for about 40 days. At the same time, about 10 days after pollination, storage protein and starch synthesis commence. Starch is deposited into amyloplasts which develop from proplastids (118), while the major portion of storage proteins starts to accumulate in suborganelle structures of the rough endoplasmic reticulum, called protein bodies (12).

With the advent of protein chemistry, fractionation techniques as well as amino acid composition analysis were used to describe and classify the most abundant proteins produced during endosperm development. Using this approach, Osborne (77) named the alcohol-soluble fraction prolamins, the fraction that is soluble in neutral pH and high salt globulins, in neutral pH and low salt albumins, and the one in high pH glutelins. Since the endosperm tissue makes up about 80% of mature kernel weight, the amino acid composition of these proteins determines the nutritional quality of maize.

The prolamins constitute the largest fraction and represent up to 60% of the total protein. The name indicates the proportions of both proline and amide nitrogen in their amino acid composition. This group of proteins is often designated by their species name; thus the prolamins of corn *(Zea mays)* are called zein, those of barley *(Hordeum vulgare)* are called hordein, and so on. In the maize kernel, they are further subdivided according to their solubility in alcohol, without and with reducing agent, and named zein-1 (z1) and zein-2 (z2). On the other hand, solubility has not always been a clear way to classify the zein proteins. The z2 fraction was originally not included because it required a reducing agent, and its larger component of Mr 27,000 is often not grouped with the z2 fraction because of its distinct properties. In fact, it is also soluble in high pH and, therefore, should be referred to as the alcohol-soluble glutelin (ASG) (33).

The Zein Protein Families

In addition to solubility, amino acid composition can be used to divide the storage proteins into different protein families. The use of the reducing agent indicates already that the z2 fraction is rich in sulfur (35). However, this fraction still contains proteins that are associated with the z1 fraction. The additional components that are only detected in the z2 fraction can be separated by SDS gel electrophoresis into Mr 10,000, 14,000, 15,000, and 27,000. The Mr 14,000 and 15,000 contain seven to eight times the number of cysteine residues that is found in the Mr 10,000 class, and the latter contains two to four times the methionine of the former. The ASG has a much higher proline content than the others, and like the Mr, 14,000 is also rich in cysteine (28).

The z1 fraction can be separated into Mr 19,000 and Mr 22,000 components, which have very similar amino acid compositions and are high in glutamine, leucine, proline, and alanine. All the zeins are low in tryptophan and lysine, two essential amino acids for the diet of monogastric animals (118). When subjected to isoelectric focusing (IEF), the z1 fraction is separated into many more components than by SDS gel electrophoresis, while the additional components in the z2 fraction do not show the same complexity (40, 114; J. Kirihara and J. Messing, unpublished). Because of the natural variation of these components between different inbred lines of maize, components have been correlated with different genetic loci which are inherited in a simple Mendelian fashion (102). More than 20 different zein genes have been identified genetically (103), indicating that the z1 fraction is encoded by a very complex multigene family, while the z2 multigene family is much simpler.

The Primary Structure of the z1 Protein Family

The charge heterogeneity within the z1 fraction has finally been resolved at the amino acid sequence level. Because of the similarity of the proteins within this family, direct amino acid sequence analysis of these proteins shows a very conserved amino terminus of the mature proteins (6). The central portion of these proteins appears to be less conserved, and direct protein sequencing analysis was impossible with the current protein separation techniques. However, the cloning of cDNAs from mRNAs encoding zein proteins (14, 32–34, 42, 46, 67, 69, 79, 112, 117) allowed the deduction of the amino acid sequence from the DNA sequence. The first complete amino acid sequence of a storage protein published (34) revealed an unusual repetitive structure subsequently observed in many storage proteins.

Sequence comparison of the z1 zein proteins can be used to divide the protein structure into at least four regions (74) (Figure 1). Region 1 contains the signal peptide that is processed during deposition into the protein bodies (12, 15), and regions 2 and 4 contain the rather conserved amino and carboxy termini, respectively. The central region 3 contains a tandem repetition of a block of 20 amino acids starting with a row of glutamines. Sequence variation within these repeats is largely the cause of the charge heterogeneity observed in the isoelectric focusing gels, while the number of repeats determine the length of the protein. Argos et al (2) have used the circular dichroism spectra of zeins to speculate on the folding of these tandem repeats. Although such spectra are not conclusive, the authors propose an interesting model of how these proteins may be organized in protein bodies. They suggest that the repeats fold back on one another in an antiparallel fashion where hydrogen bonding of the polar groups gives rise to a cylindrical molecule. While the nonpolar side chains are buried in

z1 Protein Structure

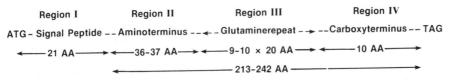

Figure 1 z1 protein structure. The z1 proteins can be divided into 4 regions (74): the signal peptide, the amino- and carboxytermini, and the glutaminerepeat. The length of each region is given in amino acid residues. For region III only one of the 9 to 10 repeats is schematically drawn. The length variation of the mature z1 zeins without the signal peptide varies between 213–242 amino acid residues.

the cylinder, the polar ones would be exposed to the surface of the molecule. The predicted rod-shaped molecule has been seen by electron microscopy (59).

It is interesting to note that the repetitive nature of the central portion of the protein leads to the amplification of certain amino acids like glutamine, leucine, proline, and alanine, and thus results in a highly hydrophobic protein with a high content of easily accessible amino groups which is the optimal form of biostorage for the dormancy period. This amplification also has important consequences for the development of the zein multigene family, as will be discussed below.

Zein Subfamilies

In addition to the deduction of the amino acid sequences of zein proteins, the cDNA clones have also been used in hybridization studies to determine their relationships. Based on the thermal stability of heteroduplexes between cDNA clones, the zein genes can be divided into different subfamilies (14, 67, 78, 112). While in the z2 fraction the molecular weight components each correspond to single subfamilies, in the z1 fraction related sequences fall into classes that not only differ significantly in their size but also in their complexity.

Using the isoelectric focusing pattern in translation experiments of hybrid-selected mRNA, cDNA clones of the z1 fraction can be divided into four subfamilies (14, 74, 78) (Table 1). All subsequent characterized z1 cDNA and genomic clones can be related to A20, A30, B49, or B59 as reference clones (74). We will call these subfamilies z1A, z1B, z1C, and z1D. Using Southern blotting (104), a relative estimate of gene copy number for each subfamily can be obtained. From these studies, it appears that the z1A subfamily is the largest gene family and has about 25 members; z1B and z1C subfamilies are somewhat smaller and have up to 20 and 15 members, respectively. The smallest subfamily is z1D with about 5 members (33, 69). These numbers should be taken as approximations because they are derived with heterologous probes, and they also do not distinguish between active and inactive genes. These estimates are twofold higher with 65 total members than the number of zein protein species

Table 1 Prolamin fraction of maize (alcohol soluble) zein multigene family[a]

Subfamily	z1 (non-reducing conditions)				z2 (reducing conditions)		
	z1A	z1B	z1C	z1D	z2A (ASG)	z2B	z2C
representing cDNA clone	A20	A30	B49	B59	B36	Z15A	—
Mr x1000	mostly 19	mostly 19 some 22	mostly 22 some 19	mostly 19	27	15	10
Locus	4L, 7S, 10L	4L, 7S	4L	?	?	?	?
Predominant amino acid	glutamine	glutamine	glutamine	glutamine	proline	cysteine	methionine
timing of expression	ca 12 dap	ca 12 dap	ca 18 dap	ca 12 dap	ca 18 dap	ca 18 dap	ca 18 dap
Transacting mutants (o6+++,f12+,Mc+)	o2+,o7+++ De*-B30+	o2+,o7++ De*-B30+	o2+++,o7+ De*-B30+++	o2+,o7++ De*-B30+	o2+,o7+ De*-B30+	o2+,o7++ De*-B30+	o2++,o7++ De*-B30+
No. of genes	<25	<20	<15	<5	2	2	?

+   reduced synthesis
++   increasingly reduced synthesis
+++ strongly reduced synthesis

[a]The prolamin fraction of maize is based on hybridization data (see text) of cDNA clones; this fraction contains proteins that fall into 7 subfamilies. Proteins of the z1 fraction are extracted only under reducing conditions. Although this does not permit the separation of proteins of the z1 fraction, the additional components have some distinct differences from the components of the z1 fraction and are listed here as the z2 protein family. Therefore, the subfamilies that belong to the z1 proteins are listed as z1A, z1B, z1C, and z1D, and the ones that belong to the z2 proteins as z2A, z2B, and z2C. The table lists a representative cDNA clone of each subfamily (14, 69, 74, 78), the molecular weight in 1000 daltons, the chromosomal location if known, the predominant amino acid residue, the time of expression in days after pollination (dap), the mutant loci and the degree (+, ++, +++) with which it reduces the synthesis of the proteins of the particular subfamily (the ones that have a general effect on the entire multigene family are listed in the first column), and the number of genes per haploid genome.

observed by 2D gel electrophoresis (41), suggesting that about half the genes are inactive. The z2 multigene family likely consists of only 6 members.

z1A, z1B, and z1D diverge from each other in sequence by 15–20%, and all show about 35% sequence divergence from z1C. The latter also encodes mostly Mr 22,000 proteins, while the z1B subfamily encodes mostly Mr 19,000 proteins and a few Mr 22,000, largely because of a duplication of a sequence of 32 amino acid residues in the central region of the protein (42, 46, 69). z1A and z1D represent mainly the small molecular weight Mr 19,000 zein component.

The structural genes of the z1 family have been located on three of the ten maize chromosomes by genetic linkage studies. The genes of the z1A subfamily are located on chromosomes 4, 7, and 10 (100, 101, 110); the ones of the z1B family on chromosomes 4 and 7 close to the z1A sequences (101, 110); and the ones of the z1C subfamily on chromosome 4 only (100). These results are also supported in part by the in situ hybridization data presented by Viotti et al (112). From genomic walking experiments, it appears that many of these genes are clustered as interdispersed repeated sequences (33).

Although the genomic organization of the genes coding for the z2 fraction are only partially characterized, they appear to occur in single chromosomal locations and only in numbers of one or two (33, 69). Therefore, only three subfamilies—z2A, z2B, and z2C—have been identified for the z2 fraction that have no significant homology to any of the four subfamilies of the z1 fraction (Table 1).

Control of Zein Synthesis

A developmental window is set for the synthesis of the zein proteins between 10 days and 50 days after pollination with a peak around day 35 that can differ for the various subfamilies (48, 68). The synthesis occurs exclusively in the inner endosperm tissue. During peak production, up to 85% of the total ribosomes in the developing endosperm are associated with the protein bodies that have been described above and thus are involved in zein synthesis (113). The majority of the mRNA associated with these ribosomes is polyadenylated and is 1100 to 1200 nucleotides long (13).

Several mutants affecting the overall deposition of zein are known (73, 76). All mutants confer an opaque phenotype to the kernel. Strains with floury-2 (f12), Mucronate (Mc), and opaque-6 (o6) genotypes have an overall reduction in their zein content, while the opaque-7 (o7) mutation preferentially reduces the amount of the Mr 19,000 and the opaque-2 (o2), and De*-B30 strains have little or no Mr 22,000 protein (102). The map positions for these loci do not coincide with any of the known structural zein genes, although o2, o7, and De*-B30 each are closely linked to a group of structural genes. These mutations often affect structural genes located elsewhere in the genome, e.g. o2 located on chromosome 7 reduces the expression of the z1C genes on chromosome 4. In each case the paucity of zein is a reflection of a reduction of zein mRNA (16, 58, 89). Double mutants show that o2 and o7 are epistatic to f12 but that the action of o2 is independent of o7 or Mc (26, 89). These results indicate that the expression of zein is regulated on several levels.

Strains with the o2 genotype deserve a special mention as they are the so-called High Lysine Corn strains (73). By lowering the contribution of zein to the total protein content, the balance is shifted in favor of proteins with a higher lysine content, thus resulting in a higher nutritional value of the protein (109). However, in addition to lowering the overall protein content, the mutation has a pleiotropic effect also causing a lower grain yield, a softer kernel, and as a consequence a higher susceptibility to various pests (111). The latter two defects result from a higher water content which is caused by a higher concentration, and thus a higher osmolarity, of free amino acids in the endosperm because of a reduced incorporation into zeins (75).

Structure and Regulation of Zein Genes in Plants

The molecular biology of the zein system is relatively well established in comparison to most other plant systems. For instance, 12 of the 57 entries of plant DNA sequences in the combined compilation of DNA sequences in the EMBO and Genbank sequence libraries are either genomic or mRNA zein sequences. Comparison of genomic and cDNA clones demonstrate one striking feature of zein genes: they belong to the rare class of eukaryotic genes that do not contain intervening sequences (46, 74, 79).

The regulation of zein transcription has three interesting aspects: 1. Transcription is tissue specific, occurring only in the endosperm. 2. The level of RNA is modulated during endosperm development, and it correlates with a tremendous increase of DNA in endosperm nuclei up to 200C (49a). 3. Whatever changes in the genome occur, they do not have to be reversible because this tissue is terminal in contrast to the embryo.

The specific increase of zein gene DNA in the nuclei of endosperm cells during midstage development would certainly explain the tremendous amplification of zein mRNA. On the other hand, no selective amplification of the genome has yet been demonstrated. So far it seems that even genes that are not expressed are also amplified.

The regulatory signals of zein genes have yet to be defined experimentally. Similar to other eukaryotic systems, a TATA box is found 30 bp upstream of the transcriptional initiation site. In some cases two of these signals are found 10 bp apart, and one case of an active second promoter preceding the regular site of initiation of transcription by 900 bp has been reported (57). The CAAT box present about 70 bp upstream of the RNA initiation site in most animal and some plant genes cannot be identified. Instead, a different consensus sequence ${}^C_T A_{2-5} {}^G_T NGA_{2-4} {}^{CC}_{TT}$ can be found around position-80 when zein genes and a number of other plant genes are compared (74). Whether this sequence has any regulatory function is not known. It might substitute for the CAAT box, or it might represent the plant version of an enhancer sequence, as it bears some resemblance to the animal core enhancer consensus sequence GTGG$^{AAA}_{TTT}$. Although the AGGA box is not found in all plant genes, it is present in a significant number of the current published sequences.

It will be interesting to see how transcription is initiated and how the mRNAs are processed. In the case of the B49 subfamily, 10 genes are arranged in tandem within 90 kilobase pairs (33); is there a position effect for any member in the cluster?

The clustering of zein genes also has some interesting implications for the evolution and conservation of their structure. Clustering is found at two levels; 20 amino acids are repeated tandemly up to nine times within each gene, and the genes themselves are closely arranged along the chromosome. The internal amplification is probably a consequence of an initial duplication which led to unequal crossing over. Evidence for the persistent occurrence of unequal cross-overs between nonequivalent repeats in the genes are the duplications found in some genes of one subfamily, e.g. ZN4 and ZG7 are one and one-half repeats longer than most of the other genes described for this subfamily.

The close linkage of groups of zein genes was shown by genetic mapping (100, 101, 110), in situ hybridization (112), and most recently, by molecular cloning (33, 106). Amplification and clustering of zein genes is not only a consequence of the plant's need to produce large amounts of zein products, but may also be needed to maintain a certain level of active zein genes. To explain this statement, it is necessary to list some peculiarities of the zein system.

First, the genes of different subfamilies are quite diverged, indicating that variation of the zein protein can be tolerated and that the genes are quite old in evolutionary terms. However, genes within one subfamily are homogenous, suggesting that selection on the protein sequence is stringent and/or that the genes originated recently or that efficient gene conversion occurs to maintain homogeneity. Even within subfamilies, cohorts of sequences can be identified: the z1B genes of the W22 inbred line fall into two classes. ZN4 and ZG7, even though slightly different from each other, share a number of significant changes that distinguish them from ZG124, 31 and 19. These sharp breaks in the level of divergence seem to indicate that amplification and diversification occurred in "quantum jumps" rather than in a continuous fashion.

Second, zein genes have a high proportion of codons (32%) that can mutate into a stop codon in one step. The high glutamine content of zein is mainly responsible for this, as glutamine is encoded by CAA and CAG. The C to U transition is the most frequently occurring mutation (61), and as a consequence, the frequency of stop codon generation in zein genes is twice as high (6.8%) for all mutations as that of average genes. As zeins store amide nitrogen in the form of glutamine, it seems that frequent gene inactivation, and consequently further amplification, is inevitable, leading to a cycle unless an efficient proofing mechanism is in place.

Introduction of a stop codon probably will not prevent synthesis of the mRNA. However, it will be translated into a shorter protein unless the stop codon is suppressed. It would seem that any suppression is very inefficient and that the mRNAs of inactivated genes are quickly degraded, as only one of the about 20 cDNA clones sequenced to date carry a nonsense mutation, while about half the genomic clones are inactivated by stop codons. This is the case in three out of the five complete genomic sequences: two have one such mutation, the other has four. However, none of the genes have experienced any further mutations which introduced codons uncharacteristic of zein genes, nor have they accumulated any deletions or insertions, as is commonly found to be the case for pseudogenes in other systems. In other words, the inactivated genes look as if they still belong to the same gene pool as the active genes subjected to the same selective pressures.

Unequal crossing over and gene conversion are mechanisms that can explain these peculiarities. While they would not eliminate inactivating stop codons completely, they would provide the means to maintain a relatively constant ratio between active and inactive genes, eliminating the need for permanent amplifications. Genes within each cluster would be constantly homogenized, and their number expanded and contracted according to the selective pressure on the gene family as a whole. This seems to be a more efficient way of counteracting the frequency with which inactivation is occurring than either reversion or rounds of amplification. Furthermore, it explains the homogeneity of genes within one subfamily or one subgroup thereof. These constitute the clusters that are homogenized by gene conversion. A mutation in any one gene in a cluster is eliminated or spreads through the cluster in a random drift fashion. This process accounts for the divergence between the different subfamilies that became separated by gross mutations such as translocation and thus are no longer involved in gene conversion or unequal crossing over with that sub-family. Gene conversion and unequal crossing over are also thought to maintain the ribosomal RNA and histone multigene families (43, 99) and the satellite DNA sequences (98).

It is instructive to speculate about the evolutionary history of the zein multigene family (Figure 2). After amplification of a sequence encoding a storage peptide, genes emerged which in turn were amplified into a multigene family in answer to the plant's need to produce large amounts of zein protein in a short time span. Once amplification had occurred the selective pressure on the individual genes was relaxed as some variation in the protein is tolerated. However, with inactivation of genes occurring frequently because of the high number of glutamine codons, the integrity of the gene family as a whole had to be safeguarded to avoid cycles of recurring gene inactivation and amplification. Clustering of the genes facilitates unequal crossing over and gene conversion, and thus may be a necessary feature of the zein system. Histone genes also need a constant proofing of one gene against the other, in this case to keep all genes homogenous, as aberrant proteins would interfere with the functioning of all histones. While the reason for the strong selection on facilitated gene conversion is different in the zein and the histones, one aspect of their structure is identical: both gene families have no intervening sequences. We postulate that the absence of intervening sequences is a reflection of the selective pressures on both systems to maintain the colinearity of nonequivalent genes to facilitate gene conversion and concerted evolution. Since the primary structure of intervening sequences is not subjected to the same selective pressure as the coding sequences, they could accumulate mutations such as insertions and deletions that would interfere with gene conversion.

Superimposed on amplification by unequal crossing over may be transposition of zein units, as it has been inferred from the structure of the inverted repeats that flank two genomic sequences (105). However, as this would lead to

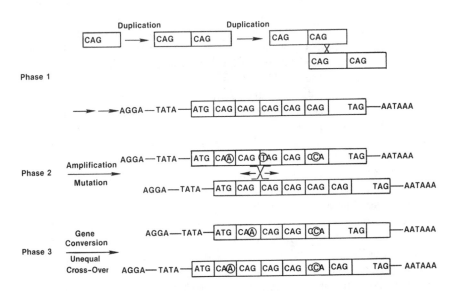

Figure 2   The origin of the z1 multigene family. The duplication of the internal glutaminerepeat has occurred in phase I by unequal crossing over as described in the text. In phase II the selective pressure has been relaxed and variations allowed to occur within single repeats without changing the glutamine content drastically. In the final phase, when inactivation of glutamine codons occurred, the clustering of genes provided the base for gene conversion as a safeguard against a cycle of inactivation and amplification.

the generation of nonclustered or "orphan" genes that would be quickly in-activated, transposition of single genes should be a rare event.

Storage Proteins in Other Plants

Recently, more information has been obtained from storage proteins in other cereals as well as legumes. In general, the major fraction of storage proteins in dicotyledons like the legumes are globulins, while those of monocotyledons like the graminae are prolamins. An exception is the storage protein of oats that is also a globulin (22). All storage proteins are high in glutamine, asparagine, and proline, consistent with their role as a nitrogen source for the germinating seedling. This bias in amino acid composition means that certain amino acids are underrepresented. As in the case of the zein they are lysine, tryptophan, and threonine in the cereal proteins and methionine, cysteine, and tryptophan for the legumes.

Two major size classes of 7 to 8S and 11 to 14S can be distinguished in the case of the globulins (22). The larger protein is composed of six subunits and each subunit consists of a basic and an acid polypeptide (25, 70, 93). These two polypeptides are generated by endoproteolytic cleavage from the same trans-lational product. The smaller protein also consists of several subunits which can be either primary translational products or processed proteins, as in the case of vicilin which is probably generated from a larger precursor (17, 25). The prolamins of other graminae besides corn are structured similarly to the zein proteins. In most cases, sulfur-rich and sulfur-poor proteins can be dis-tinguished (97), as with the z1 and z2 fractions of zein proteins.

All storage proteins investigated to date are encoded by multigene families (3, 11, 21, 29, 30, 45, 49, 62–66, 82, 85, 91, 97, 108). The structure of these families varies. In some cases, as in wheat or barley, two major subgroups can be distinguished: the α- and γ-gliadins and the B- and C-hordeins, respectively (29, 49, 82, 85). Within each subgroup, several subfamilies can be dis-tinguished. Often short repeats account for at least part of the structure of the polypeptides. These repeats constitute links through which different subfami-lies within the same species are related. Thus, the amino terminal part of B-hordeins is composed of repeats of eight amino acid which also make up most of the C-hordeins (29), and which are in turn related to repeats in A-gliadin of wheat (49). In fact, this short repeat is related to the longer one in the zein proteins. Although one must be careful not to overestimate the significance of the relatedness of these short sequences, as they might represent a case of convergent evolution, the fact that the chromosomal locations for some of the barley and wheat genes are equivalent indicates a common origin of these sequences. Similar findings have been reported for the 7S polypeptides of legumes and the 11S globulins of legumes as well as of oats and cucurbits (65, 91, 92).

In the context of the evolution of these multigene families, it is interesting to note that the overall content of proline in the storage proteins of wheat and barley is increased at the expense of glutamine (29, 49, 82). This change should result in a lower incidence of gene inactivation by mutation of a glutamine codon into a stop codon. Based on the discussion above, one would predict a lower ratio of inactive to active genes in wheat and barley than in corn. However, no pseudogenes have been described yet for these storage protein genes. So far only one example has been found outside of the cereals, namely in a member of the legumin gene family (11), where again a glutamine codon mutated to a termination codon. Otherwise, more drastic changes have occurred including deletions. It is interesting to note that these storage protein genes contain introns in contrast to the zein genes and thus may undergo a slower rate of gene conversion. In addition to the legumin genes the other legume storage protein genes have intervening sequences; examples are phaseolin and the 7S protein genes of soybeans (92, 97).

REGULATORY SEQUENCES ON PLANT mRNAs

Information Derived from Multigene Families

Multigene families are of special interest to geneticists. The patterns of divergence and conservation of the different members, as well as their chromosomal arrangement in relation to each other, provide information about the function of multigene families, and, to a certain degree, of genes in general. Features essential to the gene members and their products are revealed by the fact that they are conserved throughout the whole family.

While similarities highlight the presence and location of sequences and structures important to the functioning of the genes, small differences in these conserved sequences can elucidate the mechanism by which they operate. Although the normally limited number of changes in a natural population of sequences might not provide enough data to do so conclusively, they could serve as a basis for the design of mutagenesis experiments. For this study, we concentrated on changes in and around sequences that have the potential of being regulatory signals and are contained on the mRNA.

The Initiator Codon AUG in Plant mRNAs

Here we address in detail the changes and similarities within and surrounding recognition sequences that can also be used to further define these signals. The zein genes show a peculiarity not previously reported in other systems. We noted that 11 nucleotides surrounding the translational initiation site are duplicated in the z1B subfamily when compared to the z1A. This led us to the realization that this sequence was repeated at least three times in the 5' nontranslated region of all zein messages (see Figure 3). The repeats may

```
TCAACCATATTATTGAGACCAACAAGCAACATAGAAAGTGGAATCCAGTAGCAACAATAGAGGAACAATGG
          -----------                                -----------
             -------------                                  -----------


TCGCACATATTATTGAGACCAACTAGCAACATAGAAAGCACAATATTGT**********ACCAACAATGG
          -----------
             -------------                                  -----------
```

Nucleotide	WINDOW										
	1	2	3	4	5	6	7	8	9	10	11
A	7	–	–	7	7	–	6	5	2	1	4
C	–	3	6	–	–	7	–	–	–	2	–
G	–	4	1	–	–	–	–	–	2	4	3
U	–	–	–	–	–	–	1	2	3	–	–
Consensus	A	G/C	C	A	A	C	A	A	N	N	A/G

Figure 3　The 5' nontranslated region of z1 mRNAs. When a cDNA clone of the z1A family is compared to one of the z1B family, their sequences can be aligned as if an 11 nucleotide long deletion has occurred in the z1B cDNA clone 7 nucleotides prior the ATG initiator codon. These missing 11 nucleotides define a repeat within the 5' end of the z1 mRNA. Actually, the first 4 nucleotides of the coding region are part of the last repeat. In the preceding repeats the TG of the ATG of course are variant and prevent the presence of multiple ATGs which would lead to multiple AUG initiator codons. The repeats in the cDNA clones are underlined, the deletion is indicated by *, and the table below derives the consensus sequence of the repeat.

explain the high efficiency with which zein messages are translated: the repeats might facilitate the recruitment of ribosomes in general or of an appropriate subfraction of ribosomes, i.e. membane-bound subpopulation. Furthermore, comparison of the sequences immediately surrounding the initiation codon may also provide information about the regulation of translation.

Before attempting to identify any signal within a sequence by comparing several sequences that might fulfill the same function, it is important to realize the limitations of a purely comparative approach without verifying the conclusions experimentally. For instance, in the case of the initiator signal, several aspects govern what a specific initiator signal will look like. One aspect is that it needs to be recognized by ribosomes. This is the constant part of the signal; but, on the other hand, different mRNAs need to be translated with different efficiencies, which means that the signal sequences have to be different. A further variation might be introduced by adapting the signal to a certain subclass of ribosomes, or, to be more accurate, adapting the signal to certain aspects of the translational machinery, a term which includes different initiation factors, different location inside the cell, as well as various components of the ribosome itself. Obviously, this means that the consensus signal obtained by compilation does not necessarily represent the optimal signal, if such a signal exists at all. In this case, we define the optimal signal as that which works best in an in vitro translational system. This definition is, of course, arbitrary, but useful insofar as it is one which can be most easily tested; and, indeed, some information is already available from Kozak's data. She compiled the translational initiator sequences of higher eukaryotic messenger RNAs (51, 53) and found that the initiation of translation occurs at the first AUG triplet in the mRNAs 97% of the time. In addition, she studied the binding efficiencies of various short oligonucleotides to wheat germ ribosomes. The results of this study, in conjunction with various other experiments, led her to propose that the 40S ribosomal initiation complex attaches at the 5' cap and scans the mRNA until it finds an AUG triplet (50, 52). The efficiency of initiation is then dependent on the actual sequence that surrounds the initiator codon (51).

The initiator codon for the translation of zein is, in all cases, the first AUG triplet on the mRNA and 20 to 21 codons upstream from the start of the coding sequence for the mature zein, which agrees with the finding that the precursor is about Mr 2500 larger (15). Table 2 lists the 23 nucleotide sequences surrounding the initiator codons for the zein precursor proteins of all genomic and cDNA clones sequenced to date, and Figure 4 graphically summarizes the frequency distribution for each nucleotide at each position. The consensus of this sequence is:

5' Pu C A A Py A N Py N U A N C A A C A AUG G C N

Obviously, one would expect a higher degree of homology between members of a multigene family than between unrelated genes. The differences are minute within subfamilies; however, at positions −6, −9, and −11 each subfamily shows different nucleotides almost exclusively, and position +6 is completely fixed for a different base in each subfamily. This finding indicates that the signal does not include these positions or that members of the different subfamilies have distinct signals. It is impossible to decide between these two possibilities without further experiments as no information is as yet available for the efficiency of initiation of translation of different zein mRNAs. Conversely, the fact that divergence in this region does occur argues that the other positions are conserved by selection and thus play a role as components of the initiation recognition sequence. It is of interest that positions as far as −16 are fixed for one nucleotide, although again it is unclear whether this is the result of

Table 2 Zein gene initiator sequences[a]

Gene	Type of clone	Sequence		Ref.
ZN7	genomic	G C A A C T T C C T A A C A A C A	ATG G C T	55
ZA1	genomic	G C A A C G A C C T A A C A A C A	ATG G C T	105
Z22.1	cDNA	G C A A C G A C C T A A C A A C A	ATG G C T	67
Z22.3	cDNA	C A A C C T A A C A A C A	ATG G C T	67
Z22A1	CDNA	G C A A C G A C C T A A C A C C A	ATG G C T	69
pML1	genomic	G C A A C G A C C T A A C A A C A	ATG G C T	57
pcM1	cDNA	G C A A C G A C T T A A C A A C A	ATG G C T	105
ZN4	genomic	A C A A T A T T G T A C C A A T A	ATG G C A	46
Z19B1	cDNA	A C A A T A G T G T A C C A A C A	ATG G C A	69
ZG99	genomic	A C A A T A T T G T A C C A A T A	ATG G C A	79
Z19.1	cDNA	A C A A T A T T G T A C C A A T A	ATG G C A	79
ZG7	cDNA	A C A A T A G T G T A C C A A C A	ATG G C A	42
ZG124	cDNA	A C A A T A G T G T A C C A A C A	ATG G C A	42
ZE19	genomic	A C A A T A G T G T A C C A A C A	ATG G C A	106
ZE25	genomic	G C A A T A G T G T A G C A A C A	ATG G C A	106
Z19D1	cDNA	G C A A C A T C T T A G C A C C A	ATG G C A	69
A20	cDNA	G C A A C A A C A G A G C A A C A	ATG G C G	32
ZG14	cDNA	G C A A C A A T A G A G C A A C A	ATG G C G	ms. in prep.
ZG15	cDNA	G C A A C A A T A G A G C A A C A	ATG G C G	ms. in prep.
Z19C2	cDNA	G C A A C A A T A G A G C A A C A	ATG G C G	69
Z18C1	cDNA	G C A A C A A C A G A G C A A C A	ATG G G G	69

[a]The 5' nontranslated regions of z1 mRNAs have been grouped into three categories that are defined by the second codon. The second codon in all cases codes for alanine, but three different codons are used: GCT, GCA, and GCG. Besides the second codon, 17 nucleotides preceding the ATG are listed when available from the sequence data. Each clone has been listed with its original designation, whether it was genomic or a cDNA clone, and with its reference. A consensus sequence is presented in Table 4, and a graphic representation of the frequency distribution of each nucleotide around the ATG is given in Figure 4.

sequence conservation within the gene family or requirements of the recognition sequence. To distinguish between these two possibilities, it might be helpful to compare the zein initiation signal to those of other genes of plants and all eukaryotes in general. Table 3 lists the 23 nucleotides surrounding the initiator AUG of most plant or plant viral genes that were available at this time and for which the initiator codon could be identified. In most cases, the assignment was based on protein sequence data; in other cases where no protein data was available, the first AUG was chosen if it defined a long open reading frame, and if it could be assumed that the DNA sequence represented a complete transcription unit.

With one exception, mRNAs initiate translation at the first AUG triplet from the 5' end. The exception is the mRNA for *Phaseolus* lectin which actually contains four AUGs in all three frames, the last one being the initiator. The

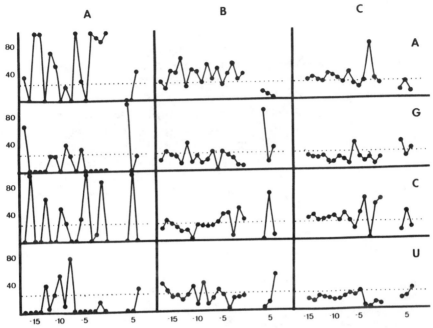

Figure 4 Graphic representation of the frequency distribution of nucleotides around the ATG initiator codon of z1 mRNAs (panel A), plant mRNAs (panel B), and animal mRNAs (panel C). The nucleotide immediately preceding the initiator codon is number −1, the nucleotides +4 to +6 represent the second codon of the coding sequence. The dotted line across each panel indicates the 25% value that would be expected on a random basis. The panels under heading A represent the z1 zein genes and their values are derived from Tables 2 and 4, those labeled B are other plant genes and their values are derived from Tables 3 and 4, and the panels under C represent the animal genes and their values are derived from the compilation prepared by Kozak (53).

preceding AUGs are in an inefficient context probably resulting in very poor recognition and thus initiation. The same was found to be the case in animal mRNAs initiating at a second or even later AUG (54, 84). This finding does not contradict Kozak's model of the scanning ribosome which recognizes the initiator AUG codon mostly on the basis of it being the first such triplet the ribosome encounters. However, while her compilation contained many more sequences, only a few were plant genes; thus it would be more correct to term a consensus derived from her compilation the animal initiator recognition sequence. The compilation of plant sequences supports her scanning ribosome model for initiation in eukaryotes in general, but it shows that there might be some subtle differences in the sequences surrounding the actual initiation codon in plants and animals. Figure 4 summarizes the data from Tables 2 and 3 and also presents Kozak's data (53) for comparison purposes. Based on these data, the consensus for the sequence surrounding the initiator AUG in plant genes is:

5'N N N N A N A/U N U/A A N N N N A N N AUG G C U

Overall, as was to be expected, more variation can be observed for plant genes in general than for zeins alone. At some positions, both consensus sequences coincide: positions −15 and −3 show a preference for A, while +4 is almost fixed for G and +5 has a high incidence of C. In addition, plant genes show a high frequency of A at position −8 and −13; overall they are rich in A and T residues in their 5' ends. This would mean that any secondary structure in the nontranslated region would be relatively unstable and could be melted easily by the scanning ribosomal 40S subunit. Exceptions are the α-amylase genes in barley which have a 5' end that is high in GC and have a GC-rich palindrome preceding the initiator codon. It is unclear whether this structure regulates the translation of the message or stabilizes the whole RNA molecule to enable it to survive through dormancy.

Table 3 Plant gene initiator sequence[a]

Gene	Sequence		Ref.
Barley amylase A	C G A C A G T A G C G C G C G C C	ATG G G G	88
Barley amylase B	A G A G A G A G C T G A A G A A C	ATG G C G	86
Barley amylase high pI	C G A C A G T A G C G C G C G C G	ATG G G G	86
Barley thiol protein	T C C G C C G G C G A A A C G A A	ATG G C C	87
Barley hordein 2-1	G T T A A C A C C A A T C C A C T	ATG A A G	85
Brassica napus 1.7	C A T A C A C G A	ATG G C G	21
Maize actin	T T A C A G A A T A G T T G A G A	ATG G C T	96
Maize ADH 1	A T T T G C G G A C G G G G G C A	ATG G C G	24
Maize ADH 2	G A G A G A G A G A G C A A G C A	ATG G C G	24
Pea thylakoid protein	G C C T C C T C T A G T A G C C C	ATG G C A	19
Pea legumin A	T C T T A G T A T C T C T C T T C	ATG G C T	63
Pea legumin B	T C T T A G T A T C T C T C T T C	ATG G C T	63
Pea legumin C	T C T T A G T A T C T C T C T T C	ATG G C T	63
Pea legumin D	A T A T A T T C T A T C C A A C T	ATG G C T	11
Pea preprovicilin	C A A T C A A A C C G T T A	ATG T T G	65
Pea RUBISCO SS	C T A A G A A A G T C A G A A A A	ATG G C T	20
Phaseolin	T T C A A T A C T A C T C T A C T	ATG A T G	97
Phaseolus lectin	A T G A A T G C A T G A T C	ATG G C T	44
Soybean actin	C T G T A G G T T T G T A A A A G	ATG G C A	95
leghemoglobin LBA	A A A A T T A A A A A G A A A T	ATG G T T	47
LBC	A A A A G T A G A A A A G A A A T	ATG G G T	47
LBG	A A A G G T A G A A A A G A A A T	ATG G G T	10
lectin	T T A G C T G A A G C A A A G C A	ATG G C T	115
RUBISCO SS	T G A G A A C T A A G A A G A A A	ATG G C T	5
Thaumatin	G G C A T C A T C A T A C A T C A	ATG G C C	27
Tobacco atp2-1	T A G C C A A A C C C T C C A C C	ATG G C T	8
Tobacco RUBISCO SS	T C T A A G T G T A A T T A A C A	ATG G C T	71
Wheat gliadin W8233	G T C A A T A C A A A T C C A C C	ATG A A G	82
Wheat cab protein	A C C A T A A G T G C A G C G C A	ATG G C G	56

Table 3   *(continued)*

Gene	Sequence		Ref.
Alfalfa RNA1	A C T G T G A A G A T T T C A C A	ATG A A T	18
35K	T C G T G A G T A A G T T G C A A	ATG G A G	81
Coat	T T T C A A A T A C T T C C A T C	ATG A G T	9
CaMV ORF1	T A A T C T T C T G T G T T G A G	ATG G A T	31
ORF2	G A T C T G T G G A G A A T A A A	ATG G A G	31
ORF3	G A A T A T T A T T G G C T G A A	ATG G C T	31
ORF4	T T C A A A G C T C C A G C A G G	ATG G C C	31
ORF5	C T C T A C A G A A G A A A G C G	ATG A T G	31
ORF6	A A A T C A G A C C T C C A A G C	ATG G A G	31
Brome RNA3	T A C T G T T T T T G T T C C C G	ATG T C T	1
Coat	G T A T T T A A T A	ATG T C G	23
CCMC Coat	G T A A T T T A T C	ATG T C T	23
CuMV RNA3	T T T A G A T T A C G A A G G T T	ATG G C T	38
Coat	T G C G T C T C A G T G T G C C T	ATG G A C	38
TMV 126 K	T A C T A T T T A C A A T T A C A	ATG G C A	36
30 K	T A G A A G T T T G T T T A T A G	ATG G C T	36
Coat	T C G G A T T C G T T T T A A A T	ATG T C T	36
TuYMV Coat	T A G C A A T C A G C C C C A A C	ATG G A A	39

[a]The 5' nontranslated regions of plant mRNAs have been divided into two sections, one for nuclear genes and one for viral genes. Besides the second codon, 17 nucleotides preceding the ATG are listed when available from the sequence data. Each clone has been listed with its reference and the name of the gene. A consensus sequence is presented in Table 4, and a graphic representation of the frequency distribution of each nucleotide around the ATG is given in Figure 4.

In general, the high content in A and T residues also means that any interaction with the rRNA of the ribosome would be weak. In the case of the animal consensus which is:

5' C C A C C AUG G C

such an interaction has been suggested to occur with the 3'GGUGG sequence found at the base of the 3'-terminal hairpin of 18S rRNA. Plant sequences in general clearly do not show a high frequency of C at positions −2 and −5.

As can be seen in Figure 4, the animal consensus does not extend beyond this region. Whether the length of the plant consensus sequence is increased to compensate for the lack of G or C residues, and thus to stabilize some hybrid structure between mRNA and rRNA, or is merely a reflection of the smaller number of sequences in the study that might lead to statistical errors, cannot be decided at this point. It is interesting that zein genes, which are probably translated very efficiently, show both the longer consensus sequence and a higher than average content of C just preceding the initiator AUG. The notion that the difference in the sequences surrounding the initiator AUG codon in plant and animal genes is more than a matter of statistical variation and is a

Table 4 Initiator codon of translation of plant mRNAs (zein genes)[a]

Nucleotide	Window																				
	−17	−16	−15	−14	−13	−12	−11	−10	−9	−8	−7	−6	−5	−4	−3	−2	−1	AUG	+4	+5	+6
A	7	20	20			15	11		5		21	6		21	19		21	AUG			9
C	20			13	13	5	5	10	6	5		7	21		2	18		AUG		21	5
G	13					5	5		8	5	7	7						G	21		
U				8	8	1	5	11	2	16						3		AUG			7
Consensus	Pu	C	A	A	Py	A	N	Py	N	U	A	N	C	A	A	C	A	AUG	G	C	N

Initiator codon of translation of plant mRNAs (nuclear genes)

Nucleotide	Window																				
	−17	−16	−15	−14	−13	−12	−11	−10	−9	−8	−7	−6	−5	−4	−3	−2	−1	AUG	+4	+5	+6
A	7	5	11	11	15	6	12	12	8	15	8	12	7	11	15	8	12	AUG	3	2	1
C	4	7	6	5	4	5	1	6	6	7	6	8	8	10	2	14	8	AUG		18	2
G	5	5	3	5	5	10	5	7	4	4	9	1	8	7	7	2	2	G	23	4	9
U	10	9	6	7	4	7	9	3	11	3	5	8	6	1	5	5	6	AUG	1	3	14
Consensus	N	N	N	A	N	N	N	N	N	A	N	N	N	A	N	N	N	AUG	G	C	T

Initiator codon of translation of plant mRNAs (viral genes)

Nucleotide	Window																				
	−17	−16	−15	−14	−13	−12	−11	−10	−9	−8	−7	−6	−5	−4	−3	−2	−1 AUG	+4	+5	+6
A	2	8	3	3	8	6	3	3	7	6	2	5	3	5	10	7	6	3	7	7
C	1	3	5	3	2	2		5	1	5	2	2	4	5	3	5	4		9	2
G	2	1	4	3	3	3	3	4	3	4	6	3	1	3	4	2	5	11	1	5
U	11	4	4	7	3	5	10	6	7	3	8	8	10	5	1	4	3	4	1	9
Consensus	T	N	N	N	N	N	T	N	N	N	N	N	T	N	A	N	N	G	C	N

[a] Consensus sequences of the region around the plant ATG initiator codon. A window of 17 nucleotides preceding the ATG initiator codon and the nucleotides of the second codon are presented. The frequency of As, Cs, Gs, and Us in each position of the corresponding mRNAs are derived from Tables 2 and 3. The frequency distribution of each nucleotide has been listed for the z1 mRNAs, the nuclear mRNAs, and the viral mRNAs in separate tables. From this frequency distribution a consensus sequence for each category has been derived. A position has been specified only if one particular nucleotide occurs more than 50%.

consequence of the still small numbers of plant gene sequences is supported by in vitro binding studies with synthetic oligonucleotides (51). According to these studies, wheat germ ribosomes showed the highest binding to sequences in which the AUG was followed by a G residue; binding was also enhanced by an A at position −3 and a C at position +5, though to a somewhat lesser extent. These results are in agreement with the plant consensus sequence, while the animal sequence suggests a higher importance for a purine, particularly an A residue, at position −3 and a good deal of freedom as to the nucleotide that follows the initiation codon, though a slight preference for guanine can be observed.

The fixation of a C at position +5, in conjunction with the prevalence of G at +4, means that the majority of the proteins encoded by the genes listed here start with the met-ala dipeptide. Animal genes also show this preference for ala as the second residue, though it is less pronounced (20%). Whether the high frequency of alanine at that position is a result of requirements at the nucleic acid or the protein level is unclear at this point, although the former seems more likely as there is no linkage between the incidence of G in position +4 and C in position +5, as would be expected if the selection was on an alanine residue as the second residue in the polypeptide. The frequency for the GC dinucleotide in both plants and animal genes is the product of the frequencies of the single nucleotides. Possibly the efficiency and stability with which the second tRNA can bind is of importance for the association of the 60S subunit, and thus also influences initiation and level of translation. The binding constant of the second tRNA would be dependent on the composition of the codon and on the concentration of the tRNA in the cell. It is interesting to note in this context that

Table 5 Poly A addition signal of plant mRNAs[a]

Nucleotide	Window						
	−3	−2	−1	U	1	2	3
A	21	47	47	—	47	47	28
C	7	—	1	—	—	—	3
G	8	—	—	—	1	—	10
U	12	1	—	—	—	1	7
Consensus	A/U 69%	A 98%	A 98%	U —	A 98%	A 98%	PU 79%

[a]From 47 plant mRNAs of which the 3' end was known by mapping or cloning, the 7 nucleotides preceding the 3' end about 15-35 nucleotides and resembling the sequence AAUAAA have been aligned. The frequency distribution of each nucleotide around the central U has been listed and a window +3 and −3 has been chosen. From this distribution a consensus sequence has been derived. Below the consensus sequence the frequency of which a nucleotide was found in that position has been expressed as a percentage.

the second codon of the zein genes, while always encoding alanine, varies at the third position for each subfamily. The availability of the different alanine-tRNAs during endosperm development could influence the expression of the various zein subfamilies.

The Polyadenylation Signal Sequence

The polyadenylation signals precede the end of the mature message by 15 to 23 nucleotides (74). Most plant genes have more than one of the consensus sequences for the polyadenylation signal (74), as do some of the animal genes (94). The role of the multiple sites and what, if anything, controls the choice of

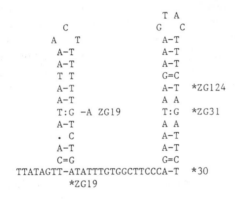

(A)

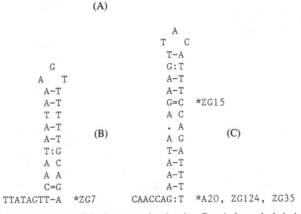

Figure 5   Secondary structures of the 3' nontranslated region. Part A shows the hairpin structures that can be formed by all z1B-like sequences except ZG7. Part B represents the hairpin in ZG7. Part C gives the distal hairpin in the z1A-like sequences. The sites for polyadenylation are indicated by *. Nucleotide substitutions are given next to the affected position. The first hairpin in structure (A) has a ΔG of +0.4 for A30 and −1.1 for ZG19; the second hairpin has a ΔG of −4.8. The ΔGs of structures (B) and (C) are +3.8 and −1.1, respectively (90).

which one is used is not yet clear. While the consensus sequence for this signal is a very conserved AAUAAA in animal systems, plant mRNA deviate from this theme frequently, as shown in Table 5. Zein genes alone show three different deviations and, in some cases, the variant sequence is preferred even when it is preceded by the canonical sequence. In most cases, the second signal is recognized, but for instance in the two cDNA clones, ZG7 and ZG19 of the A30 subfamily of the z1 zein fraction, polyadenylation does not bypass any signal sequence (42).

When these cDNA clones are compared to other cDNA clones of the same subfamily that use the second signal, sequence differences affect a potential hairpin structure around the polyadenylation signal. As shown in Figure 5, ZG7 and ZG19 stabilize the first hairpin compared to the cDNA clones ZG31, ZG124, and A30. A similar hairpin structure involvement has been proposed for the late transcripts of adenovirus (60). Site-directed mutagenesis and gene transfer experiments should help to substantiate these predictions. Although other plant transcripts yield a similar result (37, 62, 82), not all cases can be explained this way. If hairpin structures are involved in defining the 3' end of mRNAs, then alternate or modified pathways may exist, as suggested for animal mRNAs (4).

ACKNOWLEDGMENTS

We thank our colleagues for sending us yet to be published manuscripts, and especially Mike Freeling, John Gatehouse, Grantley Lycett, John Rogers, Diter von Wettstein, and others for their reprints. The computer-assisted work was supported by grants from the Department of Energy, DOE #FG02-84ER13210, and the National Institutes of Health, GM31499. Part of this work represents the thesis of G.H.

Literature Cited

1. Ahlquist, P., Luckow, V., Kaesberg, P. 1981. Complete nucleotide sequence brome mosaic virus RNA3. *J. Mol. Biol.* 153:23–28

2. Argos, P., Pedersen, K., Marks, M. D., Larkins, B. A. 1982. A structural model for maize zein proteins. *J. Biol. Chem.* 257:9984–90

3. Bartels, D., Thompson, R. D. 1983. The characterization of cDNA clones coding for wheat storage proteins. *Nucleic Acids Res.* 11:2961–77

4. Berget, S. M. 1984. Are U4 small nuclear ribonucleoproteins involved in polyadenylation? *Nature* 309:179–82

5. Berry-Lowe, S. L., McNight, T. D., Shah, D. M., Meagher, R. B. 1982. The nucleotide sequence, expression and evolution of one member of a multigene family encoding the small subunit of ribulose-1,5-biphosphate carboxylase in soybean. *J. Mol. Appl. Genet.* 1:483–98

6. Bietz, J. A., Paulis, J. W., Walls, J. S. 1979. Zein subunit homology revealed through amino terminal sequence analysis. *Cereal Chem.* 56:327–32

7. Deleted in proof

8. Boutry, M., Chua, N. H. 1985. A nuclear gene encoding the beta subunit of the mitochondrial ATP synthase in *Nicotiana plumbaginifolia. EMBO J.* In press

9. Brederode, F. T., Koper-Zwarthoff, E. C., Bol, J. F. 1980. Complete sequence of alfalfa mosaic virus RNA4. *Nucleic Acids Res.* 8:2213–23

10. Brisson, N., Verma, D. P. S. 1982. Soy-

bean leghemoglobin gene family: Normal, pseudo and truncated genes. *Proc. Natl. Acad. Sci. USA* 79:4055–59

11. Brown, D., Levasseur, M., Croy, R. R. D., Boulter, D., Gatehouse, J. A. 1985. Sequence of a pseudogene in the legumin gene family of pea (*Pisum sativum* L.) Submitted for publication

12. Burr, B., Burr, F. A. 1976. Zein synthesis in maize endosperm by polyribosomes attached to protein bodies. *Proc. Natl. Acad. Sci. USA* 73:515–19

13. Burr, B., Burr, F. A., Rubenstein, I., Simon, M. N. 1978. Purification and translation of zein messenger RNA from maize endosperm protein bodies. *Proc. Natl. Acad. Sci. USA* 75:696–700

14. Burr, B., Burr, F. A., St. John, T. P., Thomas, M., Davis, R. D. 1982. Zein storage protein gene family of maize. *J. Mol. Biol.* 154:33–49

15. Burr, F. A., Burr, B. 1981. In vitro uptake and processing of prezein and other maize preproteins by maize membranes. *J. Cell. Biol.* 90:427–34

16. Burr, F. A., Burr, B. 1982. Three mutations affecting zein accumulation. *J. Cell. Biol.* 94:201–6

17. Chrispeels, M. J., Higgins, T. J. V., Spencer, D. 1982. Assembly of storage protein oligomers in the endoplasmic reticulum and processing of the polypeptides in the protein bodies of developing pea cotyledons. *J. Cell. Biol.* 93:306–13

18. Cornelissen, J. B. C., Bol, J. F., Jaspers, E. M. J. 1983. Complete nucleotide sequence of alfalfa mosaic virus RNA1. *Nucleic Acids Res.* 11:1253–65

19. Coruzzi, G., Broglie, R., Cashmore, A., Chua, N. H. 1983. Nucleotide sequence of ribulose-1,5-diphosphate carboxylase and the major chlorophyll *a*/*b* binding thylakoid polypeptide. *J. Biol. Chem.* 258:1399–1402

20. Coruzzi, G., Broglie, R., Edwards, C., Chua, N. H. 1984. Tissue-specific and light-regulated expression of a pea nuclear gene encoding the small subunit of ribulose-1,5-bisphosphate carboxylase. *EMBO J.* 3:1671–79

21. Crouch, M., Tenberge, K., Simon, N. E., Ferl, R. 1983. Sequence of the 1.7 k storage protein of *Brassica napus*. *Mol. Appl. Genet.* 2:273–83

22. Danielsson, C. E. 1949. Seed globulins of the graminae and leguminosae. *Biochem. J.* 44:387–400

23. Dasgupta, R., Ahlquist, P., Kaesberg, P. 1982. Complete nucleotide sequences of the coat protein messenger RNAs of brome mosaic virus and cowpea chlorotic mottle virus. *Nucleic Acids Res.* 10: 703–13

24. Dennis, E. S., Sachs, M. M., Gerlach, W. L., Finnegan, E. J., Peacock, W. J. 1985. Molecular analysis of the alcohol dehydrogenase 2 (ADH2) gene of maize. *Nucleic Acids Res.* 13:727–43

25. Derbyshire, E., Wright, D. J., Boulter, D. 1976. Legumin and vicilin storage proteins of legume seeds. *Phytochemistry* 15:3–24

26. DiFonzo, N., Fornasari, E., Salamini, F., Reggiani, R., Soave, C. 1979. Interaction of the mutants floury-2, opaque-2 with opaque-2 in the synthesis of endosperm proteins. *J. Hered.* 71:397–402

27. Edens, L., Hesling, L., Klok, R., Lederboer, A. M., Maat, J., et al. 1982. Cloning of cDNA encoding the sweet tasting plant protein thaumatin and its expression in *E. coli*. *Gene* 18:1–12

28. Esen, A., Bietz, J. A., Paulis, J. W., Wall, J. S. 1981. Tandem repeats in the N-terminal sequence of a proline-rich protein from corn endosperm. *Nature* 296:678–79

29. Forde, B. G., Kreis, M., Williamson, M. S., Fry, R. P., Pywell, J., et al. 1985. Short tandem repeats shared by B- and C-hordein cDNAs suggest a common evolutionary origin for two groups of cereal storage protein. *EMBO J.* 4:9–15

30. Forde, J., Forde, B. G., Fry, R. P., Kreis, M., Shewry, P. R., Miflin, B. J. 1983. Identification of barley and wheat cDNA clones related to the high Mr polypeptides of wheat gluten. *FEBS Lett.* 162:360–66

31. Gardner, R., Howarth, A., Hahn, P., Brown-Luedi, M., Shepherd, R., Messing, J. 1981. The complete nucleotide sequence of an infectious clone of Cauliflower Mosaic Virus by M13mp7 shotgun sequencing. *Nucleic Acids Res.* 9:2871–88

32. Geraghty, D., Messing, J., Rubenstein, I. 1982. Sequence analysis and comparison of cDNAs of the zein multigene family. *EMBO J.* 1:1329–35

33. Geraghty, D., Messing, J., Rubenstein, I. In preparation

34. Geraghty, D., Peifer, M. A., Rubenstein, I., Messing, J. 1981. The primary structure of a plant storage protein: zein. *Nucleic Acids Res.* 9:5163–74

35. Gianazza, E., Viglienghi, V., Rhigetti, P. C., Salamini, F., Soave, C. 1977. Amino acid composition of zein molecular components. *Phytochemistry* 16:315–17

36. Goelet, P., Lomonosoff, G. P., Butler, P. J. G., Akam, M. E., Gait, M. J., Karn, J. 1982. Nucleotide sequence of

tobacco mosaic virus RNA. *Proc. Natl. Acad. Sci. USA* 79:5818–22

37. Goldberg, R. B., Hoschek, G., Vodkin, L. O. 1983. An insertion sequence blocks the expression of a soybean lectin gene. *Cell* 33:465–75

38. Gould, A. R., Symons, R. H. 1982. Cucumber mosaic virus RNA3: determination of the nucleotide sequence provides the amino acid sequences for protein 3A and viral coat protein. *Eur. J. Biochem.* 126:217–26

39. Guilley, H., Briand, J. P. 1978. Nucleotide sequence of turnip yellow mosaic virus coat protein mRNA. *Cell* 15:113–22

40. Hagen, G., Rubenstein, I. 1980. Two-dimensional gel analysis of the zein proteins in maize. *Plant Sci. Lett.* 19:217–23

41. Hagen, G., Rubenstein, I. 1981. Complex organization of zein genes in maize. *Gene* 13:239–49

42. Heidecker, G., Messing, J. 1983. Sequence analysis of zein cDNAs obtained by an efficient mRNA cloning method. *Nucleic Acids Res.* 11:4891–4906

43. Hentschel, C. C., Birnstiel, M. L. 1981. The organization and expression of histone gene families. *Cell* 25:301–13

44. Hoffman, L. M., Ma, Y., Barker, R. F. 1982. Molecular cloning of *Phaseolus vulgaris* lectin mRNA and use of cDNA as a probe to estimate lectin transcript levels in various tissues. *Nucleic Acids Res.* 10:7819–28

45. Hopp, H. E., Rasmussen, S. K., Brandt, A. 1983. Organization and transcription of B1 hordein genes in high lysine mutants of barley. *Carlsberg Res. Commun.* 48:201–16

46. Hu, N-T., Peifer, M. A., Heidecker, G., Messing, J., Rubenstein, I. 1982. Primary structure of a zein genomic clone. *EMBO J.* 1:1337–42

47. Hyldig-Nielsen, J. J., Jensen, E. O., Paludan, K., Wiborg, O., Garrett, R., et al. 1982. The primary sequence of two leghemoglobin genes from soybean. *Nucleic Acids Res.* 10:689–701

48. Ingle, J., Beitz, D., Hageman, R. H. 1965. Changes in composition during development and maturation of maize seeds. *Plant Physiol.* 40:835–39

49. Kasarda, D. D., Okita, T. W., Bernardin, J. E., Baecker, P. A., Nimmo, C. C., et al. 1984. DNA and amino acid sequences of alpha and gamma gliadins. *Proc. Natl. Acad. Sci. USA* 81:4712–16

49a. Kowles, R. V., Phillips, R. I. 1985. DNA amplification patterns in maize endosperm nuclei during kernel development. *Proc. Natl. Acad. Sci. USA* 82:7010–14

50. Kozak, M. 1980. Role of ATP in binding and migration of 40S ribosomal subunits. *Cell* 22:459–67

51. Kozak, M. 1981. Possible role of flanking nucleotides in recognition of the AUG initiator codon by eukaryotic ribosomes. *Nucleic Acids Res.* 9:5233–52

52. Kozak, M. 1983. Comparison of initiation of protein synthesis in procaryotes, eucaryotes and organelles. *Microbiol. Rev.* 47:1–45

53. Kozak, M. 1984. Compilation and analysis of sequences upstream from the translational start site in eukaryotic mRNAs. *Nucleic Acids Res.* 12:857–72

54. Kozak, M. 1984. Selection of initiation sites by eukaryotic ribosomes: Effect of inserting AUG triplets upstream from the coding sequence for preproinsulin. *Nucleic Acids Res.* 12:3873–93

55. Kridl, J. C., Vieira, J., Rubenstein, I., Messing, J. 1984. Nucleotide sequence analysis of a zein genomic clone with a short open reading frame. *Gene* 28:113–18

56. Lamppa, G. K., Morelli, G., Chua, N. H. 1985. Structure and developmental regulation of a wheat gene encoding the major chlorophyll *a/b*-binding polypeptide. *Mol. Cell Biol.* 5:1370–78

57. Langridge, P., Feix, G. 1983. A zein gene of maize is transcribed from two widely separated promoter regions. *Cell* 34:1015–22

58. Langridge, P., Pintor-Toro, J. A., Feix, G. 1982. Transcriptional effect of the opaque-2 mutation of *Zea mays* L. *Planta* 156:166–70

59. Larkins, B. A., Pedersen, K., Marks, M. D., Wilson, D. R. 1984. The zein proteins of maize endosperm. *Trends Biol. Sci.* 9:306–8

60. LeMoullec, J. M., Akusjaervi, G., Stalhandske, P., Pettersson, U., Chambraud, B., et al. 1983. Polyadenylic acid addition sites in adenovirus type-2 major late transcription unit. *J. Virol.* 48:127–34

61. Li, W-H. 1983. Evolution of duplicated genes and pseudogenes. In *Evolution of Genes and Proteins*, ed. M. Nei, R. K. Koehn, pp. 14–37. Sunderland, MA: Sinauer

62. Lycett, G. W., Croy, R. R. D., Shirsat, A. H., Boulter, D. 1984. The complete nucleotide sequence of a legumin gene from pea (*Pisum sativum* L.). *Nucleic Acids Res.* 12:4493–4505

63. Lycett, G. W., Croy, R. R. D., Shirsat, A. H., Richards, D. M., Boulter, D. 1985. The 5'-flanking regions of three

pea legumin genes: Comparison of the DNA sequences. *Nucleic Acids Res.* 13:6733–43

64. Lycett, G. W., Delauney, A. J., Croy, R. R. D. 1983. Are plant genes different? *FEBS Lett.* 153:43–46

65. Lycett, G. W., Delauney, A. J., Gatehouse, J. A., Gilroy, J., Croy, R. R. D., Boulter, D. 1983. The vicilin gene family of pea: A complete cDNA coding sequence for preprovicilin. *Nucleic Acids Res.* 11:2367–80

66. Lycett, G. W., Delauney, A. J., Zhao, W., Gatehouse, J. A., Croy, R. R. D., Boulter, D. 1984. Two cDNA clones coding for the legumin protein of *Pisum sativum* L. contain sequence repeats. *Plant Mol. Biol.* 3:91–96

67. Marks, M. D., Larkins, B. A. 1982. Analysis of sequence microheterogeneity among zein messenger RNAs. *J. Biol. Chem.* 257:9976–83

68. Marks, M. D., Lindell, J. S., Larkins, B. A. 1985. Quantitative analysis of the accumulation of zein mRNA during maize endosperm development. *J. Biol. Chem.* In press

69. Marks, M. D., Lindell, J. S., Larkins, B. A. 1985. Nucleotide sequence analysis of zein mRNAs from maize endosperm. *J. Biol. Chem.* In press

70. Matlashewski, G. J., Adeli, K., Altosaar, I., Shewry, P. R., Miflin, B. J. 1982. In vitro synthesis of oat globulin. *FEBS Lett.* 145:208–12

71. Mazur, B. J., Chui, C. F. 1985. Sequence of a genomic DNA clone for the small subunit of ribulose bis-phosphate carboxylase-oxygenase from tobacco. *Nucleic Acids Res.* 13:2373–86

72. McClintock, B. 1965. The control of gene action in maize. *Brookhaven Symp. Biol.* 18:162–84

73. Mertz, E. T., Bates, L. S., Nelson, O. E. 1964. Mutant gene that changes protein composition and increases lysine content of maize endosperm. *Science* 145:279–80

74. Messing, J., Geraghty, D., Heidecker, G., Hu, N-T., Kridl, J., Rubenstein, I. 1983. Plant gene structure. In *Genetic Engineering of Plants,* ed. T. Kosuge, C. P. Meredith, A. Hollaender, pp. 211–27. New York: Plenum

75. Misra, P. S., Mertz, E. T., Glover, D. V. 1975. Studies on corn proteins VIII: Free amino acid content of opaque-2 double mutants. *Cereal Chem.* 52:844–48

76. Nelson, O. E., Mertz, E. T., Bates, L. S. 1965. Second mutant affecting the amino acid pattern of maize endosperm proteins. *Science* 150:1469–70

77. Osborne, T. B. 1924. *The Vegetable Pro-*

teins. London: Longmans, Green. 154 pp. 2nd ed.

78. Park, W. D., Lewis, E., Rubenstein, I. 1980. Heterogeneity of zein mRNA and protein in maize. *Plant Physiol.* 65:98–106

79. Pedersen, K., Devereux, J., Wilson, D. R., Sheldon, E., Larkins, B. A. 1982. Cloning and sequence analysis reveal structural variation among related zein genes in maize. *Cell* 29:1015–26

80. Deleted in proof

81. Pinck, M., Fritsch, C., Ravelonandro, M., Thivent, C., Pinck, L. 1981. Binding of ribosomes to the 5′ leader sequence (n=258) of RNA3 of alfalfa mosaic virus. *Nucleic Acids Res.* 9:1087–1100

82. Rafalski, J. A., Scheets, K., Metzler, M., Peterson, D. M., Hedgcoth, C., Soll, D. G. 1984. Developmentally regulated plant genes: The nucleotide sequence of a wheat gliadin genomic clone. *EMBO J.* 3:1409–15

83. Randolph, L. F. 1936. Developmental morphology of the caryopsis in maize. *J. Agric. Res.* 53:881–916

84. Rasheed, S., Norman, G., Heidecker, G. 1983. Nucleotide sequence of the Rasheed rat sarcoma virus oncogene: New mutations. *Science* 221:155–57

85. Rasmussen, S. K., Hopp, H. E., Brandt, A. 1983. Nucleotide sequences of cDNA clones for B1 hordein polypeptides. *Carlsberg Res. Commun.* 48:187–99

86. Rogers, J. C. 1985. Two barley alpha-amylase gene families are regulated differently in aleurone cells. *J. Biol. Chem.* 260:3731–38

87. Rogers, J. C., Dean, D., Heck, G. R. 1985. Aleurain: A barley thiol protease closely related to mammalian cathepsin. *Proc. Natl. Acad. Sci. USA.* In press

88. Rogers, J. C., Milliman, C. 1983. Isolation and sequence analysis of a barley alpha-amylase cDNA clone. *J. Biol. Chem.* 258:8169–74

89. Salamini, F., DiFonzo, N., Fornasari, E., Gentinetta, E., Reggiani, R., Soave, C. 1983. Mucronate, Mc, a dominant gene of maize which interacts with opaque-2 to suppress zein synthesis. *Theor. Appl. Genet.* 65:123–28

90. Salser, W. 1978. Globin mRNA sequences: Analysis of base pairing and evolutionary implications. *Cold Spring Harbor Symp. Quant. Biol.* 42:985–1002

91. Schuler, M. A., Ladin, B. F., Pollaco, J. C., Freyer, G., Beachy, R. N. 1982. Structural sequences are conserved in the genes coding for the alpha, alpha′ and beta-subunits of the soybean 7S seed stor-

age protein. *Nucleic Acids Res.* 10:8245–61

92. Schuler, M. A., Schmitt, E. S., Beachy, R. N. 1982. Closely related genes code for the alpha and alpha' subunits of the 7S storage protein complex. *Nucleic Acids Res.* 10:8225–44

93. Sengupta, C., Deluca, V., Bailey, D., Verma, D. P. S. 1981. Posttranslational processing of 7S and 11S components of soybean storage proteins. *Plant Mol. Biol.* 257:14753–59

94. Setzer, D. R., McGrogan, M., Nunberg, J. H., Schimke, R. T. 1980. Size heterogeneity in the 3' end of dihydrofolate reductase messenger RNAs in mouse cells. *Cell* 22:361–70

95. Shah, D. M., Hightover, R. C., Meagher, R. B. 1982. Complete nucleotide sequence of a soybean actin gene. *Proc. Natl. Acad. Sci. USA* 79:1022–26

96. Shah, D. M., Hightover, R. C., Meagher, R. B. 1983. Genes encoding actin in higher plants: Intron positions are highly conserved but the coding sequences are not. *J. Mol. Appl. Genet.* 2:111–16

97. Slightom, J. L., Sun, S. M., Hall, T. C. 1983. Complete nucleotide sequence of french bean storage protein gene: phaseolin. *Proc. Natl. Acad. Sci. USA* 80:1897–1901

98. Smith, G. P. 1973. Evolution of repeated DNA sequences by unequal crossing over. *Science* 191:528–35

99. Smith, G. P. 1973. Unequal crossover and the evolution of multigene families. *Cold Spring Harbor Symp. Quant. Biol.* 38:507–13

100. Soave, C., Reggiani, R., DiFonzo, N., Salamini, F. 1982. Genes for zein subunits on maize chromosome 4. *Biochem. Genet.* 11:1027–37

101. Soave, C., Rhigetti, R., DiFonzo, N., Salamini, F. 1981. Clustering of genes for 20 k subunits in the short arm of maize chromosome 7. *Genetics* 97:363–77

102. Soave, C., Salamini, F. 1983. Genetic organization and regulation of maize storage proteins. In *Seed Proteins*, ed. J. Deussant, J. Mosse, J. Vaughan, pp. 205–18. New York: Academic

103. Soave, C., Salamini, F. 1984. The role of structural and regulatory genes in the development of maize endosperm. *Dev. Genet.* 5:1–25

104. Southern, E. 1975. Detection of specific sequences among DNA fragments separated by gel electrophoresis. *J. Mol. Biol.* 98:503–17

105. Spena, A., Viotti, A., Pirrotta, V. 1982. A homologous repetitive block structure underlies the heterogeneity in heavy and light chain zein genes. *EMBO J.* 1:1589–94

106. Spena, A., Viotti, A., Pirrotta, V. 1983. Two adjacent genomic zein sequences: structure, organization and tissue specific restriction pattern. *J. Mol. Biol.* 168:799–812

107. Steffensen, D. M. 1968. A reconstruction of cell development in shoot apex of maize. *Am. J. Bot.* 55:354–69

108. Thompson, R. D., Bartels, D., Harberd, N. P., Flavell, R. B. 1983. Characterization of the multigene family coding for HMW glutenin subunits in wheat using cDNA clones. *Theor. Appl. Genet.* 67:87–96

109. Tsai, C. Y., Huber, D. M., Warren, H. L. 1980. A proposed role of zein and glutelin as N sinks in maize. *Plant Physiol.* 66:330–33

110. Valentini, G., Soave, C., Ottaviano, E. 1979. Chromosomal location of zein genes in *Zea mays* L. *Heredity* 42:33–40

111. Vasal, S. K., Villegas, E., Bauer, R. 1979. Present status of breeding quality protein maize. Seed protein improvement in cereals and legumes. *IAEA Proc. Ser.* 2:127–48

112. Viotti, A., Abildstein, D., Pogna, N., Sala, E., Pirrotta, V. 1982. Multiplicity and diversity of cloned zein cDNA sequences and their chromosomal localization. *EMBO J.* 1:53–58

113. Viotti, A., Sala, E., Alberti, P., Soave, C. 1975. RNA metabolism and polyribosome profiles during seed development in normal and opaque-2 maize endosperms. *Maydica* 20:111–23

114. Vitale, A., Inaniotto, E., Longhi, T., Galante, E. 1982. Reduced soluble proteins associated with endosperm protein bodies. *J. Exp. Bot.* 33:349–448

115. Vodkin, L. O., Rhodes, P. R., Goldberg, R. B. 1983. A lectin gene insertion has the structural features of a transposable element. *Cell* 34:1023–31

116. Weiher, H., Konig, M., Gruss, P. 1983. Multiple point mutations affecting the simian virus-40 enhancer. *Science* 219:626–31

117. Wienand, U., Brusch, C., Feix, G. 1979. Cloning of double-stranded DNAs derived from polysomal mRNA of maize endosperm: Isolation and characterization of zein clones. *Nucleic Acids Res.* 6:2707–15

118. Wilson, C. H. 1983. Seed protein fractions of maize, sorghum and related cereals. In *Seed Proteins: Biochemistry, Genetics, Nutritive Value*, ed. W. Gottschalk, H. P. Mueller, pp. 276–99. The Hague/Boston/London: Martinus Nijhoff/Dr. W. Junk

Ann. Rev. Plant Physiol. 1986. 37:467–507

ANALYSIS OF PHOTOSYNTHESIS WITH MUTANTS OF HIGHER PLANTS AND ALGAE

C. R. Somerville

MSU-DOE Plant Research Laboratory and Department of Botany and Plant Pathology, Michigan State University, East Lansing, Michigan 48824

CONTENTS

467

0066-4294/86/0601-0467$02.00

INTRODUCTION[1]

The use of mutants as experimental tools in the analysis of photosynthetic phenomena is a well-established approach, and large collections of mutants are available in species such as *Zea mays* (41, 119), *Hordeum vulgare* (192), and *Chlamydomonas reinhardii* (102), which have been favored experimental organisms. Thus, for example, there are at least 117 separately designated loci in maize and at least 96 loci in barley that directly or indirectly affect photosynthetic pigmentation. Although many of these and related mutations have not been investigated in detail, a few have been utilized extensively for biochemical, biophysical, or structural studies. For instance, the *chlorina-f2* mutant of barley has probably figured in more than 50 research articles. Thus, in attempting to synthesize a cogent overview of the massive literature that has accumulated since the last review of the topic in this series (102), it was necessary to establish relatively narrow guidelines as to what might be most useful. In this respect, my first priority has been to describe the results obtained with several relatively new approaches to mutant isolation and characterization which have greatly expanded the potential utility of mutant analysis. Second, because it does not appear possible to formulate simple rules that explain how to exploit mutants, I present a wide range of examples of the successful and creative use of mutants with altered photosynthetic properties. Finally, I include references to observations that may merit further study and suggest possible opportunities for new initiatives.

Because of space limitations, I have largely neglected the extensive literature that deals with non-Mendelian mutations. This topic has been the subject of a recent review (30). Several other recent reviews describe in greater detail than possible here certain aspects of mutants with alterations in CO_2 fixation and photorespiration (159) or the photosynthetic membranes (57a). Also, several

---

[1]*Abbreviations:*
chl, chlorophyll
CF_1, coupling factor
GOGAT, glutamate synthase
kd, kilodalton
LDS, lithium lauryl sulfate
LHC-II, light-harvesting complex of PSII
PRK, phosphoribulokinase
PSI, photosystem I
PSII, photosystem II
RuBP, D-ribulose-1,5-bisphosphate
Rubisco, ribulosebisphosphate carboxylase/oxygenase
rbcL, gene encoding the large subunit polypeptide of Rubisco
rbcS, gene encoding small subunit polypeptide of Rubisco
SDS, sodium lauryl sulfate

chapters in a recent compendium provide useful descriptions of rationales and methods for isolating and characterizing mutants of plants and algae with alterations in photosynthesis (69). Finally, the comprehensive volume by Kirk & Tilney-Bassett (94a) provides an excellent overview of most aspects of chloroplasts.

MUTATIONS AFFECTING CO_2 FIXATION AND PHOTORESPIRATION

Mutants of Higher Plants

Very few mutations have been characterized in higher plants which specifically affect photosynthetic CO_2 fixation. The major reason is probably that mutants incapable of photoautotrophic growth are difficult to maintain and propagate in the homozygous state. However, several nonphotosynthetic mutants of higher plants have been maintained in tissue culture using carbohydrate rather than CO_2 as the carbon source (80, 88). In species such as maize, which have large endosperm reserves, nonphotosynthetic mutants can be maintained as heterozygous lines, and physiological or biochemical experiments can be performed on the homozygous mutants from self-fertilized plants. This approach has been used to characterize a collection of mutants of Zea mays which have normal photosynthetic electron transport activity but are unable to photoassimilate CO_2 (119). By default, many of these mutants are expected to have deficiencies in the enzymes required for CO_2 fixation. However, only one mutant that is deficient in NADP-ferredoxin oxidoreductase activity has been shown to have a specific defect (120).

Another method of propagating mutants of higher plants with defects in photosynthesis is, in principle, to identify leaky mutants. Such mutants might be expected to be somewhat chlorotic because of reduced photosynthetic capacity. Although many of the pigment-deficient mutants of barley apparently have reduced rates of CO_2 fixation when expressed on a leaf area or freshweight basis, the difference is not apparent when chlorophyll content is the standard for comparison (34). Thus, it seems that the pigment-deficient phenotype is not a good criterion to identify mutants that specifically affect CO_2 fixation.

There are, in principle, relatively few ways in which photosynthetic CO_2 fixation could be impaired specifically by a mutation. Most of these would involve loss of activity for an enzyme associated with the Calvin cycle or possibly in the subsequent metabolism of triose phosphate to sucrose. Although a collection of mutants with defects in these reactions would be valuable for testing our understanding of photosynthetic carbon metabolism, many of the most interesting mutants may not be those that lead to loss of Calvin cycle activity but rather those that directly or indirectly modulate the activity of the cycle. For example, the concept that leaf starch exerts a regulatory influence on the rate of photosynthesis has been a persistent theme, but it has been difficult to study because no specific inhibitors of starch biosynthesis are known. Mutants of *Arabidopsis* lacking leaf starch were isolated by simply screening a population of mutagenized plants for the presence of starch by staining small leaf samples with iodine (36). One of the mutants was characterized as lacking the chloroplast isozyme of phosphoglucomutase. This relatively innocuous mutation has a pronounced effect on photosynthetic capacity and on growth and development in response to variation in daylength. For instance, the mutant has a very poor growth rate in short days but has the same growth rate as the wild type when grown in continuous illumination. This effect has been associated with excess respiratory loss of carbon brought about by the increased accumulation of soluble sugars in the mutant (36). The mutant, which lacks starch in all tissues, promises to be useful for examining the role of starch in guard cell function, gravitropism, and photorespiration. This mutant illustrates the point that although direct screening methods usually involve some initial tedium, the approach produces biochemically defined mutations and permits the isolation of a very wide range of mutants. All that is required to isolate a specific mutation is a facile and sensitive assay.

An aspect of photosynthetic carbon metabolism that does not appear to have received any attention with respect to mutant analysis is C_4 photosynthesis. Mutants could be very useful in establishing the genetic complexity of the morphological and biochemical adaptations associated with the syndrome, and could be useful tools in studying the cell-cell interaction which is difficult to approach by conventional physiological analysis.

Algal Mutants

The most suitable experimental organisms for isolating mutants with defects in CO_2 fixation are the algae such as *Chlamydomonas reinhardii* and *Scenedesmus obliquus,* which can grow heterotrophically with acetate as a carbon source. Unlike higher plants that require light for normal chloroplast development, most strains of *Chlamydomonas* and other algae develop a normal chloroplast when grown heterotrophically in darkness. In spite of the obvious advantages of this system, which have been accentuated by the development of methods for genetically transforming the organism (145) and for isolating intact chloroplasts (14, 95), it is only recently that the full potential of this system as a tool for analyzing photosynthetic CO_2 fixation has begun to be realized. The problem was that until recently only one mutation affecting the dark reactions had been found, an acetate-requiring strain lacking phosphoribulokinase (PRK) activity (15, 127). The existence of this single mutant was, in retrospect, misleading because it suggested that other mutants with defects in photosynthetic carbon metabolism could be isolated as nonphotoautotrophic mutants capable of growth on acetate in the light.

In attempting to broaden the spectrum of available mutations, Spreitzer & Mets (178) isolated a number of acetate-requiring mutants in darkness and found that many of these mutants were unconditionally light sensitive. The first mutant to be characterized carried a chloroplast mutation in the structural gene *(rbcL)* for the large subunit polypeptide of ribulosebisphosphate carboxylase/oxygenase (Rubisco) (46, 176). On the basis of preliminary characterization, the other mutants appeared to include both uniparental and Mendelian mutants with defects in both the light reactions and in photosynthetic carbon metabolism. Several other Rubisco-deficient mutants have also been found with the same phenotype (179), and recently a light-sensitive acetate-requiring mutant has been characterized as lacking PRK activity (148). The enzyme from the mutant has been shown to have an altered isoelectric point, suggesting an alteration in the structural PRK gene.

The discrepancy between the light sensitivity of the Rubisco-deficient mutants and the PRK-deficient mutant of Salvucci & Ogren (148) and the light insensitivity of the previously isolated F60 (PRK-deficient) mutant (127) requires an explanation. Unfortunately, the basis for light sensitivity is not known. It might be caused by photoinhibition resulting from an inability to dissipate photosynthetic reductant (144). Alternatively, the recent discovery of a photosystem-linked chloroplast respiratory chain in *Chlamydomonas* (57) could be involved. Whatever the case, it seems likely that the F60 mutant line carries a second mutation that relieves the light sensitivity which would otherwise result from the deficiency of PRK activity. In support of this argument, Spreitzer & Ogren (180) have isolated nuclear mutations that suppress the photosensitivity associated with the Rubisco-deficient mutant. One of the mutants lacked PSII activity, another class lacked pigmentation, and yet another class had no obvious deleterious effect when the suppressor mutation was crossed into a wild-type strain. This latter class of mutants should provide an excellent source of material to investigate the basis for this intriguing aspect of photosynthesis which is not apparent by other criteria.

The availability of the well-characterized *rbcL* mutant provided an opportunity to reevaluate an earlier suggestion that selection for arsenate resistance in *Chlamydomonas* results in preferential recovery of nonphotosynthetic mutants (73, 155). The enrichment scheme was based on the idea that a mutation that reduced utilization of phosphate by photophosphorylation would confer arsenate resistance by increasing the size of the available phosphate pool, thereby reducing the effectiveness of arsenate as a lethal phosphate analog. In a reconstruction experiment employing the *rbcL* mutant, no obligate relationship between arsenate resistance and an acetate requirement was found (177), suggesting that the selection technique may be somewhat unreliable.

Although the green algae such as *Chlamydomonas* have a Rubisco with a ratio of RuBP carboxylase to oxygenase activity similar to that in higher plants

(91), they do not photorespire or show the O_2 inhibition of photosynthesis associated with photorespiratory phenomena. This has been attributed to the presence of an environmentally responsive mechanism for concentrating CO_2 inside the cell (10, 130). However, because of the rapidity of the HCO_3^- and CO_2 interconversion, it has been difficult to study the CO_2-concentrating mechanism by conventional techniques, and the precise biochemical mechanisms responsible for the effect have remained elusive (130).

In order to distinguish the effects of the various components believed to play a role in the CO_2-concentrating mechanism in *Chlamydomonas*, mutants that lack this activity have been isolated by screening for colonies that are unable to grow in air but that grow normally in atmospheres enriched in CO_2 (171–173). One mutant *(ca-1)* appears to be specifically deficient in carbonic anhydrase activity (172). The rate of photosynthetic CO_2 fixation by this mutant at atmospheric levels of CO_2 was greatly reduced relative to wild-type and was strongly inhibited by O_2. However, the *ca-1* mutant retained the ability to concentrate CO_2. This observation supports the concept that carbonic anhydrase is involved in the suppression of photorespiratory phenomena but clearly indicates that another component is required. Another mutation *(pmp-1)* isolated by the same criterion has normal levels of carbonic anhydrase activity but lacks the ability to concentrate CO_2 and is believed to have a lesion in a component of the CO_2-concentrating mechanism (171). The existence of this mutant is convincing evidence for a multicomponent process. Additional mutants of this kind would be useful in permitting an estimate of the biochemical complexity of the CO_2-concentrating system. The mutants may also prove useful in attempts to identify the genes and the gene products involved in this important and interesting phenomenon.

Mutations in Photorespiratory Metabolism

Mutants with defects in photorespiratory metabolism have been isolated in *Arabidopsis* and barley by exploiting the observation that the flow of carbon into the photorespiratory pathway can be prevented without any deleterious effects by placing plants in air containing high levels (1%) of CO_2 (159, 161). The basis for this effect is that CO_2 competitively inhibits RuBP oxygenase activity, the first reaction of the photorespiratory pathway (135). Thus, mutants with defects in photorespiratory metabolism were isolated as plants which were normal in appeerence when grown in air containing 1% CO_2 but which turned chlorotic within several days after being transferred to standard atmospheric conditions because of the disruption of photorespiratory metabolism. Because intermediates of the photorespiratory pathway are rapidly labeled with $^{14}CO_2$, many of the mutants were readily characterized by simply labeling the plants with $^{14}CO_2$, examining the distribution of label in the photorespiratory metabolites and performing confirming enzyme assays. This approach permitted the characterization of mutants deficient in activity for seven enzymes associated with photorespiratory metabolism (Table 1). The properties of many of these mutants have been described in several recent reviews (31, 135, 159).

Approximately 50 nuclear mutants of *Arabidopsis* with defects in one of six loci associated with photorespiratory metabolism have been characterized (159). Mutations at each of the six loci result in loss of activity for one enzyme involved in photorespiratory metabolism (Table 1). Sixty-one mutants have been identified from a similar screen of 120,000 M2 barley seedlings, and a number of these have been assigned to one of five distinct classes on the basis of biochemical characteristics (31, 68, 94, 98). All of these mutants share the property that when illuminated in atmospheric conditions that do not support significant amounts of photorespiratory metabolism (i.e. nitrogen containing 350 ul l^{-1} CO_2, 2% O_2), photosynthetic CO_2 fixation is quantitatively and qualitatively similar in mutant and wild type. By contrast, photosynthesis is rapidly inhibited following a brief period of illumination in standard atmospheric conditions. As noted below, this inhibition of photosynthesis, and the inviability of the mutants in air, is thought to result from the relatively large amounts of photorespiratory intermediates that accumulate at the site of the enzymatic lesion.

Most of the mutants have been useful as tools for physiological studies. For instance, labeling studies with the *p*-glycolate phosphatase-deficient mutant of *Arabidopsis* helped establish the precursor-product relationship between phosphoglycolate and photorespiratory glycolate (161). The absence of photorespiration in mutants lacking mitochondrial serine transhydroxymethyl-

Table 1   Mutants of higher plants and algae that require CO_2 enrichment for growth

Gene symbol	Enzyme deficiency	Species	Reference
ca-1	carbonic anhydrase	C. reinhardtii	172
cat3	catalase	H. vulgare	94
dct	dicarboxylate transport	A. thaliana	167
gluS	glutamate synthase	A. thaliana	162
		H. vulgare	31
glyD	glycine decarboxylase	A. thaliana	165
pcoA	p-glycolate phosphatase	A. thaliana	166
		H. vulgare	68
pmp	CO_2 pump	C. reinhardtii	171
rca	Rubisco activation	A. thaliana	166
sat	serine:glyoxylate amino-transferase	A. thaliana	163
stm	serine transhydroxymethyl-ase	A. thaliana	164

ase activity provided an unequivocal demonstration that glycine decarboxylation was the sole site of photorespiratory CO_2 release (164). Also, the normal growth of these mutants in permissive conditions demonstrated an important difference between C_1 metabolism in plants and yeast where mutants lacking this activity were multiply auxotrophic for compounds derived from methylene tetrahydrofolate. The implication is that C_1 metabolism of leaf cells is entirely dependent upon another isozyme of serine transhydroxymethylase.

Because the operation of the photorespiratory pathway depends on recycling of ammonia, mutants were also found with defects in several enzymes associated with the photorespiratory ammonia cycle. For instance, mutants lacking 97% of chloroplast glutamate synthase (GOGAT) activity were unable to recycle photorespiratory ammonia and therefore had a de facto block in the photorespiratory pathway at the glyoxylate amination step (31, 162). The normal growth of the mutants in high CO_2 indicated that the vast majority of GOGAT activity is required only for photorespiratory ammonia recycling. The observation that both leaf and root GOGAT activities are similarly deficient in a barley mutant indicates that the same gene is expressed in both tissues (31).

One of the most informative mutants isolated by this approach was deficient in activity for the chloroplast envelope dicarboxylate translocator (168). This mutant had greatly reduced ability to transport aspartate, glutamate, oxoglutarate, and malate into the chloroplast. The loss of activity was correlated with the absence of a normally abundant 42 kd protein in the envelopes of *Arabidopsis* chloroplasts (170), suggesting that this protein is a component of the dicarboxylate transporter. This mutant requires high CO_2 for growth because the transporter is an essential component of the mechanisms that recycle photorespiratory ammonia. Because of the defect, the mutant is functionally equivalent to the GOGAT-deficient mutants in that it is unable to cycle oxoglutarate and glutamate in and out of the chloroplast at a high enough rate to support the operation of the photorespiratory nitrogen cycle. The mutant represents a clear example of how it is possible to genetically disrupt a pathway without inactivating an enzyme per se.

The first photorespiratory mutant characterized in barley was deficient in catalase activity (94). This enzyme is located in the peroxisome where it is required to dispose of hydrogen peroxide generated during glycolate oxidation. Although two *cat* isozymes are apparent by starch gel electrophoresis of wild-type extracts, the *cat* mutant lacked both isozymes, indicating that the two forms are under the control of a single locus. It appears that the mutant is inviable in air because H_2O_2 produced during glycolate oxidation accumulates to the point that peroxidation of lipids disrupts membrane integrity and function (139). The mutant was used to reexamine a proposal that some photorespiratory CO_2 release could be the result of nonenzymatic oxidation of glyoxylate by H_2O_2 to produce CO_2 and formate. However, no difference in the rate of

^{14}C-glycolate decarboxylation was observed between mutant and wild-type barley, indicating that this is not a physiologically significant route of photorespiratory CO_2 production, even under extreme conditions (94). Analysis of the operation of other H_2O_2 scavenging systems in the barley *cat* mutant showed that following exposure of the mutant to air, the glutathione levels underwent a rapid five- to tenfold increase (158). It was suggested that oxidation of glutathione by H_2O_2 in the mutant could relieve feedback inhibition of the biosynthetic enzymes by reduced glutathione.

A mutant lacking one of the three *cat* allozymes *(cat2)* in maize was identified by screening directly for isozyme variants by starch gel electrophoresis (188). The *cat2* gene product is the only catalase produced in bundle sheath cells. The viability of this mutant may be considered confirming evidence that there is little or no flux through the photorespiratory pathway in maize resulting from the CO_2-concentrating mechanisms operating in C_4-species. However, the photosynthetic properties of this interesting mutant do not yet appear to have been examined.

Although multiple mutant alleles have been isolated for most of the photorespiratory enzymes listed in Table 1, there are many functions associated with photorespiration for which corresponding mutants have not yet been isolated. The reasons for this are not known, so it is important to attempt to understand why the available mutants are inviable and why photosynthesis is rapidly inhibited in air. In this respect, it seems probable that the *p*-glycolate phosphatase-deficient mutant is inviable in air because *p*-glycolate competitively inhibits the Calvin cycle enzyme triose phosphate isomerase. Although there is no significant short-term inhibition of photosynthesis in the *cat* mutant of barley, the accumulation of peroxide appears to lead to lethal membrane disruption (139). With all the other classes of mutants, it appears that following illumination in photorespiratory but not nonphotorespiratory atmospheric conditions, the activation level of Rubisco declines (38). A similar effect was observed in soybean leaves following inhibition of glycine decarboxylation by aminoacetonitrile (43). Since the provision of serine and ammonia was shown to delay the inhibition of photosynthesis in the *stm* mutant (169), and considering the other available information, it seems likely that a photorespiratory metabolite produced from serine is required for Rubisco activation, or that the indirect inhibition of glyoxylate metabolism (164) has an inhibitory effect on Rubisco. Glyoxylate has been shown to inhibit photosynthesis in isolated chloroplasts, but the mechanism is unknown (133).

It has also been suggested (164) that since most of the mutants are unable to recycle photorespiratory ammonia, the chlorotic phenotype could be related to glutamate deficiency which would be reflected in reduced chlorophyll content. The implication is that mutants with defects in other steps of the pathway may not be recovered simply because they do not block amino group recycling.

Whatever the precise reason, the fact that certain classes of mutants have not been recovered remains a nagging problem.

As a final point, it should be noted that if these mutants are grown in air, they appear as chl-deficient plants with gross abnormalities in chloroplast ultrastructure. This serves as a clear example of the fact that pigment-deficient mutants need have no direct connection with pigment synthesis or chloroplast biogenesis.

Mutations Affecting Rubisco

Because of the central importance of this enzyme in photosynthetic CO_2 fixation and photorespiratory metabolism, there has been substantial interest in mutations that affect the expression or activity of Rubisco. The gene for the large subunit of maize was among the first plant genes cloned, and the use of this gene as a heterologous probe has led to the isolation of Rubisco genes from a large number of photosynthetic organisms (126). With these cloned genes it is now possible to analyze mutations that affect the expression or activity of this enzyme at the molecular level.

A number of mutants deficient in Rubisco activity have been isolated in *Chlamydomonas* on the basis of a light-sensitive acetate requirement for growth (176). One of these has normal amounts of Rubisco polypeptides, but the large subunit polypeptide exhibits an altered isoelectric point. The *rbcL* gene from the mutant has been cloned and the DNA sequence compared to that of the wild type. A single nucleotide substitution was found that converted the Gly 171 residue to an Asp (46), and a revertant was shown to have a restoration of the wild-type sequence (181). This mutation permitted the first correlation between the genetic map and the physical map of *Chlamydomonas* chloroplast DNA. Several other uniparentally inherited mutants of *Chlamydomonas* lack the polypeptides associated with both subunits and also map to the *rbcL* locus (179). These have been shown by DNA sequence analysis to have nonsense mutations in the *rbcL* gene (175) which lead to premature termination of translation of the large subunit polypeptide. The absence of the small subunit polypeptide in the mutant is consistent with other observations indicating accelerated turnover of unassembled small subunit polypeptides (152).

Revertants of the mutants that do not accumulate Rubisco have also been isolated (174). These chloroplast-encoded second-site mutations appear to be maintained as stable heteroplasmic genes, in which the suppressor is lost when the revertant strains are grown in the dark (thereby removing selection). In view of the fact that these mutations are known to be nonsense mutations, it appears likely that the suppressors affect translational fidelity, thereby allowing read-through of nonsense (UGA) codons. It seems possible that a specific single-copy tRNA gene has been altered so that it recognizes UGA codons rather than the normal codon. Thus, in order for the mutant to carry out protein synthesis it must have both a wild-type and a mutant genome in the same chloroplast. If this is substantiated by further work, it will be the first instance of genetically induced informational suppression in a plant.

Several plastome mutations in *Oenothera* which result in the presence of very low levels of both subunits of Rubisco have been isolated and maintained by tissue culture (70). The mutation in one of these appears specific for Rubisco since at least four other Calvin cycle enzymes are present in the mutants in normal amounts. Translation of chloroplast mRNA in vitro demonstrated that the translation product of the *rbcL* mRNA from the mutant is only 30 kd rather than the expected 55 kd (80). The simplest explanation is that a nonsense mutation has occurred in the *rbcL* gene about 900 nucleotides from the translation start site. If confirmed by DNA sequencing, this would appear to be the first nonsense mutation characterized in a higher plant. There were normal levels of mRNA for the *rbcS* gene in the mutant, and the in vitro translation product was of normal size. Thus it seems likely that the absence of the small subunit polypeptide in the mutant is the result of rapid turnover of unassembled small subunits as noted above for the similar mutants of *Chlamydomonas*.

A mutant of *Arabidopsis*, designated *rca*, has been isolated in which Rubisco appears to be poorly activated in vivo. This *rca* mutant was recovered as a line that required high levels of atmospheric CO_2 for growth (166). In atmospheres enriched with very high levels of CO_2 this mutant is capable of CO_2-fixation rates comparable to that of the wild type. Thus, the photosynthetic apparatus and the enzyme complement of the Calvin cycle are intact. However, in standard atmospheric conditions this mutant has a very low photosynthesis rate and an abnormally elevated RuBP pool because although Rubisco is present in normal amounts in the mutant, it is not normally activated in vivo. No differences could be found in the physical or catalytic properties of Rubisco purified from the *rca* mutant as compared to that from the wild type. Recently, a reconstituted chloroplast system has been developed in which Rubisco is activated up to fivefold by illumination in the presence of RuBP, stromal extracts, and thylakoids (149). Stromal extracts of the mutant do not support this activation phenomenon and are missing two moderately abundant polypeptides. Thus, it has been suggested that the mutant lacks a protein(s) (Rubisco activase) which facilitates activation of Rubisco. The discovery of this activity may help explain why higher CO_2 concentrations are required to activate Rubisco in vitro than in vivo, and could have important consequences for any attempts to enhance photosynthetic productivity.

Several recent applications of recombinant DNA technology have created entirely new possibilities for studying the structural basis of Rubisco function. The Rubisco genes from a number of organisms, including *Rhodospirillum rubrum, Anabaena variabilis, Anacystis nidulans,* and *Z. mays,* have been expressed at high levels in *E. coli* (55, 56, 66, 160, 167). Expression of the prokaryotic genes resulted in the accumulation of a fully functional holoenzyme in *E. coli*. Several mutations have been introduced into the gene from *R. rubrum* by oligonucleotide-directed mutagenesis and the effects on the properties of the enzyme determined (51, 67). The demonstration that the heteromeric cyanobacterial enzymes assemble correctly in *E. coli* to give a hexadecameric L_8S_8 holoenzyme creates the possibility that hybrid enzymes can be created by splicing together homologous fragments of Rubisco genes from taxonomically diverse species. This may facilitate determination of the structural basis for differences in the Michaelis constants of enzymes from different species (91). Although several groups have expressed the *Z. mays rbcL* gene in *E. coli*, the product is insoluble and not catalytically active (55, 160). By analogy with the results obtained with the cyanobacterial genes, it may be possible to obtain a functional enzyme by co-expressing an appropriate *rbcS* gene in the same cell.

Because the *rbcS* gene is nuclear and the *rbcL* gene is chloroplast-encoded, there is uncertainty about the mechanisms which must coordinate the synthesis of the two genes that are present in very different copy numbers. Substantial attention has been directed to the concept that the synthesis of one subunit is coupled with the synthesis of the other. However, it appears that many experiments suggesting coupled synthesis were the result of an accelerated turnover of unassembled subunits (125, 152). Since no mutations are known that specifically regulate the amount of synthesis of Rubisco, an attempt has been made to examine gene dosage effects in the aneuploid series of wheat (89). Although differences in amount of Rubisco per cell are observed in this aneuploid series, the pleiotropic effects of these gross genetic changes precludes a simple interpretation.

MUTATIONS AFFECTING PHOTOSYNTHETIC ELECTRON TRANSPORT

Methods of Mutant Isolation

One of the most elegant and effective methods for isolating mutants with defects in photosynthesis was devised by Garnier (54a) and Bennoun & Levine (23) on the basis of chlorophyll fluorescence. The basic concept is that any lesion that inhibits electron flow through the electron transport chain will result in more of the absorbed energy being reemitted as chlorophyll fluorescence. By irradiating with light of less than 640 nm and viewing or photographing colonies through a filter that transmits only wavelengths greater than 650 nm, those colonies with defects are readily apparent. In contrast to the identification of mutants on the basis of chl-deficiency, this and related approaches permit the isolation of mutants with normal chloroplast development.

The high chlorophyll fluorescence technique was successfully adapted to higher plants by Miles & Daniel (120), who isolated a number of *hcf*-mutants of maize. Because most of the *hcf* mutants are lethal, the lines were isolated by screening segregating families and are maintained as heterozygotes (119). Homozygous mutant seedlings grow normally until starch reserves are exhausted, at which time the lethals undergo wilting and chlorosis. However, the homozygous mutant seedlings can be differentiated from the heterozygotes or wild-type siblings by high chlorophyll fluorescence long before the chlorosis is apparent, and can be used for experimental purposes at this stage.

The mutants isolated on the basis of the *hcf* phenotype fall into many classes which include mutants with defects in PSI, PSII, photophosphorylation, CO_2 fixation, and most other functions related to photosynthesis. More than 31 loci which give rise to the *hcf* phenotype have been assigned to chromosome arms of maize (122). A detailed description of the protocols and methodologies for classifying and analyzing the mutants has been presented (119). The utility of the approach may be greatly enhanced by the isolation of *hcf* mutants such as *hcf-Mu5* (87a) by transposon mutagenesis. This may eventually permit the cloning of genes by transposon tagging and could be particularly valuable in identifying the primary products of genes that may exert a regulatory or a structural role in supporting photosynthesis. A major limitation of screening for lethal *hcf* mutants of plants is the necessity of maintaining thousands of families during the screening. Also, the method depends upon the oxidation state of the primary acceptor of PSII and, therefore, cannot reveal genetic defects on the oxidizing side of PSII. Mutants of this kind have been isolated in algae on the basis of reduced fluorescence yield (48, 113), but this is not considered feasible in higher plants.

Several authors have described procedures that should permit the isolation of mutants with defects anywhere between water and the oxidizing side of PSI. One approach exploits the fact that the herbicide Diquat (1,1'-ethylene-2,2'-dipyridylium bromide) is reduced by the electron transport chain near the primary acceptor of PSI. Once reduced, it is subsequently oxidized by a series of reactions which ultimately lead to the production of destructive peroxides. Miles (118) demonstrated in a reconstruction experiment that mutants of maize that were unable to reduce Diquat were not made chlorotic by the herbicide (but were, of course, seedling lethals). Similarly, mutants of *Chlamydomonas* have been isolated by using the PSI electron acceptor metronidazole (2-methyl-5-nitro imidazole-1-ethanol) (58, 151). It is thought that the compound is reduced by ferredoxin and then autooxidized to produce toxic superoxide radical. Up to 50% of the colonies selected for resistance to this compound appeared to be photosynthetic electron transport mutants. The approach has the advantage, as compared with fluorescence techniques, of permitting selection of mutants rather than screening.

PSI-Deficient Mutants

PSI-deficient nuclear mutants of *Chlamydomonas* were originally isolated on the basis of high chlorophyll fluorescence by Levine and collaborators (102). Following the development of high-resolution gel electrophoresis techniques, several of these mutants (F1 and F14) were found to lack both a 66 kd polypeptide and the chl-protein complex CP1, suggesting that the 66 kd protein is the apoprotein of the P700-associated PSI reaction center complex (40). More recently, in what must be the most comprehensive genetic analysis of a specific thylakoid component, 25 nuclear-encoded PSI-deficient mutants of *Chlamydomonas*, isolated by four different criteria, were analyzed genetically and biochemically (58). Analysis of the thylakoid membrane composition by SDS polyacrylamide electrophoresis demonstrated that all of these mutants lacked the putative apoprotein of PSI (CP1) and six low molecular weight proteins (58). These same proteins were found to be the constituents of PSI particles prepared by detergent fractionation of thylakoid membranes. The interpretation was that the missing polypeptides are components of a multi-subunit complex in which the absence of one or more of the constituent polypeptides blocks the synthesis or assembly of other polypeptides of the same complex. Since the number of loci involved (13) was larger than the number of polypeptides affected, the majority of the genes identified probably do not encode the missing polypeptides. It is surprising that no mutants were identified in which the polypeptides were present but simply not functional. This raises the unlikely possibility that no single amino acid substitution will completely eliminate PSI activity, and that the only way of completely eliminating activity is to prevent assembly of the complex. Alternatively, turnover of the complexes might be regulated by a mechanism that is responsive to function. Thus, a mutation which eliminated function would cause automatic removal of the complex from the membrane. It is apparent that this collection of mutants offers many interesting opportunities to identify genes regulating the synthesis and assembly of PSI.

Maize mutants deficient in PSI activity have been isolated on the basis of high chl fluorescence (119). By analogy with the PSI-deficient mutants of *Chlamydomonas*, at least five mutants that lack PSI activity lack the P700 reaction center complex (CP1) (121). One of the mutants, *hcf-50*, (originally designated *hcfE1481*) had a strong reduction in the amount of a 68 kd polypeptide and at least one low molecular weight polypeptide (123). Electron microscopy of the *hcf-50* mutant revealed normal chloroplast morphology (123). Freeze-etch studies showed that the unstacked regions of the thylakoids had a reduction in the size of particles on the unstacked protoplasmic face (PFu). By contrast, no alterations were seen in the stacked regions. This observation was interpreted as evidence for physical partitioning of the PSI and PSII reaction centers into unstacked and stacked regions of the thylakoids, respectively (123). A similar conclusion has been reached from studies involving mechanical shearing of wild-type thylakoids and separation of the stroma and grana lamellae by polyethylene glycol-dextran phase partitioning methods or detergent fractionation (183).

PSI-deficient mutants have also been identified among the large collection of chl-deficient mutants in barley. Lethal mutations at five nuclear loci *(xantha-q, viridis-h, -q, -n, -zb)* give rise to a chl-deficient PSI deficiency (157). Possibly because of the criterion by which they were isolated (i.e. chl deficiency), each of these mutants exhibited some perturbation in thylakoid ultrastructure. The mutants *viridis-n*[34] and *viridis-zb*[63] have been characterized extensively with respect to thylakoid polypeptide composition, ultrastructure, and electron transport activity (81, 103, 104, 128). The mutants have reduced amounts of the PSI reaction center apoprotein, P700, and three low molecular weight proteins. The composition of the membranes was otherwise similar to wild type with respect to polypeptides, chl proteins, and cytochromes. As in the PSI-deficient mutant of maize, the major ultrastructural change in the thylakoids was a reduction in the number of large PFu particles.

Non-Mendelian mutants deficient in PSI have been isolated on the basis of chl deficiency in *Antirrhinum majus* (78) and by scoring for high chl fluorescence or by selecting for arsenate resistance in *Chlamydomonas* following 5-fluorodeoxyuridine treatment to reduce chloroplast DNA copy number (21, 155). At least three chloroplast loci in *Chlamydomonas* have been shown to cause loss of the CP1 complex from thylakoid membranes (155), indicating that several chloroplast gene products are required for assembly of CP1. No other thylakoid proteins were missing, suggesting that the normal function of these gene products is highly specific. Recent progress in correlation of the genetic and molecular maps of the chloroplast genomes of *Chlamydomonas* (50) has created the possibility that these mutations will eventually be characterized at the molecular level.

PSII-Deficient Mutants

Elucidation of the composition of PSII has proved more difficult than for any of the other complexes of the thylakoids. Whereas PSI components have been identified using both particle isolation techniques (131, 132) and mutant analysis (58, 81), it has been difficult to isolate PSII particles in the absence of contaminating thylakoid components. Thus, mutants deficient in PSII have been widely used to identify the polypeptides that are specifically associated with PSII (20, 25, 39, 44, 106, 113–115, 124, 196).

There are, in principle, two distinct classes of PSII-deficient mutants; those missing one or more of the polypeptides associated with the complex and those that are simply lacking activity. An example of the former is the normally pigmented PSII-deficient *Chlamydomonas* mutant F34, in which several chl proteins (CPIII and CPIV) are missing (44). The absence of these chl-protein complexes is accompanied by the loss of eight thylakoid polypeptides (of 50, 47, 32, 21, 18, 14, 6, 3 kd) (20). A second mutant (BF25) had reduced amounts of five of the polypeptides and was completely missing the other three (21, 18, 6 kd). In contrast to F34, which was completely deficient in PSII activity, the BF25 mutant was specifically defective in water hydrolysis, suggesting that the three missing polypeptides are required for this aspect of PSII activity. These three polypeptides were also found to be more readily extracted from wild-type thylakoids, suggesting that they are extrinsic proteins and providing a possible basis for their selective absence in BF25. A study of independently isolated PSII-deficient mutants of *Chlamydomonas* (106) showed that at least five polypeptides could be missing (50, 47, 33, 27, 19 kd). The electrophoretic system used did not resolve the low molecular weight polypeptides. The absence of a 27 kd polypeptide, not noticed by others (20), was made possible by the use of a double mutant lacking both PSII and LHC-II, which permitted scrutiny of the region of the gel normally obscured by LHC-II polypeptides. Considered together, these studies identified at least nine polypeptides associated with PSII. The involvement of these polypeptides was substantiated by the observation that isolated PSII particles from the wild type contained all of the missing polypeptides except for the 6 kd polypeptide (20).

At least two of the polypeptides (47 and 50 kd) missing in the F34 mutant are translated on chloroplast ribosomes (57a). The implication is that the product of the locus altered by the nuclear mutation in strain F34 is involved in the synthesis, assembly, or degradation of at least two chloroplast-encoded proteins. Therefore, it is particularly interesting that a mutation at another nuclear locus (SU-1) partially restores PSII activity and the missing polypeptides. In view of recent evidence emphasizing the importance of protein degradation in regulating thylakoid polypeptide composition (17, 99), a reexamination of the SU-1 mutation could prove useful in understanding this phenomenon.

The second type of PSII-deficient mutants is exemplified by a mutant of *Scenedesmus* that was isolated by scoring for low chl fluorescence. The mutant (LF-1) is unable to use water as an electron source but retains rates comparable to the wild type using alternate donor systems (113). Electrophoresis of chl-protein complexes in LDS gels at low temperature revealed that the only difference between the mutant and wild type is that the mutant is missing a nonpigmented 34 kd polypeptide but has gained a 36 kd polypeptide. The simplest interpretation is that the protein is not correctly processed in the mutant, or the translational reading frame has been extended by mutation. Whatever the precise cause, this appears to be the first identification of a nonpigmented component on the water side of PSII. The results of subsequent analysis (25) is consistent with the suggestion (116) that the 34 kd polypeptide is the manganese binding polypeptide of PSII.

One of the most extensively characterized PSII-deficient mutants in higher plants is the nuclear recessive *hcf-3* mutation of maize. This mutant, which remains green until the seed reserves are exhausted, was one of several similar mutants isolated on the basis of high chlorophyll fluorescence (101). It is completely lacking in PSII activity and lacks cytochrome b_{559}, but has normal levels of PSI activity and normal levels of plastoquinone, cytochrome f and cytochrome b_6, LHC-II and non-PSII chl (115). Electrophoresis of membrane proteins established the presence of the majority of the thylakoid polypeptides as well as the chl-protein complexes associated with PSI and LHC-II (115). However, six or seven polypeptides are missing, several of which had not previously been attributed to PSII (115). A lower chl a/b ratio also suggests the loss of a specific set of chl molecules associated with the PSII reaction center. Overall, it appears that *hcf-3* is specifically missing the entire PSII reaction center complex (114, 115).

Electron microscopy of thin sections indicated that the *hcf-3* mutant has essentially normal thylakoid morphology. However, examination of thylakoids from the mutant by freeze-fracture electron microscopy revealed a dramatic reduction in the density of the large EFs particles in both bundle sheath and mesophyll chloroplasts (7). A similar reduction in the density of EFs particles has been reported in several PSII-deficient *Chlamydomonas* mutants (i.e. F34) (196) and the barley mutants *xantha-b*^{12}, *viridis-c*^{12}, *-e*^{64}, *-zd*^{69} and *-m*^{29} (157). On the basis of these and related studies there appears to be general agreement that the large EFs particles represent the PSII reaction center complex in association with peripherally arranged subunits of LHC-II. Interestingly, the *hcf-3* mutant and some of the mutants of *Chlamydomonas* had an increase in the particle density of PF fracture faces in thylakoid membranes (7, 196). This was interpreted as indicating that LHC-II subunits, which are normally part of EFs particles in the wild type, might fractionate to the PFs face in the absence of the PSII reaction center complex.

As in the case of the F34 mutant of *Chlamydomonas*, the *hcf-3* mutant carries a recessive nuclear mutation that leads to the loss of several polypeptides which are synthesized on chloroplast ribosomes. These and related observations strongly support the concept that PSII exists as a physiological unit whose presence in the membrane is regulated independently of the other major complexes of the thylakoids (115). Trace amounts of all the polypeptides are present in *hcf-3*, indicating that synthesis occurs at a very low level or that turnover is very high. Recent evidence suggests that the loss of the 48 kd and 34.5 kd PSII reaction center polypeptides is caused by their constitutive turnover from the thylakoid following quantitatively normal synthesis and association with mutant thylakoids (99). This accelerated turnover was specific to PSII-associated polypeptides. The mechanism is not known but is thought to reflect the existence of a mechanism that accelerates the turnover of improperly or incompletely assembled proteins (16, 17, 152). A similar phenomenon involving ubiquitin has been studied in animal mitochondria (45). Accelerated turnover of unassembled polypeptides, although apparently quite common, is not always observed. Recent studies employing a monoclonal antibody specific for the 23 kd polypeptide associated with PSII have demonstrated the presence of the polypeptide in membranes of several barley mutants (i.e. vir-zd^{69}, -m^{29}) which lack the PSII complex (84a).

In addition to the accelerated turnover phenomenon, other regulatory mechanisms are probably involved in mediating the pleiotropic effects of certain mutants. Jensen & Schmidt (90) have reported that a PSII-deficient mutant of *Chlamydomonas* accumulates normal amounts of translatable mRNA for a 32–35 kd thylakoid protein, but it is not translated. They suggested that the absence of another polypeptide that is missing in the mutant may impose a negative translational control over a group of proteins. This would be particularly interesting if substantiated because instances of translational control are relatively rare.

The pleiotropism associated with mutants such as *hcf-3* is not confined to protein components of the thylakoid chl-protein complexes and is useful in defining other complex-specific components which may be difficult to assign because of a labile association. For example, the *viridis*-zd^{69} of barley is devoid of PSII activity and lacks several chl *a* proteins. This mutant also has reduced levels of neoxanthin, violaxanthin, and β-carotene, suggesting that these three carotenoids are normally associated with the missing complexes. By contrast, lutein levels were normal, suggesting that this carotenoid is not normally bound to PSII (77).

Coupling Factor

In contrast to the many mutations that specifically affect PSI or PSII, few mutants have been shown to have specific defects in coupling factor. Levine (102) described several mutants of *Chlamydomonas* deficient in photophosphorylation, but these do not appear to have been characterized in any detail. However, non-Mendelian acetate-requiring mutants of *Chlamydomonas* with apparent defects in CF_1 have been isolated by several groups (21, 24, 155). The mutants are characterized by a high amount of delayed luminescence which is thought to result from the inability of the mutants to dissipate the electric field generated in the light across the thylakoid membrane (24). Mutants at distinct chloroplast loci have been isolated which appear to be blocked in the synthesis, assembly, or integration of CF_1 into the thylakoid membrane (155). The physical basis of these mutations has not yet been established, but if they were found to map in appropriate locations, a detailed analysis should be possible because the genes for the three chloroplast-encoded subunits (i.e. α, β, ε) have been cloned (79).

Some plant species are resistant to the fungal toxin Tentoxin where others are not. Tentoxin binds noncompetitively to CF_1 and inhibits activity in sensitive but not resistant species (184). Thus, it is inferred that the resistant species have an altered gene for a CF_1 subunit. In support of this, sexual and nonsexual hybridizations between resistant and susceptible species of tobacco have shown that the resistance is inherited in a non-Mendelian manner (53). If substantiated by subsequent work, such a mutation might prove very useful as a selectable marker in future attempts to improve the efficiency of chloroplast transformation.

Mutants Lacking Chlorophyll b

Perhaps the most intensively utilized mutant affecting photosynthesis is the spontaneous *chlorina f2* mutant of barley which has normal PSI and PSII activities but lacks chl *b* (28). Similar mutants have been reported in *Arabidopsis thaliana* (82), *Zea Mays* (121), *Melilotus alba* (105), *Triticum aestivum* (47), *Pisum sativum* (154), and *Chlamydomonas* (117), but have been much less widely utilized. Early studies of the barley mutant revealed that chl *b* lacked an essential photochemical role but was involved in increasing the efficiency of light capture (28). Thus, chl *b*-deficient mutants have photosynthesis rates comparable to the wild type at high light intensities when expressed on a leaf area or fresh weight basis, but have much higher rates when expressed on a chl basis. At low light intensity the mutants generally have relatively poor photosynthetic rates because they are not light-saturated.

The precise biochemical defect in the *chlorina f2* mutant or the other chl *b*-deficient mutants is not known because of uncertainty concerning the pathway of chl *b* biosynthesis (36) and because the chl *b*-less mutants also lack several polypeptides associated with chl-protein complexes that contain the chl *b* (16, 32, 150, 186). The *chlorina f2* mutant lacks the 27 kd polypeptide, contains only trace amounts of the 25 kd polypeptide, and has about 50% of the 24 kd polypeptide present in the LHC-II complex from wild-type membranes (16). It also lacks the three polypeptides in the 20–22 kd range which are constituents of the PSI peripheral antenna complex (16, 132), suggesting that these polypeptides also normally bind chl *b*. This has recently been substantiated by the demonstration that the PSI antenna complex contains chl *b* (76). Thus, in principle, an alteration that prevents binding of chl to the LHC or that prevents uptake of the cytoplasmically synthesized protein into chloroplast membranes could prevent the accumulation of chl *b*. However, chloroplasts of the *chlorina f2* mutant exhibit normal transport into the chloroplast of an LHC-II polypeptide that is missing in the mutant (16, 150). Also, because at least one of the LHC-II polypeptides is encoded by a multigene family (42), it is unlikely that a mutation in one of the genes would abolish chl binding by the products of the others. It seems more likely that the *chlorina-f2* mutant has a defect in chl *b* biosynthesis, and the missing LHC-II polypeptides do not accumulate because they are not stabilized by bound chl *b* (16, 17). In support of this, normal levels of translatable polysome-associated mRNA for the missing polypeptides were found in both the barley and pea mutants (6, 16, 153). Thus, it is inferred that the mRNA is translated at normal rates but that the products turn over more quickly. It is not known if chl *a* is converted to chl *b* before or after it binds to the protein, or whether chl *a* can bind to the protein in the absence of chl *b*. The availability of cloned genes for the LHC-II polypeptides should permit novel approaches to this question.

The response to loss of chl *b* in *Chlamydomonas* appears to be somewhat different to that in higher plants. A mutant *(pg113)* lacking chl *b* lacks the chl-protein complex CPII (i.e. LHCP[3]) in SDS-polyacrylamide gels, but contains all of the proteins that comprise CPII (117). Also, the quantum yield of the mutant is very similar to the wild type, suggesting the existence of a functional antenna. The absence of CPII may be caused by greater detergent lability of the chl *b*-less LHC-II. If substantiated, this might be interpreted as evidence that the first step in LHC-II assembly is the binding of chl *a*. The fact that the chl *b*-less LHC-II accumulates suggests a fundamental difference in the mechanisms that regulate protein turnover in *Chlamydomonas* as compared to higher plants.

Current theories to explain the mechanisms of thylakoid stacking and cation-induced changes in the distribution of excitation energy between PSI and PSII invoke LHC-II as the mediator of these effects (18, 183). This model is based in part upon the observation that the *chlorina-f2* mutant of barley has a quantitatively altered cation requirement for functional reorganization of the antennae associated with changes in variable fluorescence (29, 32). Because the mutants are not completely lacking in granal stacking or cation effects, it is inferred that these residual effects are mediated by the LHC-II components that remain in the membranes of the mutants (32, 146). These membrane reorganizations are also associated with the activity of a thylakoid protein kinase (18). Although the barley mutant has normal levels of this kinase activity in vitro (19, 105), it lacks or has substantially reduced protein kinase activity in vivo (19). This interesting effect may be explained by the unusual distribution of chl between the photosystems in the mutant which maintains a disproportionately rapid turnover of PSI. Thus, the plastoquinone pool is highly oxidized and this, in turn, inhibits the kinase activity (19). The observation that the chl *b*-less mutant *(pg113)* of *Chlamydomonas* has normal levels of phosphorylation of the LHC-II apoproteins is consistent with this interpretation (117). Thus, the fact that chloroplasts of the *chlorina-f2* mutant do not exhibit state I/state II transitions in phosphorylating conditions (75) is consistent with the theory that phosphorylation of LHC-II mediates a functional reorganization of the thylakoid membrane (19, 183).

The chl *b*-less mutants of both barley and maize have been useful in helping to identify the chl-protein complexes which are resolved by electrophoresis in SDS or LDS-containing gels (104, 121). It has been proposed that several of the chl *b*-containing bands represent multimeric forms of a common monomeric unit. However, the loss of the putative multimeric forms without the concomitant loss of the monomeric forms in several mutants of maize (121) raises the possibility that the situation may be more complex. The mutants could lack an additional (colorless) component that is required for multimer organization. Indeed, loss of the multimers but not the monomer has also been observed under certain conditions in a mutant of *Arabidopsis* that lacks the chloroplast-specific lipid acyl group *trans*-hexadecenoic acid (110). In this case it was proposed that the absence of the lipid simply made the LHC-II complex more accessible to detergent-mediated dissociation. Because of possible changes in the protein and pigment-to-lipid ratio in the maize mutants, a similar explanation may account for these observations.

Thylakoid Ultrastructure

The predominant ultrastructural characteristic of thylakoid membranes from higher plants and many green algae is the presence of granal stacks. A substantial body of evidence supports a model in which stacking is mediated by direct LHC-II adhesion (183). Another well-developed model attributes stacking to differences in the surface charge density of thylakoid surfaces (11). In this case membranes are proposed to stack in regions where repulsive electrostatic forces between adjacent membranes are low. In both models, phosphorylation of LHC-II is proposed to play an important role in unstacking the membranes.

Many mutants have been reported which alter thylakoid ultrastructure in subtle or dramatic ways. The properties of several of these mutants bear directly on the problems associated with determining the mechanism of thylakoid stacking. In particular, a pale-green nuclear recessive mutant of *Chlamydomonas* (strain *ac-31*) that is capable of photoautotrophic growth and has a chl *a/b* ratio of about 3 completely lacks thylakoid stacks (60). Although this mutant does not appear to have been studied since the advent of high-resolution electrophoretic methods for analysis of thylakoid membrane composition, the presence of chl *b* implies the presence of LHC-II. If LHC-II is present, this mutant could prove useful in identifying one or more components required for stacking.

One of the most frequently cited lines of evidence for the role of LHC-II in thylakoid membrane adhesion is the absence of normal stacking in the *chlorina-f2* mutant of barley, which lacks or has greatly reduced amounts of several polypeptides associated with LHC-II and LHC-I (16). However, even a cursory glance at the original micrographs of the membranes from this mutant (59), or one of several subsequent reexaminations (156), indicates that this mutant exhibits very well-developed thylakoid stacks. Similarly, a chl *b*-deficient mutant of pea that lacks LHC-II also exhibits well-developed granal stacks (153). Thus, rather than supporting the model that depends on LHC-II, the properties of the mutants must be considered evidence against a determinative role for LHC-II in mediating membrane stacking. This is also suggested by the analysis of a PSII-deficient mutant of tobacco in which the thylakoid membranes are almost completely unstacked (124). Although this mutant appears to lack the PSII reaction center, it contains LHC-II. In attempting to explain the stacking evident in the *chlorina-f2* mutant, Ryrie (146) used antibodies to show that the 23 and 23.5 kd polypeptides are present but in greatly reduced amounts. He proposed that these polypeptides may simply be required in very small amounts to permit stacking. Although this seems reasonable, it might be interesting to survey the polypeptide composition of several mutants that do not form thylakoid stacks to determine if other gene products may be involved.

Analysis of Fluorescence Phenomena

Measurement of fluorescence induction kinetics and low-temperature fluorescence emission spectra has been one of the most widely utilized methods for the analysis of mutants with defects in photosynthetic electron transport (i.e. 22, 119, 157). In one of the most ambitious applications of this approach, 42 *viridis* mutants of barley were analyzed (157). However, because of the unexpected complexities of the fluorescence emission spectra from many of the mutants, it was concluded that the approach has limited utility in identifying the primary lesion. For example, each of the four classes of PSII-deficient mutants examined had a distinctly different low-temperature fluorescence emission spectrum. The implication is that the emission spectra observed in wild-type membranes is an average of many different pigment-protein complexes that can be partially disrupted in many ways. Thus, the approach seems most useful when the effects of the mutation lead to complete loss of a complex, as in the PSI-deficient mutant of *Chlamydomonas* in which the loss of PSI activity was accompanied by the loss of the long wavelength fluorescence emission maximum (22).

Although fluorescence may not be widely useful in mutant analysis, mutants may be useful in understanding fluorescence phenomena. Recently, several relatively well-defined mutants have been exploited to analyze the decay kinetics of room temperature fluorescence. In an attempt to attribute the three exponential decay components to PSI, PSII, or LHC-II, the fluorescence decay kinetics of the PSII-deficient *hcf-3* mutant of maize, the PSI-deficient *hcf-50* mutant of maize, and the LHC-deficient *chlorina-f2* mutant of barley were compared with the corresponding wild-type lines (63). The fact that the mutants retained the complex decay kinetics ruled out a simple origin for any component.

A widely accepted model to explain light absorbtion and utilization within the two photosystems and their antennae predicts that the fluorescence emission bands of higher plants F686, F696, and F735 (F718 in *Scenedesmus*) arise from LHC-II, CPa, and CP1, respectively (33). However, mutants of *Scenedesmus* lacking CP1 retain the long-wavelength low-temperature fluorescence emission (26), indicating that CP1 is not the exclusive emitter. The situation is further complicated by the observation that a mutant lacking β-carotene but having PSI activity lacks the long wavelength emission. A similar conclusion concerning the origin of long-wavelength fluorescence was reached following analysis of the fluorescence properties of the *viridis-zb*[63] mutant of barley which is devoid of PSI activity, lacks CP1, and lacks three low molecular weight polypeptides believed to carry iron sulfur centers of PSI (81). Leaves of the *viridis-zb*[63] mutant retain a high ratio of long/short wavelength fluorescence which differs from that of the wild type by having a 6 nm blue shift in the emission maximum (81). Therefore, it is likely that the 735 nm fluorescence has multiple origins and that a minor component becomes predominant when CP1 is missing. Finally, analysis of the *chlorina-f2* mutant of barley showed a 10 nm blue shift in the long wavelength fluorescence rather than the expected shift at 686 nm (132). Thus, the evidence from mutant analysis appears inconsistent with the tripartite model. In view of the broad support for this model by other criteria, the mutant studies should be extended to resolve the anomalies.

Recently evidence has accumulated suggesting the presence of two physically distinct types of PSII designated $PSII_\alpha$ and $PSII_\beta$ (111). The $PSII_\beta$ centers, originally proposed to explain a slow component in fluorescence induction kinetics, have been shown by membrane fractionation techniques to be located primarily in the stromal lamellae whereas the $PSII_\alpha$ centers are located in the granal lamellae. The observation that the chl-deficient *Su/su* mutant of tobacco had a decreased ratio of $PSII_\alpha/PSII_\beta$ was important in establishing that a physical basis existed for the fluorescence phenomena. Recently, it has been inferred that the $PSII_\beta$ centers may represent PSII reaction centers that are in the process of being assembled and have altered fluorescence induction kinetics because they lack the LHC-II antenna (111). A dramatically altered ratio of α/β centers in the *chlorina-f2* mutant of barley is consistent with this model.

Chlorophyll-Deficient Mutants

The study of mutants of *Chlorella* with defects in chl biosynthesis was instrumental in the elucidation of the chl biosynthetic pathway (61). By contrast, in spite of the abundance of pigment-deficient mutants of maize (41) and barley (192), relatively few studies of this kind have been carried out in higher plants because mutants with defects in chl biosynthesis generally do not accumulate significant quantities of precursors. The discovery that detached dark-grown shoots that were provided with exogenous δ-aminolevulinate accumulated chl precursors (37) opened the way to mutant investigations in higher plants, many of which have focused on the complex mechanisms that regulate chl biosynthesis (37).

Of the more than 96 loci that affect pigmentation in barley (192), only five *(xantha-f,g,h,l,u)* are thought to directly affect structural genes for chl biosynthesis (61, 191). The evidence for this is largely the result of metabolite analysis, and a specific enzymatic lesion has not been demonstrated by enzyme assay in most cases. Several mutants of maize have been similarly characterized as having specific defects in chl biosynthesis (108). One class of (yellow) mutants was defective in the conversion of protoporphorin IX to Mg-protoporphorin, and another class appears defective in the conversion of Mg-protoporphorin monomethyl ester to protochlorophyllide.

One of the few recent studies of chl biosynthesis per se in higher plants has employed a partially chl-deficient mutant of maize designated *olive necrotic 8147,* which has been useful in revealing heterogeneity in the oxidation state of the C4 side chain of chl and precursors (12, 13, 197). Chl extracted from the mutant has a red-shifted Soret excitation maximum. On the basis of NMR and mass spectroscopy, both the chl *a* and chl *b* from the mutant have been shown to have a vinyl rather than an ethyl group at position 4 of the chl macrocycle (197). Although there is uncertainty about the physiological significance and the biosynthetic origin (37), it seems likely that the mutant lacks activity for an enzyme which reduces the vinyl group at position 4 to an ethyl group. The existence of the mutant may be useful in evaluating a model that invokes specific modifications of chl as a regulatory mechanism for partitioning chl molecules to different pigment-protein complexes.

The *tigrina* mutants of barley are both dramatic and informative. When grown in continuous light these mutants are viable and uniformly green, although some are partially chl-deficient (35). However, when grown in an alternating light/dark cycle the mutants develop alternate transverse green and necrotic bands on the leaves and are inviable. It appears that several of these mutants *(tig-d, tig-b, tig-o)* lack a control mechanism which operates normally in the dark to restrict δ-aminolevulinate formation and thereby prevent excessive protochlorophyllide accumulation in darkness and subsequent bleaching by light (62, 191).

The simplest of the *tigrina* mutants is the *tig-d* mutant in which the only effect of the mutation appears to be constitutive δ-aminolevulinate formation. For this reason the *tig-d* gene product was postulated to be a specific repressor of the activity or synthesis of the δ-aminolevulinate synthase (191). The *tig-b,f,m,n,o* mutants have partial blocks in carotenoid biosynthesis (35) and unaccountably accumulate an unusual form of chl *a* (chl a_{743}) which is also found in leaves treated with an inhibitor of carotenoid biosynthesis. Analysis of the *tig-o* and *tig-b* mutants suggests that these mutants are defective in the synthesis of heme.

Presumably they are carotenoid deficient because of the involvement of heme-containing enzymes required for carotenoid biosynthesis. They bleach in light because the δ-aminolevulinate synthase is no longer inhibited by heme (62), and as a result, they accumulate protochlorophyllide in darkness and bleach during the subsequent light period. A similar connection between chl and carotenoid biosynthesis is apparent in the case of the *xan-u* mutant which is defective in both porphyrin synthesis and the synthesis of β-carotene in darkness. None of the other nine classes of *tigrina* mutants appear to have defects in carotenoids, and the basis for the phenotype is unknown. It might be interesting to evaluate the operation of the thioredoxin-based light activation system in these mutants.

A regulatory effect of metabolites on chl biosynthesis was also revealed in a study of the *xantha-f*[10] mutant (191). This locus is thought to code for a protein that participates in the insertion of magnesium into protoporphorin IX. Unlike the wild type, induction of δ-aminolevulinate synthesis does not take place in the *xan-f* mutant, suggesting that photoreduction of protochlorophyllide to chlorophyllide *a* is a prerequisite for the induction of δ-aminolevulinate synthesis by light (i.e. a positive feedback loop). Construction of the double mutant *tig-d/tig-d, xan-f/xan-f* caused constitutive protoporphorin accumulation consistent with previous work indicating the regulatory effects of the *tig-d* mutation (191). This suggests that chlorophyllide may normally be required to inactivate the *tig-d* gene product.

Wild-type *Chlamydomonas* synthesizes chl in both the light and the dark. The spontaneous mutant *y-1* does not synthesize chl in darkness and forms a relatively undifferentiated plastid. When transferred to light, chl rapidly accumulates to wild-type levels. Thus, the *y-1* mutation is thought to block the enzymatic step which converts protochlorophyllide to chlorophyllide in darkness (57). Because light-activated chl synthesis in *y-1* is accompanied by synthesis of membrane proteins and the formation of thylakoid membranes, the mutant has been extremely useful for the study of chloroplast biogenesis (57, 136, 195). Oddly, the mutant will grow indefinitely in very low light (0.5 lux) (194) but dies after about nine generations in complete darkness (136, 194). The physiological basis for the cell-death phenomenon is particularly intriguing, as is the high reversion frequency of the mutation which is suggestive of the properties expected of a transposon-induced mutation.

Carotenoid-Deficient Mutants

A number of mutants of maize (3), sunflower (193), *Chlamydomonas* (147), and *Scenedesmus* (26) are partially or completely pigmented in low light but are completely bleached by high light because of defects in carotenoid biosynthesis. The basis for this effect is that when a chl molecule absorbs a quantum of light energy it enters an excited singlet state (^{1}chl). Most of the excitation energy is passed via resonance transfer to P680 and P700 or lost as fluorescence. If energy is not dissipated in this way, the ^{1}chl molecule may convert to the relatively stable triplet state (^{3}chl) which, by virtue of triplet-triplet interactions with molecular O_2, yields toxic singlet oxygen ($^1O_2$). The major protective mechanism in higher plants is afforded by carotenoids of which there is approximately one molecule to five of chl (71). Carotenoids are thought to protect chl against photobleaching by quenching of ^{3}chl by singlet or triplet carotenoid. In contrast to chl, triplet carotenoid does not interact directly with O_2 but dissipates energy by radiationless decay.

The primary evidence for the protective role of carotenoids in plants has come from studies of mutants of maize such as *white-3 (w-3)* (3). This mutant, which is defective in the conversion of phytofluene to γ-carotene (129), is bleached by growth in normal conditions of illumination, but has normal chloroplast ultrastructure and about 50% as much chl as the wild type when grown in very low light (3, 9). When the plants grown in low light are shifted to high light, they bleach within one hour in aerobic but not anaerobic conditions.

Unlike mutants with defects in chl biosynthesis, at least some of the maize carotenoid-deficient mutants accumulate the carotenoid precursor at the site of the lesion (3). The fact that not all of the mutants do this may be useful in elucidating the mechanism of regulation of the carotenoid pathway. The *w-3* and a number of other carotenoid-deficient mutants of maize are viviparous (i.e. *vp-2, vp-5, vp-7, vp-9, y-3, and y-9*). This presumably reflects the fact that some or all abcissic acid is derived from a carotenoid precursor (117a). The metabolic blocks in the *vp-5* and *vp-7* mutants have recently been established (129). It should be noted that because the mutants have been isolated as extreme chl-deficient mutants, all of the carotenoid-deficient mutants appear to have defects in early steps of the pathway which are common to all of the carotenoids. Presumably mutants with a less extreme phenotype can be found which are deficient in only some of the carotenoid species but have normal levels of the others. Such mutants might be useful in investigating the specific roles of the various species of carotenoids.

The albescent (*al* or *y-3*) mutant of maize deserves special mention. When grown in low-intensity light, *al* mutants are indistinguishable from the wild type in chl and carotenoid content. However, when germinated in darkness, the etiolated seedlings completely lack carotenoids and rapidly bleach when transferred to high light. Although the basis for this effect was not established, it was suggested that there may be two pathways for carotenoid synthesis: one active in darkness and another active in light. This hypothesis does not appear to have been pursued but may merit reexamination. The mutant might be particularly useful for determining the action spectrum of carotenoid synthesis (187). Also, as with some of the other carotenoid-deficient maize mutants (9), proplastids of the *al* mutant lacked plastoglobuli, suggesting that carotenoids are essential components of these uncharacterized plastid components.

In contrast to photosynthetic bacteria, little is known about the organization of carotenoids in higher plants. However, fluorescence emission spectra of PSI and PSII protein-chl complexes demonstrate that carotenoids contribute to the chl *a* fluorescence, implying that carotenoids are integral components of each complex. On this basis it has been suggested that all of the photosynthetic pigments of higher plants are complexed to protein in vivo (4). In support of this, analysis of the carotenoid content of chl-protein complexes isolated from acrylamide gels indicated that CP1, CP1a, and CPa are enriched in β-carotene. By contrast, the LHC-II contains primarily lutein and neoxanthin. The chl *b*-deficient *chlorina f2* mutant of barley has a 50% reduction in the amount of neoxanthin and lutein (65, 186). Thus, it seems that if the normal protein carrier is absent, as in the *chlorina f2* mutant, carotenoid fails to accumulate because it is either not synthesized at the same rate as in the wild type or turns over more quickly. Similarly, the *viridis zb*[63] mutant of barley, which lacks the chl-containing polypeptides associated with PSI, had less β-carotene.

In addition to a role in energy transduction and photoreduction, carotenoids may have a role in enhancing the efficiency of reaction center photochemistry. Extraction of β-carotene from nonaqueous chloroplasts appears to greatly reduce PSI activity (154). On this basis it was suggested that β-carotene bound at the reaction center may decrease the rate of transfer of excitation energy away from the reaction center. However, the β-carotene-deficient mutant C-6E of *Scenedesmus* apparently shows high levels of in vivo and in vitro PSI activity (26). A comparable study of the carotenoid-deficient maize mutants does not appear to have been performed.

Chlorophyll is not a requirement for the accumulation of mRNA for LHC-II polypeptides because etiolated barley seedlings exposed to red light accumulate mRNA for at least one of the polypeptides but do not accumulate chl (5). Thus, the accumulation of normal amounts of mRNA for an LHC-II polypeptide in chl-deficient (yellow) mutants of maize is not extraordinary. However, the finding that carotenoid-deficient seedlings contain greatly reduced levels of this mRNA but normal levels of Rubisco mRNA is unusual (72, 107). The observation that several different nuclear mutants have the same effect suggests that the carotenoid deficiencies and the failure to accumulate specific mRNAs are directly related. However, it is not possible to postulate a mechanism at present.

Stoichiometry of the Photosynthetic Apparatus

A relatively simple but important observation that has arisen from the analysis of chl-deficient mutants is that, in general, decreases in chl content result in decreases in photosynthetic unit size rather than the number of photosynthetic units. For example, in the chl-deficient doubly heterozygous mutant line of tobacco *Su/su, Var/var,* the photosynthetic unit size is about 12% of the wild type (137). Presumably this reflects the fact that in the mutant, the available chl is evenly distributed to the reaction centers rather than preferentially distributed to one of the chl-binding complexes. It also suggests that whatever mechanisms regulate the number of reaction centers are largely insensitive to chl content.

Many chl-deficient mutants exhibit substantially higher rates of photosynthesis per unit of chl than the wild type at saturating but not subsaturating light intensities because of reduced photosynthetic unit size (1, 27, 137). An exception to this general rule is the y_{11} mutant of soybean (92, 93). Homozygous mutant plants are almost completely lacking in chl and are inviable. However, heterozygotes, which have about 50% as much chl as the wild type, have the same photosynthesis rate as the wild type on a leaf area basis and a much higher rate on a chl basis at all light intensities. Because of a somewhat reduced carotenoid content, the mutant bleaches when grown under very high light intensities (96) but can be grown to maturity in the field. An extensive analysis of the composition and functional properties of this mutant indicated that it has 3 to 5 times as much plastoquinone per unit chl, and the half-time for plastoquinone oxidation was much higher in the mutant. This could relieve a rate limitation in electron transport, which may account for the increased photosynthetic efficiency of the mutant.

Unfortunately, the mutant exhibits a number of differences in thylakoid composition that precludes a simple interpretation. One intriguing difference is that the amount of monogalactosyl diglyceride in the mutant is strongly correlated with the amount of chl. Since this correlation is not normally seen in other chl-deficient mutants, it raises the possibility that the primary defect is in the galactosyltransferase and that the reduced amount of this lipid limits thylakoid development. It is noteworthy that the amounts of several other chloroplast lipids was not affected. Whatever the precise defect in the y_{11} mutant, the normal growth and productivity of this mutant may be considered evidence that chl is normally present in excess.

Plants adapted to low light exhibit, among many other characteristics, large granal stacks and low chl *a/b* ratios. Under high light intensities, reduced grana stacking and high chl *a/b* ratios are evident (27), but the photosynthetic unit size (chl/P_{700}) does not change significantly. However, Hopkins et al (86) have characterized a dominant mutant of maize *(Oy-yg)* that is sensitive to light intensity. Under low light (ca 40 $\mu E\ m^{-2}s^{-1}$) chl composition of the mutant is normal. At higher light intensities the chl content of the mutant is reduced to about 20% of normal, the chl *a/b* ratio increases, and stacking is virtually absent. It is possible that the wild-type allele at this locus may be involved in mediating normal adaptation mechanisms and that this mutation represents an extreme expression of that control (86). However, the possibility of a leaky mutation in carotenoid biosynthesis or some related function might also explain the phenotype.

Unfortunately, few mutants have been quantitatively analyzed to determine the effects of missing components on the amounts of other components. One informative exception is the *hcf-3* mutant of maize. This mutant, which lacks the PSII recation center complex, has normal levels of plastoquinone, cytochrome f, cytochrome b_6, PSI, LHC-II, and non-PSII chl (115). Thus, whatever regulates the amounts of these components is apparently unrelated to the presence of the PSII components or function. By contrast, a mutant of tobacco that has reduced amounts of LHC-II has an increased ratio of PSII/PSI reaction centers, possibly as a compensatory mechanism (112). This is one of the few instances in which there is apparent regulation of a component in response to altered capacity brought about by a mutation. The seemingly different behavior of the monocot and dicot in this respect may warrant further attention.

In addition to regulation of thylakoid components, it seems likely that the number of chloroplasts per cell is under genetic regulation. This is apparent from the fact that the number of chloroplasts per cell ˙ ;anges as leaf expansion occurs. In general, partially chl-deficient mutants do not compensate for loss of photosynthetic capacity by increasing the number of chloroplasts (92, 93). Thus, whatever the mechanism that regulates plastid number, it does not appear to be responsive to demand for photosynthate or reducing equivalents. No mutation of higher plants that results in altered plastid number has been reported.

Mutations Affecting Chloroplast Development

A common genetic variant affecting pigmentation of higher plants is the class known as virescent in which the seedlings first appear as pigment-deficient but gradually intensify in color and may eventually be indistinguishable from the normally green plant (1). Such mutants are known for many species including barley (97), maize (87), cotton, peanuts, and soybeans (1, 2). In most virescent mutants, greening appears to be retarded at lower temperatures, but at higher temperatures the level of pigmentation may approach that of the wild type. In a systematic study of the temperature effect, it was found that four of five maize mutants with lesions at different loci had a specific threshold temperature below which they would not develop normal pigmentation (87). Recently it has been discovered that all of these lines are deficient in 70S chloroplast ribosomes (85), apparently because of a problem in assembly. In this respect these mutants are similar to the *ram* mutants of *E. coli*, in which ribosome assembly is reduced at low temperature because a change in a structural ribosome protein increases the threshold temperature for ribosome assembly (52). It was suggested that the low number of ribosomes in the virescent mutants causes a delay in development by simply delaying the accumulation of chloroplast-encoded proteins required for normal chloroplast development. Electrophoretic analysis of chloroplast ribosomes might permit a definitive demonstration that the ribosomes are the primary cause.

A virescent mutant of maize *v*-424* which exhibits precocious expression of LHC-II was isolated by screening a collection of virescent mutants for exceptional individuals that exhibit high chlorophyll fluorescence (143). The goal was to identify mutants with an altered sequence of developmentally regulated gene expression. During normal chloroplast development, the assembly of LHC-II is delayed until the reaction centers are assembled. The fact that *v*-424* has precocious LHC-II development could be explained by (*a*) defective protein synthesis in plastids, (*b*) altered turnover of LHC-II, and (*c*) perturbation of developmental timing in assembly of thylakoid complexes. Although the basis of the mutant phenotype is not known, the selection of this potentially interesting mutant is an elegant example of the utility of mutants as tools for selecting new classes of mutants that otherwise would not be identifiable.

Mutations Conferring Herbicide Resistance

One of the most active areas in chloroplast genetics and molecular biology during the past several years has concerned the functional and molecular analysis of spontaneous and induced genetic resistance to triazine herbicides (64, 83). Atrazine and related triazines inhibit PSII activity by displacing a bound plastoquinone that functions as the secondary acceptor on the reducing side of the PSII reaction (189). PSII activity of thylakoid membranes from resistant biotypes is not inhibited by the herbicide in vitro at levels which completely inhibit activity by thylakoid membranes from the sensitive lines (8). Comparisons of radioactive Atrazine binding to isolated thylakoid membranes from resistant biotypes showed the complete loss of high-affinity triazine binding sites in the resistant lines (140). These studies led to the conclusion that the resistant biotypes had an altered binding site on a protein designated Q_B, which resulted in selective loss of triazine affinity without causing a loss of function as an electron carrier.

The synthesis of a functional photoaffinity triazine (azidoatrazine) permitted the identification of the Q_B-binding protein as a 34 to 32 kd thylakoid protein. By demonstrating that the comparable protein in the resistant biotype was not labeled with $[^{14}C]$-azidoatrazine, the identity of the 32 kd protein as the Q_B protein was confirmed (141). The role of the 32 kd polypeptide in PSII was substantiated by the observation that the loss of PSII activity in the *hcf-3* mutant of maize was accompanied by the loss of atrazine binding sites (100).

The observation that triazine resistance was maternally inherited (170a) focused attention on the plastid genome. A maize chloroplast gene (designated *psbA*) for a 32 kd polypeptide which had previously been cloned in the course of studies on chloroplast development was used as a heterologous probe to isolate the corresponding *psbA* genes from both triazine-resistant and susceptible biotypes of several species (185). The subsequent comparison of the DNA sequences from Atrazine-resistant and sensitive biotypes of *Amaranthus hybridus* demonstrated three nucleotide substitutions between the two genes. Two of these were neutral, but one resulted in a serine-to-glycine change at residue 264 (84). Subsequently, the same change was found in an Atrazine-resistant *Solanum nigrum* biotype (83). Similarly, a uniparentally inherited mutation conferring Diuron (DCMU) resistance in *Chlamydomonas* was found to be caused by a single serine-to-alanine change at position 264 of the *psbA* gene(49).

Although the initial work raised the possibility that only one specific mutation could confer resistance to triazine herbicides, recent studies in which mutants of *Chlamydomonas* were selected for resistance to Atrazine, Diuron, Bromacil, or related compounds have indicated that there are other possibilities (54). Some classes of mutants show cross-resistance to the various herbicides where others do not, and the level of resistance varies from one mutant to another.

There has been some question as to whether herbicide resistance is coupled to a deleterious reduction in photosynthetic capactity. The triazine-resistant biotypes of *Amaranthus hybridus* had reduced rates of CO_2 fixation relative to the sensitive types, and had a 23% reduction in quantum yield because of slower transfer of electrons from the primary to the secondary acceptor. However, this step does not appear to limit photosynthesis because thylakoids from both resistant and sensitive lines had equivalent rates of whole-chain photosynthetic electron transport (138). Thus, the effects on CO_2 fixation may be unrelated to the herbicide resistance. Less ambiguous evidence was obtained with herbicide-resistant mutants of *Chlamydomonas* which did not exhibit a difference in growth rate as compared to the wild type (54). Thus, the resistant phenotype need not be deleterious. However, some of the highly resistant mutants did exhibit a depressed growth rate, suggesting that some resistance mutations are deleterious.

Unlike higher plants that carry only one copy of the *psbA* gene, the *psbA* gene in *Chlamydomonas* is located in the inverted repeat region of the chloroplast genome (49). From analysis of deletions in the repeat, it had previously been suggested that a gene conversion mechanism is operative in *Chlamydomonas* chloroplasts (134). This is supported by the observation that both copies of the *psbA* gene in the DCMU-resistant mutants have the same mutation (49).

CONCLUDING REMARKS

Mutants with alterations in photosynthesis have been widely embraced as experimental tools and have been very useful in many aspects of research. This is remarkable in view of the fact that the biochemical basis for most of the mutations that have been employed for functional studies is not known. In this respect, the relatively recent development of methods for gene isolation and characterization, and for genetically transforming plants and algae, will certainly enhance the utility of mutants as a means of marking genes for which the function of the gene product is not known. Thus, the molecular basis of interesting phenotypes which cannot be solved by direct biochemical or biophysical analyses may eventually be elucidated by isolating the gene and working backward from gene product to function. The isolation of photosynthesis-deficient *hcf* mutants of maize by transposon mutagenesis (87a) is a very promising first step in this direction. Other techniques such as chromosome walking and cloning-by-complementation may eventually be useful in some organisms. Thus, serious consideration should be given to the characteristics of an experimental organism before undertaking the isolation of new mutants.

Although mutations that eliminate gene function have many uses, there are a number of interesting questions which can only be answered by creating quantitative rather than qualitative variation. In particular, we know very little about what regulates the expression of genes and what effects variation in the amount or kind of gene products may have on photosynthetic functions. The unexpectedly high photosynthesis rate of the y_{11} soybean mutant may be considered a case in point. It seems likely that we may soon be able to create quantitative variation in the amount of expression of a particular gene by introducing multiple copies of cloned genes into plants or algae or by in vitro mutagenesis of the region of the gene which affects gene transcription or targets the gene product or the mRNA for degradation. Similarly, we may be able to evaluate the effects of quantitative variation in the kinetic constants or the optima of a particular enzyme by introducing the homologous gene from a divergent species, or by altering the gene by site-directed mutagenesis in vitro. Although these approaches are vastly more powerful than conventional approaches, they may depend upon first being able to eliminate the endogenous gene by the isolation of a conventional null mutation. Thus, in order to fully exploit the enormous potential of molecular biology, we must develop more sophisticated approaches to mutant isolation and characterization.

ACKNOWLEDGMENTS

I thank Don Miles and Sylvia Darr for helpful suggestions, W. L. Ogren and R. Spreitzer for communicating unpublished results, and R. Dilley, G. Schmidt and W. Hopkins for providing useful information.

Literature Cited

1. Alberte, R. S., Hesketh, J. D., Hofstra, G., Thornber, J. P., Naylor, A. W., et al. 1974. Composition and activity of the photosynthetic apparatus in temperature-sensitive mutants of higher plants. *Proc. Natl. Acad. Sci. USA* 71:2414–18

2. Alberte, R. S., Hesketh, J. D., Kirby, J. S. 1976. Comparisons of photosynthetic activity and lamellar characteristics of virescent and normal green peanut leaves. *Z. Pflanzenphysiol.* 77:152–59

3. Anderson, I. C., Robertson, D. S. 1960. Role of carotenoids in protecting chlorophyll from photodestruction. *Plant Physiol.* 35:531–34

4. Anderson, J. M., Waldron, J. C., Thorne, S. W. 1978. Chlorophyll-protein complexes of spinach and barley thylakoids. Spectral characteristics of six complexes resolved by an improved electrophoretic procedure. *FEBS Lett.* 92:227–33

5. Apel, K. 1979. Phytochrome-induced appearance of mRNA activity for the apoprotein of the light-harvesting chlorophyll a/b protein of barley *(Hordeum vulgare). Eur. J. Biochem.* 97:183–88

6. Apel, K., Kloppstech, K. 1980. The effect of light on the biosynthesis of light-harvesting chlorophyll a/b protein. *Planta* 150:426–30

7. Armond, P. A., Staehelin, L. A., Arntzen, C. J. 1977. Spatial relationship of photosystem I, photosystem II, and the light-harvesting complex in chloroplast membranes. *J. Cell Biol.* 73:400–18

8. Arntzen, C. J., Ditto, C. L., Brewer, P. E. 1979. Chloroplast membrane alterations in triazine-resistant *Amaranthus retroflexus* biotypes. *Proc. Natl. Acad. Sci. USA* 76:278–82

9. Bachmann, M. D., Robertson, D. S., Bowen, C. C., Anderson, I. C. 1973. Chloroplast ultrastructure in pigment-deficient mutants of *Zea mays* under reduced light. *J. Ultrastruct. Res.* 45:384–406

10. Badger, M. R., Kaplan, A., Berry, J. A. 1980. Internal inorganic carbon pool of *Chlamydomonas reinhardtii.* Evidence for a carbon dioxide-concentrating mechanism. *Plant Physiol.* 66:407–13

11. Barber, J. 1982. Influence of surface charges on thylakoid structure and function. *Ann. Rev. Plant Physiol.* 33:261–95

12. Bazzaz, M. B. 1981. New chlorophyll chromophores isolated from a chlorophyll-deficient mutant of maize. *Photobiochem. Photobiophys.* 2:199–207

13. Bazzaz, M. B., Govindjee, Paolillo, D. J. Jr. 1984. Biochemical, spectral and structural studies of olive necrotic 8147 mutant of *Zea Mays* (L.). *Z. Pflanzenphysiol.* 72:181–92

14. Belknap, W. R. 1983. Partial purification of intact chloroplasts from *Chlamydomonas reinhardii. Plant Physiol.* 72:1130–32

15. Belknap, W. R., Togasaki, R. K. 1982. The effects of cyanide and azide on the photoreduction of 3-phosphoglycerate and oxalocacetate by wild-type and two reductive pentose phosphate cycle mutants of *Chlamydomonas reinhardtii. Plant Physiol.* 70:469–75

16. Bellemare, G., Bartlett, S. G., Chua, N. H. 1982. Biosynthesis of chlorophyll a/b binding polypeptides in wild-type and the chlorina F2 mutant of barley. *J. Biol. Chem.* 257:7762–67

17. Bennett, J. 1981. Biosynthesis of the light-harvesting chl a/b protein. Polypeptide turnover in darkness. *Eur. J. Biochem.* 118:61–70

18. Bennett, J. 1983. Regulation of photosynthesis by reversible phosphorylation of the light-harvesting chlorophyll a/b protein. *Biochem. J.* 212:1–13

19. Bennett, J., Williams, R., Jones, E. 1984. Chlorophyll-protein complexes of higher plants: Protein phosphorylation and preparation of monoclonal antibodies. In *Advances in Photosynthesis Research,* ed. C. Sybesma, 3:99–106. The Hague: Nijhoff

20. Bennoun, P., Diner, B., Wollman, F., Schmidt, G., Chua, N. H. 1981. Thylakoid polypeptides associated with photosystem II in *Chlamydomonas reinhardii:* Comparison of system II mutants and particles. In *Structure and Molecular Organization of the Photosynthetic Apparatus,* ed. G. Akoyunaglou, 3:839–49. Philadelphia: Balban Int. Sci. Serv.

21. Bennoun, P., Girard, J., Chua, N. H. 1977. A uniparental mutant of *Chlamydomonas reinhardtii* deficient in chlorophyll protein complex CPI. *Mol. Gen. Genet.* 153:343–48

22. Bennoun, P., Jupin, H. 1976. Spectral properties of system I deficient mutants of *Chlamydomonas reinhardii. Biochim. Biophys. Acta* 440:122–30

23. Bennoun, P., Levine, R. P. 1967. Detecting mutants that have impaired photosynthesis by their increased level of fluorescence. *Plant Physiol.* 42:1284–87

24. Bennoun, P., Masson, A., Delosme, M. 1980. A method for complementation

analysis of nuclear and chloroplast mutants of photosynthesis in *Chlamydomonas*. *Genetics* 95:39–47

25. Bishop, N. I. 1984. Further assessment through mutational analysis of chloroplast membrane polypeptides required in water photolysis. In *Advances in Photosynthesis Research*, ed. C. Sybesma, 1:321–32. The Hague: Nijhoff

26. Bishop, N. I., Oquist, G. 1980. Correlation of the photosystem I and II reaction center chloroplast-protein complexes, CP-$_{aI}$ and CP-$_{aII}$ with photosystem activity and low temperature fluorescence emission properties in mutants of *Scenedesmus*. *Physiol. Plant.* 49:477–86

27. Boardman, N. K. 1977. Comparative photosynthesis of sun and shade plants. *Ann. Rev. Plant Physiol.* 28:355–77

28. Boardman, N. K., Highkin, H. R. 1966. Studies on a barley mutant lacking chlorophyll *b*. I. Photochemical activity of isolated chloroplasts. *Biochim. Biophys. Acta* 126:189–99

29. Boardman, N. K., Thorne, S. W. 1976. Cation effects on light-induced chlorophyll *a* fluorescence in chloroplasts lacking both chlorophyll *b* and chlorophyll-protein complex II. *Plant Sci. Lett.* 7:219–24

30. Borner, T., Sears, B. B. 1986. Plastome mutants. *Curr. Genet.* In press

31. Bright, S. W., Lea, P. J., Arruda, P., Hall, N. P., Kendall, A. C., et al. 1984. Manipulation of key pathways in photorespiration and amino acid metabolism by mutation and selection. In *The Genetic Manipulation of Plants and its Application to Agriculture*, ed. P. J. Lea, G. R. Stewart, pp. 141–69. Oxford: Univ. Press

32. Burke, J. J., Steinback, K. E., Arntzen, C. J. 1979. Analysis of the light-harvesting pigment-protein complex of wild type and a chlorophyll-*b*-less mutant of barley. *Plant Physiol.* 63:237–43

33. Butler, W. L. 1978. Energy distribution in the photochemical apparatus of photosynthesis. *Ann. Rev. Plant Physiol.* 29: 345–78

34. Carlsen, B. 1977. Barley mutants with defects in photosynthetic carbon dioxide fixation. *Carlsberg Res. Commun.* 42: 199–209

35. Casadoro, G., Hoyer-Hansen, G., Gaminikannangara, C., Gough, S. P. 1983. An analysis of temperature and light sensitivity in *tigrina* mutants of barley. *Carlsberg Res. Commun.* 48:95–129

36. Caspar, T., Huber, S. C., Somerville, C. R. 1985. Alterations in growth, photosynthesis and respiration in a starchless mutant of *Arabidopsis* deficient in chloroplast phosphoglucomutase activity. *Plant Physiol.* 79:11–17

37. Castelfranco, P. A., Beale, S. I. 1983. Chlorophyll biosynthesis: Recent advances and areas of current interest. *Ann. Rev. Plant Physiol.* 34:241–78

38. Chastain, C. J., Ogren, W. L. 1985. Photorespiration-induced reduction of ribulose bisphosphate carboxylase activation level. *Plant Physiol.* 77:851–56

39. Chua, N.-H., Bennoun, P. 1975. Thylakoid membrane polypeptides of *Chlamydomonas reinhardtii*: Wild-type and mutant strains deficient in photosystem II reaction center. *Proc. Natl. Acad. Sci. USA* 72:2175–79

40. Chua, N. H., Matlin, K., Bennoun, P. 1975. A chlorophyll-protein complex lacking in photosystem I mutants of *Chlamydomonas reinhardtii*. *J. Cell Biol.* 67:361–77

41. Coe, E. H., Neuffer, M. G. 1977. The genetics of corn. In *Corn and Corn Improvement*, ed. G. F. Sprague, pp. 148–55. Madison: Am. Soc. Agron.

42. Coruzzi, G., Broglie, R., Cashmore, A., Chua, N. H. 1983. Nucleotide sequences of two pea cDNA clones encoding the small subunit of ribulose 1,5-bisphosphate carboxylase and the major chlorophyll *a/b* binding thylakoid polypeptide. *J. Biol. Chem.* 258:1399–1402

43. Creach, E., Stewart, C. R. 1982. Effects of aminoacetonitrile on net photosynthesis, ribulose 1,5-bisphosphate levels and glycolate pathway intermediates. *Plant Physiol.* 70:1444–48

44. Delepelaire, P., Chua, N. H. 1979. Lithium dodecyl sulfate/polyacrylamide gel electrophoresis of thylakoid membranes at 4°C: Characterizations of two additional chl *a*-protein complexes. *Proc. Natl. Acad. Sci. USA* 76:111–15

45. Desautels, M., Goldberg, A. L. 1982. Liver mitochondria contain an ATP-dependent, vanadate-sensitive pathway for the degradation of proteins. *Proc. Natl. Acad. Sci. USA* 79:1869–73

46. Dron, M., Rahire, M., Rochaix, J. D., Mets, L. 1983. First DNA sequence of a chloroplast mutation: A missense alteration in the ribulose bisphosphate carboxylase large subunit gene. *Plasmid* 9:321–24

47. Duysen, M. E., Freeman, T. P., Williams, N. D., Olson, L. L. 1984. Regulation of excitation energy in a wheat mutant deficient in light-harvesting pigment protein complex. *Plant Physiol.* 76:561–66

48. Epel, B. L., Butler, W. L., Levine, R. P. 1979. A spectroscopic analysis of low

fluorescence mutants of *Chlamydomonas reinhardtii* blocked in their water splitting oxygen evolving apparatus. *Biochim. Biophys. Acta* 274:395–400

49. Erickson, J. M., Rahire, M., Bennoun, P., Delepelaire, P., Diner, B., Rochaix, J. D. 1984. Herbicide resistance in *Chlamydomonas reinhardtii* results from a mutation in the chloroplast gene for the 32-kilodalton protein of photosystem II. *Proc. Natl. Acad. Sci. USA* 81: 3617–21

50. Erickson, J. M., Schneider, M., Vallet, J. M., Dron, M., Bennoun, P., Rochaix, J. D. 1984. Chloroplast gene function, combined genetic and molecular approach in *Chlamydomonas reinhardii*. In *Advances in Photosynthesis Research*, ed. C. Sybesma, 4:491–99. The Hague: Nijhoff

51. Estelle, M., Hanks, J., McIntosh, L., Somerville, C. R. 1985. Site-specific mutagenesis of ribulose-1,5-bisphosphate carboxylase/oxygenase. *J. Biol. Chem.* 260:9523–26

52. Feunteun, J., Monier, R., Vola, C., Roset, R. 1974. Ribosomal assembly defective mutants of *Escherichia coli*. *Nucleic Acids Res.* 1:149–69

53. Flick, C. E., Evans, D. A. 1982. Evaluation of cytoplasmic segregation in somatic hybrids of *Nicotiana:* tentoxin sensitivity. *J. Hered.* 73:264–66

54. Galloway, R. E., Mets, L. J. 1984. Atrazine, Bromacil, and Diuron resistance in *Chlamydomonas*. A single nomendelian genetic locus controls the structure of the thylakoid binding site. *Plant Physiol.* 74:469–74

54a. Garnier, J. 1967. Une méthode dédetection, par photographie, de souches d'Algues vertes émettant in vitro une fluorescence anormale. *C. R. Acad. Sci. Paris Ser. D* 265:874–77

55. Gatenby, A. A. 1984. The properties of the large subunit of maize ribulose bisphosphate carboxylase/oxygenase synthesized in *E. coli*. *Eur. J. Biochem.* 144:361–66

56. Gatenby, A. A., van der Vies, S., Roadley, D. 1985. Assembly in *E. coli* of a functional multi-subunit ribulose bisphosphate carboxylase from a blue-green alga. *Nature* 314:617–20

57. Gfeller, R. P., Gibbs, M. 1985. Fermentative metabolism of *Chlamydomonas reinhardtii*. *Plant Physiol.* 77:509–11

57a. Gillham, N. W., Boynton, J. E., Chua, N. H. 1978. Genetic control of the photosynthetic membrane. *Curr. Top. Bioenerg.* 8:209–59

58. Girard, J., Chua, N. H., Bennoun, P.,

Schmidt, G., Delosme, M. 1980. Studies on mutants deficient in the photosystem I reaction centers in *Chlamydomonas reinhardii*. *Curr. Genet.* 2:215–21

59. Goodchild, D. J., Highkin, H. R., Boardman, N. K. 1966. The fine structure of chloroplasts in a barley mutant lacking chlorophyll *b*. *Exp. Cell Res.* 43:684–88

60. Goodenough, U. W., Armstrong, J. J., Levine, R. P. 1969. Photosynthetic properties of ac-31, a mutant strain of *Chlamydomonas reinhardii* devoid of chloroplast membrane stacking. *Plant Physiol.* 44:1001–12

61. Gough, S. 1972. Defective synthesis of porphyrins in barley plastids caused by mutations in nuclear genes. *Biochim. Biophys. Acta* 286:36–54

62. Gough, S. P., Kannangara, C. G. 1979. Biosynthesis of δ-aminolevulinate in *tigrina* mutants of barley. *Carlsberg Res. Commun.* 44:403–16

63. Green, B. R., Karukstis, K. K., Sauer, K. 1984. Fluorescence decay kinetics of mutants of corn deficient in photosystem I and photosystem II. *Biochim. Biophys. Acta* 767:574–81

64. Gressel, J. 1985. The molecular anatomy of resistance to photosystem II herbicide. *Oxford Surveys Plant Mol. Cell Biol.* 2:321–28

65. Grumbach, K. H. 1985. Pigment and quinone content of two photosynthetic barley mutants. *Physiol. Plant* 63:265–68

66. Gurevitz, M., Somerville, C. R., McIntosh, L. 1985. Pathway of assembly of ribulosebisphosphate carboxylase/oxygenase from *Anabaena* 7120 expressed in *Escherichia coli*. *Proc. Natl. Acad. Sci. USA* 82:6546–50

67. Gutteridge, S., Sigal, I., Thomas, B., Arentzen, R., Cordova, A., Lorimer, G. 1984. A site-specific mutation within the active site of ribulose-1,5-bisphosphate carboxylase of *Rhodospirillum rubrum*. *EMBO J.* 3:2737–43

68. Hall, N. D., Kendall, A. C., Turner, J. C. 1985. Characteristics of a photorespiratory mutant of barley deficient in phosphoglycolate phosphatase. *Plant Physiol.* 77S:25

69. Hallick, R., Edelman, M., Chua, N. H., eds. 1982. *Methods in Chloroplast Molecular Biology*. New York: Elsevier. 1140 pp.

70. Hallier, V. W., Schmitt, J. M., Heber, U., Chaianova, S. S., Volodarsky, A. D. 1978. Ribulose-1,5-disphosphate carboxylase-deficient plastome mutants of *Oenothera*. *Biochim. Biophys. Acta* 504: 67–83

71. Halliwell, B. 1981. *Chloroplast Metabolism.* Oxford: Oxford Univ. Press
72. Harpster, M. H., Mayfield, S. P., Taylor, W. C. 1984. Effects of pigment-deficient mutants on the accumulation of photosynthetic proteins in maize. *Plant Mol. Biol.* 3:59–71
73. Harris, E. H., Boynton, J. E., Gillham, N. W. 1982. Induction of nuclear and chloroplast mutations which affect the chloroplast in *Chlamydomonas reinhardtii.* In *Methods in Chloroplast Molecular Biology,* ed. R. Hallide, M. Edelman, N. Chua, pp. 3–24. New York: Elsevier
74. Hayden, D. B., Hopkins, W. G. 1977. A second distinct chl *a* protein complex in maize mesophyll chloroplasts. *Can J. Bot.* 55:2525–29
75. Haworth, P., Kyle, D., Arntzen, C. J. 1982. Protein phosphorylation and excitation energy distribution in normal intermitten-light-grown, and a chlorophyll *b*-less mutant of barley. *Arch. Biochem. Biophys.* 218:199–206
76. Haworth, P., Watson, J. L., Arntzen, C. J. 1983. The detection, isolation and characterization of a light-harvesting complex which is specifically associated with PS I. *Biochim. Biophys. Acta.* 724:151–58
77. Henry, L. E. A., Mikkelsen, J. D., Moller, B. L. 1983. Pigment and acyl lipid composition of photosystem I and II vesicles and of photosynthetic mutants in barley. *Carlsberg. Res. Commun.* 48:131–48
78. Herrmann, F. 1971. Genetic control of pigment-protein complexes I and Ia of the plastid mutant En: Alba-I of *Antirrhinum majus. FEBS Lett.* 19:267–69
79. Herrmann, R. G., Westhoff, P., Alt, J., Winter, P., Tittgen, J., et al. 1982. Identification and characterization of genes for polypeptides of the thylakoid membrane. In *Structure and Function of Plant Genomes,* ed. O. Ciferri, L. Dure, pp. 143–55. New York: Plenum
80. Hildebrandt, J., Bottomley, W., Moser, J., Herrmann, R. G. 1984. A plastome mutant of *Oenothera hookeri* has a lesion in the gene for the large subunit of ribulose-1,5-bisphosphate carboxylase/oxygenase. *Biochim. Biophys. Acta* 783:67–73
81. Hiller, R. G., Moller, B. L., Hoyer-Hansen, G. 1980. Characterization of six putative PS I mutants in barley. *Carlsberg. Res. Commun.* 45:315–28
82. Hirono, Y., Redei, G. P. 1963. Multiple allelic control of chlorophyll *b* level in *Arabidopsis thaliana. Nature* 197:1324–25

83. Hirschberg, J., Bleecker, A., Kyle, D. J., McIntosh, L., Arntzen, C. J. 1984. The molecular basis of triazine-herbicide resistance in higher plant chloroplast. *Z. Naturforsch. Teil C* 39:412–20
84. Hirschberg, J., McIntosh, L. 1983. The molecular basis of herbicide resistance in *Amaranthus hybridus. Science* 222: 1346–49
84a. Honberg, L. S. 1984. Probing barley mutants with a monoclonal antibody to a polypeptide involved in photosynthetic oxygen evolution. *Carlsberg Res. Commun.* 49:703–19
85. Hopkins, W. G., Elfman, B. 1984. Temperature-induced chloroplast ribosome deficiency in virescent maize. *J. Hered.* 75:207–11
86. Hopkins, W. G., Hayden, D. B., Neuffer, M. G. 1980. A light-sensitive mutant in maize (*Zea mays* L.) I. chlorophyll, chlorophyll-protein and ultrastructure studies. *Z. Pflanzen physiol.* 99:417–26
87. Hopkins, W. G., Walden, D. B. 1977. Temperature sensitivity of virescent mutants of maize. *J. Hered.* 68:283–86
87a. Hunt, M., Miles, D. 1984. A mutator-induced photosynthesis mutant. *Maize Genet. Coop Newsl.* 58:68–70
88. Izhar, S., Davey, M. R., Gatenby, A. A. 1979. A light-sensitive mutant in *Petunia.* Growth ultrastructure and regulation of chlorophyll and ribulose-1,5-bisphosphate carboxylase content in the leaves. *Plant Sci. Lett.* 15:75–82
89. Jellings, A. J., Leese, B. M., Leech, R. M. 1983. Location of chromosomal control of ribulose bisphosphate carboxylase amounts in wheat. *Mol. Gen. Genet.* 192:272–74
90. Jensen, K. H., Schmidt, G. W. 1984. Translational arrest of chloroplast-synthesized 32-34 kDa polypeptide in a photosystem II mutant of *Chlamydomonas. Plant Physiol.* 75S:173
91. Jordan, D., Ogren, W. L. 1981. Species variation in the specificity of ribulose diphosphate carboxylase/oxygenase. *Nature* 291:513–15
92. Keck, R. W., Dilley, R. A., Allen, C. F., Biggs, S. 1970. Chloroplast composition and structure differences in a soybean mutant. *Plant Physiol.* 46:692–98
93. Keck, R. W., Dilley, R. A., Ke, B. 1970. Photochemical characteristics in a soybean mutant. *Plant Physiol.* 46:699–704
94. Kendall, A. C., Keys, A. J., Turner, J. C., Lea, P. J., Miflin, B. J. 1983. The isolation and characterization of a catalase-deficient mutant of barley (*Hordeum vulgare* L.). *Planta* 159:505–11

94a. Kirk, J. T. O., Tilney-Bassett, R. A. E. 1978. *The Plastids: Their Chemistry, Structure, Growth and Inheritance.* Amsterdam: Elsevier. 960 pp. 2nd ed.

95. Klein, U., Chen, C., Gibbs, M., Platt-Aloia, K. 1983. Cellular fractionation of *Chlamydomonas reinhardii* with emphasis on the isolation of the chloroplast. *Plant Physiol.* 72:481–87

96. Koller, H. R., Dilley, R. A. 1974. Light intensity during leaf growth affects chlorophyll concentration and CO$_2$ assimilation of a soybean chlorophyll mutant. *Crop Sci.* 14:779–82

97. Kyle, D. J., Zalik, S. 1982. Photosystem II activity, plastoquinone A levels, and fluorescence characterization of a virescens mutant of barley. *Plant Physiol.* 70:1026–31

98. Lea, P. J., Hall, N. P., Kendall, A. C., Keys, A. J., Turner, J. C. 1984. The isolation of a photorespiratory mutant of barley unable to convert glycine to serine. *Plant Physiol.* 75S:155

99. Leto, K. J., Bell, E., McIntosh, L. 1985. Nuclear mutation leads to an accelerated turnover of chloroplast-encoded 48 kd and 34.5 kd polypeptides in thylakoids lacking photosystem II. *EMBO J.* 4: 1645–53

100. Leto, K. J., Keresztes, A., Arntzen, C. J. 1982. Nuclear involvement in the appearance of a chloroplast-encoded 32,000 dalton thylakoid membrane polypeptide integral to the photosystem II complex. *Plant Physiol.* 69:1450–58

101. Leto, K. J., Miles, C. D. 1980. Characterization of three photosystem II mutants in *Zea mays* lacking a 32,000 dalton lamellar polypeptide. *Plant Physiol.* 66:18–24

102. Levine, R. P. 1969. The analysis of photosynthesis using mutant strains of algae and higher plants. *Ann. Rev. Plant Physiol.* 20:523–40

103. Machold, O., Hoyer-Hansen, G. 1976. Polypeptide composition of thylakoid from viridis and xantha mutants in barley. *Carlsberg Res. Commun.* 41:360–66

104. Machold, O., Simpson, D. J., Moller, B. L. 1979. Chlorophyll-proteins of thylakoids from wild-type and mutants of barley (*Hordeum vulgare* L.). *Carlsberg Res. Commun.* 44:235–54

105. Markwell, J. P., Webber, A. N., Lake, B. 1985. Mutants of sweetclover (*Melilotus alba*) lacking chlorophyll *b*. *Plant Physiol.* 77:948–51

106. Maroc, J., Guyon, D., Garnier, J. 1983. Characterization of new strains of nonphotosynthetic mutants of *Chlamydomonas reinhardii*. III. Photosystem II-related thylakoid proteins in five mutants and double mutants. *Plant Cell Physiol.* 24:1217–30

107. Mayfield, S. P., Taylor, W. C. 1984. Carotenoid-deficient maize seedlings fail to accumulate light-harvesting chlorophyll *a/b* binding protein (LHCP) mRNA. *Eur. J. Biochem.* 144:79–84

108. Mascia, P. 1978. An analysis of precursors accumulated by several chlorophyll biosynthetic mutants of maize. *Mol. Gen. Genet.* 161:237–44

109. Deleted in proof

110. McCourt, P., Browse, J., Watson, J., Arntzen, C. J., Somerville, C. R. 1985. Photosynthetic antenna function in a mutant of *Arabidopsis* lacking *trans*-hexadecenoic acid. *Plant Physiol.* 78: 853–58

111. Melis, A. 1985. Functional properties of photosystem II$_B$ in spinach chloroplasts. *Biochim. Biophys. Acta* 808:334–42

112. Melis, A., Brown, J. S. 1980. Stoichiometry of system I and system II reaction centers end of plastoquinone in different photosynthetic membranes. *Proc. Natl. Acad. Sci. USA* 77:4712–16

113. Metz, J., Bishop, N. I. 1980. Identification of a chloroplast membrane polypeptide associated with the oxidizing side of photosystem II by the use of select low-fluorescent mutants of *Scenedesmus*. *Biochem. Biophys. Res. Commun.* 94:560–66

114. Metz, J. G., Krueger, R. W., Miles, D. 1984. Chlorophyll-protein complexes of a photosystem II mutant of maize. *Plant Physiol.* 75:238–41

115. Metz, J. G., Miles, D. 1982. Use of a nuclear mutant of maize to identify components of photosystem II. *Biochim. Biophys. Acta* 681:95–102

116. Metz, J. G., Wong, J., Bishop, N. I. 1980. Changes in electrophoretic mobility of a chloroplast membrane polypeptide associated with the loss of the oxidizing side of photosystem II in low fluorescent mutants of *Scenedesmus*. *FEBS Lett.* 114:61–66

117. Michel, H., Tellenbach, M., Boschetti, A. 1983. A chlorophyll *b*-less mutant of *Chlamydomonas reinhardii* lacking in the light-harvesting chlorophyll *a/b*-protein complex but not in its apoproteins. *Biochim. Biophys. Acta* 725:417–24

117a. Milborrow, B. V. 1983. Pathways to and from abscissic acid. In *Abscisic Acid*, ed. F. Addicott, pp. 79–111. New York: Praeger Scientific

118. Miles, C. D. 1976. Selection of diquat resistance photosynthesis mutants from maize. *Plant Physiol.* 57:284–85

119. Miles, C. D. 1982. The use of mutations to probe photosynthesis in higher plants. In *Methods in Chloroplast Molecular Biology,* ed. R. Hallick, M. Edelman, N. Chua, pp. 75–109. New York: Elsevier

120. Miles, C. D., Daniel, D. J. 1974. Chloroplast reactions of photosynthetic mutants in *Zea mays. Plant Physiol.* 53:589–95

121. Miles, C. D., Markwell, J. P., Thornber, J. P. 1979. Effect of nuclear mutation in maize on photosynthetic activity and content of chlorophyll-protein complexes. *Plant Physiol.* 64:690–94

122. Miles, D. M., Leto, K. J., Neuffer, M. G., Polacco, M., Hanks, S., Hunt, M. A. 1985. Chromosome arm location of photosynthesis mutants in *Zea mays* using B-A translocations. In *Molecular Biology of the Photosynthetic Apparatus.* Cold Spring Harbor, NY: Cold Spring Harbor Lab. In press

123. Miller, K. R. 1980. A chloroplast membrane lacking photosystem I changes in unstacked membrane regions. *Biochim. Biophys. Acta* 592:143–52

124. Miller, K. R., Cushman, R. A. 1979. A chloroplast membrane lacking photosystem II. *Biochim. Biophys. Acta* 546:481–97

125. Mishkind, M. L., Schmidt, G. W. 1983. Post transcriptional regulation of ribulose-1,5-bisphosphate carboxylase small subunit accumulation in *Chlamydomonas reinhardtii. Plant Physiol.* 72:847–54

126. Miziorko, H. M., Lorimer, G. H. 1983. Ribulose-1,5-bisphosphate carboxylase-oxygenase. *Ann. Rev. Biochem.* 52:507–35

127. Moll, B., Levine, R. P. 1970. Characterization of a photosynthetic mutant strain of *Chlamydomonas reinhardii* deficient in phosphoribulokinase activity. *Plant Physiol.* 46:576–80

128. Moller, B. L., Smillie, R. M., Hoyer-Hansen, G. 1980. A photosystem I mutant in barley. *Carlsberg Res. Commun.* 45:87–99

129. Moore, R., Smith, J. D. 1985. Graviresponsiveness and abscisic-acid content of roots of carotenoid-deficient mutants of *Zea mays* L. *Planta* 164:126–28

130. Moroney, J. V., Tolbert, N. E. 1985. Inorganic carbon uptake by *Chlamydomonas reinhardtii. Plant Physiol.* 77:253–58

131. Mullet, J. E., Burke, J. J., Arntzen, C. J. 1980. Photosystem I chlorophyll-proteins. *Plant Physiol.* 65:814–27

132. Mullet, J. E., Burke, J. J., Arntzen, C. J. 1980. A developmental study of photosystem I peripheral chlorophyll proteins. *Plant Physiol.* 65:823–27

133. Mulligan, R. M., Wilson, B., Tolbert, N. E. 1983. Effects of glyoxylate on photosynthesis by isolated chloroplast. *Plant Physiol.* 72:415–19

134. Myers, A. M., Grant, D. M., Rabert, D. K., Harris, E. H., Boynton, J. E., Gillham, N. W. 1982. Mutants of *Chlamydomonas reinhardtii* with physical alterations in their chloroplast DNA. *Plasmid* 7:133–51

135. Ogren, W. L. 1984. Photorespiration: Pathways, regulation, and modification. *Ann. Rev. Plant Physiol.* 35:415–42

136. Ohad, I. 1975. Biogenesis of chloroplast membranes. In *Membrane Biogenesis,* ed. A. Tzagoloff, pp. 279–350. New York: Plenum

137. Okabe, K., Schmid, G. H., Straub, J. 1977. Genetic characterization and high efficiency photosynthesis of an aurea mutant of tobacco. *Plant Physiol.* 60:150–56

138. Ort, D. R., Ahrens, W. H., Martin, B., Stoller, E. W. 1983. Comparison of photosynthetic performance in triazine-resistant and susceptible biotypes of *Amaranthus hybridus. Plant Physiol.* 72:925–30

139. Parker, M. L., Lea, P. J. 1983. Ultrastructure of the mesophyll cells of leaves of a catalase-deficient mutant of barley (*Hordeum vulgare* L.). *Planta* 159:512–17

140. Pfister, K., Radosevich, S. R., Arntzen, C. J. 1979. Modifications of herbicide binding to photosystem II in two biotypes of *Senecio vulgaris* L. *Plant Physiol.* 64:995–99

141. Pfister, K., Steinback, K. E., Gardner, G., Arntzen, C. J. 1981. Photoaffinity labeling of an herbicide receptor protein in chloroplast membranes. *Proc. Natl. Acad. Sci. USA* 78:981–85

142. Deleted in proof

143. Polacco, M. L., Chang, M. T., Neuffer, M. G. 1985. Nuclear virescent mutants of *Zea mays* (L.) with high levels of chlorophyll (*a/b*) light harvesting complex during thylakoid assembly. *Plant Physiol.* 77:795–800

144. Powles, S. B. 1984. Photoinhibition of photosynthesis induced by visible light. *Ann. Rev. Plant Physiol.* 35:15–44

145. Rochaix, J. D., van Dillewijn, J. 1982. Transformation of the green alga

Chlamydomonas reinhardii with yeast DNA. *Nature* 296:70–72

146. Ryrie, J. J. 1983. Immunological evidence for apoproteins of the light-harvesting chlorophyll-protein complex in a mutant of barley lacking chl *b*. *Eur. J. Biochem.* 121:149–55

147. Sager, R., Zalokar, M. 1958. Pigments and photosynthesis in a carotenoid-deficient mutant of *Chlamydomonas*. *Nature* 182:98–100

148. Salvucci, M. E., Ogren, W. L. 1985. A *Chlamydomonas reinhardii* mutant with catalytically and structurally altered ribulose-5-phosphate kinase. *Planta* 165:340–47

149. Salvucci, M. E., Portis, A. R., Ogren, W. L. 1985. A soluble chloroplast protein catalyzes ribulose bisphosphate carboxylase/oxygenase activation in vivo. *Photosynth. Res.* In press

150. Schmidt, G. W., Bartlett, S. G., Grossman, A. R., Cashmore, A. R., Chua, N. H. 1981. Biosynthetic pathways of two polypeptide subunits of the light-harvesting chl *a/b* protein complex. *J. Cell Biol.* 91:468–78

151. Schmidt, G. W., Matlin, K. S., Chua, N. H. 1977. A rapid procedure for selective enrichment of photosynthetic electron transport mutants. *Proc. Natl. Acad. Sci. USA* 74:610–14

152. Schmidt, G. W., Mishkind, M. L. 1983. Rapid degradation of unassembled ribulose 1,5-bisphosphate carboxylase small subunits in chloroplasts. *Proc. Natl. Acad. Sci. USA* 80:2632–36

153. Schwarz, H. P., Kloppstech, K. 1982. Effects of nuclear gene mutations on the structure and function of plastids in pea. The light harvesting chl *a/b* protein. *Planta* 155:116–23

154. Searle, G. F. W., Wessels, J. S. C. 1978. Role of β-carotene in the reaction centers of photosystem I and II of spinach chloroplasts prepared in nonpolar solvents. *Biochim. Biophys. Acta* 504:84–99

155. Shepherd, H. S., Boynton, J. E., Gillham, N. W. 1979. Mutations in nine chloroplast loci of *Chlamydomonas* affecting different photosynthetic functions. *Proc. Natl. Acad. Sci. USA* 76:1353–57

156. Simpson, D. J. 1979. Freeze-fracture studies on barley plastid membranes III. Location of the light-harvesting chlorophyll-protein. *Carlsberg Res. Commun.* 44:305–36

157. Simpson, D. J., von Wettstein, D. 1980. Macromolecular physiology of plastids XIV. Viridis mutants in barley: genetic fluoroscopic and ultrastructural charac-

terization. *Carlsberg Res. Commun.* 45:283–314

158. Smith, I. K., Kendall, A. C., Keys, A. J., Turner, J. C., Lea, P. J. 1984. Increased levels of glutathione in a catalase-deficient mutant of barley. *Plant Sci. Lett.* 37:29–33

159. Somerville, C. R. 1984. The analysis of photosynthetic carbon dioxide fixation and photorespiration by mutant selection. *Oxford Surveys Plant Mol. Cell Biol.* 1:103–31

160. Somerville, C. R., McIntosh, L., Fitchen, J., Gurevitz, M. 1985. The cloning and expression in *E. coli* of RuBP carboxylase/oxygenase large subunit genes. *Methods Enzymol.* 118:419–33

161. Somerville, C. R., Ogren, W. L. 1979. A phosphoglycolate phosphatase deficient mutant of *Arabidopsis*. *Nature* 280:833–36

162. Somerville, C. R., Ogren, W. L. 1980. Inhibition of photosynthesis in *Arabidopsis* mutants lacking leaf glutamate synthase activity. *Nature* 286:257–59

163. Somerville, C. R., Ogren, W. L. 1980. Photorespiration mutants of *Arabidopsis* deficient in serine: Glyoxylate aminotransferase activity. *Proc. Natl. Acad. Sci. USA* 77:2684–87

164. Somerville, C. R., Ogren, W. L. 1981. Photorespiration-deficient mutants of *Arabidopsis thaliana* lacking serine transhydroxymethylase activity. *Plant Physiol.* 67:666–71

165. Somerville, C. R., Ogren, W. L. 1982. Mutants of the cruciferous plant *Arabidopsis thaliana* lacking glycine decarboxylase activity. *Biochem. J.* 202:373–80

166. Somerville, C. R., Portis, A. R., Ogren, W. L. 1982. A mutant of *Arabidopsis thaliana* which lacks activation of RuBP carboxylase in vivo. *Plant Physiol.* 70:381–87

167. Somerville, C. R., Somerville, S. C. 1984. Cloning and expression of the *Rhodospirillum rubrum* ribulose bisphosphate carboxylase gene in *E. coli*. *Mol. Gen. Genet.* 193:214–19

168. Somerville, S. C., Ogren, W. L. 1983. An *Arabidopsis thaliana* mutant defective in chloroplast dicarboxylate transport. *Proc. Natl. Acad. Sci. USA* 80:1290–94

169. Somerville, S. C., Somerville, C. R. 1983. Effect of oxygen and carbon dioxide on photorespiratory flux determined from glycine accumulation in a mutant of *Arabidopsis*. *J. Exp. Bot.* 34:415–24

170. Somerville, S. C., Somerville, C. R.

1984. A mutant of *Arabidopsis* deficient in chloroplast dicarboxylate transport is missing an envelope protein. *Plant Sci. Lett.* 37:217–20

170a. Souza Machado, U., Bandeen, J. D., Stephenson, G. R., Lavigne, P. 1978. Uniparental inheritance of chloroplast atrazine tolerance in *Brassica campestris*. *Can. J. Plant Sci.* 58:977–81

171. Spalding, M. H., Spreitzer, R. J., Ogren, W. L. 1983. Reduced inorganic carbon transport in a CO$_2$-requiring mutant of *Chlamydomonas reinhardii*. *Plant Physiol.* 73:273–76

172. Spalding, M. H., Spreitzer, R. J., Ogren, W. L. 1983. Carbonic anhydrase deficient mutant of *Chlamydomonas* requires elevated carbon dioxide concentration for photoautotrophic growth. *Plant Physiol.* 73:268–72

173. Spalding, M. H., Spreitzer, R. J., Ogren, W. L. 1983. Genetic and physiological analysis of the CO$_2$-concentrating system of *Chlamydomonas reinhardii*. *Planta* 159:261–66

174. Spreitzer, R. J., Chastain, C. J., Ogren, W. L. 1984. Chloroplast gene suppression of defective ribulose bisphosphate carboxylase/oxygenase in *Chlamydomonas reinhardii:* Evidence for stable heteroplasmic genes. *Curr. Genet.* 9:83–89

175. Spreitzer, R. J., Goldschmidt-Clermont, M., Rahire, M., Rochaix, J. D. 1985. Nonsense mutations in the *Chlamydomonas* gene that codes for the large subunit of ribulose bisphosphate carboxylase/oxygenase. *Proc. Natl. Acad. Sci. USA* 82:5460–64

176. Spreitzer, R. J., Mets, L. J. 1980. Nonmendelian mutation affecting ribulose-1,5-bisphosphate carboxylase structure and activity. *Nature* 285:114–15

177. Spreitzer, R. J., Mets, L. J. 1981. An assessment of arsenate selection as a method for obtaining nonphotosynthetic mutants of *Chlamydomonas*. *Genetics* 100:417–25

178. Spreitzer, R. J., Mets, L. 1981. Photosynthesis deficient mutants of *Chlamydomonas reinhardii* with associated light-sensitive phenotypes. *Plant Physiol.* 67:565–69

179. Spreitzer, R. J., Ogren, W. L. 1983. Rapid recovery of chloroplast mutations affecting ribulose bisphosphate carboxylase/oxygenase in *Chlamydomonas reinhardii*. *Proc. Natl. Acad. Sci. USA* 80:6293–97

180. Spreitzer, R. J., Ogren, W. L. 1983. Nuclear suppressors of the photosensitivity associated with defective photosyn-

thesis in *Chlamydomonas reinhardii*. *Plant Physiol.* 71:35–39

181. Spreitzer, R. J., Rahire, M., Rochaix, J. D. 1985. True reversion of a mutation in the chloroplast gene encoding the large subunit of ribulosebisphosphate carboxylase/oxygenase in *Chlamydomonas*. *Curr. Genet.* 9:229–31

182. Deleted in proof

183. Staehelin, L. A., Arntzen, C. J. 1983. Regulation of chloroplast membrane function: Protein phosphorylation changes the spatial organization of membrane components. *J. Cell Biol.* 97:1327–37

184. Steele, J. A., Uchytil, T. F., Durbin, R. D., Bhatnager, P., Rich, D. H. 1976. Chloroplast coupling factor-1: A species specific receptor for tentoxin. *Proc. Natl. Acad. Sci. USA* 73:2245–48

185. Steinback, K. E., McIntosh, L., Bogorad, L., Arntzen, C. J. 1981. Identification of the triazine receptor protein as a chloroplast gene product. *Proc. Natl. Acad. Sci. USA* 78:7463–67

186. Thornber, J. P., Highkin, H. R. 1974. Composition of the photosynthetic apparatus of normal barley leaves and a mutant lacking chlorophyll *b*. *Eur. J. Biochem.* 41:109–16

187. Troxler, R. F., Lester, R., Craft, F. O., Albright, J. T. 1969. Plastid development in maize. *Plant Physiol.* 44:1609–18

188. Tsaftaris, A. S., Bosabalidis, A. M., Scandalios, J. G. 1983. Cell-type-specific gene expression and acatalasemic peroxisomes in a null cat2 catalase mutant of maize. *Proc. Natl. Acad. Sci. USA* 80:4455–59

189. Velthuys, B. R. 1981. Electron-dependent competition between plastoquinone and inhibitors for binding to photosystem II. *FEBS Lett.* 126:277–81

190. Vernotte, C., Briantais, J. M., Remy, R. 1976. Light-harvesting pigment protein complex requirement for spill-over changes induced by cations. *Plant Sci. Lett.* 6:135–41

191. von Wettstein, D., Kahn, A., Nielsen, O. F., Gough, S. 1974. Genetic regulation of chlorophyll synthesis analysed with mutants in barley. *Science* 184:800–2

192. von Wettstein, D., Kristiansen, K. 1973. Stock list for nuclear gene mutants affecting the chloroplast. *Barley Genet. Newslett.* 3:113–17

193. Walles, B. 1965. Plastid structures of carotenoid-deficient mutants of sunflower (*Helianthus annus* L.). *Hereditas* 53:247–56

194. Wang, W. Y. 1978. Effect of dim light on the y-1 mutant of *Chlamydomonas reinhardtii*. *Plant Physiol.* 61:842–46

195. Wang, W. Y., Boynton, J. E., Gillham, N. W. 1977. Genetic control of chlorophyll biosynthesis: Effect of increased δ-aminolevulinic acid synthesis on the phenotype of the Y-1 mutant of *Chlamydomonas*. *Mol. Gen. Genet.* 152:7–12

196. Wollman, F. A., Olive, J., Bennoun, P., Recouvreur, M. 1980. Organization of PS II centers and their associated antennae in the thylakoid membranes: A comparative ultrastructural, biochemical and biophysical study of *Chlamydomonas* wild type and mutants lacking in PS II reaction centers. *J. Cell Biol.* 87:728–35

197. Wu, S. M., Rebeiz, C. A. 1985. Chloroplast biogenesis. Molecular structure of chlorophyll *b*. *J. Biol. Chem.* 260:3632–34

Ann. Rev. Plant Physiol. 1986. 37:509–38

GENES SPECIFYING AUXIN AND CYTOKININ BIOSYNTHESIS IN PHYTOPATHOGENS

Roy O. Morris

Department of Agricultural Chemistry, Oregon State University, Corvallis, Oregon 97331

CONTENTS

INTRODUCTION

This review is concerned primarily with recent evidence that two dissimilar phytopathogenic bacteria, *Agrobacterium tumefaciens* and *Pseudomonas syringae* pv. *savastanoi,* contain genes that specify the biosynthesis of cytoki-

509

0066-4294/86/0601-0509$02.00

nins and indoleacetic acid.[1] The first organism, *A. tumefaciens,* causes crown gall, a neoplastic disease of dicotyledonous plants. It does so by transferring a fragment (the T-DNA) of a large endogenous plasmid (the Ti plasmid) to its hosts. The T-DNA is subsequently integrated into the host genome where its expression causes the tumor phenotype. The second organism, *P. savastanoi,* causes olive knot disease, a hyperplastic condition of woody species including olive, oleander, and privet. It does not appear to transfer any of its DNA to its host, but rather exerts its pathogenic effect, at least in part, by secreting high levels of auxins and cytokinins.

By 1983, strong circumstantial evidence indicated that *Agrobacterium* T-DNA was responsible for control of auxin and cytokinin levels in crown gall tumors. Nevertheless, it was not clear whether control was direct, via insertion of T-DNA-encoded phytohormone biosynthetic genes, or indirect, via insertion of a regulatory element which, in turn, controlled the biosynthetic activities of endogenous host genes (128). Recent work has answered the question. In summary, the T-DNA of wide host range *A. tumefaciens* Ti plasmids carries structural genes (*ipt, tms1* and *tms2*) that specify, respectively, a cytokinin biosynthetic prenyl transferase and two IAA biosynthetic enzymes: tryptophan monooxygenase and indoleacetamide hydrolase. The direct hypothesis is therefore correct. William of Ockham would have been pleased.

Phytohormone biosynthetic genes are present also in other parts of the Ti plasmid. Nopaline-catabolizing *A. tumefaciens* Ti plasmids contain a cytokinin

[1]*Abbreviations used:*

DHZ	dihydrozeatin
DHZMP	dihydrozeatin riboside 5'-monophosphate
DHZR	dihydrozeatin riboside
DHZOG	dihydrozeatin-O-glucoside
DMAPP	dimethylallylpyrophosphate
IAA	indole-3-acetic acid
IAM	indole-3-acetamide
iP	*iso*-pentenyladenine
iPA	*iso*-pentenyladenosine
iPMP	*iso*-pentenyladenosine 5'-monophosphate
iP7G	*iso*-pentenyladenine-7-glucoside
Me 1″-DHZR	1″-methyl-dihydrozeatin riboside
Me 1″-Z	1″-methyl-zeatin
Me 1″-ZR	1″-methyl-zeatin riboside
RIA	radioimmunoassay
Z	zeatin
ZMP	zeatin riboside 5'-monophosphate
ZOG	zeatin-O-glucoside
ZR	zeatin riboside
ZROG	zeatin riboside-O-glucoside
Z7G	zeatin-7-glucoside
Z9G	zeatin-9-glucoside

biosynthetic prenyl transferase gene *(tzs)* close to the *vir* region. There is also evidence for a *vir*-region gene that modulates bacterial IAA biosynthesis. Finally, *P. savastanoi* has been found to possess a set of phytohormone biosynthetic genes comparable to those of *A. tumefaciens* both in structure and function.

The molecular biology of plant transformation by *A. tumefaciens* has been reviewed frequently and in depth (19, 62, 108). This review therefore describes only the salient molecular aspects in order to provide a structure against which phytohormone-related aspects may be considered. Recent reviews related to phytohormones and crown gall are available (16, 107), as are reviews of auxin (36) and cytokinin (80) biosynthesis in untransformed plants. The literature survey was completed on November 15, 1985.

MOLECULAR BIOLOGY OF TRANSFORMATION BY *A. TUMEFACIENS*

A recent review (107) has summarized the genetic and molecular aspects of tumor induction and growth and should be consulted for further details.

A. tumefaciens is classified among the Rhizobiaceae, and is considered to be a close relative of the nitrogen-fixing *Rhizobia*. It has recently been assigned to the alpha-2 subgroup of the purple bacteria (159), and on the basis of rRNA homology, it is thought to be closely related to the Rickettsiae (155) and to plant mitochondria (163).

All virulent *A. tumefaciens* strains harbor several large plasmids. The Ti plasmids, first identified by Zaenen et al (165), are responsible for tumor formation. There are several types of Ti plasmid, usually categorized by the nature of the opine (agropine, octopine, or nopaline) produced upon transfer to the plant. Plasmids that carry T-DNA genes for the production of a particular opine in the plant usually carry genes elsewhere in the plasmid for the catabolism of that opine (107). T-regions differ in size among plasmids of different origin, but they are functionally defined by possessing consensus direct repeats at their extremities (161) which are necessary for integration into the host genome.

T-DNA from different *A. tumefaciens* (and *A. rhizogenes*) strains differs in its insertion patterns (107). Octopine strains usually integrate a single copy of the leftward T_L-DNA, and may or may not integrate single or multiple copies of a rightward T_R-DNA. T_R-DNA is not required for maintenance of tumor growth. Other strains, notably those having limited host range, may integrate a split T_L-DNA.

The complete nucleotide sequence of T_L-DNA from two octopine plasmids has been determined (14, 49). It encodes (157) 7 transcripts (Figure 1). Nopaline strains encode 13 (156) transcripts, 6 of which are homologous to

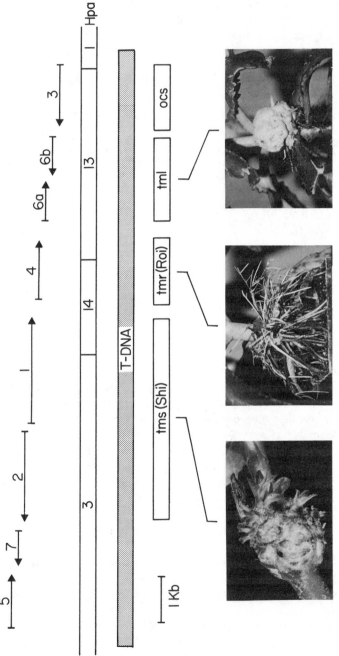

Figure 1 Physical, genetic, and transcriptional map of typical octopine T$_L$-DNA. The figure represents a synthesis of data from (48, 156). *Tms* (*Shi*), *tmr* (*Roi*), and *tml* are loci which when mutated cause the overproduction of roots, shoots, or increase in tumor size, respectively. *Ocs* is the octopine synthase locus. Transcripts 1 and 2 (*tms1* and *tms2*) encode auxin biosynthetic enzymes. Transcript 4 (*tmr*, *ipt*) encodes a cytokinin biosynthetic enzyme. The insets are photographs of tumors incited by a *tms* mutant (on tobacco) and *tmr* and *tml* mutants on *Kalanchoe*.

those found in octopine T_L-DNA. A large region of the Ti plasmid *(vir)*, located at some distance from the T-region, is involved in the transfer of T-DNA to the plant but is not believed to be transferred itself.

EVIDENCE FOR PHYTOHORMONE PARTICIPATION IN CROWN GALL TUMOR GROWTH

An extensive body of evidence indicates that auxin and cytokinin overproduction, encoded by T-DNA genes, is central to the maintenance of the transformed state. The initial description of auxin-like growth effects in plants bearing crown gall tumors (42), the growth of axenic tumor tissue in culture in the absence of added auxin (25), the stimulation of growth of normal (untransformed) tissue in culture by diffusable factors secreted by adjacent tumor tissue (42), and the demonstration of elevated auxin (57) and cytokinin (99) levels have been reviewed recently (107). Braun (24) suggested that growth of axenic crown gall tissue in culture was the result of its ability to produce sufficient endogenous auxin and cell division factors. Subsequently, a theoretical model explaining crown gall growth as the consequence of overproduction of auxins and cytokinins within the same tissue locus (in contrast to the spatial separation of auxin and cytokinin production in untransformed plants) was used to rationalize tumor phenotype (127).

Studies of tumor morphology lend support to the idea that hormones play a central role in maintaining the tumor phenotype. *A. tumefaciens* strains may produce either shooty teratomata or unorganized tumors (90), indicating that differential synthesis of auxin and/or cytokinin may determine tumor morphology in the same way that normal tobacco callus morphology is determined in culture by the application of exogenous auxins or cytokinins (137).

Fine structure genetic analysis of the T-DNA of octopine plasmids is possible by the introduction and integration of transposable elements such as Tn5 (47, 48, 78, 79, 111). Mutation of the T-DNA causes startling changes in tumor morphology (Figure 1). Three loci have been defined as a result of these studies: the *tms* (or *Shi*) locus, which when inactivated causes tumors to overproduce shoots; the *tmr (Roi)* locus, which when inactivated causes tumors to overproduce roots; and the *tml* locus, which when inactivated causes increase in tumor size. Lesions in either *tmr* or *tms* can be complemented by co-infection with strains bearing mutations in the other locus (149).

Growth of root-overproducing tumor mutants in culture is enhanced by the addition of cytokinins, whereas growth of shoot-overproducing mutants requires the addition of auxins (67, 111). Endogenous cytokinin levels are depressed in root-bearing mutant tumors and elevated in shoot-bearing tumors (3). Conversely, IAA levels are elevated in root-bearing tumors and depressed in shoot-bearing tumors.

The host range of Ti plasmids is determined in part by the nature of the genes carried by the T-DNA. Some strains are limited in host range to certain plant species (75, 164) because they do not apparently carry a functional copy of the *tmr* locus (27). Their host range can be extended to other plants if *tmr* is introduced into the T-DNA (28, 60).

GENES CODING FOR CYTOKININ BIOSYNTHESIS

Because many of the studies on cytokinin involvement in tumorigenesis have relied on the availability of improved analytical tools, this area will be briefly reviewed. The proposed pathway of cytokinin biosynthesis in untransformed plants and the phenomenon of cytokinin habituation will also be summarized.

Advances in Cytokinin Analysis

Since 1979, major advances have been made in methods for measurement of cytokinin levels, estimation of biosynthesis, and assay of related enzymes. At the high-technological end of the analytical scale, the use of isotopically labeled (^{2}H or ^{15}N) internal standards coupled with GC/MS multiple ion quantitation provides precise measurements of cytokinin levels (131, 140). The technique does require expensive equipment and is not suited to the rapid processing of multiple samples. At the opposite extreme, the introduction of economical and rapid RIA and ELISA assays (55, 153) has allowed analysis of hormones in multiple samples. Sub-picomolar amounts can be detected with reasonable precision. Because currently available antibodies do not distinguish between closely related cytokinins such as Z, ZR, or Z9G; RIA of unfractionated tissue extracts may give uninterpretable data. The problem is partially alleviated by combining HPLC with RIA (8, 85, 86), thus allowing mixtures of cytokinins containing, for example, Z, ZR, DHZ, DHZR, iP, and iPA to be analyzed. The approach must be used with caution, however, on samples containing excessive organic contaminants (86) because retention times shift. The introduction of immunoaffinity chromatography on immobilized anticytokinin antibodies as a prepurification step has provided a solution (40, 86), and in most cases gives samples that are essentially analytically homogeneous. With the availability of monoclonal anticytokinin antibodies (146), the assay of cytokinin levels and biosynthesis rates has become practical. One useful application of immunoaffinity chromatography has been the development of a rapid assay for the cytokinin biosynthetic prenyl transferase (61).

Cytokinin Biosynthesis and Metabolism in Untransformed Plants

Cytokinin biosynthesis and metabolism have been reviewed recently (80) and are illustrated in Figure 2. Synthesis of cytokinins in vitro was first reported by

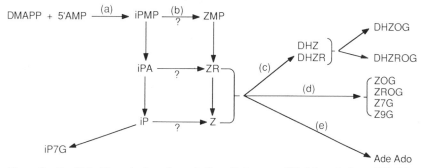

Figure 2 Cytokinin biosynthesis and metabolism. Pathway modified from Letham (80). (a) *iso*-pentenyl transferase; (b) cytokinin hydroxylase; (c,d) cytokinin glucosylation pathways; (e) cytokinin oxidase (oxidative degradation).

Chen & Melitz (32) who described a prenyl transferase from habituated tobacco callus (Figure 2a) which transfers the isoprenoid side chain from dimethyl-allylpyrophosphate to 5'-AMP to give iPMP. The 5'AMP was an obligatory substrate; neither adenine nor adenosine were suitable. Hydroxylation of the cytokinin side chain (Figure 2b) to give the corresponding *trans*-zeatin derivative was observed in vivo (44, 112) and in vitro (31), but it is not yet clear whether it occurs at the nucleotide, nucleoside, or the free base level. The oxidase preparation from cauliflower (31) *trans*-hydroxylated iP to Z and iPA to ZR in vitro. Inhibitor studies indicated that the system was cytochrome P-450 dependent. Although there have been numerous reports that *cis*-Z and *cis*-ZR occur as free cytokinins in normal (see 80) and transformed plant cells (52), recent studies (143) of normal tobacco shoots failed to find any evidence for them, and they appear to occur only in plant (and some bacterial) tRNA.

Extensive metabolic studies (80) indicate that when many plant tissues are presented with an excess of either synthetic or natural cytokinins, they detoxify the excess by oxidative side chain cleavage (Figure 2e) or they glucosylate and sequester the hormones (Figure 2c,d). Glucosylation thus may provide a mechanism whereby the active cytokinin pool is reduced to physiologically acceptable levels.

Cytokinin Habituation

One of the difficulties in interpretation of data on cytokinin content in trans-formed tissue is that the plant has already available to it a complement of genes capable of cytokinin biosynthesis. Such genes are presumably inactive in tissue cultures of *Nicotiana tabacum* because hormones must be added for growth to occur. Habituation to a cytokinin autotrophy can occur, and Meins and his coworkers have provided fascinating insight into the epigenetic and genetic changes associated with habituation (91–96).

Nicotiana tabacum cv. Havana 425 can be habituated readily in culture to cytokinin autotrophy (91). Pith tissue contained two cell types: readily habituatable (C^+) and resistant to habituation (C^-) (95, 96). Habituation to the C^+ phenotype was achieved by maintenance of C^- cultures at 35°C. Upon regeneration of plants from C^+ cells, pith tissues regained the C^- phenotype. The shift was termed epigenetic. It is characterized by high frequency ($>10^{-3}$/ cell generation) and involves no permanent change in the genome. By definition, it is not retained through meiosis.

Similar epigenetic shifts between the C^+ and C^- states were observed in the developing plant (91). Leaf lamina cells are normally C^- whereas stem cortical cells are C^+. Regeneration of plants from either C^- lamina sources or C^+ cortical sources gave progeny that exhibited the C^- phenotype in leaf laminae and the C^+ phenotype in stem cortex.

Recently, a permanent genetic shift in habituation was observed (93). It was first noted in a single plant, and is characterized by permanent alteration of leaf lamina cells of regenerated plants to C^+. The gene governing this phenotype segregated as a partially dominant Mendelian factor, *habituated leaf* (Hl). Subsequent experiments (92, 94) allowed direct selection for Hl by growth of C^- cells on a low-cytokinin medium. Regenerated plants exhibited the habituated leaf lamina C^+ trait (Hl) as a single dominant Mendelian factor.

Transformation of either hl/hl or Hl/Hl tobacco lines with a pTiT37 plasmid containing either a fully active *ipt* gene (cyt$^+$) (see below) or an insertionally inactivated *ipt* gene (cyt$^-$) indicated that Hl complemented the *ipt* locus (94). Transformants having the hl/hl *cyt$^+$* genotype would not regenerate because of the presence of the functional *ipt* locus (21). Neither could Hl/Hl *cyt$^-$* transformants be regenerated. Further, Hl/Hl *cyt$^-$* transformants formed tumors when transplanted into intact plants. The data were interpreted to mean that the Hl locus has an oncogenic function comparable to that of *ipt*.

Cytokinin Biosynthetic Genes Within A. tumefaciens T-DNA

The evidence that *A. tumefaciens* T-DNA encodes a cytokinin biosynthetic prenyl transferase falls into several categories: elevated tumor cytokinin content (99, 132), enhanced cytokinin biosynthesis (114), unchanged oxidative catabolism of cytokinins (135), correlation of tumor morphology with cytokinin content (3, 101), requirement of *tmr* mutants for cytokinins (21, 111), presence of a cytokinin biosynthetic prenyl transferase in tumor tissue but not in untransformed tissue (1, 2, 101), association of the cytokinin biosynthetic enzyme with a specific T-DNA locus (101), and finally, cloning of this locus into *E. coli* and expression of cytokinin biosynthetic activity (2, 15, 29).

CYTOKININ BIOSYNTHESIS AND METABOLISM IN CROWN GALL TUMORS
There have been numerous reports (summarized in 107), of the identifica-

tion of active cytokinins, usually Z and ZR, from different crown gall lines. In addition to the active cytokinins, crown gall tumors also contain cytokinin glucosides. The major cytokinin glucosides (ZOG, ZROG, and Z9G) were, in fact, first isolated from crown gall tumors (100, 118, 133, 134).

Probably the most precise measurements of cytokinins in crown gall were made for *Datura* and *Nicotiana* tumors, illustrated in Table 1. Levels of free cytokinins and cytokinin glucosides were markedly elevated over those of untransformed tissue, although there were some differences in the types of glucoside synthesized by each tumor.

Most cytokinin measurements have been made at a single time point during the exponential phase of tumor growth. Recently, the time course of cytokinin accumulation during the culture growth cycle has been determined (54, 106, 147, 160). Cytokinin levels generally increase to a maximum during the exponential phase of cell growth and then decrease precipitously. Thus, for three cloned sibling tumor lines derived from a single T37 transformation (54), cytokinins peaked during exponential growth and then dropped as tissue growth ceased. Peak cytokinin levels of each clone were quantitatively different (ZR, 30–230 pmol/g; iP 21–77 pmol/g) but were always greater than the amounts present in untransformed tissue (<1 pmol/g). Two cloned tobacco tumors possessing different phenotypes (160) (one readily greening in the light, the other remaining pale yellow) accumulated ZR at the same rate but to very different levels (136 and 35 pmol/g, respectively). Cytokinin levels in untransformed tissues were much lower. A somewhat different time profile of cytokinin accumulation was noted by Weiler & Spanier (154). Zeatin accumulated late in the growth phase and reached a maximum in stationary phase. These studies were carried out, however, upon unfractionated extracts, and it is possible that such components as zeatin-7-glucoside were being measured and that changes in Z and ZR were obscured.

Precursor feeding studies indicate that crown gall contains an active cytokinin biosynthetic pathway. Adenine is rapidly incorporated into ZR and Z in *V. rosea* crown gall (139), and when precautions are taken to reduce phosphatase activity during extraction, much of the incorporation is into ZMP (114). No incorporation of label into cytokinin is observed in untransformed *V. rosea* callus grown on cytokinin and 2,4-D.

Active hydroxylation pathways are also present. Labeled iP is *trans*-specifically hydroxylated by *V. rosea* crown gall to Z, ZR, and ZMP in vivo (112). Indirect evidence for *trans*-specific hydroxylation was also obtained from study of the tRNA-bound cytokinins in *Vinca rosea* crown gall (113). Transfer RNA from untransformed *Vinca rosea* contains primarily *cis*-ZR. Transfer RNA from crown gall tumors contains substantial amounts of *trans*-ZR. Although the data could be interpreted as evidence for tRNA-mediated cytokinin biosynthesis, as has been suggested for a number of years (80), they

Table 1 Cytokinin levels in *Datura*[a] and *Nicotiana*[b] crown gall

Cytokinin	Cytokinin (pmol/g)	
	Datura	Nicotiana
iP	5.4	—
iPA	1.5	5.5
iPMP	2.6	—
Z	123	34
ZR	223	130
ZMP	108	807
DHZ	16	7
DHZR	19	22
DHZMP	17	37
ZOG	50	16
ZROG	52	24
Z7G	518	203
Z9G	656	—
DHZOG	19	—
DHZROG	7.7	—
DHZ9G	339	—
DHZ7G	—	28

[a]Data taken from (115).
[b]Data taken from (132).

are also consistent with the induction of a *trans*-specific hydroxylation pathway as a consequence of transformation. *Trans*-hydroxylation of tRNA-bound iPA could be a byproduct of excess free cytokinin synthesis.

Oxidative catabolism of cytokinins in tumors is no different than that in untransformed tissues (135). Given the biosynthetic pathway proposed by Letham (80), the data indicate extensive de novo biosynthesis of cytokinins in crown gall tissue and considerable activity in the isopentenyl transferase and hydroxylation pathways.

TUMOR MORPHOLOGY MUTANTS: CYTOKININ RELATIONSHIPS   Studies of the auxin and cytokinin content of unorganized and shoot teratomata (5, 6) indicated that endogenous hormone levels favorable to shoot formation (high cytokinin:auxin ratios) were characteristic of teratoma lines, whereas unorganized tumors exhibited lower cytokinin:auxin ratios. Addition of exogenous cytokinin to unorganized tumor lines tended to produce shoots. Stronger circumstantial evidence implicating T-DNA in control of tumor cytokinin levels was obtained from studies of tumors bearing T-DNA in which specific loci had been inactivated by transposon insertion (Figure 1). Tumors in which the *tmr* locus had been inactivated required an external supply of cytokinin for growth in culture (21, 67).

Studies of the hormone content of morphological mutants also indicated that the *tmr* locus was involved in cytokinin biosynthesis (3). Zeatin riboside levels in primary tumors incited by *A. tumefaciens* in which the *tms, tmr,* or *tml* loci had been inactivated by insertion were: normal tobacco stem tissue, <1 pmol/g; parental (unmutagenized) crown gall 50 pmol/g; shoot-bearing tumors *(tms),* >1200 pmol/g; root-bearing tumors *(tmr)* <1 pmol/g; and large tumors *(tml),* 40 pmol/g. A functional *tmr* locus is therefore necessary for elevated ZR levels. When ZR levels are very high, iPA levels are also elevated, indicating that control of biosynthesis is at the level of isopentenyl transferase.

ISOPENTENYL TRANSFERASE IN CROWN GALL TUMORS   Isopentenyl transferase activity is exhibited by a number of crown gall lines. Cloned multicopy T-DNA tumor lines (1) exhibit high activity, whereas untransformed tobacco callus contains no detectable activity. Measurements of isopentenyl transferase activity in cloned cell lines derived from morphological mutants indicated that inactivation of *tmr* abolished activity (2, 101), whereas inactivation of *tms* increased it (2).

EXPRESSION OF THE IPT GENE   The *tmr* locus has been sequenced (50, 56, 81) and overexpressed behind the lambda P_R promoter (136). It encodes a 27 kd protein. Good evidence that this protein is isopentenyl transferase (Figure 2a) has been obtained by three groups (2, 15, 29). A DNA fragment encompassing the *tmr* region was cloned and expressed in *E. coli.* Cells bearing *tmr* express a 27 kd protein together with isopentenyl transferase activity. Attempted expression of *tmr* containing a Tn5 insertion results in no enzyme activity. High performance liquid chromatography and RIA indicate that iPMP, iPA, and iP are produced in vitro (2). *Iso*-pentenyladenosine identity was confirmed by mass spectrometry (29). The *tmr* gene product expressed in *E. coli* is similar or identical to the isopentenyl transferase expressed in crown gall tumors because an antibody raised against a synthetic decapeptide derived from the coding region of *tmr* cross-reacts with the tumor enzyme. The enzyme could not be detected, nor was cross-reacting material seen in extracts from habituated (untransformed) cell lines.

Cytokinin Biosynthetic Genes Outside A. tumefaciens T-DNA

Not only does *A. tumefaciens* incite tumors that contain an excess of cytokinins as a result of the activity of the T-DNA *ipt* gene, there is also evidence that free-living bacteria secrete cytokinins. Rigorous identification of iP and Z production by the nopaline strain C58 was obtained (35, 68) by GC/MS. A direct association between Z production and the presence of nopaline Ti plasmids was established (125). Both octopine and nopaline strains secreted iP (ca 500 ng/L), but only wild-type nopaline strains or exconjugants containing nopaline plasmids in octopine or nopaline chromosomal backgrounds secreted Z (ca 200–800 ng/L). A nopaline Ti plasmid locus *(tzs)* specifying *trans*-zeatin synthesis or secretion was postulated. *Tzs* has now been mapped to a fragment outside the T-region close to *vir* (4, 17). Cloning and expression of *tzs* in *E. coli* results in the secretion of zeatin. The sequence has been reported both for C58 (107) and the closely related strain T37 (4), and it has been shown to have extensive homology with the *ipt* gene from the T-DNA. Finally, *tzs* has been shown to code for an isopentenyl transferase, apparently similar in function to that coded for by *ipt* (4, 17, 102).

Cytokinin Biosynthesis In Pseudomonas savastanoi

Gall formation by *P. savastanoi* is characterized both by cell enlargement and cell division, and it includes differentiation into abnormally arranged tracheary elements and sieve tubes. On the basis of the morphology, it was suggested that two agents, one inducing hypertrophy and one inducing hyperplasia, might be involved (158). The initial discovery that cytokinins are produced by *P. savastanoi* was made by Surico and his coworkers (142). TLC fractionation of ethyl acetate extracts of culture filtrates of *P. savastanoi* followed by bioassay gave cytokinin-active species similar to iPA and iP. Recently this work was extended and a novel biologically active cytokinin was identified (141). It was shown by mass spectrometry and proton NMR to be 1″-methyl-*trans*-ribosylzeatin (Figure 3).

Identity of the cytokinin was recently confirmed and a series of other methylated derivatives were reported, including 1″-methyl-Z and 1″-methyl-DHZR (87, 102). Studies of other *P. savastanoi* strains have indicated that there is a diversity of cytokinins being synthesized. Some strains (E. M. S. MacDonald and E. H. Terhune, personal communication) produce primarily iP and iPA, while others produce zeatin, and in some cases 1″-methylated zeatins. The gene(s) responsible for 1″-methylation has yet to be isolated. A recent report (121b) indicates that the 1″ hydroxymethyl zeatin is present in immature wheat seeds. Wild-type *P. savastanoi* strains contain multiple plasmids (39), and production of zeatin by a number of strains is correlated with plasmid complement. *P. savastanoi* strain 1006 contains two plasmids, 105 and 81 Kb, and secretes Z, ZR and Me 1″-ZR (87). Deletion of part of the 105 Kb plasmid abolishes almost all secretion. Similarly, strain 213-2 has five plasmids (64, 56, 48, 42, and 38 Kb), and secretes primarily Z and ZR. Deletion of the 42 Kb plasmid abolishes Z and ZR secretion.

Construction of a library of fragments from the plasmids of *P. savastanoi* 1006 has allowed identification of a plasmid-borne gene, *ptz*, encoding cytokinin biosynthesis (102). Expression of *ptz* in *E. coli* causes secretion of Z, ZR, iP, and iPA. Cell lysates of *E. coli* bearing *ptz* contain isopentenyl transferase

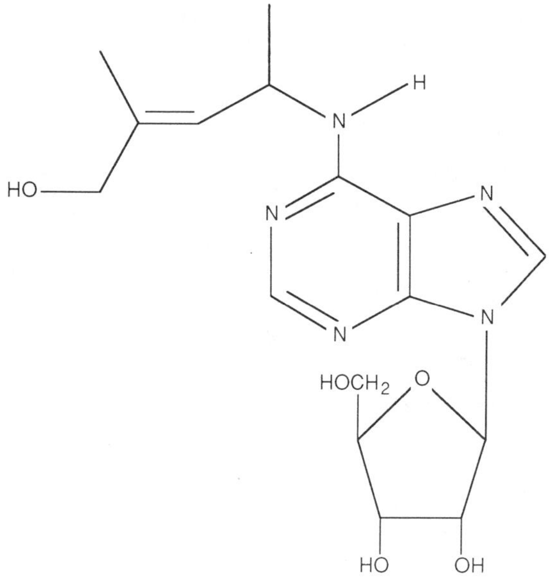

Figure 3   Structure of 1″-methyl-*trans*-zeatin riboside (141).

activity (102), indicating that *ptz* has considerable functional homology with the *tzs* and *ipt* genes of *A. tumefaciens*.

Homology Between Cytokinin Biosynthetic Genes of *A. tumefaciens* and *P. savastanoi*

The nucleotide sequences of the coding regions of *ipt* from *A. tumefaciens* strains Ach5, T37, and 15955 (50, 56, 81), *ptz* from *P. savastanoi* strain 1006, and *tzs* from *A. tumefaciens* C58 and T37 (4, 102, 120) have been determined. A comparison of the deduced amino acid sequences of these coding regions is illustrated in Figure 4.

Overall amino acid homology is at least 50%, and an examination of the nucleotide sequence homologies indicates that conservation of the first and second bases occurs. Homology is confined to the coding regions. The *ipt* gene of *A. tumefaciens* has associated with it a typical eukaryotic promoter sequence (14), whereas the *ptz* gene of *P. savastanoi* has a promoter region and a ribosome binding site almost identical to the consensus sequences of *E. coli* (120). *A. tumefaciens tzs* has a ribosome binding site but does not have the −10 and −35 *E. coli* promoter consensus sequences (17). At the 3' end of the coding region, the *ipt* gene has a polyadenylation site whereas *tzs* and *ptz* have fairly

```
              10        20        30        40        50
Ptz   MKIYLIWGATCTGKTEHSIKLSKSTGWPVIVLDRVQCCFDIATGSGRPHP
Tzs   .LLH..Y.P..S...DMA.QIAQE.....VA.......PQ........LE
Ipt   .DLH..F.P......TTA.A.AQQ..L..LS.......PQLS......TV

              60        70        80        90       100
Ptz   EELQSTRRIYLDNRRISEGVISAEEANDRLKLEVNKHIDSGGVILEGGSI
Tzs   S...........S.PLT..ILD..S.HR..IF..DWRKSEE.L.......
Ipt   ...KG.T.L...D.PLV..I.A.KQ.HH..IE..YN.EAN..L......T

             110       120       130       140       150
Ptz   SLLKLISKDPYWCDRFIWSQHRMRLQDTDVFMDKAKARVRRMLVGSTETT
Tzs   ...NCMA.S.F.RSG.Q.HVK.L..G.S.A.LTR..Q..AE.FAIREDRP
Ipt   ...NCMARNS..SAD.R.HII.HK.P.QET..KA.....KQ..HPAAG-H

             160       170       180       190       200
Ptz   GLLDELVAAQSDLNAKLAIQDIDGYRYIMNYAQARRLSITQLLNVMTGDM
Tzs   S..E..AELWNYPA.RPILE......CAIRF.RKHD.A.S..P.IDA.-R
Ipt   SIIQ...YLWNEPRLRPILKE......A.LF.SQNQITADM..QLDAN-.


             210       220       230       240
Ptz   KEELINGIALEYYEHAKWQERDFPAEWLAERSTR
Tzs   HV...EA..N..L...LS......QWPEDGAGQPVCPVTLTRIR
Ipt   EGK......Q..FI..RQ..QK..QVNA.AFDGFEGHPFGMY
```

Figure 4 Deduced amino acid sequence homologies between *ptz*, *tzs*, and *ipt*. Single letter amino acid codes are used. Gaps have been inserted in *tzs* (position 199) and *ipt* (positions 149 and 198) to maximize homology. The *tzs* sequence is from pTiC58 (17) and the *ipt* sequence is from the T-DNA of the octopine-type plasmid pTiAch5 (56). The *ptz* sequence is from *P. savastanoi* strain 1006 (120). Amino acids matching those from *tzs* are indicated by dots. Functionally conserved amino acid changes in *tzs* and *ipt* are underlined.

well-defined inverted repeats associated with transcription termination (17, 120).

Cytokinin Production by Rhizobium

The evidence that *Rhizobium* strains produce cytokinins or that cytokinins are involved in nodulation is fragmentary and conflicting. Early studies indicated that log-phase cultures of *R. japonicum* and *R. leguminosarum* (109, 119) produce zeatin-like cytokinins in culture (ca 1 μg kinetin equivalent/L). No iP was detected. A more recent study (26) reported similar data. An exhaustive study of *R. leguminosarum* (152) has failed to confirm these findings. Bioassay and chemical ionization GC-mass spectrometry were used to show that iP was the only cytokinin secreted and that it was present at <1% of the amount produced by *A. tumefaciens* strain B6. Because most *A. tumefaciens* strains produce about 500 ng of iP/L (125), production by *R. leguminosarum* is exceedingly modest.

Studies of nodule cytokinin levels are in a similarly confused state. New-comb et al (109) identified iP and Z in pea root nodules incited by *R. legumino-sarum*. Twelve days after infection, levels were high (2 μg equivalent/100 g) but declined significantly by 24 days. A new cytokinin was produced late in the nodule development cycle. A subsequent study was unable to confirm these data. Pea root nodules contained no more cytokinins than those present in uninfected roots at the same stage of development (152). While it was not possible (121a) to detect significant amounts of cytokinin in either *Phaseolus vulgaris* roots in culture or in culture filtrates of *R. phaseoli*, culture filtrates in which nodulated roots had been grown contained substantial levels of Z and iP.

A study of cytokinin metabolism in infected pea root nodules lent little support for the concept of bacterial cytokinin synthesis (10). Zeatin riboside supplied to the root tips of nodulated peas accumulated rapidly in the nodules and was metabolized rapidly to Z, ZOG, ZROG, ZMP and DHZ derivatives. No major differences in cytokinin metabolism were noted between effective and ineffective nodules, and nodule metabolism resembled that of unnodulated roots.

One report (53) suggests that part of the T-DNA is homologous to *Rhizobium* chromosomal DNA. Whether this homology relates to genes specifying phyto-hormone synthesis remains to be determined. *A. tumefaciens* carrying *nod* genes from *R. meliloti* can form pseudonodules on alfalfa (58, 59), and the Ri plasmid from *A. rhizogenes* can enhance nodulation by *R. meliloti* (138), but again, the role of phytohormones, if any, remains to be established.

GENES CODING FOR AUXIN BIOSYNTHESIS

Auxin Biosynthetic Genes in P. savastanoi

AUXIN LEVELS AND BIOSYNTHESIS   Early studies (18, 88) established the presence of IAA and other growth-promoting substances in the culture filtrates of *P. savastanoi* and in the galls incited on host plants. Kosuge and his coworkers (65, 88) proposed an IAA biosynthetic pathway in the bacterium that requires a two-step conversion of tryptophan to IAA via indoleacetamide (IAM):

$$\text{Trp} + O_2 \xrightarrow[\text{monooxygenase}]{\text{tryptophan}} \text{IAM} + CO_2 + H_2O$$

$$\text{IAM} + H_2O \xrightarrow[\text{hydrolase}]{\text{indoleacetamide}} \text{IAA} + NH_4^+$$

Both tryptophan monooxygenase and indoleacetamide hydrolase may be assayed in semipurified bacterial extracts; genes encoding the two enzymes were termed *iaaM* and *iaaH* (38, 76). *P. savastanoi* strains resistant to 5-methyltryptophan were defective in IAA production and in virulence. Other mutants overproduced IAA and exhibited enhanced pathogenesis (38).

PLASMID STATUS AND AUXIN BIOSYNTHESIS   *P. savastanoi* strain 2009 has a complement of four plasmids: 58, 52, 41, and 34 Kb (37). Mutants that have lost the 52 Kb (PIAA1) plasmid [either by treatment with acridine orange or by selection for resistance to 5-methyltryptophan (38)] lose also the ability to synthesize IAA and IAM. Reintroduction of pIAA1 into IAA⁻ mutants of *P. savastanoi* causes reacquisition of tryptophan monooxygenase, indolacetamide hydrolase and virulence. A parallel study of plasmid loss and IAA production was carried out in strain 213 (39). Here, *iaaH* and *iaaM* were present on a 73 Kb plasmid (pIAA2) which showed significant homology to pIAA1. IAA⁻ mutants derived from other *P. savastanoi* strains exhibited no deletions or loss of plasmids. It was concluded that in those organisms, IAA biosynthesis was probably chromosomally encoded.

TRYPTOPHAN MONOOXYGENASE: THE iaaM GENE   The *iaaM* gene was located on pIAA1 by DNA fragment transfer to *E. coli* mediated by the plasmid RSF1010. *E. coli* bearing the appropriate EcoRI fragment produces IAM but not IAA, and expresses tryptophan monooxygenase but not indoleacetamide hydrolase (38). The *iaaM* gene has now been isolated and sequenced (162). It contains an open reading frame coding for 557 amino acids and should produce a protein of molecular weight of 61,789. In vitro transcription-translation studies gave a translation product having a molecular weight of approximately 62 kd. The sequence is characterized by the presence of strong homology at amino acids 42–66 with the amino acid sequence 5–29 of 4-hydroxybenzoate hydroxylase from *Pseudomonas fluorescens*. This region is known to define an FAD binding site. The data therefore indicated that tryptophan monooxygenase was a FAD-requiring enzyme.

Tryptophan monooxygenase has now been purified to homogeneity (64). Its molecular weight is 62 kd; it utilizes FAD as a cofactor, and it converts tryptophan to IAM. It is subject to feedback inhibition by IAM ($K_i = 7$ μM) and IAA ($K_i = 225$ μM), indicating that it regulates the biosynthetic pathway. The sequence of an N-terminal peptide from the enzyme was identical to that specified by the open reading frame of *iaaM*.

INDOLEACETAMIDE HYDROLASE: THE iaaH GENE   The sequence of the *iaaH* gene has also been determined (162). It is located adjacent to the *iaaM* locus and was cloned on a 4 Kb fragment that contained both genes. In vitro transcription-translation experiments produced two proteins, 62 kd and 47 kd. The 62 kd protein corresponds to tryptophan monooxygenase, the 47 kd protein to indoleacetamide hydrolase. The nucleotide sequence indicated the presence of an open reading frame specifying a 455 amino acid protein with a deduced molecular weight of 48,515, in good agreement with the observed value. Cell-free extracts of *E. coli* carrying the two genes contain both tryptophan monooxygenase and indoleacetamide hydrolase, whereas *iaaM* alone codes only for the oxygenase. Culture filtrates of *E. coli* strains bearing both genes accumulated IAA to approx. 58 ug/mg cell dry weight. Neither tryptophan monooxygenase nor indoleacetamide hydrolase activities nor significant levels of IAA secretion were noted in cell-free preparations of *E. coli* containing the vector alone. There is significant homology between the deduced amino acid sequence of *iaaH* and that of *tms2* from *A. tumefaciens* (see below).

Auxin Biosynthetic Genes Within A. tumefaciens T-DNA

The evidence that T-DNA codes for genes responsible for IAA biosynthesis falls into several categories: elevated tumor IAA content (5, 34, 57, 82, 116, 154), a different profile of IAA accumulation during tumor growth in culture (7, 104, 106, 117, 147), increased metabolism of tryptophan to IAA (124), differential metabolism of tryptophan to indoleacetamide (iaaM) (124), the presence of indoleacetamide in tumors but not in untransformed tissues (148), reduction of tumor IAA content in tumors incited with mutated *tms* loci (3, 147), an IAA requirement for growth of mutant tumors (67), sequence homology of *tms1* with *iaaM* from *P. savastanoi* (162), sequence homology of *tms2* with the *P. savastanoi iaaH* gene (162), and the expression of *tms2* in *E. coli* to give indoleacetamide hydrolase (73, 144).

TUMOR IAA CONTENT   Early studies on IAA in tumors (57, 82) have been reviewed recently (107). Subsequent work (5, 104, 116, 117, 154) has provided further confirmation that IAA levels are elevated over those of untransformed tissues. However, there is little agreement on the time course of IAA concentration change during tumor growth. IAA levels have been reported to peak prior to log phase in *Nicotiana tabacum* tumors (106, 117, 147), during mid to late log phase in *Helianthus, Kalanchoë, Parthenocissus,* and *Daucus* tumors (104), or just prior to the cessation of tumor growth (7, 154). One report found little change in IAA levels during growth of soybean tumors (160).

Because the assay techniques have been various, and because cloned and uncloned tumor lines (arising from different plant species) have been analyzed, perhaps the difference in these results is not surprising. A rigorous evaluation of IAA levels by GC/MS or electrochemical means would be of interest.

Transformation of *Nicotiana tabacum* with *tms1* or with *tml* (genes 6a and 6b, Figure 1) gave cell lines containing significantly different levels of IAM

(148). Levels were 1000-fold greater in plants bearing *tms1* than in uninfected plants or in plants bearing *tml* genes. Significantly, IAA levels were not different, suggesting that there is no endogenous indoleacetamide hydrolase activity in untransformed tissue in culture.

Ethylene biosynthesis (30) and 1-amino-cyclopropane-1-carboxylic acid (98) levels were increased in crown gall tumors, and major differences in ability to synthesize ethylene were noted between A6 and A66 tumor lines. The phenomenon was felt to be host-dependent.

METABOLISM OF IAA IN TUMORS   A study of auxin metabolism (124) in potato crown gall tumors and wounded untransformed tissues defined a significant alteration in tryptophan metabolism. Both tissues converted L-tryptophan to indoleacetaldehyde, indole-3-ethanol, and indoleacetamide, although conversion was more efficient by crown gall. Indoleacetoxime was formed only in crown gall tissues.

TUMOR MORPHOLOGY MUTANTS: AUXIN RELATIONSHIPS   Major differences in peak IAA concentrations and changes with time were noted on comparing shoot teratomata and unorganized tumors (117). Spontaneous T-DNA mutants were found to have an auxin requirement for growth in planta (111). A mutant of Ach5 in which the IS element IS60 was inserted into the leftmost part of the T-DNA produced shoot-bearing tumors on tobacco. It was avirulent on *Kalanchoë* leaves, and produced only small tumors on tomato. However, these tomato tumors grew to full size if supplied with naphthaleneacetic acid.

A similar mutant, A66 (22), gives tumors on *Nicotiana tabacum* which are characterized by shoot proliferation and tumors on tomato which grow very slowly. Cloned mutant tobacco tumors exhibit an auxin requirement for growth in culture that is overcome if the tumor is allowed to produce shoots. Presumably the shoots act as an auxin source. The phenotype was found to depend on the host plant. Tumors on *Nicotiana glauca* and *Nicotiana glutinosa* exhibited unorganized growth and were capable of growth on hormone-free media. It was concluded that *N. glauca* and *N. glutinosa* were able to supply enough auxin for growth.

An examination of engineered inserts and deletions in pTiC58 has further defined the role of the *tms (Shi)* locus with respect to auxin requirement (67). *Shi* mutants were avirulent on potato but tumors were produced if naphthaleneacetic acid was added. Mutants defective in either *tms1* or *tms2* required napthaleneacetic acid in order to produce tumors (66). Defects in *tms1* could be overcome by addition of naphthaleneacetamide but defects in *tms2* could not. It was suggested that *tms2* was capable of converting naphthaleneacetamide to free napthaleneacetic acid and that *tms1* was active in the synthesis of an amide-like IAA precursor.

Analyses of primary tumors incited by pTiA6NC carrying mutations in the T-DNA locus (3) demonstrated a significant decrease in IAA levels if the *tms* locus was inactivated. Similar data were reported for cloned tobacco crown gall lines grown in culture (147). The IAA levels in parental (pTiA6NC) tumors reached 400 pmol/g of tissue, whereas tumors with inactivated *tms* loci had significantly lower levels of IAA (approximately 75 pmol/g).

THE TMS1 GENE: PUTATIVE TRYPTOPHAN MONOOXYGENASE *Tms1* has been sequenced (74) and shown to encode an 83,769 dalton protein. The deduced amino acid sequence displayed significant homology with that of 4-hydroxybenzoate hydroxylase from *Pseudomonas fluorescens*. Specifically, residues 239 to 263 of *tms1* were almost identical in sequence to residues 5–29 of 4-hydroxybenzoate hydroxylase. This amino acid sequence had been shown by crystallographic studies to be the adenine binding pocket responsible for FAD attachment. It was considered likely therefore that the *tms1* gene product bound FAD or adenine.

Although *tms1* has not yet been expressed in *E. coli* to give an enzyme of known function, it seems probable that it encodes tryptophan monooxygenase. The *iaaM* gene of *P. savastanoi*, known to specify tryptophan monooxygenase, was found to have over 50% sequence homology at the deduced amino acid level with *tms1* (162). Homology was very high in the putative FAD-binding region. The major difference between the genes is the presence in *tms1* of an extra 198 amino acids at the N-terminus.

THE TMS2 GENE: INDOLEACETAMIDE HYDROLASE The sequence of *tms2* has been published (130). An open reading frame is present that encodes a 49 kd protein. The promoter region contains sequences typical of eukaryotic genes. A HindIII fragment from pTiAch5 containing the complete coding region of *tms2* was cloned into an *E. coli* vector and expressed (129). It specified the formation of a 49 kd protein, in good agreement with that encoded by the open reading frame. Bacterial extracts were able to convert IAM to IAA if *tms2* was present, but not if it was absent. Similarly, crown gall tissues were able to hydrolyze IAM but habituated (untransformed) tissues were not. The product of hydrolysis was shown by GC/MS to be IAA.

Partial purification of the enzyme has since been reported (73). It has a high pH optimum (8.5–9.5), a K_m of 1.2 μM and, although fairly stable at 0°C, has resisted further purification. It hydrolyzes a wide range of substrate analogs, including indoleacetonitrile, IAA glucosyl ester, IAA myoinositol, naphthaleneacetamide, and phenylacetamide. It cannot, however, hydrolyze IAA aspartate or IAA conjugates of alanine, glutamic acid, or glycine. Comparison of the deduced amino acid sequence with that of *iaaH* from *P. savastanoi* shows significant homology (162).

Auxin Biosynthetic Genes Outside *A. tumefaciens* T-DNA

Free-living strains of *A. tumefaciens* have been reported to secrete IAA (83), and secretion has been proposed to influence tumorigenesis (33, 70, 84). *A. tumefaciens* strain C58 (83) secretes IAA at 1.4 mg/L (measured colorimetrically), but Ti plasmid-cured strains secrete 50% less (0.7 mg/L). Reintroduction of pTiC58 restores secretion of IAA to its original level. *A. tumefaciens* strain 15955 (octopine) shows concomitant decreases in IAA secretion and virulence if the *incW* R-plasmids pSa or R388 are introduced (33), although there is no apparent loss of the Ti plasmid. Changes in IAA concentrations were modest, and avirulent strains still secreted over 20 μg IAA/mg cell protein. Whether these high IAA levels reflect reality remains to be determined by application of rigorous analytical techniques. If valid, they indicate at least the presence of a chromosomal biosynthetic pathway.

Mutants of pTiC58 carrying Tn5 insertions were reported to have altered host range, and one of these mutants could be complemented by the addition of naphthaleneacetic acid (70). The mutation mapped outside of the T-DNA. Another avirulent mutant of pTiC58 induced by Tn5 transposition exhibited decreased IAA secretion (84). Changes in IAA secretion were quite small and were observed only when the medium was supplemented with tryptophan. The locus responsible mapped outside of the T-DNA and was designated *iaaP*. It has recently been reported (89) that binding of *A. tumefaciens* to carrot cells is inhibited by presence of the pSa plasmid, but binding can be restored by the addition of auxin.

It is difficult at this stage to draw any convincing conclusions because of the complications introduced by the production of high levels of endogenous IAA presumably coded for by the bacterial chromosome. It is not clear that IAA is always being measured because GLC and colorimetric techniques were employed. An unequivocal demonstration of the presence of IAA by GC/MS and a rigorous quantitation of its levels by stable isotope dilution or electrochemical means would be welcome.

In summary, therefore, there is evidence that auxins may play a role in the process of tumor induction. The numerous reports (23, 63, 107) of auxin stimulation of the frequency of tumor induction and the finding that virulence is dependent upon the mode of infection (126) cannot be discounted. Because the transformation event and the subsequent growth of tumor cells are conceptually separable (103), it should be possible to utilize the introduction of other genes as transformation markers in order to probe the role of phytohormones in the transformation event itself.

Auxin Production by Rhizobium

In 1960, Kefford (72) determined that a number of *Rhizobium* species produce IAA in culture. Production was confirmed by many others and has been found to be enhanced by tryptophan or 2-ketoglutaric acid (see 20). Bacteroids, as opposed to free-living *Rhizobium*, also produce significant amounts of IAA and indole-3-lactic acid (145).

Most early studies utilized bioassays or colorimetric measurement of IAA levels. Unequivocal demonstration (by GC/MS) of IAA secretion was provided for both *R. trifolii* and *R. leguminosarum* (11, 12, 151). Growth on media containing tryptophan, glutamic acid, or phenylalanine increased IAA levels tenfold (to about 1 mg/L). When nod^+ and nod^- bacterial mutants were compared, three out of four nod^- strains produced less IAA at stationary phase than the corresponding nod^+ strains. However, detailed measurements of IAA levels in log phase cultures of *R. trifolii* nod^+ and nod^- pairs showed no differences. Similar results were obtained for *R. leguminosarum* nod^+ and nod^- strains; in the absence of tryptophan, no differences in IAA production were seen. Tryptophan stimulation of IAA production, however, was greater in wild-type strains (nod^+) than in nod^- mutants.

Detailed mass spectrometric analysis of IAA metabolism was reported for the same set of *R. leguminosarum* and *R. trifolii* strains (9, 13). Indole-3-aldehyde, tryptophol, indole-3-carboxylic acid, indole-3-glycolic acid, indole-3-lactic acid, indole-3-glyoxylic acid, *N*-acetyl-L-tryptophan, and indole-3-pyruvic acid were identified. While there were some differences in the levels and types of metabolites produced by nod^+ and nod^- strains, there was no obvious correlation between nodulation ability and IAA metabolism. Because of the complexity of nodulation (43), existing mutations in the nodulation loci might well not affect IAA levels or metabolism. A study of IAA$^-$ mutants having concomitant lesions in nodulating ability would be of interest.

CYTOKININ PRODUCTION BY OTHER MICROORGANISMS

In her exhaustive review, Greene (51) summarized the literature on cytokinin production by microorganisms through about 1978. Significant production was documented for a number of bacteria including *Corynebacterium fascians*, leaf nodule endophytes *(Chromobacterium lividum)*, rhizosphere bacteria including *Arthrobacter* sp. and *Azobacter* sp., ectomycorrhizal fungi including *Rhizopogon roseolus, Amanita rubescens,* and *Boletus edulis,* and by a large number of fungal pathogens including those that cause leaf curl and fasciation (*Taphrina* sp.), canker (*Nectria* sp.), and club root disease *(Plasmodiophora brassicae)*. A number of rusts and mildews produce cytokinin-like symptoms upon infection, notably the "green island" symptom on senescing leaves.

There has not been a great deal of activity in the further examination of cytokinin production by many of these organisms. The phytohormone status of *C. fascians* has been reviewed recently (108) and will not be re-reviewed here. The work of Murai et al (105) provided good evidence for the existence of a plasmid-coded cytokinin biosynthetic pathway, although virulence has been found not to depend on the presence of a 78 Md plasmid (77). Recently, hybridization between the *ipt* gene of *A. tumefaciens* and chromosomal DNA sequences from a number of *C. fascians* strains has been demonstrated (97). The extent to which *Corynebacterium* cytokinin biosynthetic genes are related to those of *A. tumefaciens* remains to be determined.

Part of the difficulty in studying cytokinin biosynthesis in a number of the fungi is that they are obligate parasites and not amenable to axenic culture. Among those that are not, two are of interest. Ectomycorrhizae associated with *Pinus sylvestris* and with *Pinus radiata* synthesize cytokinins (69, 110) having chromatographic and biological properties similar to those of Z, ZR, and iP. Isotope dilution-mass spectrometry has also been applied to the analysis of IAA production by a number of mycorrhizae (45). Some species, notably *Pisolithus tinctorius,* produce extremely high levels (up to 18 μg/mg of mycelial dry weight).

Several reports (41, 71, 122, 123) have appeared on auxins and cytokinins in tissues infected by *P. brassicae.* A recent study (41) of *Brassica campestris* infected with *P. brassicae* showed that Z and ZR levels were significantly elevated in infected tissues. A novel cytokinin, apparently a conjugate of Z and glucose-6-phosphate linked through the side chain hydroxyl group, was also detected. Whether this is a normal metabolite or is formed in response to fungal infection is not known.

Finally, in an intriguing developmental situation reviewed recently (46), cytokinins have been reported in gall-forming wasp larvae. At present it is not clear whether the larvae are secreting cytokinins or accumulating them from the host plant. Van Staden (150) studied cytokinins of *Erythrina latissima* galls and found that although galls had lower overall cytokinin levels than leaf laminae, most cytokinin was located in the larvae. Zeatin and ZR were present in the larvae whereas leaves contained primarily Z glucosides. A recent study (K. Anderson, personal communication) has shown that *Andricus* larvae contain substantial levels of iP. Because the galls themselves contain mostly Z and ZR, it seems that the larvae may be a cytokinin source.

SUMMARY

The characterization of bacterial genes coding for the biosynthesis of two major plant hormones is of great interest. Not only should it now be possible to introduce these genes into plants under suitable control and observe the developmental consequences, it should also be possible to compare the bacterial and plant biosynthetic pathways. The question as to the origin of the genes is also clearly of interest. Although untransformed plants apparently do not synthesize IAA by a pathway similar to that demonstrated for *A. tumefaciens* and *P. savastanoi* (36), it is difficult to imagine a cytokinin biosynthetic pathway that differs from the bacterial route. There is obviously much to be learned in the near future from a study of normal phytohormone biosynthetic pathways.

For additional data see *Note added in proof,* p. 538.

ACKNOWLEDGMENTS

I wish to thank E. Larsen, E. M. S. MacDonald, and J. W. Morris for untiring assistance in the preparation of the literature review and manuscript. I also thank G. K. Powell for critical comments on the manuscript. The work of this laboratory is supported by grants from the National Science Foundation (Grant PCM 83-03371) and the U.S. Department of Agriculture (Grant 83-CRCR-1-1249).

Literature Cited

1. Akiyoshi, D. E. 1984, thesis. Oregon State Univ., Corvallis
2. Akiyoshi, D. E., Klee, H., Amasino, R. M., Nester, E. W., Gordon, M. P. 1984. T-DNA of *Agrobacterium tumefaciens* encodes an enzyme of cytokinin biosynthesis. *Proc. Natl. Acad. Sci. USA* 81:5994–98
3. Akiyoshi, D. E., Morris, R. O., Hinz, R., Mischke, B. S., Kosuge, T., et al. 1983. Cytokinin-auxin balance in crown gall tumors is regulated by specific loci in the T-DNA. *Proc. Natl. Acad. Sci. USA* 80:407–11
4. Akiyoshi, D. E., Regier, D. A., Jen, G., Gordon, M. P. 1985. Cloning and nucleotide sequence of the *tzs* gene from *Agrobacterium tumefaciens* strain T37. *Nucleic Acids Res.* 13:2773–88
5. Amasino, R. M., Miller, C. O. 1982. Hormonal control of tobacco crown gall tumor morphology. *Plant Physiol.* 69:389–92
6. Amasino, R. M., Miller, C. O. 1983. Effect of temperature on the morphology and cytokinin levels of tobacco crown gall teratoma tissues. *Plant Sci. Lett.* 28:245–53
7. Atsumi, S. 1980. Relation between auxin autotrophy and tryptophan content in sunflower crown gall cells in culture. *Plant Cell Physiol.* 21:1031–39
8. Badenoch-Jones, J., Letham, D. S., Parker, C. W., Rolfe, B. G. 1984. Quantitation of cytokinins in biological samples using antibodies against zeatin riboside. *Plant Physiol.* 75:1117–25
9. Badenoch-Jones, J., Rolfe, B. G., Letham, D. S. 1983. Phytohormones, *Rhizobium* mutants, and nodulation in legumes. III. Auxin metabolism in effective and ineffective pea root nodules. *Plant Physiol.* 73:347–52
10. Badenoch-Jones, J., Rolfe, B. G., Letham, D. S. 1984. Phytohormones, *Rhizobium* mutants, and nodulation in legumes. VI. Metabolism of zeatin riboside applied via the tips of nodulated pea roots. *J. Plant Growth Regul.* 3:41–49
11. Badenoch-Jones, J., Summons, R. E., Djordjevic, M. A., Shine, J., Letham, D. S., Rolfe, B. G. 1982. Mass spectrometric quantification of indole-3-acetic acid in *Rhizobium* culture supernatants: Relation to root hair curling and nodule initiation. *Appl. Environ. Microbiol.* 44:275–80
12. Badenoch-Jones, J., Summons, R. E., Entsch, B., Rolfe, B. G., Parker, C. W., Letham, D. S. 1982. Mass spectrometric identification of indole compounds produced by *Rhizobium* strains. *Biomed. Mass Spectrom.* 9:429–37
13. Badenoch-Jones, J., Summons, R. E., Rolfe, B. G., Letham, D. S. 1984. Phytohormones, *Rhizobium* mutants, and nodulation in legumes. IV. Auxin metabolites in pea root nodules. *J. Plant Growth Regul.* 3:23–39
14. Barker, R. F., Idler, K. B., Thompson,

D. V., Kemp, J. D. 1983. Nucleotide sequence of the T-DNA from *Agrobacterium tumefaciens* octopine Ti plasmid pTi15955. *Plant Mol. Biol.* 2:335–50

15. Barry, G. F., Rogers, S. G., Fraley, R. T., Brand, L. 1984. Identification of a cloned cytokinin biosynthetic gene. *Proc. Natl. Acad. Sci. USA* 81:4776–80

16. Bayer, M. H. 1977. Phytohormone und pflanzliche tumorgenese. *Beitr. Biol. Pflanz.* 53:1–54

17. Beaty, J. S., Powell, G. K., Lica, L., Regier, D. A., MacDonald, E. M. S., et al. 1985. *Tzs*, a nopaline Ti plasmid gene from *Agrobacterium tumefaciens* associated with *trans*-zeatin biosynthesis. *Mol. Gen. Genet.* In press

18. Beltra, R. 1959. El acido beta-indoleacetic y los tumores vegetales de origen bacteriano. *Revista Latinoam. Microbiol.* 2:23–32

19. Bevan, M. W., Chilton, M. D. 1982. T-DNA of the *Agrobacterium* Ti and Ri plasmids. *Ann. Rev. Genet.* 16:357–84

20. Bhowmick, P. K., Basu, P. S. 1984. Contents of hormones and indoleacetic acid metabolism in root nodules of *Erythrina indica* LAMK., *Sesbania grandiflora* PERS. and *Pterocarpus santalinus* LINN. *Biochem. Physiol. Pflanz.* 179:455–62

21. Binns, A. N. 1983. Host and T-DNA determinants of cytokinin autonomy in tobacco cells transformed by *Agrobacterium tumefaciens. Planta* 158:272–79

22. Binns, A. N., Sciaky, D., Wood, H. N. 1982. Variation in hormone autonomy and regenerative potential of cells transformed by strain A66 of *Agrobacterium tumefaciens. Cell* 31:605–12

23. Bouckaert, U. A. M., Vendrig, J. C. 1981. The influence of plant growth regulators on crown-gall initiation on cotyledonary leaves of *Helianthus giganteus* L. *in vitro. Z. Pflanzenphysiol.* 103:75–81

24. Braun, A. C. 1958. A physiological basis for the autonomous growth of the crown gall tumor cell. *Proc. Natl. Acad. Sci. USA* 44:344–49

25. Braun, A. C., White, P. R. 1943. Bacteriological sterility of tissues derived from secondary crown gall tumors. *Phytopathology* 33:85–100

26. Brun, G. A., Sabel'Nikova, V. I. 1982. Capacity of nodule bacteria to synthesize cytokinin-type substances. *Izv. Akad. Nauk mold. SSR. Ser. Biol. Khim. Nauk* P43–45

27. Buchholz, W. G., Thomashow, M. F. 1984. Comparison of T-DNA oncogene complements of *Agrobacterium tumefaciens* tumor-inducing plasmids with lim-

ited and wide host ranges. *J. Bacteriol.* 160:319–26

28. Buchholz, W. G., Thomashow, M. F. 1984. Host range encoded by the *Agrobacterium tumefaciens* tumor-inducing plasmid pTiAg-63 can be expanded by modification of its T-DNA oncogene complement. *J. Bacteriol.* 160:327–32

29. Buchmann, I., Marner, F. J., Schröder, G., Waffenschmidt, S., Schröder, J. 1985. Tumor genes in plants: T-DNA encoded cytokinin biosynthesis. *EMBO J.* 4:853–59

30. Canfield, M. L., Moore, L. W. 1983. Production of ethylene by *Daucus carota* inoculated with *Agrobacterium tumefaciens* and *Agrobacterium rhizogenes. Z. Pflanzenphysiol.* 112:471–74

31. Chen, C. M., Leisner, S. M. 1984. Modification of cytokinins by cauliflower microsomal enzymes. *Plant. Physiol.* 75:442–46

32. Chen, C. M., Melitz, D. K. 1979. Cytokinin biosynthesis in a cell-free system from cytokinin-autotrophic tobacco tissue cultures. *FEBS Lett.* 107:15–20

33. Chernin, L. S., Lobanok, E. V., Fomicheva, V. V., Kartel, N. A. 1984. Crown gall-suppressive IncW R plasmids cause a decrease in auxin production in *Agrobacterium tumefaciens. Mol. Gen. Genet.* 195:195–99

34. Chirek, Z. 1984. Changes in auxin level in the course of growth of a sunflower crown-gall suspension in culture. *Acta Soc. Bot. Pol.* 52:285–94

35. Claeys, M., Messens, E., Van Montagu, M., Schell, J. 1978. GC/MS determination of cytokinins in *Agrobacterium tumefaciens* cultures. *Fresenius' Z. Anal. Chem.* 290:125–26

36. Cohen, J. D., Bialek, K. 1984. Biosynthesis of indole-3-acetic acid in higher plants. In *The Biosynthesis and Metabolism of Plant Hormones,* ed. A. Crozier, J. R. Hillman. *Soc. Exp. Biol. Sem. Ser.* 23:165–81. Cambridge Univ. Press

37. Comai, L., Kosuge, T. 1980. Involvement of plasmid deoxyribonucleic acid in indoleacetic acid synthesis in *Pseudomonas savastanoi. J. Bacteriol.* 143:950–57

38. Comai, L., Kosuge, T. 1982. Cloning and characterization of *iaaM,* a virulence determinant of *Pseudomonas savastanoi. J. Bacteriol.* 149:40–46

39. Comai, L., Surico, G., Kosuge, T. 1982. Relation of plasmid DNA to indoleacetic acid production in different strains of *Pseudomonas syringae* pv. *savastanoi. J. Gen. Microbiol.* 128:2157–63

40. Davis, G. C., Hein, M. B., Neely, B. C., Sharp, C. R., Carnes, M. G. 1985. Strat-

egies for the determination of plant hormones. *Anal. Chem.* 57:638A
41. Dekhuijzen, H. M. 1980. The occurrence of free and bound cytokinins in clubroots and *Plasmodiaphora brassicae* infected turnip tissue cultures. *Physiol. Plant.* 49:169–76
42. De Ropp, R. S. 1951. The crown gall problem. *Bot. Rev.* 17:629–70
43. Downie, J. A., Hombrecher, G., Ma, Q. S., Knight, C. D., Wells, B., Johnston, A. W. B. 1983. Cloned nodulation genes of *Rhizobium leguminosarum* determine host-range specificity. *Mol. Gen. Genet.* 190:359–65
44. Einset, J. W. 1984. Conversion of N_6-isopentenyladenine to zeatin by *Actinidia* tissues. *Biochem. Biophys. Res. Commun.* 124:470–74
45. Ek, M., Ljungquist, P. O., Stenström, E. 1983. Indole-3-acetic acid production by mycorrhizal fungi determined by gas chromatography-mass spectrometry. *New Phytol.* 94:401–7
46. Elzen, G. W. 1983. Cytokinins and insect galls. *Comp. Biochem. Physiol.* A76:17–19
47. Garfinkel, D. J., Nester, E. W. 1980. *Agrobacterium tumefaciens* mutants affected in crown gall tumorigenesis and octopine catabolism. *J. Bacteriol.* 144:732–43
48. Garfinkel, D. J., Simpson, R. B., Ream, L. W., White, F. F., Gordon, M. P., Nester, E. W. 1981. Genetic analysis of crown gall: Fine structure map of the T-DNA by site-directed mutagenesis. *Cell* 27:143–53
49. Gielen, J., De Beuckeleer, M., Seurinck, J., De Boeck, F., De Greve, H., et al. 1984. The complete nucleotide sequence of the T_L DNA of the *Agrobacterium tumefaciens* plasmid pTiAch5. *EMBO J.* 3:835–46
50. Goldberg, S. B., Flick, J. S., Rogers, S. G. 1984. Nucleotide sequence of the *tmr* locus of *Agrobacterium tumefaciens* pTiT37 T-DNA. *Nucleic Acids Res.* 12:4665–77
51. Greene, E. M. 1980. Cytokinin production by microorganisms. *Bot. Rev.* 46:25–74
52. Guerin, B., Kahlem, G., Teller, G., Durand, B. 1984. Evidence for host genome involvement in cytokinin metabolism by male and female cells of *Mercurialis annua* transformed by strain 15,955 of *Agrobacterium tumefaciens*. *Plant Physiol.* 74:139–45
53. Hadley, R. G., Szalay, A. A. 1982. DNA sequences homologous to the T-DNA region of *Agrobacterium tumefaciens* are present in diverse *Rhizobium* species. *Mol. Gen. Genet.* 188:361–69
54. Hansen, C. E., Meins, F. Jr., Milani, A. 1985. Clonal and physiological variation in the cytokinin content of tobacco-cell lines differing in cytokinin requirement and capacity for neoplastic growth. *Differentiation* 29:1–6
55. Hansen, C. E., Wenzler, H., Meins, F. Jr. 1984. Concentration gradients of *trans*-zeatin riboside and *trans*-zeatin in the maize stem. *Plant Physiol.* 75:959–63
56. Heidekamp, F., Dirkse, W. G., Hille, J., van Ormondt, H. 1983. Nucleotide sequence of the *Agrobacterium tumefaciens* octopine Ti plasmid-encoded *tmr* gene. *Nucleic Acids Res.* 11:6211–23
57. Henderson, J. H. M., Bonner, J. 1952. Auxin metabolism in normal and crown gall tumor tissue of sunflower. *Am. J. Bot.* 39:444–51
58. Hirsch, A. M., Drake, D., Jacobs, T. W., Long, S. R. 1985. Nodules are induced on alfalfa roots by *Agrobacterium tumefaciens* and *Rhizobium trifolii* containing small segments of the *Rhizobium meliloti* nodulation region. *J. Bacteriol.* 161:223–30
59. Hirsch, A. M., Wilson, K. J., Jones, J. D. G., Bang, M., Walker, V. V., Ausubel, F. M. 1984. *Rhizobium meliloti* nodulation genes allow *Agrobacterium tumefaciens* and *Escherichia coli* to form pseudonodules on alfalfa. *J. Bacteriol.* 158:1133–43
60. Hoekema, A., de Pater, B. S., Fellinger, A. J., Hooykaas, P. J. J., Schilperoort, R. A. 1984. The limited host range of an *Agrobacterium tumefaciens* strain extended by a cytokinin gene from a wide host range T-region. *EMBO J.* 3:3043–47
61. Hommes, N. G., Akiyoshi, D. E., Morris, R. O. 1985. Assay and partial purification of the cytokinin biosynthetic enzyme dimethylallylpyrophosphate: 5'AMP transferase. *Methods Enzymol.* 110:340–47
62. Hooykaas, P. J. J., Schilperoort, R. A. 1984. The molecular genetics of crown gall tumorigenesis. *Adv. Genet.* 22:209–83
63. Hrouda, M., Ondrej, M. 1983. The effect of plant growth regulators on formation of crown gall tumors on potato tuber disks. *Biol. Plant.* 25:28–32
64. Hutcheson, S., Kosuge, T. 1985. Regulation of 3-indoleacetic acid production in *Pseudomonas syringae* pv. *savastanoi*. Purification and properties of tryptophan-2-monooxygenase. *J. Biol. Chem.* 260:6281–87

65. Hutzinger, O., Kosuge, T. 1967. Microbial synthesis and degradation of indole-3-acetic acid. *Biochim. Biophys. Acta* 136:389–91

66. Inzé, D., Follin, A., Van Lijsebettens, M., Simoens, C., Genetello, C., et al. 1984. Genetic analysis of the individual T-DNA genes of *Agrobacterium tumefaciens;* Further evidence that two genes are involved in indole-3-acetic acid synthesis. *Mol. Gen. Genet.* 194:265–74

67. Joos, H., Inzé, D., Caplan, A., Sormann, M., Van Montagu, M., Schell, J. 1983. Genetic analysis of T-DNA transcripts in nopaline crown galls. *Cell* 32:1057–67

68. Kaiss-Chapman, R. W., Morris, R. O. 1977. *Trans*-zeatin in culture filtrates of *Agrobacterium tumefaciens*. *Biochem. Biophys. Res. Commun.* 76:453–59

69. Kampert, M., Strzelczyk, E. 1978. Production of cytokinins by mycorrhizal fungi of pine (*Pinus sylvestrus* L.). *Bull. Acad. Pol. Sci.* 26:499–503

70. Kao, J. C., Perry, K. L., Kado, C. I. 1982. Indoleacetic acid complementation and its relation to host range specifying genes on the Ti plasmid of *Agrobacterium tumefaciens*. *Mol. Gen. Genet.* 188:425–32

71. Kavanagh, J. A., Williams, P. H. 1981. Indole auxins in *Plasmodiaphora* infected cabbage roots and hypocotyls. *Trans. Br. Mycol. Soc.* 77:125–30

72. Kefford, N. P., Brockwell, J., Zwar, J. A. 1960. The symbiotic synthesis of auxin by legumes and nodule bacteria and its role in nodule development. *Aust. J. Biol. Sci.* 13:456–67

73. Kemper, E., Waffenschmidt, S., Weiler, E. W., Rausch, T., Schröder, J. 1985. T-DNA-encoded auxin formation in crown-gall cells. *Planta* 163:257–62

74. Klee, H., Montoya, A., Horodyski, F., Lichtenstein, C., Garfinkel, D., et al. 1984. Nucleotide sequence of the *tms* genes of the pTiA6NC octopine Ti plasmid: 2 gene products involved in plant tumorigenesis. *Proc. Natl. Acad. Sci. USA* 81:1728–32

75. Knauf, V. C., Panagopoulos, C. G., Nester, E. W. 1983. Comparison of Ti plasmids from three different biotypes of *Agrobacterium tumefaciens* isolated from grape vines. *J. Bacteriol.* 153:1535–42

76. Kosuge, T., Heskett, M. G., Wilson, E. E. 1966. Microbial synthesis and degradation of indole-3-acetic acid. I. The conversion of L-tryptophan to indole-3-acetamide by an enzyme system from *Pseudomonas savastanoi*. *J. Biol. Chem.* 241:3738–44

77. Lawson, E. N., Gantotti, B. V., Starr, M. P. 1982. A 78-megadalton plasmid occurs in avirulent strains as well as virulent strains of *Corynebacterium fascians*. *Curr. Microbiol.* 7:327–32

78. Leemans, J., Deblaere, R., Willmitzer, L., De Greve, H., Hernalsteens, J. P., et al. 1982. Genetic identification of functions of T_L-DNA transcripts in octopine crown galls. *EMBO J.* 1:147–52

79. Leemans, J., Shaw, C., Deblaere, R., De Greve, H., Hernalsteens, J. P., et al. 1981. Site-specific mutagenesis of *Agrobacterium* Ti plasmids and transfer of genes to plant cells. *J. Mol. Appl. Genet.* 1:149–64

80. Letham, D. S., Palni, L. M. S. 1983. The biosynthesis and metabolism of cytokinins. *Ann. Rev. Plant Physiol.* 34:163–97

81. Lichtenstein, C., Klee, H., Montoya, A., Garfinkel, D., Fuller, S., et al. 1984. Nucleotide sequence and transcript mapping of the *tmr* gene of the pTiA6NC octopine Ti plasmid bacterial gene involved in plant tumorigenesis. *J. Mol. Appl. Genet.* 2:354–62

82. Link, G. K. K., Eggers, V. 1941. Hyperauxiny in crown gall of tomato. *Bot. Gaz.* 103:87–96

83. Liu, S. T., Kado, C. I. 1979. Indoleacetic acid production: A plasmid function of *Agrobacterium tumefaciens* C58. *Biochem. Biophys. Res. Commun.* 90:171–78

84. Liu, S. T., Perry, K. L., Schardl, C. L., Kado, C. I. 1982. *Agrobacterium* Ti plasmid indoleacetic acid gene is required for crown gall oncogenesis. *Proc. Natl. Acad. Sci. USA* 79:2812–16

85. MacDonald, E. M. S., Akiyoshi, D. E., Morris, R. O. 1981. Combined high-performance liquid chromatography-radioimmunoassay for cytokinins. *J. Chromatogr.* 214:101–9

86. MacDonald, E. M. S., Morris, R. O. 1985. Isolation of cytokinins by immunoaffinity chromatography and isolation by HPLC-radioimmunoassay. *Methods Enzymol.* 110:347–58

87. MacDonald, E. M. S., Powell, G. K., Regier, D. A., Glass, L., Kosuge, T., Morris, R. O. 1986. Secretion of zeatin, ribosylzeatin and ribosyl-1″-methylzeatin by *Pseudomonas savastanoi:* Plasmid-coded cytokinin biosynthesis. Submitted for publication

88. Magie, A. R., Wilson, E. E., Kosuge, T. 1963. Indoleacetamide as an intermediate in the synthesis of indoleacetic acid in *Pseudomonas savastanoi*. *Science* 141:1281–82

89. Matthysse, A. G. 1985. Auxin-de-

pendent attachment of an *Agrobacterium tumefaciens* strain containing PSA to carrot suspension culture cells. *85th Ann. Meet. Am. Soc. Microbiol. Abstr.* 26

90. Meins, F. Jr. 1971. Regulation of phenotypic expression in crown gall teratoma tissues of tobacco. *Dev. Biol.* 24:287–300

91. Meins, F. Jr., Binns, A. N. 1982. Rapid reversion of cell-division factor habituated cells in culture. *Differentiation* 23:10–12

92. Meins, F. Jr., Foster, R. 1985. Reversible, cell-heritable changes during the development of tobacco pith tissues. *Dev. Biol.* 108:1–5

93. Meins, F. Jr., Foster, R., Lutz, J. D. 1983. Evidence for a Mendelian factor controlling the cytokinin requirement of cultured tobacco cells. *Dev. Genet.* 4: 129–41

94. Meins, F. Jr., Hansen, C. E. 1986. Epigenetic and genetic factors regulating the cytokinin requirement of cultured cells. In *Plant Growth Substances*, ed. M. Bopp. Berlin: Springer-Verlag. In press

95. Meins, F. Jr., Lutz, J. D. 1980. The induction of cytokinin habituation in primary pith explants of tobacco. *Planta* 149:402–7

96. Meins, F. Jr., Lutz, J. D., Binns, A. N. 1980. Variation in the competence of tobacco pith cells for cytokinin habituation in culture. *Differentiation* 16:71–75

97. Mellano, M. A., Cooksey, D. A. 1985. Detection of a cytokinin gene homologue in *Corynebacterium fascians*. *Phytopathology*. In press

98. Miller, A. R., Pengelly, W. L. 1984. Ethylene production by shoot-forming and unorganized crown gall tumor tissues of *Nicotiana* and *Lycopersicon* cultured *in-vitro*. *Planta* 161:418–24

99. Miller, C. O. 1974. Ribosyl-*trans*-zeatin, a major cytokinin produced by crown gall tumor tissue. *Proc. Natl. Acad. Sci. USA* 71:334–38

100. Morris, R. O. 1977. Mass spectroscopic identification of cytokinins. Glucosyl zeatin and glucosyl ribosylzeatin from *Vinca rosea* crown gall. *Plant. Physiol.* 59:1029–33

101. Morris, R. O., Akiyoshi, D. E., MacDonald, E. M. S., Morris, J. W., Regier, D. A., Zaerr, J. B. 1982. Cytokinin metabolism in relation to tumor induction by *Agrobacterium tumefaciens*. In *Plant Growth Substances*, ed. P. F. Wareing. London: Academic

102. Morris, R. O., Powell, G. K., Beaty, J.

S., Durley, R. C., Hommes, N. G., et al. 1986. Cytokinin biosynthetic genes and enzymes from *Agrobacterium tumefaciens* and other plant-associated prokaryotes. See Ref. 94

103. Mousdale, D. M. A. 1981. Endogenous indolyl-3-acetic acid and pathogen-induced plant growth disorders: Distinction between hyperplasia and neoplastic development. *Experientia* 37:972–73

104. Mousdale, D. M. 1982. Endogenous indolyl-3-acetic acid and the growth of auxin-dependent and auxin-autotrophic (crown gall) plant tissue cultures. *Biochem. Physiol. Pflanz.* 177:9–17

105. Murai, N., Skoog, F., Doyle, M. E., Hanson, R. S. 1980. Relationships between cytokinin production, presence of plasmids, and fasciation caused by strains of *Corynebacterium fascians*. *Proc. Natl. Acad. Sci. USA* 77:619–23

106. Nakajima, H., Yokota, T., Takahashi, N., Matsumoto, T., Noguchi, M. 1981. Changes in endogenous ribosyl-*trans*-zeatin and IAA levels in relation to the proliferation of tobacco crown gall cells. *Plant Cell Physiol.* 22:1405–10

107. Nester, E. W., Gordon, M. P., Amasino, R. M., Yanofsky, M. F. 1984. Crown gall: A molecular and physiological analysis. *Ann. Rev. Plant Physiol.* 35:387–413

108. Nester, E. W., Kosuge, T. 1981. Plasmids specifying plant hyperplasias. *Ann. Rev. Microbiol.* 35:531–65

109. Newcomb, W., Syono, K., Torrey, J. G. 1977. Development of an ineffective pea root nodule: Morphogenesis, fine structure, and cytokinin biosynthesis. *Can. J. Bot.* 55:1891–1907

110. Ng, P. P., Cole, A. L. J., Jameson, P. E., McWha, J. A. 1982. Cytokinin production by ectomycorrhizal fungi. *New Phytol.* 91:57–62

111. Ooms, G., Hooykaas, P. J. J., Moolenaar, G., Schilperoort, R. A. 1981. Crown gall plant tumors of abnormal morphology induced by *Agrobacterium tumefaciens* carrying mutated octopine Ti plasmids; analysis of T-region DNA functions. *Gene* 14:33–50

112. Palni, L. M. S., Horgan, R. 1983. Cytokinin biosynthesis in crown gall tissue of *Vinca rosea* L.: Metabolism of *iso*-pentenyl adenine. *Phytochemistry* 22: 1597–1602

113. Palni, L. M. S., Horgan, R. 1983. Cytokinins in transfer RNA of normal and crown-gall tissue of *Vinca rosea* L. *Planta* 159:178–81

114. Palni, L. M. S., Horgan, R., Darrall, N. M., Stuchbury, T., Wareing, P. F. 1983. Cytokinin biosynthesis in crown gall tis-

sue of *Vinca rosea* L. The significance of nucleotides. *Planta* 159:50–59

115. Palni, L. M. S., Summons, R. E., Letham, D. S. 1983. Mass spectrometric analysis of cytokinins in plant tissues. V. Identification of the cytokinin complex of *Datura innoxia* crown gall tissue. *Plant Physiol.* 72:858–63

116. Pengelly, W. L., Meins, F. Jr. 1982. The relationship of indole-3-acetic acid content and growth of crown-gall tumor tissues of tobacco in culture. *Differentiation* 21:27–31

117. Pengelly, W. L., Meins, F. Jr. 1983. Growth, auxin requirement, and indole-3-acetic acid content of cultured crown-gall and habituated tissues of tobacco. *Differentiation* 25:101–5

118. Peterson, J., Miller, C. O. 1977. Glucosyl zeatin and glucosyl ribosyl-zeatin from *Vinca rosea* L. crown gall tumor tissue. *Plant Physiol.* 59:1026–28

119. Phillips, D. A., Torrey, J. G. 1972. Studies on cytokinin production by *Rhizobium*. *Plant Physiol.* 49:11–15

120. Powell, G. K., Morris, R. O. 1985. Nucleotide sequence and expression of a *Pseudomonas savastanoi* cytokinin biosynthetic gene: Homology with *Agrobacterium tumefaciens tmr* and *tzs* loci. *Nucleic Acids Res.* In press

121a. Puppo, A., Rigaud, J. 1978. Cytokinins and morphological aspects of frenchbean roots in the presence of *Rhizobium*. *Physiol. Plant.* 42:202–6

121b. Rademacher, W., Graebe, J. E. 1984. Hormonal changes in developing kernels of two spring wheat varieties differing in storage capacity. *Ber. Dtsch. Bot. Ges.* 97:167–81

122. Rausch, T., Butcher, D. N., Hilgenberg, W. 1981. Nitrilase activity in clubroot diseased plants. *Physiol. Plant.* 52:467–70

123. Rausch, T., Butcher, D. N., Hilgenberg, W. 1983. Indole-3-methylglucosinolate biosynthesis and metabolism in clubroot diseased plants. *Physiol. Plant.* 58:93–100

124. Rausch, T., Minocha, S. C., Hilgenberg, W., Kahl, G. 1985. L-tryptophan metabolism in wound-activated and *Agrobacterium tumefaciens*-transformed potato tuber cells. *Physiol. Plant.* 63:335–44

125. Regier, D. A., Morris, R. O. 1982. Secretion of *trans*-zeatin by *Agrobacterium tumefaciens:* A function determined by the nopaline Ti plasmid. *Biochem. Biophys. Res. Commun.* 104:1560–66

126. Ryder, M. H., Tate, M. E., Kerr, A. 1985. Virulence properties of strains of *Agrobacterium* on the apical and basal surfaces of carrot root disks. *Plant Physiol.* 77:215–21

127. Sachs, T. 1975. Plant tumors resulting from unregulated hormone synthesis. *J. Theor. Biol.* 55:445–53

128. Schell, J., Van Montagu, M., Schröder, J., Schröder, G., Inze, D., et al. 1984. Genes involved in development and differentiation control in plants. *Horm. Cell Regul.* 8:245–54

129. Schröder, G., Waffenschmidt, S., Weiler, E. W., Schröder, J. 1984. The T-region of Ti plasmids codes for an enzyme synthesizing indole-3-acetic acid. *Eur. J. Biochem.* 138:387–91

130. Sciaky, D., Thomashow, M. F. 1984. The sequence of the *tms* transcript 2 locus of the *A. tumefaciens* plasmid pTiA6 and characterization of the mutation in pTiA66 that is responsible for auxin attenuation. *Nucleic Acids Res.* 12:1447–61

131. Scott, I. M., Horgan, R. 1980. Quantification of cytokinins by selected ion monitoring using ^{15}N labelled internal standards. *Biomed. Mass Spectrosc.* 7:446–49

132. Scott, I. M., Horgan, R. 1984. Mass-spectrometric quantification of cytokinin nucleotides and glycosides in tobacco-crown gall tissue. *Planta* 161:345–54

133. Scott, I. M., Horgan, R., McGaw, B. A. 1980. Zeatin-9-glucoside, a major endogenous cytokinin of *Vinca rosea* L. crown gall tissue. *Planta* 149:472–75

134. Scott, I. M., Martin, G. C., Horgan, R., Heald, J. K. 1982. Mass spectrometric measurement of zeatin glycoside levels in *Vinca rosea* L. crown gall tissue. *Planta* 154:273–76

135. Scott, I. M., McGaw, B. A., Horgan, R., Williams, P. E. 1982. Biochemical studies on cytokinins in *Vinca rosea* crown gall tissue. See Ref. 101, pp. 165–74

136. Sibold, L., Guiso, N., De Buckeleer, M., Van Montagu, M. 1984. Cloning and expression in *Escherichia coli* of the T_L-DNA gene 4 of *Agrobacterium tumefaciens* under the control of the PR promoter of bacteriophage lambda. *Biochimie* 66:547–56

137. Skoog, F., Miller, C. O. 1957. Chemical regulation of growth and organ formation in plant tissues cultured *in vitro*. *Symp. Soc. Exp. Biol.* 11:118–31

138. Strobel, G. A., Lam, B., Harrison, L., Hess, B. M., Lam, S. 1985. Introduction of the hairy root plasmid into *Rhizobium meliloti* results in increased nodulation on its host. *J. Gen. Microbiol.* 131:355–61

139. Stuchbury, T., Palni, L. M., Horgan, R., Wareing, P. F. 1979. The biosynthesis of cytokinins in crown-gall tissue of *Vinca rosea* L. *Planta* 147:97–102
140. Summons, R. E., Duke, C. C., Eichholzer, J. V., Entsch, B., Letham, D. S., et al. 1979. Mass spectrometric analysis of cytokinins in plant tissues. II. Quantitation of cytokinins in *Zea mays* kernels using deuterium labelled standards. *Biomed. Mass Spectrosc.* 6:407–13
141. Surico, G., Evidente, A., Iacobellis, N. S., Randazzo, G. 1985. A new cytokinin from the culture filtrate of *Pseudomonas syringae* pv. *savastanoi*. *Phytochemistry* 24:1499–1502
142. Surico, G., Sparapano, L., Lerario, P., Durbin, R. D., Iacobellis, N. 1975. Cytokinin-like activity in extracts from culture filtrates of *Pseudomonas savastanoi*. *Experientia* 31:929–30
143. Tay, S. A. B., MacLeod, J. K., Palni, L. M. S. 1986. On the reported occurrence of *cis*-zeatin riboside as a free cytokinin in tobacco shoots. *Plant Sci.* In press
144. Thomashow, L. S., Reeves, S., Thomashow, M. F. 1984. Crown gall oncogenesis: Evidence that a T-DNA gene from the *Agrobacterium* Ti plasmid pTiA6 encodes an enzyme that catalyzes synthesis of indole-acetic acid. *Proc. Natl. Acad. Sci. USA* 81:5071–75
145. Trinchant, J. C., Rigaud, J. 1977. Synthesis of indole-3-lactic acid and indole-3-acetic acid by bacteroid extracts from *Phaseolus vulgaris* L. nodules and its relation with atmospheric nitrogen fixation. *C. R. Acad. Sci. Ser. D* 284:301–3
146. Trione, E. J., Krygier, B. B., Banowetz, G. M., Kathrein, J. M. 1985. The development of monoclonal antibodies against the cytokinin zeatin riboside. *J. Plant Growth Regul.* 4:101–9
147. Van Onckelen, H., Rüdelsheim, P., Hermans, R., Horemans, S., Messens, E., et al. 1984. Kinetics of endogenous cytokinin, IAA and ABA levels in relation to the growth and morphology of tobacco crown gall tissue. *Plant Cell Physiol.* 25:1017–25
148. Van Onckelen, H., Rüdelsheim, P., Inzé, D., Follin, A., Messens, E., et al. 1985. Tobacco plants transformed with the *Agrobacterium* T-DNA gene 1 contain high amounts of indole-3-acetamide. *FEBS Lett.* 181:373–76
149. Van Slogteren, G. M. S., Hooykaas, P. J. J., Schilperoort, R. A. 1984. Tumor formation on plants by mixtures of attenuated *Agrobacterium tumefaciens* T-DNA mutants. *Plant Mol. Biol.* 3:337–44

150. Van Staden, J., Davey, J. E. 1978. Endogenous cytokinins in the larvae and galls of *Erythrina latissima* leaves. *Bot. Gaz.* 139:36–41
151. Wang, T. L., Wood, E. A., Brewin, N. J. 1982. Growth regulators, *Rhizobium* and nodulation in peas. Indole-3-acetic acid from the culture medium of nodulating and non-nodulating strains of *R. leguminosarum*. *Planta* 155:345–49
152. Wang, T. L., Wood, E. A., Brewin, N. J. 1982. Growth regulators, *Rhizobium* and nodulation in peas. The cytokinin content of a wild-type and a Ti-plasmid-containing strain of *R. Leguminosarum*. *Planta* 155:350–55
153. Weiler, E. W. 1980. Radioimmunoassay for *trans*-zeatin and related cytokinins. *Planta* 149:155–62
154. Weiler, E. W., Spanier, K. 1981. Phytohormones in the formation of crown gall tumors. *Planta* 153:326–37
155. Weisburg, W. G., Woese, C. R., Dobson, M. E., Weiss, E. 1985. A common origin of Rickettsiae and certain plant pathogens. *Science* 230:556–58
156. Willmitzer, L., Dhaese, P., Schreier, P. H., Schmalenbach, W., Van Montagu, M., Schell, J. 1983. Size, location and polarity of T-DNA-encoded transcripts in nopaline crown gall tumors; common transcripts in octopine and nopaline tumors. *Cell* 32:1045–56
157. Willmitzer, L., Simons, G., Schell, J. 1982. The TL-DNA in octopine crown-gall tumours codes for seven well-defined polyadenylated transcripts. *EMBO J.* 1:139–46
158. Wilson, E. E. 1935. The olive knot disease: its inception, development and control. *Hilgardia* 9:231–64
159. Woese, C. R., Stackebrandt, E., Weisburg, W. G., Paster, B. J., Madigan, M. T., et al. 1984. The phylogeny of purple bacteria: The alpha subdivision. *Syst. Appl. Microbiol.* 5:315–26
160. Wyndaele, R., Van Onckelen, H., Chritiansen, J., Rüdelsheim, P., Hermans, R., De Greef, J. 1985. Dynamics of endogenous IAA and cytokinins during the growth cycle of soybean crown gall and untransformed cells. *Plant Cell Physiol.* 26:1147–54
161. Yadav, N. S., Vanderleyden, J., Bennett, D. R., Barnes, W. M., Chilton, M.-D. 1982. Short direct repeats flank the T-DNA on a nopaline Ti-plasmid. *Proc. Natl. Acad. Sci. USA* 79:6322–26
162. Yamada, T., Palm, C. J., Brooks, B., Kosuge, T. 1985. Nucleotide sequences of the *Pseudomonas savastanoi* indoleacetic acid genes show homology

with *Agrobacterium tumefaciens* T-DNA. *Proc. Natl. Acad. Sci. USA* 82:6522–26

163. Yang, D., Oyaizu, Y., Oyaizu, H., Olsen, G. J., Woese, C. R. 1985. Mitochondrial origins. *Proc. Natl. Acad. Sci. USA* 82:4443–47

164. Yanofsky, M., Montoya, A., Knauf, V., Lowe, B., Gordon, M., Nester, E. 1985.

Limited-host-range plasmid of *Agrobacterium tumefaciens:* Molecular and genetic analyses of transferred DNA. *J. Bacteriol.* 163:341–48

165. Zaenen, I., Van Larebeke, N., Teuchy, H., Van Montagu, M., Schell, J. 1974. Supercoiled circular DNA in crown gall inducing *Agrobacterium* strains. *J. Mol. Biol.* 86:109–27

NOTE ADDED IN PROOF

Conclusive evidence that the *tms1* locus from *Agrobacterium tumefaciens* encodes tryptophan monooxygenase was recently obtained by two groups: Thomashow, M. F., Hugly, W. G., Buchholz, L. S., Thomashow, L. S. 1986. Molecular basis for the auxin independent phenotype of crown gall tumor tissues. *Science* 231:616–18; and Van Oncklen, H., Prinsen, E., Inze, D., Rudelsheim, P., Follin, A., et al. 1986. *Agrobacterium* T-DNA gene 1 codes for tryptophan-2-monooxygenase activity in tobacco crown gall cells. *FEBS Lett.* In press.

Ann. Rev. Plant Physiol. 1986. 37:539-74

PRODUCTS OF BIOLOGICAL NITROGEN FIXATION IN HIGHER PLANTS: Synthesis, Transport, and Metabolism

Karel R. Schubert

Biology Department, Washington University, St. Louis, Missouri 63130

CONTENTS

INTRODUCTION

Nitrogen (N) is the nutrient most limiting plant growth, especially in agricultural systems (34, 47). Plants normally acquire N from the soil in inorganic form (nitrate or ammonium). In the absence of an adequate supply of available soil N, certain leguminous and actinorhizal species are capable of forming a

0066-4294/86/0601-0539$02.00

symbiotic association with N_2-fixing microorganisms. Through this symbiosis, the plant is able to obtain part or all of the N required for plant growth from its symbiotic partner. In return, the plant provides photoassimilates to support the growth and function of the nodule, the symbiotic organ of the plant containing the N_2-fixing bacteria.

This review, which is not intended to be comprehensive, focuses on the synthesis, transport, and metabolism of the products of N_2 fixation. We begin with an overview of this exchange of reduced carbon and nitrogen from a whole plant perspective.

THE EXCHANGE OF REDUCED CARBON AND NITROGEN: AN OVERVIEW

The formation and function of the symbiotic association between legumes and *Rhizobium* or actinorhizal species and the actinomycete *Frankia* involve a complex set of adaptations that lead to an intertwining of metabolism and a need for elaborate genetic and biochemical control mechanisms. Because of the importance of N_2-fixing plants to agriculture, forestry, and the global N cycle, efforts have been focused on unraveling these complex, sometimes subtle, metabolic and regulatory interrelationships.

The symbiosis depends on an exchange of C and N between the host plant and the nodule (Figure 1). Prior to the onset of N_2 fixation, the host plant provides C and N substrates to support the growth of the developing nodule which contains plant cells infected with the endophyte. These substrates are derived from seed storage reserves or from the reallocation and remobilization of the plant's resources. Once N_2 fixation begins, the nodule and its bacterial inhabitants maintain their dependency upon the host plant for both C and N.

The reducing potential and ATP generated through photosynthesis are used to drive the carbon reduction cycle. Photoassimilates produced are subsequently allocated to the various sinks within the plant. The nodule and its subtending root system represent one of the strongest sinks, receiving an estimated 15 to 30% or more of the net photosynthate of the plant (64, 75, 84, 100, 116, 121). This supply of photosynthate transported via the phloem is used for energy-yielding substrates and carbon skeletons to support (*a*) the growth and maintenance of the nodule tissue; (*b*) the energy-consuming reactions associated with the reduction of N_2 in the endophyte and the assimilation of the NH_4^+ produced in the host cytosol; and (*c*) the synthesis of N-containing organic compounds for export from the nodule.

These nitrogenous solutes transported via the xylem (82, 83) to the vegetative and reproductive structures of the plant contribute to the soluble N pool (3, 48). As N is needed, these compounds are withdrawn from this pool and degraded, releasing NH_4^+ and other byproducts. The products of catabolism

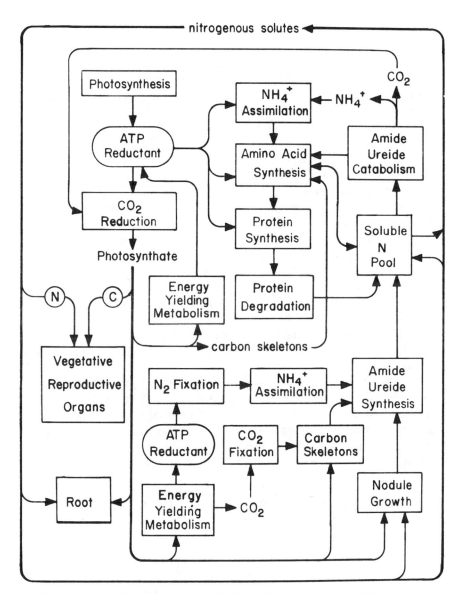

Figure 1   An overview of the exchange of carbon and nitrogen in nitrogen-fixing plants.

are subsequently reassimilated, producing the specific complement of amino acids needed for the synthesis of other metabolites (e.g. nucleic acids, secondary products) and macromolecules (e.g. protein, DNA, RNA). In the light, the energy-consuming reactions of NH_4^+ assimilation, amino acid biosynthesis, and macromolecular synthesis are directly coupled to the energy-generating reactions of photosynthesis while in the dark or in nonphotosynthetic tissues, photosynthate is consumed to produce the ATP and reductant necessary to drive these processes.

The insoluble N pool serves as a reserve. As N is required for vegetative or reproductive growth, this reserve is mobilized. Turnover of macromolecules, especially prominent during reproductive growth, feeds nitrogenous compounds into the soluble N pool for recycling. This recycling process frequently involves catabolism, reassimilation, and resynthesis of N-containing metabolites (58, 82).

Organic N for sinks such as newly expanding leaves arrives primarily in the xylem stream (82, 83, 87), whereas N for seed growth is delivered to a great extent through the phloem (3, 4, 82, 83, 87). Once again the nitrogenous solutes reaching a given sink are often metabolized and the products reassimilated and used for the synthesis of required metabolites.

Although recently fixed N_2 is used to meet the N requirements of the nodule (81, 82), approximately half of the N for nodule and root growth is obtained from the soluble N pool of the shoot and delivered through the phloem (56, 82, 84; D. A. Polayes and K. R. Schubert, unpublished). Excess N is apparently returned to the shoot in the xylem (82). In addition to providing general control over the utilization of the plant's N resources, provision of N for nodule growth from reserves in the shoot may serve as another mechanism of regulating nodule growth and activity, preventing what otherwise might develop into a parasitic relationship.

Estimates of the general carbon costs associated with N_2 fixation have been considered in various reviews (64, 75, 84, 100, 116, 121). Pate and others (48, 56, 74, 84) have constructed C and N budgets for several annual legumes which provide extensive data on the transport of C and N. This review will focus on the synthesis and metabolism of the major products of N_2 fixation exported from the nodule and the carbon costs associated with these processes.

MAJOR PRODUCTS OF RECENTLY FIXED N_2 EXPORTED FROM NODULES

Bergersen (10) and Kennedy (52) demonstrated that NH_4^+ is the first stable product of N_2 fixation in legume nodules. The NH_4^+ produced is excreted by *Rhizobium* bacteroids (12) into the host cell cytoplasm where it is assimilated and used in the synthesis of organic N for transport. Nitrogen-fixing plants can be classified as amide exporters or ureide exporters based on the composition of the xylem fluid collected from excised nodules or nodulated root systems (Table 1). The amide exporters transport asparagine (ASN), glutamine (GLN) or 4-methyleneglutamine (MeGln) while ureide exporters transport either allantoin (ALN) and allantoic acid (ALC) or citrulline (CIT).

Table 1   Nitrogenous constituents of xylem sap of nitrogen-fixing plants

Plant species	Major nitrogenous constituents[a]	References
Legumes		
Lathyrus cicera	ASN, GLN	96
Lathyrus sativa	ASN, GLN	96
Lens esculenta	ASN, GLN	96
Lupinus albus	ASN, GLN	96
Lupinus angustifolius	ASN, GLN	96
Lupinus mutabilis	ASN, GLN	96
Pisum arvense	ASN, GLN	88
Pisum sativum	ASN, GLN	44
Trifolium repens	ASN	82
Vicia calcarata	ASN, GLN	96
Vicia ervilia	ASN, GLN	96
Vicia faba	ASN, GLN	86
Vicia sativa	ASN, GLN	96
Arachis hypogaea	ASN, ALN, ALC	36, 96
	4-MeGln	148
Albizzia lophantha	CIT, ALN, ALC	23
Cajanus cajan	ASN, ALN, ALC	96
Cicer arientinum	ASN, ALN, ALC	96
Cyamopsis tetragonoloba	ALN, ALC	96
Glycine max	ALN, ALC	68, 130
	(ASN)	(133, 151)
Phaseolus vulgaris	ALN, ALC	33
Psophocarpus tetragonolobus	ALN, ALC	85
Macrotyloma uniflorum	ALN, ALC	96
Vigna angularis	ALN, ALC	96
Vigna mungo	ALN, ALC	96
Vigna radiata	ALN, ALC	96
Vigna unguiculata	ALN, ALC	48, 85
Actinorhizal plants		
Casuarina equisetifolia	ASN	23
	(CIT)	146a, (b)
Coriaria ruscifolia	GLN, ASN (CIT)	23
Myrica gale	ASN	60, (b)
Myrica cerifera	ASN, GLN	(b)
Comptonia spp	ASN	(b)
Datisca glomerata	GLN, ASN	(b)
Parasponia rigida	ASN	(b)
Eleagnus umbellata	ASN	(b)
Eleagnus angustifolia	ASN	(b)
Ceanothus americana	ASN	(b)
Alnus glutinosa	CIT	23, 73
Alnus rubra	CIT	(b)
Alnus crispa	CIT	(b)

[a]Abbreviations: ASN, asparagine; GLN, glutamine; MeGln, 4-methyleneglutamine; CIT, citrulline; ALN, allantoin, ALC, allantoic acid.
[b]P. R. McClure, G. T. Coker III, J. D. Tjepkema, K. R. Schubert, unpublished results.

Legumes of the tribes Vicieae, Genisteae, and Trifolieae are generally amide exporters. These tribes of more temperate origin include pea, lupin, broad bean, alfalfa, and clover. The majority of the actinorhizal symbioses examined also export the amide ASN (Table 1). Based on results of $^{15}N_2$-labeling studies, asparagine is synthesized from the products of recent N_2 fixation in nodules of amide-exporting symbioses (1, 52, 60).

One tropical legume, peanut *(Arachis hypogaea)*, a member of the tribe Aeschynomenaea, is reportedly an amide exporter. Fowden (35, 36) discovered an unusual nonprotein amino acid amide 4-methyleneglutamine (MeGln) in peanuts. This amide occurs in high concentrations in the cotyledons and leaves of peanut. Winter et al (148) recently reported that MeGln accounted for 90% of the N in the root bleeding sap of nodulated peanut. Rainbird's (96) observation of high levels of ASN, ALN, and ALC in peanut xylem sap is an as yet unresolved contradiction to the findings above. Fowden and coworkers (35, 36) concluded that MeGln plays a major role in N transport in this species. The role of MeGln in the transport of recently fixed N_2, however, is still open. Short-term labeling studies will be required to substantiate the proposed role of MeGln in the transport of fixed N.

Tropical legumes of the tribe Phaseoleae synthesize and transport the ureides allantoin and allantoic acid from recently fixed N_2. The occurrence of ALN and ALC in legumes was noted in the 1930s [see reviews by Tracey and others (140)]. These two compounds account for 60 to 90% of the total N in the xylem sap of soybeans (68, 133), cowpeas (48, 85), garden beans (33, 81), and other legumes (82, 85, 96) grown symbiotically. The role of these compounds in the transport of recently fixed N_2 was confirmed using $[^{15}N]N_2$ (48, 66, 80).

The other major ureide found in N_2-fixing plants is citrulline which was first detected in leaves, roots, and nodules of *Alnus* (alder) by Miettinen & Virtanen (73). Leaf, Gardner & Bond (59) subsequently demonstrated that CIT was rapidly labeled after exposure to ^{15}N-labeled N_2 and exported to the shoot (43).

Although these are the major nitrogenous solutes in the xylem stream of N_2-fixing plants, lesser amounts of many other protein and nonprotein amino acids are present (32, 33, 35, 36, 48, 68, 81, 82, 96, 130, 133). Low levels of ALN and ALC occur in xylem sap of amide-exporting legumes, and amides are normally present in sap from the tropical legumes. The absolute amounts and relative proportion of these primary and secondary nitrogenous solutes vary with developmental and environmental factors (50, 65, 85, 89).

The presence of inorganic N (NO_3^- or NH_4^+) in the soil has the most dramatic effect on this distribution (33, 50, 65, 70, 85). Increasing levels of soil N decrease the relative proportion of organic N in the xylem fluid. In many ureide-exporting legumes, a shift from ureide synthesis to amide synthesis occurs when NO_3^- or NH_4^+ are present in the rooting media (50, 65, 85, 89,

136). Similar shifts have not been noted for amide-exporting legumes or actinorhizal species grown on inorganic N.

The "plant's rationale" for exporting amides or ureides is the subject of continued research and speculation. Using short-term amino acid feeding studies, McNeil et al (69) demonstrated that in lupin particular amino acids were selectively removed from the xylem stream and accumulated. Likewise, certain nitrogenous solutes such as ASN are more readily transferred to the phloem than others, for example, GLN (4, 81–83, 87, 90, 98). Thus, control of the biochemical composition of xylem and phloem fluids may provide a mechanism to control the selective partitioning of N into various sinks.

The ureides are less soluble than the amides. Sprent (128) has suggested that this may be one reason why high levels of ureides are found only in tropical legumes. The growth pattern of nodules of ureide-exporting legumes is essentially determinant, with vascular strands fused apically to form a series of closed loops connected to the main vascular network at the base of the nodule. The reduced resistance to water flow in this closed loop system provides for a higher flux of transpirational water through the system to flush the products of fixation from the nodule.

In contrast, the nodule growth pattern for the amide-exporting members of the Vicieae and Trifolieae is indeterminant with an open, branched vascular system with restricted water flow (128). Gunning et al (44, 86) have observed special transfer cells in pea nodules which apparently actively secrete nitrogenous solutes into the xylem. As a net result of the reduced flux of water and the active loading of solutes, the molar concentration of amides in the xylem sap is severalfold higher than the corresponding concentration of ureides in the tropical legumes (44, 83, 86, 133, 151). Thus, the water use efficiency for N export should be higher for amide-exporting symbioses. The higher solubility of amides would be an advantage in temperate climates or in more arid environments. In a similar fashion, this may explain why the export of ureides is restricted to tropical legumes.

The ureides may be more efficient forms of N transport both in terms of the moles of C used for carbon skeletons and the energetics of synthesis. The increased production and transport of ureides in other tissues or plant species under conditions in which carbon may be limiting (77, 94) is consistent with the theory that ureides are a more efficient form of N transport. In fact, the ureide-producing legumes are apparently more frugal in the use of photoassimilates when compared to amide exporters (82, 84). Pate et al (84) have shown that cowpeas use 5.4 g C per g N fixed while lupins consume 6.9 g C/g N. These differences in apparent efficiency cannot be attributed arbitrarily to differences between the cost of producing ureides or amides. Many other factors (e.g. reduced rates of H_2 evolution, H_2 recycling, PEP carboxylase) may contribute to this overall improvement in efficiency (84).

SYNTHESIS OF AMIDES AND UREIDES

Ammonium Assimilation

The enzymes of NH_4^+ assimilation are repressed in bacteroids (or N_2-fixing cultures) of *Rhizobium* (25) and induced in host plant tissue (18, 108, 109). Since isolated bacteroids excrete NH_4^+ (12), Robertson et al (108, 109) suggested that NH_4^+ is excreted from bacteroids in vivo and assimilated in the host cell cytoplasm.

Within the host cytosol, NH_4^+ is assimilated via the combined activities of glutamine synthetase and glutamate synthase. This conclusion is based on a variety of enzymological and tracer studies. Although initial workers using $^{15}N_2$ suggested that NH_4^+ was first assimilated in the reaction catalyzed by glutamate dehydrogenase (1, 52, 59), additional findings do not support this conclusion. More recent studies by Meeks et al (72) using $[^{13}N]N_2$, a short-lived radioactive isotope of N, demonstrated that NH_4^+ is first incorporated into the amide position of GLN in the reaction catalyzed by GLN synthetase. This incorporation is blocked by methionine sulfoximine, an inhibitor of GLN synthetase (72). The amide group is subsequently transferred to the 2-carbon of oxoglutarate in the reductive amination reaction carried out by GLU synthase (72). The latter is blocked by azaserine, a general inhibitor of amide-group transfer reactions. Similar results have been obtained by Japanese scientists using $[^{15}N]N_2$ (80). Through these two reactions, the GLU and GLN necessary for the synthesis of other N-carrying molecules are produced.

Both enzymes have been isolated and purified from legume nodules (16, 33a, 43a, 71). Glutamine synthetase is localized totally in the cytoplasm while GLU synthase exists in the cytoplasm and in the plastid (8, 19, 123a, 124). Both enzymes are induced during nodule development (7, 18, 43b, 105, 108, 109, 115).

The pathway of assimilation of NH_4^+ is apparently the same in amide-producing and ureide-producing legume nodules. Although this is assumed for nodules of actinorhizal species, some differences are apparent. To begin, unlike *Rhizobium*, N_2-fixing cultures of *Frankia* can grow on the N fixed (for further details, see chapter by Tjepkema et al in this volume). Leaf et al (59) concluded that GLU was the first organic product of N_2 fixation and suggested that GLU dehydrogenase was responsible for the initial assimilation of NH_4^+, based on their studies with $[^{15}N]N_2$. Using $[^{13}N]$-labeled NH_4^+, Schubert et al (118, 120) demonstrated that NH_4^+ was incorporated initially into the amide group of GLN. The observed pattern of assimilation, however, was different from that observed by Meeks et al (72) for soybean nodules. The results for alder were not consistent with a major role for GLU synthase in the assimilatory process. These differences have not been resolved by biochemical studies. Glutamine synthetase and GLU dehydrogenase activities were detected in extracts of alder nodules (13a). Glutamine synthetase has been purified from this source (48a) and apparently is located in the host cytoplasm (13a, 48a). Blom et al (13a) were unable to detect GLU synthase activity in nodule extracts. They suggested that the enzyme may require ferredoxin for activity but did not address this experimentally. One must be cautious, however, when interpreting these findings because the site and pathway of assimilation of tracer levels of NH_4^+ may be distinctly different from those for NH_4^+ produced via N_2 fixation; unfortunately, the biochemical studies were incomplete. Definitive biochemical and tracer experiments using labeled N_2 must be carried out to resolve these differences.

Asparagine Synthesis

The pathway of ASN synthesis in plants has been elusive. Synthesis can occur by several different routes. The first involves the incorporation of HCN via the β-cyanoalanine pathway (14). The second is the direct synthesis of ASN from aspartate (ASP) in the GLN-dependent ASN synthetase reaction. Although labeled HCN is converted into ASN in cotyledons and leaves of legume seedlings, this does not appear to be the primary pathway (62, 131).

Asparagine synthetase activity has been detected in cotyledons of lupin (110) and soybean (129, 131), and subsequently the enzyme has been purified from lupin (57, 111). Scott et al (122) reported finding ASP aminotransferase and ASN synthetase activities in the cytosol of lupin nodules. The former has been further purified from this source (106, 107).

On the basis of early reports which indicated that ASN was the predominant amino acid in the xylem sap of soybeans (130, 151), much of the work on ASN synthesis has been carried out on soybean nodules. Although ureides are the major export product in symbiotically dependent soybeans, the concentration of ASN in nodules is severalfold higher than the concentration of ureides, and detectable amounts of ASN are exported from the nodulated root system (68, 115, 130), especially during early nodule development when ASN may be imported from the phloem for nodule growth. The rates of ASN synthesis (119) and the concentration of ASN in the xylem increase when inorganic N is added to the growth medium (33, 50, 70, 83, 119).

Attempts to demonstrate ASN biosynthesis using labeled precursors, however, have been complicated by the rapid metabolism of ASP (131) and by the presence of an active asparaginase in soybean nodules (132). Label from $^{14}CO_2$, however, is incorporated into ASP and subsequently ASN through the action of PEP carboxylase. This has been used to probe C metabolism associated with N assimilation in soybean roots and nodules (29, 31, 32, 118, 119). Results of these studies support the synthesis of ASN by ASN synthetase in vivo. Fujihara & Yamaguchi (42) have confirmed this using $^{15}NH_4^+$ and [amide-^{15}N]GLN. Asparagine synthetase activity has been detected in soybean nodule extracts and purified (49).

In lupin nodules, ASP aminotransferase (18, 107) and ASN synthetase (122) along with GLN synthetase activities increase with the onset of N_2 fixation. Aspartate aminotransferase exists in two forms in lupin (106, 107) and soybean (113) nodules, one of which is specifically induced in response to N_2 fixation and the export of fixed N. This isozyme of ASP aminotransferase and ASN synthetase are compartmentalized within the plastid in soybean nodules (19). Shelp & Atkins (123a) found that the distribution of ASP aminotransferase was similar in lupin nodules. In contrast to results with soybean nodules, ASN synthetase was apparently cytosolic.

The synthesis and export of ASN requires a continuous supply of carbon skeletons for ASP synthesis. Of the photosynthate received by the nodulated lupin root, 25 to 30% is returned to the shoot primarily as nitrogenous solutes (48, 56, 82, 84). Dark CO_2 fixation catalyzed by PEP carboxylase can function as an anaplerotic pathway to supply part of the carbon skeletons needed. Legume nodules exhibit high levels of PEP carboxylase activity (27, 32, 33, 54, 56, 74, 145). PEP carboxylase activity is highly correlated with N_2-fixing activity in lupin (27) and alfalfa (145) nodules. Although dark CO_2 fixation may serve other roles in the nodule (32, 119), this activity is important in replenishing 2-oxo acids used in amino acid biosynthesis.

Aspartate and ASN are labeled after exposure of nodules or roots to $^{14}CO_2$ (31, 119). CO_2 fixation accounts for an estimated 20 to 25% of the C returned to the shoot in lupin (84) and alfalfa (145). PEP carboxylase has been purified from soybean (92) and alfalfa (144) nodules and is cytoplasmic. (Note: See Figure 4 for diagram showing pathway and subcellular organization of ASN synthesis in soybean nodules.)

Synthesis of 4-Methyleneglutamine

Fowden (35, 36) suggested that MeGln might be synthesized in a manner similar to GLN synthesis. Although glutamine synthetase will catalyze the transfer of NH_4^+ to 4-methyleneglutamate (MeGlu), the K_m for MeGlu is tenfold higher than for GLU (149). Recently, Winter et al (150) isolated a specific MeGln synthetase from peanut cotyledons. This enzyme catalyzes the reaction analogous to that catalyzed by GLN synthetase but using MeGlu as an acceptor and forming AMP + PP_i. To date, the occurrence of MeGln synthesis or the presence of the MeGln synthetase in peanut nodules have not been documented.

Fowden (36) suggested that MeGlu was formed either by reductive amination or transamination of 4-methylene-2-oxoglutarate. Fowden & Webb (38) reported that this compound was present in extracts of peanut seedlings, and a mechanism for synthesis involving the condensation of two molecules of pyruvate was proposed (38). Experimental evidence supporting this proposal was obtained by feeding labeled substrates including pyruvate, which was rapidly incorporated into MeGlu and MeGln in cotyledons, the main site of synthesis (39). Nodule extracts, however, were not tested for activity. Powell & Dekker (95) have proposed a scheme in which MeGlu functions as an N-carrier (Figure 2). According to this scheme, MeGlu is synthesized initially in the cotyledons and transported to the nodulated root system where MeGln synthesis occurs. MeGln is exported to the shoot and acted upon by a specific amidohydrolase releasing NH$_4^+$ and MeGlu, the latter being cycled back to the nodule or metabolized. The presence of deamidase activity in extracts of peanut was reported by Fowden (37), and the enzyme was recently purified (95). This proposal is intriguing but currently lacks scientific support. Evidence supporting the role of MeGln in the transport of recently fixed N$_2$ and biochemical and tracer studies to support the synthesis of MeGln in the nodule are essential. The hypothetical model for a MeGln-MeGlu cycle in peanuts (Figure 2) is presented here not as fact, but to stimulate critical experimentation.

Synthesis of Citrulline

Although CIT is the major nitrogenous solute in nodules, roots, stems, leaves, and xylem sap of nodulated and non-nodulated alder (73) and *Casuarina* (146a), very few biochemical studies have been conducted to elucidate the pathway of synthesis. Gardner & Leaf (43) have shown that alder nodules readily fix CO$_2$, and labeled products after exposure to ^{14}CO$_2$ include CIT, amino acids, and organic acids. To examine the synthesis of CIT and the role of CO$_2$ fixation, McClure et al (67) extended the studies of Gardner and Leaf.

Citrulline is normally synthesized from ornithine (ORN) and carbamoyl phosphate, the ORN required being produced from glutamate via the ORN cycle (see 139). Thus, the synthesis and export of CIT requires an anaplerotic mechanism for the production of oxoglutarate. After exposure of alder nodules to ^{14}CO$_2$, GLU was labeled in the C1 position (67). Labeling patterns were consistent with CIT synthesis from GLU with 10 to 20% of the ^{14}C incorporated into CIT present in the C1 position. The remaining 80 to 90% of the ^{14}C in CIT was located in the carbamoyl carbon. Although ORN was labeled, ORN pools were very low.

Both PEP carboxylase and carbamoyl phosphate synthetase are responsible for dark CO$_2$ fixation in alder nodules (67). PEP carboxylase produces oxalacetate (OAA) which can enter the tricarboxylic acid cycle with subsequent conversion into oxoglutarate labeled in the C1 position. The observation that citrate and oxoglutarate are labeled rapidly and that label accumulates in GLU support the conclusions above. The labeling pattern in pulse-chase experiments suggests that at least two pools of malate exist. The largest pool is relatively inert and possibly located in the vacuole. The second pool, representing a third of the label in the organic acid fraction, is rapidly chased into the amino acid pool.

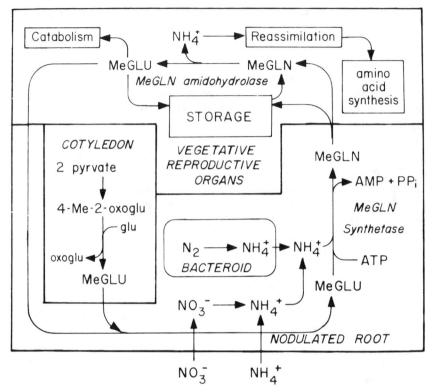

Figure 2 Hypothetical model for the role of 4-methyleneglutamate as a nitrogen carrier in peanuts.

Very little has been done to study the localization or properties of the enzymes involved in CIT synthesis. Using immunocytochemistry, Perrot-Rechenmann et al (91) have shown that PEP carboxylase is localized primarily in the host cytoplasm in alder nodules. Based on studies in *Neurospora,* Weiss & Davis (146b) concluded that carbamoyl phosphate synthetase and all of the enzymes of the ORN cycle, except acetylglutamate phosphotransferase, were located in the mitochondria. In contrast, Shargool et al (122b) reported that carbamoylphosphate synthetase and ORN carbamoyltransferase were located in the plastids of cultured soybean cells. Similarly, these two enzymes along with acetylornithine: GLU acetyltransferase have been localized in the chloroplast of pea leaves (135a). Gardner and coworkers (42a, 122a) used cytochemical techniques to examine the location of ORN carbamoyltransferase in alder nodules. They found that the enzyme was mitochondrial and concluded that this organelle was the intracellular site of CIT synthesis. This enzyme recently has been purified from alder nodules (64a). The intracellular site of this enzyme must be confirmed and the other enzymes of the ORN cycle must be characterized and localized.

The following (Figure 3) is presented as a model for the synthesis of CIT and essential precursors in alder nodules. The model is consistent with the limited knowledge available and presented to highlight the many gaps in our understanding of N and C metabolism in the actinorhizal symbiosis and to encourage investigation of this important topic.

Biogenesis of Allantoin and Allantoic Acid

Reinbothe and Mothes (77, 101) proposed that ALN and ALC could be formed by either of two routes: 1. the condensation of urea and a two-carbon compound such as glyoxylate or glycine; or 2. the oxidative catabolism of purines. Although in 1953 Mothes suggested that the latter pathway was unlikely, evidence supporting the condensation pathway has been limited to labeling studies in certain fungi and banana (see 24).

The fact that enzymes of purine degradation (uricase, allantoinase, allantoicase, and urease) occur in plants including legumes has been known for more than 50 years (see 140). The catabolic route to ureide biogenesis is supported by the following observations: 1. high levels of xanthine (XAN) dehydrogenase, uricase, and allantoinase activity are present in nodules of cowpea (2, 5, 7), soybean (40, 41, 115, 135), and garden bean (102, 105, 136, 137); 2. the levels of these enzymes are normally low in nodules of amide-exporting legumes such as lupin (102, 105) and pea (28); 3. the levels of these enzymes increase in response to the onset of N_2 fixation, NH_4^+ assimilatory activity, and ureide export within the nodule (2, 5, 7, 105, 115); and 4. the addition of allopurinol, an irreversible inhibitor of XAN dehydrogenase, results in a decrease in the levels of ureides in nodules and ureide export in the xylem and an increase in XAN in the nodule (2, 5, 40, 142). The ability of cell-free extracts of nodules to convert intermediates such as IMP, XMP, XAN, and hypoxanthine into ureides and for allopurinol to block this conversion (142, 152, 153) substantiate the competency of nodules to form ureides via purine oxidation.

Purines for ureide biogenesis may arise via de novo synthesis or by turnover of RNA and DNA. Fujihara & Yamaguchi (40) suggested that the salvage pathway may be important in ureide biogenesis. Although this mechanism does seem to be involved in ureide biogenesis in certain tissues such as developing soybean seedlings (94), the salvage pathway does not seem to be important for ureide formation in nodules.

In nodules, purines apparently are synthesized de novo. This is not unexpected when one considers how rapidly ureides are labeled when nodules are exposed to labeled N_2 (80, 118) and the large flux of N exported as ureides from the nodule. A number of lines of evidence support this conclusion.

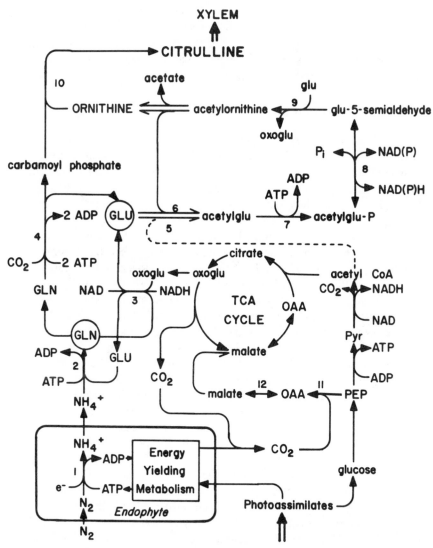

Figure 3 Proposed pathways of carbon and nitrogen metabolism involved in the synthesis of citrulline in root nodules of *Alnus*. Numbered reactions: 1. nitrogenase; 2. glutamine synthetase; 3. glutamate synthase; 4. carbamoylphosphate synthetase; 5. acetyl CoA: glutamate N-acetyltransferase; 6. N^2-acetylornithine: glutamate N-acetyltransferase; 7. N-acetylglutamate 5-phosphotransferase; 8. N-acetylglutamyl-5-phosphate: $NAD(P)^+$ oxidoreductase; 9. N^2-acetylornithine: 2-oxoglutarate aminotransferase; 10. ornithine carbamoyltransferase; 11. PEP carboxylase; 12. malate dehydrogenase.

De novo purine synthesis proceeds by a series of reactions requiring phosphoribosylpyrophosphate (PRPP), GLN, ASP, CO$_2$, glycine (GLY), methenyl tetrahydrafolate (methenyl FH$_4$), and formyl FH$_4$. Levels of enzymes involved in purine synthesis (PRPP amidotransferase) and the synthesis of purine precursors (PRPP synthetase, phosphoglycerate [PGA] dehydrogenase, serine [SER] hydroxymethylase, and methylene FH$_4$ dehydrogenase) are elevated severalfold in nodules of ureide-producing legumes over those found in nodules of amide-producing plants (7, 102, 105, 118). Activity of these enzymes along with GLN synthetase, GLU synthase, and ASP aminotransferase are induced as N$_2$ fixation and ureide export begin (6, 7, 102, 105, 118).

Label from [^{14}C]GLY, a purine precursor, was incorporated into ureides (152, 153) in cell-free extracts of cowpea nodules. These results are consistent with the role of de novo purine synthesis but do not discount a condensation mechanism of synthesis. Subsequently, Boland & Schubert (20) confirmed that purines were being synthesized de novo and oxidized in vivo. During the synthesis of the purine ring, carbon from CO$_2$ is incorporated at C6. If de novo purine synthesis is occurring, purines should be labeled with ^{14}C after exposure to [14]CO$_2$. Since this carbon is lost as CO$_2$ through the action of uricase, label would not accumulate in ureides. By blocking the complete oxidation of the purine ring with allopurinol, Boland and Schubert have shown that xanthine labeled specifically in the 6-position accumulates. The presence of ^{14}C in XAN was detectable even in the absence of the inhibitor while the levels of ^{14}C in IMP, the end product of purine synthesis, or in other purines were very low. These results confirm that purines synthesized de novo are rapidly oxidized in nodules as a part of ureide biogenesis.

Several enzymes involved in ureide biogenesis have been purified and characterized from nodule sources. These include PRPP amidotransferase (99, 103), PGA dehydrogenase (22), SER hydroxymethylase (76), IMP dehydrogenase (6a, 123), XAN dehydrogenase (15, 17, 143), uricase (63, 97), ASP aminotransferase (113), and phosphoserine aminotransferase (101a). Boland et al (19) fractionated soybean nodule extracts on sucrose gradients in order to localize the enzymes involved in NH$_4^+$ assimilation, the synthesis of purine precursors, and de novo purine biosynthesis. Based on these results, the authors suggested that GLN synthetase is located in the cytoplasm while GLU synthase occurs in the cytoplasm and the plastid. PRPP synthetase activity was not stable during the isolation, and the low level of activity remaining was mostly soluble. Some activity, however, was associated with the organelle and bacteroid fractions. PRPP amidotransferase, PGA dehydrogenase, SER hydroxymethylase, and methylene FH$_4$ dehydrogenase were all found in the plastid fraction. Using gel electrophoretic techniques, the isozyme of ASP aminotransferase associated with N$_2$ fixation was localized in the plastid.

Glutamate-dependent phosphoserine aminotransferase (101a) and a phosphoserine-specific phosphatase (20a) are also present in the plastid fraction. On this basis, the authors suggested that de novo purine synthesis and the reactions associated with the synthesis of ASP, GLY, methenyl FH_4, and formyl FH_4 occur in the plastid.

Boland & Schubert (21) confirmed this hypothesis. Plastids, bacteroids, and mitochondria were separated on sucrose step gradients, and each fraction was incubated with labeled GLY (or SER) and other purine precursors. Label was incorporated into IMP by the plastid fraction. While this fraction was contaminated with some mitochondria, the mitochondrial fraction did not synthesize purines. Contrary to previous suggestions, the bacteroid fraction did not actively synthesize purines. The residual activity present in this fraction could be accounted for by plastid contamination. The level of purine synthesis in the soluble fraction was less than expected from plastid breakage based on release of marker enzymes. The decreased stability of the isolated enzymes may account for the reduced activity in the soluble fraction. Subsequently, Shelp et al (124) found similar results after fractionation of cowpea nodules on sucrose gradients, thus confirming the conclusions of Boland and Schubert.

Because of the low level of PRPP synthetase activity remaining, the location of this enzyme was ambiguous (19). This activity does appear to be associated with the plastid. Isolated plastids that were incubated with ribose-5-phosphate and ATP, substrates for PRPP synthesis, incorporated ^{14}C from labeled GLY into IMP at two-thirds the rate for incubation mixtures containing excess PRPP (21).

On the basis of enzymological and tracer studies, SER apparently gives rise to GLY and methenyl FH_4. The latter is readily interconverted into formyl FH_4. Serine can be synthesized in plants from glycolytic intermediates, either glycerate or 3-PGA, via a nonphosphorylated or phosphorylated pathway, respectively. The latter is normally present in nonphotosynthetic tissues (126). The presence of PGA dehydrogenase (19), a GLU-dependent phosphoserine aminotransferase (101a), a specific phosphoserine phosphatase (20a), and phosphoserine in nodule extracts (102) indicate that the phosphorylated pathway is responsible for the production of SER for purine synthesis. Theoretically, this mechanism of synthesis of GLY and C1 groups would lead to excess production of GLY. Oxidation and/or resynthesis of SER from GLY would eliminate excess GLY. The presence of SER hydroxymethylase, normally found in the mitochondria as part of the photorespiratory cycle, within the plastid presents many interesting questions concerning organelle development.

Although nodules of ureide exporters contain an active PEP carboxylase, the role of this enzyme in ureide synthesis is not clear. Cookson et al (33) found that ^{14}C from $^{14}CO_2$ was incorporated into SER and GLY in French bean nodules. Although ureides were not labeled extensively in nodules, labeled ureides accounted for about 20% of the total ^{14}C in the xylem sap. They proposed that label was incorporated into GLY via the glyoxylate cycle. Johnson et al (51) previously found no evidence for an intact glyoxylate cycle in soybean nodules. For this reason, Coker and Schubert (32, 119) investigated the role of dark CO_2 fixation in soybean nodules. Dark CO_2 fixation declined as nodules began to actively export ureides. Labeled SER, GLY, and ureides accounted for 3, 1, and 3% of the total radioactivity, respectively, after a 60 min exposure to $^{14}CO_2$ (29, 119). These levels of radioactivity are similar to those reported for beans (33). The incorporation of label into these compounds can also be accounted for by the synthesis of 3-PGA from OAA via the action of phosphoenolpyruvate carboxykinase, enolase, and phosphoglyceromutase (29). Further work must be carried out to determine the exact route by which $^{14}CO_2$ is incorporated into SER, GLY, and ureides in these symbioses.

Early investigators presumed that hypoxanthine was formed from IMP and then oxidized by XAN dehydrogenase to XAN and ultimately uric acid. This is not consistent with the observed accumulation of XAN (and not hypoxanthine) in extracts of nodules treated with allopurinol. Boland & Schubert (20) were unable to detect labeled hypoxanthine after long-term exposure of nodules to $^{14}CO_2$. For this reason, they suggested that alternative pathways might be responsible for IMP catabolism (20, 104). The most likely of these involves IMP dehydrogenase which converts IMP into XMP. Early attempts to detect this activity were unsuccessful (see discussion in 20, 104). Using [^{14}C]IMP, Boland & Schubert (21) detected IMP dehydrogenase activity in plastids. Shelp & Atkins (123) have found a cytoplasmic IMP dehydrogenase activity and isolated the enzyme from cowpea nodules (6a). The levels of plastid breakage and contamination were not reported for the cowpea enzyme characterization (123). Thus, the differences in apparent localization have not been resolved. The plastid fraction also contains nucleosidase and nucleotidase activity based on the synthesis of small amounts of XMP, xanthosine, and XAN (21). These observations are consistent with, but not proof for, the synthesis of XAN within plastids. Compartmentalization of the enzymes of purine synthesis and IMP catabolism would help shuttle purines into ureide biogenesis. The final steps of ureide formation are also compartmentalized. Hanks et al (46) demonstrated that XAN dehydrogenase, uricase, and allantoinase were located in the cytoplasm, peroxisome, and endoplasmic reticulum, respectively.

The following model (Figure 4) is proposed for the pathway and subcellular localization of amide synthesis and ureide biogenesis in nodules of ureide-producing legumes. The model is based on current knowledge combined with some speculation on certain aspects as noted in the text. As such, the model requires further confirmation.

The infected zone of soybean (11) and cowpea (124) nodules contains both infected and uninfected cells in approximately equal numbers. Newcomb &

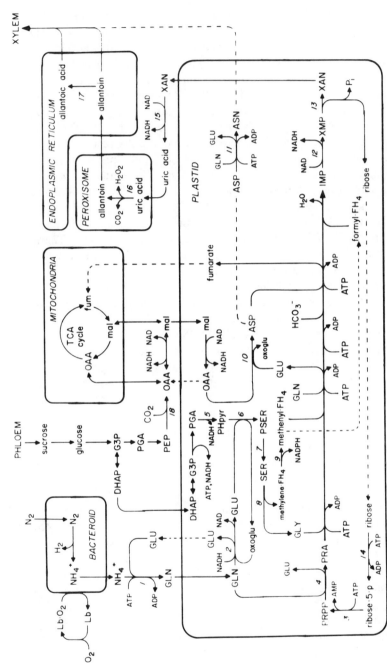

Figure 4 Proposed pathways and subcellular organization of the reactions involved in amide synthesis and ureide biogenesis in ureide-exporting legume nodules. Numbered reactions: 1. glutamine synthetase; 2. glutamate synthase; 3. PRPP synthetase; 4. PRPP amidotransferase; 5. PGA dehydrogenase; 6. phosphoserine aminotransferase; 7. phosphoserine phosphatase; 8. serine hydroxymethylase; 9. methylene FH_4 dehydrogenase; 10. aspartate aminotransferase; 11. asparagine synthetase; 12. IMP dehydrogenase; 13. nucleotidase, nucleosidase; 14. ribokinase; 15. xanthine dehydrogenase; 16. uricase; 17. allantoinase; 18. PEP carboxylase.

Tandon (79) observed that the development of peroxisomes was restricted in infected cells of soybean but proceeded normally in uninfected cells. In addition, there was a proliferation of smooth endoplasmic reticulum in the uninfected cells while both cells contained plastids. They suggested that these uninfected cells played an important role in ureide biogenesis. Hanks et al (45) fractionated protoplasts of infected and uninfected cells on sucrose step gradients and determined the levels of enzymatic activity for enzymes involved in ureide biogenesis for each fraction. Uricase and allantoinase were predominantly in the uninfected cell fraction. Results for PRPP amidotransferase were inconclusive while total PGA dehydrogenase and ASP aminotransferase activities were split between the two fractions. Based on these findings, it was difficult to determine conclusively the primary location of plastids responsible for the synthesis of purines to be used for ureide production.

Shelp et al (124) completed a similar separation on protoplasts from cowpea nodules. Their results were qualitatively similar to those of Hanks et al (45). They observed that the majority of the activity for enzymes associated with NH$_4^+$ assimilation, synthesis of purine precursors and de novo purine synthesis was in the infected cell fraction. Although activity of these enzymes was present in the uninfected cell fraction, the activity was attributed to contamination of 1 to 2% except for GLU synthase, de novo purine synthesis, and uricase. Upon analysis of their data, the actual level of contamination is closer to 6 to 11% based on leghemoglobin analyses and cell counts, respectively (Table 2). The use of hydroxybutyrate dehydrogenase as a marker is suspect because of poor breakage of bacteroids under these assay conditions (K. R. Schubert, unpublished). Shelp et al (124) proposed that the initial assimilation of NH$_4^+$

Table 2   Enzymes of ureide biogenesis in infected and uninfected cells from cowpea nodules[a]

Assay	Percentage total activity recovered in two fractions	
	Infected cell fraction	Uninfected cell fraction
Number infected cells[b]	89 (2.5×10^6)	11 (0.3×10^6)
Number uninfected cells[b]	10 (0.4×10^6)	90 (3.5×10^6)
Leghemoglobin	94	6
Hydroxybutyrate dehydrogenase	98	2
PGA dehydrogenase	95	5
Aspartate aminotransferase	93	7
Glutamine synthetase	98	2
Glutamate synthase	83	17
De novo purine synthesis	84	16
Methylene FH$_4$ dehydrogenase	95	5

[a]Calculated from data presented by Shelp et al (124).
[b]Cell numbers in parentheses.

occurs in the infected cells and that all of the reactions of ureide biogenesis from that point are located in both cell types.

In fact, their data strongly support the model presented by Schubert & Boland (117) for the cellular organization of ureide biogenesis. The percentage of the total GLU synthase activity (17%) and de novo purine synthesis (16%) in the uninfected cells is not greatly different from the level of contamination of this fraction with infected cells (Table 2). Presumably, low levels of GLU synthase and purine biosynthetic enzymes are required for normal function of all cells. This as well as the cytoplasmic location of GLU synthase may account for the activities found in the uninfected cells. This does not explain the distribution of the other plastid enzymes involved in the synthesis of purine precursors which differed from that observed for de novo purine synthesis. These enzymes were restricted almost totally to infected cells (greater than 95% in the infected cell fraction).

The high level of uricase in infected cells is difficult to explain. This result differs substantially from that obtained with soybean. Recently, Bergmann et al (13) used immunocytochemical techniques to confirm that uricase was present only in the uninfected cells of soybean nodules. Triplett (141) used a histological procedure to examine the localization of XAN dehydrogenase in soybean nodules and reported that this enzyme is in infected cells only. Unfortunately, Shelp et al (124) did not measure XAN dehydrogenase and allantoinase activities in their cell preparations. Figure 5 presents a model for the cellular organization of ureide biogenesis consistent with our current knowledge. According to this model, NH_4^+ assimilation, de novo purine synthesis and associated reactions, and purine oxidation through XAN dehydrogenase occur in infected cells. The final steps of purine oxidation and ureide formation take place in the peroxisomes and endoplasmic reticulum of the uninfected cells.

The model can be rationalized on the basis of known properties of the enzymes involved. The enzymes of purine synthesis as well as some of the intermediates in the pathway are oxygen labile (112). These enzymes would be protected by the low oxygen concentration in the infected cells that contain leghemoglobin. On the other hand, uricase has a K_m for O_2 of approximately 30μM (63, 97). Localization of uricase in the uninfected cells circumvents this apparent O_2 limitation. Confirmation of the model, however, will require biochemical and immunological studies to substantiate the results and speculation presented previously.

METABOLISM OF AMIDES AND UREIDES

Although appreciable amounts of amides and ureides are stored in the shoot (3, 48, 82), Pate (82) has estimated that 75 to 90% of the imported amide or ureide is rapidly metabolized. The products of catabolism are reassimilated and used

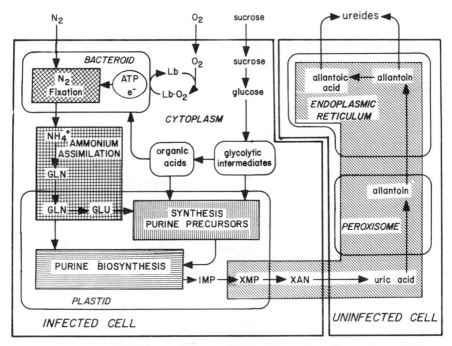

Figure 5 Proposed model for the cellular organization of the reactions of ammonium assimilation, de novo purine synthesis, and ureide biogenesis in nodules of ureide-producing legumes.

in the synthesis of the insoluble organic fraction. The primary routes of amide and ureide catabolism are still not known with certainty.

Asparagine is rapidly metabolized in shoots and reproductive organs. Joy and associates (9, 49a, 62a, 134) found that 75 and 45% of the [^{14}C]ASN supplied to pea leaves was metabolized within 3.5 h in the light and dark, respectively. Although ASP and other amino acids were labeled, the major product that accumulated was 2-hydroxysuccinamate. The latter was metabolized slowly in the light and more rapidly in the dark (62a). Likewise, 80% of the amide-N of ASN was lost in 2 h (9, 134).

During seed development, over 80% of the [U-^{14}C, ^{15}N (amide)]ASN supplied to lupin fruits was broken down with ^{15}N appearing in the amide N of GLN and ASN and in 13 other amino acids (4). Similar results were obtained using [^{13}C, ^{15}N]ASN and cross-polarization NMR (114). These observations demonstrate that ASN is degraded extensively in shoots and reproductive organs, and the products of catabolism are used in the synthesis of other amino acids.

There is evidence to support several pathways of ASN catabolism (Figure 6). Asparaginase, which catalyzes the removal of the amide-N, has been found in

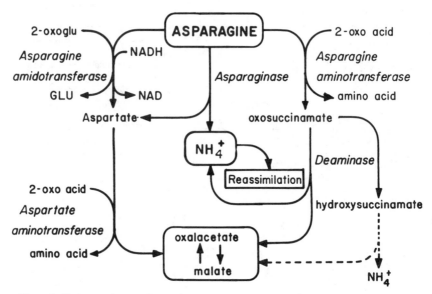

Figure 6 Pathways of asparagine catabolism in vegetative and reproductive tissues of amide- and ureide-producing legumes.

roots and nodules (26, 61, 122, 132) and developing seeds (127). Fowler et al (see 58) suggested that ASN could act directly as an amide-group donor, similar in role to GLN. Miflin and Lea reported finding ASN amidotransferase activity but proved later that the activity was an artifact (see 58). There is strong evidence for ASN aminotransferase which transfers the amino N of ASN to a 2-oxo acid and forms 2-oxosuccinamate (49a, 62a, 131). This activity is present in leaves but does not seem to be important in seeds. Thus, there are multiple paths by which ASN may be degraded.

The metabolism of MeGln is open. An active amidohydrolase, first detected by Fowden (37), has been purified (95) from peanut cotyledons. The enzyme is specific for MeGln and does not hydrolyze GLN. The fate of the MeGlu produced is uncertain.

Likewise, nothing is known about the fate of CIT in N_2-fixing plants. Citrulline is a precursor for arginine in many plant tissues. Arginine may then be hydrolyzed by arginase to urea and ORN (see Figure 7). Alternatively, CIT may be metabolized by reversing the synthetic reaction catalyzed by ORN carbamoyltransferase or through the action of CIT hydrolase (see 67, 139). Two forms of ORN carbamoyltransferase exist in bacteria and plants. Eid et al (35a) have suggested that the cytosolic form is important in CIT catabolism while the smaller mitochondrial enzyme is involved in CIT synthesis. Carbamoyl phosphate hydrolysis can be coupled to ATP synthesis, recouping some of

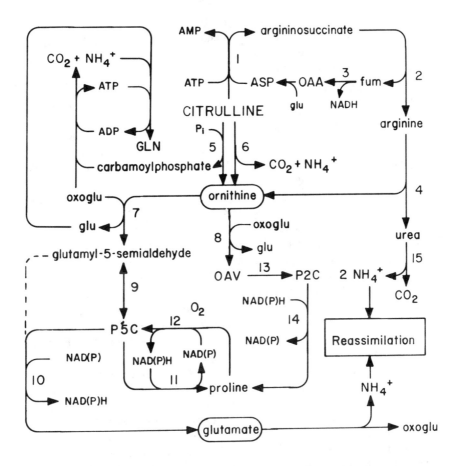

Figure 7 Possible pathways for citrulline catabolism in plants which transport reduced nitrogen as citrulline. Numbered reactions: 1. argininosuccinate synthetase; 2. argininosuccinate lyase; 3. fumarase and malate dehydrogenase; 4. arginase; 5. ornithine carbamoyltransferase; 6. citrulline hydrolase; 7. ornithine: 5-oxoglutarate aminotransferase; 8. ornithine: 2-oxoglutarate aminotransferase; 9. nonenzymatic; 10. Δ^1-pyrroline-5-carboxylic acid: NAD(P)$^+$ oxidoreductase; 11. proline: NAD(P)$^+$ 5-oxidoreductase; 12. proline oxidase; 13. nonenzymatic; 14. proline: NAD(P)$^+$ 2-oxidoreductase; 15. urease. Abbreviations: 2-oxo-5-aminovaleric acid, OAV; Δ^1-pyrroline-5-carboxylic acid, P-5-C; Δ^1-pyrroline-2-carboxylic acid, P-2-C.

the cost of synthesis. The amino group of ORN can be transferred to oxoglutarate to produce GLU, giving rise to glutamyl-5-semialdehyde or 2-oxo-5-aminovaleric acid. Glutamate or proline (which can be catabolized to form GLU) can be produced from the latter two compounds. Proline catabolism proceeds through the action of proline oxidoreductase. Several forms of this enzyme have been detected in plants, using either molecular oxygen or NAD(P)$^+$ as electron acceptors. The product of proline oxidation Δ^1-pyrroline-

5-carboxylic acid (P-5-C), which is in equilibrium with glutamyl-5-semialdehyde (G-5-SA). The final oxidation step of G-5-SA (or P-5-C) is coupled to $NAD(P)^+$ reduction.

The presence of allantoinase, "allantoicase," and urease activity in plants was noted in the early 1900s (24, 77, 140). Allantoinase is present at high levels in nodules, stems, leaves, and germinating seedlings of soybean (115, 135, 137, 138), cowpea (5, 48), and other legumes. Urease is abundant in seeds of many legumes. Although Bollard (24) indicated that urease was not present in tissues that metabolized urea, a second urease was recently identified which is present in seeds and leaves (53, 93). "Allantoicase" is generally defined as an enzyme that degrades ALC. In fact, there are several possible enzymes that can carry out this reaction, as discussed by Vogels & van der Drift (146). These alternative paths are shown in Figure 8.

Allantoicase produces urea and ureidoglycolate. The latter is further metabolized to urea and glyoxylate via the action of allantoicase or ureidoglycolase. On the other hand, allantoic acid amidohydrolase produces ureidoglycine, NH_4^+, and CO_2. In bacteria, ureidoglycine can be deaminated or donate its amino group to a 2-oxo acid producing an amino acid (i.e. glycine) and oxalurate. Oxalurate can be converted to ureidoglycolate through a NAD^+-linked oxidation. In *Streptococcus allantoicase* and other bacteria, oxamate and carbamoyl phosphate are produced from oxalurate (see 146). To date, there is no evidence for the formation of oxamate in legumes.

Several different approaches have been tried to examine ALC metabolism with the hope of distinguishing between alternate pathways of catabolism. Coker & Schaefer (30) fed $[^{13}C, {}^{15}N]$ALN to cultured soybean cotyledons and then examined ureide metabolism using solid state NMR. The ^{13}C-^{15}N bonds of $[5-^{13}C; 1-^{15}N]$ and $[2-^{13}C; 1,3-^{15}N]$ALN were broken and C2 and half of C5 were lost, presumably as CO_2. These findings were consistent with degradation via either allantoicase or amidohydrolase. Atkins et al (3a) incubated cowpea leaf tissue slices with $[2-^{14}C]$ALN or $[^{14}C]$urea. This tissue degraded the labeled substrates releasing $^{14}CO_2$. Shelp & Ireland (125) repeated this type of experiment on soybean leaf tissue using the urease inhibitor, acetohydroxamate. Release of $^{14}CO_2$ and NH_4^+ was almost completely blocked in the presence of inhibitor. In the absence of inhibitor, the stoichiometry of release was four NH_4^+ to two CO_2. These results are consistent with, but do not prove, that degradation occurs through the action of allantoicase. Blevins and coworkers (147) recently reported that acetohydroxamate inhibits ALC-degrading activity in soybeans. Using another urease inhibitor, phenylphosphordiamidate, they obtained results which suggest that ALC is degraded by an amidohydrolase. Thus, the pathway by which ALC is degraded is still uncertain.

Both ureides and amides are transported to developing legume seeds (4, 55,

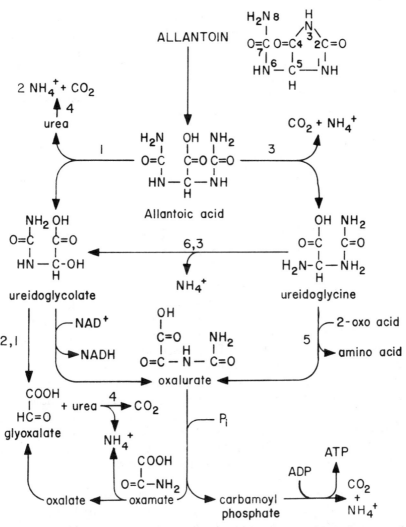

Figure 8   Alternative pathways for ureide catabolism in vegetative and reproductive organs of nitrogen-fixing plants. Numbered reactions: 1. Allantoicase; 2. ureidoglycolase; 3. allantoate amidohydrolase; 4. urease; 5. aminotransferase; 6. deaminase.

83, 87). Rainbird et al (98) have found that the growth rates of cultured soybean embryos were highest when GLN was supplied in the media. Growth with ALN as an N source was slow. Developing seeds import nitrogenous solutes from the phloem. These solutes are transported through the seed coat to feed the embryo. Glutamine and ASN accounted for 52 and 19% of the N present in soybean seed coat exudate, respectively, while ureide levels were negligible. Peoples et al (90) obtained similar results with cowpea fruits. In addition, they have shown that the levels and distribution of enzymes of NH_4^+ assimilation, amino acid synthesis, and amide and ureide catabolism changed during seed development (also see 78). Enzymes of ureide catabolism (allantoinase and urease) were present only in the seed coat until synthesis of these "storage" proteins occurred in the embryo. Asparagine catabolism was restricted to the seed coats in the immature seed, but as the seed developed, the embryo acquired the ability to degrade ASN. Throughout the period of time studied, both the seed coat and embryo were capable of assimilating NH_4^+ and synthesizing amino acids. These findings suggest that the capacity of the embryo to use ureides for growth is limited by the restricted export of ureides from the seed coat and the absence of ureide-degrading enzymes in the embryo.

ENERGETICS OF SYNTHESIS AND METABOLISM OF THE PRODUCTS OF N_2 FIXATION

A number of groups have presented and compared theoretical estimates of the costs of synthesizing amides and ureides (75, 82, 84, 100, 115). The costs associated with metabolism of these compounds, however, have not been included. The following is an analysis of the costs of synthesis and metabolism for the major products of N_2 fixation exported from nodules. The synthesis cost estimates include the cost of N_2 reduction, ATP and reductant used in the synthesis of the amide and amino-group donors (i.e. GLN and GLU), and the ATP, reductant, and carbon skeletons used in amide and ureide synthesis. The actual costs of transport are not included in these estimates. Pate and others have assumed that the cost of transport of a molecule across a membrane is equivalent to one ATP. This component of the cost could be much higher than normally assumed based on the complicated patterns of metabolic compartmentalization in nodules.

The cost of the carbon skeleton was estimated on the basis of the equivalent cost of the carbon skeleton synthesized from 3-PGA. The cost is expressed in ATP-equivalents, the amount of ATP that could be generated upon the complete oxidation of the carbon skeleton through the tricarboxylic acid cycle. Others have assumed that each carbon is equivalent to 6 ATP based on 36 ATP produced per glucose oxidized. This introduces some inaccuracy since oxalacetate is certainly not equivalent, energetically, to succinate. The cost of carbon skeletons is a very large component of the total cost. This artificially inflates the net cost because these carbon skeletons may be "salvaged" upon catabolism. Some have suggested that dark CO_2 fixation decreases the cost of exporting fixed N$_2$ (33, 56, 82, 84, 145). CO$_2$ fixation does decrease the total carbon required from the host but energetically the PEP consumed is essentially equivalent to OAA in ATP equivalents. The costs are summarized in Table 3. In cases in which NAD(P)H is produced or consumed, the net cost of synthesis is reduced or increased by 3 ATP, respectively. These estimates do not include the cost of the enzymatic machinery necessary for the synthesis of these compounds. These values are difficult to estimate without measurements of enzyme turnover rates.

Allantoin and ALC are by far the least expensive to produce. The lowered cost results from the plant's ability to recapture much of the energy costs associated with purine synthesis by coupling purine oxidation and other associated reactions to the reduction of NAD(P)$^+$ (Figure 4). In addition, the carbon skeleton associated with these ureides is already highly oxidized and therefore the costs of these compounds are quite low. Asparagine and CIT are intermediate in synthesis cost while MeGln is apparently the most expensive. However, if the cyclic mechanism of MeGln transport (95) operates in peanuts, then the net cost of transporting one N is reduced to two ATP equivalents, making it the least expensive.

Examination of the total cost of synthesis of certain amino acids such as ASP or GLU provides a rationale for why N is exported from the nodule as amides and ureides with subsequent synthesis of required amino acids coupled to photosynthesis in the shoot. Synthesis in the nodule would be much more costly in terms of the C budget of the plant.

The costs of metabolism are more difficult to estimate because the catabolic fate of these compounds is not known with certainty. The energy value of ALN and ALC is very low. At best only a few ATP could be generated upon the complete oxidation of ALC (Figure 8). The energy value of the carbon skeleton of ASN (Figure 6) and MeGln is conserved based on known or proposed routes of metabolism. Thus, the total energy cost of using ASN, for example, should be decreased by the cost of the carbon skeleton (i.e. net synthesis cost = 7 ATP). Potentially, citrulline metabolism may be the most efficient, assuming catabolism proceeds in the order CIT to ORN to GLU (Figure 7). Most, if not all, of the synthesis costs may be recouped through the metabolism of CIT and its metabolites.

The cost of reassimilation is the same for all products that release NH$_4$$^+$ upon degradation. The net cost of reassimilation can be reduced by utilizing compounds such as ASN and CIT which produce metabolites that serve as amino-group donors. Lea & Miflin (58) have estimated that a nitrogenous solute may be catabolized and reassimilated five or more times before ending up in the seed. Thus, the theoretical costs of reassimilation and resynthesis may be much greater than the original cost of synthesis. However, the actual cost to the plant

Table 3 Theoretical estimates of the costs of synthesis of amino acids, amides, and ureides

Product	N$_2$ Fixation[a]	NH$_4$ Assimilation[b]	Synthesis[c]	Carbon skeleton	Total[d]	Cost[e]/N
					Cost estimates in ATP-equivalents	
Glutamate	12	4	—	27	31(43)	31 (4)
Amino-N	12	4	—	—	4(12)	— —
Glutamine	24	5	—	27	32(56)	16 (2.5)
Amide-N	12	1	—	—	1(12)	— —
Aspartate	12	4	—	18	22(34)	22 (4)
Asparagine	24	5	2	18	25(49)	12.5(3.5)
Alanine	12	4	—	15	19(31)	19 (4)
4-Methyleneglutamate	12	4	—	30	34(46)	34 (4)
4-Methyleneglutamine	24	6	—	30	36(60)	18 (3)
Ornithine	24	8	4	27	39(63)	13 (6)
Carbamoyl phosphate	12	1	2	—	3(15)	3 (3)
Citrulline	36	9	6	27	42(78)	14 (5)
Arginine	48	13	-1(2-3)	27	39(87)	10 (3)
Serine	12	4	—	13	17(29)	17 (4)
Glycine	12	4	—	8	12(24)	12 (4)
C1	—	—	-3	5	2	— —
Allantoin	48	10	-8(7-15)[f]	18	20(68)	5 (0.5)
(Allantoic acid)			[-9.5(7-16.5)][g]	[19.5]		

[a]Estimated cost of N$_2$ fixation assuming 2 ATP hydrolyzed per e$^-$ transferred to substrates, N$_2$ reduction equal to H$_2$ evolution, and P/O = 2 for bacteroids.

[b]Cost associated with synthesis of glutamine and glutamate used in other synthetic reactions.

[c]Specific costs associated with synthesis (i.e. ATP or reductant).

[d]Total cost of synthesis (NH$_4^+$ assimilation, synthesis, and carbon skeleton costs). Values in parentheses include cost of N$_2$ fixation.

[e]Synthesis cost per N excluding cost of N$_2$ fixation. Values in parentheses include assimilatory and synthesis cost per N.

[f]Cost of purine synthesis is equivalent to 7 ATP. During synthesis of purine precursors and purine oxidation, 5 NAD(P)H are produced giving a net cost of −8 ATP.

[g]Based on the proposed pathway of glycine and C1 synthesis, excess glycine would be produced. This glycine could be used to resynthesize serine with the production of CO$_2$, NH$_4^+$, and NAD(P)H. reactions catalyzed by GLY decarboxylase and SER hydroxymethylase or GLY synthase. Net utilization of 1.5 SER with production of 5.5 NAD(P)H and 0.5 CO$_2$.

may be negligible when these processes take place in the light where excess ATP and reductant are available to drive the assimilatory processes (see Figure 1). Reassimilation in the dark or in nonphotosynthetic tissues, which must be coupled to photosynthesis indirectly through utilization of reduced carbon, is an inherently less efficient process. This may help explain the reduced rates of metabolism in the dark (62a).

SUMMARY

N$_2$-fixing plants have evolved a complex system for the exchange of reduced C and N. Although there are still many unknowns, our knowledge of the reactions involved in NH$_4^+$ assimilation and the synthesis of asparagine, allantoin, and allantoic acid has increased dramatically over the last 10 years. In contrast, our understanding of other areas such as the pathways of citrulline synthesis, the role of 4-methyleneglutamine, and the mechanisms of ureide and amide metabolism is still at the embryonic stage. One area of enlightenment that has emerged recently is the complicated pattern of cellular and subcellular compartmentalization within nodules and plants in general. We have not fully comprehended the impact of this compartmentalization. In the future, knowledge of the biochemical and regulatory role of compartmentalization as well as the biochemical and molecular mechanisms of regulating the enzymes of NH$_4^+$ assimilation and amide and ureide synthesis and catabolism will be essential to our understanding of how plants grow. Finally, although the theoretical costs of synthesis and metabolism for ureides and amides differ, the impact of these differences on plant growth and yield is not known. We have yet to determine the real costs of using fixed nitrogen.

ACKNOWLEDGMENTS

I would like to thank my associates and collaborators, including M. J. Boland, J. F. Hanks, N. E. Tolbert, G. T. Coker, P. R. McClure, D. A. Polayes, G. DeShone, P. H. S. Reynolds, and D. G. Blevins, who have contributed greatly to the work discussed herein; my colleagues J. S. Pate, C. A. Atkins, J. G. Robertson, K. J. F. Farnden, B. J. Miffin, P. J. Lea, E. Newcomb, and their coworkers, who have contributed enormously to our overall understanding of C and N metabolism in legumes; and my wife Karen, who assisted in the preparation of this manuscript. The author's work has been supported by grants from the United States Department of Agriculture Competitive Research Grants Program, the National Science Foundation, and the Michigan Agricultural Experiment Station.

Literature Cited

1. Aprison, M. H., Magee, W. E., Burris, R. H. 1954. Nitrogen fixation by excised soy bean root nodules. *J. Biol. Chem.* 208:29–39
2. Atkins, C. A. 1981. Metabolism of purine nucleotides to form ureides in nitrogen-fixing nodules of cowpea (*Vigna unguiculata* L. Walp). *FEBS Lett.* 125:89–93
3. Atkins, C. A., Pate, J. S., McNeil, D. L. 1980. Phloem loading and metabolism of xylem-borne amino compounds in fruiting shoots of a legume. *J. Exp. Bot.* 31:1509–20
3a. Atkins, C. A., Pate, J. S., Ritchie, A., Peoples, M. B. 1982. Metabolism and translocation of allantoin in ureide-producing grain legumes. *Plant Physiol.* 70:476–82
4. Atkins, C. A., Pate, J. S., Sharkey, P. J. 1975. Asparagine metabolism—Key to the nitrogen nutrition of developing legume seeds. *Plant Physiol.* 56:807–12
5. Atkins, C. A., Rainbird, R. M., Pate, J. S. 1980. Evidence for a purine pathway of ureide synthesis in N_2-fixing nodules of cowpea. *Z. Pflanzenphysiol.* 97:249–60
6. Atkins, C. A., Ritchie, A., Rowe, P. B., McCairns, E., Sauer, D. 1982. *De novo* purine synthesis in nitrogen-fixing nodules of cowpea (*Vigna unguiculata* L. Walp) and soybean (*Glycine max* L. Merr). *Plant Physiol.* 70:55–60
6a. Atkins, C. A., Shelp, B. J., Storer, P. J. 1985. Purification and properties of inosine monophosphate oxidoreductase from nitrogen-fixing nodules of cowpea (*Vigna unguiculata* L. Walp). *Arch. Biochem. Biophys.* 236:807–14
7. Atkins, C. A., Shelp, B. J., Storer, P. J., Pate, J. S. 1984. Nitrogen nutrition and the development of biochemical functions associated with nitrogen fixation and ammonia assimilation of nodules on cowpea seedlings. *Planta* 162:327–33
8. Awonaike, K. O., Lea, P. J., Miflin, B. J. 1981. The location of the enzymes of ammonia assimilation in root nodules of *Phaseolus vulgaris* L. *Plant Sci. Lett.* 23:189–95
9. Bauer, A., Urquhart, A. A., Joy, K. W. 1977. Amino acid metabolism of pea leaves. Diurnal changes and amino acid synthesis from ^{15}N-nitrate. *Plant Physiol.* 59:915–19
10. Bergersen, F. J. 1965. Ammonia—an early stable product of nitrogen fixation by soybean root nodules. *Aust. J. Biol. Sci.* 18:1–9
11. Bergersen, F. J., Goodchild, D. J. 1973. Aeration pathways in soybean root nodules. *Aust. J. Biol. Sci.* 26:729–40
12. Bergersen, F. J., Turner, G. L. 1967. Nitrogen fixation by the bacteroid fraction of breis of soybean root nodules. *Biochim. Biophys. Acta* 141:507–15
13. Bergmann, H., Preddie, E., Verma, D. P. S. 1983. Nodulin-35: A subunit of specific uricase (uricase II) induced and localized in the uninfected cells of soybean nodules. *EMBO J.* 2:2333–39
13a. Blom, J., Roelofsen, W., Akkermans, A. D. L. 1981. Assimilation of nitrogen in root nodules of alder (*Alnus glutinosa*). *New Phytol.* 89:321–26
14. Blumenthal-Goldschmidt, S., Butler, G. W., Conn, E. E. 1963. Incorporation of hydrocyanic acid labelled with carbon-14 into asparagine in seedlings. *Nature* 197:718–19
15. Boland, M. J. 1981. NAD: xanthine dehydrogenase from nodules of navy beans: Partial purification and properties. *Biochem. Int.* 2:567–74
16. Boland, M. J., Benny, A. G. 1977. Enzymes of nitrogen metabolism in legume nodules. Purification and properties of NADH-dependent glutamate synthase from lupin nodules. *Eur. J. Biochem.* 79:355–62
17. Boland, M. J., Blevins, D. G., Randall, D. D. 1983. Soybean nodule xanthine dehydrogenase: A kinetic study. *Arch. Biochem. Biophys.* 222:435–41
18. Boland, M. J., Fordyce, A. M., Greenwood, R. M. 1978. Enzymes of nitrogen metabolism in legume nodules: A comparative study. *Aust. J. Plant Physiol.* 5:553–59
19. Boland, M. J., Hanks, J. F., Reynolds, P. H. S., Blevins, D. G., Tolbert, N. E., Schubert, K. R. 1982. Subcellular organization of ureide biogenesis from glycolytic intermediates and ammonium in nitrogen-fixing soybean nodules. *Planta* 155:45–51
20. Boland, M. J., Schubert, K. R. 1982. Purine biosynthesis and catabolism in soybean root nodules: Incorporation of $^{14}CO_2$ into xanthine. *Arch. Biochem. Biophys.* 213:486–91
20a. Boland, M. J., Schubert, K. R. 1982. The biosynthesis of glycine and methenyl tetrahydrofolate. Precursors for ureide synthesis in soybean nodules. *Plant Physiol.* 69S:112
21. Boland, M. J., Schubert, K. R. 1983. Biosynthesis of purines by a proplastid fraction from soybean nodules. *Arch. Biochem. Biophys.* 220:179–87
22. Boland, M. J., Schubert, K. R. 1983.

Phosphoglycerate dehydrogenase from soybean nodules. Partial purification and some kinetic properties. *Plant Physiol.* 71:658–61

23. Bollard, E. G. 1957. Translocation of organic nitrogen in the xylem. *Aust. J. Biol. Sci.* 10:292–301

24. Bollard, E. G. 1959. Urease, urea and ureides in plants. In *Utilization of Nitrogen and Its Compounds by Plants*, ed. H. K. Porter. *Symp. Soc. Exp. Biol.*, 13:304–29. London: Cambridge Univ. Press

25. Brown, C. M., Dilworth, M. J. 1975. Ammonia assimilation by *Rhizobium* cultures and bacteroids. *J. Gen. Microbiol.* 86:39–48

26. Chang, K. S., Farnden, K. J. F. 1981. Purification and properties of asparaginase from *Lupinus arboreus* and *Lupinus angustifolius*. *Arch. Biochem. Biophys.* 208:49–58

27. Christeller, J. T., Laing, W. A., Sutton, W. D. 1977. Carbon dioxide fixation by lupin nodules. I. Characterization, association with phosphoenolpyruvate carboxylase, and correlation with nitrogen fixation during nodule development. *Plant Physiol.* 60:47–50

28. Christensen, T. M. I. E., Jochimsen, B. U. 1983. Enzymes of ureide synthesis in pea and soybean. *Plant Physiol.* 72:56–59

29. Coker, G. T. III. 1982. *Dark CO₂ fixation and amino acid metabolism in symbiotic N₂-fixing systems. Labeling studies with ¹⁴C and ¹³N-labeled tracers.* PhD thesis. Mich. State Univ., East Lansing

30. Coker, G. T. III, Schaefer, J. 1985. ¹⁵N and ¹³C NMR determinations of allantoin metabolism in developing soybean cotyledons. *Plant Physiol.* 77:129–35

31. Coker, G. T. III, Schubert, K. R. 1980. Asparagine synthesis in soybean roots and nodules. *Plant Physiol.* 65(S):111

32. Coker, G. T. III, Schubert, K. R. 1981. Carbon dioxide fixation in soybean roots and nodules. I. Characterization and comparison with N₂ fixation and composition of xylem exudate during early nodule development. *Plant Physiol.* 67:691–96

33. Cookson, C., Hughes, H., Coombs, J. 1980. Effects of combined nitrogen on anapleurotic carbon assimilation and bleeding sap composition in *Phaseolus vulgaris* L. *Planta* 148:338–45

33a. Cullimore, J. V., Lara, M., Lea, P. J., Miflin, B. J. 1983. Purification and properties of two forms of glutamine synthetase from the plant fraction of *Phaseolus* root nodules. *Planta* 157:245–53

34. Date, R. A. 1973. Nitrogen, a major limitation in the productivity of natural communities, crops and pastures in the Pacific area. *Soil Biol. Biochem.* 5:5–18

35. Done, J., Fowden, L. 1951. A new amino-acid amide in the groundnut plant *(Arachis hypogaea):* Evidence of the occurrence of γ-methyleneglutamine and γ-methyleneglutamic acid. *Biochem. J.* 51:451–58

35a. Eid, S., Waly, Y., Abdelal, T. 1974. Separation and properties of two ornithine carbamoyltransferases from *Pisum sativum* seedlings. *Phytochemistry* 13:99–102

36. Fowden, L. 1954. The nitrogen metabolism of groundnut plants: The role of γ-methyleneglutamine and γ-methyleneglutamic acid. *Ann. Bot.* (NS) 18:417–40

37. Fowden, L. 1955. The deamidase of groundnut plants *(Arachis hypogaea).* *J. Exp. Bot.* 6:362–70

38. Fowden, L., Webb, J. A. 1955. Evidence for the occurrence of γ-methylene-α-oxoglutaric acid in groundnut plants *(Arachis hypogaea).* *Biochem. J.* 59:228–34

39. Fowden, L., Webb, J. A. 1958. The incorporation of ¹⁴C-labelled substrates into the amino-acids of groundnut plants *(Arachis hypogaea).* *Ann. Bot.* (NS) 22:73–93

40. Fujihara, S., Yamaguchi, M. 1978. Effects of allopurinol [4-hydroxypyrazolo-(3,4-d)-pyrimidine] on the metabolism of allantoin in soybean plants. *Plant Physiol.* 62:134–38

41. Fujihara, S., Yamaguchi, M. 1978. Probable site of allantoin formation in nodulating soybean plants. *Phytochemistry* 17:1239–43

42. Fujihara, S., Yamaguchi, M. 1980. Asparagine formation in soybean nodules. *Plant Physiol.* 66:139–41

42a. Gardner, I. C. 1976. Ultrastructural studies of non-leguminous root nodules. In *Symbiotic Nitrogen Fixation in Plants*, ed. P. S. Nutman, pp. 485–95. Cambridge: Cambridge Univ. Press

43. Gardner, I. C., Leaf, G. 1960. Translocation of citrulline in *Alnus glutinosa.* *Plant Physiol.* 35:948–50

43a. Groat, R. G., Schrader, L. E. 1982. Isolation and immunochemical characterization of plant glutamine synthetase in alfalfa *(Medicago sativa* L.) nodules. *Plant Physiol.* 70:1759–61

43b. Groat, R. G., Vance, C. P. 1981. Root nodule enzymes of ammonia assimilation in alfalfa *(Medicago sativa* L.). Developmental pattern and response to applied nitrogen. *Plant Physiol.* 67:1198–1203

44. Gunning, B. E. S., Pate, J. S., Minchin, F. R., Marks, I. 1974. Quantitative aspects of transfer cell structure in relation to vein loading in leaves and solute transport in legume nodules. In *Transport at the Cellular Level. Symp. Soc. Exp. Biol.* 28:87–125

45. Hanks, J. F., Schubert, K. R., Tolbert, N. E. 1983. Isolation and characterization of infected and uninfected cells from soybean nodules. Role of uninfected cells in ureide synthesis. *Plant Physiol.* 71:869–73

46. Hanks, J. F., Tolbert, N. E., Schubert, K. R. 1981. Localization of enzymes of ureide biosynthesis in peroxisomes and microsomes of nodules. *Plant Physiol.* 68:65–69

47. Hardy, R. W. F., Havelka, U. D. 1975. Nitrogen fixation research: A key to world food. *Science* 188:633–43

48. Herridge, D. F., Atkins, C. A., Pate, J. S., Rainbird, R. M. 1978. Allantoin and allantoic acid in the nitrogen economy of the cowpea (*Vigna unguiculata* (L.) Walp). *Plant Physiol.* 62:495–98

48a. Hirel, B., Perrot-Rechenmann, C., Maudinas, B., Gadal, P. 1982. Glutamine synthetase in alder *(Alnus glutinosa)* root nodules. Purification, properties and cytoimmunochemical localization. *Physiol. Plant.* 55:197–203

49. Huber, T. A., Streeter, J. G. 1984. Asparagine biosynthesis in soybean nodules. *Plant Physiol.* 74:605–10

49a. Ireland, R. J., Joy, K. W. 1981. Two routes of asparagine metabolism in *Pisum sativum* L. *Planta* 151:289–92

50. Israel, D. W., McClure, P. R. 1980. Nitrogen translocation in the xylem of soybeans. In *World Soybean Research Conference II: Proceedings*, ed. F. T. Corbin, pp. 111–27. Boulder: Westview

51. Johnson, G. V., Evans, H. J., Ching, T. M. 1966. Enzymes of the glyoxylate cycle in *Rhizobia* and in nodules of legumes. *Plant Physiol.* 41:1330–36

52. Kennedy, I. R. 1966. Primary products of symbiotic nitrogen fixation I. Short-term exposures of serradella nodules to $^{15}N_2$. *Biochim. Biophys. Acta* 130:285–94

53. Kerr, P. S., Blevins, D. G., Rapp, B., Randall, D. D. 1983. Soybean leaf urease: Comparison with seed urease. *Physiol. Plant* 57:339–45

54. Lawrie, A. C., Wheeler, C. T. 1975. Nitrogen fixation in the root nodules of *Vicia faba* L. in relation to the assimilation of carbon II. The dark fixation of carbon dioxide. *New Phytol.* 74:437–45

55. Layzell, D. B., LaRue, T. A. 1982. Modeling C and N transport to developing soybean fruits. *Plant Physiol.* 70:1290–98

56. Layzell, D. B., Rainbird, R. M., Atkins, C. A., Pate, J. S. 1979. Economy of photosynthate use in nitrogen-fixing legume nodules. *Plant Physiol.* 64:888–91

57. Lea, P. J., Fowden, L. 1975. The purification and properties of glutamine-dependent asparagine synthetase isolated from *Lupinus albus*. *Proc. R. Soc. London Ser. B* 192:13–26

58. Lea, P. J., Miflin, B. J. 1980. Transport and metabolism of asparagine and other nitrogen compounds within the plant. In *The Biochemistry of Plants*, ed. B. J. Miflin, 5:569–608. New York: Academic

59. Leaf, G., Gardner, I. C., Bond, G. 1958. Observations on the composition and metabolism of the nitrogen-fixing root nodules of *Alnus*. *J. Exp. Bot.* 9:320–31

60. Leaf, G., Gardner, I. C., Bond, G. 1959. Observations on the composition and metabolism of the nitrogen-fixing root nodules of *Myrica*. *Biochem. J.* 72:662–67

61. Lees, E. M., Blakeney, A. B. 1970. The distribution of asparaginase activity in legumes. *Biochim. Biophys. Acta* 215:145–51

62. Lever, M., Butler, G. W. 1971. The relationship between cyanide assimilation and asparagine biosynthesis in lupins. *J. Exp. Bot.* 22:285–90

62a. Lloyd, N. D. H., Joy, K. W. 1978. 2-Hydroxysuccinamic acid: A product of asparagine metabolism in plants. *Biochem. Biophys. Res. Commun.* 81:186–92

63. Lucas, K., Boland, M. J., Schubert, K. R. 1983. Uricase from soybean root nodules: Purification, properties, and comparison with the enzyme from cowpea. *Arch. Biochem. Biophys.* 226:190–97

64. Mahon, J. D. 1983. Energy relationships. In *Nitrogen Fixation*, ed. W. J. Broughton, 3:299–325. Oxford: Oxford Univ. Press

64a. Martin, F., Hirel, B., Gadal, P. 1983. Purification and properties of ornithine carbamoyl transferase 1 from *Alnus glutinosa* root nodules. *Z. Pflanzenphysiol.* 111:413–22

65. Matsumoto, T., Yatazawa, M., Yamamoto, Y. 1977. Effects of exogenous nitrogen-compounds on the concentrations of allantoin and various constituents in several organs of soybean plants. *Plant Cell Physiol.* 18:613–24

66. Matsumoto, T., Yatazawa, M., Yamamoto, Y. 1977. Incorporation of ^{15}N into allantoin in nodulated soybean plants supplied with ^{15}N$_2$. *Plant Cell Physiol.* 18:459–62

67. McClure, P. R., Coker, G. T. III, Schubert, K. R. 1983. Carbon dioxide fixation in roots and nodules of *Alnus glutinosa* I. Role of phosphoenolpyruvate carboxylase and carbomyl phosphate synthetase in dark CO$_2$ fixation, citrulline synthesis, and N$_2$ fixation. *Plant Physiol.* 71:652–57

68. McClure, P. R., Israel, D. W. 1979. Transport of nitrogen in the xylem of soybean plants. *Plant Physiol.* 64:411–16

69. McNeil, D. L., Atkins, C. A., Pate, J. S. 1979. Uptake and utilization of xylemborne amino compounds by shoot organs of a legume. *Plant Physiol.* 63:1076–81

70. McNeil, D. L., LaRue, T. A. 1984. Effect of nitrogen source on ureides in soybean. *Plant Physiol.* 66:720–25

71. McParland, R. H., Guevara, J. G., Becker, R. R., Evans, H. J. 1976. The purification and properties of the glutamine synthetase from the cytosol of soyabean root nodules. *Biochem. J.* 153:597–606

72. Meeks, J. C., Wolk, C. P., Schilling, N., Shaffer, P. W., Avissar, Y., Chien, W.-S. 1978. Initial organic products of fixation of [^{13}N]dinitrogen by root nodules of soybean *(Glycine max)*. *Plant Physiol.* 61:980–83

73. Miettinen, J. K., Virtanen, A. I. 1952. The free amino acids in the leaves, roots and root nodules of the alder *(Alnus)*. *Physiol. Plant.* 5:540–57

74. Minchin, F. R., Pate, J. S. 1973. The carbon balance of a legume and the functional economy of its root nodules. *J. Exp. Bot.* 24:295–308

75. Minchin, F. R., Summerfield, R. J., Hadley, P., Roberts, E. H., Rawsthorne, S. 1981. Carbon and nitrogen nutrition of nodulated roots of grain legumes. *Plant Cell Environ.* 4:5–26

76. Mitchell, M. K., Reynolds, P. H. S., Blevins, D. G. 1986. Serine hydroxymethylase from soybean root nodules: Purification and kinetic properties. *Plant Physiol.* In press

77. Mothes, K. 1961. The metabolism of urea and ureides. *Can. J. Bot.* 39:1785–1807

78. Murray, D. R., Kennedy, I. R. 1980. Changes in activities of enzymes of nitrogen metabolism in seedcoats and cotyledons during embryo development in pea seeds. *Plant Physiol.* 66:782–86

79. Newcomb, E. H., Tandon, S. K. 1981.

Uninfected cells of soybean root nodules: Ultrastructure suggests key role in ureide production. *Science* 212:1394–96

80. Ohyama, T., Kumazawa, K. 1978. Incorporation of ^{15}N into various nitrogenous compounds in intact soybean nodules after exposure to ^{15}N$_2$ gas. *Soil Sci. Plant Nutr.* 24:525–33

81. Pate, J. S. 1973. Uptake, assimilation and transport of nitrogen compounds by plants. *Soil Biol. Biochem.* 5:109–19

82. Pate, J. S., Atkins, C. A. 1983. Nitrogen uptake, transport, and utilization. In *Nitrogen Fixation*, ed. W. J. Broughton, 3:245–98. Oxford: Oxford Univ. Press

83. Pate, J. S., Atkins, C. A., Hamel, K., McNeil, D. L., Layzell, D. B. 1979. Transport of organic solutes in phloem and xylem of a nodulated legume. *Plant Physiol.* 63:1082–88

84. Pate, J. S., Atkins, C. A., Rainbird, R. M. 1981. Theoretical and experimental costing of nitrogen fixation and related processes in nodules of legumes. In *Current Perspectives in Nitrogen Fixation*, ed. A. H. Gibson, W. E. Newton, pp. 105–16. Canberra: Aust. Acad. Sci.

85. Pate, J. S., Atkins, C. A., White, S. T., Rainbird, R. M., Woo, K. C. 1980. Nitrogen nutrition and xylem transport of nitrogen in ureide-producing grain legumes. *Plant Physiol.* 65:961–65

86. Pate, J. S., Gunning, B. E. S., Briarty, L. G. 1969. Ultrastructure and functioning of the transport system of the leguminous root nodule. *Planta* 85:11–34

87. Pate, J. S., Sharkey, P. J., Atkins, C. A. 1977. Nutrition of a developing legume fruit. *Plant Physiol.* 59:506–10

88. Pate, J. S., Wallace, W. 1964. Movement of assimilated nitrogen from the root system of the field pea (*Pisum arvense* L.). *Ann. Bot.* 28:83–89

89. Patterson, T. G., LaRue, T. A. 1983. N$_2$ fixation (C$_2$H$_2$) and ureide content of soybeans: Environmental effects and source sink manipulations. *Crop Sci.* 23:819–24

90. Peoples, M. B., Atkins, C. A., Pate, J. S., Murray, D. R. 1985. Nitrogen nutrition and metabolic interconversions of nitrogenous solutes in developing cowpea fruits. *Plant Physiol.* 77:382–88

91. Perrot-Rechenmann, C., Vidal, J., Maudinas, B., Gadal, P. 1981. Immunocytochemical study of phosphoenolpyruvate carboxylase in nodulated *Alnus glutinosa*. *Planta* 153:14–17

92. Peterson, J. B., Evans, H. J. 1979. Phosphoenolpyruvate carboxylase from soybean nodule cytosol: Evidence for isozymes and kinetics of the most active

component. *Biochim. Biophys. Acta* 567:445–52

93. Polacco, J. C., Winkler, R. G. 1984. Soybean leaf urease: A seed enzyme? *Plant Physiol.* 74:800–3

94. Polayes, D. A., Schubert, K. R. 1984. Purine synthesis and catabolism in soybean seedlings. The biogenesis of ureides. *Plant Physiol.* 75:1104–10

95. Powell, G. K., Dekker, E. E. 1983. Purification and properties of a 4-methylene-L-glutamine amidohydrolase from peanut leaves. *J. Biol. Chem.* 258:8677–83

96. Rainbird, R. M. 1983. *Elements in the cost of nitrogen fixation with special reference to the legume cowpea (Vigna unguiculata (L.) Walp. cv. Caloona).* PhD thesis. Univ. West. Australia, Nedlands

97. Rainbird, R. M., Atkins, C. A. 1981. Purification and some properties of urate oxidase from nitrogen-fixing nodules of cowpea. *Biochim. Biophys. Acta* 659:-132–40

98. Rainbird, R. M., Thorne, J. H., Hardy, R. W. F. 1984. The role of amides, amino acids, and ureides in the nutrition of developing soybean seeds. *Plant Physiol.* 74:329–34

99. Rao, K. P., Blevins, D. G., Randall, D. D. 1980. Glutamine-phosphoribosylpyrophosphate amidotransferase from soybean nodules. *Plant Physiol.* 65(S):110

100. Rawsthorne, S., Minchin, F. R., Summerfield, R. J., Cookson, C., Coombs, J. 1980. Carbon and nitrogen metabolism in legume root nodules. *Phytochemistry* 19:341–55

101. Reinbothe, H., Mothes, K. 1962. Urea, ureides and guanidines in plants. *Ann. Rev. Plant Physiol.* 13:129–50

101a. Reynolds, P. H. S., Blevins, D. G. 1986. Phosphoserine aminotransferase in soybean nodules: Demonstration and localization. *Plant Physiol.* In press

102. Reynolds, P. H. S., Blevins, D. G., Boland, M. J., Schubert, K. R., Randall, D. D. 1982. Enzymes of ammonia assimilation in legume nodules: A comparison between ureide- and amide-transporting plants. *Physiol. Plant* 55:255–60

103. Reynolds, P. H. S., Blevins, D. G., Randall, D. D. 1984. 5-Phosphoribosylpyrophosphate amidotransferase from soybean root nodules: Kinetic and regulatory properties. *Arch. Biochem. Biophys.* 229:623–31

104. Reynolds, P. H. S., Boland, M. J., Blevins, D. G., Randall, D. D., Schubert, K. R. 1982. Ureide biogenesis in leguminous plants. *Trends Biochem. Sci.* 7:366–68

105. Reynolds, P. H. S., Boland, M. J., Blevins, D. G., Schubert, K. R., Randall, D. D. 1982. Enzymes of amide and ureide biogenesis in developing soybean nodules. *Plant Physiol.* 69:1334–38

106. Reynolds, P. H. S., Boland, M. J., Farnden, K. J. F. 1981. Enzymes of nitrogen metabolism in legume nodules: Partial purification and properties of the aspartate aminotransferase from lupine nodules. *Arch. Biochem. Biophys.* 209:-524–33

107. Reynolds, P. H. S., Farnden, K. J. F. 1979. The involvement of aspartate aminotransferases in ammonium assimilation in lupin nodules. *Phytochemistry* 18:1625–30

108. Robertson, J. G., Farnden, K. J. F., Warburton, M., Banks, J. M. 1975. Induction of glutamine synthetase during nodule development in lupin. *Aust. J. Plant Physiol.* 2:265–72

109. Robertson, J. G., Warburton, M., Farnden, K. J. F. 1975. Induction of glutamate synthase during nodule development in lupin. *FEBS Lett.* 55:33–37

110. Rognes, S. E. 1970. A glutamine-dependent asparagine synthetase from yellow lupine seedlings. *FEBS Lett.* 10:62–66

111. Rognes, S. E. 1975. Glutamine-dependent asparagine synthetase from *Lupinus luteus. Phytochemistry* 14:1975–82

112. Rowe, P. B., McCairns, E., Madsen, G., Sauer, D., Elliott, H. 1978. *De novo* purine synthesis in avian liver. Copurification of the enzymes and properties of the pathway. *J. Biol. Chem.* 253:7711–21

113. Ryan, E., Bodley, F., Fotrell, P. F. 1972. Purification and characterization of aspartate aminotransferases from soybean root nodules and *Rhizobium japonicum. Phytochemistry* 11:957–63

114. Schaefer, J., Skokut, T. A., Stejskal, E. O., McKay, R. A., Varner, J. E. 1981. Asparagine amide metabolism in developing cotyledons of soybean. *Proc. Natl. Acad. Sci. USA* 78:5978–82

115. Schubert, K. R. 1981. Enzymes of purine biosynthesis and catabolism in *Glycine max:* Comparison of activities with N₂ fixation and composition of xylem exudate during nodule development. *Plant Physiol.* 68:1115–1122

116. Schubert, K. R. 1982. *The Energetics of Biological Nitrogen Fixation.* Workshop Summaries—I. *Am. Soc. Plant Physiol.*, pp. 1–30

117. Schubert, K. R., Boland, M. J. 1984. The cellular and intracellular organization of the reactions of ureide biogenesis

in nodules of tropical legumes. In *Advances in Nitrogen Fixation Research*, ed. C. Veeger, W. E. Newton, pp. 445–51. The Hague: Nijoff/Junk

118. Schubert, K. R., Coker, G. T. III. 1982. Studies of nitrogen and carbon assimilation in N₂-fixing plants: Short-term studies using [¹³N] and [¹¹C]. In *Recent Developments in Biological and Chemical Research with Short-lived Radioisotopes*, ed. J. W. Root, K. A. Krohn. *Adv. Chem. Ser.* 197:317–39. Washington, DC: Am. Chem. Soc.

119. Schubert, K. R., Coker, G. T. III. 1985. Carbon metabolism in soybean roots and nodules: Role of dark CO₂ fixation. *World Soybean Res. Conf. III Proc.*, ed. R. Shibles, pp. 815–23. Boulder: Westview

120. Schubert, K. R., Coker, G. T. III, Firestone, R. B. 1981. Ammonia assimilation in *Alnus glutinosa* and *Glycine max*: Short-term studies using ¹³NH₄⁺. *Plant Physiol.* 67:662–65

121. Schubert, K. R., Ryle, G. J. A. 1980. The energy requirements for N₂ fixation in nodulated legumes. In *Advances in Legume Science*, ed. R. J. Summerfield, A. H. Bunting, pp. 85–96. London: Kew Botanical Gardens

122. Scott, D. B., Farnden, K. J. F., Robertson, J. G. 1976. Ammonia assimilation in lupin nodules. *Nature* 263:703–5

122a. Scott, A., Gardner, I. C., McNally, S. F. 1981. Localization of citrulline synthesis in the alder root nodule and its implication in nitrogen fixation. *Plant Cell Rep.* 1:21–22

122b. Shargool, P. D., Steeves, T., Weaver, M., Russell, M. 1978. The localization within plant cells of enzymes involved in arginine biosynthesis. *Can. J. Biochem.* 56:273–79

123. Shelp, B. J., Atkins, C. A. 1983. Role of inosine monophosphate oxidoreductase in the formation of ureides in nitrogen-fixing nodules of cowpea (*Vigna unguiculata* L. Walp.). *Plant Physiol.* 72:1029–34

123a. Shelp, B. J., Atkins, C. A. 1984. Subcellular location of enzymes of ammonia assimilation and asparagine synthesis in root nodules of *Lupinus albus* L. *Plant Sci. Lett.* 36:225–30

124. Shelp, B. J., Atkins, C. A., Storer, P. J., Canvin, D. T. 1983. Cellular and subcellular organization of pathways of ammonia assimilation and ureide synthesis in nodules of cowpea. (*Vigna unguiculata* L. Walp.). *Arch. Biochem. Biophys.* 224:429–41

125. Shelp, B. J., Ireland, R. J. 1985. Ureide metabolism in leaves of nitrogen-fixing

soybean plants. *Plant Physiol.* 77:779–83

126. Slaughter, J. C., Davies, D. D. 1968. Isolation of 3-phosphoglycerate dehydrogenase from peas. *Biochem. J.* 109:743–48

127. Sodek, L., Lea, P. J., Miflin, B. J. 1980. Distribution and properties of a potassium-dependent asparaginase isolated from developing seeds of *Pisum sativum* and other plants. *Plant Physiol.* 65:22–26

128. Sprent, J. I. 1980. Root nodule anatomy, type of export product and evolutionary origin in some Leguminosae. *Plant Cell Environ.* 3:35–43

129. Streeter, J. G. 1970. Asparagine biosynthesis in higher plants: Evidence for a reaction involving glutamine as the nitrogen donor. *Plant Physiol.* 46:44

130. Streeter, J. G. 1972. Nitrogen nutrition of field-grown soybean plants. I. Seasonal variations in soil nitrogen and nitrogen composition of stem exudate. *Agron. J.* 64:311–14

131. Streeter, J. G. 1973. *In vivo* and *in vitro* studies on asparagine biosynthesis in soybean seedlings. *Arch. Biochem. Biophys.* 157:613–24

132. Streeter, J. G. 1977. Asparaginase and asparagine transaminase in soybean leaves and root nodules. *Plant Physiol.* 60:235–39

133. Streeter, J. G. 1979. Allantoin and allantoic acid in tissues and stem exudate from field-grown soybean plants. *Plant Physiol.* 63:478–80

134. Ta, T. C., Joy, K. W., Ireland, R. J. 1984. Amino acid metabolism in pea leaves. Utilization of nitrogen from amide and amino groups of [¹⁵N] asparagine. *Plant Physiol.* 74:822–26

135. Tajima, S., Yamamoto, Y. 1975. Enzymes of purine catabolism in soybean plants. *Plant Cell Physiol.* 16:271–82

135a. Taylor, A. A., Stewart, G. R. 1981. Tissue and subcellular localization of enzymes of arginine metabolism in *Pisum sativum*. *Biochem. Biophys. Res. Commun.* 101:1281–89

136. Thomas, R. J., Feller, U., Erismann, K. H. 1980. Ureide metabolism in non-nodulated *Phaseolus vulgaris* L. *J. Exp. Bot.* 31:409–17

137. Thomas, R. J., Schrader, L. E. 1981. Ureide metabolism in higher plants. *Phytochemistry* 20:409–17

138. Thomas, R. J., Schrader, L. E. 1981. The assimilation of ureides in shoot tissues of soybeans. I. Changes in allantoinase activity and ureide contents of leaves and fruits. *Plant Physiol.* 67:973–76

139. Thompson, J. F. 1980. Arginine synthesis, proline synthesis, and related processes. In *The Biochemistry of Plants*, ed. B. J. Miflin, 5:375–402. New York: Academic

140. Tracey, M. V. 1955. Urea and ureides. In *Modern Methods of Plant Analysis*, ed. K. Paech, M. V. Tracey, 4:119–41. Berlin: Springer-Verlag

141. Triplett, E. W. 1985. Intercellular nodule localization and nodule specificity of xanthine dehydrogenase in soybean. *Plant Physiol.* 77:1004–9

142. Triplett, E. W., Blevins, D. G., Randall, D. D. 1980. Allantoic acid synthesis in soybean root nodule cytosol via xanthine dehydrogenase. *Plant Physiol.* 65:1203–6

143. Triplett, E. W., Blevins, D. G., Randall, D. D. 1982. Purification and properties of soybean nodule xanthine dehydrogenase. *Arch. Biochem. Biophys.* 219:39–46

144. Vance, C. P., Stade, S. 1984. Alfalfa root nodule carbon dioxide fixation II. Partial purification and characterization of root nodule phosphoenolpyruvate carboxylase. *Plant Physiol.* 75:261–64

145. Vance, C. P., Stade, S., Maxwell, C. A. 1983. Alfalfa root nodule carbon dioxide fixation I. Association with nitrogen fixation and incorporation into amino acids. *Plant Physiol.* 72:469–73

146. Vogels, G. D., van der Drift, C. 1976. Degradation of purines and pyrimidines by microorganisms. *Bacteriol. Rev.* 40:403–68

146a. Walsh, K. B., Ng, B. H., Chandler, G. E. 1984. Effects of nitrogen nutrition on xylem sap composition of Casuarinaceae. *Plant Soil* 81:291–93

146b. Weiss, R. L., Davis, R. H. 1973. Intracellular localization of enzymes of arginine metabolism in *Neurospora*. *J. Biol. Chem.* 248:5403–8

147. Winkler, R. G., Polacco, J. C., Blevins, D. G., Randall, D. D. 1985. Enzymic degradation of allantoate in developing soybeans. *Plant Physiol.* 79:787–93

148. Winter, H. C., Powell, G. K., Dekker, E. E. 1981. 4-Methyleneglutamine in peanut plants: Dynamics of formation, levels, turnover in relation to other free amino acids. *Plant Physiol.* 68:588–93

149. Winter, H. C., Powell, G. K., Dekker, E. E. 1982. Glutamine synthetase of germinating peanuts. Properties of two chromatographically distinct forms and their activity toward 4-methyleneglutamic acid. *Plant Physiol.* 69:41–47

150. Winter, H. C., Su, T.-Z., Dekker, E. E. 1983. 4-methyleneglutamine synthetase: A new amide synthetase present in germinating peanuts. *Biochem. Biophys. Res. Commun.* 111:484–89

151. Wong, P. P., Evans, H. J. 1971. Poly-β-hydroxybutyrate utilization by soybean (*Glycine max* Merr.) nodules and assessment of its role in maintenance of nitrogenase activity. *Plant Physiol.* 47:750–55

152. Woo, K. C., Atkins, C. A., Pate, J. S. 1980. Biosynthesis of ureides from purines in a cell-free system from nodule extracts of cowpea [*Vigna unguiculata* (L.) Walp.]. *Plant Physiol.* 66:735–39

153. Woo, K. C., Atkins, C. A., Pate, J. S. 1981. Ureide synthesis in a cell-free system from cowpea (*Vigna unguiculata* [L.] Walp.) nodules. *Plant Physiol.* 67:1156–60

AUTHOR INDEX

586   AUTHOR INDEX

SUBJECT INDEX

A

Abscisic acid, 260-62, 266-68, 369, 372
β-N-Acetylglucosaminidase, 141
Actinorhizal nodules physiology, 209-32
conclusions, 225-26
Frankia
carbon metabolism, 214-16
hyphae, 212-13
isolation techniques, 211
morphology and development, 212-14
nitrogen and hydrogen metabolism, 216-18
nodule initiation, 214
response to oxygen, 218-19
sporangia and spores, 213
spore plus and minus strains, 224-25
taxonomy and host specificity, 211-12
vesicles, 213-14
introduction 209-11
host plants, 210
plant habitats, 210
nitrogen fixation and oxygen
hemoglobins and oxygen diffusion, 219-20
nodule response to soil pO₂, 220-21
nitrogen fixation control
spore plus nodules, 225
seasonal patterns of nodule activity, 222-23
Acyl lipids, 97-98
S-Adenosyl-l-methionine-C-24-methyltransferase
inhibition, 294-96
Aerobactin, 193-95, 197
Agrobacterium tumefaciens, 509-11, 522-23, 530
auxin biosynthetic genes
outside T-DNA, 528
within T-DNA, 528
cytokinin biosynthetic genes
homology with *Pseudomonas* genes, 521-22
outside T-DNA, 519-20
within T-DNA, 516-19
tumor induction, 511, 513
T-DNA, 511-13
zeatin, 519
Agrobactin, 197-98
structure, 197
Albomycin, 189
Alcohol dehydrogenase, 364

Allantoic acid, 542, 544, 562, 565
biogenesis, 551, 553-58
Allantoic acid amidohydrolase, 562
Allantoicase, 562
Allantoin, 542, 544, 563, 565
biogenesis, 551, 553-58
Allantoinase, 551, 557-58, 562-63
Allopurinol, 551
Amide
exporters, 542, 544-45
metabolism, 558-64
synthesis
ammonium assimilation, 546-47
pathways, 56
δ-Aminolevulinate, 489-91
δ-Aminolevulinate synthase, 491
Ammonia, 216-17, 475
Ammonium
assimilation
amides and ureides synthesis, 546-47
AMO 1618, 294
α-Amylase, 85-86, 370
β-Amyrin, 294
Amytal, 316
Anaerobic stress
gene expression change, 364-67
Andreae, Wolf, 7
Anisomycin, 425
Antennapedia complex, 163-64
Apigenin, 153
Arabinoxylans, 179-80
Arginine
active transport, 150
Arthobactin, 194
Ascorbic acid, 56
Asparaginase, 559
Asparagine, 542, 544-45, 563, 565
catabolism pathways, 560
metabolism, 559
synthesis, 547-48
Asparagine synthetase, 547-48
Aspartate, 547
Aspartate aminotransferase, 548, 553, 557
ATP-dependent proton pumping
vacuoles, 147-48
ATP synthetase, 97, 113, 123, 125
light regulation 111
ATPase
tonoplast, 145, 153

activity enrichment, 146-47
inhibitors, 146
substrate specificity, 146
Atrazine, 496-97
Auxin, 40
see also Gene regulation by auxin; Phytopathogen genes, auxin and cytokinin biosynthesis
Avenic acid, 201, 203
25-Azacycloartanol, 295, 301
2-Aza-2,3-dihydrosqualene, 293, 301
15-Aza-24-methylene-D-homocholesta-8,14-dien-3β-ol, 297
Azaserine, 546
Azide, 55, 63, 146

B

Beadle, G. W., 5
Benzoate, 56
Benzoic acid, 424
N-Benzyl-8-aza-4α,10-dimethyl-*trans*-decal-3β-ol, 297
Bioluminescence, 50-51
Black, Joseph, 2
Brant, Joseph, 2
Brown, Allan, 5-6
Butylated hydroxyanisole, 59

C

C₄ plants
sucrose formation compartmentation, 242
Cadmium, 372
Calcium, 154, 183, 314
bridges
pectin solubilization, 178
cellular activities regulation, 77
cross-linking, 179
gradient in tip-growing cells, 32
ions
cell growth, 26
plasma membrane and cell polarity, 35-39
NADH dehydrogenase requirement, 313-14
Calmodulin, 32, 38
CAM plants
organic acids accumulation and mobilization, 153-54
Carbamoyl phosphate synthetase, 549-50

CUMULATIVE INDEXES

CONTRIBUTING AUTHORS, VOLUMES 29–37

A

Abeles, F. B., 37:49–72
Albersheim, P., 35:243–75
Amasino, R. M., 35:387–413
Amesz, J., 29:47–66
Andersen, K., 29:263–76
Anderson, J. M., 37:93–136
Appleby, C. A., 35:443–78
Akazawa, T., 36:441–72

B

Badger, M. R., 36:27–53
Bandurski, R. S., 33:403–30
Barber, J., 33:261–95
Bauer, W. D., 32:407–49
Beale, S. I., 29:95–120; 34:241–78
Bedbrook, J. R., 30:593–620
Beevers, H., 30:159–93
Bell, A. A., 32:21–81
Benedict, C. R., 29:67–93
Benson, D. R., 37:209–32
Benveniste, P., 37:275–308
Bernstam, V. A., 29:25–46
Berry, J., 31:491–543
Bewley, J. D., 30:195–238
Biale, J. B., 29:1–23
Bickel-Sandkötter, S., 35:97–120
Björkman, O., 31:491–543
Boller, T., 37:137–64
Bottomley, W., 34:279–310
Boyer, J. S., 36:473–516
Brenner, M. L., 32:511–38
Buchanan, B. B., 31:341–74
Burnell, J. N., 36:255–86
Butler, W. L., 29:345–78

C

Castelfranco, P. A., 34:241–78
Chaney, R. L., 29:511–66
Chapman, D. J., 31:639–78
Clarke, A. E., 34:47–70
Clarkson, D. T., 31:239–98; 36:77–115

Cogdell, R. J., 34:21–45
Cohen, J. D., 33:403–30
Conn, E. E., 31:433–51
Cosgrove, D., 37:377–405
Craigie, J. S., 30:41–53
Cronshaw, J., 32:465–84
Cullis, C. A., 36:367–96

D

Darvill, A. G., 35:243–75
Davies, D. D., 30:131–58
Dennis, D. T., 33:27–50
Diener, T. O., 32:313–25
Digby, J., 31:131–48

E

Edwards, G. E., 36:255–86
Eisbrenner, G., 34:105–36
Eisinger, W., 34:225–40
Elbein, A. D., 30:239–72
Ellis, R. J., 32:111–37
Elstner, E. F., 33:73–96
Etzler, M. E., 36:209–34
Evenari, M., 36:1–25
Evans, H. J., 34:105–36
Evans, L. T., 32:485–509

F

Farquhar, G. D., 33:317–45
Feldman, J. F., 33:583–608
Feldman, L. J., 35:223–42
Fincher, G. B., 34:47–70
Firn, R. D., 31:131–48
Fischer, R. A., 29:277–317
Flavell, R., 31:569–96
Fork, D. C., 37:335–61
Foy, C. D., 29:511–66
Freeling, M., 35:277–98
French, C. S., 30:1–26
Fry, S. C., 37:165–86
Furuya, M., 35:349–73

G

Galston, A. W., 32:83–110
Galun, E., 32:237–66

Gantt, E., 32:327–47
Giaquinta, R. T., 34:347–87
Gifford, E. M. Jr., 34:419–40
Gifford, R. M., 32:485–509
Glass, A. D. M., 34:311–26
Good, N. E., 37:1–22
Goodwin, T. W., 30:369–404
Gordon, M. P., 35:387–413
Graham, D., 33:347–72
Gray, M. W., 33:373–402
Green, P. B., 31:51–82
Greenway, H., 31:149–90
Grisebach, H., 30:105–30
Guerrero, M. G., 32:169–204
Gunning, B. E. S., 33:651–98

H

Haehnel, W., 35:659–93
Hahlbrock, K., 30:105–30
Halperin, W., 29:239–62
Hanson, A. D., 33:163–203
Hanson, J. B., 31:239–98
Hara-Nishimura, I., 36:441–72
Hardham, A. R., 33:651–98
Harding, R. W., 31:217–38
Harris, N., 37:73–92
Haselkorn, R., 29:319–44
Hatch, M. D., 36:255–86
Haupt, W., 33:205–33
Heath, R. L., 31:395–431
Heber, U., 32:139–68
Hedden, P., 29:149–92
Heidecker, G., 37:439–66
Heldt, H. W., 32:139–68
Hepler, P. K., 36:397–439
Higgins, T. J. V., 35:191–221
Hirel, B., 36:345–65
Hitz, W. D., 33:163–203
Ho, T.-H. D., 37:363–76
Hoffman, N. E., 35:55–89
Hotta, Y., 29:415–36
Howell, S. H., 33:609–50
Huber, S. C., 37:233–46

J

Jackson, M. B., 36:145–74

607

CHAPTER TITLES, VOLUMES 29–37

ORGANELLES AND CELLS